C0-AJY-753

Signals in Linear Circuits

José B. Cruz, Jr.
University of Illinois

M. E. Van Valkenburg
Princeton University

Houghton Mifflin Company · Boston
Atlanta · Dallas · Geneva, Illinois · Hopewell, New Jersey
Palo Alto · London

Printed in the U.S.A.

Library of Congress Catalog Card Number: 73-8076

ISBN: 0-395-16971-2

Contents

Preface

Some of the guidelines we have followed in the preparation of Signals in Linear Circuits are the following: (1) The approach should be general, though elementary. We recognize that the circuits which model physical systems are becoming more complicated due to the solid-state device revolution. Further, solutions are frequently obtained by computer simulation so that simplifying assumptions are no longer imperative. Thus it is important that the treatment of subjects within the book be as general as possible, stressing assumptions and limitations. (2) The sequence of topics should be ordered to present simple ideas first, followed by more complicated situations, and then the most general cases. (3) It no longer makes sense, as it once might have, to teach the subject of circuits while paying little or no attention to the properties of signals. A circuit is primarily a signal processor, even though it is true that there are important special cases in which the properties of the signal are of less importance—the study of the electric power system, for example.

In keeping with this plan, the topics we have selected are treated in sufficient detail to arouse student interest, but not in such depth that he or she will lose sight of the end objectives. We believe this kind of treatment will give the student an accurate picture of the objectives and methods of circuit theory. With this background, the student will be more effective both in subsequent studies of circuit theory per se and in many applications of circuit theory to the topics of electrical engineering.

It is our experience that students must have extensive drill in their first course in circuits and that this drill must make it possible to first understand the principle and then to extend it to the solution of more complicated problems. Thus there is a need for both easy and more difficult problems. We have organized the text so that drill exercises, with answers in an appendix, are provided at the end of each section. These will permit the student to test his knowledge and the subject.

Problems are given at the end of the chapter, some of equal difficulty and others of increased difficulty. Those which are true challenges and therefore should be assigned with care are marked with a star.

In stressing concept over specific facts, we have chosen to delay the study of methods for the efficient solution of simultaneous algebraic equations until Chapter 14. There will be no serious difficulty, of course, if this material is covered earlier, even before Chapter 1. Each instructor should decide the point at which this information has maximum usefulness. His decision will be influenced by the degree to which the computer will be used as an adjunct to the course. Similarly, topics in graph theory in Chapters 12 and 13 may be studied first for courses requiring this more general foundation. Finally, the topics of the Fourier analysis, Fourier integral and transform, and the Laplace transform of Chapters 16 and 17 may be studied immediately after the sinusoidal steady-state analysis of Chapter 10 for courses with this topic sequence.

We have attempted to increase the flexibility available to the instructor using this textbook through appendices. Appendix A contains an introduction to mechanical circuits, stressing the mechanical-electrical analogy. This material can be inserted in Chapter 4 if the course outline specifies these topics. Many courses are planned with a collateral software laboratory, serving the roles of practice in network analysis, an introduction to numerical methods through the use of the digital computer, and also the use of the computer to perform tasks relating to signal and circuit analysis. Appendix B is intended to assist the instructor in planning his software laboratory, selecting appropriate exercises, and finding tabulations of subroutines if these are not available in his computer center. Appendix C is an introduction to the Laplace transform. In the preparation of the textbook, we envisioned that the subject of the Laplace transform would receive only an introduction, leaving the subject to later courses for a more complete and detailed study. We recognize that many classes have course outlines that require a more complete treatment in the introductory course, and for this purpose a more detailed coverage is provided in Appendix C.

We are gratefully indebted to Houghton Mifflin Company for making this book part of their Electrical Engineering series. It represents a substantial revision and enlargement of our earlier book *Introductory Signals and Circuits* published originally in 1967 with Blaisdell Publishing Company. The Blaisdell list was later transferred to Xerox College Publishing, who having decided to deemphasize engineering publishing returned the rights for the new edition to the authors.

We are indebted to many people who have assisted in the writing of the book, the most important of these have been our students over the years who provided the motivation for writing and who participated in the testing of much of the material. In preparing this revision we are especially indebted to Professor William R. Perkins with whom we have discussed both technical and pedagogical problems.

The material in *Introductory Signals and Circuits* was developed over a number of years in classes at the University of Illinois. We acknowledge our indebtedness to the community of scholars in the circuits and systems area at Illinois who provided a stimulating and congenial milieu in which such a book could develop. These include William R. Perkins, Donald A. Calahan, Leon O. Chua, Franklin F. Kuo, S. Louis Hakimi, Wataru Mayeda, James A. Resh, Ronald A. Rohrer, the late Sundaram Seshu, Manoel Sobral, Jr., Timothy N. Trick, Nelson Wax, and James R. Young. We benefited from discussions with Charles A. Desoer, Ernest S. Kuh, Benjamin J. Leon, and Leon O. Chua concerning conventions and symbols that should be used. We express our special appreciation to Professor John P. Gordon of Northern Arizona University who while at the University of Illinois assisted in the preparation of the manuscript in many ways, including the completion of the *Solutions Manual*.

Additionally we gratefully acknowledge the assistance of the many reviewers who aided us in the preparation of this book. They are: Professor Henry Meadows, Columbia University; Professor George T. Etzweiler, Pennsylvania State University; Professor S. L. Hakimi, Northwestern University; Professor Leonard S. Bobrow, University of Massachusetts; Professor Nirmal K. Bose, University of Pittsburgh; and Professor Johnny Andersen, University of Washington.

Finally, it is a pleasure to acknowledge our indebtedness to our wives, Pat Cruz and Evelyn Van Valkenburg, for their support, encouragement, and cooperation during the writing of the book.

José B. Cruz, Jr.

M. E. Van Valkenburg

Urbana, Illinois
Princeton, New Jersey

List of Frequently Used Symbols

Symbol	Meaning
a, b	General constants
A, B, C, D	Transmission parameters
a_n, b_n, c_n	Fourier-series coefficients
B	Susceptance in mhos, the imaginary part of the admittance
C	Capacitance in farads
D	Damping coefficient, Nsec/m (newton-seconds/meter)
e	Naperian logarithm base, 2.718 ...
$\det A$	Determinant of A
$e^{j\theta}$	Unit phasor with angle θ, equivalent to $\underline{/\theta}$
$e^{j\omega t}$	Unit rotating phasor, rotating at ω radians/sec
E^2	Mean-square error
f	Frequency in Hz (hertz) or cycles/sec
$F(j\omega)$	Fourier transform
F_n	Exponential Fourier-series coefficients
G, g	Conductance, real part of admittance, mhos
g_m	Incremental transconductance of electronic device
g_{xy}	Two-port g-parameters
$g(t)$	Approximating function
h_{xy}	Two-port h-parameters
$h(t)$	Impulse response
I, i	Current, usually I is a constant or rms value and i is a function of time
$\mathcal{I}$	Column matrix of currents
Im	Imaginary part of a complex function
j	$\sqrt{-1}$
J	Moment of inertia in Nmsec^2 (newton-meters-sec^2)
k	Real constant
K	Compliance, m/N
KCL	Kirchhoff's current law
KVL	Kirchhoff's voltage law
l	Length or distance in complex plane
L	Inductance in henrys
$\mathcal{L}[\ ]$	Laplace transformation of [ ]
M	Mutual inductance in henrys
M	Mass in kilograms

N_v	Number of vertices (nodes) in network
$p(s)$	Polynomial in s
$p(t)$	Power in watts or va (volt-amperes)
p	Differential operator, d/dt
P_{av}	Average power in watts
p.f.	Power factor
q, Q	Charge in coulombs
Q	Quality factor in coil
Q	Reactive power in vars
$q(s)$	Polynomial in s
r	Incremental resistance of electronic device
R	Resistance in ohms
Re	Real part of a complex function
s	Complex frequency, $\sigma + j\omega$
S	Complex power in va (volt-amperes)
t	Time, sec
T	Time constant, period in sec, or a constant
$u(t)$	Unit step function
V, v	Voltage in volts, usually with V as a constant or rms value and v as a time-varying quantity
$\bar{v}$	Average value of a signal
$\hat{v}$	Velocity in m/sec
$V(s)$	Laplace transform of $v(t)$
$\mathscr{V}$	Column matrix of voltages
V_{rms}	Rms or effective value of a signal; phasor representation of a signal of magnitude $V_m/\sqrt{2}$, where V_m is the maximum value of the sinusoidal $v(t)$
w, W	Energy in joules or watt-sec
x	General variable, dummy variable
x	Displacement, m (meters)
X	Reactance, imaginary part of impedance
y	General variable
y_{xz}	Two-port y-functions
Y	Admittance in mhos, $1/Z$
$\mathscr{Y}$	Node-admittance matrix
z_{xy}	Two-port impedance parameters
Z	Impedance in ohms
$\mathscr{Z}$	Loop-impedance matrix
α	Current gain, ratio of output to input

$\delta(t)$	Impulse function
Δ	An increment, a determinant, or a constant
ζ	Damping ratio
θ	Angle of a phasor with respect to a reference
μ	A constant
ξ	Dummy variable
ϕ	Angle of a phasor with respect to a reference
σ	Damping, real part of s
τ	Time constant, dummy variable
τ	Torque in Nm (newton-meters)
ψ	Flux linkages in volt-sec
ω	Frequency in radians/sec
ω_0	Fundamental frequency, radians/sec
$n\omega_0$	nth harmonic frequency, radians/sec
ω_r	Resonant frequency, radians/sec
ω	Rotational angular frequency, radians/sec
$\|C\|$	Magnitude of complex C
C^*	Complex conjugate of C
$\underline{/\theta}$	Unit phasor with angle θ, equivalent to $e^{j\theta}$
$\triangleq$	Equal by definition

Prefixes for Units

Prefix	Factor	Symbol
giga-	10^9	G
mega-	10^6	M
kilo-	10^3	k
deci-	10^{-1}	d
centi-	10^{-2}	c
milli-	10^{-3}	m
micro-	10^{-6}	μ
nano-	10^{-9}	n
pico-	10^{-12}	p

Chapter 1

Signal Sources and Signal Processing

1.1 Reference conventions for signals

We use the word *signal* to mean a time history of voltage or current in an electric network. The voltage or current variation with time may represent a message, music, a television picture, etc. In our elementary study of signals, we will include the sinusoidal waveform used to transmit energy. We exclude signals which must be described by statistical properties, leaving this important subject for later study. All signals we study may be described by real numbers which may be either positive or negative (and zero). It is important that we first understand the meaning of the positive or negative sign identified with the signal.

Consider a battery, an electric-energy source, which has its own exclusive symbol, that of Figure 1.1(a). The terminal of the battery with an excess of positive charge is called the *anode*, that with an excess of negative charge the *cathode*. It is conventional that the anode be distinguished by a plus sign and the cathode by a minus sign, as shown in the figure.

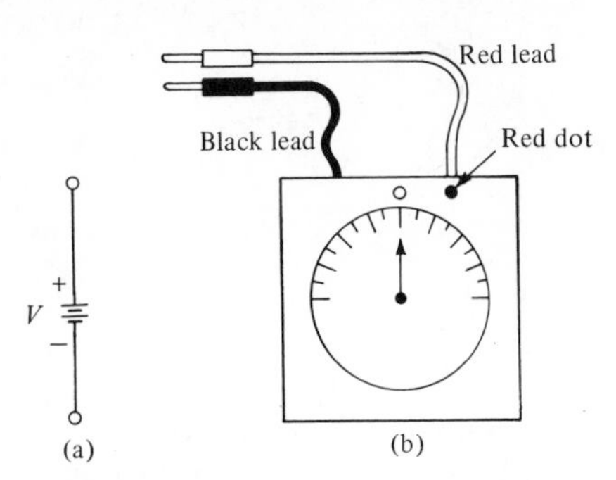

Figure 1.1 (a) Symbol for a battery. (b) Representation of an ideal voltmeter.

The battery voltage may be measured by a voltmeter. An idealized voltmeter or galvanometer suited to measurements of our interest measures instantaneous voltage, has a zero center scale, and is capable of deflecting to both the left and the right as in Figure 1.1(b). Reversal of voltmeter leads reverses the direction of deflection. It is usual to call one direction of deflection positive and the other negative. This is the *second* use to which we have put the words positive and negative, but this need not be a source of confusion if we make clear which use we intend.

Consider the experiment in which the two leads of the voltmeter are connected to the battery and the deflection is positive. The lead connected to the plus terminal or anode is designated as the positive reference lead (red); the other is the negative reference lead (black).* With these identifications defining the voltmeter deflections, we consider a general electric source. This source may reverse polarity with the passage of time. How can we describe this polarity variation?

We first postulate that the output terminals of the source have reference marks painted on them, one a plus and one a minus, these assignments having been made *arbitrarily* at the time the source left the factory. To this source we connect

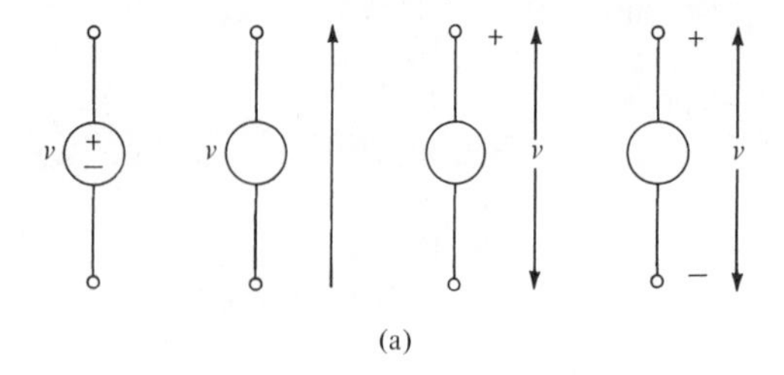

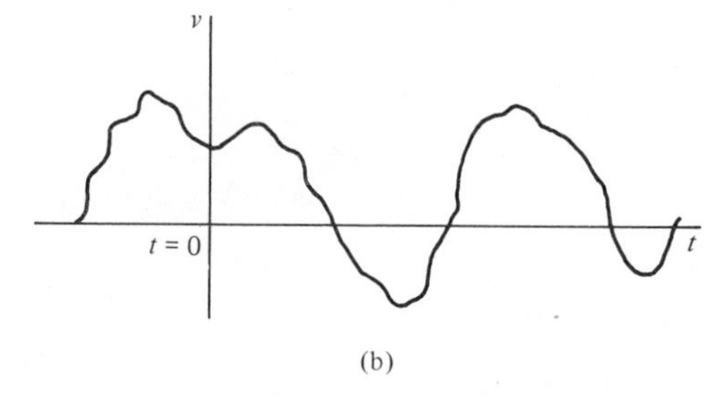

Figure 1.2 (a) Several alternative representations of voltage sources and their associated reference marks. (b) A plot of voltage as a function of time.

* In the laboratory, our red lead will often be called the "live" or "hot" lead, and the black one the "ground" lead.

our calibrated voltmeter with the red lead of the voltmeter connected to the plus terminal of the source, and the black lead to the minus terminal. With this connection, a positive deflection of the voltmeter (or its counterpart in the form of an oscillograph) implies that the plus terminal of the source has an excess of positive charge and the source polarity is like that of the battery. Similarly, a negative deflection implies a source polarity opposite to that of the battery. Thus we see that the plus mark on the general source has a different meaning than the plus mark on the battery. In the case of a battery, the plus mark implies positive polarity; for the source, the plus mark is a *reference* in terms of which the polarity variation with time may be described. For this reason, the plus mark on the general source could just as well be a sign of the Zodiac. We can make the identifications "like the battery" and "opposite to the battery" only in terms of the reference marks on the source. Plots like those in Figure 1.2(b) have meaning only in terms of the reference marks.

For example, if the equation of a recorded voltage is $v(t) = V \sin t$, where V is a positive number, then the plus terminal of the source has a positive polarity from $t = 0$ to $t = \pi$, a negative polarity from $t = \pi$ to $t = 2\pi$, etc. There are a number of schemes which can be substituted for our plus and minus designation; three of the most common alternatives are shown in Figure 1.2(a).

The need for reference marks is further illustrated by the connection of two sources shown in Figure 1.3(a). Without reference marks on the sources, we do not know the relationship of v_1, v_2, and v. With the reference marks of Figure 1.3(b), we see that $v = v_1 + v_2$; with those of Figure 1.3(c), we have $v = v_1 - v_2$.

Statements similar to those given for the voltage reference apply to the current reference. The symbol used to indicate reference direction* is shown in Figure 1.4(a). The ammeter deflects in a positive direction when current is in the reference direction, negative when in the opposite of the reference direction. The marks of

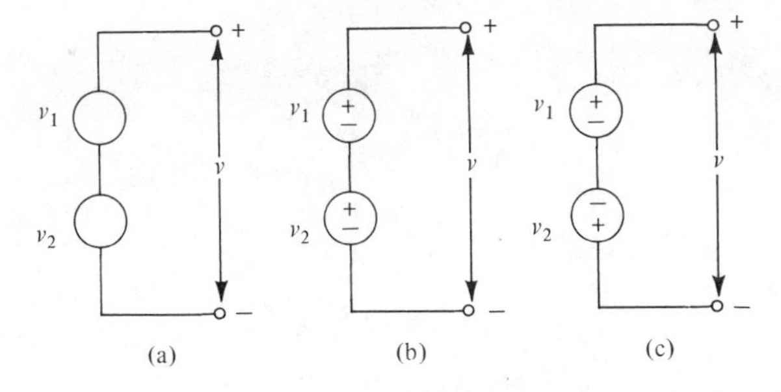

gure 1.3 Two voltage sources connected in series, illustrating the need for reference marks in describing the voltage of the combination.

* Ammeters found in the laboratory are marked with plus and minus signs. The plus goes with the tail of the arrow, the minus with the head.

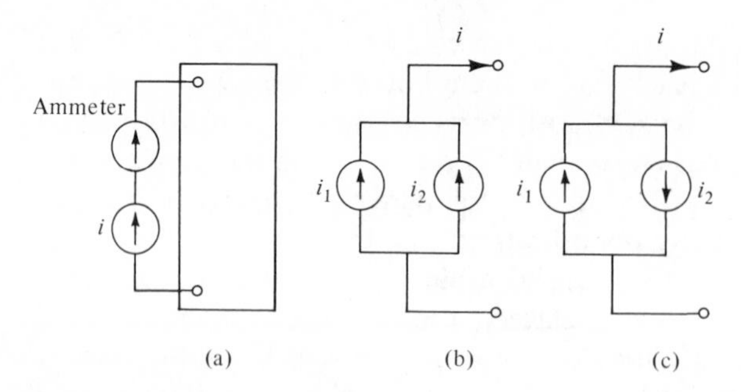

Figure 1.4 Conventional reference marks for a current source. Clearly, for (b) $i = i_1 + i_2$ while for (c) $i = i_1 - i_2$.

Figure 1.4(b) imply that $i = i_1 + i_2$, while those of Figure 1.4(c) imply that $i = i_1 - i_2$.

In our discussion of source notation, we have used a single-subscript symbol like v_1 together with reference marks (plus and minus) to describe sources. Another notation commonly used in circuit theory is known as the *double-subscript notation.* If we use this notation, the voltage v_{jk} is *defined* to be the voltage between j and k with the plus mark implied at j. Thus we read v_{12} as the potential at 1 with respect to 2, meaning the voltmeter reading with the red lead connected to 1 and the black lead to 2.

In describing current sources by double subscripts, we see that i_{jk} is the current in the path from j to k with the reference arrow from j to k. Thus i_{12} is a current with the reference direction such that the tail of the arrow is at 1, the head at 2.

The equivalences for the two forms of notation for voltage and current are illustrated in Figure 1.5. Observe that for both cases, the reversal of subscripts implies a change of sign; thus we have

$$v_{12} = -v_{21}, \qquad i_{ab} = -i_{ba}, \qquad \text{etc.}$$

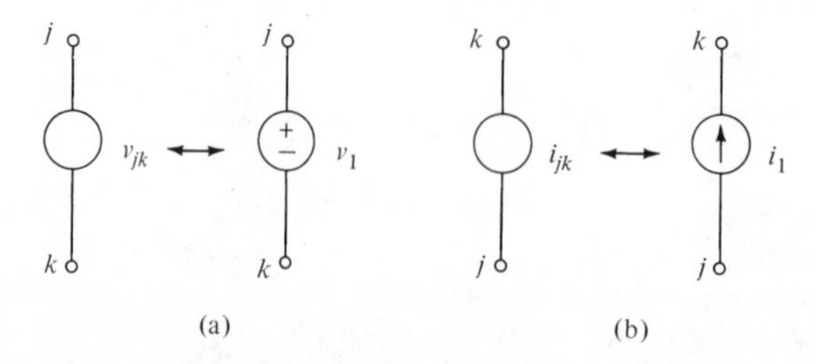

Figure 1.5 Equivalent representations for double-subscript and single-subscript notations.

XERCISES

1.1.1 The voltage signal from a source is described by the equation $v_1(t) = 5 \sin (2t + \pi/6)$ volts. When $t = \pi/2$ sec, determine the terminal voltage and its polarity with respect to the reference marks of the source.

1.1.2 The current in a circuit is $i_{12} = 5 \sin (2t + \pi/6)$ amperes. At $t = \pi/2$ sec, what is the currect direction? Repeat for $t = 0$.

1.1.3 Two identical ideal voltmeters are connected across a time-varying source. The red lead of one meter and the black lead of the other are connected to the same terminal of the source. The remaining leads are connected to the other terminal. What is the algebraic sum of the readings of the meters at all instants of time? How do the readings of the two meters compare?

.2 Models of signal sources and notation

The devices associated with electric networks are commonplace: coils, capacitors, resistors, transistors, vacuum tubes. In studying circuit theory or network theory, we are not concerned with these devices per se but with abstractions or idealizations of them known as *models*. Good models are required to be simple and yet must accurately represent the device (or system) under specified conditions. In general, the choice of a model represents a compromise between simplicity and accuracy requirements.

Devices which are important sources of electric energy include the battery, electromechanical generators, and electronic generators. These and other electric energy sources may be represented by two models: the voltage source and the current source.*

Let a source of electric energy be connected to an arbitrary network as in Figure 1.6(a) and let the resulting current be denoted by $i_1(t)$ and the voltage across the terminal pair by $v_1(t)$. Suppose that instead of using Network No. 1,

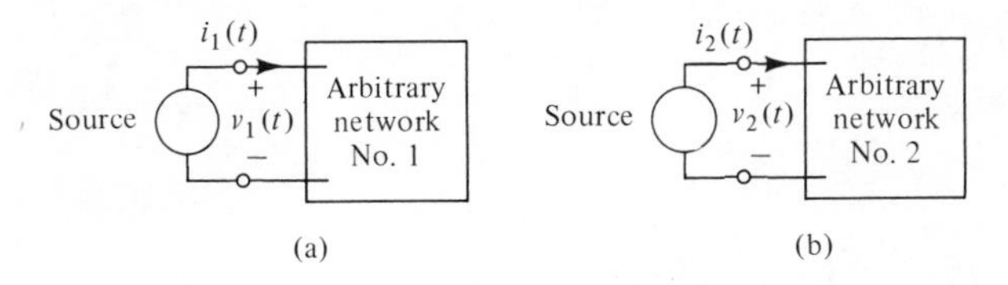

Figure 1.6 Arbitrary networks connected to identical sources of electrical energy. If the sources are voltage sources, $v_1(t) = v_2(t)$. If the sources are current sources, $i_1(t) = i_2(t)$.

* More elaborate source models will be introduced later.

we had used some other arbitrary network as in Figure 1.6(b). In general, the resulting $i_2(t)$ would be different from $i_1(t)$ and $v_2(t)$ would be different from $v_1(t)$. However, if $v_1(t) = v_2(t)$ *no matter what arbitrary network we use,* then we define the source as an *ideal voltage source* or simply a *voltage source.* The voltage-source model is a source of electric energy with a prescribed voltage across its terminals. The resulting current depends not only on $v(t)$ but also the nature of the connected network.

If in Figure 1.6, $i_1(t) = i_2(t)$ *no matter what arbitrary network we use,* then we define the source as an *ideal current source* or simply a *current source.* This time, the source has a prescribed current through its terminals. Both the network and the current source $i(t)$ determine the resulting voltage $v(t)$. To distinguish between voltage and current sources, we will adopt the symbols in Figure 1.7.

When the voltage of a voltage source or the current of a current source is controlled by or depends on another voltage or current, the resulting model is distinguished by the name *controlled source* or *dependent source.* The symbols that we will use for the controlled voltage source and the controlled current source are shown in Figures 1.8(a) and 1.8(b), respectively. An example of a controlled voltage source is one which depends on a voltage $v_1(t)$

$$v_2(t) = \mu v_1(t), \qquad \mu = \text{real constant}, \tag{1.1}$$

which is a special case of the more general relationship, $v_2(t) = f[v_1(t)]$. The voltage $v_1(t)$ does not have to be a voltage source. It may be a voltage resulting from electric sources somewhere in the system. Such a controlled voltage source is called a voltage-controlled voltage source. An example of a controlled current source is one which depends on a current $i_1(t)$ such that

$$i_2(t) = \alpha i_1(t), \tag{1.2}$$

where $i_1(t)$ is a current in some part of a network and α is a real constant. This example generalizes to the form $i_2(t) = f[i_1(t)]$ where i_1 is the current from some source or in some element. Two other common controlled sources are the current-controlled voltage source and the voltage-controlled current source.

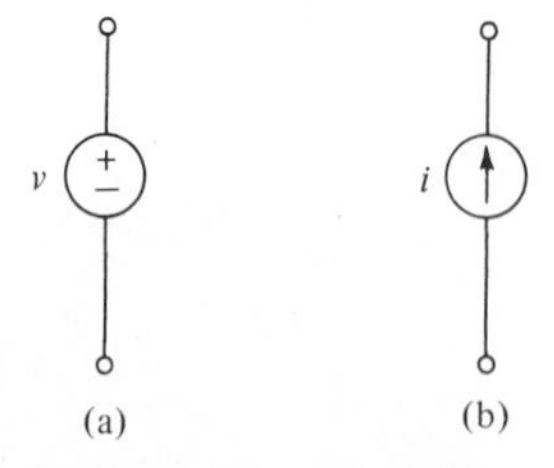

Figure 1.7 (a) Standard symbol for a voltage source. (b) Standard symbol for a current source.

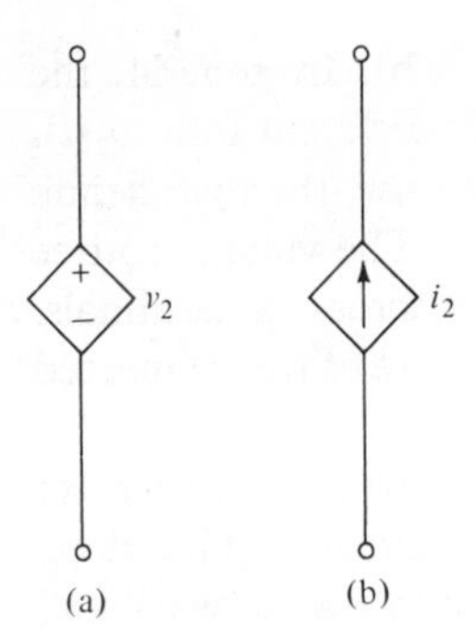

Figure 1.8 (a) Standard symbol for a controlled (or dependent) voltage source. (b) Standard symbol for a controlled (or dependent) current source.

A number of conventions have been introduced in the foregoing discussion, as well as in the preceding section. Throughout the book we will continue to use lower-case letters for time-varying quantities. This will be explicit when we write $v_1(t)$ or implied when we write simply v_1. Upper-case letters will be reserved for time-invariant quantities: real constants and later complex numbers.

In writing $v_1(t)$, we have assumed that the student is familiar with the usual function notation by which it is read as *the value of* v_1 *at* t. Here v_1 will be a voltage or, as determined by the context of the discussion, a general signal; t is time, usually in seconds. We usually think of $v_1(t)$ as given by an equation (in closed form) although it need not be. A function will sometimes be expressed by two sets of numbers occurring in ordered pairs as in Table 1.1.

Table 1.1

t	v_1
0	1
1	3
2	3.4
...	...

A graph of a function is made by plotting the two numbers of an ordered pair such that v_1 is the *ordinate* and t is the *abscissa*. Such a graph of a signal as a function of time is called a *waveform*.

The time t is reckoned with respect to an arbitrary reference. A common reference is the closing of a switch; another is the firing of a gun that starts the race. This reference time is $t = 0$; $t = 1$ sec is the instant 1 sec after the closing of the switch; $t = -1$ sec refers to the instant of time 1 sec before the closing of the switch.

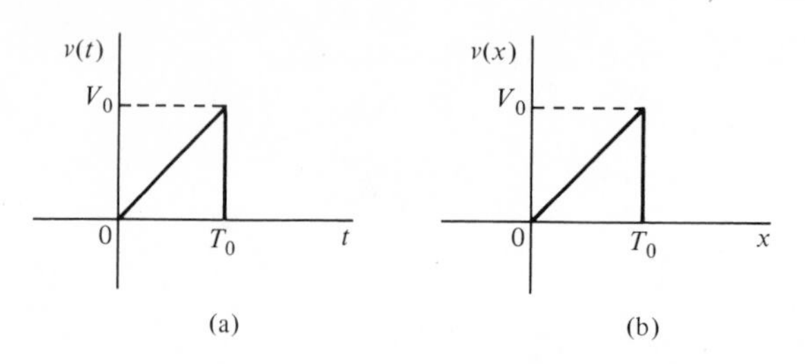

Figure 1.9 Given a function $v(t)$ displayed in (a) with t as abscissa, the graph of $v(x)$ using x as abscissa is shown in (b).

Another set of names used in describing $v_1(t)$ are *dependent variable* for v_1 and *independent variable* for t. Time is the usual independent variable, but in some mathematical operations in circuit theory it is necessary to replace t by another variable, say x. It is important that we understand that when t is replaced by x in the function $v(t)$, then $v(x)$ plotted against x looks exactly like $v(t)$ plotted against t. This statement is illustrated in Figure 1.9. All of this sounds ridiculously obvious,* but it may prove to be a stumbling block later.

To carry this discussion a bit further, let t be replaced by $x + T$ where T is a constant. Clearly, if $v(x + T)$ is plotted against $x + T$, we get one more plot identical to the two of Figure 1.9. But if $v(x + T)$ is plotted against x, then things become a little different. This results in a *shift* or *translation* of the graph with respect to the abscissa. To illustrate, consider the signal waveform of Figure 1.9(b) which is described by the equations

$$v(x) = \begin{cases} 0, & x < 0, \\ \dfrac{V_0}{T_0}x, & 0 \le x \le T_0, \\ 0, & x > T_0. \end{cases} \tag{1.3}$$

Replacing x by $x + T$, we have

$$v(x + T) = \begin{cases} 0, & x + T < 0, \\ \dfrac{V_0}{T_0}(x + T), & 0 \le x + T \le T_0, \\ 0, & x + T > T_0. \end{cases} \tag{1.4}$$

* Any fact is *obvious* once we understand it.

Rearranging the inequalities in the intervals of the piecewise description of $v(x + T)$, we find that

$$v(x + T) = \begin{cases} 0, & x < -T, \\ \dfrac{V_0}{T_0}(x + T), & -T \leq x \leq T_0 - T, \\ 0, & x > T_0 - T. \end{cases} \tag{1.5}$$

This equation is plotted in Figure 1.10(a) for T positive and in Figure 1.10(b) for T negative. In the second figure, note that since T is negative, $-T$ is positive. Such a shifting operation will occur frequently in our study.

Another common transformation on $v(t)$ is to replace t by Kx or Kt to form $v(Kx)$ or $v(Kt)$ and plot this with respect to x or t. This is called *time scaling*. It finds use in the selection of an appropriate time unit; for example, replacing time in seconds by time in microseconds. The effect on the graph is either stretching or compressing the abscissa. More on this in a later chapter.

An operation we will encounter often is the *integration* of a signal $v(t)$ to give another signal $f(t)$. From our discussion on change of variable, it is clear that

$$\int_{t_1}^{t_2} v(t)\, dt = \int_{t_1}^{t_2} v(x)\, dx. \tag{1.6}$$

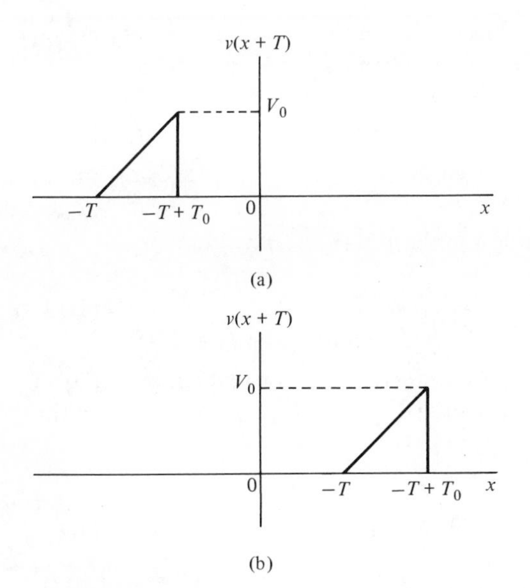

Figure 1.10 The signal waveform of Figure 1.9 is shown shifted as specified by Equation 1.5.

For that matter, we may use any other letter in the English, Greek, Cyrillic, or any other alphabet. The value of the integral will be the same in all cases. Because of the invariance and independence of the value of the integral with respect to the choice for our independent variable, t or x in Equation 1.6 is called a *dummy variable*. To illustrate the use of a dummy variable, suppose that the signal

$$v(t) = Ke^{-at} \tag{1.7}$$

is to be integrated beginning at $t = 0$ to give the new signal $f(t)$. It is confusing to write

$$f(t) = \int_0^t v(t)\, dt \tag{1.8}$$

as the value of f at time t for t is *both* a variable and a limit of integration. Instead, we make use of the dummy variable x as

$$f(t) = \int_0^t v(x)\, dx\,. \tag{1.9}$$

Carrying out these operations, we have

$$f(t) = K\int_0^t e^{-ax}\, dx = \frac{-K}{a} e^{-ax}\Big|_0^t = \frac{K}{a}(1 - e^{-at}). \tag{1.10}$$

Another application of the dummy variable is used when writing an expression for the *energy* that a source has supplied between time t_1 and time t. If the voltage of the source is $v(t)$ and the current $i(t)$, then the *energy* is

$$w(t) = \int_{t_1}^t v(x)i(x)\, dx\,. \tag{1.11}$$

The time rate of energy transfer at time t is the *power*

$$p(t) = \frac{dw(t)}{dt} = v(t)i(t)\,. \tag{1.12}$$

We shall often express the integrand by different functions which are valid in different ranges. For example, the waveform of Figure 1.11(a) is expressed by the equations

$$v_1(t) = \begin{cases} 0, & t < 0, \\ V_0, & 0 \le t \le T_0, \\ 0, & t > T_0. \end{cases} \tag{1.13}$$

Then depending on the value of t, the proper expression in this set of equations is

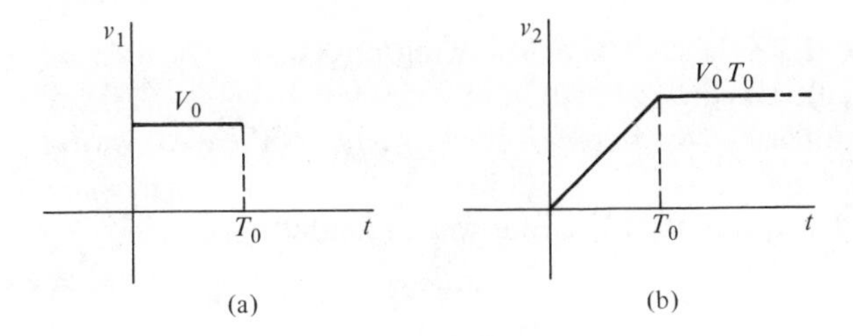

Figure 1.11 The integral of the signal of (a) is that shown in (b).

selected for the range being considered. Let v_2 be related to the v_1 of Equations 1.13 by the equation

$$v_2(t) = \int_{-\infty}^{t} v_1(x)\, dx. \tag{1.14}$$

Substituting Equations 1.13 into this integral, we have

$$v_2 = \begin{cases} 0, & \text{for } t < 0, \\ \int_0^t V_0\, dx = V_0 t, & \text{for } 0 \le t < T_0, \\ \int_0^{T_0} V_0\, dx + \int_{T_0}^{t} 0\, dx = V_0 T_0, & \text{for } t \ge T_0. \end{cases} \tag{1.15}$$

The waveform for v_2 is shown in Figure 1.11(b). Integration is readily visualized as the adding of increments of area. In this example, we see that the area is zero for $t < 0$ and is a constant value for $t > T_0$ as shown in the figure.

EXERCISES

1.2.1 If $f(t) = e^{-at} \sin bt$, write $f(x)$, $f(\xi)$, and $f(\zeta + 1)$.

1.2.2 Consider the signal

$$v(t) = \begin{cases} 0, & t < 0, \\ \sin t, & 0 \le t \le 2\pi, \\ 0, & t > 2\pi. \end{cases}$$

Sketch $v(x + 2\pi)$ and $v(x - \pi)$ against x.

1.2.3 Let $v(t) = \sin t$ and $i(t) = \frac{1}{2} \sin t$. Find $w(t)$ using Equation 1.11 with $t_1 = 0$. What is the expression for $p(t)$?

1.2.4 If the signal $v(t)$ is that shown in Figure 1.9(a) with $V_0 = 5$ and $T_0 = 10$, sketch the following: (a) $v(x + 5)$ vs. x; (b) $v(t + 5)$ vs. t; (c) $v(x - 5)$ vs. x; (d) $v(t - 5)$ vs. t.

1.2.5 For $v(t)$ of Exercise 1.2.4, sketch the following: (a) $v(10x)$ vs. x; (b) $v(10t)$ vs. t; (c) $v(x/10)$ vs. x; (d) $v(t/10)$ vs. t.

1.2.6 For $v(t)$ of Exercise 1.2.4, sketch the following: (a) $v(10t - 5)$ vs. t; (b) $v(t/10 + 5)$ vs. t.

1.2.7 For $v(t)$ of Exercise 1.2.4, calculate the following integrals:

(a) $I_1 = \int_0^{10} v(t)\, dt\,,$

(b) $I_2 = \int_0^{10} v(x)\, dx\,,$

(c) $I_3 = \int_{-10}^{0} v(x + 10)\, dx\,,$

(d) $I_4 = \int_5^{15} v(x - 5)\, dx\,.$

Are these integrals expected to be equal to each other?

1.2.8 For $v(t)$ of Exercise 1.2.4, calculate

$$I = \int_0^t v(\tau)\, d\tau$$

and sketch I vs. t for $-1 \leq t \leq 20$.

1.2.9 The waveform shown in Figure 1.2(b) is observed on a cathode-ray oscilloscope connected to the output of the device described by Equation 1.14. With the oscilloscope leads reversed, draw the waveform that will be observed on the oscilloscope screen.

1.3 Nature of signal processing in networks

In the chapters to follow, we will be concerned with the determination of the characteristics of signals throughout a network of electrical elements. As illustrated by Figure 1.12, a signal source will be connected to the network at a pair of terminals (or *nodes*). The network may alter the waveform of the signal at each output pair of nodes that we identify and thus produce a modified or processed signal. The characteristics of the modified signal will be determined through network analysis employing laws and mathematical techniques which we will develop. As an example, the input signal may be a time function $f(t)$, and the desired ideal output signal may be a constant K multiplied by the same time function $f(t)$. For this example, the network is commonly called an *amplifier* for K

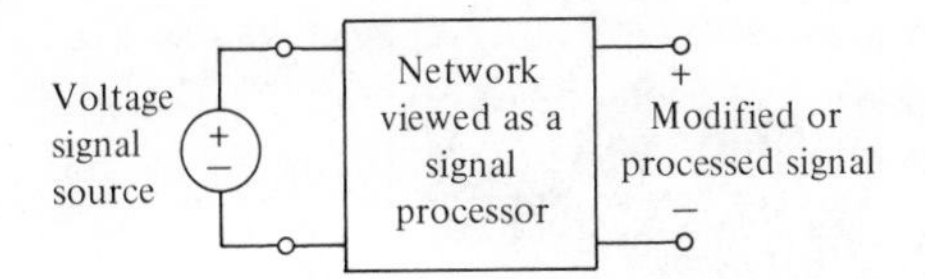

Figure 1.12 An excitation — the input signal — is applied to the network. The network modifies the signal as measured at another pair of terminals.

greater than 1 and an *attenuator* for K less than 1. As another example, the input signal may be of the form $K_1 + K_2 \sin K_3 t$, where K_1, K_2, and K_3 are constants, and the desired output may be of the form $K_1 + K_4 K_2 \sin K_3 t$, where K_4 is very much less than 1. The network may be descriptively called a *filter*, since it tends to suppress and discriminate against a component of the input.

There are several possibilities for voltage–current relationships in the elements we will consider in our study which may be discussed with the aid of Figure 1.13 and the reference directions assigned. We will find that for the elements,

1. v may be linearly related to i,
2. v may be the time integral of i or vice versa,
3. v may be the time derivative of i or vice versa,
4. v may be proportional to the square of i, etc.,
5. the roles of v and i may be interchanged in the above.

These possibilities for specific electrical elements will be explored fully in Chapter 4.

The interaction and interrelationships among various signals such as those identified as input and output in Figure 1.12 are governed by two sets of fundamental constraints. One set is the *voltage–current relationships* we have just mentioned. The other is the *Kirchhoff laws* which we will now discuss briefly and explore in depth in Chapter 5. These laws were first stated by Gustav Kirchhoff in 1848 and are the consequences of the laws of conservation of energy and

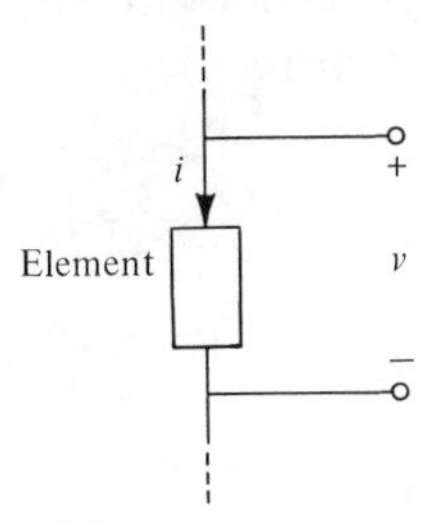

Figure 1.13 Voltage and current reference directions for any electric element.

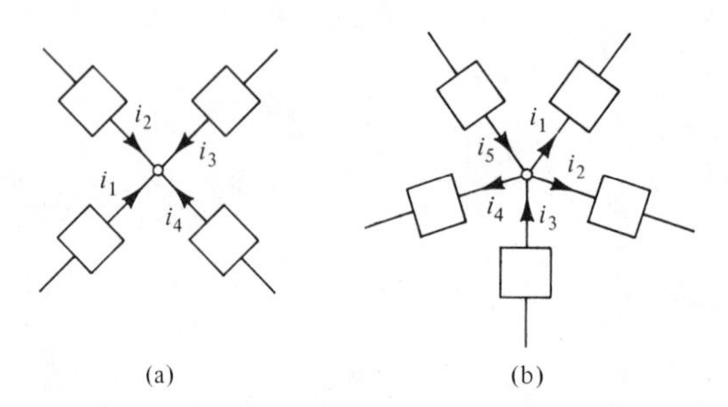

Figure 1.14 Four branches in (a) and five in (b) are connected at a node as shown. Kirchhoff's current law is applied at the node.

conservation of charge. The *Kirchhoff current law* states that the sum of currents at a *node* (or junction) in a network must equal zero, meaning that the current entering must exactly equal the current leaving for each instant of time, for no current may be stored nor lost in the node if charge is conserved. If reference directions are assigned for each branch of the network, as in Figure 1.14(a), then it is necessary that

$$-i_1 - i_2 - i_3 - i_4 = 0. \tag{1.16}$$

The particular form of this equation is arrived at by assigning a positive sign to currents with a reference direction leaving the node and a negative sign for those entering it. (The arbitrariness of this choice is seen from the fact that Equation 1.16 may be multiplied by -1 to change all signs to be positive. It is important that some choice be made and then used consistently!) For the network of Figure 1.14(b),

$$i_1 + i_2 - i_3 + i_4 - i_5 = 0. \tag{1.17}$$

At any instant of time, the currents will be either in the reference direction or in the direction opposite to the reference direction. Hence they may be represented by either a positive number or negative number, respectively, in such a way that the law is satisfied. For any number of branches j, the Kirchhoff current law says simply that at every node

$$\sum i_j = 0, \tag{1.18}$$

where a positive sign is associated with i_j for a reference direction out of the node, negative otherwise.

To introduce the Kirchhoff voltage law, we will make use of the network, only partially shown, of Figure 1.15(a). Here the voltage references are assigned, and a path around the loop from node a to node b to c to d and return to a is identified.

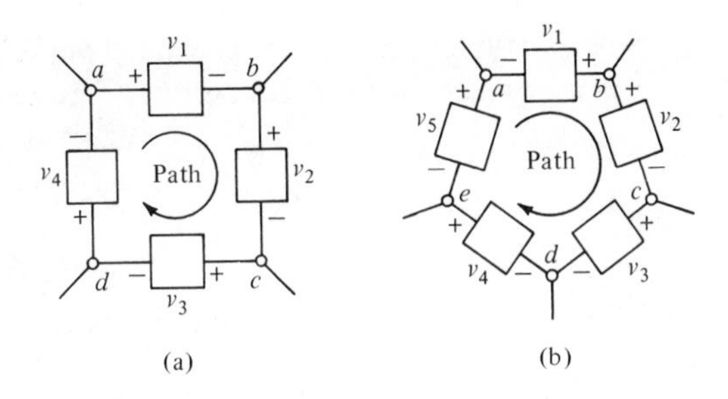

Figure 1.15 Closed paths through the elements as well as the voltages of each element are identified. Kirchhoff's voltage law is applied around the loop.

The Kirchhoff voltage law states that around any such closed path the voltages of the elements in that path must sum to zero. If we assign a positive sign to the reference direction from + to − as the loop is traversed and a negative sign for the direction − to +, then

$$v_1 + v_2 + v_3 + v_4 = 0\,. \tag{1.18a}$$

The assignment of reference directions need not be uniform with respect to the path traversed in closing the loop. In Figure 1.15(b), for example, the Kirchhoff voltage law requires that

$$-v_1 + v_2 + v_3 - v_4 - v_5 = 0\,. \tag{1.19}$$

So much for the time being on the mechanics of writing the equations; it will become clear and routine only with considerable experience. The justification of the law stems from the law of conservation of energy. If we take a point charge in an electric field and move it from node a to node b, then energy will be either lost or gained. If we move the charge around any path and eventually return to the point of origination, the energy lost must exactly equal the energy gained since energy is conserved. Now energy per unit charge is voltage. As long as we use a proper reference scheme to indicate energy lost or gained, we can apply the law by simply summing the voltages around the closed path. If we traverse the path in the direction we have selected, then a positive sign is indicated for voltage if we encounter the + and then the −, as we have illustrated in Equation 1.18(a); the − to + sequence results in assigning a negative sign. In writing Equation 1.18(a) for the network of Figure 1.15(a), all reference directions were such that the signs of the voltages were positive. At each instant of time, the element voltages will agree with the reference directions or be reversed with respect to them in such a way that the algebraic sum is zero.

It should be noted that these laws are analogous to Newton's and D'Alembert's laws in mechanics which require that the algebraic sum of all forces on a body be zero. Kirchhoff's laws govern the equilibrium of voltages and currents in a network just as Newton's laws govern the equilibrium of mechanical forces and velocities.

EXERCISES

1.3.1 An idealized version of a circuit known as a full-wave rectifier has the following equilibrium equations:

$$v_2(t) = \begin{cases} v_{\text{in}}(t), & \text{for } v_{\text{in}}(t) \geq 0, \\ -v_{\text{in}}(t), & \text{for } v_{\text{in}}(t) < 0, \end{cases}$$

$$v_{\text{out}}(t) = \frac{1}{T}\int_0^T v_2(\tau)\, d\tau, \qquad T = \text{constant}.$$

Express the output signal in terms of the input signal.

1.3.2 A simple network is described by the following set of equations:

$$av_1(t) + bv_2(t) = v_3(t),$$
$$cv_1(t) + dv_2(t) = v_4(t).$$

Where $v_3(t)$ and $v_4(t)$ are input signals, $v_1(t)$ is the output signal, and a, b, c, and d are constants such that $ad - bc \neq 0$. Express the output signal in terms of the input signals.

1.3.3 The equations describing a network are as follows:

(a) $L\dfrac{d}{dt}i(t) + Ri(t) + \dfrac{1}{C}\displaystyle\int_0^t i(\tau)\, d\tau = v_1(t),$

(b) $v_2(t) = \dfrac{1}{C}\displaystyle\int_0^t i(\tau)\, d\tau,$

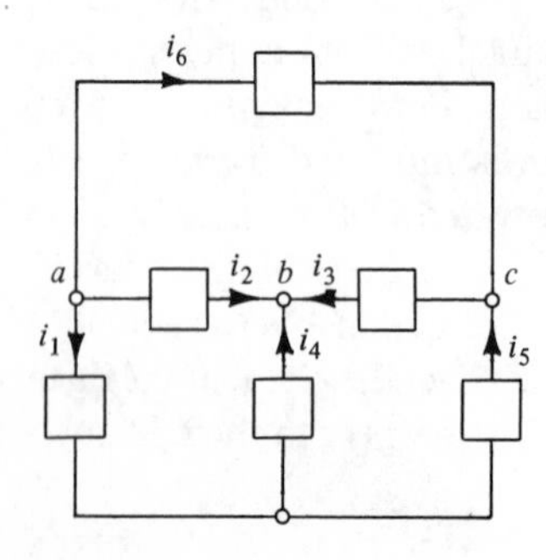

Figure 1.16

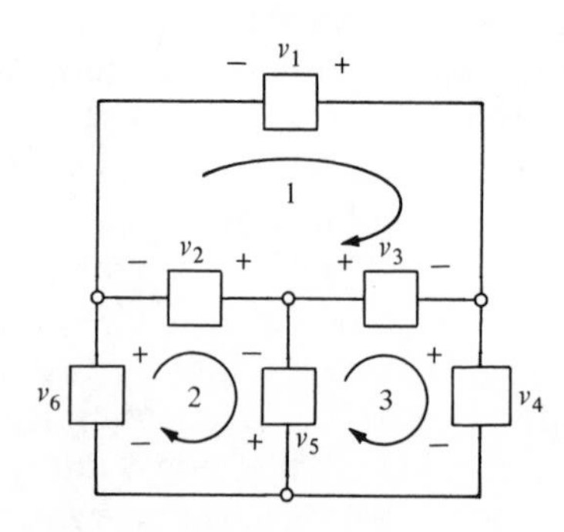

Figure 1.17

where $v_1(t)$ is the input, and $v_2(t)$ is the output. Obtain a differential equation relating $v_1(t)$ to $v_2(t)$.

1.3.4 For the network shown in Figure 1.16, apply the Kirchhoff current law to each labeled node and thereby write a set of equations for the currents of the network.

1.3.5 For the network of Figure 1.17, apply the Kirchhoff voltage law to each loop that is identified, thus obtaining a set of equations for the voltages of the network.

.4 Summary

1. The standard symbol for the independent voltage source is a circle with a + and a − inside indicating reference directions. Here v_1 is maintained for all values of current from the source. The standard symbol for the independent current source is a circle with an arrow inside indicating the reference direction. Here i_1 is maintained for all values of terminal voltage from the source (Figure 1.18).
2. The double-subscript notation may be used in lieu of polarity marks if, in v_{jk}, j is identified with the positive terminal and k with the negative, and in i_{jk}, j is the "into" terminal and k is the "out of" terminal (Figure 1.19).
3. Network descriptions also require dependent sources in which v or i is a function of some other v or i in the network. Such sources are required to model or represent many solid-state devices, for example, and are distinguished by a diamond shape.
4. Given a signal waveform $v(t)$, then $v(t - T)$ has the same shape as $v(t)$ but is shifted T units of time to the right for T positive or to the left for T negative.
5. (a) The Kirchhoff voltage law (KVL) states that the sum of voltages of the branches around a closed path (loop) is zero if each voltage v_j is given a sign determined by the order of the + and − polarity marks with respect to the direction of traversal. Namely a + sign is associated with v_j when the + is encountered before the − and a − sign when the − precedes the +.

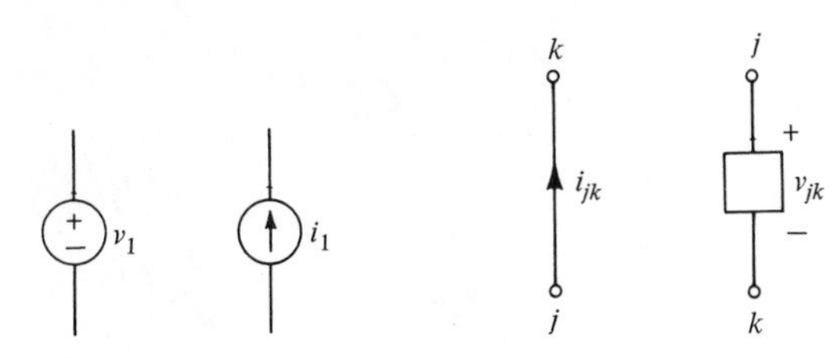

Figure 1.18

Figure 1.19

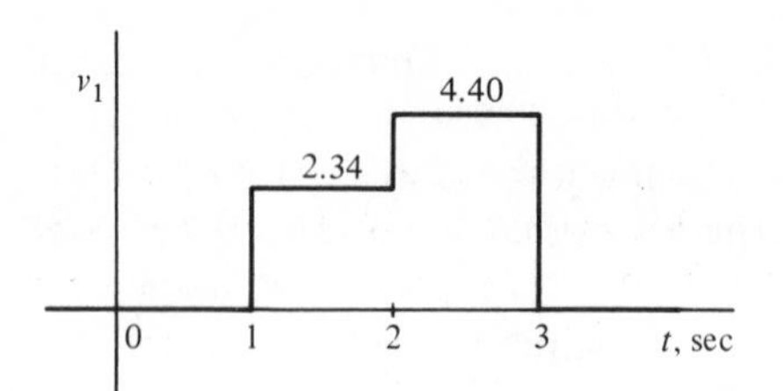

Figure 1.20

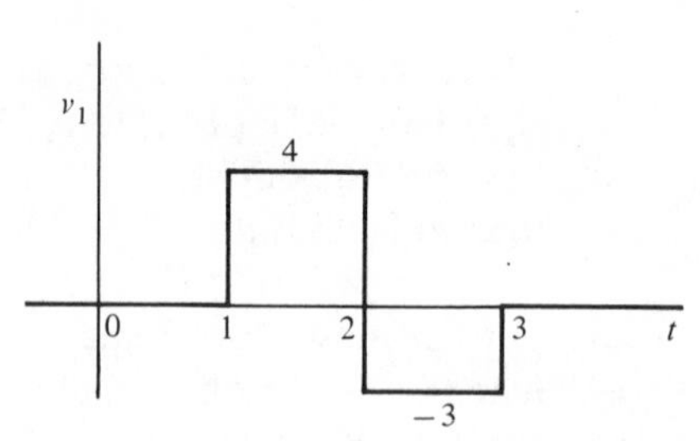

Figure 1.21

(b) The Kirchhoff current law (KCL) states that the sum of currents associated with the branches at a node is zero if each current i_j has a sign determined by the reference direction assigned to the current. Namely, a + sign is associated with i_j if the reference direction is out of the node and a − sign if directed into the node.

PROBLEMS

1.1 Suppose that the source labeled as v_1 in Figure 1.5(a) is such that $v_{kj}(t) = \cos 100t$. Determine v_1 at $t = 0$.

1.2 A system is characterized by the equations

$$v(t) = \int_0^t v_1(\tau)\, d\tau, \qquad v_2(t) = 0.02v(t).$$

If the input $v_1(t)$ is the staircase function shown in Figure 1.20, sketch the output waveform $v_2(t)$.

1.3 Repeat Problem 1.2 for the waveform of Figure 1.21.

1.4 If $v_1(t) = V_1 \sin \omega t$ is the signal passed through a device with an input-output characteristic like that of Figure 1.22, plot the output, $v_2(t)$. This model represents some rectifier devices.

1.5 Repeat Problem 1.4 for the different input-output characteristic of Figure 1.23.

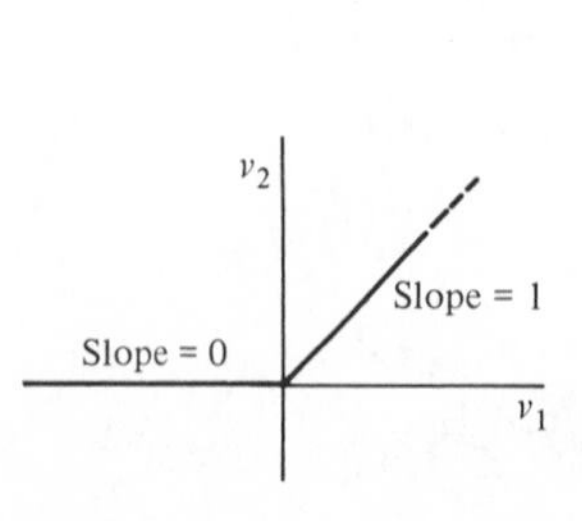

Figure 1.22

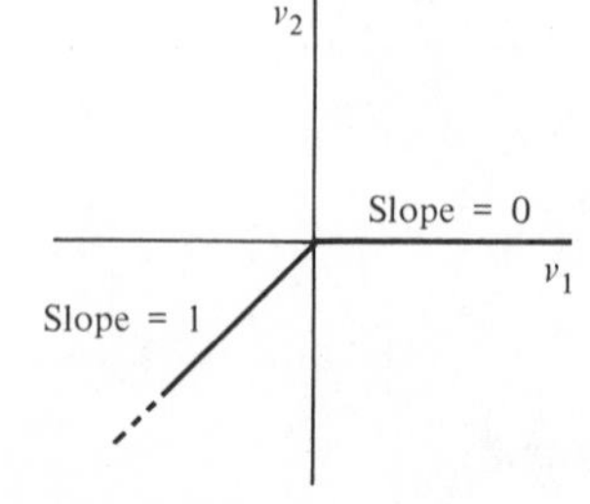

Figure 1.23

v_2
Slope = 1
v_1
Slope = 0.1

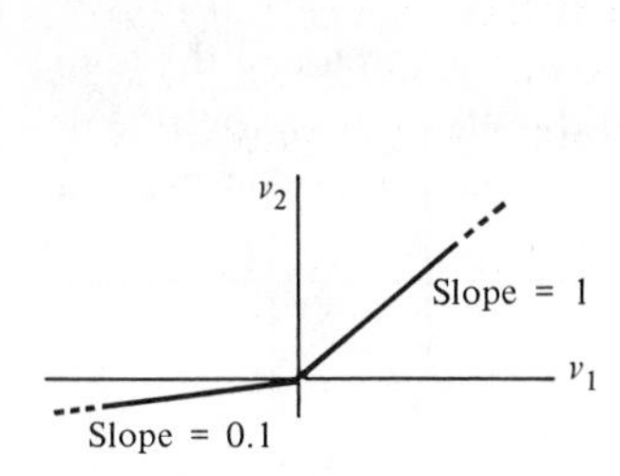

Figure 1.24

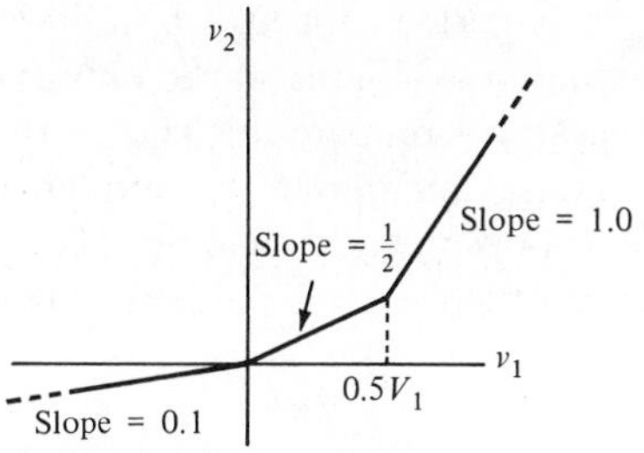

v_2
Slope = 1.0
Slope = $\frac{1}{2}$
v_1
$0.5V_1$
Slope = 0.1

Figure 1.25

1.6 Repeat Problem 1.4 for the different input-output characteristic of Figure 1.24.

1.7 Repeat Problem 1.4 for the different input-output characteristic of Figure 1.25.

1.8 If $i(t)$ is the integral in Exercise 1.2.8, calculate di/dt. Is di/dt equal to $v(t)$? The result of this exercise is an example of a more general rule for differentiating an integral with respect to the upper limit (check your calculus book). Thus we have

$$\frac{d}{dt}\int_0^t v(\tau)\,d\tau = v(t),$$

for *any* integrable $v(t)$. This result is important in circuit theory and will come up often in future discussions.

1.9 Let $v_1(t)$ be the signal in Figure 1.9(a) and define

$$v(t) = v_1(t) + v_1(2T_0 - t).$$

Calculate

$$v_0(t) = \int_{-\infty}^t v(x)\,dx.$$

v_2
Slope = 0
1
Slope = 1
v_1
1

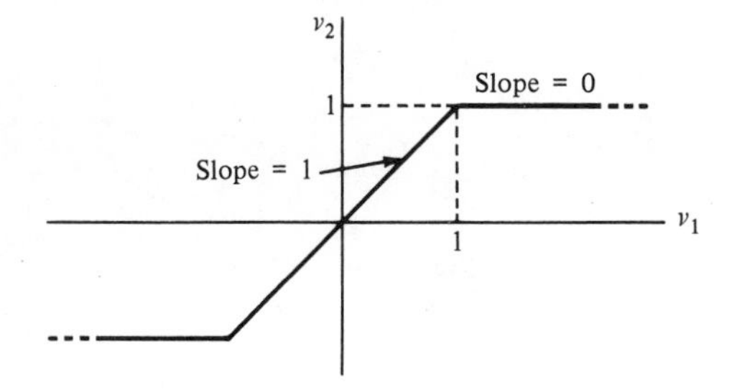

Figure 1.26

1.10 A signal is passsed through a limiter (which prevents the output from exceeding a fixed value) and the output of the limiter is then passed through an integrator. Compare this system to that where the order of operations of limiting and integration are interchanged. Substantiate your comparison with appropriate equations. See Figure 1.26.

1.11 Repeat Problem 1.10 if the limiter is replaced by a differentiator.

Chapter 2

Signal Waveforms

For both analysis and design, it is necessary that waveforms produced by generators be represented by mathematical *abstractions* (formulas). These formulas are approximations to the actual waveforms in the same sense that models are idealizations of actual generators. In this chapter, we will study waveforms of importance in electrical engineering. As we will see, many of these waveforms may be expressed as linear combinations of exponential functions such as*

$$v(t) = 3e^{-t} + 2e^{-3t} + 5e^{jt} + 5e^{-jt}. \tag{2.1}$$

This equation may be written in the compact form

$$v(t) = \sum_{k=0}^{n} a_k e^{s_k t}. \tag{2.2}$$

The $s_0, s_1, \ldots$ may be real, imaginary, or complex.

2.1 The exponential function

The function

$$v(t) = Ke^{\sigma t} \tag{2.3}$$

with σ negative real is known as an *exponentially decreasing waveform* and an *exponentially increasing waveform* for σ positive real. The exponential waveform is shown in Figure 2.1. We define $T = 1/|\sigma|$, $\sigma < 0$, as the *time constant* of $v(t)$ in Equation 2.3. Observe from Equation 2.3 that for $\sigma < 0$, $v(T) = K/e$ which is

* The j in this equation is the same as the i used by mathematicians and is equal to $\sqrt{-1}$. This is to avoid confusion with i as a symbol for current.

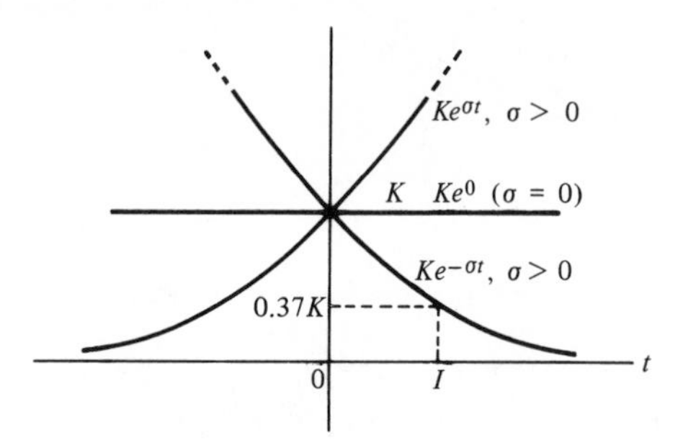

Figure 2.1 Exponential signal waveform for $\sigma > 0$, $\sigma = 0$, and $\sigma < 0$.

approximately $0.37K$. Thus in one time constant, the exponential signal decays to 37 percent of its $t = 0$ or *initial value*. For $\sigma < 0$, $|\sigma|$ is also known as *damping*.

The exponential signal has a number of interesting properties. For example, the derivative of $v(t)$ in Equation 2.3 is

$$\frac{d}{dt} Ke^{\sigma t} = \sigma K e^{\sigma t}, \tag{2.4}$$

which is the same waveform as v except for a scale factor. Similarly, if we integrate v from t_1 to t, we have

$$\int_{t_1}^{t} Ke^{\sigma\tau}\, d\tau = \frac{K}{\sigma} e^{\sigma t} + K_1 . \tag{2.5}$$

Thus the passage of an exponential signal through an amplifier, integrator, or differentiator changes only the signal magnitude, except for a possible additional constant corresponding to the initial value of the integrator output. For example, if a system is characterized by the equation

$$v_3 = \frac{3dv_1}{dt} - 2v_1, \tag{2.6}$$

where v_1 is the input and v_3 is the output, then for a signal $v_1 = e^{-t}$ at the input, the output is found to be

$$v_3 = -3e^{-t} - 2e^{-t} = -5e^{-t}. \tag{2.7}$$

Consider next the equation

$$i = 3e^{-t} + 10e^{-2t}, \qquad t > 0, \tag{2.8}$$

which is a *linear combination* of exponentials. Observe that the time constant of the first term is 2 times longer than that of the second term. A plot of the total signal has the characteristic shape shown in Figure 2.2. Such waveforms are encountered in the electronic recording of the radiation from radioactive materials in which two decay processes are present.

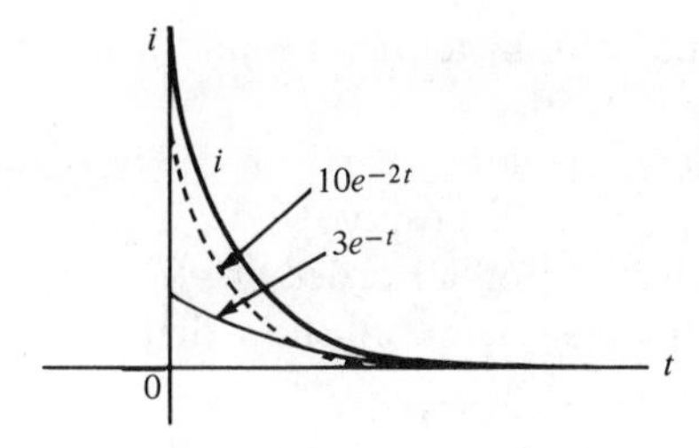

Figure 2.2 A signal waveform composed of two exponential terms.

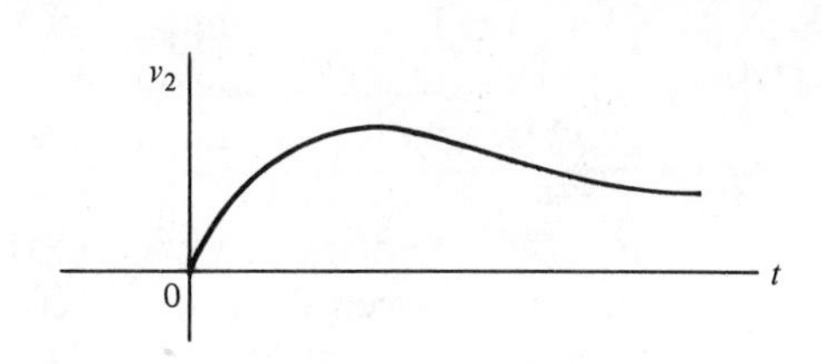

Figure 2.3 A linear combination of exponential signals with a "long tail."

Another linear combination of exponentials,

$$v_2 = \tfrac{1}{2} - e^{-t} + \tfrac{1}{2}e^{-0.1t}, \qquad t > 0, \tag{2.9}$$

has a characteristic form illustrated by Figure 2.3. Observe that the signal waveform begins at zero value, rapidly assumes its maximum value, and then decays slowly. This slow decay, described by the colorful name *long tail*, is due to the third term in Equation 2.9, or the long duration component of the signal. The final value of the signal, $v_2(\infty)$, is $\frac{1}{2}$.

EXERCISES

2.1.1 A signal is described by the equation

$$v_1(t) = 5e^{-t} - 5e^{-3t}.$$

Does this signal have a long tail like Equation 2.9? Sketch v_1 for the range, $0 \leq t \leq 3$ sec.

2.1.2 The signal of Exercise 2.1.1 enters the system described by Equation 2.6. Determine the output of this system.

2.1.3 A signal of the form $v(t) = K(e^{-\alpha t} - e^{-\beta t})$ enters the limiter shown in

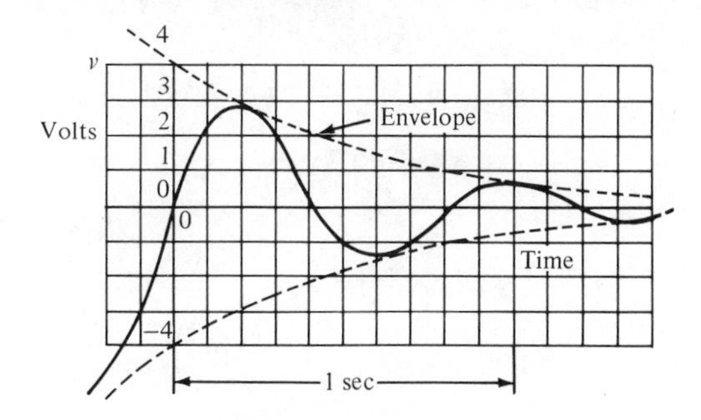

Figure 2.4

Figure 1.26 (p. 19). Find the largest value of K in terms of α and β for which the input and output waveforms will be identical.

2.1.4 (a) Show that the plot of an exponential function like $v(t) = Ke^{\sigma t}$ on semilogarithmic coordinates is a straight line for K positive and σ either positive or negative. (b) Show that derivatives of all order of $v(t)$ are straight lines parallel to $v(t)$ in this coordinate system. (c) Show that the spacing between the lines described in (b) is constant and determine the value of the constant.

2.1.5 The signal in Figure 2.4 may be approximated by the equation $v(t) = Ke^{-\sigma t} \sin bt$. Determine appropriate numerical values for K, σ, and b.

2.2 The step, ramp, and impulse

At time $t = 0$, the reference time at which we usually begin our experiments, the switch shown in Figure 2.5 is thrown from position 1 to position 2. This ideal switch acts in zero time connecting a 1-volt time-invariant source to an ideal recording meter (a model for instruments like the voltmeter, oscillograph, recording penmotor, etc.) which records or indicates instantaneous values. The waveform produced by the switching action is known as a *unit step function*, having been so named by the English engineer Oliver Heaviside (1850–1925). The *step* nature of the waveform is shown in Figure 2.6. We represent this step function by the symbol $u(t)$, where we have

$$u(t) = \begin{cases} 1, & \text{for} \quad t \geq 0, \\ 0, & \text{for} \quad t < 0. \end{cases} \tag{2.10}$$

If the 1-volt source is replaced by a source of K volts, then clearly the new voltage is $Ku(t)$.

Let the unit step voltage $u(t)$ be applied at the input of an ideal integrator. The output voltage v_2 is

$$v_2(t) = \int_0^t 1\, dx + v_2(0), \qquad \text{for} \quad t \geq 0. \tag{2.11}$$

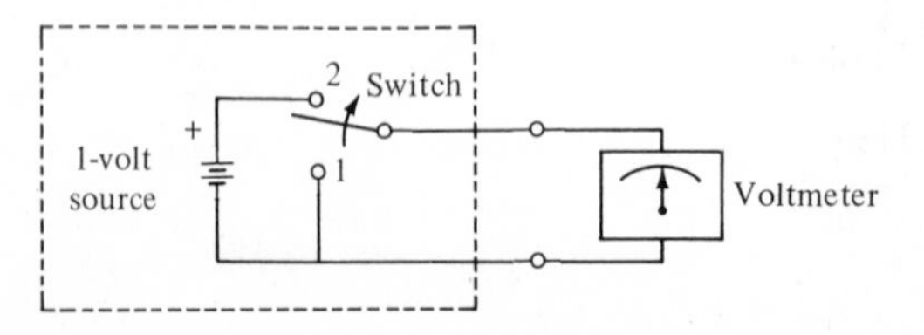

Figure 2.5 A simple step function generator.

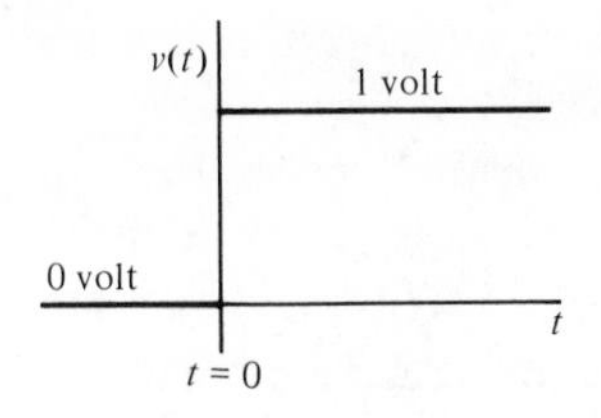

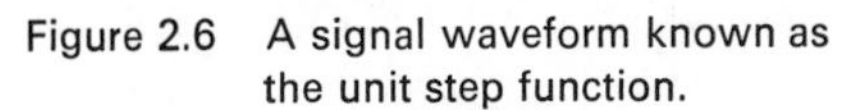

Figure 2.6 A signal waveform known as the unit step function.

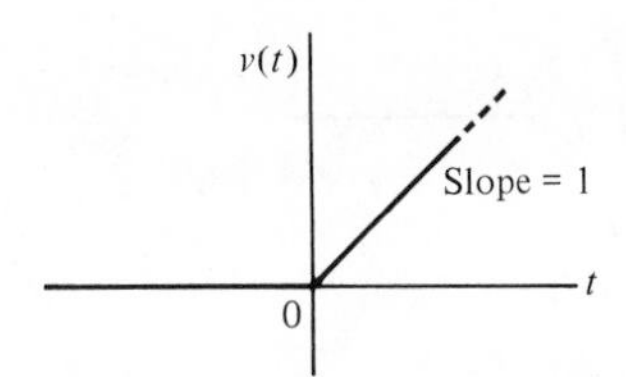

Figure 2.7 A signal waveform identified as the unit ramp function.

If the initial condition $v_2(0)$ at the integrator output is zero and if there is no other input before $t = 0$, then $v_2(t)$ is zero for negative time. The output waveform v_2 is described by the equation

$$v_2(t) = \begin{Bmatrix} t, & \text{for} \quad t \geq 0 \\ 0, & \text{for} \quad t < 0 \end{Bmatrix} \equiv r(t). \tag{2.12}$$

Because of the shape of this response, shown in Figure 2.7, $r(t)$ is known as a *ramp function*. When the ramp function has unit slope, it is known as a *unit ramp*. Should we both integrate and also multiply by a constant K, then for $t \geq 0$, $v_2 = Kt = Kr(t)$, which is described as a ramp of slope K. Observe that the slope of the ramp equals the magnitude of the step. In summary, we see that

$$\int_{-\infty}^{t} u(x)\, dx = r(t) \qquad \text{and} \qquad \frac{d}{dt} r(t) = u(t). \tag{2.13}$$

Notice that we are using the symbol $-\infty$ (minus infinity) to denote integration over the entire past history of the signal.

We next turn to a discussion of the derivative of the unit step function. The step function is discontinuous at $t = 0$, and the derivative is therefore not ordinarily defined at that point. A function generated by passing the unit step function through an ideal differentiator is known as a unit impulse function or a unit Dirac delta function, $\delta(t)$. To follow the philosophy of Equation 2.13, we require that

$$\frac{d}{dt} u(t) = \delta(t) \qquad \text{and} \qquad \int_{-\infty}^{t} \delta(x)\, dx = u(t), \tag{2.14}$$

or simply that when we integrate and then differentiate that we get back to where we started. The plausibility of the integration of Equation 2.14 being reasonable is suggested by Figure 2.8(a) which shows a waveform similar to the step function but with a finite slope at its "leading edge." The *pulse* waveform obtained by differentiating the modified step is shown in Figure 2.8(b). This pulse has width a and height $1/a$ such that the area of the pulse is $a \times 1/a = 1$. In the limit as a approaches zero, the height of the pulse approaches infinity and the width

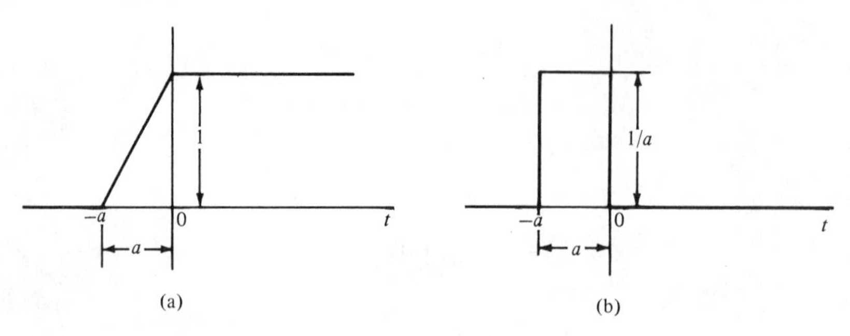

Figure 2.8 Waveforms which are useful in describing the impulse function as a limit of another waveform.

approaches zero, but the area remains equal to 1.* Several steps in approaching this limit are shown in Figure 2.9. Starting with the pulse v_1, we halve the width and double the height thus obtaining v_2, repeat to get v_3, and continue this process observing that as the width approaches zero, the height approaches an infinite value keeping the area under the pulse unity. Since the integral of Equation 2.14 is the area of $\delta(t)$, we see that a property of the unit impulse is

$$\int_{-\infty}^{t} \delta(x)\, dx = \begin{cases} 1, & t \geq 0, \\ 0, & t < 0. \end{cases} \tag{2.15}$$

More generally, $\delta(x)$ has the following defining properties:

(a) $$\delta(x) = 0, \quad \text{for} \quad x \neq 0; \tag{2.16}$$

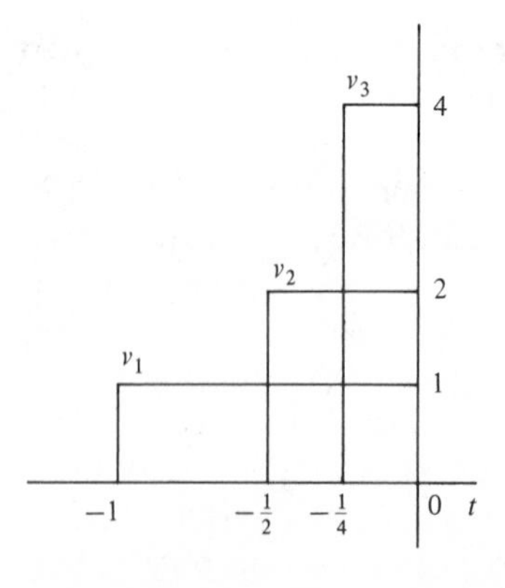

Figure 2.9 A set of pulse functions whose limit defines the impulse function.

* This heuristic argument is supported by reference to the properties of Schwarz distributions in advanced treatises.

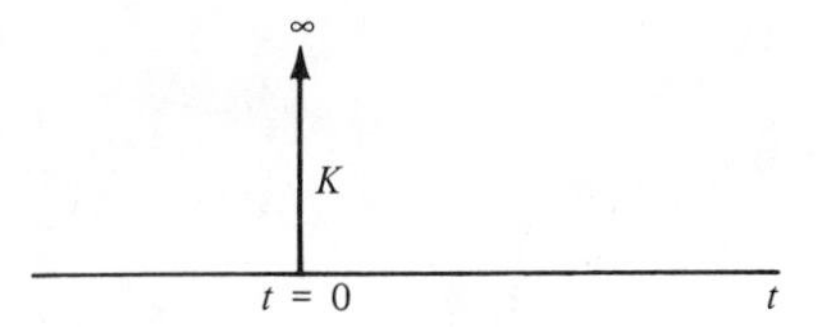

Figure 2.10 Waveform representation of a unit-impulse function $K\delta(t)$. We sometimes draw the arrow length proportional to K.

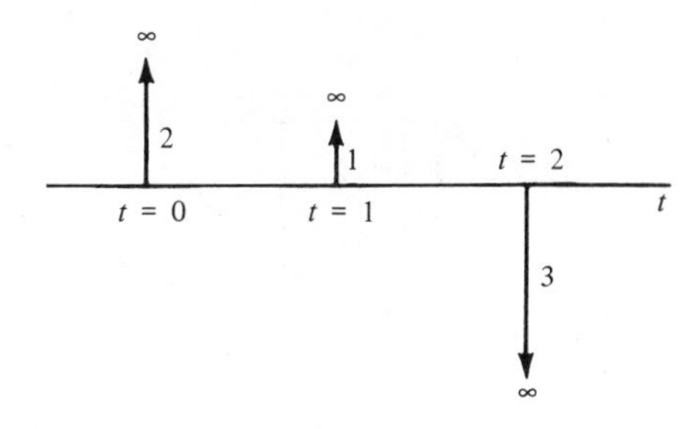

Figure 2.11 A train of impulse functions.

(b) for any function $f(x)$ which is continuous at $x = 0$, we have

$$\int_b^a f(x)\delta(x)\,dx = f(0), \qquad \text{for} \quad a \geq 0 \quad \text{and} \quad b < 0. \tag{2.17}$$

The sampling property expressed in Equation 2.17 states that if a continuous function $f(x)$ is multiplied by a delta function $\delta(x)$ and integrated over a range which includes $x = 0$, the value of the integral is $f(0)$, which is a sample of $f(x)$ at $x = 0$. For example, if $f(x)$ is $\sin(x + \pi/4)$, $a = -1$, and $b = 1$, then the value of the integral in Equation 2.16 is $\sin(\pi/4) = 0.707$.

If we differentiate $Ku(t)$ rather than $u(t)$, then we obtain $K\delta(t)$, where K represents the area under the curve of the impulse function. In order to represent the impulse function and the corresponding value of K, the symbol shown in Figure 2.10 is used. The number beside the arrow and the arrow length represent K; a negatively directed arrow implies a negative value of K. Using this convention, we see that a number of impulses may be conveniently represented as shown in Figure 2.11. Impulses, steps, ramps, etc., may be shifted away from the origin. For the signal in Figure 2.11, the equation for $v(t)$ is

$$v(t) = 2\delta(t) + \delta(t - 1) - 3\delta(t - 2).$$

From the defining properties of the delta function we see that the integral of $\delta(x - T)$ is $u(x - T)$, that is, the integral of a delayed impulse is a delayed step with the same delay T. The height of the step is equal to the area of the impulse. For example, the integral of $K\delta(x - T)$ is $Ku(x - T)$ where K is an arbitrary real number. With this in mind, the integral of the signal in Figure 2.10 is

$$v_1(t) = 2u(t) + u(t - 1) - 3u(t - 2).$$

This integral is shown in Figure 2.12. Notice the jumps of height 2 at $t = 0$, height 1 at $t = 1$, and height -3 at $t = 2$. If the signal in Figure 2.12 is passed through a differentiator, the signal in Figure 2.11 is obtained. At every jump discontinuity of $v_1(t)$, $v(t)$ has an impulse whose area equals the height of the jump of $v_1(t)$. This result is worth remembering, and it applies to the derivative

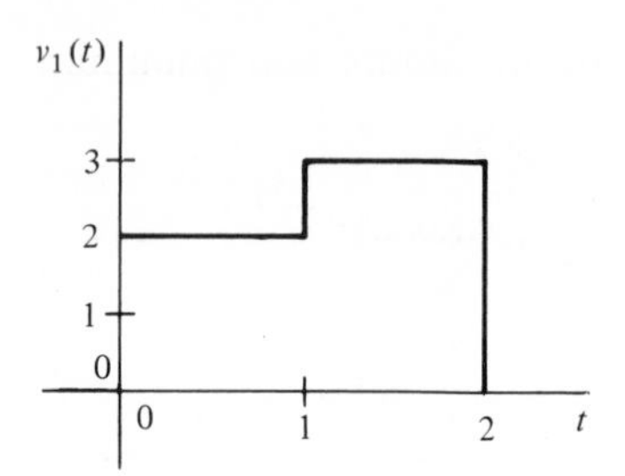

Figure 2.12 The integral of the signal of Figure 2.11.

of any function which has finite jump discontinuities but otherwise differentiable in the ordinary sense. The generalized derivative will contain an impulse at every point of discontinuity of the original function, and the area of the impulse will equal the height of the jump of the original function.

The step, ramp, and impulse are, of course, *abstractions* of signal waveforms produced by practical generators. The impulse function is an approximation for pulses of relatively short duration of large magnitude. The most common form of electronic ramp generator produces a repetitive ramp or a "sawtooth."

EXERCISES

2.2.1 If the input v_1 has the dimension of volts and if t has the dimension time, what are the units of the output v_2 for each of the following: (a) ideal integrator; (b) ideal differentiator.

2.2.2 The ramp function shown in Figure 2.13 is applied to an ideal differentiator. Find the resulting $v_2(t)$ in the form of a sketch.

2.2.3 A step function of amplitude 3 volts as shown in Figure 2.14 is applied to an ideal differentiator. Describe $v_2(t)$.

2.2.4 The signal shown in Figure 1.11(a) (p. 11) is the input v_1 for the system described by

$$v_2(t) = -v_1(t) + \int_0^t v_1(\tau)\, d\tau .$$

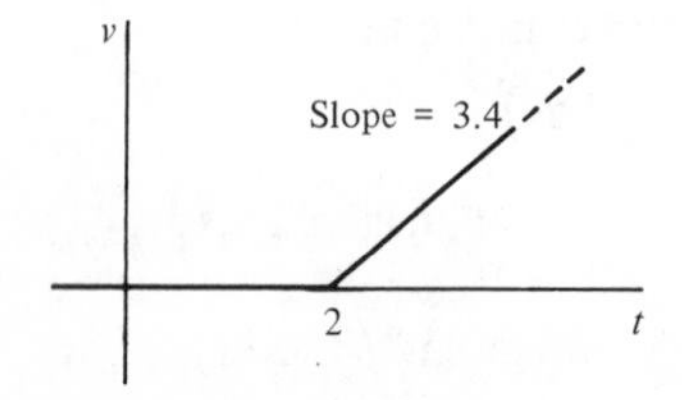

Figure 2.13

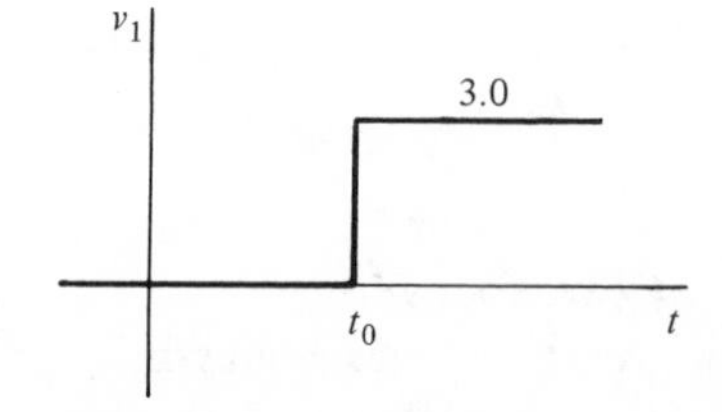

Figure 2.14

Draw the signal waveform of $v_2(t)$ indicating maximum and minimum signal values.

2.2.5 Figure 2.11 shows a *train* of impulses. Find the output that results when this impulse train is applied to an ideal integrator. Sketch $v_2(t)$.

2.2.6 The waveform shown in Figure 1.20 (p. 18) is the input to an ideal differentiator. Find the corresponding output.

2.2.7 The American electrical engineer E. A. Guillemin (1898–1970) uses the symbol $u_0(t)$ for a unit impulse and then describes other singular functions through the recursion relationship

$$u_{n+1}(t) = \frac{du_n(t)}{dt}.$$

By this system, what symbol should be used to describe a unit step? A unit ramp?

3 The sinusoid

The circular functions sine and cosine are familiar from studies in trigonometry as the vertical and horizontal projections of a point on a unit circle as functions of the angular displacement of a line through this point and the origin. The corresponding sinusoidal time-varying signal is

$$v_1(t) = V_1 \sin \omega t, \tag{2.18}$$

where ω is the constant of proportionality between time displacement and angular displacement (a scale factor). This constant is defined as the *frequency* of $v_1(t)$. Here V_1 is a real, positive number which is the *maximum magnitude* of the sine wave. The sine wave is shown in Figure 2.15. If the sinusoidal signal is shifted such

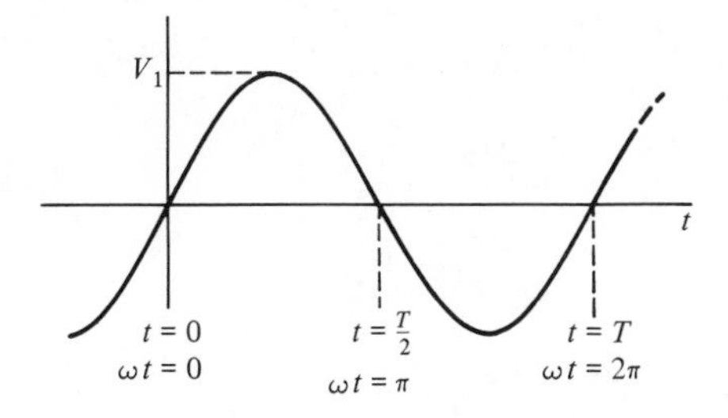

ure 2.15 A sinusoidal signal of amplitude V_1, frequency ω, and period T.

that it does not have zero value at $t = 0$, this shift is described by a *phase angle* ϕ in the equation

$$v_2(t) = V_2 \sin(\omega t + \phi_2), \tag{2.19}$$

or in the alternative representation

$$v_2(t) = V_c \cos(\omega t + \phi_3). \tag{2.20}$$

In comparing Equation 2.19 with Equation 2.18, we say that v_2 *leads* v_1 if ϕ_2 is positive or *lags* v_1 for negative ϕ_2. Shifted sine waves are shown in Figure 2.16.

Since the sine and cosine are related by identities like

$$\cos(\theta - \tfrac{1}{2}\pi) = \sin\theta \tag{2.21}$$

and

$$\sin(\theta + \tfrac{1}{2}\pi) = \cos\theta, \tag{2.22}$$

it is evident that a sine function can be expressed by an equivalent cosine function with a different phase angle and *vice versa*. We use the term *sinusoid* to mean a signal waveform which may be expressed as either a sine or cosine function with any phase angle.

Sinusoids may be expressed in exponential form. Thus* we have

$$\sin\omega t = \frac{1}{2j}(e^{j\omega t} - e^{-j\omega t}) \tag{2.23}$$

and

$$\cos\omega t = \tfrac{1}{2}(e^{j\omega t} + e^{-j\omega t}). \tag{2.24}$$

The unit of the quantity ωt in the equations of this section has the dimension of *radians*, although it is common in electrical engineering to use the older unit, the *degree* (which dates back to the Babylonians in 2000 B.C.). When ωt is in radians and t is in seconds, the unit for the frequency ω is the radian/sec.

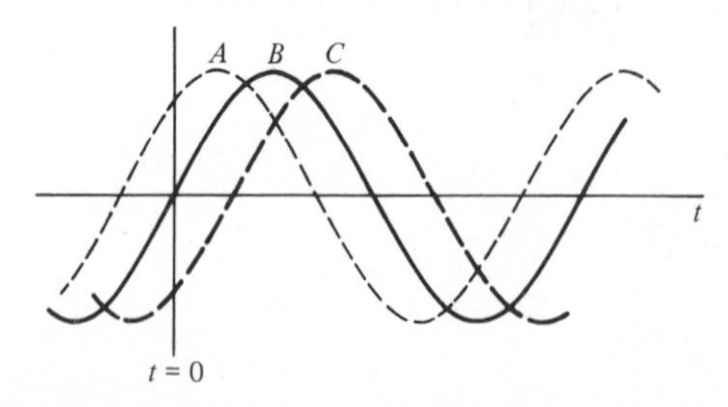

Figure 2.16 With sinusoid B as the reference, A leads B and C lags B.

* These two equations follow from the familiar *Euler formula* $e^{\pm jx} = \cos x \pm j \sin x$.

We next seek the smallest positive number T such that

$$\sin \omega(t + T) = \sin \omega t. \tag{2.25}$$

By an inspection of Figure 2.15, it is clear that if we shift the curve (either to the left or to the right) by an amount $\omega t = \omega T = 2\pi$, the new curve will coincide with the old. No amount of shift less than 2π will cause such a complete overlapping. Alternatively, we may expand the left-hand side of Equation 2.25 to get

$$\sin \omega t \cos \omega T + \cos \omega t \sin \omega T = \sin \omega t. \tag{2.26}$$

For this equation to hold for all values of t, $\cos \omega T$ must be equal to 1 and $\sin \omega T = 0$. These conditions require that

$$T = \frac{2k\pi}{\omega}, \qquad k = \text{any integer}. \tag{2.27}$$

The smallest positive value of T corresponds to $k = 1$ which agrees with our previous discussion. This value is known as the *period* of the sine wave. During T seconds, *one cycle* of the sine wave $\sin [(2\pi/T)t]$ is described, and another cycle during the subsequent T sec. The number of cycles of sine wave per second is seen to be $1/T$, and since $\omega T = 2\pi$, we have

$$\frac{1}{T} = \frac{\omega}{2\pi} \equiv f, \tag{2.28}$$

where f is the frequency in the unit of cycles/sec or Hz (hertz). Other related units for f are kilohertz, megahertz, and gigahertz (10^3, 10^6, and 10^9 cycles/sec or Hz).

What happens to a sinusoidal signal when it passes through an ideal differentiator or ideal integrator? Since

$$\frac{d}{dt} \sin \omega t = \omega \cos \omega t = \omega \sin \left(\omega t + \frac{\pi}{2}\right) \tag{2.29}$$

and

$$\int_{t_1}^{t} \sin \omega\tau \, d\tau = -\frac{1}{\omega} \cos \omega t + \frac{1}{\omega} \cos \omega t_1 = \frac{1}{\omega} \sin \left(\omega t - \frac{\pi}{2}\right) + K, \tag{2.30}$$

we see that except for a possible integrator initial condition, these operations *change the magnitude and phase of the sinusoid but do not change its frequency.* We shall exploit this important conclusion in later chapters in the study of networks excited exclusively by sinusoidal signals.

Another signal waveform of importance is the *damped sinusoid* which is given by the equation

$$v_3(t) = Ae^{-\sigma t} \sin \omega t, \qquad \sigma > 0, \tag{2.31}$$

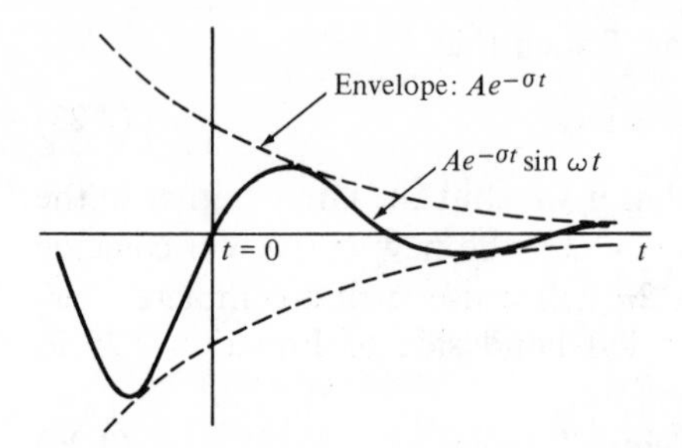

Figure 2.17 A damped sinusoidal signal and its envelope.

where A is a real, positive constant. Expressing the sine function in exponential form and combining with $e^{-\sigma t}$, we have

$$v_3(t) = \frac{A}{2j}(e^{(-\sigma+j\omega)t} - e^{(-\sigma-j\omega)t}). \tag{2.32}$$

From this form of the equation, we may interpret the damped sinusoid as a special linear combination of exponentials with *complex* time constants.

This waveform is shown in Figure 2.17. Observe that the function $\pm Ae^{-\sigma t}$ is the bound of the oscillations and is therefore known as the *envelope* of the waveform. The time constant of the envelope is $T = 1/\sigma$.

The product of two sine waves of different frequencies like

$$v_4(t) = A \sin \omega_1 t \sin \omega_2 t \tag{2.33}$$

is another signal waveform of interest which arises in the study of modulation. Using the identity

$$2 \sin \omega_1 t \sin \omega_2 t = \cos (\omega_1 - \omega_2)t - \cos (\omega_1 + \omega_2)t, \tag{2.34}$$

we see that v_4 is equivalent to the sum of two sinusoidal functions whose fre-

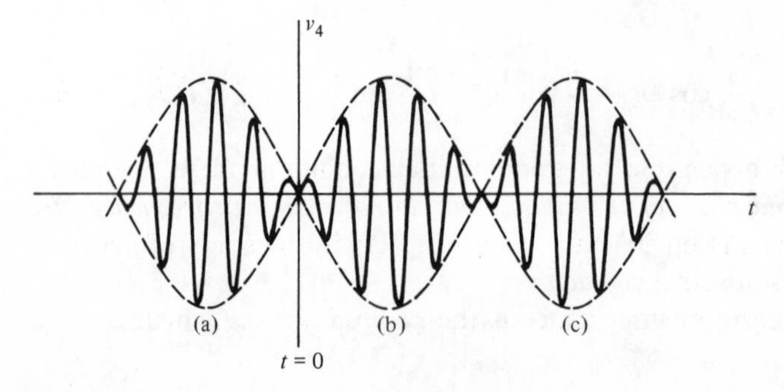

Figure 2.18 A segment of a modulated signal waveform.

quencies are the sum and difference of ω_1 and ω_2. A typical signal of this form is shown in Figure 2.18.

ERCISES

2.3.1 Figure 2.19 shows three sinusoidal waveforms which are recorded by means of an oscillograph for different voltages and currents in a given network. From these records, determine the following: (a) the frequency ω; (b) the period T; (c) the number of Hz; (d) the phase relationship of the three waveforms; (e) write an equation for each waveform using the numerical values found previously. State any assumptions made in writing these equations.

2.3.2 A signal is known to be a damped sinusoid. When $t = \frac{1}{2}$ sec, the envelope has decayed to 37 percent of its initial value, and one full cycle of oscillation is completed. It is also known that $v_1(0) = 0$, and $v_1(\frac{1}{8}) = 2$ volts. Write an expression for the signal $v_1(t)$, determining as many numerical values for the parameters as you can.

2.3.3 The inputs to a multiplier consist of two sine waves, one of maximum magnitude $V_1 = 2$ and frequency $\omega_1 = 3$, the other $V_2 = 1$ and $\omega_2 = 2$. Sketch the product waveform.

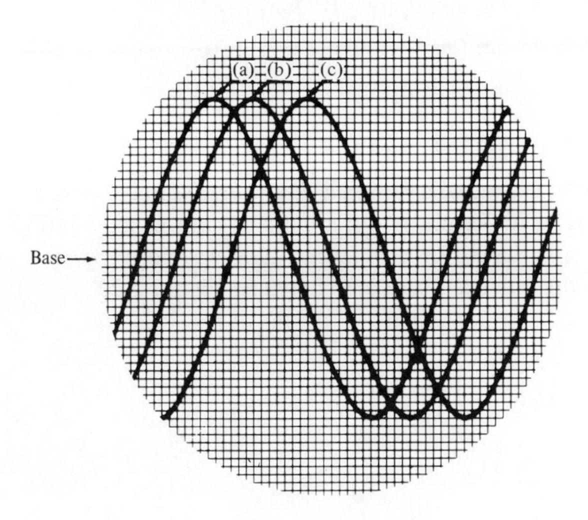

Figure 2.19 Horizontal sensitivity: 0.416 millisec per square. Vertical sensitivities: (*a*), 0.38 ampere per square; (*b*), 8.5 volts per square; (*c*), 0.6 ampere per square.

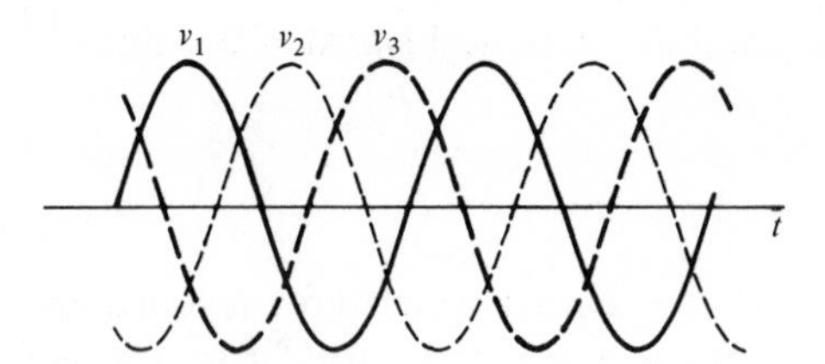

Figure 2.20 Balanced three-phase sinusoidal voltages.

2.3.4 For the system represented by the equation $v_2(t) = (dv_1/dt) - v_1$, determine $v_2(t)$ when $v_1(t) = \sin 2t$. Write $v_2(t)$ in the form $V_2 \sin(\omega t + \phi_2)$ and determine numerical values for V_2, ω, and ϕ_2. [*Note:* V_2 should be chosen to be real and *positive*.]

2.4 Three-phase sinusoidal sources

Three voltage sources displaced in phase by 120° and described by the equations

$$v_1(t) = V_1 \sin \omega t,$$
$$v_2(t) = V_2 \sin(\omega t - 120°), \tag{2.35}$$

and

$$v_3(t) = V_3 \sin(\omega t - 240°),$$

where $V_1 = V_2 = V_3$, constitute a set called a *balanced three-phase* voltage source. If the peak values are not equal or if the phase differences are not exactly 120° and 240°, the three-phase source is said to be *unbalanced*. The waveforms corresponding to Equations 2.35 for the balanced case are shown in Figure 2.20.

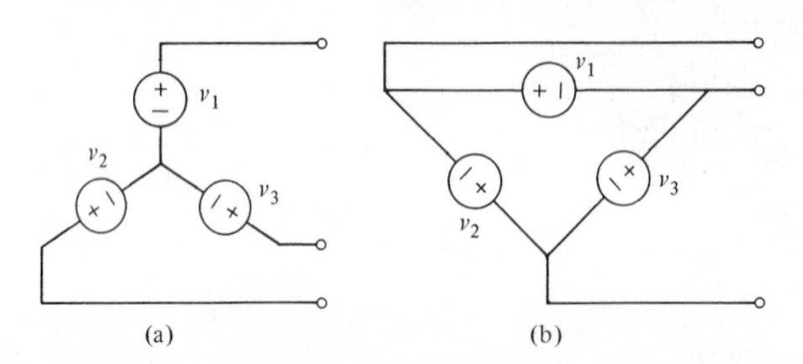

Figure 2.21 The two common methods of connecting three-phase sinusoidal sources together: (a) is known as a *Y*-connection, and (b) as a delta, Δ.

The balanced three-phase voltages (similar definitions may be made for currents) are produced by commercial *alternators* and are of great importance in the power-generation and distribution field. In the alternator, these voltages are constrained to be of precisely the same frequency. The two common methods for connecting the three sources together are illustrated in Figure 2.21. These are the Y (or wye) and the Δ (delta) connections.

5 Sampled signals and quantized signals

The operation of some systems containing a digital computer as one of the parts (an automatic control system in a high-speed aircraft or missile, for example) is based on *sampling* of a continuous input. In such a system, known as a *sampled data system*, the signal is not a continuous function of time but rather a train of sample values of the form shown in Figure 2.22. The sampling concept applies to waveforms like those studied earlier in this chapter and also to general signal waveforms like that of Figure 2.22.

The sampled output signal may be *restored* approximately from the train of signals to a continuous signal by employing an electronic device known as a *hold circuit* to *clamp* the signal at the last sampled value until the next sample arrives. An example of a partially restored waveform resulting from clamping is shown in Figure 2.23. Observe that a better approximation to a smooth curve is obtained if the sampling rate is high.

The operation of sampling is closely related to *quantizing*. In a *quantized signal*, the values of the signal at different instants of time are interpreted to be any one of a specified set of numbers. For example, a signal $v(t)$ may be sampled at $t = 0, 1, 2, 3, \ldots$ as in the second column of Table 2.1.

Suppose that the quantized signal is allowed to have values among only the following set of numbers: 0, 0.5, 1, 1.5, 2, 2.5, 3.0. Then to produce the quantized signal, each sampled value is replaced or approximated by a closest available value

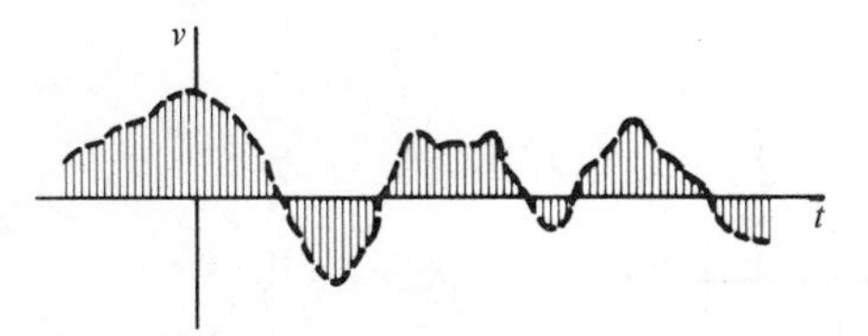

ure 2.22 A sampled signal representation. Vertical lines represent the signal at discrete instants of time. The continuous signal is shown as a dashed line.

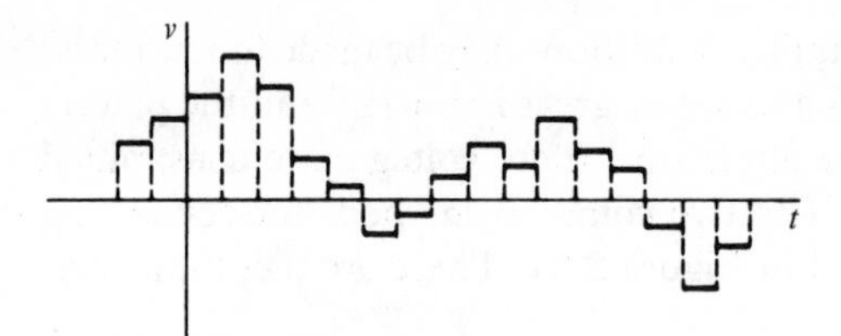

Figure 2.23 A partially restored sample signal.

allowable in the quantized signal. In this example, 0.45 is replaced by 0.50, 1.20 by 1.00, etc. as is shown in Table 2.1 and illustrated by Figure 2.24.

In systems using sampling, quantizing is also usually used right after sampling. This is the case in a digital computer in a control system. Quantizing a sampled signal is analogous to the "rounding off" of empirical data. (See Problem 2.14.)

Table 2.1

t	*Sampled value*	*Quantized value*
0	0.45	0.50
1	1.20	1.00
2	1.60	1.50
3	1.40	1.50
4	1.35	1.50
5	2.10	2.00

2.6 Signals with time delay

The notation which we shall use in describing signals which are shifted in time is easily described in terms of the operation of a magnetic tape recorder. We shall specify that our special tape recorder have two playback heads, identified as No. 1

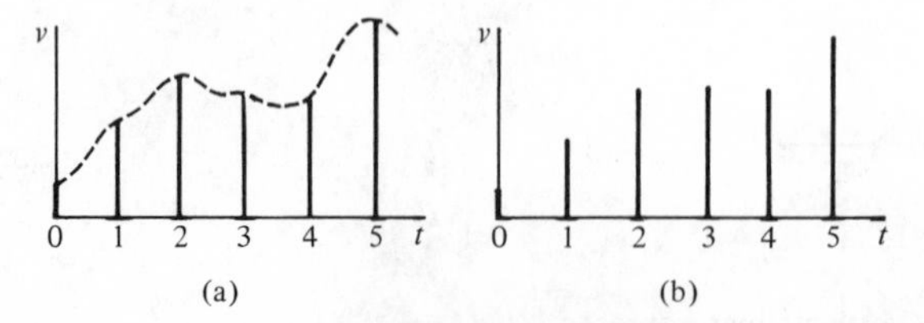

Figure 2.24 An illustration of the dual operations of first sampling (a) and then quantizing (rounding off) in (b).

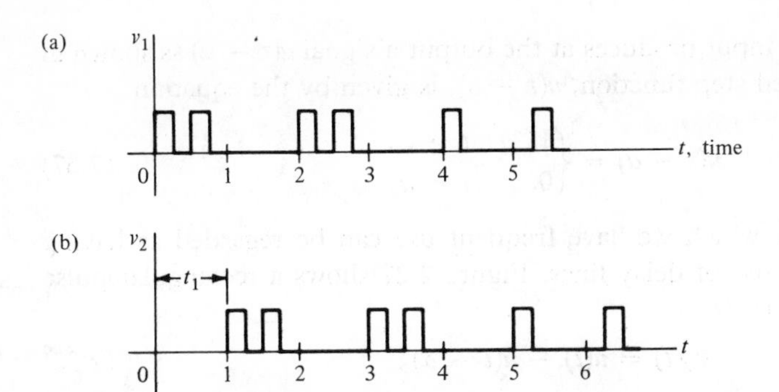

ure 2.25 Signal waveform which is delayed by t_1; $v_2(t)$ is delayed t_1 sec compared to $v_1(t)$.

and No. 2. With the tape moving forward at normal speed, there is a time interval t_1 in the passage of a given point on the tape through the two heads. Unless otherwise specified, assume that the two playback heads are set at the same gain level.

At the reference time $t = 0$, the tape begins to move with constant speed and head No. 1 plays back a recorded signal. Let $v_1(t)$ be the value of v_1 at time t having the waveform of Figure 2.25(a). The signal corresponding to head No. 2 is shown in Figure 2.25(b). How might we describe $v_2(t)$? Observe that the signal from the second head is *shifted* t_1 units of time to the right with respect to the signal from head No. 1. Clearly, the second signal waveform is described in terms of the first by the equation

$$v_2(t) = v_1(t - t_1). \tag{2.36}$$

The tape recorder is one example of a device capable of delaying a signal in time. Others are the transmission line and quartz crystal used as an acoustical delay line. All such devices may be represented by a model which will be identified as the *ideal delay line*. If the input signal to the ideal delay line is $v_1(t)$, then the output is $v_1(t - a)$, where a is the *delay*. For example, the application of a unit

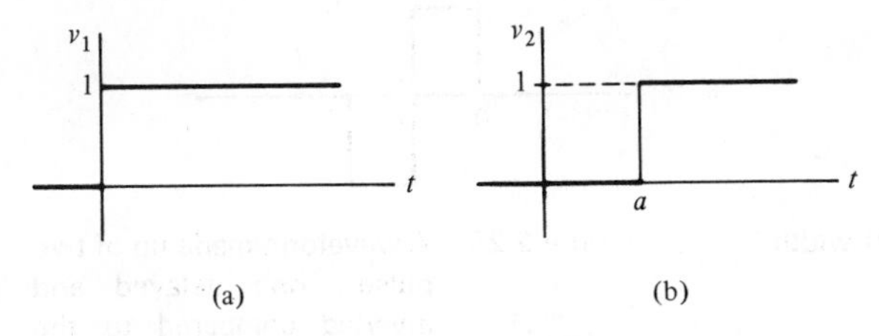

gure 2.26 Input and output waveforms for a delay line.

step signal, $u(t)$, at the input produces at the output a signal $u(t - a)$ as shown in Figure 2.26. The shifted step function, $u(t - a)$, is given by the equation

$$u(t - a) = \begin{cases} 1, & t \geq a, \\ 0, & t < a. \end{cases} \tag{2.37}$$

Some waveforms for which we have frequent use can be regarded as having been generated by the use of delay lines. Figure 2.27 shows a rectangular pulse which may be written as

$$v_2(t) = u(t) - u(t - a), \tag{2.38}$$

which is a pulse of width a. The linear combination of the two step functions, one positive, the other negative and delayed in time, together form a pulse. A model for a system which generates such a pulse may be represented by the set of equations

$$\begin{aligned} v_2(t) &= v_1(t) + v_3(t), \\ v_3(t) &= v_4(t - a), \\ v_4(t) &= -v_1(t), \end{aligned} \tag{2.39}$$

where $v_1(t)$ is the input, $v_2(t)$ is the output, and the pulse is obtained when the input is a unit step function.

A more complicated pulse is shown in Figure 2.28. This pulse may be represented by the equation

$$v_2(t) = u(t) - 2u(t - a) + u(t - 2a). \tag{2.40}$$

This scheme for representing waveforms may be extended to describe a "square wave,"

$$v_2(t) = u(t) + \sum_{k=1}^{\infty} (-1)^k 2u(t - ka). \tag{2.41}$$

as shown in Figure 2.29.

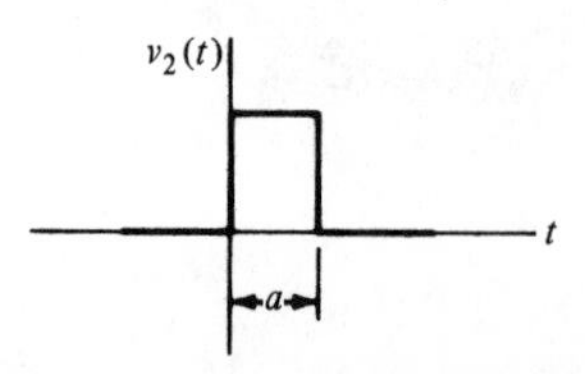

Figure 2.27 A rectangular pulse of width a and amplitude 1.

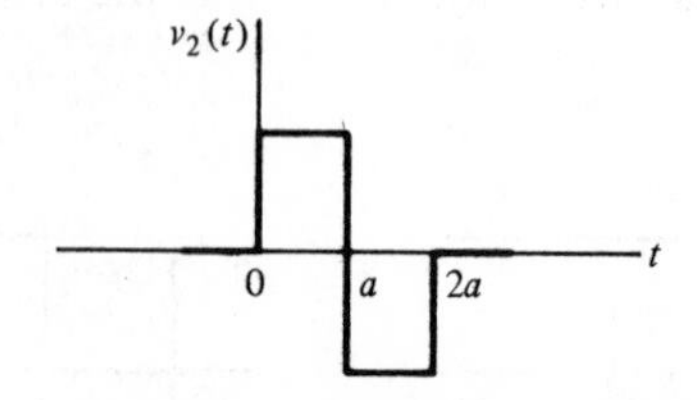

Figure 2.28 A waveform made up of two pulses, one delayed and inverted compared to the other.

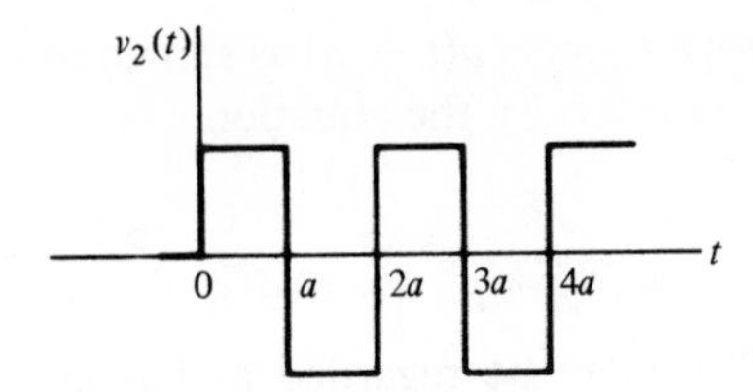

ure 2.29 A signal waveform known as a square wave.

ERCISES

2.6.1 Let v_2 of Equation 2.36 be recorded from head No. 2 with $t_1 = 4$. Sketch $v_2(t)$ if v_1 has the waveform of Figure 2.25(a).

2.6.2 Let v_1 be the input to an ideal delay line of delay a, $v_1(t) = K_1 e^{-\sigma_1 t} \sin(\omega_1 t + \phi_1) u(t)$. Determine $v_2(t)$.

2.6.3 Repeat Exercise 2.6.2 for $v_1 = A_2 \sin(\omega_2 t + \phi_2)$. Show that the derivative of the negative of the phase shift caused by the passage of a sine wave through an ideal delay line with respect to the angular frequency ω_2 is the time delay.

2.6.4 A system is described by the equation $v_2(t) = v_1(t) - 2v_1(t - a/2)$. Determine the output $v_2(t)$ when the input is a rectangular pulse of duration a. Sketch the waveform of the output signal.

2.6.5 Repeat Exercise 2.6.4 for the system described by $v_2(t) = v_1(t) - v_1(t - a)$.

2.6.6 A system is described by

$$v_2(t) = \int_0^t v_1(\tau)\, d\tau - v_1(t - 1)$$

as schematically represented in Figure 2.30. Sketch the output for the given input.

2.6.7 A system is described by $v_2(t) = 2v_1(t) + v_1(t - \pi/2)$. Sketch $v_2(t)$ for $v_1(t) = \sin 2t$. See Figure 2.31.

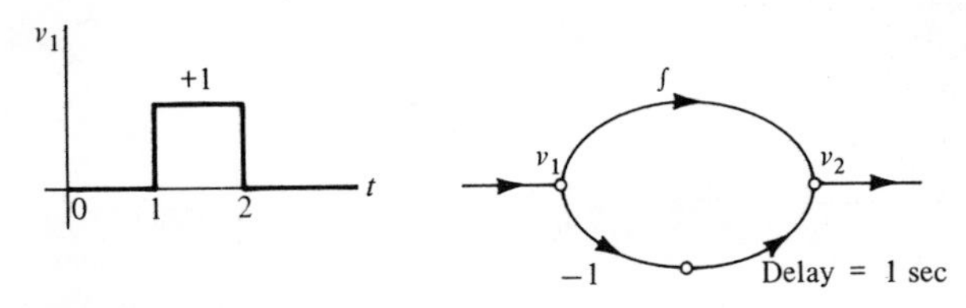

Figure 2.30

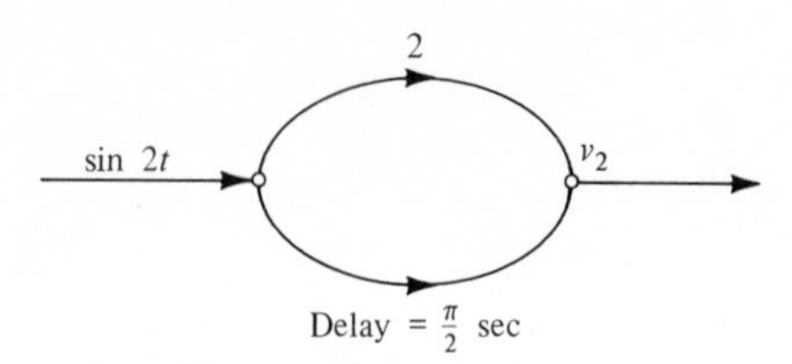

Figure 2.31

2.6.8 Given the following signal

$$v_1(t) = \begin{cases} 0, & \text{for } t < 0, \\ v_0 t, & \text{for } 0 \le t \le 1, \\ 0, & \text{for } t > 1, \end{cases}$$

sketch $v_2(t) = v_1(t-2) + v_1(t-3)$.

2.6.9 The output of a system is the sweep voltage shown in Figure 2.32. Express the output in terms of delayed functions.

2.6.10 Repeat for $v_2(t)$ as in Figure 2.33.

2.7 Summary

1. The exponential function is defined as $v(t) = Ke^{\sigma t}$. For $\sigma < 0$, $T = 1/|\sigma|$ is the time constant. When $\sigma < 0$, for each lapse of T sec, the exponential decreases by a factor of $1/e \approx 0.37$. We deal with linear combinations of exponential functions, $\sum K_i e^{\sigma_i t}$. For such functions, differentiation or integration changes only the magnitude of the component functions except for the possible addition of a constant in the case of integration.

2. The step function is defined as

$$u(t) = 1, \qquad t \ge 0; \quad u(t) = 0 \text{ otherwise}.$$

The unit ramp function is

$$r(t) = tu(t).$$

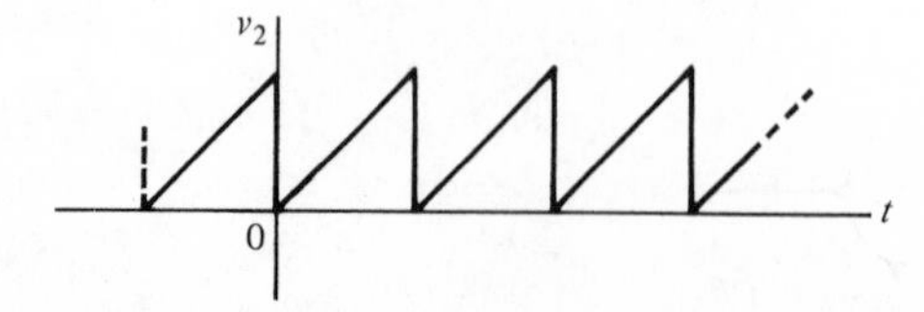

Figure 2.32

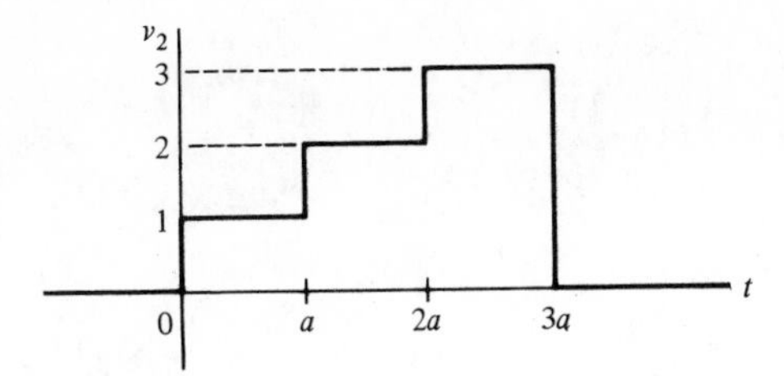

Figure 2.33

The unit-impulse function, $\delta(t)$, has infinite value at $t = 0$ and is zero otherwise. It has the property

$$\int_{-a}^{b} \delta(\tau)\, d\tau = 1$$

for any positive a and b.

All functions may be scaled by K, where K is the *magnitude* of the step, the *slope* of the ramp, and the *strength* of the impulse. Linear combinations of impulses are called *impulse trains*.

3. The unit step, unit ramp, and unit-impulse functions are related by differentiation or integration.

$$\frac{d}{dt}\,\text{ramp} = \text{step}\,; \qquad \frac{d}{dt}\,\text{step} = \text{impulse}\,.$$

$$\int \text{step}\; dt = \text{ramp}\,; \qquad \int \text{impulse}\; dt = \text{step}\,.$$

4. The word *sinusoid* is used to denote the sine or the cosine or linear combinations of both at the same frequency. One general form of the sinusoid is

$$v(t) = V \sin(\omega t + \phi), \qquad \text{all } t,$$

where V is the magnitude, a positive number; ω is the frequency (radians/sec) which is equal to $2\pi f$ where f is frequency in Hz; $T = 1/f$ is the period; and ϕ is the phase angle.

5. The *damped sinusoid* is

$$v(t) = Ae^{\sigma t} \sin(\omega t + \phi), \qquad \sigma < 0,$$

where A is real and positive. The function $\pm Ae^{\sigma t}$ is called the *envelope*.

6. Balanced three-phase voltage sources have equal voltage magnitude but phase displacement of 120°. Two common connections for three sources (or the corresponding loads) are the Y- and the Δ-connections.

7. Sampled signals are defined only at discrete values of time. Quantized signals are permitted to have only certain values of magnitude.

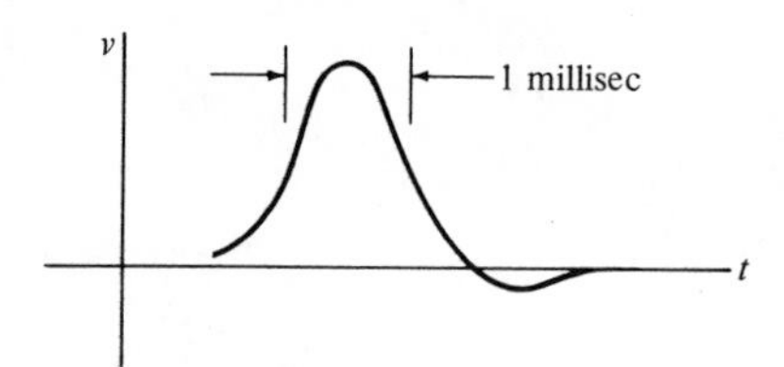

Figure 2.34

8. If $v_2(t) = v_1(t - t_1)$, then for $t_1 > 0$, $v_2(t)$ is said to be *delayed* t_1 sec with respect to $v_1(t)$.

PROBLEMS

2.1 One of the most interesting signals is that which propagates along a nerve fiber in the human body when a neuron is properly excited (say by a sock on the jaw). The signal is in the form of a pulse as shown in Figure 2.34 with a time duration of about 1 millisec and an amplitude of about 200 millivolts. Write an equation, a mathematical abstraction, of this signal waveform, evaluating all constants in your equation.

2.2 (a) Sketch the function

$$f(t) = \begin{cases} \frac{1}{\sigma} e^{-t/\sigma}, & t \geq 0, \\ 0, & t < 0, \end{cases}$$

as a function of t for $\sigma = 1, 0.5, 0.1$.

(b) Calculate the area under the curve for this function.

(c) The limit of $f(t)$ as σ approaches zero may be used as a definition for the delta function. Does this definition have advantages over the definition based on a pulse?

2.3 Repeat Problem 2.2 for the function

$$f(t) = \frac{1}{\sqrt{2\pi}\,\sigma} e^{-t^2/(2\sigma^2)}.$$

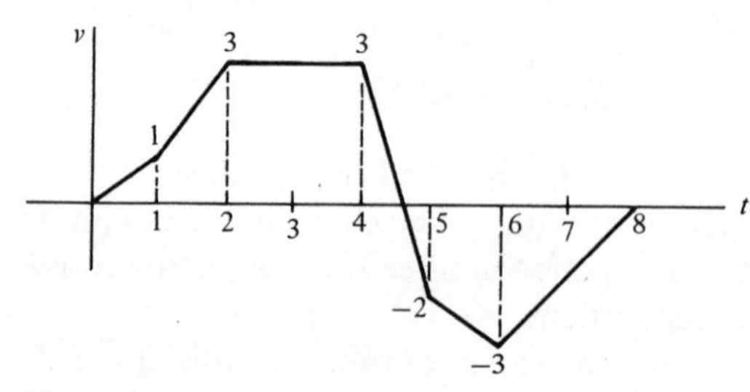

Figure 2.35

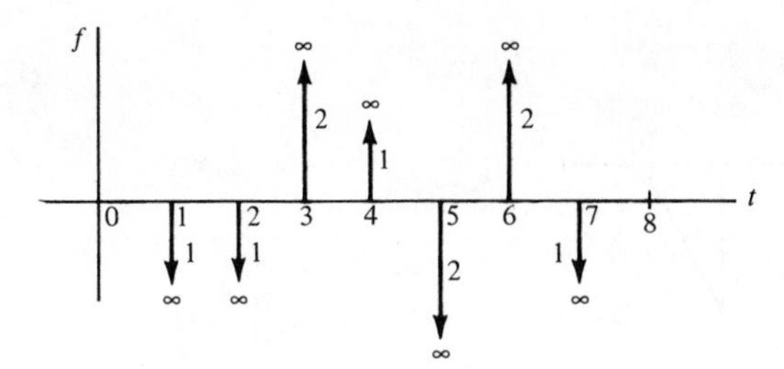

Figure 2.36

Calculate the integral of this function from $-\infty$ to $+\infty$. This function is known as a Gaussian or normal distribution.

2.4 Repeat Problem 2.3 if t is replaced by $t - T_0$ where T_0 is a constant.

2.5 Sketch the function

$$v(t) = \begin{cases} \dfrac{\pi}{2\epsilon} \sin\left[\dfrac{\pi}{\epsilon}(t+\epsilon)\right], & -\epsilon \le t \le 0, \\ 0, \quad t < -\epsilon \text{ and } t > 0. \end{cases}$$

Show that the integral of $v(t)$ approaches a unit step as $\epsilon \to 0$.

2.6 A new function is defined by the equation

$$\int_{-\infty}^{t} r(\tau)\, d\tau = p(t),$$

where $r(t)$ is the unit ramp. Sketch $p(t)$. What name for $p(t)$ do you think is appropriate?

2.7 A waveform is approximated by a number of straight-line segments in Figure 2.35. For this "piecewise linear" waveform, plot dv/dt and d^2v/dt^2.

2.8 The impulse train shown in Figure 2.36 represents the second derivative of a function, $f(t)$. Determine $f(t)$. State any assumptions made in arriving at your answer.

2.9 The input signal to the system represented by the signal-flow schematic of Figure 2.37 or the equation $v_2(t) = v_1(t) + dv_1(t)/dt$ is $v_1 = 2 \sin 2t$.

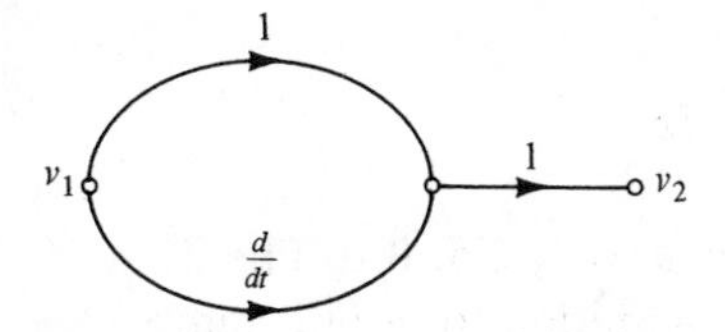

Figure 2.37

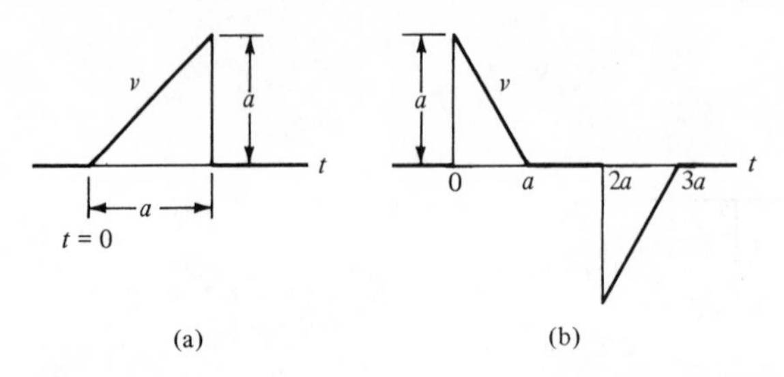

Figure 2.38

Show that the output signal, $v_2(t)$, has the form $v_2 = A \sin(2t + \phi)$ and determine A and ϕ.

2.10 (a) Express the waveform in Figure 2.38(a) in terms of steps, ramps, delayed steps, delayed ramps, etc.
(b) Repeat (a) for the waveform in Figure 2.38(b).
(c) Specify a system (by means of its equations) where input is a unit step and where output is the waveform in Figure 2.38(a).
(d) Repeat (c) for the waveform in Figure 2.38(b).

2.11 Prepare sketches of the waveforms described by the following equations:

(a) $r(t-3)$,

(b) $\sum_{n=0}^{10} \delta(t - nT), \qquad \text{for} \quad T = 1,$

(c) $u(t) \sin \frac{2\pi t}{T}, \qquad u\left(t - \frac{T}{2}\right) \sin \frac{2\pi t}{T}, \qquad u(t - T) \sin \frac{2\pi t}{T},$

(d) $r(t)u(t-1), \qquad r(t) - r(t-1) - u(t-1).$

2.12 Sketch the function

$$g(t) = \int_{-\infty}^{t} f(x)\, dx$$

vs. t, where $f(t)$ is given in Problem 2.2 for $\sigma = 1, 0.5, 0.1$. Discuss the limiting plot for $\sigma \to 0$.

2.13 Sketch the function

$$r_1(t) = \int_{-\infty}^{t} g(x)\, dx$$

vs. t, where $g(t)$ is given in Problem 2.12 for $\sigma = 1, 0.5, 0.1$. The limit of $r_1(t)$ as σ approaches zero may be used as a definition for a unit ramp.

2.14 In this and the previous chapter, we have introduced two similarly related

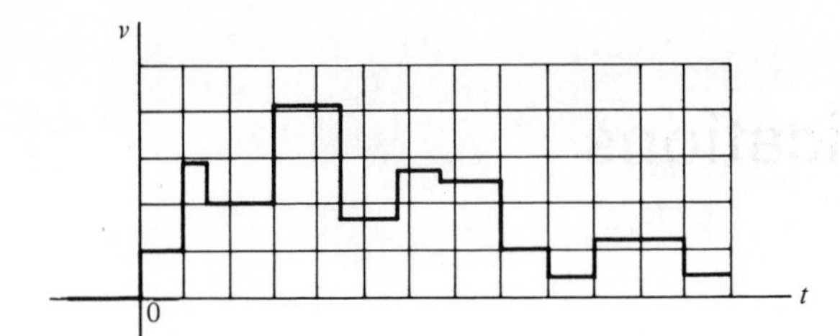

Figure 2.39

descriptions: (1) devices and models; and (2) device waveforms and their mathematical abstractions. In working problems and in equipment design, we make use of *exact numbers* and *approximate numbers*. In distinguishing these two kinds of numbers, the following expressions are useful: (a) roundoff of a number, roundoff error; (b) number of significant digits; and (c) margin of error. Discuss the meaning of these expressions with respect to the actual and idealized quantities listed under (1) and (2).

2.15 Show that any staircase signal such as that shown in Figure 2.39 can be represented as a sum of step functions of various magnitudes and delays, that is

$$v(t) = \sum_{k=0}^{N} a_k u(t - \tau_k).$$

2.16 Specify the equations of a system where input is a unit impulse and where output is as shown in Figure 2.35. Your system may contain amplifiers, integrators, ideal delay lines, and adders.

2.17 A special tape recorder (No. 1) contains a tape with a recorded signal $v_1(t)$ on it, while another similar recorder (No. 2) is loaded with a blank tape. The output from No. 1 is to be recorded on No. 2. Both recorders are turned on simultaneously with the tapes beginning to travel in the same forward direction. After t_a sec, the tapes are stopped. A switch for reversing the direction of travel of tape No. 1 is closed. The tapes are started again, with tape No. 2 traveling forward and tape No. 1 backward, for a period of t_b sec. After t_b sec, the tapes are stopped and the switch is thrown so that tape No. 1 may run forward again. The tapes are then started and are permitted to run forward until they run out completely. Determine an equation for the signal $v_2(t)$ in terms of $v_1(t)$.

Chapter **3**

Partial Signal Specifications

In Chapters 1 and 2, we discussed signals that were specified either in equation form or by tabular or graphic display. For many purposes, it is not necessary to give such *complete* specification of a signal. If, for example, we are interested in knowing if a device or component such as a capacitor will fail due to overvoltage, then we need know only the maximum or peak value of the magnitude of the signal voltage. For other purposes, we find frequent use for such specifications as average value, root-mean-square value, rise time, overshoot, and settling time. These quantities are *partial specifications* of a waveform; they represent digested information about some interesting property of a signal. These characterizing partial specifications are considered in this chapter.

3.1 Average value of a signal

The average (or arithmetic mean) of n numbers is found from the familiar relationship

$$\bar{a} = \frac{a_1 + a_2 + \cdots + a_n}{n}. \tag{3.1}$$

What do we mean by the average or mean value of a continuous function rather than n numbers? If the function $v(t)$ can be integrated over an interval from t_1 to t_2, then the average value $\bar{v}$ multiplied by the interval is equal to the area under $v(t)$ from t_1 to t_2 as shown in Figure 3.1. Thus we have

$$(t_2 - t_1)\bar{v} = \int_{t_1}^{t_2} v(t)\, dt. \tag{3.2}$$

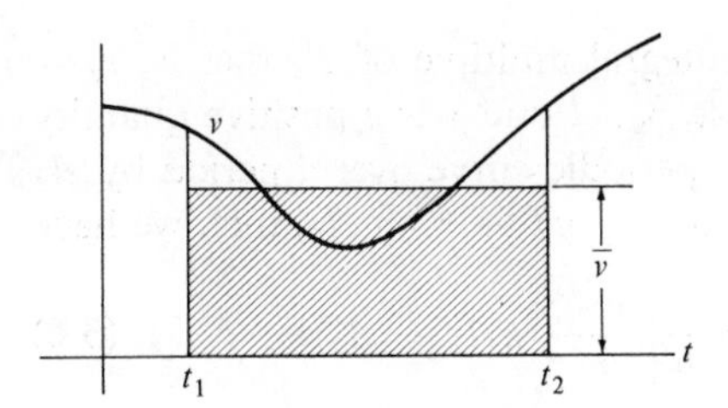

ure 3.1 $\bar{v}$ is the average or mean value of $v(t)$ over the interval from t_1 to t_2.

In other words, the *average value* of $v(t)$ over the range t_1 to t_2 is

$$\bar{v} = \frac{1}{t_2 - t_2} \int_{t_1}^{t_2} v(t)\, dt. \tag{3.3}$$

We consider now an important class of signals, the set of periodic signals, in which the signal waveform repeats itself. More precisely, $v(t)$ is periodic if there is a quantity T such that

$$v(t) = v(t + T) \tag{3.4}$$

for all t. The smallest T that satisfies the condition in Equation 3.4 is called the *period* as defined in the previous chapter for sinusoidal signals. If $t_2 - t_1 = T$, then the average value $\bar{v}$ is

$$\bar{v} = \frac{1}{T} \int_{t_1}^{t_1+T} v(t)\, dt. \tag{3.5}$$

Notice that the area under a periodic curve over a range $t_2 - t_1 = T$ is the same for all t_1. This is illustrated in Figure 3.2. We should always choose t_1 for the most convenient numerical evaluation in a specific situation. If $t_2 - t_1 = 2T$, we see that the total area under the curve will be twice what it would be if $t_2 - t_1$ were just T. Hence the average value $\bar{v}$ is the same as that for $t_2 - t_1 = T$. It is easy to see that the average value of a periodic signal over an integral multiple of a complete period is equal to the average value over one complete period.

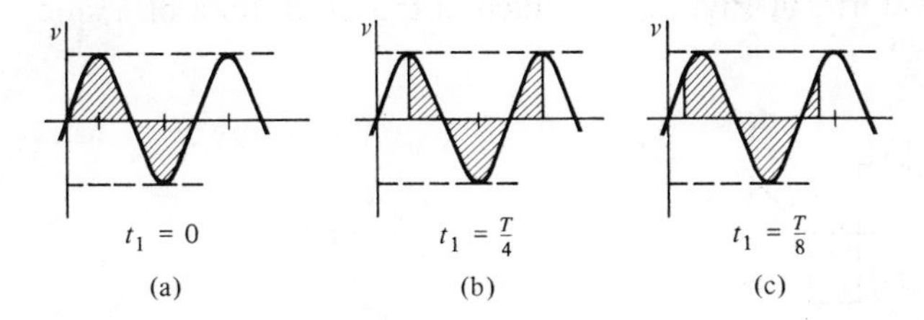

ure 3.2 The area under a periodic waveform over the range $t_2 - t_1 = T$ is the same regardless of the value of t_1.

Suppose however that $t_2 - t_1$ is not an integral multiple of T; that is, $t_2 = t_1 + nT + \tau$, where n is 0 or any integer 1, 2, 3, . . . and τ is a positive quantity less than T. If we denote the area under the periodic curve over a period by A_1 and the area under the curve from $t_1 + nT$ to $t_1 + nT + \tau$ by A_0, then we have

$$\bar{v} = \frac{1}{nT + \tau} \int_{t_1}^{t_1 + nT + \tau} v(t)\, dt = \frac{1}{nT + \tau} (nA_1 + A_0). \tag{3.6}$$

We see that provided A_0 is finite, then as n becomes large, $\bar{v}$ becomes almost equal to A_1/T. In practical devices for measuring average value, $t_2 - t_1$ is often not an integral multiple of T, but $t_2 - t_1$ is often relatively much larger than T and A_0 is finite in all practical cases so that the error involved in assuming that it is the value given in Equation 3.5 is small. We shall then take the average value of a periodic signal as that which corresponds to one complete period.

An interesting special case of Equation 3.5 occurs when the periodic function of period T has symmetry. This leads to simplification in the calculation of the area under the curve.

EXAMPLE 3.1.1

For the periodic signal shown in Figure 3.3, we see that the waveform repeats itself after every 10 sec so that $T = 10$. The average value is seen by inspection to be

$$\bar{v} = \frac{1}{10}\left(\frac{1 \times 4}{2} + 1 \times 3 - 1 \times 3\right) = 0.2. \tag{3.7}$$

EXAMPLE 3.1.2

For the sine wave $v(t) = V \sin [(2\pi/T)t]$, the area under the curve from 0 to $T/2$ is the negative of the area under the curve from $T/2$ to T. Hence the average value of a sine wave over any number of periods is zero.

EXAMPLE 3.1.3

Consider the periodic waveform of Figure 3.4 which is the magnitude of a sine

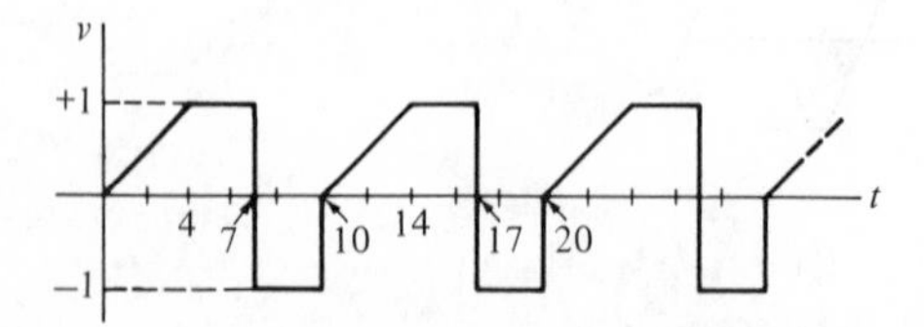

Figure 3.3 Signal for Example 3.1.1.

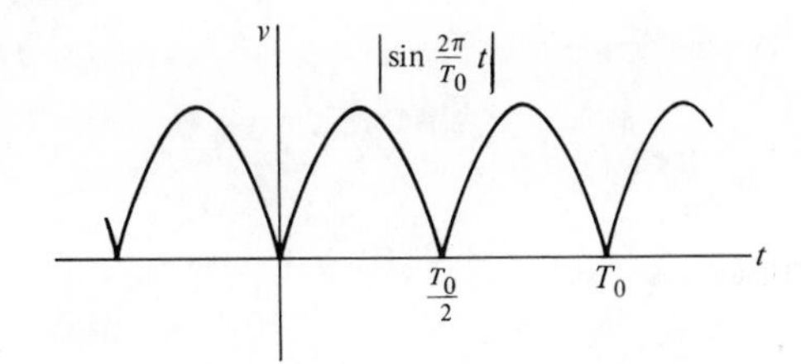

ure 3.4 A waveform of a signal known as a full-wave-rectified sine wave.

function,

$$v(t) = \left|\sin \frac{2\pi}{T_0} t\right|. \tag{3.8}$$

This wave is known as a *full-wave-rectified sine wave.* By inspection of the waveform, we see that the period is $T_0/2$. Note also that the area under the curve from 0 to $T_0/4$ is equal to that from $T_0/4$ to $T_0/2$. This symmetry allows us to write

$$\bar{v} = \frac{1}{T_0/2} \cdot 2 \cdot \int_0^{T_0/4} \sin \frac{2\pi}{T_0} t \, dt = \frac{2}{\pi}. \tag{3.9}$$

ERCISES

3.1.1 What is the period of the signal $v(t)$ in Figure 3.5?. This waveform is known as a half-wave-rectified sine wave.

3.1:2 What is the average value of $v(t)$ for Exercise 3.1.1?

3.1.3 Is the waveform shown in Figure 3.6 periodic? Give a reason or reasons for your answer.

3.1.4 What is the average value of the signal of Figure 3.14? The waveshape is shown for 1 cycle.

3.1.5 Determine the average value of the waveform of Figure 3.7 as a function of a and K.

3.1.6 If the average value of a periodic signal $v_{12}(t)$ is V_0, what is the average value of $v_{21}(t)$?

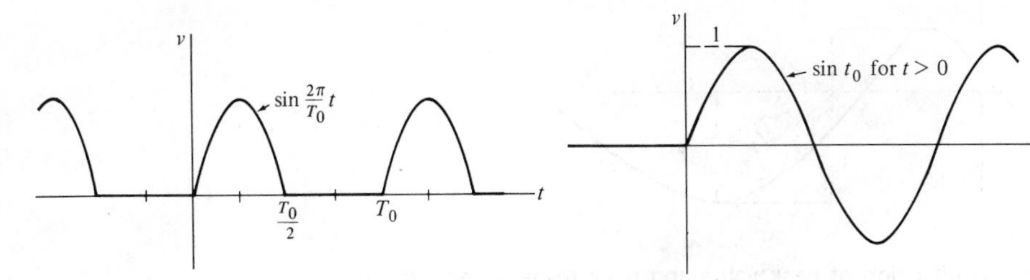

Figure 3.5

Figure 3.6

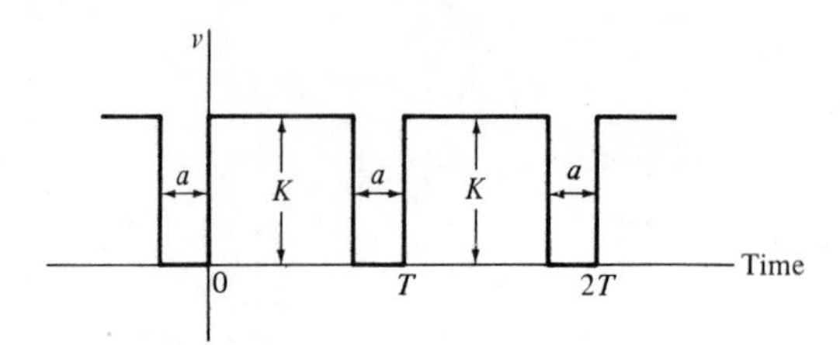

Figure 3.7

3.2 Peak value of a signal

The *peak value* of a signal is its maximum absolute value. This definition is general, applying to both periodic and aperiodic (nonperiodic) signal waveforms. Applying this definition to the waveforms of Figures 3.3 and 3.4, we see that each has a peak value of one. For periodic signals, especially sinusoidal signals, we will use capital letters with the subscript *m* or *max*. Thus, if $v(t)$ is sinusoidal, we designate its peak value V_m or V_{max} in the equation, $v(t) = V_m \sin(\omega t + \phi)$. In the introduction to the chapter, we mentioned that peak values of current are similarly important in establishing the rating of many electric devices. In vacuum-tube and semiconductor diodes, the allowable peak inverse voltage is a critical design and operating quantity.

A quantity useful in the calibration and use of a cathode-ray oscillograph is the *peak-to-peak* value of a signal. We define this value to be the algebraic difference of the maximum signal value and the minimum signal value. In the case of a sinusoidal signal of maximum value V_m, the peak-to-peak value is clearly $2V_m$. For a general signal, it is as represented in Figure 3.8.

EXERCISES

3.2.1 Find the peak value and the peak-to-peak value of the signal waveforms shown in Figure 3.9.

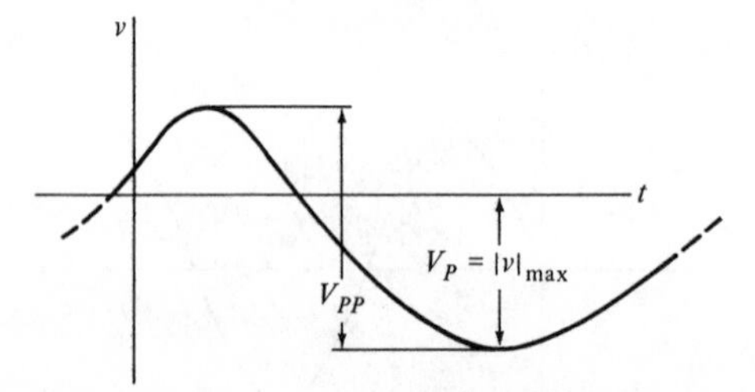

Figure 3.8 The identification of peak value and peak-to-peak value for an arbitrary signal.

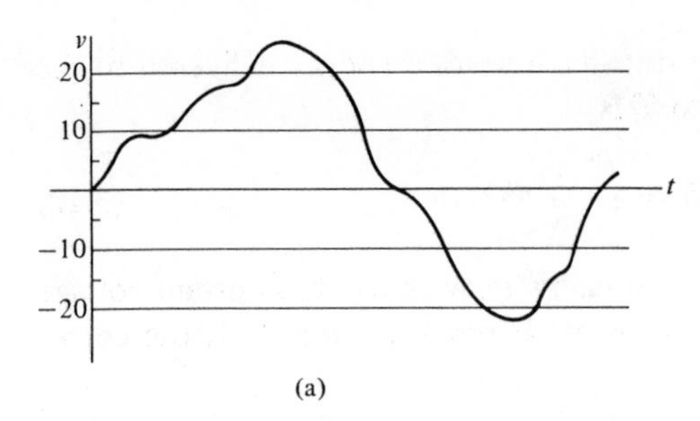

(a)

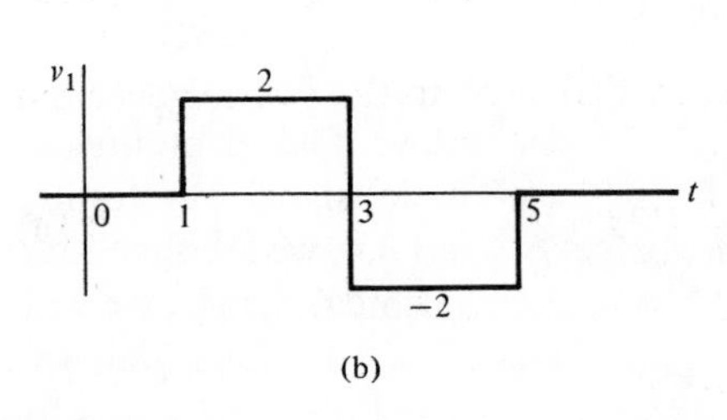

(b)

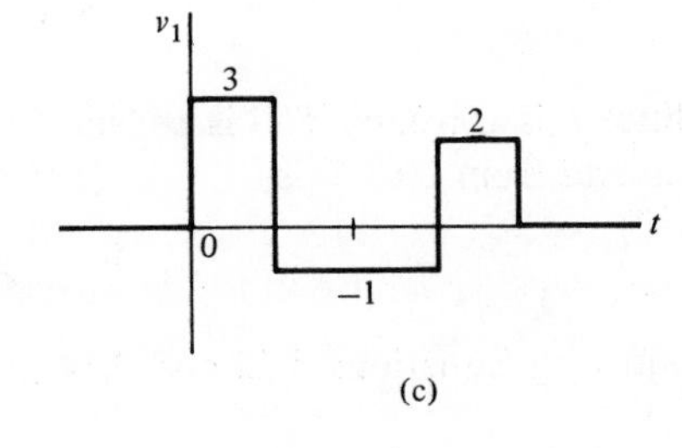

(c)

Figure 3.9

3 Root-mean-square value of a signal

A periodic voltage $v(t)$ is impressed across the resistor shown in Figure 3.10. The current through the resistor is related to this voltage by Ohm's law,

$$i(t) = \frac{1}{R} v(t). \tag{3.10}$$

The instantaneous power is

$$p(t) = v(t)i(t) = \frac{1}{R} v^2(t). \tag{3.11}$$

Our objective is to characterize $v(t)$ under the constraint that the energy delivered to the resistor is equal to that of an equivalent time-invariant voltage source.

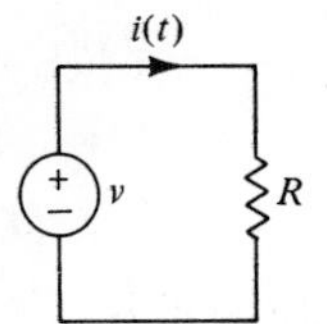

Figure 3.10 A periodic voltage source connected to a resistor, used in the definition of rms value.

Since power is the time rate of energy transfer, the total energy delivered to the resistor over the time interval from 0 to T is

$$W = \int_0^T p(t)\,dt = \int_0^T \frac{1}{R} v^2(t)\,dt. \tag{3.12}$$

Suppose now that instead of the periodic voltage $v(t)$, we have the constant voltage of an ideal battery V connected to the resistor. From Equation 3.11, the corresponding instantaneous power is

$$p = VI = \frac{1}{R} V^2. \tag{3.13}$$

Since p is constant if V is constant, the energy delivered to the resistor over the interval from 0 to T is

$$W = pT = \frac{1}{R} V^2 T. \tag{3.14}$$

Equating Equations 3.12 and 3.14 gives

$$\frac{1}{R} V^2 T = \frac{1}{R} \int_0^T v^2(t)\,dt, \tag{3.15}$$

$$V = \sqrt{\frac{1}{T} \int_0^T v^2(t)\,dt}. \tag{3.16}$$

This characteristic value is known as the *effective value* or *root-mean-square* (abbreviated rms) *value* of the periodic voltage $v(t)$. The rms value concept applies only to *periodic* signals. It is the effective voltage in the sense that it results in the same heating of the resistor (heat and energy being proportional) as would result from a time-invariant source of the same value. This quantity will be distinguished by the subscripts *eff* or *rms*. This same definition applies to a periodic current and to other physical quantities. As for the average value studied in the first section of this chapter, the integration may be performed starting at any time t_a and continuing for one period to $t_a + T$. Thus, Equation 3.16 becomes

$$V_{\mathrm{rms}} = \sqrt{\frac{1}{T} \int_{t_a}^{t_a+T} v^2(t)\,dt}. \tag{3.17}$$

In any given situation, t_a may be chosen to make computation easy.

EXAMPLE 3.3.1

Consider the sine wave $v(t) = V_m \sin \omega t$. The effective value of this signal is

$$V_{\mathrm{rms}} = \sqrt{\frac{1}{T} \int_{t_a}^{t_a+T} V_m^2 \sin^2 \omega t\,dt} \tag{3.18}$$

$$V_{\text{rms}} = \sqrt{\frac{V_m^2}{T}\int_0^T \frac{1 - \cos 2\omega t}{2}\,dt} \tag{3.19}$$

$$= \sqrt{\frac{V_m^2}{T}\left(\frac{1}{2}t - \frac{1}{4\omega}\sin 2\omega t\right)\Bigg|_0^T} \tag{3.20}$$

$$= \sqrt{\frac{V_m^2}{T}\left(\frac{1}{2}T - \frac{1}{4\omega}\sin 2\omega T\right)}. \tag{3.21}$$

Now, since $\omega T = 2\pi$, $\sin 2\omega T = 0$, and we have

$$v_{\text{rms}} = \frac{V_m}{\sqrt{2}} \approx 0.707 V_m. \tag{3.22}$$

This result will be used repeatedly in working with sinusoidal signals. You should memorize it.

XERCISES

3.3.1 Calculate the rms value of the signal shown in Figure 3.11.

3.3.2 The sine waves or parts of sine waves shown in Figure 3.12 all have the same maximum amplitude. If the rms value of the signal shown as (a) is 0.648 volts, find the rms value of the other signals.

3.3.3 The periodic waveform shown in (a) of the figure causes a power dissipation of 1 watt in a resistor. Find the power dissipation in the same resistor produced by the other voltage waveforms of Figure 3.13.

3.3.4 Determine the rms value of the signal given in Figure 3.3 (p. 48).

3.3.5 Repeat Exercise 3.3.4 for the signal of Figure 3.4 (p. 49).

3.3.6 Calculate the rms value of the signal $v(t) = V_m \cos(\omega t + \phi)$.

3.3.7 Find the rms value of the signal represented by the equation $v(t) = 5 + 2 \sin 3t$.

3.3.8 Determine the rms value of the signal $v(t) = 5 \sin \omega t + 4 \cos(\omega t + \pi/6)$.

3.3.9 If the rms value of $v_{12}(t)$ is V, what is the rms value of $v_{21}(t)$?

3.3.10 The portion of a sine wave shown in Figure 3.14 represents the current

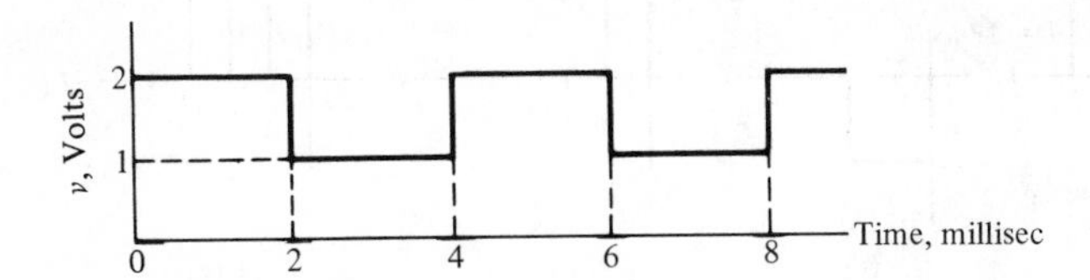

Figure 3.11

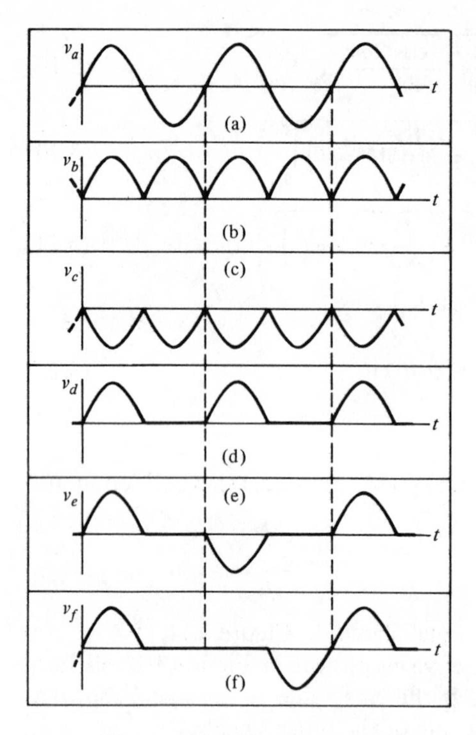

Figure 3.12 All waveforms are periodic.

in a grid-controlled Thyratron. In 1 cycle, the current is sinusoidal from $\omega t = 135°$ to $180°$ but otherwise zero. If the signal has a maximum amplitude of one ampere, find the rms value of the current.

3.4 Specification of the response to a step input

The step function described in Chapter 2 is an important signal in the testing of a network or a system. With a step applied at the input, the output or response of

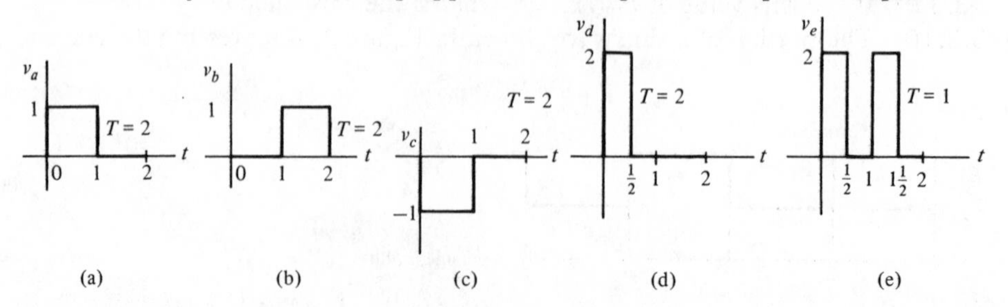

Figure 3.13

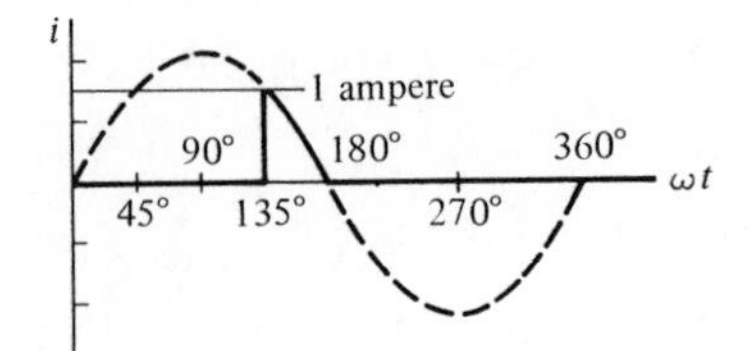

Figure 3.14

many systems has the waveform similar to that shown in Figure 3.15. In comparison to the input step, the output overshoots, undershoots, and continues to oscillate with decreasing amplitude. The overshoot is one characteristic by which the response may be described. If the steady-state value of the response $v(t)$ is $v(\infty)$ and if $v(\infty)$ is positive, then the *overshoot* is

$$\text{Overshoot} = v_{\max} - v(\infty), \tag{3.23}$$

where $v_{\max}$ is the peak value. We sometimes speak of *percent overshoot* by which we mean

$$\text{Percent overshoot} = \frac{v_{\max} - v(\infty)}{v(\infty)} \times 100\%. \tag{3.24}$$

Percent overshoot is not defined when $v(\infty) = 0$. The signals of Figure 3.16, for example, are such that overshoot is zero and undefined, respectively.

Rise time is defined for the same response signal as overshoot. Referring to Figure 3.15, we find that the rise time is defined as

$$\text{Rise time} = t_2 - t_1, \tag{3.25}$$

where t_1 is the time at which $v(t) = 0.1v(\infty)$ and t_2 is the time at which $v(t) = 0.9v(\infty)$ as shown in the figure. The numbers 0.1 and 0.9 are most commonly used but other conventions are in existence. Rise time is an important specification

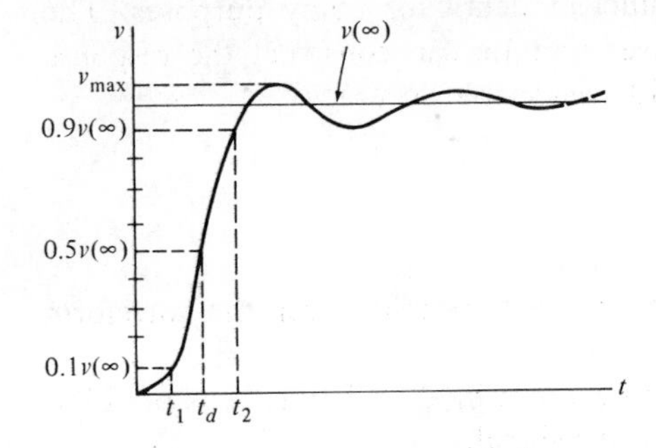

Figure 3.15 The form of signal waveform for which overshoot, rise time, time delay, and settling time are defined.

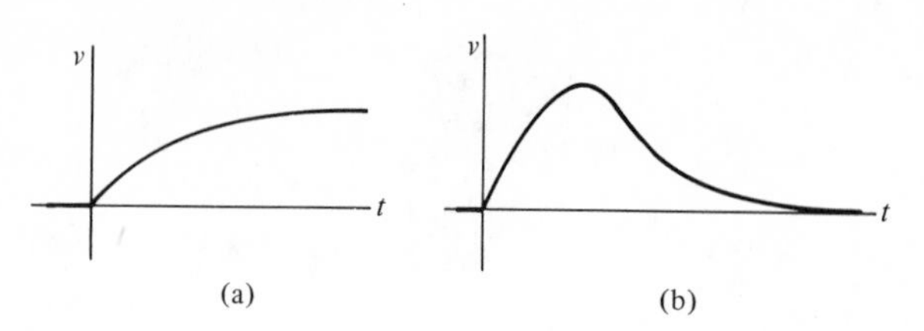

Figure 3.16 Signal waveform of (a) has zero overshoot, while for (b) overshoot is not defined.

when speed of response is important and thus applies to components of digital computers, nuclear counters, pulse generators and test equipment like cathode-ray oscillographs. Rise times range from fractions of a nanosecond* in electronic equipment to several seconds in the case of large electromechanical servomechanisms.

Rise time gives one measure of the speed of a system in reacting to a step input. Another is *time delay* (which we must distinguish from delay time in a delay line as studied in Chapter 2). For some responses, the time elapsed from the value $t = 0$ to the time at which $v(t) = 0.1v(\infty)$ may be large compared to the rise time. For such a situation, an additional specification is useful. The *time delay* is defined as the time interval from $t = 0$ until the time at which $v(t) = 0.5v(\infty)$. Roughly speaking, it is the time required for an "appreciable" response to appear. Knowing both the rise time and the time delay gives a more accurate description of the response waveform than either of the two specifications alone.

The signal $v(t)$ of Figure 3.15 has an oscillatory behavior as it approaches its steady value $v(\infty)$. A rough measure of the decay of the oscillation is called the *settling time*. It is defined as the time elapsed between the application of the step input ($t = 0$) and the time at which the oscillation is "negligible." By negligible, we usually mean that $[v(t) - v(\infty)]$ is not greater than 5 percent of the steady value $v(\infty)$.

The four characteristics, overshoot, rise time, time delay, and settling time, together describe a step response in sufficient detail for many purposes. The student will find in working the exercises that he can construct the response waveform from the four specifications with reasonable accuracy.

EXERCISES

3.4.1 Determine the overshoot and the percent overshoot for the waveform given in Figure 3.17.

3.4.2 A response signal is given by the equation $v_1(t) = 1 - e^{-t} \cos \omega t$. Find an expression for the overshoot of this signal.

* 1 nanosec = 10^{-9} sec.

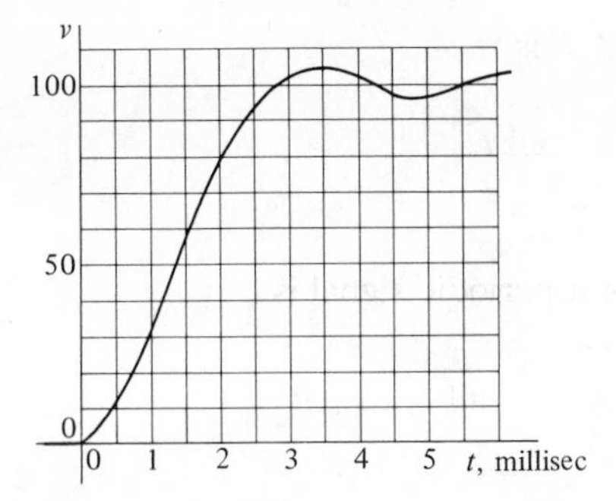

Figure 3.17

3.4.3 Given the response signal

$$v(t) = K(1 - e^{-\sigma_1 t}), \qquad t \geq 0 \text{ and } \sigma_1 > 0,$$

show that the rise time is approximately $2.2/\sigma_1$.

3.4.4 Determine the rise time and time delay of the waveform shown in Figure 3.17.

3.4.5 For the signal described in Exercise 3.4.3, find an expression for the time delay in terms of σ_1.

3.4.6 Find the settling time for the signal of Exercise 3.4.2 which $\omega = 1$.

3.5 The measurement of signal characteristics

Practical meters and instruments measure many of the characteristics described in this chapter. For example, D'Arsonval-type meters measure average values of current and voltage, dynamometer- and thermocouple-type meters measure effective or root-mean-square values of voltage and current. Other instruments have been designed to measure peak values of voltage or current.

The step-response characteristics described in Section 3.4 are not ordinarily directly measured by an instrument. From a recording of such a waveform or the waveform displayed on a cathode-ray oscillograph, it is an easy matter to measure the four quantities that have been described.

3.6 Summary

1. The average value of a signal between times t_1 and t_2 is

$$v = \frac{1}{t_2 - t_1} \int_{t_1}^{t_2} v(t)\, dt\,.$$

The average value of a *periodic* signal of period T is

$$\bar{v} = \frac{1}{T}\int_{t_a}^{t_a+T} v(t)\, dt, \qquad \text{all } t_a.$$

For a sinusoid, the average value is zero.

2. The root-mean-square or effective value of a periodic signal is

$$V_{\text{rms}} = \sqrt{\frac{1}{T}\int_{t_1}^{t_1+T} v^2(t)\, dt}, \qquad \text{all } t_1.$$

For the sinusoidal signal

$$V_{\text{rms}} = 0.707 V_{\max}.$$

3. Specifications for the time response of a system to a step input are most often made in terms of (a) overshoot, (b) rise time, (c) time delay, and (d) settling time.

PROBLEMS

3.1 This problem is concerned with the properties of periodic functions. (a) Is the product of two sine waves, $v(t) = \sin \omega_1 t \sin \omega_2 t$, always periodic? (b) Is $v(t) = \sin \omega t + \sin 2\omega t$ periodic? How about $v_1(t) = \sin \omega t + \sin \sqrt{2}\omega t$? In general, is the sum of any number of sine waves always periodic? Explain. (d) Is the product of two periodic functions always periodic? Explain.

3.2 Show that the average value of the sum of N periodic functions with identical periods is equal to the sum of the average values of the N periodic functions.

3.3 A sinusoidal signal is passed through a limiter which clips off the top and bottom of the sinusoid. The clipping level is set at one half the peak value of the signal. Calculate the rms value of the clipped signal. Assume an arbitrary peak value V_0 and frequency ω.

3.4 Show that if $v_1(t)$ and $v_2(t)$ are periodic and if their periods are rationally related ($nT_1 = mT_2$, where T_1 and T_2 are the periods and n and m are integers), then the rms value of $v(t) = v_1(t) + v_2(t)$ is given by

$$V_{\text{rms}}^2 = V_{1\text{rms}}^2 + V_{2\text{rms}}^2,$$

provided further that

$$\int_0^T v_1(t)v_2(t)\, dt = 0,$$

where T is the period of $v(t)$.

3.5 Extend the result in Problem 3.4 to the sum of N functions.

3.6 Show that in general the rms value of the sum of periodic waves is not equal to the sum of the rms values of the periodic waves.

3.7 For the signal response resulting from a unit-step input $v_1(t) =$

$K_1(1 - e^{-\sigma_1 t} \cos \omega t)$, determine an expression for the rise time, the time delay, and the overshoot. Let $\sigma_1 = \frac{1}{2}\omega$.

3.8 If $v(t)$ is a periodic signal with an average value V_{av} and an rms value of V_{rms}, show that $V_{av} \leq V_{rms}$. [*Hint:* Let $v(t) = V_{av} + v(t) - V_{av}$.]

3.9 If $\bar{v}_1$ is the average value of a periodic signal $v_1(t)$, what is the average value of $v_2(t) = v_1(t + \tau)$, where τ is a constant? How about the average value of $v_3(t) = Kv_1(t + \tau)$, where K and τ are constants?

3.10 If V_1 is the rms value of a periodic signal $v_1(t)$, what is the rms value of $v_2(t) = v_1(t + \tau)$, where τ is a constant? How about the rms value of $v_3(t) = Kv_1(t + \tau)$, where K and τ are constants.

3.11 If V_m is the peak value of a signal $v_1(t)$, what is the peak value of $v_1(t + \tau)$, where τ is a constant?

3.12 Show that if $v(t) = V_0 + v_1(t)$, where $v_1(t)$ is periodic with zero average value and V_0 is constant, then

$$V_{rms} = \sqrt{V_0^2 + V_{1\,rms}^2},$$

where $V_{1\,rms}$ is the rms value of v_1.

Chapter **4**

Network Models for Devices

In the preceding chapters, we have introduced idealized operations such as differentiation, integration, addition, and multiplication for the processing of signals. In this chapter we shall associate some of these operations with models for the physical components or devices. An interconnection of these components constitutes a *network*. The word *circuit* is sometimes used interchangeably for network, but usually, circuit refers to a relatively simple (perhaps one-loop) network consisting of a few components. Another term which we have used frequently in the preceding chapters is *system*. The meaning we attach to it is a set of things or parts forming a whole. Thus a network may or may not be a system depending on whether it is the whole or simply a part. A physical system may contain components which are electrical, mechanical, hydraulic, pneumatic, etc. In this chapter, we shall describe the most commonly used models for electrical and mechanical components. These models are called *network elements*.

4.1 Terminal properties of *R*, *L*, and *C* elements

A complete understanding of the electric behavior of electric components and devices requires an understanding of the interaction of materials with electric, magnetic, and electromagnetic fields. However, in studying the total behavior of networks — interconnections of components — it is often sufficient to characterize the components by their external behavior. For example, the effect of a component on the overall behavior of an interconnection of components may be adequately assessed by knowing *what* the terminal voltage–terminal current relationship is for the particular component, rather than knowing *how* or *why* such a component exhibits its characteristics. The main concern of network analysis is to determine

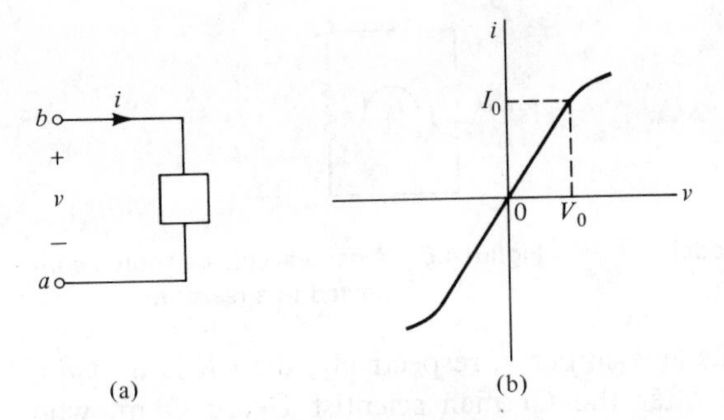

ure 4.1 An example of the terminal characteristics of a resistor model.

not only *what* but *how* and *why* a total network characteristic is caused by individual component behaviors and characteristics. The components are represented by models whose external behaviors approximate those of the actual components.

The first model we consider is that of a *resistor*, a two-terminal component or device which has relatively negligible electric or magnetic-field energy-storage capability. Its external behavior is approximately described by a function relationship between voltage and current at the terminals. The relationship may be given graphically rather than analytically. An example of such a relationship is shown in Figure 4.1. The graph specifies what the current will be for any voltage v across the resistor model. Suppose that the resistor model of Figure 4.1 is connected to a voltage source such as an ideal battery as shown in Figure 4.2. Since the battery voltage $v_{cd} = V_0$ is impressed across terminals b and a, we see that $v_{ba} = V_0$. Then according to Figure 4.1(b), the resulting current will be I_0. A simpler model is shown in Figure 4.3. In the latter case, v and i are linearly related by

$$v = Ri, \tag{4.1}$$

where R is a *constant*. The quantity R is called the *resistance* of the model and $G = 1/R$ is called *conductance*. The symbol for the resistor model is shown in

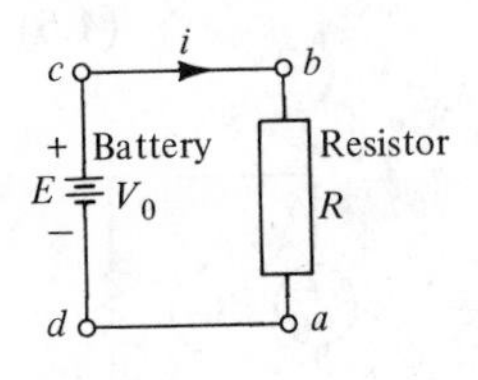

Figure 4.2 A series connection of a resistor and a battery.

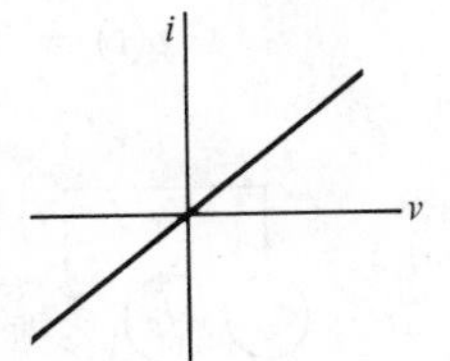

Figure 4.3 The v–i characteristic for the constant linear resistor model. The slope of the straight line is $G = 1/R$.

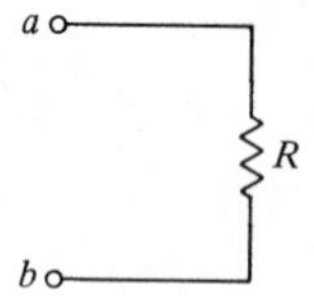

Figure 4.4 Symbol for the resistor model.

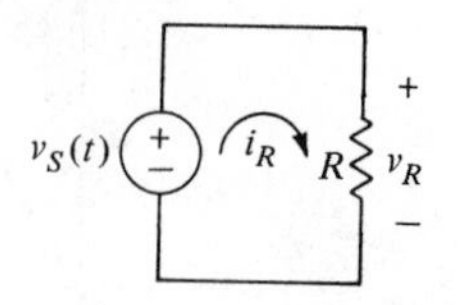

Figure 4.5 A general voltage source connected to a resistor.

Figure 4.4. If v and i are in volts and amperes, respectively, then R is in *ohms*, and G is in *mhos*, units named after the German scientist Georg Ohm, who postulated Equation 4.1. Equation 4.1 is also known as Ohm's law. Instead of a battery, suppose we connect a general voltage source with time-varying magnitude and polarity, as in Figure 4.5. If the reference direction for the current is taken in the clockwise direction shown in Figure 4.5 and the voltage polarity marks are such that the plus terminals of the source and resistor are superimposed, then comparison with the situation of Figures 4.1 and 4.2 shows that this circuit is described by Ohm's law in the form

$$v_R(t) = Ri_R(t). \tag{4.2}$$

If either the current reference or the voltage reference is reversed, then Ohm's law becomes

$$v_R(t) = -Ri_R(t). \tag{4.3}$$

For the reference possibilities of Figure 4.6, we see that (a) and (b) are described by Equation 4.2, while those of (c) and (d) require the use of Equation 4.3.

The resistor R is said to be linear and constant (or time-invariant) whenever v and i are related by a linear equation such as in Equation 4.2 and where R is not a function of time. If R is an explicit function of time, the resistor is said to be time-varying. If R in Equation 4.1 depends on i or v, then the resistor is described as nonlinear. Thus a resistor represented by the equation

$$v(t) = 22i(t) \tag{4.4}$$

is linear time-invariant; one represented by

$$v(t) = 33(1 + \tfrac{1}{2}e^{-t})i(t) \tag{4.5}$$

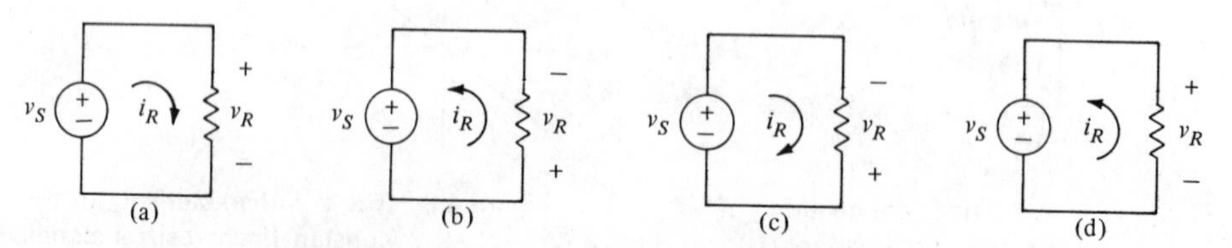

Figure 4.6 The effect of the choice of a reference direction: for (a) and (b), Ohm's law is Equation 4.2, while for (c) and (d), it is Equation 4.3.

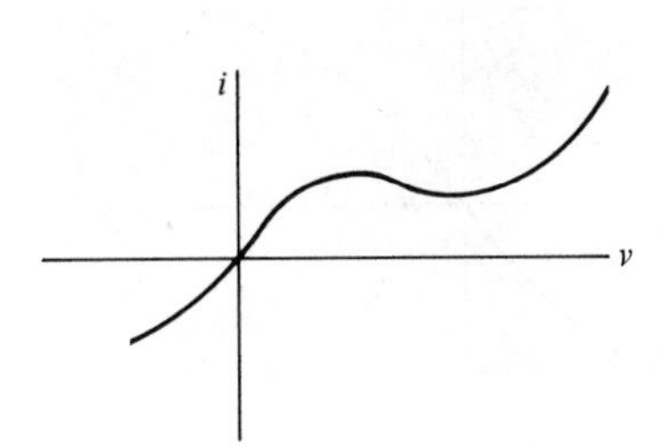

Figure 4.7 A voltage-controlled resistor characteristic.

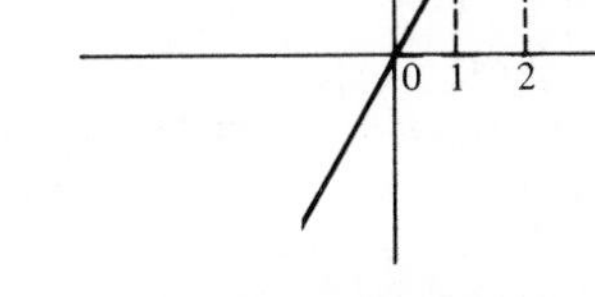

Figure 4.8 A current-controlled resistor characteristic.

is linear time-varying; one represented by

$$v(t) = 47i^2(t) = [47i(t)]i(t) \tag{4.6}$$

is nonlinear; and one represented by

$$v(t) = (100 + 10 \sin t)\sqrt{i(t)} \tag{4.7}$$

is nonlinear. For nonlinear resistors, resistance may be defined either as the ratio v/i or the slope dv/di, depending on the application. For linear resistors, the two definitions yield the same quantity. For nonlinear resistors, the voltage corresponding to a given current or the current corresponding to a given voltage may not be unique. For instance, in Figure 4.7, there are some current values for which there are three possible voltage values. In Figure 4.8, for $v = 1$, there correspond infinitely many possible values between $i = 1$ and $i = 2$ for the current. However, in Figure 4.7, v uniquely defines i, that is to say, for any given v there is only one i. This characteristic is said to be *voltage-controlled.* Similarly in Figure 4.8, i uniquely defines v and the characteristic is said to be *current-controlled.* Thus we speak of a current-controlled resistor or a voltage-controlled resistor. The characteristic in Figure 4.1 is both voltage-controlled and current-controlled. Linear resistors are always current-controlled as well as voltage-controlled.

For the nonlinear resistor, the equation relating v and i is nonlinear. Considering only voltage-controlled or current-controlled resistors, if the relationship between v and i does not contain time t explicitly, the resistor is said to be time-invariant. Otherwise, it is time-varying. For example, the relation expressed by Equation 4.6 does not involve t explicitly so that the nonlinear resistor is time-invariant; the resistor associated with Equation 4.7 is time-varying as well as nonlinear because time t appears explicitly.

The second network element we consider is the model for a component known as a *capacitor.* It is a component or device which has relatively negligible magnetic-field storage capability, negligible dissipation of energy, and so behaves as an electric-field energy-storage device. The capacitor device is approximated by a capacitor model and associated equation which relates the charge and voltage

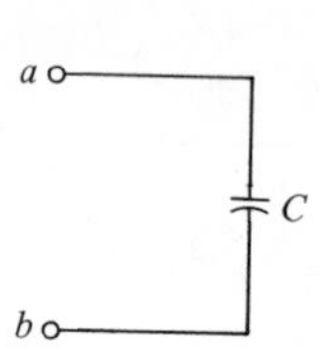

Figure 4.9 The symbol for the capacitor model.

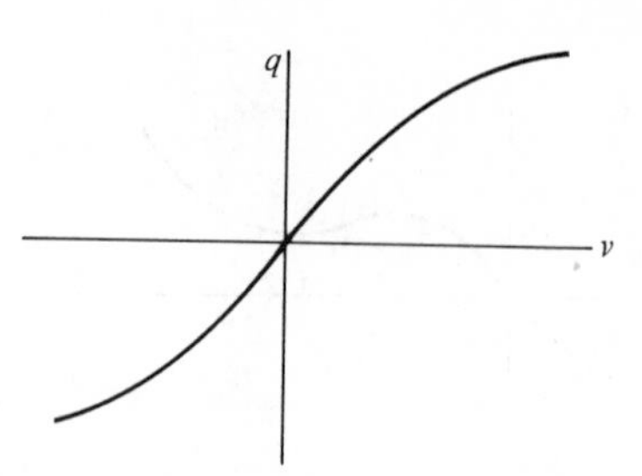

Figure 4.10 Example of a charge–voltage relationship for a capacitor model.

across the two terminals. Charge is the time integral of current, or current is the time rate of charge transfer. Figure 4.9 shows the symbol for the capacitor model, and Figures 4.10 and 4.11 show possible q–v characteristics. As in the discussion of the resistor model, we have linear, nonlinear, time-invariant, and time-varying capacitor models. If q and v are linearly related as in Figure 4.11, then we may write q as

$$q(t) = C(t)v(t), \tag{4.8}$$

where the proportionality factor C is defined as the *capacitance* corresponding to the capacitor model. If q is in coulombs (integral of current in amperes with respect to time in seconds) and v is in volts, then C is in *farads* (F). In Equation 4.8, C does not depend on v but it may depend on time t. If C is also independent of t, then C is said to be time-invariant and C is simply a constant. This constant is the slope of the q–v line as shown in Figure 4.11. For the linear time-invariant case, a differentiation of Equation 4.8 yields

$$i(t) = \frac{dq(t)}{dt} = C\frac{dv(t)}{dt}, \tag{4.9}$$

which is sometimes taken as the defining equation for a linear time-invariant capacitor model. Integrating Equation 4.9 from 0 to t, we have

$$\int_0^t i(\tau)\,d\tau = C[v(t) - v(0)] \tag{4.10}$$

or

$$\int_0^t i(\tau)\,d\tau + Cv(0) = Cv(t). \tag{4.11}$$

The integral may be interpreted as the net change in charge between 0 and t, and $Cv(0)$ may be interpreted as the initial charge at $t = 0$, giving an algebraic sum which is the total charge at time t. This is, of course, equivalent to Equation 4.8. Equations 4.8 and 4.9 are forms of Coulomb's law.

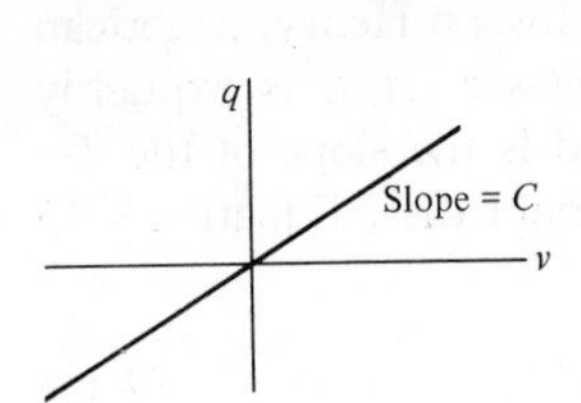

Figure 4.11 Charge–voltage relationship for a linear constant capacitor model. The slope of the line is C.

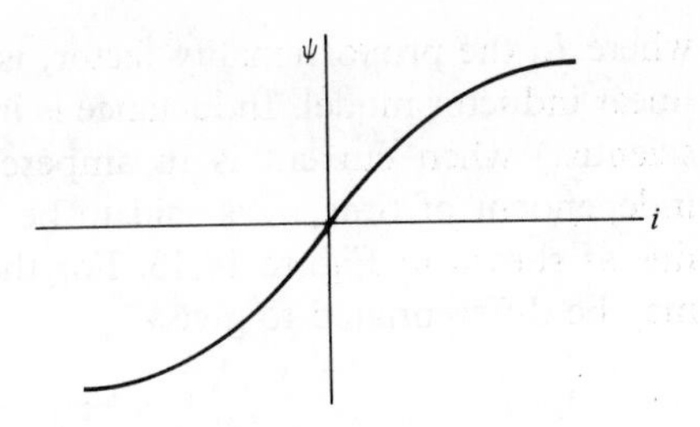

Figure 4.12 Flux-linkage–current relationship for an inductor model.

Analogous to the situation in resistor models, we may have a *charge-controlled* capacitor model or a *voltage-controlled* capacitor model or both. Again, linear models are always both charge-controlled and voltage-controlled.

A third network element model is the *inductor*. The inductor is used to represent a magnetic-field energy-storage device in which other effects may be neglected. The model used to idealize the inductor assumes that the time integral of voltage, called *magnetic flux linkages*,* or simply *flux linkage*, is related to the current by means of a function. We shall denote flux linkage by the symbol $\psi(i)$ so that we have

$$v(t) = \frac{d}{dt}\psi(i). \tag{4.12}$$

Figures 4.12 and 4.13 show possible ψ–i characteristics for an inductor and Figure 4.14 shows the symbol for the model. When ψ is directly proportional to i the model is said to be linear, and ψ may be written as

$$\psi(t) = L(t)i(t), \tag{4.13}$$

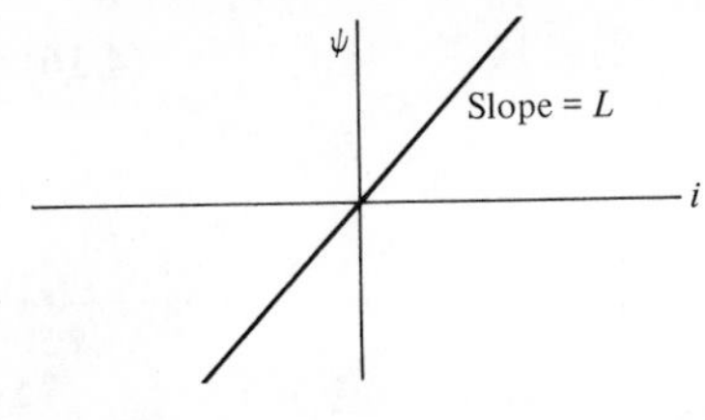

Figure 4.13 Flux-linkage–current relationship for a linear constant inductor model. The slope of the straight line is L.

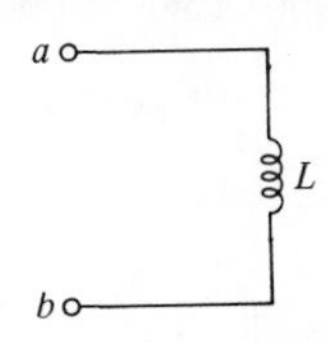

Figure 4.14 Symbol for the inductor model.

* In a coil, flux linkage is equal to $N\phi$ where N is the effective number of turns and ϕ is the magnetic flux. Since N is a fixed number, we shall use the expressions flux-linkage-controlled and flux-controlled interchangeably.

where L, the proportionality factor, is called the *inductance* corresponding to the linear inductor model. Inductance is in *henrys* (H) (after Joseph Henry, American scientist) when current is in amperes, and ψ is in volt-sec. If L is explicitly independent of time, L is said to be time-invariant, and is the slope of the ψ–i line as shown in Figure 14.13. For the linear time-invariant case, Equation 4.13 may be differentiated to give

$$v(t) = L\frac{di(t)}{dt}. \tag{4.14}$$

Analogous to Equation 4.9, we may integrate Equation 4.14 and obtain

$$\int_0^t v(\tau)\,d\tau + Li(0) = Li(t). \tag{4.15}$$

The left-hand side of Equation 4.15 is, of course, the total flux linkage at time t, the integral is the net change in flux linkage from 0 to t, and $Li(0)$ is the initial flux at time $t = 0$.

For the nonlinear inductor model, the meaning of a current-controlled and flux-controlled inductor should be evident. Again, the linear model is always flux-controlled as well as current-controlled.

It is conceivable that for any of the three models, the terminal characteristics remain unchanged even if the terminals are interchanged, as in Figure 4.15. Note that in Figure 4.15 the voltage and current reference directions are maintained, but the terminals are physically interchanged. Where network behavior is not affected by interchanges in terminal connections of the elements, the elements are defined as *bilateral*. It is readily seen that linear elements are always bilateral. Nonlinear elements are bilateral if the terminal equations ($v - i$ for resistor, $q - v$ for capacitor, and $\psi - i$ for inductor) are of the form $y = f(x)$, where $f(x)$ is an odd function. A function $y = f(x)$ is *odd* if for every x

$$-y = f(-x). \tag{4.16}$$

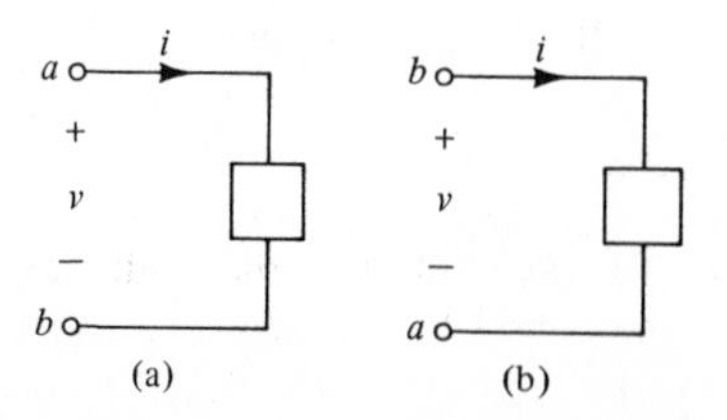

Figure 4.15 If the terminal characteristics of the element are unchanged when the terminals are interchanged, then the element is said to be bilateral.

Table 4.1 Summary of *R*, *L*, *C* description

Element and principal unit	*Symbol and reference directions for variables in equations*	*Terminal characteristics* — *Linear*	*Time-invariant*	*General*
Resistor, ohm (Ω)	i, +, v, −, R	$v = Ri$ or $i = Gv$ If time-invariant R is constant	v = function of i only (current-controlled) or i = function of v only (voltage-controlled)	v = function of i and t (current-controlled) or i = function of v and t (voltage-controlled)
Capacitor, farad (F)	i, +, v, −, C	$q = Cv$ or $v = (1/C)q$ $i = dq/dt$ If time-invariant also, C is constant	q = function of v only (v-controlled) or v = function of q only (q-controlled)	q = function of v and t (v-controlled) or v = function of q and t (q-controlled)
Inductor, henry (H)	i, +, v, −, L	$\psi = Li$ or $i = (1/L)\psi$ $v = d\psi/dt$ If time-invariant also, L is constant	ψ = function of i only (i-controlled) or i = function of ψ only (ψ-controlled)	ψ = function of i and t (i-controlled) or i = function of ψ and t (ψ-controlled)

For example, a voltage-controlled capacitor represented by the terminal or branch equation

$$q = v^3 \tag{4.17}$$

is a bilateral element. A current-controlled resistor represented by the branch equation

$$v = 3i + 5i^3 \tag{4.18}$$

is bilateral.

If the equation relating flux and current associated with an inductor does not involve time t explicitly, the inductor is said to be time-invariant. Otherwise it is time-varying. For example, if the flux-linkage–current relationship is $\psi = i^3$, then the associated inductor model is said to be time-invariant. Of course it is nonlinear. Likewise, a capacitor is said to be time-invariant if the charge–voltage relationship does not involve time t explicitly. For example, the charge–voltage relationship in Equation 4.17 is associated with a time-invariant but nonlinear capacitor.

A summary of the description of the three network elements discussed in this section is given in Table 4.1 for reference.

All the above elements are called *lumped* elements. This is because in the applications for which the above models are useful, it is reasonable to assume that the electrical properties of the elements are concentrated or "lumped" at points in space. Such an assumption leads to total derivatives in the defining equations. In applications where space variables must be considered, the defining equations for the elements involve partial derivatives. These latter elements are called *distributed* elements.*

EXERCISES

4.1.1 In most physics textbooks, the relationship of v_L and i_L for the inductor is given as $v_L = -L(di_L/dt)$. Reconcile the form of this equation with Equation 4.14. (Check with a physics textbook to find the meaning of the minus sign.)

4.1.2 A component (or part) in a radio set is marked 1 microfarad (μF) (1 $\mu\text{F} = 10^{-6}$ farad). What significance does the word *model* have with respect to this component? What are the meanings of the words *capacitor* and *capacitance* in terms of this component?

4.1.3 Give an example of a resistor model which is not bilateral.

4.1.4 Give an example of an inductor model which is nonlinear but bilateral.

4.1.5 Suppose that a nonlinear resistor as described by Equation 4.18 is connected across a constant current source I. Draw the network and label the terminals of the current source a and b. Show that the voltage v_{ab} remains the same regardless of how the resistor is connected to the current source; that is, verify that the resistor model is bilateral.

4.1.6 Give an example of a charge-controlled capacitor which is not voltage-controlled.

4.1.7 Give an example of a voltage-controlled capacitor which is not charge-controlled.

4.1.8 Give an example of a nonlinear capacitor which is both charge-controlled and voltage-controlled.

* Lumped elements and distributed elements are both mathematical models and hence represent approximations to the actual devices and components. The lumped models are cruder approximations compared to the distributed models but the resulting theory for lumped models is also much simpler. The lumped approximation is useful for signals with frequency components up to that which has a wavelength of the same order of magnitude as the physical dimensions of the components. Typically, this is up to several megahertz.

4.1.9 Give an example of a linear but time-varying inductor.
4.1.10 Obtain a differential equation describing a linear time-varying capacitor.

Voltage–current relationships for magnetically coupled coils

The lumped circuit idealizations considered in this section and the next differ from the resistor, inductor, and capacitor of Section 4.1 in that two pairs of terminals are involved rather than one. We first consider coils which are magnetically coupled as shown by Figure 4.16; the magnetic path may be air or it may be some ferromagnetic material. For a linear time-invariant model, the voltage induced in the windings marked 2–2′ and 3–3′ due to the time rate of change of current in winding 1–1′ is related by a constant known as the coefficient of mutual inductance, M_{jk}, and has a polarity which depends on the winding sense of the coil. Thus, in the magnetic system of Figure 4.16, the polarity of the voltage induced in winding 2–2′ with the terminal 2′ as the reference is opposite that of the voltage induced in winding 3–3′ with terminal 3′ as the reference. The voltage–current relationships for coupled coils depend on both the coefficient M_{jk} and on the winding sense of the coils.

It is conventional in circuit theory to indicate winding sense by a pair of dots or polarity marks placed near one of the terminals for each coil. These polarity marks may be established by a procedure which will be described with reference to Figure 4.17. There a battery is connected to coil 1 by the closing of a switch. The polarity mark of coil 1 is chosen to be that terminal which is connected to the plus terminal of the battery. Now the closing of the switch will cause the current in coil 1 to increase so that di_1/dt is positive. The polarity mark for coil 2 is assigned to the terminal which is positive as indicated by a positive deflection of a voltmeter with the plus reference terminal of the voltmeter connected to that terminal.

When both polarity marks and the coefficient M_{jk} are given, the voltage–current relationships are easily written. If the plus reference mark for the voltage v_2

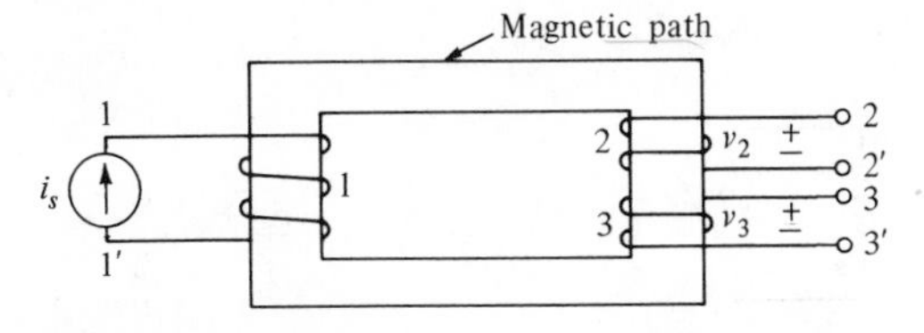

ure 4.16 A system of coupled coils with the magnetic path through which they are coupled.

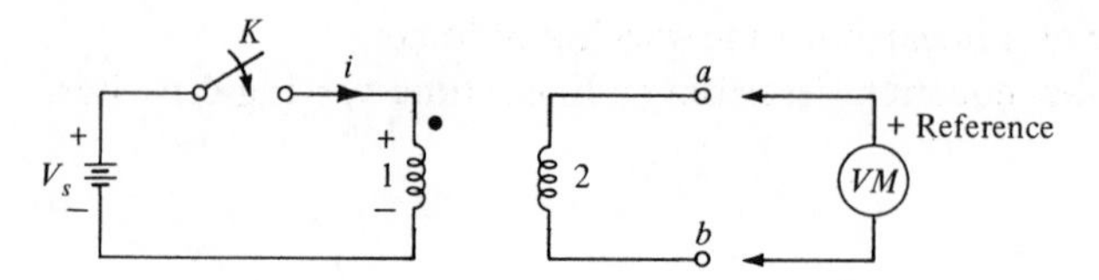

Figure 4.17 The experimental arrangement employed to establish polarity marks for two coupled coils.

coincides with the dotted terminal, the situation depicted in Figure 4.18(a), then we have

$$v_2(t) = M_{12}\frac{di_1(t)}{dt}, \qquad M_{12} > 0. \tag{4.19}$$

If the plus reference mark for v_2 is at the undotted terminal as shown in Figure 4.18(b), then we have

$$v_2(t) = -M_{12}\frac{di_1(t)}{dt}, \qquad M_{12} > 0. \tag{4.20}$$

With current in L_2, the expression for v_2 will contain an additional term $L_2(di_2/dt)$. For instance, if in Figure 4.18(a) the reference direction for i_2 is toward the dotted terminal (from right to left in the top lead), then the total $v_2(t)$ is

$$v_2(t) = L_2\frac{di_2(t)}{dt} + M_{12}\frac{di_1(t)}{dt}. \tag{4.21}$$

In Figure 4.18(b), if the reference direction for i_2 is also toward the dotted terminal (from right to left on the bottom lead), then the total $v_2(t)$ is

$$v_2(t) = -L_2\frac{di_2(t)}{dt} - M_{12}\frac{di_1(t)}{dt}. \tag{4.22}$$

Similarly, for the assigned reference directions of i_2 in both figures, the total voltage $v_1(t)$ for Figure 4.18(a) is

$$v_1(t) = L_1\frac{di_1(t)}{dt} + M_{21}\frac{di_2(t)}{dt}, \tag{4.23}$$

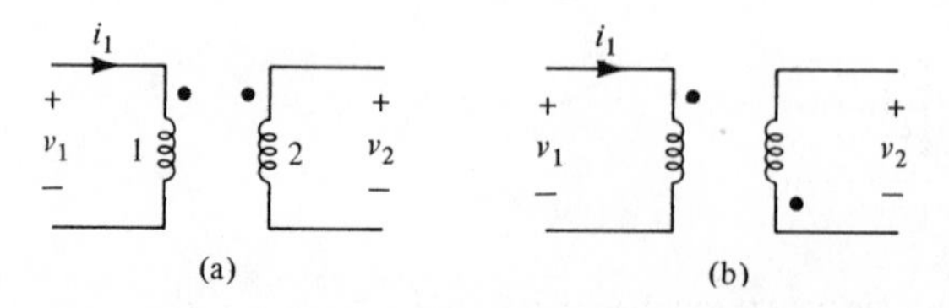

Figure 4.18 Equation 4.19 applies for (a), Equation 4.20 for (b).

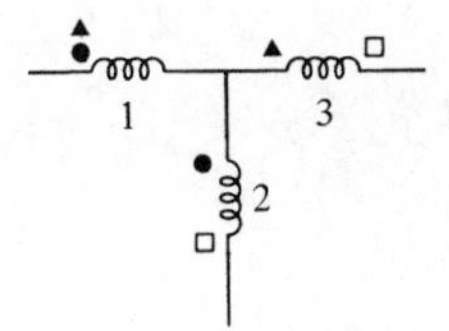

re 4.19 For more complicated magnetic systems, several shapes of polarity marks are used.

and for Figure 4.18(b) is

$$v_1(t) = L_1 \frac{di_1(t)}{dt} + M_{21} \frac{di_2(t)}{dt}. \tag{4.24}$$

We postulate here that $M_{12} = M_{21}$. That such is the case can be proved using energy considerations.*

For complicated magnetic coupling situations, it is sometimes necessary to use a number of pairs of dots. It is conventional to choose the sets of dots to have different identifying shapes as is illustrated in Figure 4.19. The circular symbols specify the coupling between coils 1 and 2, the triangular symbols specify the coupling between coils 1 and 3 and the square symbols specify the coupling between coils 2 and 3.

For nonlinear models, the fluxes and currents are no longer related by linear equations. For instance, for current-controlled nonlinear coupled coils, the terminal relationships have the general form

$$\psi_1 = f_1(i_1, i_2), \tag{4.25}$$

$$\psi_2 = f_2(i_1, i_2), \tag{4.26}$$

where ψ_1 and ψ_2 are the fluxes at the two-terminal pairs. Of course, for the time-varying case, the functions f_1 and f_2 also depend explicitly on time t.

ERCISES

4.2.1 For the magnetic circuit of Figure 4.16, establish a system of polarity marks using Lenz's law (after Lenz, German scientist). (Consult your physics textbook for Lenz's law.)

4.2.2 For the circuit shown in Figure 4.20(a), write an equation for the part of $v_2(t)$ due to the current in coil 1.

4.2.3 Repeat Exercise 4.2.2 for the circuit of Figure 4.20(b).

4.2.4 Repeat Exercise 4.2.3 for Figure 4.16.

* See N. Balabanian, *Fundamentals of Circuit Theory* (Boston: Allyn and Bacon, 1961).

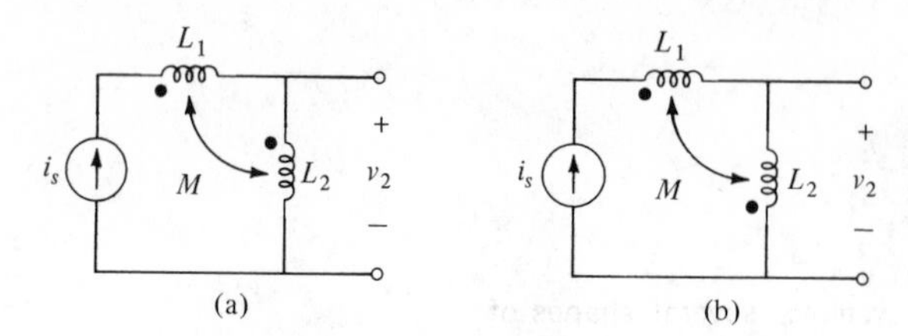

Figure 4.20

4.2.5 For the circuit in Figure 4.18(a), replace $i_1(t)$ by $i_a(t)$ which is opposite to i_1, and assume i_b as the current in coil 2 directed toward the dot. Write the equations for v_1 and v_2 in terms of L_1, L_2, M, i_a, and i_b.

4.2.6 Replace v_1 in Figure 4.18(a) by v_a which has a reference mark opposite to that of v_1. Write the equations for v_a and v_2 in terms of L_1, L_2, M, i_1, and i_2, where i_2 is directed toward the dotted terminal.

4.2.7 Repeat Exercise 4.2.6 for Figure 4.18(b) where i_2 is directed toward the undotted terminal.

4.3 Ideal converters

The ideal converters to be discussed in this section are lumped models of electrical devices which have two terminal pairs. The *ideal transformer* represented in Figure 4.21(a) is a model useful in connection with the study of actual transformers which are used in power and communications applications, and is also used often to accomplish isolation in electronic circuits. The two voltages and two currents for the ideal transformer are related by the equations

$$v_2 = nv_1 \tag{4.27}$$

and

$$i_2 = \frac{-1}{n} i_1, \tag{4.28}$$

where n is a real number. This idealization is useful not only because it is accurate for some purposes but also because it may be combined with ideal R, L, C elements to represent practical transformers.

If a linear (possibly time-varying) resistor is connected to terminals 2–2′, then v_2 and i_2 are related by Ohm's law as follows:

$$v_2 = -Ri_2. \tag{4.29}$$

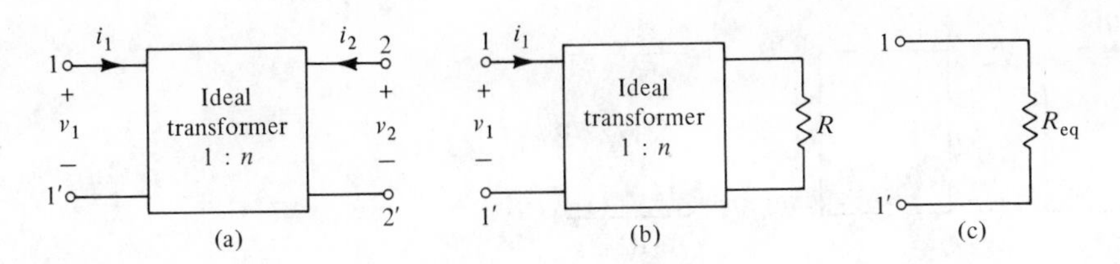

Figure 4.21 Schematic representations of the ideal transformer.

Substituting this equation into Equations 4.27 and 4.28 and solving for v_1 in terms of i_1, we have

$$v_1 = \frac{1}{n} v_2 = \frac{-1}{n} Ri_2 = \frac{1}{n^2} Ri_1 . \tag{4.30}$$

Indeed, Equation 4.30 is the defining characteristic of a linear resistor. At terminals 1–1′, the network appears as a resistor of value

$$R_{eq} = \frac{1}{n^2} R, \tag{4.31}$$

and an equivalent representation of the network of Figure 4.21(b) is that shown in (c) of the figure. Thus the ideal transformer acts to adjust the magnitude of resistance in terms of measurements made at the input terminals of the ideal transformer.

Another useful converter model is the *gyrator* first introduced by Tellegen (Dutch engineer, 1900–   ) in 1948. The gyrator is a model for a number of physical devices including *Hall-effect* semiconductor devices and also certain microwave structures. It is represented by the special symbol shown in Figure 4.22. For the voltages and currents of the figure, it is required that

$$v_1 = Ki_2 \tag{4.32}$$

and

$$v_2 = -Ki_1, \tag{4.33}$$

where K is a real constant which may be either positive or negative. Thus the

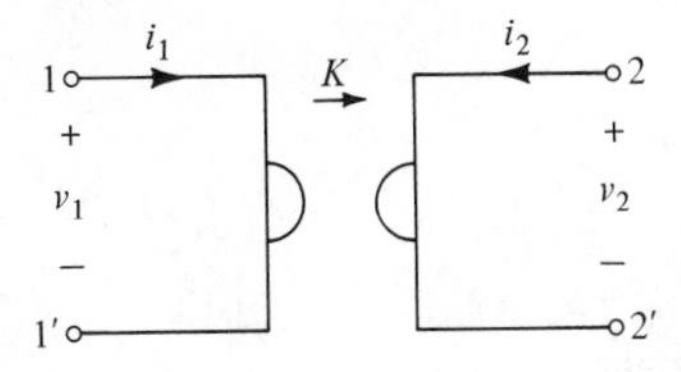

Figure 4.22 Conventional representation of the Tellegen gyrator.

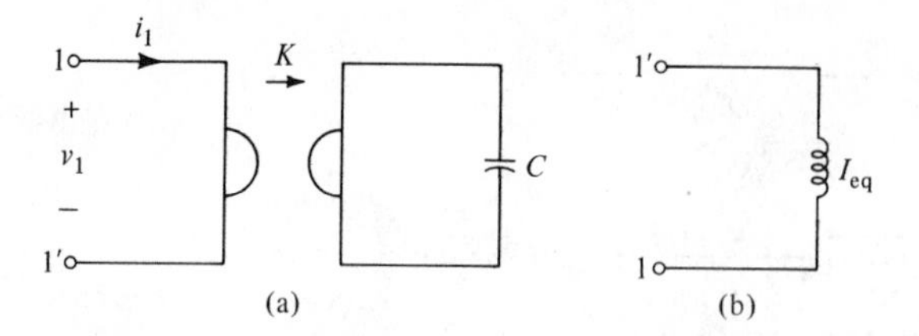

Figure 4.23 A gyrator terminated in a capacitor (a) is equivalent to an inductor (b).

gyrator relates the current corresponding to a terminal pair to the voltage corresponding to another terminal pair. (In Hall-effect devices, electric-field effects are transformed into equivalent magnetic-field effects.)

An interesting behavior of the gyrator is exhibited by studying the two-terminal equivalent of the gyrator terminated in a linear (possibly time-varying) capacitor as shown in Figure 4.23(a). For this network, we have

$$v_1 = Ki_2 = K\left(-\frac{d}{dt}Cv_2\right). \tag{4.34}$$

Now, since $v_2 = -Ki_1$, this equation becomes

$$v_1 = K\frac{d}{dt}(CKi_1) = \frac{d}{dt}(K^2Ci_1) = \frac{d}{dt}(L_{eq}i_1), \tag{4.35}$$

where the equivalent inductance has the value $L_{eq} = K^2C$. By this demonstration, we see that a gyrator, together with a linear capacitor, is equivalent to a linear inductor. Hence, if we have resistors, linear capacitors, and gyrators, we can always construct (in theory) networks which are equivalent to those containing resistors, linear inductors, and linear capacitors.

Another useful converter is the *negative converter* which is a model for devices employing transistors or vacuum tubes which are common in the telephone industry. For the negative converter as shown in Figure 4.24, we have either

$$v_1 = -kv_2 \tag{4.36}$$

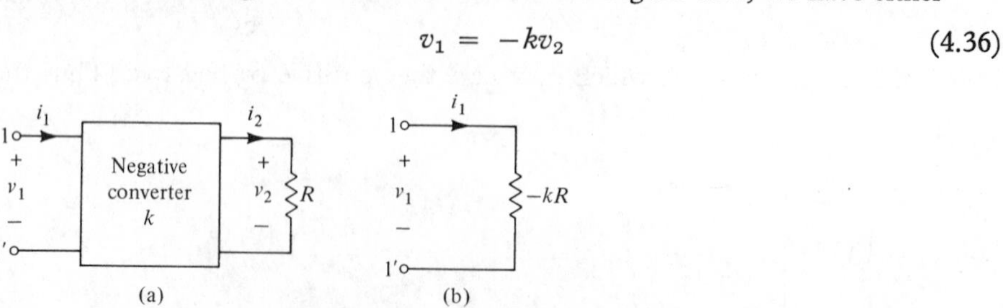

Figure 4.24 The negative converter terminated in R is equivalent to $-R$ when $k = 1$.

and

$$i_1 = i_2, \tag{4.37}$$

or

$$i_1 = -ki_2 \tag{4.38}$$

and

$$v_1 = v_2, \tag{4.39}$$

where k is a positive real constant. Equations 4.36 and 4.37 define a *voltage-inversion* type of negative converter. Equations 4.38 and 4.39 define a *current-inversion* type of negative converter. The negative converter has the property that when it is terminated in a resistor (possible nonlinear) at terminals 2–2′, the two-terminal behavior at 1–1′ is that of a resistor which is the negative of that at 2–2′. Similar statements apply to the inductor and the capacitor. Practical realizations of negative converters of high precision have been manufactured by the Bell Telephone Laboratories.*

ERCISES

4.3.1 Find the two-terminal equivalent of an ideal transformer terminated in (a) a linear inductor, (b) a linear capacitor.

4.3.2 Find the two-terminal equivalent of a gyrator terminated at 2–2′ in (a) linear R, and (b) linear L. Repeat for the R or L termination (linear) at terminals 1–1′ and find the two-terminal equivalent network at 2–2′.

4.3.3 Show that the two-terminal equivalent of a negative converter terminated in a linear L is $-L$ when $k = 1$. Show also that when the termination is a linear C, the equivalent network is $-C$ when $k = 1$.

4.3.4 A new device is constructed by connecting two gyrators in cascade (or tandem). Show that this device is equivalent to an ideal transformer.

4 Operational amplifiers

An *operational amplifier* is a multiterminal device of substantial practical significance. Advances in integrated circuit technology have made it possible to produce commercial versions which are reliable and inexpensive, with the result that it finds application in numerous electronic systems. The symbol for the

* J. G. Linvill, "Transistor Negative Impedance Converters," *Proc. IRE*, **41**, pp. 725–729, June, 1953; see also F. H. Blecher, "Application of Synthesis Techniques to Electronic Circuit Design," *IRE Trans. on Circuit Theory*, **CT-7**, Special Supplement, August, 1960, pp. 79–91. Negative converters used in industry operate up to about 1 megahertz (MHz).

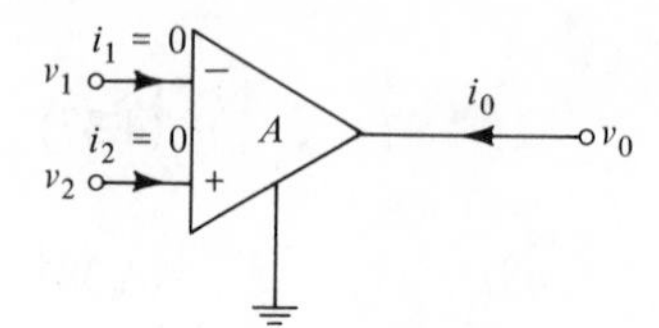

Figure 4.25 The standard symbol for an operational amplifier.

operational amplifier (or *op-amp*) is shown in Figure 4.25. It has two input voltages, v_1 and v_2, and one output voltage, v_0. Note that these voltages are measured with respect to a common ground terminal. They are related by

$$v_0 = A(v_2 - v_1) \tag{4.40}$$

where A is a large positive constant (typically 20,000). Furthermore, the input currents are zero:

$$i_1 = 0, \qquad i_2 = 0. \tag{4.41}$$

However, the output current i_0 is not constrained by the model. Of course i_0 and v_0 must satisfy the Kirchhoff's laws and the v–i relationships imposed by other connections. For example, if a resistor R is connected between ground and the output, then

$$v_0 = -Ri_0. \tag{4.42}$$

(Note the reference direction assigned to i_0.)

Integrated-circuit fabrications of operational amplifiers involve the equivalent of several transistors and field-effect transistors (FETs). Typically, the linear relationship of Equation 4.40 is valid for $v_2 - v_1$ of about $\frac{1}{2}$ millivolt or less. For larger input voltages, the amplifier saturates. Thus for linear operation, v_0 is limited to 10 volts or less, if A is at least 20,000. Early commercial versions of op-amps had only one input, with the v_2 input not available. A very useful simplifying approximation in the analysis of networks containing operational amplifiers is that $v_1 - v_2$ is zero since A is very large and v_0 is finite.

Operational amplifiers are useful components in analog computers as well as many networks such as active filters. We will illustrate some applications with examples.

EXAMPLE 4.4.1

Figure 4.26 shows a realization of an *isolator* using an op-amp. We will examine an interesting property of this network. From KVL applied to the outer loop, we obtain

$$v_{12} + v_{\text{in}} - v_0 = 0. \tag{4.43}$$

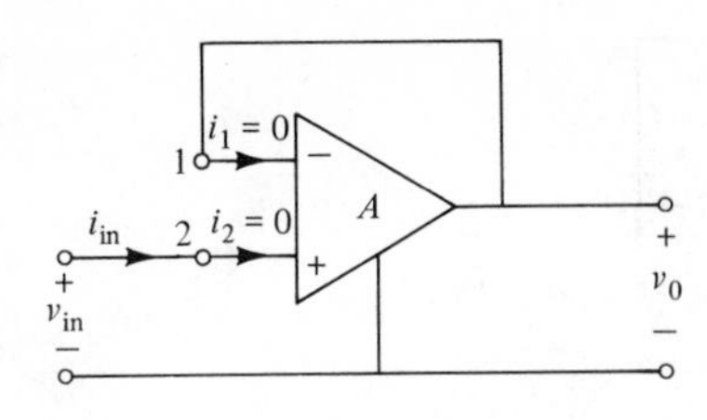

Figure 4.26 The realization of an isolator using an op-amp.

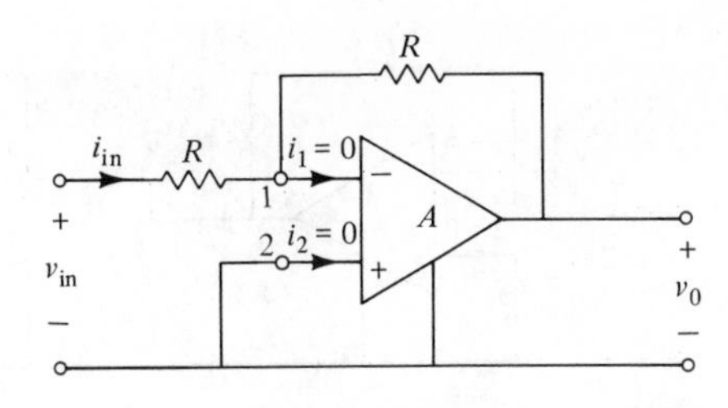

Figure 4.27 The realization of a phase inverter.

Since $v_{12} = 0$,

$$v_{in} = v_0 \tag{4.44}$$

and furthermore

$$i_{in} = i_2 = 0. \tag{4.45}$$

Hence, if v_{in} is a voltage across a terminal pair of a network to the left of the isolator, the isolator will appear to be an open circuit; that is, $i_{in} = 0$. The voltage at the output of the isolator, however, will "track" v_{in}. This network isolates two networks in the sense that the output network will not load the input network.

AMPLE 4.4.2

The network shown in Figure 4.27 is a realization of a *phase inverter*. We will show that

$$v_0 = -v_{in}. \tag{4.46}$$

Since terminal 2 is connected to the reference node (ground), and since $v_{12} = 0$, then $v_1 = 0$. Applying KCL to node 1, we have

$$\frac{v_{in}}{R} + \frac{v_0}{R} = 0. \tag{4.47}$$

Thus, for any R, Equation 4.46 holds. Furthermore, the input current

$$i_{in} = \frac{v_{in}}{R} \tag{4.48}$$

can be made sufficiently small by choosing R to be large. Typically, $R = 1$ megohm. $= 10^6$ ohms.

AMPLE 4.4.3

Figure 4.28 shows a realization of a *summer*. We will demonstrate that

$$v_0 = v_a + v_b. \tag{4.49}$$

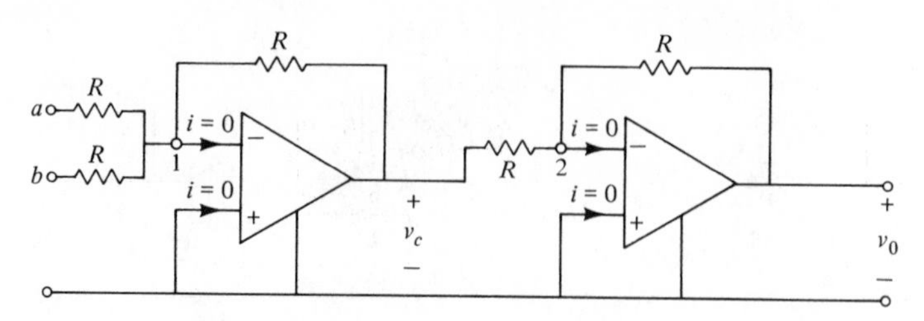

Figure 4.28 The realization of a summer: $v_0 = v_a + v_b$.

Observing that the input currents to the op-amps are zero and that $v_1 = 0$, we find that KCL applied to node 1 yields

$$\frac{v_a}{R} + \frac{v_b}{R} + \frac{v_c}{R} = 0. \tag{4.50}$$

Similarly, KCL applied to node 2 gives

$$\frac{v_c}{R} + \frac{v_0}{R} = 0. \tag{4.51}$$

Multiplying Equations 4.50 and 4.51 by R and eliminating v_c, we obtain Equation 4.49 as required. To make the input currents small, R is chosen to be large. Additional input terminals may be connected through additional resistors R to add three or more input voltages.

EXAMPLE 4.4.4

The network in Figure 4.29 is a realization of an *integrator* whose output is an integral of the input. The analysis is again made simple by the approximation that the voltages across the input terminals of the op-amps are zero. Using KCL, we have

$$\frac{v_{\text{in}}}{R_1} + \frac{v_a}{R_2} = 0 \tag{4.52}$$

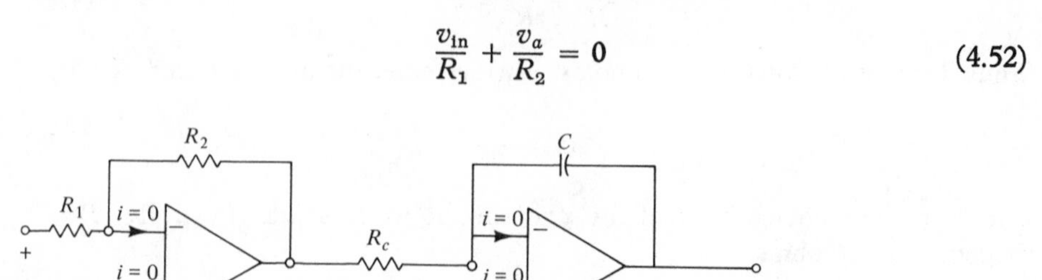

Figure 4.29 The realization of an integrator.

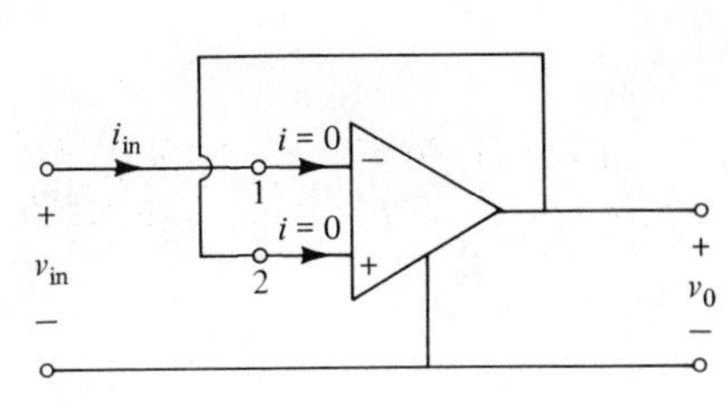

Figure 4.30

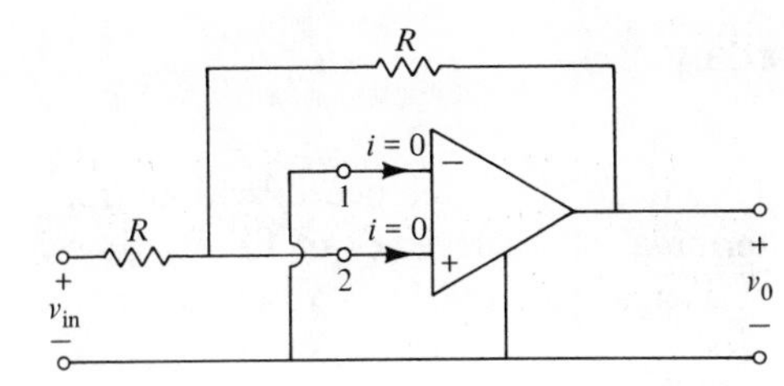

Figure 4.31

and

$$\frac{v_a}{R_c} + C\frac{dv_0}{dt} = 0. \tag{4.53}$$

Thus

$$\frac{-R_2}{R_1R_c}v_{in} + C\frac{dv_0}{dt} = 0 \tag{4.54}$$

or, finally

$$v_0(t) = v_0(0) + \frac{R_2}{R_1R_cC}\int_0^t v_{in}(\tau)\,d\tau. \tag{4.55}$$

The elements R_1, R_2, R_c, and C may be chosen to realize any desired ratio $R_2/(R_1R_cC)$. For example, by choosing the ratio $R_2/(R_1R_cC) = 1$, Equation 4.55 has an especially simple form.

ERCISES

4.4.1 Determine the terminal v–i characteristics of the network of Figure 4.30. Compare your result with the isolator of Example 4.4.1.

4.4.2 Determine the terminal v–i characteristics of the network of Figure 4.31. Compare the result with the phase inverter of Example 4.4.2.

4.4.3 Determine the terminal v–i characteristics of the network of Figure 4.32.

4.4.4 Connect a phase inverter as in Figure 4.27 to the output of the network shown in Figure 4.32. Determine the terminal v–i characteristics of the composite network. Compare your results with the summer of Figure 4.28.

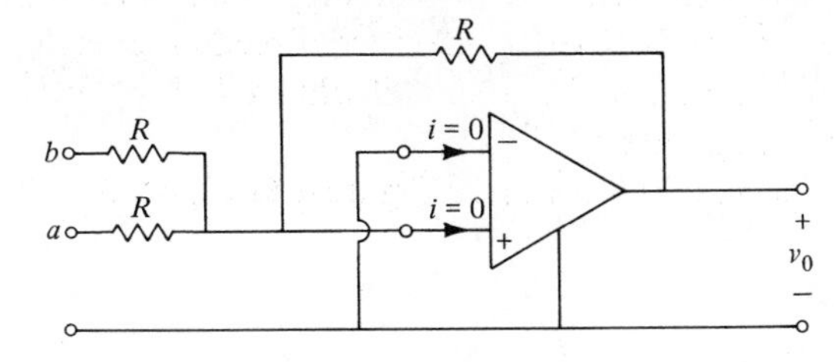

Figure 4.32

4.5 Summary

1. Given the reference directions for v and i as shown in Figure 4.15(a), the terminal characteristics of linear, time-invariant models for resistors, inductors, and capacitors are:

Resistors:

$$v(t) = Ri(t) \qquad \text{or} \qquad i(t) = Gv(t) = \frac{1}{R}v(t)$$

Inductors:

$$v(t) = L\frac{di(t)}{dt} \qquad \text{or} \qquad i(t) = i(0) + \frac{1}{L}\int_0^t v(\tau)\,d\tau$$

Capacitors:

$$i(t) = C\frac{dv(t)}{dt} \qquad \text{or} \qquad v(t) = v(0) + \frac{1}{C}\int_0^t i(\tau)\,d\tau$$

where R, G, L, and C are constants called resistance, conductance, inductance, and capacitance, respectively. When v is in volts and i in amperes, R will be in ohms, G in mhos, L in henrys, and C in farads. For time-varying and nonlinear characteristics, see Table 4.1. These elements are called two-terminal elements.

2. When two coils are magnetically coupled, in addition to voltage induced by a time rate of change of current through the coil as summarized in 1 above, there is a voltage induced by a time rate of change of current through the other coil. With reference directions as indicated in Figure 4.18,

$$v_1 = L_1\frac{di_1}{dt} + M_{12}\frac{di_2}{dt}, \qquad v_2 = M_{21}\frac{di_1}{dt} + L_2\frac{di_2}{dt},$$

where L_1 and L_2 are called self-inductances of coils 1 and 2, respectively, and M_{12} and M_{21} are mutual inductances. It is always true that $M_{12} = M_{21}$ for linear elements. The mutual inductance terms may be positive or negative depending on the winding senses. Winding polarities are usually indicated by marking the terminals with dots differing in shape for each pair of windings. The following rule applies to these dots: If a current is assigned a reference direction entering the dot of one coil, then the dotted terminal of the other coil will have + polarity for the voltage induced by this changing current.

3. An ideal transformer is a four-terminal network [see Figure 4.21(a)] described by the equations

$$v_2 = nv_1, \qquad i_2 = \frac{-1}{n}i_2.$$

4. A gyrator (see Figure 4.22) is a four-terminal network characterized by

$$v_1 = Ki_2, \qquad v_2 = -Ki_1.$$

5. A voltage-inversion type of negative converter [see Figure 4.24(a)] is a four-terminal network described by

$$v_1 = -Kv_2, \qquad K \text{ is a real, positive constant}$$
$$i_1 = i_2.$$

6. A current-inversion type of negative converter [see Figure 4.24(a)] is a four-terminal network described by

$$i_1 = -Ki_2, \qquad K \text{ is a real, positive constant}$$
$$v_1 = v_2.$$

7. An operational amplifier (see Figure 4.25) is a four-terminal network characterized by

$$v_0 = A(v_2 - v_1), \qquad A = \text{large real positive constant}$$
$$i_1 = 0, \qquad i_2 = 0.$$

)BLEMS

4.1 Find the corresponding $i(t)$ or $v(t)$ for the linear time-invariant positive elements and signal waveforms shown in Figure 4.33.

4.2 Find the two-terminal equivalent of a gyrator terminated in (a) a voltage source, and (b) a current source.

4.3 For the two cascade connections of ideal converters shown in Figure 4.34, each one terminated in a linear capacitor C_2, find the single element equivalent at terminals 1–1′.

4.4 For the circuits in Figure 4.20, write equations for v_2 in terms of i_s. [*Hint:* Both the time rate of change of the current through L_1 and that of the current through L_2 affect v_2.]

4.5 Apply KVL to the two loops of the network shown in Figure 4.35 and obtain differential equations for i_1 and i_2.

4.6 Reverse the reference direction of i_2 in Figure 4.35 and repeat Problem 4.5.

4.7 The simple network shown in Figure 4.36 contains a voltage-controlled voltage source. Show that it is equivalent to a voltage-inversion-type negative converter, with $K = 1$.

4.8 The simple network shown in Figure 4.37 contains a current-controlled current source. Show that it is equivalent to a current-inversion-type negative converter, with $K = 1$.

4.9 The simple network shown in Figure 4.38 contains two current-controlled voltage sources. Show that it is equivalent to a gyrator.

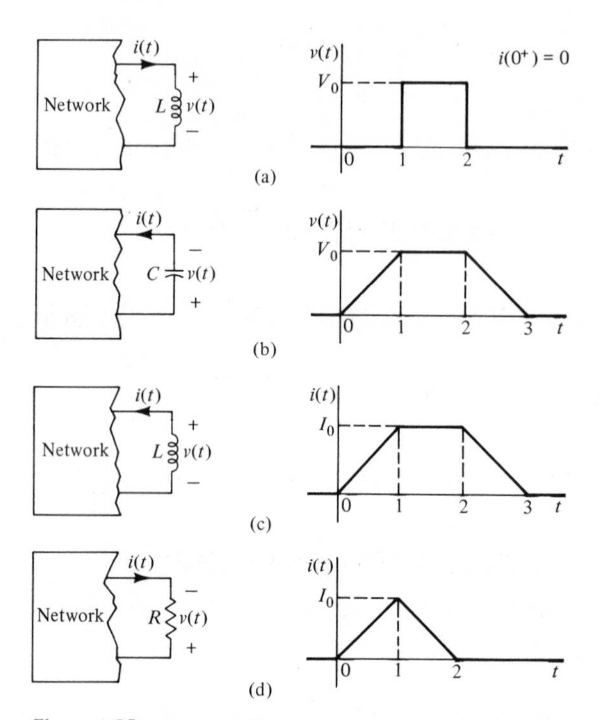

Figure 4.33

4.10 Suppose that a voltage-controlled resistor with v and i conventions as in Figure 4.1(a) is described by $i = 3v + v^2$. The resistor is connected to a voltage source as in Figure 4.6(a) whose terminal b is on top. If $v_s = 10$ volts, compute i. Suppose the resistor connection is reversed so that b is at the bottom. Compute i, for the same v_s. Is the resistor bilateral?

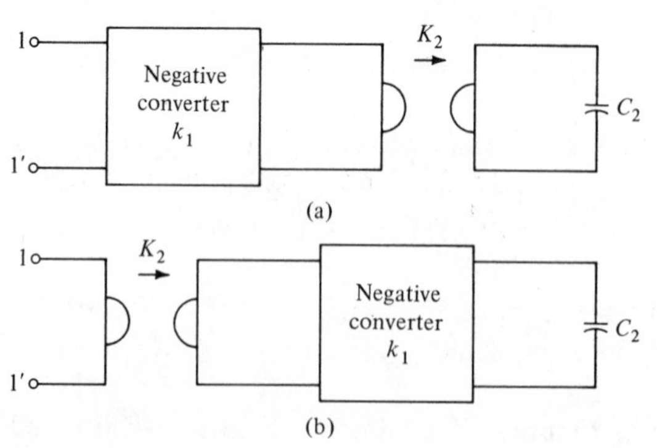

Figure 4.34

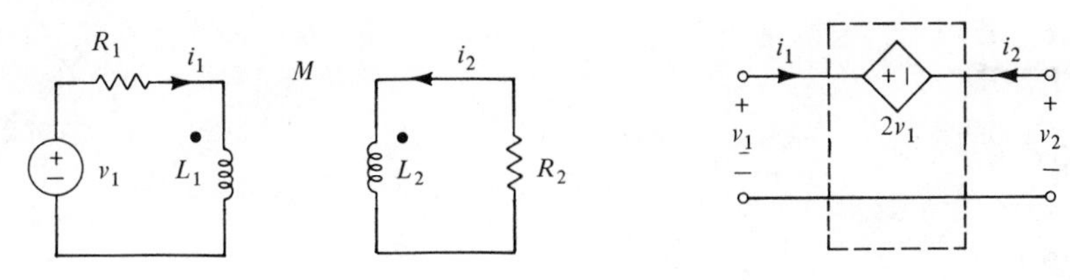

Figure 4.35

Figure 4.36

4.11 A nonlinear voltage-controlled capacitor has a model described by $q = 3v^3$. Compute the current if a voltage source with a waveform as in Figure 4.33(b) is impressed across it.

4.12 Repeat Problem 4.11 for the waveform in Figure 4.33(a).

4.13 Repeat Exercise 4.3.1 for (a) a nonlinear inductor $\psi = i^2$, reference convention as in Table 4.1, and for (b) a nonlinear capacitor $v = q^{1/2}$, reference convention as in Table 4.1.

4.14 Repeat Exercise 4.3.3 for a nonlinear current-controlled inductor as a termination.

4.15 Repeat Exercise 4.3.3 for a nonlinear flux-controlled inductor as a termination.

4.16 Repeat Exercise 4.3.3 for (a) a voltage-controlled nonlinear capacitor as a termination, and for (b) a charge-controlled nonlinear capacitor as a termination.

4.17 The network shown in Figure 4.39 contains two voltage-controlled voltage sources. Show that it is equivalent to an operational amplifier.

4.18 In Figure 4.27, replace the input resistor by R_1 and replace the feedback resistor by R_2. Determine the relationship between v_o and v_{in} for the modified network.

4.19 For the phase inverter of Figure 4.27, suppose that $A = 20{,}000$. Without making the simplifying approximation of $v_{12} = 0$, determine the relationship between v_o and v_{in}. Compute the percentage error caused by the approximation used in Section 4.4.

4.20 For the isolator of Figure 4.26, suppose that $A = 20{,}000$. Determine the

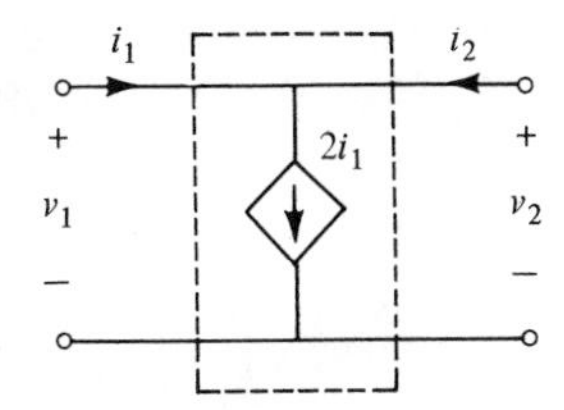

Figure 4.37

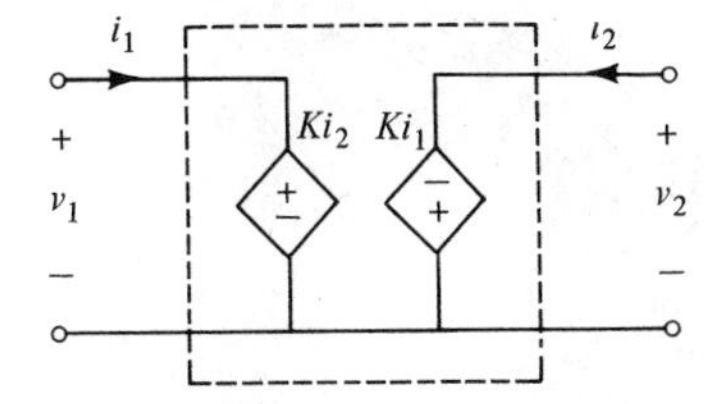

Figure 4.38

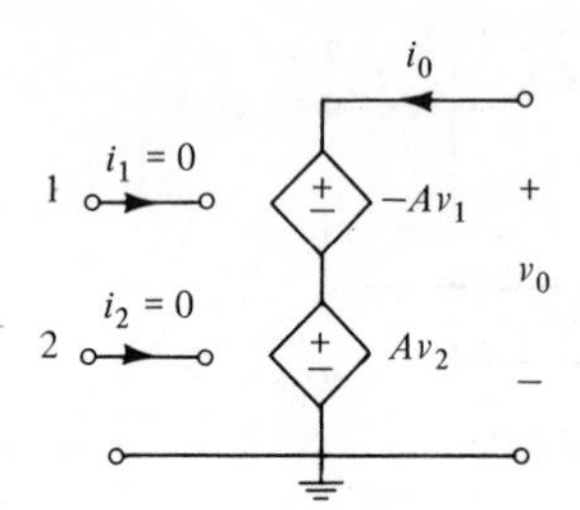

Figure 4.39

relationship between v_0 and v_{in} without using the approximation that $v_{12} = 0$ in Section 4.4.

4.21 Using only one voltage-controlled voltage source, determine a network equivalent to an isolator.

4.22 Repeat Problem 4.21 for a phase inverter.

Chapter 5

rchhoff's Laws

Network terminology

We have already introduced the words *element*, *network*, and *terminal* in previous chapters. We shall have need for additional definitions to describe certain aspects of a network. The word *branch* is used synonymously with two-terminal *element*. In the literature, however, *branch* is sometimes used in a wider sense to mean a string of elements joined end to end or even a two-terminal subnetwork. In the network of Figure 5.1, elements *c* and *d* may be considered separate branches or they may be considered parts of a single branch *cd*. Usually, the precise meaning will be clear from the context. The junction of two or more branches is called a *node*. In Figure 5.1, node 5 joins branches *d*, *e*, *f*, and *g*. If only two branches are involved, the node is called *simple*. Thus, in Figure 5.1 node 4 is a *simple node*. A *terminal* of a network is a node to which an input source may be connected, or from which an output signal may be observed, or to which another network terminal may be connected. An end-to-end connection of branches such as *c* and *d* in Figure 5.1 is called a series connection. Other series connections in the figure are *ak* and *gh*. A parallel connection of branches implies the joining of one terminal from each branch to form a node, and the remaining terminals joining to form another node. Thus branches *b* and *m* are in parallel. Branch *f* is also in parallel with branch *gh*. The word *path* implies a specified train of branches. There are several kinds of paths. One path in Figure 5.1 starts at node 3, continues to node 4, and thence to node 5. Such a path is called an *open path*. Note that for an open path there is a starting node to which only one branch is connected, and an ending node to which only one branch is connected. If the starting and ending are one and the same, we have a *closed path*. One closed path is *cdfb*. Another is *ambefk*. If each node in the closed path connects exactly two branches, then the closed path is called *simple*. Simple closed paths are also called *topological circuits* or *loops*. In Figure 5.1, *cde* is an example of a loop.

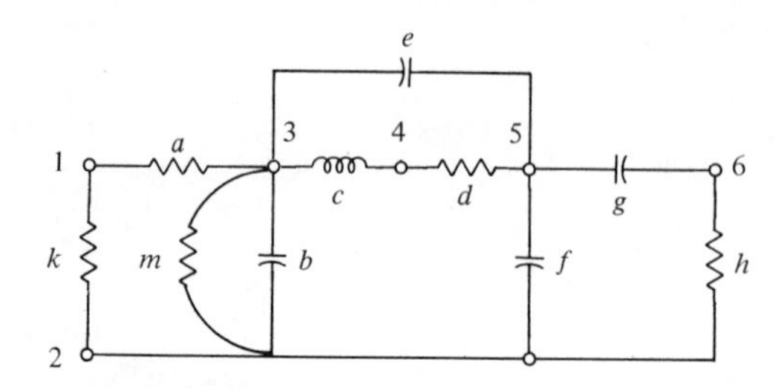

Figure 5.1 Network for illustrating the definitions for branches, nodes, paths, and circuits.

EXERCISES

5.1.1 Find all paths (a) between nodes 1 and 3 in Figure 5.2, and (b) between nodes 1 and 5 in Figure 5.1.

5.1.2 Find as many loops as you can in Figures 5.2 and 5.1. Identify these loops in terms of the node or element labels.

5.2 Kirchhoff's current law

We introduced Kirchhoff's current law (abbreviated KCL) in Chapter 1, and applied it to elementary examples there and in later chapters. In this section, we continue our study of this fundamental law in a more general setting. First, we briefly review the application of KCL to elementary examples. In Figure 5.3, consider node 1 and the three branches connected to it to be a portion of a complete network. For the given reference directions, Kirchhoff's current law requires that

$$-i_a - i_b - i_c = 0. \tag{5.1}$$

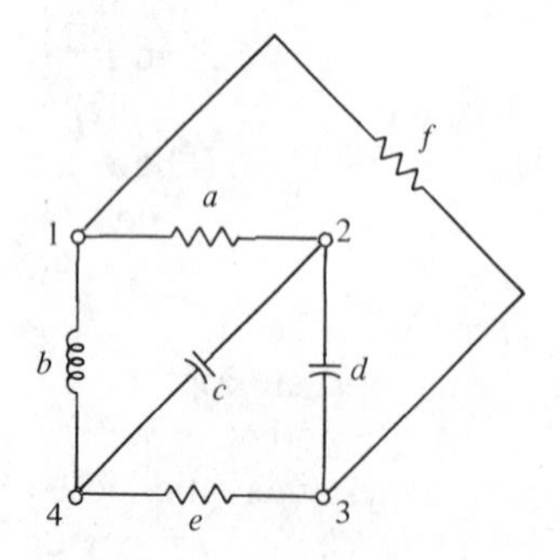

Figure 5.2

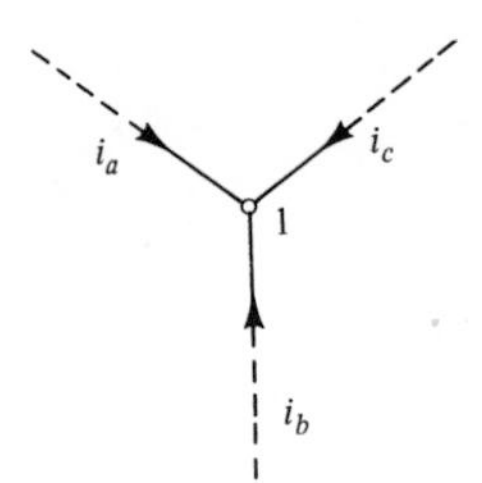

Figure 5.3 An isolated node used for discussing Kirchhoff's current law.

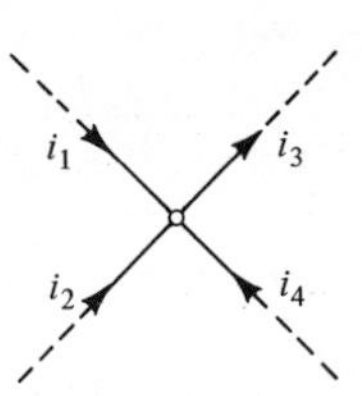

Figure 5.4 A node to which four branches are connected.

Thus, if $i_a = 5$ and $i_b = 2$, KCL requires that i_c be -7. As a second example, consider Figure 5.4. Applying KCL, we get $-i_1 - i_2 + i_3 - i_4 = 0$. Thus, if $i_1 = 5 \sin \omega t$, $i_2 = 3 \sin \omega t$, $i_3 = 6 \cos \omega t$, then i_4 must be

$$i_4 = -i_1 - i_2 + i_3 = -8 \sin \omega t + 6 \cos \omega t = 10 \sin \left(\omega t + \tan^{-1} \frac{6}{-8}\right). \quad (5.2)$$

The last step is accomplished by letting $-8 = A \cos \theta$ and $6 = A \sin \theta$, and then using the double-angle relationship

$$A \sin \omega t \cos \theta + A \cos \omega t \sin \theta = A \sin (\omega t + \theta).$$

A and θ are thus determined. The current law of Kirchhoff may be thought of as a consequence of some other assumption or postulate such as the conservation of charge. The latter states that the net charge flowing into a node is equal to the net charge flowing out of a node. The current law then follows directly by differentiation with respect to time.

In applying KCL to a given node, all branch currents with the reference arrows pointing toward the node in question, have the same sign. The reference arrows out of the node require the opposite sign for these corresponding terms in the KCL equation.

In general, if there are N_v nodes and N_b branches, then Kirchhoff's current law may be written in the very compact form

$$\sum_{j=1}^{N_b} a_{kj} i_j = 0, \qquad k = 1, \ldots, N_v, \quad (5.3)$$

where $a_{kj} = +1$ if branch j is connected to node k and the current reference for i_j is away from node k; $a_{kj} = -1$ if branch j is connected to node k and the current reference for i_j is toward node k; and $a_{kj} = 0$ if branch j is not connected to node k.

The current law may be applied to several nodes and the resulting KCL equations added. This simple addition of KCL equations leads to a generalized

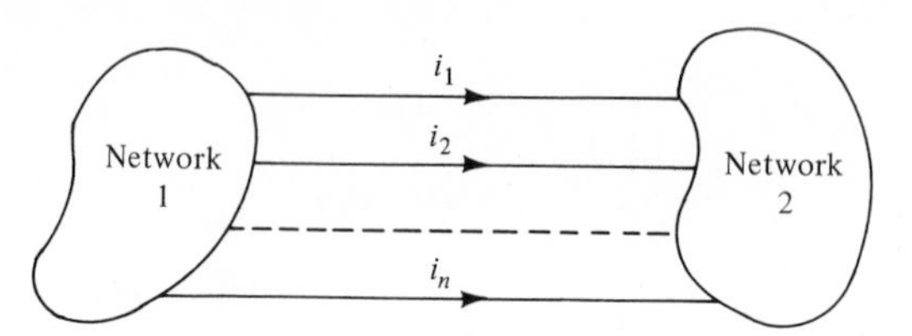

Figure 5.5 Generalized Kirchhoff current law: The algebraic sum of currents flowing from Network 1 to Network 2 is zero.

KCL as follows. Consider a network as in Figure 5.5 which is arbitrarily pulled apart so as to expose the connections. Also the pulling apart is such that if all the exposed connections were to be cut, the network would be completely separated into two parts. Now apply KCL to each node of one part labeled Network 2 in Figure 5.5. Each branch current in each KCL equation for Network 2 is either one of the currents in the exposed wires ($i_1, i_2, \ldots, i_n$) or a current for a branch connecting two nodes inside Network 2. Let j and k be any two of the nodes inside Network 2. Then, for KCL at node j, we have i_{kj} entering the node and, for KCL at node k, we have i_{jk} entering the node. If we add the KCL equations for all the nodes in Network 2, $i_{jk} + i_{kj}$ will appear in the expression. Since $i_{jk} = -i_{kj}$, we see that $i_{jk} + i_{kj} = 0$. Each such branch will be involved in exactly two node equations resulting in a cancellation. Hence, the only currents that are left are the ones pertaining to the exposed wires. Therefore, we have

$$i_1 + i_2 + \cdots + i_n = 0. \tag{5.4}$$

The above conclusion may be restated as a generalized Kirchhoff's current law: The sum of the currents entering minus the sum of the currents leaving one side of a hypothetical plane dividing a network into two parts is zero. The wires cut by the plane constitute a *cut set*.

It should be emphasized that Kirchhoff's current law is a topological constraint on the currents at a node (or through a hypothetical plane). Nothing is said about the branch characteristics. The law holds for networks containing nonlinear as well as linear elements. Regardless of how the currents come about, they have to satisfy Kirchhoff's current law at each instant of time.

EXERCISES

5.2.1 For the network of Figure 5.6, apply KCL at nodes 1, 2, 3, and 4, using single-subscript notation and the reference directions shown.

5.2.2 Repeat Exercise 5.2.1 using double-subscript notation.

5.2.3 In the network of Figure 5.6, suppose that nodes 1 and 2 are grouped together for Subnetwork 1 and nodes 3 and 4 for Subnetwork 2. Verify that $i_f + i_b - i_c + i_d = 0$ by applying KCL to nodes only.

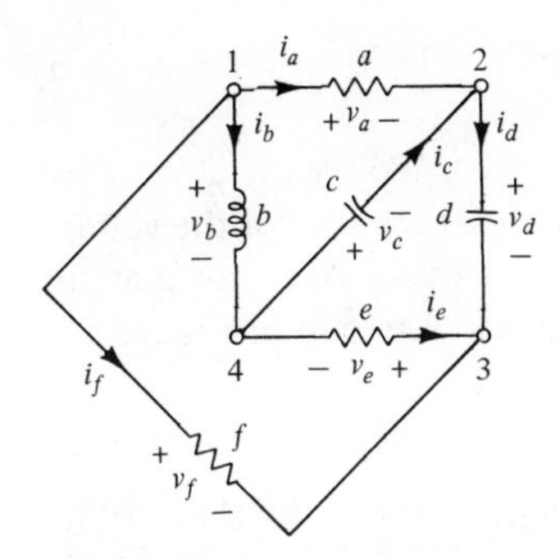

Figure 5.6

5.2.4 Suppose that nodes 1 and 3 of the network in Figure 5.6 are in Subnetwork 1 and nodes 2 and 4 are in Subnetwork 2. If the subnetworks are pulled apart, what branch currents correspond to the exposed connecting wires? Apply KCL to the hypothetical separating plane.

5.2.5 Repeat Exercise 5.2.4 for nodes 1 and 4 in Network 1 and nodes 2 and 3 in Network 2.

5.2.6 Show that for the network in Figure 5.6 the KCL equation at node 1 is obtainable from the sum of the KCL equations at the remaining nodes.

5.2.7 Repeat Exercise 5.2.6 for the KCL at node 2.

5.2.8 Repeat Exercise 5.2.6 for the KCL at node 3.

5.2.9 Repeat Exercise 5.2.6 for the KCL at node 4.

3 Kirchhoff's voltage law

We next continue our study of Kirchhoff's voltage law (abbreviated KVL) in a more general form than that previously given in Chapter 1. Recall that this law is concerned with the summation of voltages in a closed path or loop. Figure 5.7 shows a portion of a network which contains a loop. An arbitrary orientation is assigned to a loop, say clockwise, such as the loop 12341. In tracing around the loop with its specified orientation or direction which in our example is clockwise, we enter branch b at its + reference terminal and leave it at its − reference terminal. We say that in going from node 1 to node 2, we have a *voltage drop* of v_b. The variable v_b may or may not be positive, so that the drop in voltage refers to the reference polarity of the voltage only. Similarly, in going from node 3 to node 4, we enter branch d at the − reference terminal and leave it at the + reference terminal. We say that in tracing through branch d from 3 to 4, we have a *voltage rise* of v_d. Again the rise in voltage refers to the reference polarity of the voltage v_d whose value may or may not be positive.

Kirchhoff's voltage law states that in tracing around a loop with a specified orientation, the sum of the voltage drops must exactly equal the sum of the voltage rises. If different signs are assigned to rise and drop, then we may say that the sum of voltages around a loop is zero. Thus, for the clockwise loop in Figure 5.7, KVL yields

$$v_b + v_c - v_d - v_a = 0. \tag{5.5}$$

Instead of the + and − reference convention for voltage, we may use the double-subscript notation. The portion of the network in Figure 5.7 is shown again in Figure 5.8 without the + and − references. If we choose a clockwise orientation, then in going from 1 to 2, we encounter a voltage drop v_{12}. This is also equivalent to saying that in going from 1 to 2, we encounter a voltage rise of v_{21} since v_{21} is the negative of v_{12}. So applying KVL to the loop 12341 of Figure 5.8, using double-subscript notation, we have

$$v_{12} + v_{23} + v_{34} + v_{41} = 0. \tag{5.6}$$

If $v_{12} = 2$, $v_{23} = -3$, and $v_{34} = 5$, then v_{41} must be equal to -4.

We note that if we are using the double-subscript notation for voltages, the application of KVL around a circuit yields positive signs for the terms of the summation if the second subscript of every term is the first subscript of the succeeding term. Equation 5.6 may, of course, be rewritten as

$$v_{12} - v_{32} + v_{34} - v_{14} = 0. \tag{5.7}$$

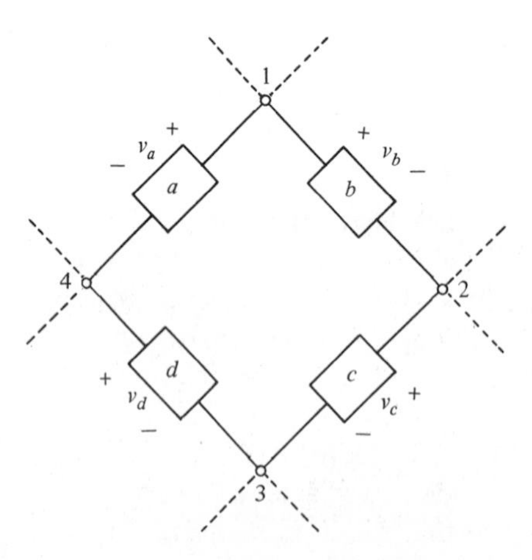

Figure 5.7 An isolated loop used for illustration of the Kirchhoff voltage law.

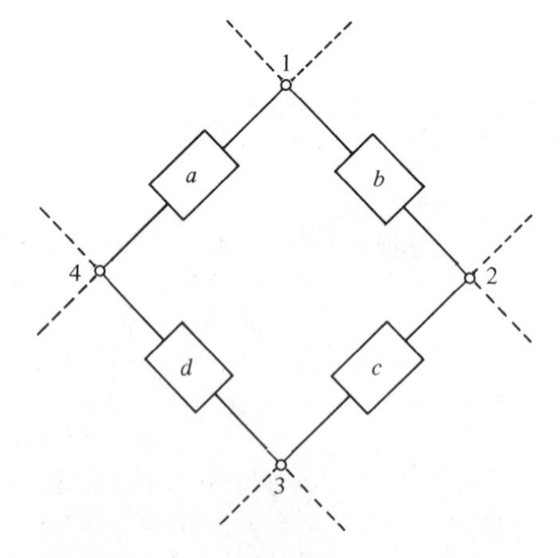

Figure 5.8 A loop embedded in a network.

If the reference convention for the voltages is the polarity mark reference, we may adopt the following convenient rule: In tracing through a loop, if we enter a branch at its + reference node, the sign of the corresponding term in the KVL equation is positive. Otherwise, the corresponding term in the KVL equation is negative. Equation 5.5 is obtained this way by tracing the loop in Figure 5.7 in the clockwise direction. Reversing the direction by which we trace a circuit changes the sign of each term in the equation formulated from KVL. In the network of Figure 5.7, tracing the loop in the counterclockwise direction, we obtain

$$-v_b + v_a + v_d - v_c = 0. \tag{5.8}$$

This equation reduces to Equation 5.5 when multiplied by -1. The above discussion leads to the following general statement of Kirchhoff's voltage law in compact form:

$$\sum_{j=1}^{N_b} b_{kj} v_j = 0, \qquad k = 1, \ldots, N_l, \tag{5.9}$$

where N_b is the number of branches and N_l is the number of loops. In tracing around a loop (the choice of direction is arbitrary), say loop k, if branch j is not in loop k, then $b_{kj} = 0$. If branch j is in loop k and in tracing through branch j the + reference of branch j is encountered first, then $b_{kj} = +1$. Otherwise $b_{kj} = -1$. Simple.

Like Kirchhoff's current law, Kirchhoff's voltage law holds for networks containing nonlinear as well as time-varying elements. KVL constrains the voltage in all loops to sum to zero for all instants of time.

ERCISES

5.3.1 For Figure 5.6, apply KVL to the following loops: (a) 12341, (b) 1241, (c) 2342, (d) 1431. Use single-subscript notation as indicated on the figure.

5.3.2 Repeat Exercise 5.3.1 using double-subscript notation.

5.3.3 For Figure 5.1, apply KVL to the following loops: (a) 1321, (b) 3453, (c) 34523, (d) 5625. Use double-subscript notation.

5.3.4 Given $v_a = -5$, $v_b = 3$, and $v_d = -6$ in Figure 5.7, determine v_c.

5.3.5 Given $v_{13} = 6$, $v_{42} = 5$, and $v_{12} = 2$ in Figure 5.8, determine v_{14}, v_{43}, and v_{32}.

4 Applications to simple networks

Consider the series-connected network in Figure 5.9. By a straightforward application of KCL, we see that $i_{12} = i_{23} = i_{34} = i_{45}$. Thus the branch currents in a

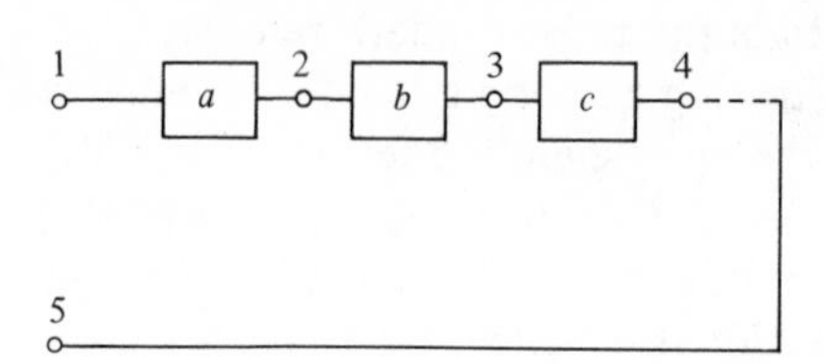

Figure 5.9 A series-connected network.

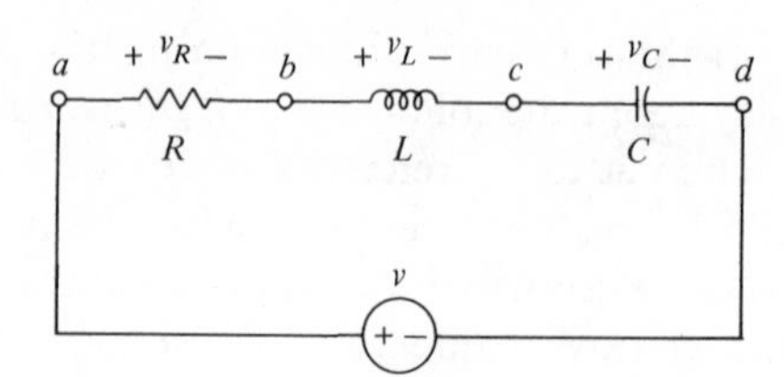

Figure 5.10 A series-connected *RLC* network.

series network are equal. A specific example of a series network is shown in Figure 5.10. Application of KVL gives

$$v_R + v_L + v_C - v = 0$$

or

$$v_R + v_L + v_C = v. \tag{5.10}$$

If the elements are both linear and time-invariant, we know that for the voltage reference conventions indicated in the figure we have

$$v_R = Ri_{ab}, \qquad v_L = L\frac{di_{bc}}{dt}, \qquad v_C = v_C(t_1) + \frac{1}{C}\int_{t_1}^{t} i_{cd}(\tau)\, d\tau. \tag{5.11}$$

Since the currents in series-connected branches are equal, that is,

$$i_{ab} = i_{bc} = i_{cd}, \tag{5.12}$$

then from Equations 5.11 and 5.12, Equation 5.10 can be written as

$$v = Ri + L\frac{di}{dt} + v_C(t_1) + \frac{1}{C}\int_{t_1}^{t} i(\tau)\, d\tau, \tag{5.13}$$

where t_1 is an arbitrary instant of time and i is the common current. If t_1 is our reference time, $v_C(t_1)$ is called the *initial voltage* across the capacitor. Usually, $t_1 = 0$ and Equation 5.13 becomes

$$v = Ri + L\frac{di}{dt} + \frac{1}{C}\int_{0}^{t} i(\tau)\, d\tau + v_C(0). \tag{5.14}$$

Equation 5.13 or 5.14 is the *integro-differential equation* relating the current i to the voltage v of the *RLC* series circuit in Figure 5.10. The right-hand side of Equation 5.14 specifies the processing of i necessary to obtain the left-hand side v.

Instead of the current $i(t)$, we may use the charge variable $q(t)$ which is related to $i(t)$ by

$$q(t) = \int_{0}^{t} i(\tau)\, d\tau + q(0). \tag{5.15}$$

Since $q(0) = Cv(0)$, Equation 5.14 may be rewritten as

$$v(t) = L\frac{d^2q(t)}{dt^2} + R\frac{dq(t)}{dt} + \frac{1}{C}q(t). \tag{5.16}$$

Equation 5.16 is the differential equation that relates the charge $q(t)$ to the voltage $v(t)$ if the elements in the network of Figure 5.10 are linear and time-invariant.

Suppose that instead of the linear branch constraints of Equation 5.11 we have the nonlinear characteristics

$$v_R = f_R(i_{ab}), \qquad \psi_L = f_L(i_{bc}), \qquad v_C = f_C(q), \tag{5.17}$$

where f_R, f_L, and f_C are nonlinear functions. Recalling that $v_L(t)$ is the time derivative of the flux ψ_L and since the series branch currents are all equal and equal to the time derivative of charge $q(t)$, Equation 5.10 becomes

$$f_R\left[\frac{dq(t)}{dt}\right] + \frac{d}{dt}\left\{f_L\left[\frac{dq(t)}{dt}\right]\right\} + f_C[q(t)] = v(t). \tag{5.18}$$

Equation 5.18 is a nonlinear differential equation relating charge $q(t)$ to voltage $v(t)$ written for the nonlinear branch characteristics of Equation 5.17.

Consider now a network with parallel branches as in Figure 5.11. A simple application of KVL shows that $v_{12} + v_{43} = 0$ or $v_{12} = v_{34}$. Similarly, $v_{34} + v_{65} = 0$ or $v_{34} = v_{56}$. In other words, the branch voltages of branches in parallel are all equal. A specific simple network consisting entirely of branches in parallel is shown in Figure 5.12. Applying KCL at node a, we have

$$-i + i_R + i_L + i_C = 0 \tag{5.19}$$

or

$$i = i_R + i_L + i_C. \tag{5.20}$$

Assuming linear time-invariant branch characteristics for the R, L, and C elements, we may write Equation 5.20 as

$$i = \frac{1}{R}v + \left[\frac{1}{L}\int_{t_1}^{t} v(\tau)\,d\tau + i_L(t_1)\right] + C\frac{dv}{dt}. \tag{5.21}$$

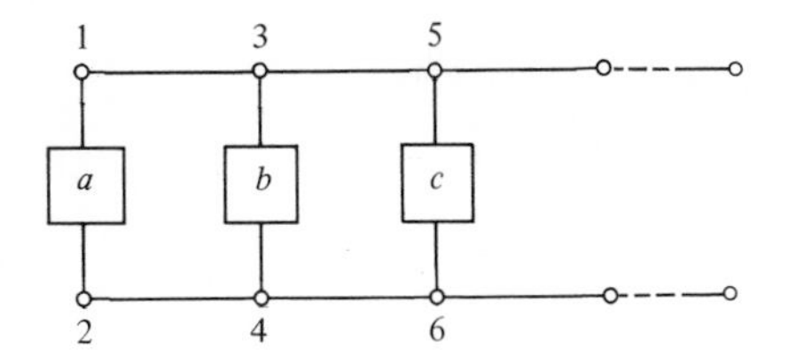

Figure 5.11 A network of branches connected in parallel.

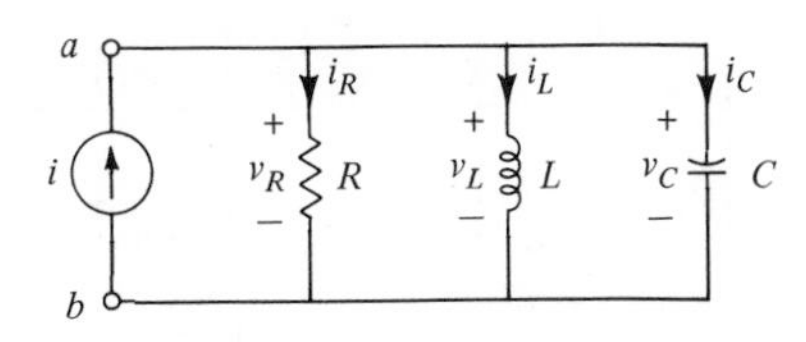

Figure 5.12 A parallel *RLC* network.

At this point, we note the similarity of this equation and Equation 5.14. If in Equation 5.14, v is replaced by i, i by v, R by G, L by C, and C by L, we obtain Equation 5.21. Insofar as the mathematical problem is concerned, the two are identical and the same type of integro-differential equation is involved in both cases. The only difference is the physical significance of the variables and the constants involved. These two networks are said to be *topological duals*. If the corresponding coefficients are numerically equal, then the solution of the mathematical problem is the same.

Equation 5.21 may be rewritten in terms of flux $\psi(t)$. Since $v = d\psi/dt$, Equation 5.21 becomes

$$i = \frac{1}{R}\frac{d\psi(t)}{dt} + \frac{1}{L}\psi(t) + C\frac{d^2\psi(t)}{dt^2}. \tag{5.22}$$

This equation is similar to Equation 5.16 so that $q(t)$ and $\psi(t)$ are dual variables also. Furthermore, if the branches have the nonlinear characteristics

$$i_R = f_R(v_R), \qquad i_L = f_L(\psi), \qquad q_C = f_C(v_C), \tag{5.23}$$

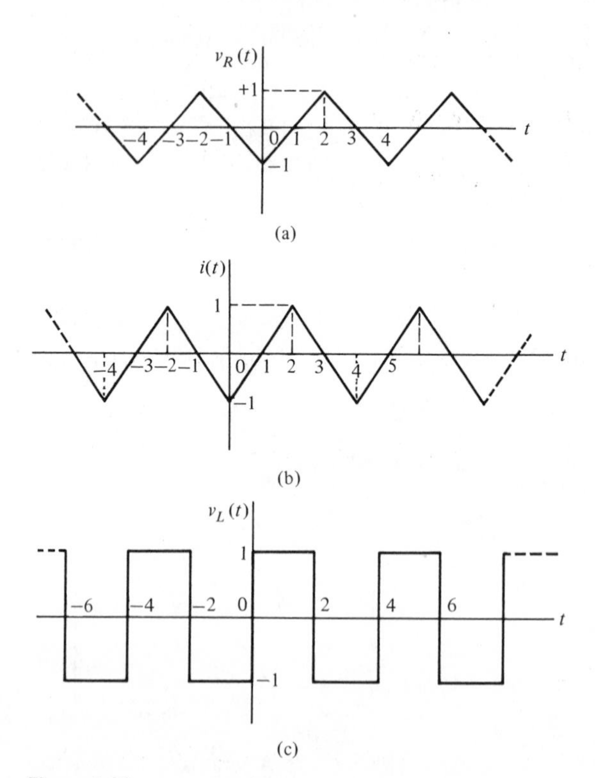

Figure 5.13

then Equation 5.20 may be rewritten as

$$i = f_R\left[\frac{d\psi(t)}{dt}\right] + f_L[\psi(t)] + \frac{d}{dt}\left\{f_C\left[\frac{d\psi(t)}{dt}\right]\right\}. \tag{5.24}$$

If the nonlinear functions in Equation 5.23 are equal to the corresponding ones in Equation 5.17 (note that f_C in Equation 5.23 corresponds to f_L in Equation 5.17), then Equation 5.24 is essentially the same differential equation as in Equation 5.18. The only difference is the use of the function $i(t)$ instead of $v(t)$ and $\psi(t)$ instead of $q(t)$.

In general, the mathematical analysis of networks is based on the solution of simultaneous equations found from KCL and KVL. The KCL equations relate currents at a node, and the KVL equations relate voltages around loops. The other equations needed specify relationships of voltage and current for the network elements, as discussed in the previous chapter. These equations are called branch equations. The topological equations and the branch equations together completely describe any network.

ERCISES

5.4.1 For the network in Figure 5.10, assume that R, L, and C are all equal to unity for simplicity. If $v_R(t)$ has the waveform as in Figure 5.13(a), verify that the waveform for $i(t)$ is as shown in Figure 5.13(b), and that the waveform for $v_L(t)$ is as shown in Figure 5.13(c). The reference direction of i is from a to b through R.

5.4.2 Given the data in Exercise 5.4.1, and $v_C(0) = 0$, show that $v_C(t)$ has the waveform as in Figure 5.14(a) for t between 0 and 2. Determine the

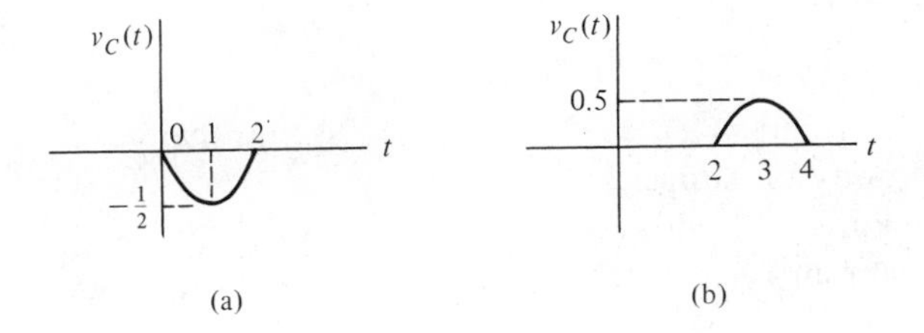

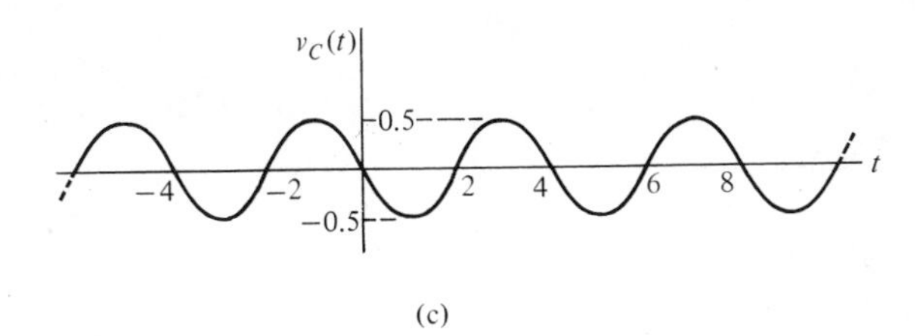

Figure 5.14

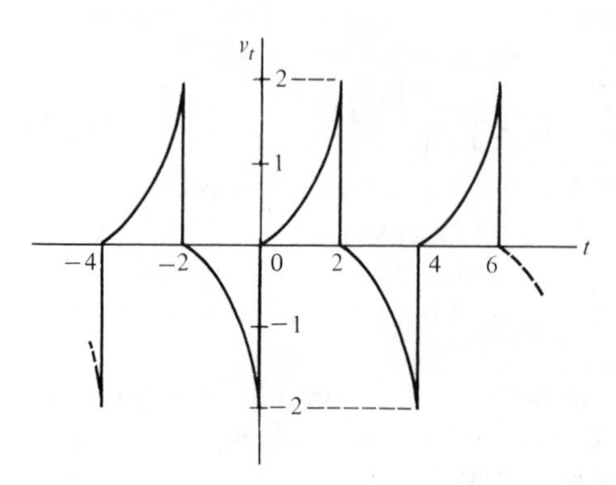

Figure 5.15

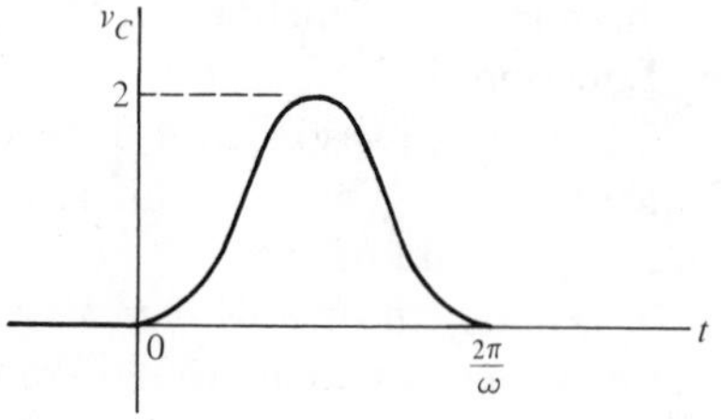

Figure 5.16

equation for $v_C(t)$ in the interval 0 to 2. Similarly, show that $v_C(t)$ has the waveform as in Figure 5.14(b) for $2 \leq t \leq 4$ and determine the equation for $v_C(t)$ in the interval. Also, since $v_C(4) = 0 = v_C(0)$ and since the waveshape of $i(t)$ in $4 \leq t \leq 8$ is the same as that in $0 \leq t \leq 4$, what can you say of the waveshape of $v_C(t)$ in $4 \leq t \leq 8$? Can you conclude that the complete waveform for $v_C(t)$ is as in Figure 5.14(c)? Give reasons.

5.4.3 For the network in Figure 5.10, the waveforms for $v_R(t)$, $v_L(t)$, and $v_C(t)$ are shown in Figures 5.13(a) and 5.13(b), and 5.14(c), respectively. Verify that the waveform for $v(t)$ is as shown in Figure 5.15.

5.4.4 If $v_C(t)$ of Figure 5.10 is defined as

$$v_C(t) = \begin{cases} 1 - \cos 5t, & 0 \leq t \leq \dfrac{2\pi}{5}, \\ 0, & \text{elsewhere}, \end{cases}$$

as shown in Figure 5.16, find $v_R(t)$, $v_L(t)$, and $v(t)$. Assume that R, L, and C are all equal to unity for simplicity.

5.4.5 For the network in Figure 5.12 with $R = 1$ ohm, $L = 0.1$ henry, $C = \frac{1}{2}$ farad, and $v_{ab}(t) = u(t) \sin 10t$ volts as shown in Figure 5.17, plot $i(t)$, given that $i_L(0) = 0$.

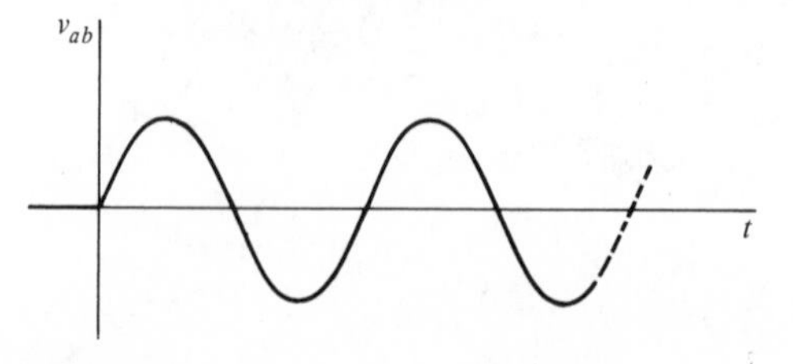

Figure 5.17

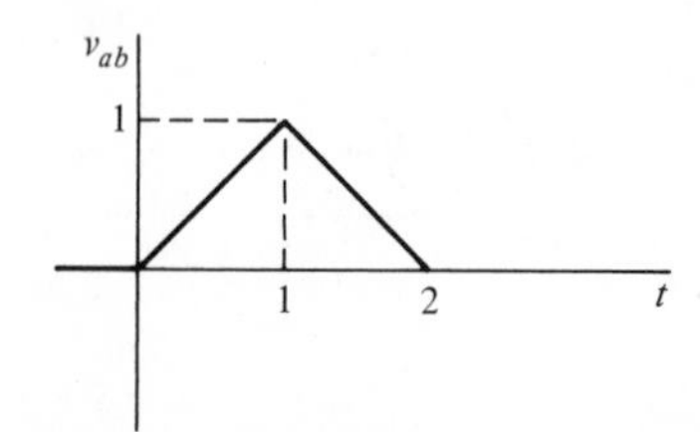

Figure 5.18

5.4.6 Repeat Exercise 5.4.5 if $v_{ab}(t)$ is as shown in Figure 5.18. The initial current $i_L(0)$ is also zero.

Choice of network variables

Thus far we have stressed procedures for applying Kirchhoff's current law at a node and Kirchhoff's voltage law around a loop. There remains the problem of the proper choice of variables that will ensure that the equations which are written from the Kirchhoff laws have a solution which satisfies Kirchhoff's laws for all nodes and for all loops. This topic is considered in detail in Chapter 12 in terms of topological considerations. The principal results will be given in this section for our use until then.

First, we should note that the choice of variables is critical for complicated networks, but such problems do not arise for simple networks. To illustrate, consider the network shown in Figure 5.19. For this network, v_1 is specified and it is required that we determine v_2, the output voltage. Now v_1 is the voltage of node 1 with respect to node 3, and v_2 is the voltage of node 2 with respect to node 3. There are no other nodes in the network and no other voltage variables, other than the voltage from node 1 to node 2 which is determined once v_1 and v_2 are known. At node 2,

$$i_C + i_{R_2} + i_{R_1} = 0$$

or

$$C\frac{dv_2}{dt} + \frac{1}{R_2}v_2 + \frac{1}{R_1}(v_2 - v_1) = 0. \tag{5.25}$$

Rearranging and dividing by C give

$$\frac{dv_2}{dt} + \frac{1}{C}\left(\frac{1}{R_1} + \frac{1}{R_2}\right)v_2 = \frac{1}{R_1C}v_1. \tag{5.26}$$

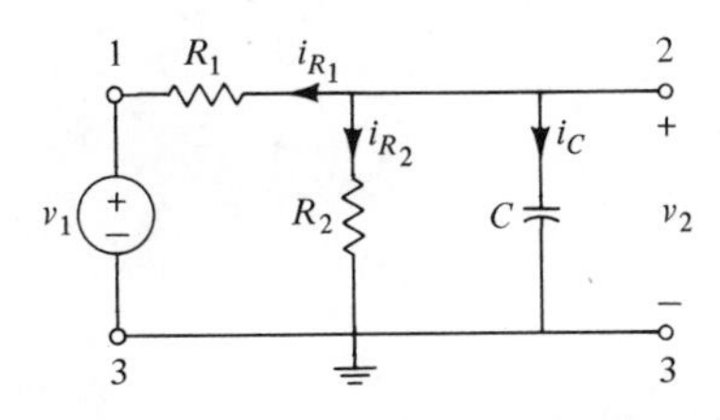

Figure 5.19 Simple RC network for which v_2 is the unknown voltage and v_1 is specified. The Kirchhoff current law is applied at node 2.

The solution of this equation will permit v_2 to be found once v_1 and the initial conditions for v_2 are specified.

The network of Figure 5.20 is complicated compared to the one we have just considered, and there clearly is some problem of choice of variable. We may select either node-to-node voltages or node-to-datum voltages as variables; it is not clear how many such voltages should be chosen. We normally wish to know a node voltage with respect to the datum (or ground), and it will turn out that this is a good choice of variables (although clearly not the only choice). It will be shown in Chapter 12 that:

(a) The node-to-datum voltage variables constitute an independent set.

(b) Eliminating known node voltages, thus leaving n nodes, and selecting one as the reference (datum), there remain $n - 1$ node-to-datum voltages. If Kirchhoff's current law is applied to each of these nodes, then the resulting equations are independent and a correct solution for each variable exists.

(c) If there are dependent sources or magnetic coupling, then these constraints may further reduce the number of equations from the $n - 1$ found through application of (a) and (b).

Returning to the network of Figure 5.20, we see that there are seven nodes, with node 7 selected as the datum. If the $n - 1 = 6$ node voltages (with respect to the datum) are chosen, then the resulting equations will permit these six unknowns to be found. The form of each of these equations will be similar to that for node 3 which is

$$C_1 \frac{d}{dt}(v_3 - v_2) + \frac{1}{R_6}(v_3 - v_6) + C_2 \frac{d(v_3 - v_4)}{dt} + \frac{1}{R_1} v_3 = 0. \quad (5.27)$$

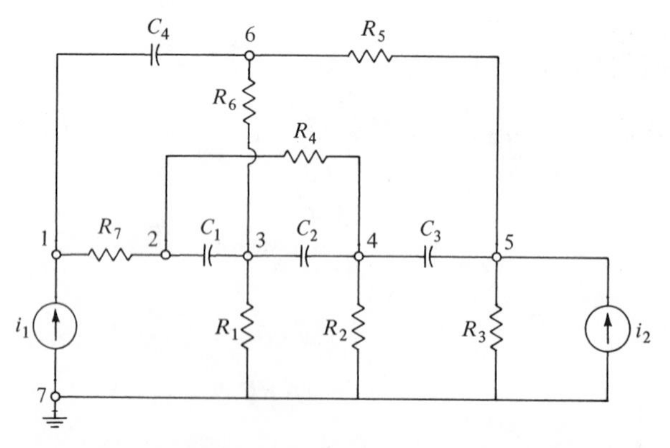

Figure 5.20 Complicated *RC* network used to illustrate the possibilities for choosing voltage variables and the complexities of selecting loops.

For the loop-current variables used in the Kirchhoff voltage law, the rules are more complicated and require that we introduce new concepts. Consider a network with n nodes and b branches, and for the time being assume that there is no magnetic coupling nor controlled sources. For this network, let us first remove all sources and then continue to remove branches one at a time until no closed paths (loops) remain. Those branches that have been removed will be called *links* (or *chords*); the circuitless structure that remains after all of the removals will be known as a *tree*. The tree will contain all nodes and $n - 1$ branches. Since there were b branches to begin with and $n - 1$ remain in the tree, clearly there are

$$b - (n - 1) = b - n + 1 \text{ links}. \tag{5.28}$$

To define an independent set of loop currents, follow these steps:

(a) Starting from the tree, replace one link and so produce a closed path.

(b) Assign a loop current to this closed path, and follow this path in the application of Kirchhoff's voltage law.

(c) Remove the link of (a), and repeat the process by replacing another link. Continue until all links have been used. There will then be $b - n + 1$ loop currents assigned.

(d) Apply Kirchhoff's voltage law to the $b - n + 1$ loops. The resulting equations will be linearly independent and so may be solved.

(e) If the network contains controlled sources or magnetic coupling, the constraining equations describing these effects may further reduce the number of equations to be solved.

If we are interested in branch currents, then each may be expressed as a linear combination of the loop currents. Once the loop currents are found, the branch currents are easily determined.

To illustrate, the simple network of Figure 5.21 may be used. The two necessary loop currents are shown as i_1 and i_2, and the Kirchhoff voltage law gives*

$$R_1 i_1 + \frac{1}{C}\int_{-\infty}^{t} (i_1 - i_2)\, d\tau - v_1(t) = 0 \tag{5.29}$$

$$\frac{1}{C}\int_{-\infty}^{t} (i_2 - i_1)\, d\tau + L\frac{di_2}{dt} + R_2 i_2 = 0. \tag{5.30}$$

Note that the branch currents may be determined from the loop currents. The equations are

$$i_C = i_1 - i_2, \qquad i_{R_1} = i_1, \qquad i_L = i_{R_2} = i_2. \tag{5.31}$$

A more complicated network is shown in Figure 5.22. One choice of a tree

* Question: What have we used as the tree?

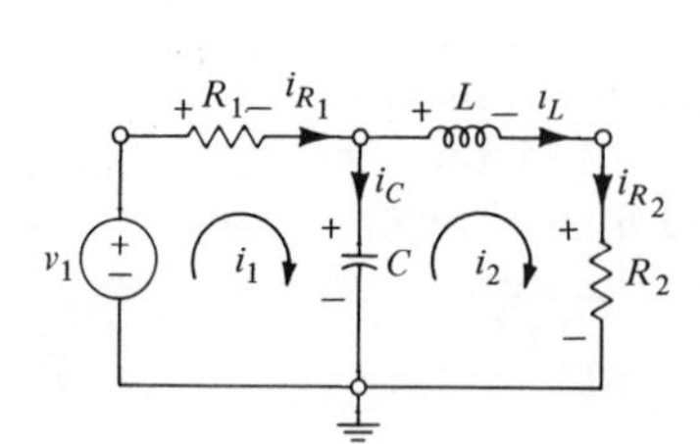

Figure 5.21 Two-loop network which is described by two equations formulated from the Kirchhoff voltage law.

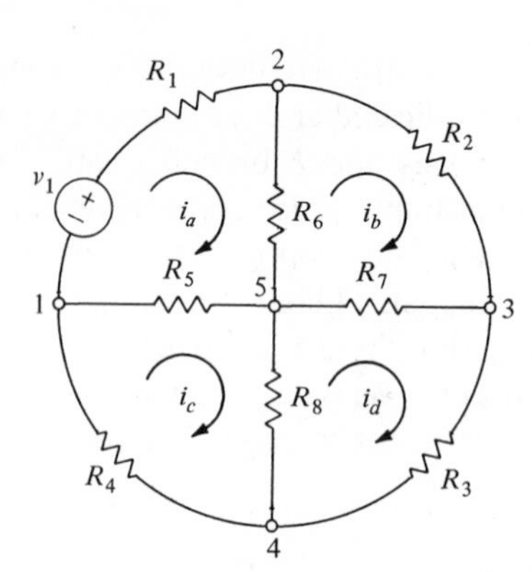

Figure 5.22 Four-loop network used to illustrate the procedure for assigning loop currents, and for Exercise 5.5.4.

(which can include many possibilities, since a tree is not unique for a network, except for networks with no loops) consists of the cross formed by the four resistors R_5–R_6–R_7–R_8. There are 8 branches (with the voltage source removed) and 5 nodes, meaning $b - n + 1 = 4$ links which are the peripheral elements R_1, R_2, R_3, and R_4. The required four loop currents are shown as i_a, i_b, i_c, and i_d. The form of the equations that are now written is illustrated by that for the loop containing i_c, which is

$$R_5(i_c - i_a) + R_8(i_c - i_d) + R_4 i_c = 0. \tag{5.32}$$

For networks which are planar, like those of our two examples of Figures 5.21 and 5.22, a relatively simple rule may be used which is that each *window pane* has a loop current assigned to it. This permits the tree and link identification to be bypassed. However, if it should be required to analyze a nonplanar network such as that of Figure 5.20, then the use of trees and links becomes necessary to ensure that the equations which are written to describe the network may be solved.

EXERCISES

5.5.1 Equation 5.27 results from the application of KCL to node 3 of Figure 5.20. Apply KCL to the other five nodes of the network.

5.5.2 For the network of Figure 5.22, select node 4 as the datum, and apply KCL to the remaining four nodes.

5.5.3 Equation 5.32 resulted from the application of KVL around loop c of Figure 5.22. Apply KVL to the other three loops of the network.

5.5.4 Consider the network of Figure 5.22, and for simplicity replace each resistor by a line segment (thus forming the *graph* of the network).

Identify four trees other than the tree formed by $R_5R_6R_7R_8$, and assign loop currents for each.

5.5.5 For the network of Figure 5.20, identify a tree and for this tree identify the proper loop currents to be used in applying KVL.

5.6 Summary

1. Kirchhoff's current law pertains to currents entering and leaving a node. If a reference convention is established (the usual one is that currents out of the node are positive, those in are negative), then

$$\sum_{j=1}^{N_b} a_{kj} i_j = 0, \qquad k = 1, \ldots, N_v$$

which is Equation 5.3. See the discussion of Equation 5.3 on page 87 for definitions of terms. KCL similarly applies to a cut set.

2. In the application of Kirchhoff's current law at a node, the branch currents may be expressed in terms of branch voltages and flux-linkages using the v–i relationships. For R, L, and C, these are

$$C\frac{dv}{dt}, \qquad \frac{1}{R}v, \qquad \text{and} \qquad \frac{1}{L}\int_{-\infty}^{t} v(\tau)\, d\tau;$$

$$C\frac{d^2\psi}{dt^2}, \qquad \frac{1}{R}\frac{d\psi}{dt}, \qquad \text{and} \qquad \frac{1}{L}\psi.$$

3. Kirchhoff's voltage law pertains to voltages which exist around a closed path (loop). With a reference convention established which relates voltage polarity to current direction, Equation 5.9 is

$$\sum_{j=1}^{N_b} b_{kj} v_j = 0, \qquad k = 1, \ldots, N_l.$$

(See discussion of Equation 5.9 on page 91 for definition of terms.)

4. In the application of Kirchhoff's voltage law around a loop, the branch currents may be expressed in terms of branch voltages using v–i relationships. For R, L, and C, the expressions are

$$L\frac{di}{dt}, \qquad Ri, \qquad \text{and} \qquad \frac{1}{C}\int_{-\infty}^{t} i(\tau)\, d\tau;$$

$$L\frac{d^2q}{dt^2}, \qquad R\frac{dq}{dt}, \qquad \text{and} \qquad \frac{1}{C}q.$$

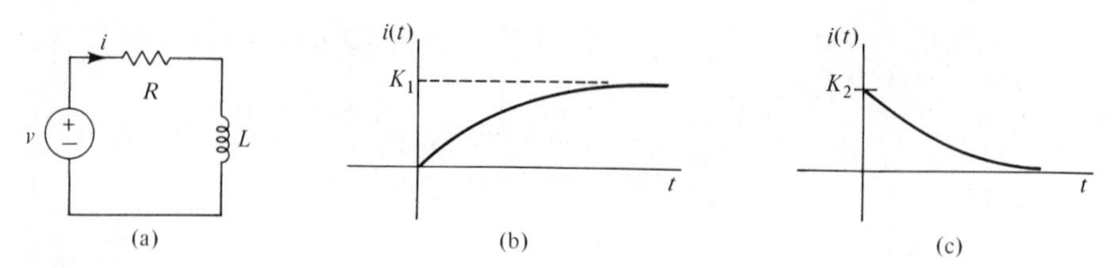

Figure 5.23

5. If we exclude known node voltages, leaving N_v nodes in the network, then $N_v - 1$ node-to-datum voltage equations must be written to describe the network.
6. For planar networks, the *window-pane* method may be used in which a loop current is assigned to each window pane. Equation formulation is usually simpler if one direction (say clockwise) is used uniformly.
7. To determine the loops and loop currents for more complicated networks, use the *tree-link* approach.

(a) Identify a tree (nonunique).

(b) Replace the links one at a time, so identifying loops and loop currents.

(c) Assign $b - n + 1$ loop currents and apply Kirchhoff's voltage law around these loops.

Additional constraints associated with controlled sources or magnetic coupling may reduce the number of equations required.

PROBLEMS

5.1 Write the differential equation relating $i(t)$ to $v(t)$ for the network in Figure 5.23(a). Assume that the elements are linear and time-invariant. If the current is

$$i(t) = K_1(1 - e^{-(R/L)t})u(t)$$

as shown in Figure 5.23(b), where K_1 is a positive constant and $u(t)$ is the unit step function, determine $v(t)$. Repeat the above for

$$i(t) = K_2 e^{-(R/L)t}u(t), \qquad K_2 > 0,$$

as shown in Figure 5.23(c).

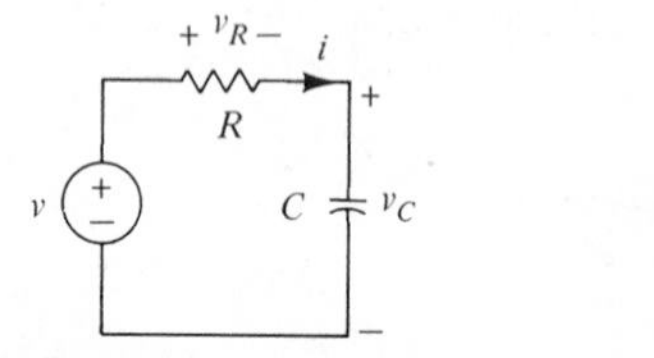

Figure 5.24

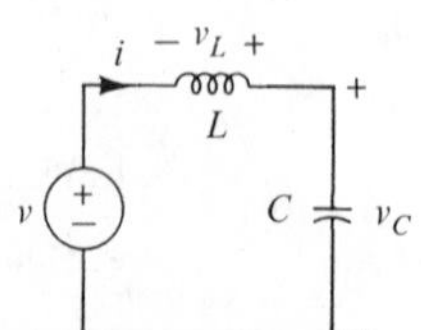

Figure 5.25

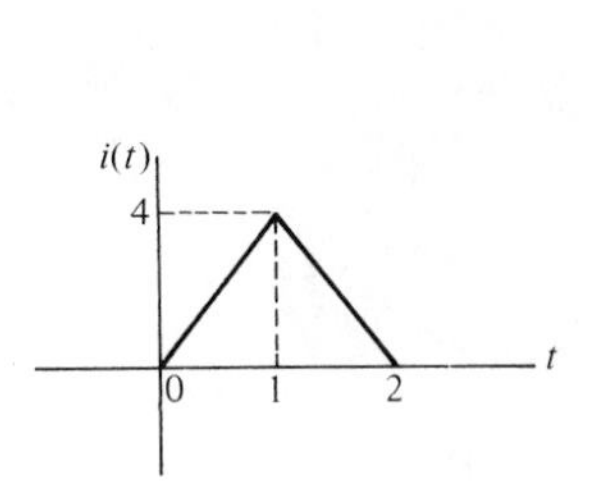

Figure 5.26

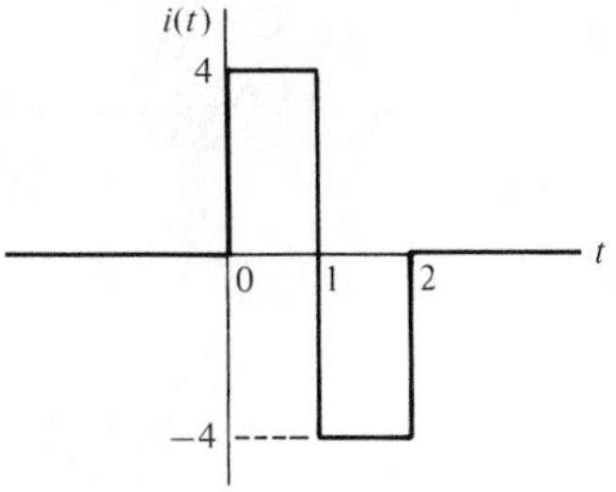

Figure 5.27

5.2 For the network in Figure 5.24, if R and C are constant and

$$i(t) = K_1 e^{-t/(RC)} u(t), \qquad K_1 > 0,$$

sketch $v_C(t)$ if $v_C(0) = 0$. Determine $v(t)$.

5.3 Write the differential equation relating $i(t)$ to $v(t)$ for the network in Figure 5.25, if L and C are linear but time-varying.

5.4 For the network in Figure 5.24, solve for $v(t)$ if $R = 8$ ohms, $C = \frac{1}{3}$ farad, and $i(t)$ is the triangular pulse shown in Figure 5.26. The initial voltage $v_C(0)$ is zero.

5.5 Repeat Problem 5.4 for the current waveform as shown in Figure 5.27.

5.6 In the network of Figure 5.28, K is closed at $t = 0$, and the current is

$$i(t) = \begin{cases} 2e^{-t}, & t \geq 0, \\ 0, & t < 0. \end{cases}$$

Capacitor C_1 is initially charged corresponding to a voltage V_0 with polarity as shown in the figure, and C_2 is initially uncharged. Determine the voltages across the two capacitors as a function of time using numerical values where possible. Determine the final values of the two capacitor voltages, i.e., the values for $t \to \infty$.

5.7 In the network shown in Figure 5.29, the switch is moved from a to b at $t = 0$. Find the differential equation for $i(t)$ for $t > 0$, if R_1 and R_2 are

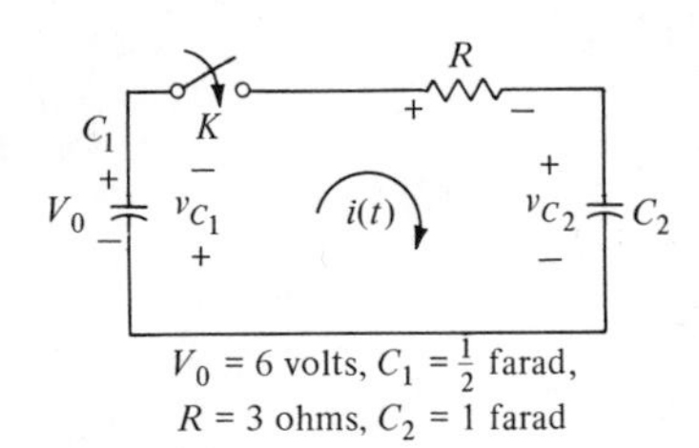

Figure 5.28

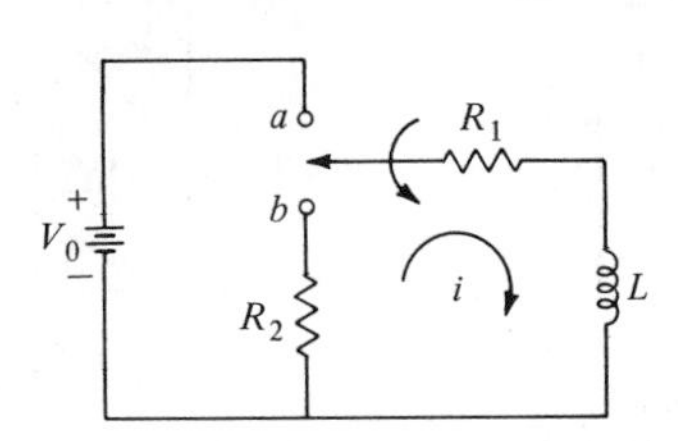

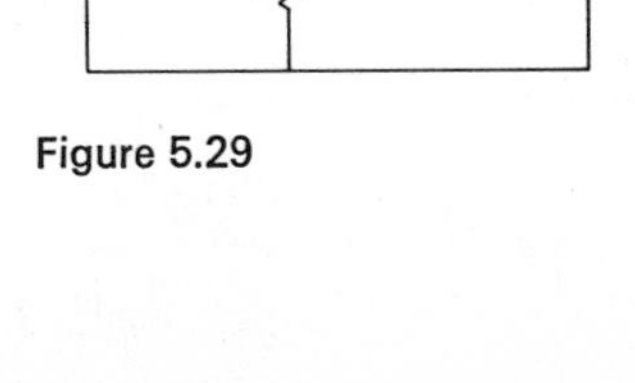

Figure 5.29

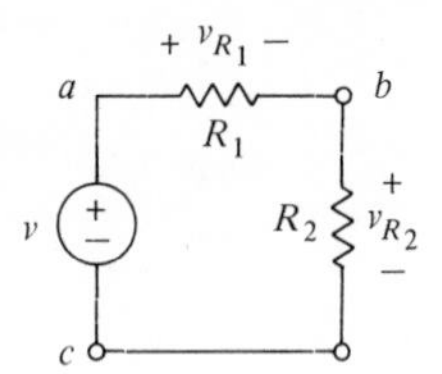

Figure 5.30

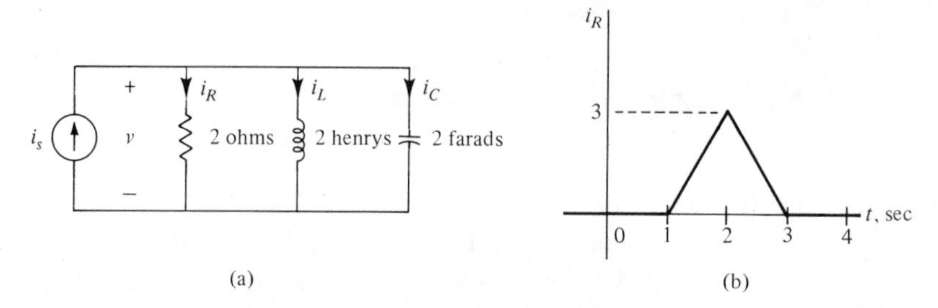

Figure 5.31

positive constants and L is nonlinear described by $\psi = Ki^2$, where K is a positive constant.

5.8 Two nonlinear current-controlled resistors are connected in series with a voltage source as shown in Figure 5.30. If the resistor characteristics are described by $v_{R_1} = K_1 f(i_{R_1})$ and $v_{R_2} = K_2 f(i_{R_2})$, where f is a specified function and K_1 and K_2 are constants, show that the resulting v_{R_1} and v_{R_2} in the network do not depend on the function f.

5.9 In the network of Figure 5.31(a), four currents and one voltage are identified. In Figure 5.31(b) is shown a signal waveform.

(1) If i_R has the waveform of (b), find the waveform for all other voltages and currents.

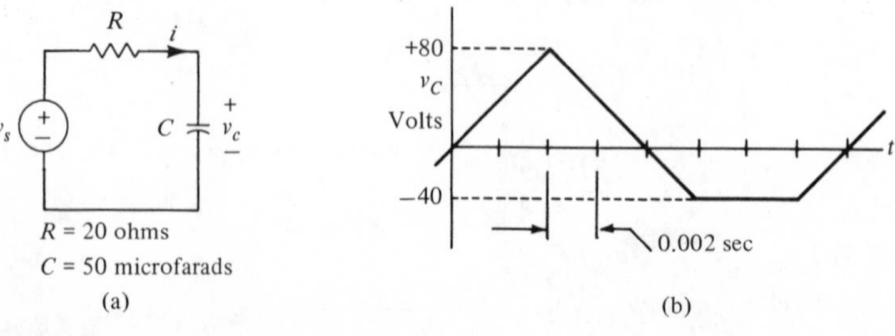

Figure 5.32

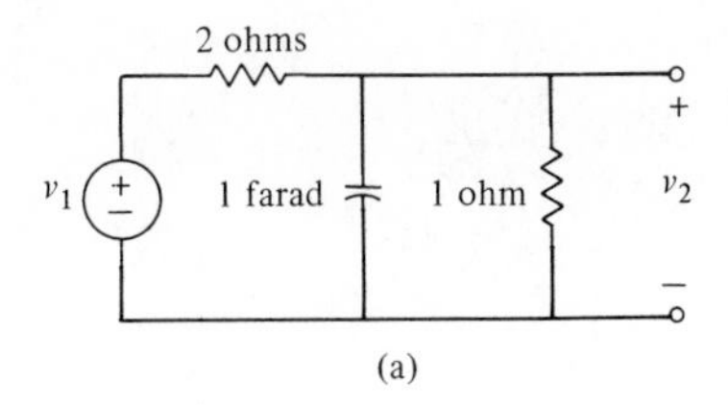

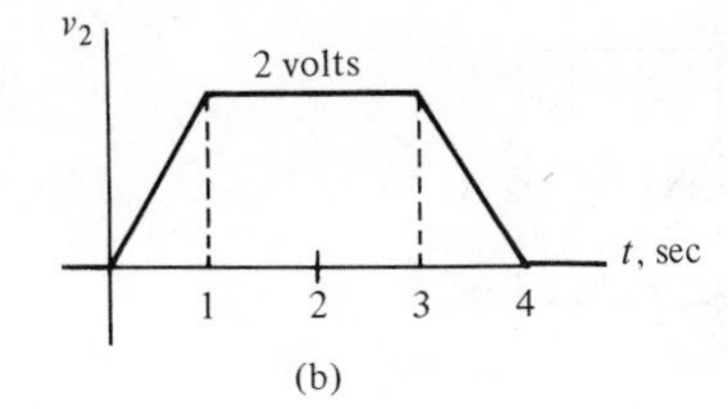

Figure 5.33

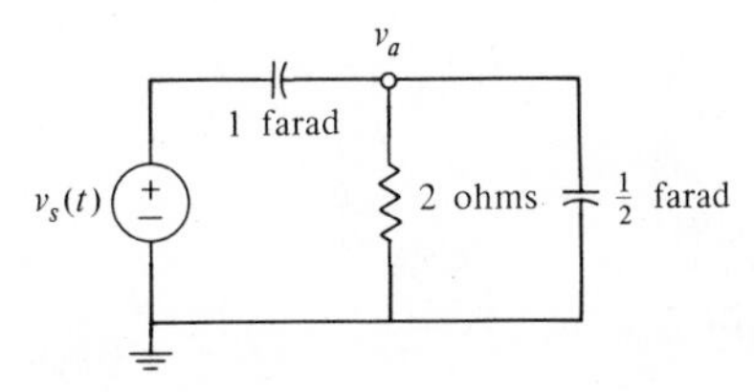

Figure 5.34

(2) Repeat (1) if i_L instead of i_R has the waveform of (b).
(3) Repeat (1) if i_C instead of i_R has the waveform of (b).
(4) Repeat (1) if v instead of i_R has the waveform of (b).

5.10 The waveform of the voltage across the capacitor of Figure 5.32(a) is shown in Figure 5.32(b). Determine the voltage generated by the source. Repeat if $v_C(t) = 100\sqrt{2} \sin [2\pi(60)t]$, writing the result in the form $v_s(t) = V_{sm} \sin [2\pi(60)t + \phi]$.

5.11 For the waveform shown in Figure 5.33 for $v_2(t)$, find the source voltage waveform $v_1(t)$ using numerical values for maximum and minimum value, slopes, etc.

5.12 (a) For the network shown in Figure 5.34, write a differential equation relating the node-to-datum voltage, v_a, to the driving voltage, v_S.
(b) If $v_a(t)$ is a unit ramp, and the capacitors are initially uncharged, determine $v_S(t)$.

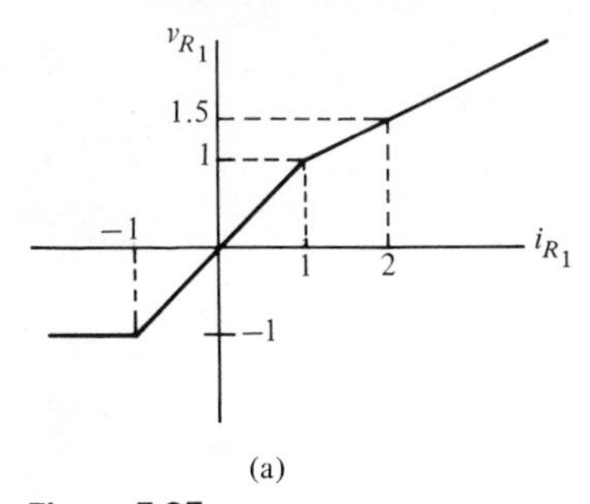

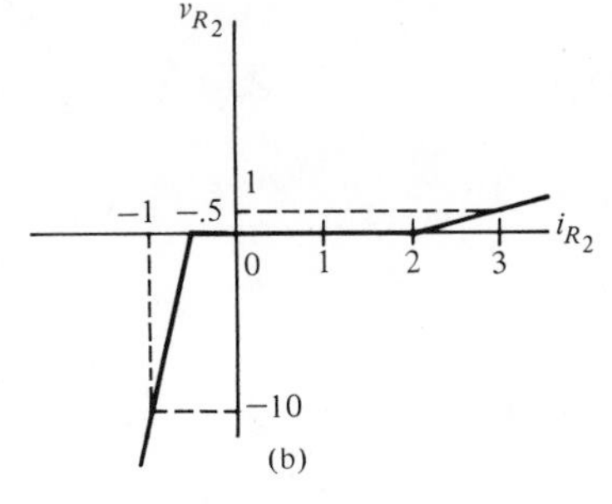

Figure 5.35

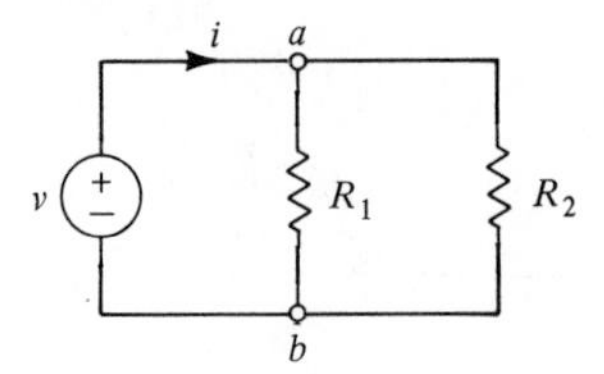

Figure 5.36

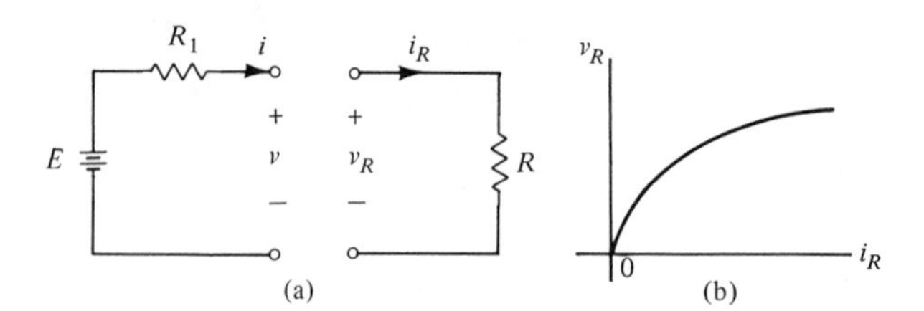

Figure 5.37

5.13 Suppose that the nonlinear resistors in Figure 5.30 have characteristics as shown in Figure 5.35. For $v = 2$ volts, determine i. [*Hint:* Obtain a composite v–i characteristic for the two series resistors first.]

5.14 Determine a composite v–i characteristic for the two resistors in parallel in Figure 5.36, if the individual characteristics are as shown in Figure 5.35. If $i = 2$, find v.

5.15 For the network in Figure 5.37(a), sketch the characteristic for v vs. i, for E and R_1 positive constants. Suppose the resistor R is connected in series with E and R_1, where R has the characteristic shown in Figure 5.37(b). Relate i to i_R and v to v_R by means of Kirchhoff's laws. How would you determine the actual v and the actual i?

Chapter **6**

nergy and Power

1 Energy dissipation and energy storage in network elements

It was mentioned in Chapter 1 that the *instantaneous power* or time rate of energy flow associated with a two-terminal or 1-port network is the product vi, where v and i are the terminal voltage and current, respectively. Unless indicated to the contrary, positive energy flow is flow *into* the network. For such an energy flow direction, the power p is equal to vi if v and i have references as shown in Figure 6.1(a). In Figure 6.1(b), *power in* is $p = -vi$. Thus the reference sign for p depends entirely on the relative references chosen for v and i. If in Figure 6.1(a) v and i are both positive or both negative at a given instant of time t_1, then $p(t_1)$ is positive, and the energy flow into the network is increasing. On the other hand, if at a given instant of time t_2, v and i in Figure 6.1(a) have opposite signs, then $p(t_2)$ is negative, and energy flow into the network is decreasing or energy flow from the network is increasing. The energy flow into the network of Figure 6.1(a) from the instant t_1 to the instant t_2 is the time integral of $p(t)$ from t_1 to t_2.

For a linear resistor, with references as in Figure 6.1(a), the instantaneous power of R is

$$p(t) = v(t)i(t) = R(t)i^2(t). \tag{6.1}$$

If the resistor is time-invariant, $R(t)$ becomes a constant. We see that so long as $R(t)$ is not negative, $p(t)$ is not negative, and energy flows into R. For $R(t) \neq 0$, we may write Equation 6.1 as

$$p(t) = v(t)\frac{1}{R(t)}v(t) = \frac{1}{R(t)}v^2(t) = G(t)v^2(t), \tag{6.2}$$

where $G(t)$, the conductance, is the reciprocal of $R(t)$. For a current-controlled

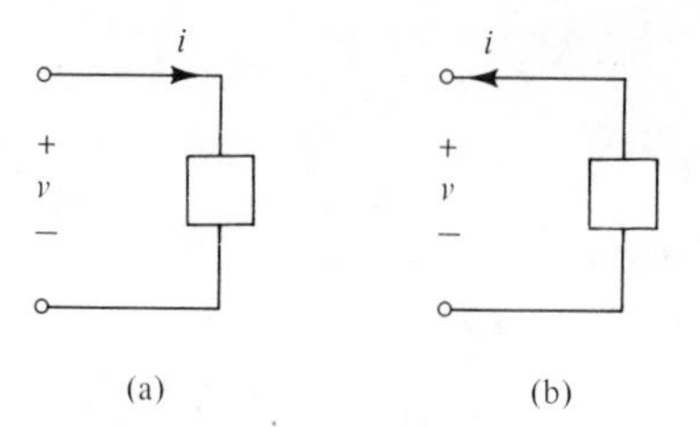

Figure 6.1 In (a), vi is the instantaneous time rate of energy flow *into* the network. In (b), vi is the instantaneous time rate of energy flow *from* the network.

nonlinear resistor characteristic such as in Figure 6.2(a), with reference directions as in Figure 6.1(a), the power of R is

$$p(t) = vi = R(i)i, \tag{6.3}$$

where $R(i)$ is the function of i representing v. Similarly, for a voltage-controlled resistor characteristic as in Figure 6.2(b), p for the network is

$$p = vi = vG(v). \tag{6.4}$$

If the characteristic is neither voltage-controlled nor current-controlled, it is no longer possible to express v as an ordinary function of i nor i as an ordinary function of v. This situation will not be treated in this book.

EXAMPLE 6.1.1

Suppose that at a certain instant t_1, the voltage and current associated with a network, with reference direction shown in Figure 6.1(a), are 5 volts and 3 amperes. Then the power is

$$p = vi = 5(3) = 15 \text{ watts}. \tag{6.5}$$

This means that the energy flow into the network is increasing at the rate of 15 watts. If the reference directions for voltage and current for the same network are

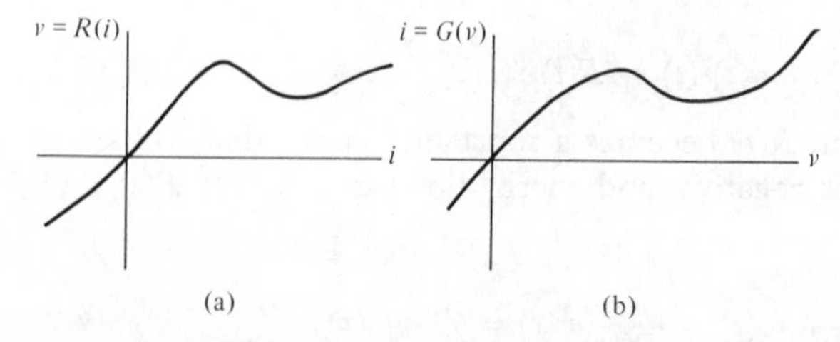

Figure 6.2 (a) Current-controlled resistance characteristic: v is a function of i. (b) Voltage-controlled characteristic: i is a function of v.

as shown in Figure 6.1(b), then v would be the same as before but i would be -3 amperes. The power is

$$p = -vi = -(5)(-3) = 15 \text{ watts}, \tag{6.6}$$

which is the same as before, as it should be.

AMPLE 6.1.2

Consider another network, with reference directions as in Figure 6.1(b), for which $v = 100 \sin 377t$ and $i = 2 \sin (377t + 3\pi/4)$. Then the power is

$$p = -vi = -200 \sin 377t \sin (377t + 3\pi/4) \tag{6.7}$$

for any instant t.

ERCISES

6.1.1 Suppose that in Figure 6.1(b), the reference polarity for v is reversed so that the plus reference mark is at the bottom. Write the expression for the instantaneous power of the network.

6.1.2 Repeat Exercise 6.1.1 for Figure 6.1(a).

6.1.3 Suppose that the reference directions for both v and i are reversed in Figure 6.1(a). Write the expression for the instantaneous power of the network.

6.1.4 Repeat Exercise 6.1.3 for Figure 6.1(b).

2 Average power for periodic signals

When $v(t)$ and $i(t)$ are periodic and of the same period, then $v(t)i(t)$ will be periodic, and if T is the period of $v(t)$ and $i(t)$, then $p(t) = p(t + T)$ for all t. However $T_1 = T$ may not be the smallest value satisfying $p(t) = p(t + T_1)$ for all t. For example, suppose $v(t) = V_0 \sin \omega_0 t$, and suppose we wish to examine the instantaneous power into a constant positive resistor R. Then $i(t) = (V_0/R) \sin \omega_0 t$. So

$$p(t) = \frac{V_0^2}{R} \sin^2 \omega_0 t = \frac{V_0^2}{2R}(1 - \cos 2\omega_0 t). \tag{6.8}$$

Thus the period of $p(t)$ is π/ω_0, but the period of $v(t)$ is $2\pi/\omega_0$. Recalling the definition for average value of Equation 3.5, we have the average power

$$P_{av} = \frac{1}{T_1} \int_{t_1}^{t_1 + T_1} p(t)\, dt, \tag{6.9}$$

where T_1 is the period of $p(t)$ and t_1 is arbitrary, and it is chosen for convenience.

For the example above,

$$P_{\text{av}} = \frac{1}{\pi/\omega_0}\int_0^{\pi/\omega_0} \frac{V_0^2}{2R}(1 - \cos 2\omega_0 t)\,dt = \frac{V_0^2}{2R} = \frac{V_{\text{rms}}^2}{R} = I_{\text{rms}}^2 R, \quad (6.10)$$

where V_{rms} is the rms value of $v(t)$ and I_{rms} is the rms value of $i(t)$. The result in Equation 6.10 applies to any constant resistor and for any periodic $v(t)$. In general, for a time-varying current-controlled resistor we have

$$P_{\text{av}} = \frac{1}{T_1}\int_0^{T_1} i(t)v[i(t), t]\,dt, \quad (6.11)$$

and for a time-varying voltage-controlled resistor we have

$$P_{\text{av}} = \frac{1}{T_1}\int_0^{T_1} v(t)i[v(t), t]\,dt, \quad (6.12)$$

provided $p(t)$ is periodic with period T_1. For a linear time-invariant capacitor, let us show that the average power is zero. Recall that for a linear time-invariant capacitor, v and i are related by

$$i = C\frac{dv}{dt}. \quad (6.13)$$

In Equation 6.13, it is assumed that i and v have relative references as in Figure 6.1(a). Using Equation 6.13 in the definition of Equation 6.9 we have

$$P_{\text{av}} = \frac{1}{T_1}\int_{t_1}^{t_1+T_1}\left(C\frac{dv}{dt}\right)v\,dt = \frac{1}{T_1}\left(\int_{v(t_1)}^{v(t_1+T_1)} Cv\,dv\right)$$

$$= \frac{1}{2T_1}Cv^2\Big|_{v(t_1)}^{v(t_1+T_1)} = \frac{C}{2T_1}[v^2(t_1 + T_1) - v^2(t_1)]. \quad (6.14)$$

Since v is periodic of period T, then $v(t_1 + T_1) = v(t_1)$ so that $P_{\text{av}} = 0$.

It is equally straightforward to demonstrate that the average power of a linear time-invariant inductor is zero. Instead of Equation 6.13 we have

$$v = L\frac{di}{dt} \quad (6.15)$$

for the reference conventions in Figure 6.1(a). Using Equation 6.15 in Equation 6.9, we have

$$P_{\text{av}} = \frac{1}{T_1}\int_{t_1}^{t_1+T}\left(L\frac{di}{dt}\right)i\,dt = \frac{1}{T_1}\left[\frac{1}{2}Li^2\right]_{i(t_1)}^{i(t_1+T_1)} \quad (6.16)$$

Since $i(t_1) = i(t_1 + T_1)$ because i is periodic of period T_1, then $P_{\text{av}} = 0$.

It will be demonstrated in Section 6.6 that even for nonlinear inductors and capacitors, provided these are time-invariant and they satisfy some mild restrictions on the type of nonlinearities, the average power is zero. For now, let us simply verify a specific example. Let the charge–voltage characteristic of a charge-controlled capacitor be given by

$$v = q^3 . \tag{6.17}$$

Note that since time does not appear explicitly, the capacitor is time-invariant. However, it is nonlinear. The average power is

$$P_{av} = \frac{1}{T_1} \int_{t_1}^{t_1 + T_1} vi \, dt = \frac{1}{T_1} \int_{q(t_1)}^{q(t_1 + T_1)} v \, dq . \tag{6.18}$$

Substituting Equation 6.17 into Equation 6.18, we obtain

$$P_{av} = \frac{1}{T_1} \left[\frac{1}{4} q^4 \right]_{q(t_1)}^{q(t_1 + T_1)} . \tag{6.19}$$

Since v and i (and, therefore, q) are periodic of period T_1, then $q(t_1) = q(t_1 + T_1)$ so that $P_{av} = 0$.

EXERCISES

6.2.1 For a linear time-invariant inductor of L henrys, and $i(t) = I_0 \sin \omega_0 t$, verify that the average power is 0 by solving for $v(t)$, and taking the time average of $v(t)i(t)$.

6.2.2 For a nonlinear time-invariant capacitor described by $q(t) = v^3(t)$ and $v(t) = V_0 \sin \omega_0 t$, verify that P_{av} is zero by solving for $i(t)$ and taking the average of $v(t)i(t)$.

6.2.3 For a time-varying linear inductor described by $\psi = (2 + \sin \omega_0 t)i$, compute the average power P_{av} into L if $i(t) = I_0 \cos \omega_0 t$. Is the inductor dissipating, generating, or simply storing energy?

6.2.4 Repeat Exercise 6.2.3 for $\psi = (2 - \sin \omega_0 t)i$.

6.2.5 Suppose we have a time-varying voltage-controlled resistor. For a periodic $v(t)$, will $p(t)$ be necessarily periodic?

6.2.6 Suppose we have a time-varying current-controlled inductor. For a periodic current, will the instantaneous power into L necessarily be periodic?

6.2.7 For a linear time-invariant resistor $R = 10$ ohms, compute the average power if $i(t) = I_1 \sin(\omega_0 t + \alpha_1) + I_2 \sin(\omega_0 t + \alpha_2)$.

6.3 Passive, active, and lossless networks

We begin this section by considering an arbitrary network with n pairs of terminals, called an *n-port* network, as shown in Figure 6.3. We connect the n-ports to n

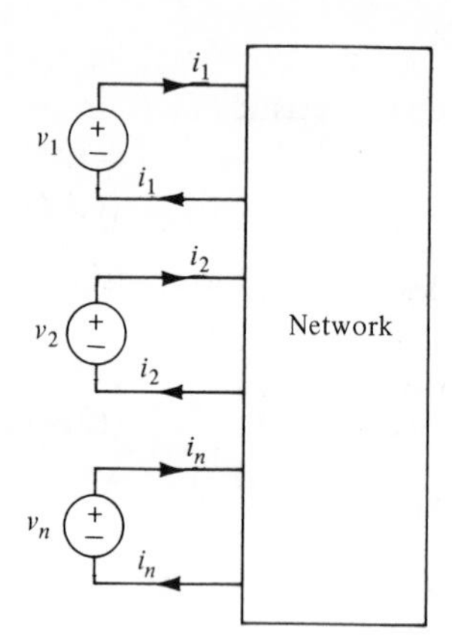

Figure 6.3 An n-port or n terminal-pair network.

sources as shown. Suppose that at the instant t_1 when the sources are connected the initial charges on capacitors are all zero and the initial fluxes in inductors are all zero. Then the energy flow into the network between t_1 and t is

$$W(t) = \sum_{k=1}^{n} \int_{t_1}^{t} v_k(\tau) i_k(\tau)\, d\tau. \tag{6.20}$$

If for any t and t_1 ($t > t_1$) and for any set of arbitrary sources, $W(t)$ is nonnegative, then the n-port is said to be *passive*. Otherwise, if $W(t)$ is negative for some value of t and some set of sources, then the n-port is said to be *active*. The definitions just given apply not only to arbitrary n-ports but, of course, to simple elements as well. We now consider a number of special cases and reserve Equation 6.20 for later use.

First consider the resistor. The energy flow into R is

$$W(t) = \int_{t_1}^{t} v(\tau) i(\tau)\, d\tau, \tag{6.21}$$

where either v is expressed in terms of i or i is expressed in terms of v. If $W(t)$ must be positive or zero for arbitrary values of t, $v(t)i(t)$ must never be negative for any time interval. For if it were, then by choosing either i or v to be zero outside the interval, we would be left with the integral of a negative quantity which would be negative. Hence, it is necessary as well as sufficient for $v(t)i(t)$ to be nonnegative in order for $W(t)$ to be nonnegative. That is, the v–i characteristic must lie in the first and third quadrant in order for a resistor to be passive.

For a linear time-invariant positive capacitor or inductor, $W(t)$ is

$$W(t) = \tfrac{1}{2} C v^2(t) \tag{6.22}$$

for a capacitor and

$$W(t) = \tfrac{1}{2}Li^2(t) \tag{6.23}$$

for an inductor, and since $C > 0$ and $L > 0$, $W(t)$ is nonnegative and hence the element is passive. Similarly, a linear time-invariant negative C or L is found to be an active element.

We have already seen that linear time-invariant capacitors and inductors are conservative elements in the sense that the integral of the periodic signal p_C or p_L over one period is zero. Conservative elements are also called *lossless* elements. In general, a time-invariant n-port is lossless if for periodic signals which are otherwise arbitrary the average power into the n-port is zero. We shall not define the lossless concept for time-varying networks.

ERCISES

6.3.1 Show that a time-invariant charge-controlled capacitor is necessarily passive if its v–q characteristic is in the first and third quadrant.

6.3.2 Give an example of a time-invariant charge-controlled passive capacitor whose characteristic is not restricted to the first and third quadrant.

6.3.3 Show that a time-invariant flux-controlled inductor is active if its i–ψ characteristic lies in the second and fourth quadrant.

6.3.4 Determine whether the ideal transformer as defined in Chapter 4 is lossless.

6.3.5 Determine whether the ideal transformer as defined in Chapter 4 is passive or active.

6.3.6 Is it possible for a network comprising a passive n-port to be an active network? Explain briefly if it is not possible or give an example if it is possible.

4 Conservation of energy

Conservation of energy is a universal physical law which may be stated in various ways. One way is as follows: Given a system or network, the total energy at any instant t_2 is equal to the sum of the total energy at any previous instant t_1 and the energy added to the system during the time interval between t_1 and t_2. The energy added to the system may be negative, i.e., positive energy may have been taken out of the system resulting in a decrease of energy at t_2. In applications where nuclear reactions may be involved, total energy must be interpreted to include the equivalent energy of mass in accordance with Einstein's mass–energy equation. In an

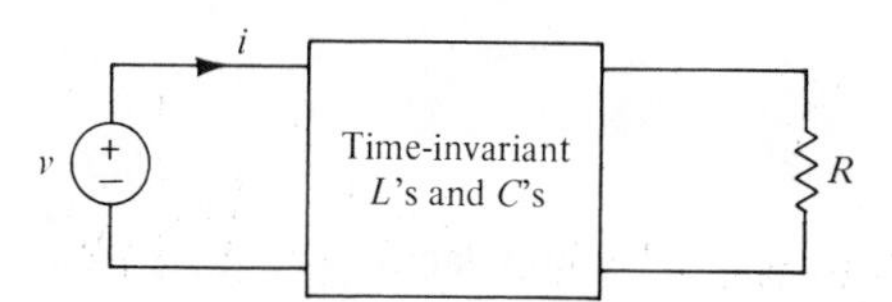

Figure 6.4 By the conservation-of-energy principle, average power into the network from the source is equal to the average power into R.

ordinary electric network where no such mass-to-energy conversion takes place, the net electric energy added to the network is equal to the net increase in the electric energy stored plus the net energy converted to other forms of energy such as heat. That is,

$$\text{Net energy added} = \text{net increase in stored energy} \\ + \text{net increase in energy converted}. \qquad (6.24)$$

Of course, any of these terms may be negative, but the equality will hold. The energy in any network or system is equal to the algebraic sum of the energies in all the parts of the network or system. Thus it is possible to determine total energy by adding the energies corresponding to all the branches. In some problems it may be convenient to partition a network into several subnetworks in which case the total energy is equal to the algebraic sum of energies in the various subnetworks.

For periodic signals, we may consider average power instead of energy so that the total average power into a network is equal to the algebraic sum of the average powers in all the subnetworks. In particular, we note that in a time-invariant network, the average power into time-invariant capacitors and inductors is zero. Hence, the total average power in this case is equal to the algebraic sum of the average powers into all the other elements. For example, for the special network

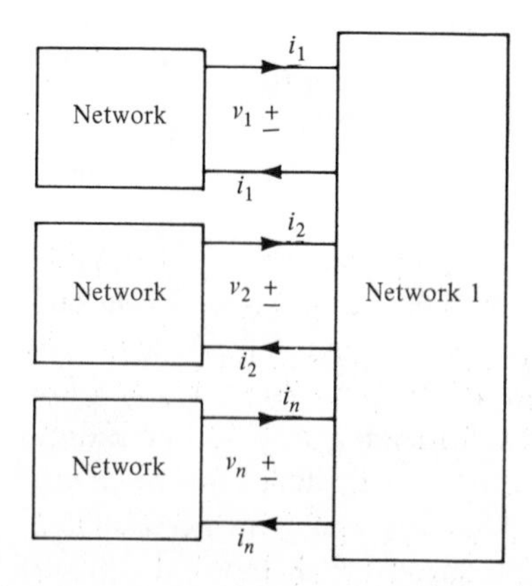

Figure 6.5 Energy flow into Network 1 is given by Equation 6.20.

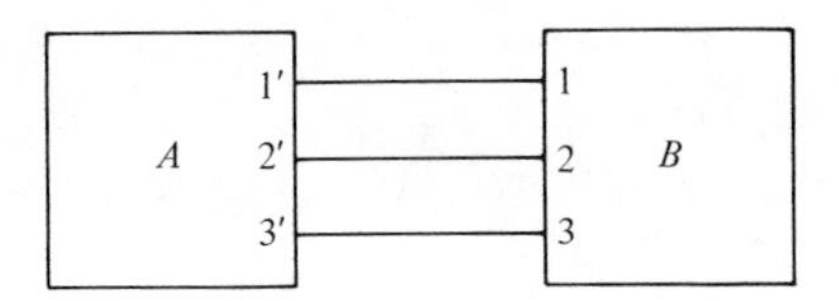

Figure 6.6 A three-terminal network connected to another three-terminal network.

in Figure 6.4, the average power from the source is equal to the average power into R.

Similarly, since Equation 6.24 holds for all t, differentiating both sides of the equation with respect to t yields a principle which may be called conservation of instantaneous power. By this we mean that the instantaneous power to the network equals the sum of the instantaneous rate of energy dissipation and the instantaneous rate of stored energy increase in all the elements. Alternatively, we may say that the instantaneous power input to the entire network is equal to the sum of the instantaneous power inputs to all the individual elements.

In a general network represented as the n-port network in Figure 6.3, the total energy into the n-port is the sum of the energies from the individual sources as in Equation 6.20. The sources do not have to be as simple as those shown in Figure 6.3. To illustrate this statement, the n-port labeled Network 1 in Figure 6.5 is connected to a set of networks. It is clear that the energy flow into Network 1 is given by Equation 6.20. When the ports are not identifiable, the situation is slightly different. Consider the three-terminal Network B in Figure 6.6. We can no longer say that terminals 1 and 2 constitute a port, for $i_{2'2}$ is not necessarily $-i_{1'1}$. All we can say is that by KCL, we have

$$i_{1'1} + i_{2'2} + i_{3'3} = 0, \tag{6.25}$$

and from KVL we have

$$v_{12} + v_{23} + v_{31} = 0. \tag{6.26}$$

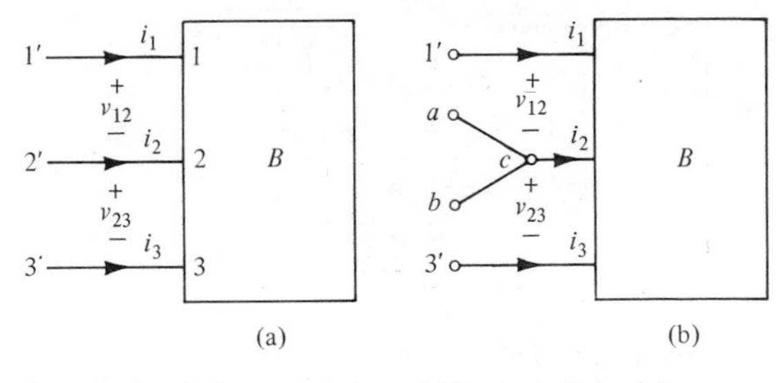

Figure 6.7 Terminal v–i characteristics of Network B in (a) are unchanged by splitting line 2 into two lines, as shown in (b).

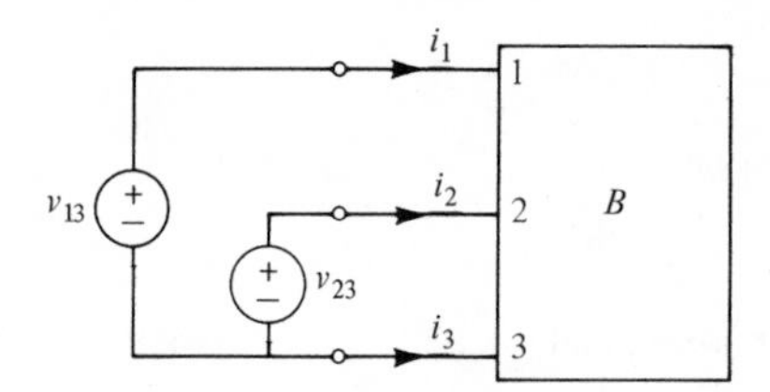

Figure 6.8 Interpretation of energy-flow calculation of Equation 6.30.

Let us develop a formula for the energy into Network B. The Network B with the terminal voltages and currents labeled is again shown in Figure 6.7(a). It is further redrawn in Figure 6.7(b). As long as $i_{ac} + i_{bc} = i_2$, it is clear that the terminal characteristics of Network B are unchanged. In particular, if it is possible to make $i_{ac} = -i_1$ and $i_{bc} = -i_3$, then the situation is analogous to a 2-port network. But from the KCL equation in Equation 6.25, we have

$$i_2 = -i_1 - i_3\,. \tag{6.27}$$

Hence, if we make $i_{ac} = -i_1$ and $i_{bc} = -i_3$, i_2 will indeed turn out to be $-i_1 - i_3$ as desired. Now it is clear that the energy flow into Network B, which is the sum of the energy flows into the ports $1'a$ and $b3'$, is

$$\int_{t_1}^{t} (v_{1'a}i_1 + v_{b3'}i_{bc})\,d\tau = \int_{t_1}^{t} [v_{12}i_1 + v_{23}(-i_3)]\,d\tau\,. \tag{6.28}$$

Since $v_{12} = v_{13} - v_{23}$ and $-i_3 = i_1 + i_2$, Equation 6.28 may be rewritten as

$$\text{Energy flow into } B = \int_{t_1}^{t} [(v_{13} - v_{23})i_1 + (i_1 + i_2)v_{23}]\,d\tau \tag{6.29}$$

or

$$W_B = \int_{t_1}^{t} (v_{13}i_1 + v_{23}i_2)\,d\tau\,. \tag{6.30}$$

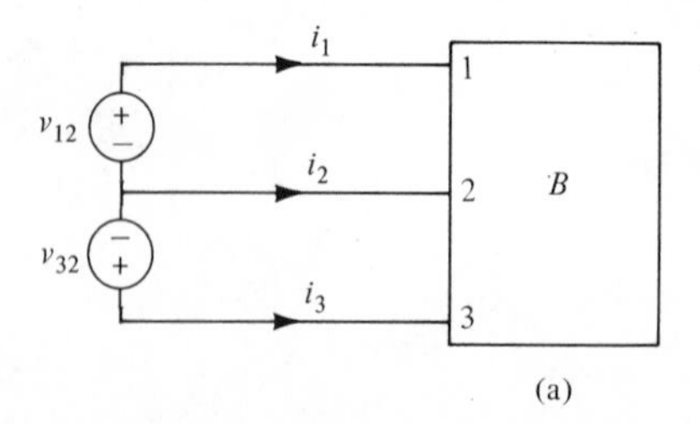

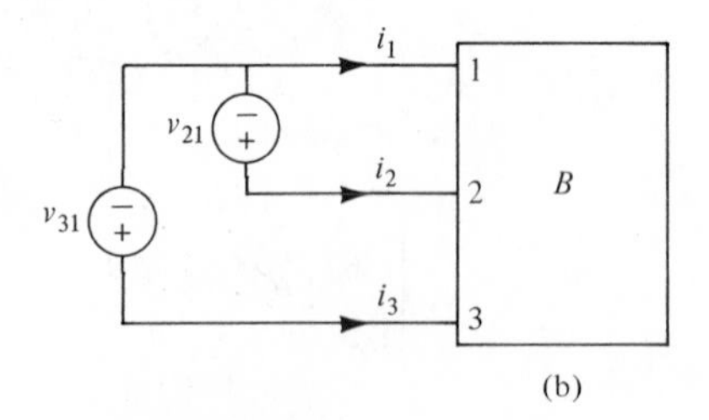

Figure 6.9 Total energy flow from sources in (a) is equal to the total energy flow from sources in (b), and also equal to total energy flow from sources in Figure 6.8.

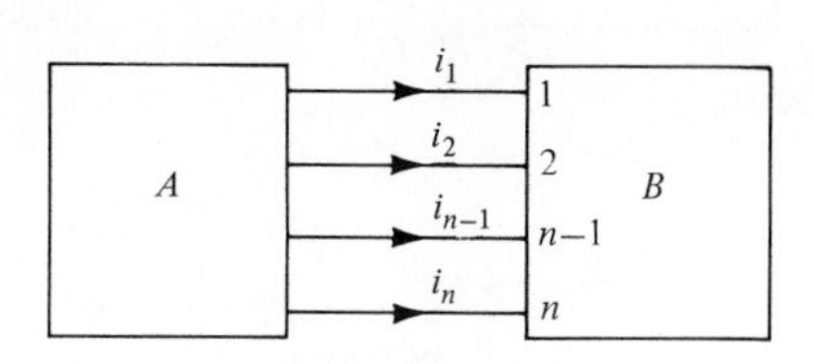

Figure 6.10 Total energy flow into Network B is equal to that in Figure 6.11.

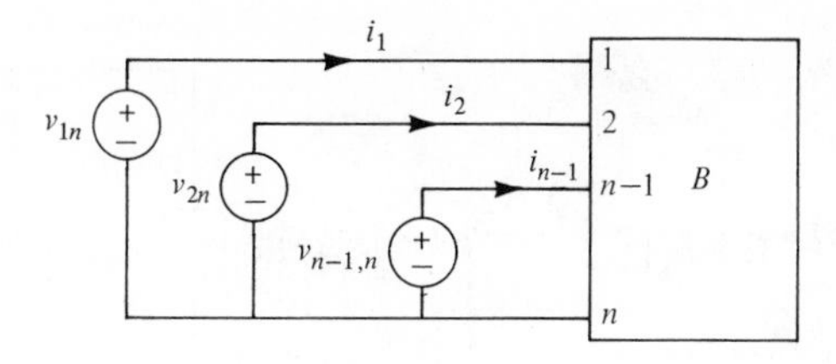

Figure 6.11 Total energy flow from the $n - 1$ sources is equal to the total energy flow into Network B in Figure 6.10.

Equation 6.30 has a particularly simple interpretation shown in Figure 6.8. Of course, the labeling of the terminals as 1, 2, and 3 is completely arbitrary. Thus the energy flow into Network B, which is the energy flow from the sources in Figure 6.8, is equal to the energy flow from the sources in Figure 6.9(a) and also equal to the energy flow from the sources in Figure 6.9(b).

Similarly, it is a fairly straightforward proposition to show (see Problem 6.8) that the energy flow into the n-terminal Network B of Figure 6.10 is equal to the energy flow from the $n - 1$ sources in Figure 6.11. The energy is

$$W_B = \int_{t_1}^{t} \sum_{k=1}^{n-1} v_{kn} i_k \, d\tau . \tag{6.31}$$

Again it should be emphasized that the terminal labeling is arbitrary and that terminals other than the nth may be used as the common return terminal.

ERCISES

6.4.1 In the network of Figure 6.12, suppose that the source v_{13} is a constant 10 volts, element a is $R_1 = 10$ ohms, and element b is $R_2 = 2$ ohms. Verify that the energy flow from the source for any time interval is equal to the sum of energy flows into the resistors.

6.4.2 Repeat Exercise 6.4.1 for element b, a nonlinear resistor specified by $v_{23} = 2i_{23}^2$.

6.4.3 Repeat Exercise 6.4.1 for $R_2 = -2$ ohms.

6.4.4 In the network of Figure 6.12, suppose element a is a resistor $R = 10$ ohms and element b is an inductor $L = 2$ henrys. If $i_{12}(0) = 0$ and $i_{12}(t)$ is

$$i_{12}(t) = 1 - e^{-(R/L)t} \qquad \text{for} \quad t \geq 0,$$

compute the energy delivered by the source from $t = 0$ to any arbitrary instant t and compare with the energy stored in L plus the energy dissipated in R.

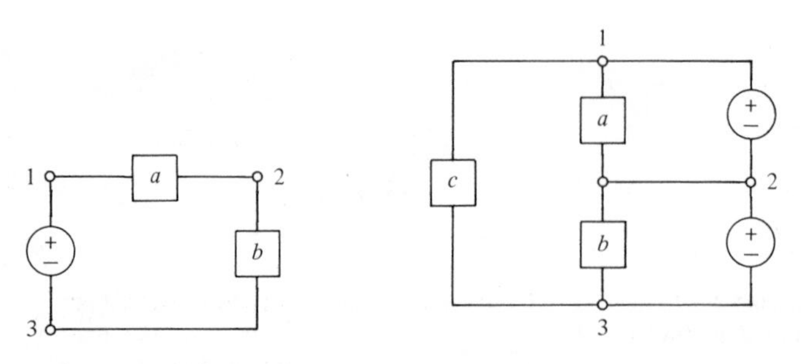

Figure 6.12 Figure 6.13

6.4.5 For the network in Figure 6.12 and element values as given in Exercise 6.4.4, compute the average power from the source if $i_{12}(t) = \sin 5t$ and compare with the sum of average powers into R and into L.

6.4.6 In the network of Figure 6.13, the elements are resistors all equal to 2 ohms. For $v_{12} = 3$ and $v_{23} = 5$, verify that the total power from the sources is equal to the sum of the powers in each branch of the network.

6.5 Tellegen's theorem

The law of conservation of energy discussed in the previous section is not an additional postulate in network theory. It is implied by Kirchhoff's laws being satisfied. This is very reassuring. Specifically, it is an immediate consequence of Tellegen's theorem* which is next proved using Kirchhoff's laws.

Tellegen's theorem pertains to a network which is an arbitrary interconnection of n_k-port subnetworks where the number of ports n_k of the kth subnetwork as well as the number of subnetworks is arbitrary. The allowable interconnections among the subnetworks are restricted to those which involve the ports only. Let the sum of the number of ports of all the subnetworks be denoted by N. Assign subnetwork port voltage- and current-reference directions such that the current-reference arrow enters the port at the terminal with the + voltage reference. Let $i_1, i_2, \ldots, i_N$ be values of currents associated with the N ports such that KCL is satisfied. Let $v_1, v_2, \ldots, v_N$ be values of voltages associated with the N ports such that KVL is satisfied. In terms of the above notation, Tellegen's theorem states that

$$\sum_{k=1}^{N} v_k i_k = 0. \tag{6.32}$$

* B. D. H. Tellegen, "A General Network Theorem, with Applications," *Philips Research Reports*, **7** (1952), 259–269. Tellegen (1900–) is at Philips Research Laboratories, Eindhoven, The Netherlands.

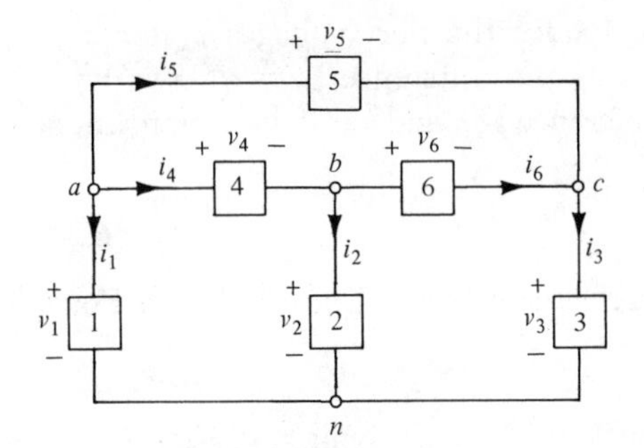

ure 6.14 Network for illustrating Tellegen's theorem.

Notice that there is no mention of the nature of branch characteristics. Hence, v_k and i_k do not have to be related by the kth branch characteristic. The set $v_1, v_2, \ldots, v_N$ is any set of voltages which satisfies KVL for the given network. Likewise, $i_1, i_2, \ldots, i_N$ is any set of currents which satisfies KCL for the given network. Before we prove the theorem, let us consider a simple example, as shown on Figure 6.14.

AMPLE 6.5.1

The network in Figure 6.14 may be viewed as an interconnection of six 1-ports. The nature of the branches is immaterial and some of the branches may be sources or nonlinear elements. For some instant of time, let us pick values for $i_1, i_2, \ldots, i_6$ which conform with KCL. For instance, let $i_1 = 3$, and $i_2 = 4$. Since KCL must be satisfied at node n, i_3 must be $-(3 + 4)$ so that $i_3 = -7$. Let $i_4 = 5$. KCL at node a yields $i_5 = -(i_1 + i_4) = -8$. KCL at node c yields $i_6 = i_3 - i_5 = 1$. KCL is satisfied at node b since $i_4 - i_2 - i_6 = 5 - 4 - 1 = 0$. Hence, the six current values satisfy the conditions of the theorem. Similarly, we may assign any voltage values to $v_1, \ldots, v_6$ so long as KVL is satisfied. For example, $v_1 = 1$, $v_4 = 4$, and $v_2 = -3$ satisfy KVL around loop *abna*. Let $v_6 = 6$. Then for KVL to be satisfied in loop *bcnb*, v_3 must be -9. Finally, for KVL to be satisfied in loop *abca*, $v_4 + v_6 - v_5 = 0$ or $v_5 = 10$. Note that KVL is satisfied for any other loop of the network so that the voltage values chosen satisfy the theorem. Now we form $\sum v_k i_k$ and obtain

$$(1)(3) + (-3)(4) + (-9)(-7) + (4)(5) + (10)(-8) + (6)(1) = 0, \quad (6.33)$$

which is predicted by the theorem.

To prove the theorem, we label the terminals of the ports of the subnetworks by $a, b, c, \ldots,$ and denote the voltages between the points $a, b, c, \ldots,$ and a fixed reference or datum node by $v_a, v_b, v_c, \ldots$. Since the port voltages v_k for $k = 1, \ldots, N$ satisfy KVL, we may express v_k as the difference between two node voltages. For instance, if v_1 has a positive reference at a and a minus reference at b, $v_1 = v_a - v_b$ and

$$v_1 i_1 = v_a i_1 - v_b i_1 . \tag{6.34}$$

Thus each $v_k i_k$ may be written as a difference similar to Equation 6.34. Form $\sum v_k i_k$ and collect terms as follows:

$$\begin{aligned}\sum_{k=1}^{N} v_k i_k = v_a &\text{ (terms involving current entering or leaving node } a) \\ &+ v_b \text{ (terms involving current entering or leaving node } b) \\ &+ v_c \text{ (terms involving current entering or leaving node } c) \\ &+ \cdots .\end{aligned} \tag{6.35}$$

For any parentheses above, say the factor multiplying v_a, the expression is simply the sum of currents leaving node a minus the sum of currents entering node a. But by KCL each pair of parentheses has a value of zero. Hence we have

$$\sum_{k=1}^{N} v_k i_k = 0,$$

which proves Tellegen's theorem.

Suppose that not only do the v_k and i_k satisfy KVL and KCL but also the branch characteristics so that the v_k and i_k constitute the network solution. Then $v_k i_k$ is the instantaneous power at the kth port and

$$\sum_{k=1}^{N} v_k i_k = \text{sum of instantaneous power at all ports of the network} = 0. \tag{6.36}$$

Since the network is the whole system, Equation 6.36 when integrated with respect to time gives one form of the law of conservation of energy. It states that the total instantaneous energy of the system does not change. Alternatively, we may say that the instantaneous power *out of* a portion of the network is equal to the instantaneous power *into* the rest of the network.

Even though this theorem might appear to be abstract and rather complex, it turns out to have many important applications in electrical engineering, one of which will be presented in Chapter 15.

EXERCISES

6.5.1 (a) For the network in Figure 6.15, apply Tellegen's theorem.
(b) Suppose that v_1, v_2, i_1, and i_2 satisfy the branch equations.

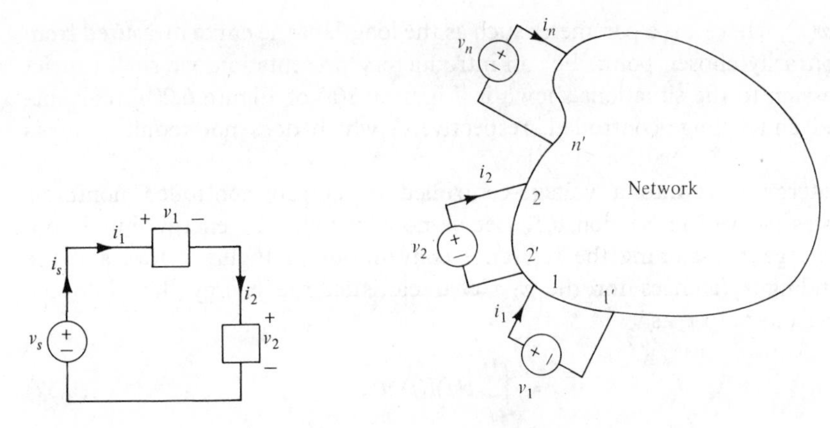

Figure 6.15

Figure 6.16

What is the significance of v_1i_1? v_2i_2? v_si_s? Verify that Tellegen's theorem implies conservation of energy for this network.

6.5.2 For the network of Figure 6.15, give an example of $v_1(t)$, $v_2(t)$, $v_s(t)$, $i_1(t)$, $i_2(t)$, and $i_s(t)$ which satisfies Tellegen's theorem.

6.5.3 For the network in Figure 6.16, show that the total power out of the n sources equals the total power into the n-port.

.6 Energy considerations in a network with nonlinear elements

The key to the theory necessary to treat the general case is due to Chua* and involves converting a curve, such as in Figure 6.17, into two ordinary functions

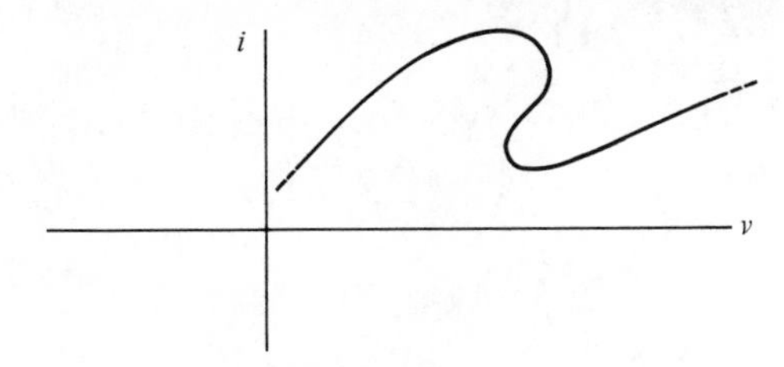

ure 6.17 A resistor characteristic which is neither current-controlled nor voltage-controlled.

* L. O. Chua, *Introduction to Nonlinear Network Theory* (New York: McGraw-Hill Book Company, 1969). See also C. A. Desoer and E. S. Kuh, *Basic Circuit Theory* (New York: McGraw-Hill Book Company, 1969), especially Chapter 13.

$i(x)$ and $v(x)$, where x is a parameter such as the length of the curve measured from some arbitrarily chosen point. For an introductory presentation, we shall restrict the discussion to the situations shown in Figure 6.2(a) or Figure 6.2(b) (current-controlled and voltage-controlled, respectively) which does not require Chua's approach.

The energy flow into a voltage-controlled or current-controlled nonlinear resistor was derived in Section 6.1. Let us now examine the energy flow into a capacitor. Again, assuming the reference conventions in Figure 6.1(a), and the corresponding references for the v–q characteristic, the energy flow into the capacitor from t_0 to t_1 is

$$W_c = \int_{t_0}^{t_1} v(t)i(t)\, dt\,. \tag{6.37}$$

For a charge-controlled capacitor, noting that $i(t)\, dt = dq(t)$ and in general that $v = v[q(t), t]$, a general function of q and t, we see that Equation 6.37 may be rewritten as

$$W_c = \int v[q(t), t]\, dq(t)\,. \tag{6.38}$$

If the capacitor is time-invariant, then Equation 6.38 reduces to

$$W_c = \int_{q_0}^{q_1} v(q)\, dq\,, \tag{6.39}$$

which has a simple geometrical interpretation, as shown in Figure 6.18, as the area under the curve between $q = q_0$ and $q = q_1$. Furthermore, if the capacitor is linear and time-invariant, then we have

$$v = \frac{1}{C} q\,, \tag{6.40}$$

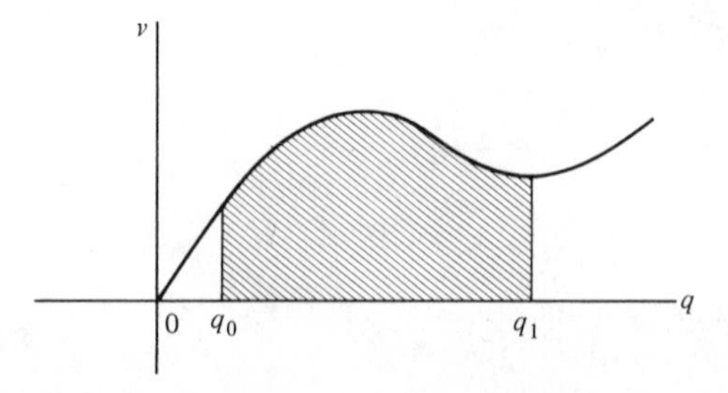

Figure 6.18 The cross-hatched area is the energy flow into a charge-controlled capacitor.

where C is a constant and

$$W_c = \int_{q_0}^{q_1} \frac{1}{C} q\, dq = \frac{1}{2C}(q_1^2 - q_0^2) \tag{6.41}$$

or

$$W_c = \tfrac{1}{2}C(v_1^2 - v_0^2), \tag{6.42}$$

where $q_1 = Cv_1$ and $q_0 = Cv_0$. Note that for the linear time-invariant case, W_c depends only on the initial q_0 (or v_0), the final q_1 (or v_1), and C. The manner in which q changes with time in going from $q(t_0)$ to $q(t_1)$ is immaterial. Likewise, for the nonlinear time-invariant capacitor, the manner in which q changes with time is immaterial. Only the initial and final values of q are of consequence. In addition, W_c depends on the particular nonlinear dependence of v on q as shown in Figure 6.18.

The calculation of energy flow into a voltage-controlled capacitor is slightly more involved. An example of a voltage-controlled capacitor characteristic is shown in Figure 6.19. We may still use the formulas as in Equations 6.37, 6.38, and 6.39. However, since v may not be a single-valued function of q, the integral symbolism has to be clarified further. Suppose that for the characteristic in Figure 6.19 we wish to determine the energy flow into the capacitor from t_0 when $v = v(t_0)$ and $q = q(t_0)$ to t_1, when $v = v(t_1)$, $q = q(t_1)$. As v changes from v_0 to v_1, q changes from q_0 to q_2 and back to q_1 as indicated in the figure. We may compute the energy in steps, first corresponding to the energy when v changes from v_0 to v_2, and next corresponding to the energy when v changes from v_2 to v_1. For the first part, the energy is equal to the area under the lower part of the characteristic curve from q_0 to q_2. On the figure, using the indicated labels, this is $A_1 + A_2$. Next, as v

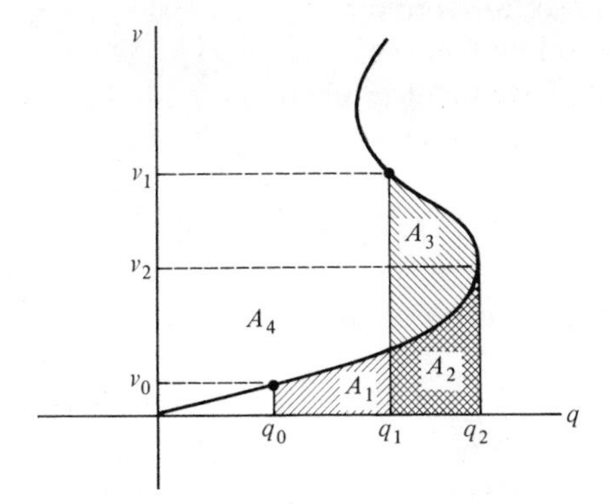

Figure 6.19 A voltage-controlled capacitor characteristic. The energy flow into the capacitor is A_1–A_3.

changes from v_2 to v_1, q changes from q_2 to q_1. Since the increment dq is negative, the energy is $-(A_2 + A_3)$. Hence, the energy from v_0 to v_1 is

$$W_c = \int_{t_0}^{t_1} v(t)i(t)\,dt = \int_{q_0,v_0}^{q_1,v_1} v\,dq = A_1 + A_2 - (A_2 + A_3) = A_1 - A_3. \quad (6.43)$$

An alternative interpretation is provided by the integration-by-parts rule. Proceeding formally, we have

$$\int_{q_0,v_0}^{q_1,v_1} v\,dq = vq\Big|_{q_0,v_0}^{q_1,v_1} - \int_{q_0,v_0}^{q_1,v_1} q\,dv = v_1q_1 - v_0q_0 - \int_{v_0}^{v_1} q\,dv. \quad (6.44)$$

We note that since q is an ordinary function of v, the last integral in Equation 6.44 is an ordinary one. From the notation in Figure 6.19, we see that

$$\int_{v_0}^{v_1} q\,dv = A_4 + A_3. \quad (6.45)$$

We also note that the rectangular area v_1q_1 minus the rectangular area v_0q_0 is equal to

$$v_1q_1 - v_0q_0 = A_4 + A_1. \quad (6.46)$$

Therefore, we have

$$\int_{q_0,v_0}^{q_1,v_1} v\,dq = (A_4 + A_1) - (A_4 + A_3) = A_1 - A_3, \quad (6.47)$$

which is the same answer as in Equation 6.43. Equation 6.44 is taken as the definition for the integral of $v\,dq$ for a voltage-controlled capacitor. The right-hand side expression in Equation 6.44 defines what the integral means. If the capacitor is neither voltage-controlled nor charge-controlled, such as the curve in Figure 6.17 if the ordinate i is replaced by q, the procedure outlined above breaks down. This more general situation has been studied by Chua.

For an inductor model, the general energy formula of Equation 6.37 still holds. This time we note that $v\,dt$ is $d\psi$, so that we may write

$$W_L = \int_{t_0}^{t_1} v(t)i(t)\,dt = \int_{t_0}^{t_1} i[\psi(t), t]\,d\psi(t). \quad (6.48)$$

Analogous to Equation 6.39, the time-invariant case reduces to

$$W_L = \int_{t_0}^{t_1} i[\psi(t)]\,d\psi(t) = \int_{(i_0,\psi_0)}^{(i_1,\psi_1)} i\,d\psi. \quad (6.49)$$

For a flux-controlled time-invariant inductor as shown in Figure 6.20, the energy flow into L from t_0 when $\psi = \psi(t_0) = \psi_0$ to t_1 when $\psi = \psi(t_1) = \psi_1$ is equal to the lined area.

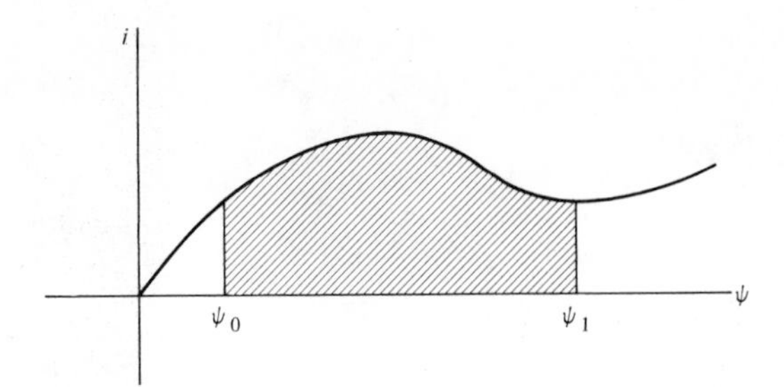

re 6.20 The cross-hatched area is the energy flow into a flux-controlled inductor.

The linear time-invariant L has a characteristic

$$i = \frac{1}{L}\psi, \tag{6.50}$$

where L is constant. Equation 6.49 reduces to

$$W_L = \int_{\psi_0}^{\psi_1} \frac{1}{L}\psi\, d\psi = \frac{1}{2L}(\psi_1^2 - \psi_0^2). \tag{6.51}$$

Since $Li_1 = \psi_1$ and $Li_0 = \psi_0$, Equation 6.51 may be written as

$$W_L = \tfrac{1}{2}L(i_1^2 - i_0^2). \tag{6.52}$$

The situation for the calculation of W_L into a current-controlled L is exactly analogous to the situation of calculating W_c into a voltage-controlled capacitor. The same integration-by-parts rule is used to define the integral of $i\, d\psi$ as

$$\int_{(i_0,\psi_0)}^{(i_1,\psi_1)} i\, d\psi = i_1\psi_1 - i_0\psi_0 - \int_{i_0}^{i_1} \psi\, di. \tag{6.53}$$

For a current-controlled inductor, the integral of $\psi\, di$ is well-defined, as shown in the example of Figure 6.21. The energy flow W_L into the inductor from $t = t_0$ corresponding to $i = i(t_0) = i_0$ and $\psi = \psi(t_0) = \psi_0$ to $t = t_1$ corresponding to $i = i(t_1) = i_1$ and $\psi = \psi(t_1) = \psi_1$ is equal to the rectangular area $i_1\psi_1$ minus the rectangular area $i_0\psi_0$ minus the lined area.

From the foregoing discussion, we note that the energy W_L does not depend on how ψ (or i) varies with respect to time, provided only that L is time-invariant and either flux-controlled or current-controlled. The only pertinent data are the initial and final values of i and ψ and the particular nonlinear but time-invariant dependence between i and ψ (the inductor characteristic). For the linear time-invariant case, W_L depends on i_0, i_1, and L. In particular, if $i_0(t_0) = i_1(t_1)$, and $\psi(t_0) = \psi(t_1)$, then no matter how i varies in time between t_0 and t_1, $W_L = 0$ (see Equations 6.51 and 6.52). That is, no matter how much energy flows into L, the same amount of

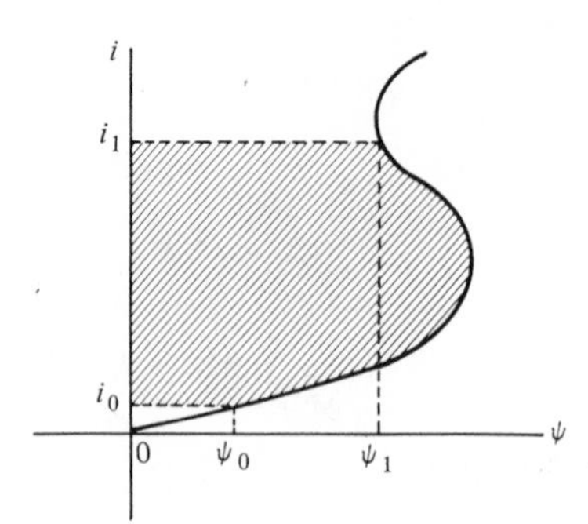

Figure 6.21 A current-controlled inductor characteristic.

energy eventually flows out within the time interval $[t_0, t_1]$, provided the initial current and initial flux are equal to the final current and final flux, respectively. Similarly, for a time-invariant nonlinear inductor which is flux-controlled, Equation 6.49 shows that if $\psi_0 = \psi_1$, then $W_L = 0$. And if a time-invariant nonlinear inductor is current-controlled, then from Equation 6.53, we note that if $i_0 = i_1$, then the right-hand side integral is zero. Furthermore, since ψ is a single-valued function of i, $\psi_0 = \psi_1$ when $i_0 = i_1$. Hence, $W_L = 0$.

If the signals are periodic, then the currents (or fluxes) separated by one period in time are equal, so that the energy over one period is zero. This means that the average power is zero. We conclude that a time-invariant inductor which is either flux-controlled or current-controlled is *lossless*. This result applies to arbitrary nonlinearities: The only restrictions are that the element must be time-invariant and either flux-controlled or current-controlled. That is, either i must be expressible as a function of ψ, or ψ must be expressible as a function of i, and t must not appear explicitly. Because the time-invariant flux-controlled (or current-controlled) inductor model is lossless, it is said to be an *energy storage element* or a *conservative element*.

Likewise for a time-invariant capacitor which is either charge-controlled or voltage-controlled, Equations 6.39 and 6.44 show that if $q_0 = q_1$ or $v_0 = v_1$, then $W_C = 0$. Thus if q and v are periodic, the net energy flow into the capacitor over one period is zero so that the average power is zero. For this reason, a time-invariant charge-controlled (or voltage-controlled) capacitor is a *lossless* element. It is also an *energy storage element*. This result applies to arbitrary nonlinearities provided only that v can be expressed as a single-valued function of q or that q can be expressed as a single-valued function of v, and t does not appear explicitly.

However, the same cannot be said of the resistor model. For example, if R is a positive constant, then $p(t)$ is never negative and for $i \neq 0$, W_R is positive even if $i(t_0) = i(t_1)$ and $v(t_0) = v(t_1)$. Furthermore, the value of W_R depends on how v and i vary with respect to time during the interval. The most common type of resistor model is a positive constant R, and for this model W_R is positive whenever $i \neq 0$; that is, positive energy is always flowing into R. This is why a resistor is commonly called a *dissipative* or nonconservative element. For the general case, W_R is not always positive. For the same resistor W_R may be positive or negative depending on the time variation of i or v. For the negative constant R, W_R is always negative and thus behaves like a source of energy. We will call any resistor dissipative, keeping in mind that dissipation may be negative. The electric-energy dissipation is converted into heat energy.

For the time-varying capacitor and time-varying inductor, whether linear or nonlinear, there may be dissipation or energy generation. For instance, take the linear time-varying capacitor $C(t)$. Since

$$q(t) = C(t)v(t), \tag{6.54}$$

we have

$$W_C = \int_{t_0}^{t_1} \frac{1}{C(t)} q(t)\, dq(t), \qquad C \neq 0. \tag{6.55}$$

The integration in Equation 6.55 cannot be carried out until the exact dependence of $q(t)$ on t is known. Moreover, even for the same $q(t_0)$ and $q(t_1)$, different variations of $q(t)$ with respect to t will result in different values of W_C. For example, for $C(t) = 1 - 2t$, $q(t) = t - t^2$, $t_0 = 0$, $t_1 = 1$, we have

$$W_C = \int_0^1 \frac{1}{1 - 2t} (t - t^2)(1 - 2t)\, dt = (\tfrac{1}{2}t^2 - \tfrac{1}{3}t^3)\big|_0^1 = \tfrac{1}{6}. \tag{6.56}$$

Note that $q(0) = q(1) = 0$ and $v(0) = v(1) = 0$ and yet $W_C \neq 0$. A similar situation obtains for the nonlinear time-varying capacitor. As an example, take

$$v(t) = \frac{1}{C(t)} q^2(t), \tag{6.57}$$

where $C(t) = 1 - 2t$, $q(t) = t - t^2$, $t_0 = 0$, $t_1 = 1$. Then we have

$$\begin{aligned} W_C &= \int_0^1 \frac{1}{1 - 2t} (t - t^2)^2(1 - 2t)\, dt \\ &= \int_0^1 (t^2 - 2t^3 + t^4)\, dt = (\tfrac{1}{3}t^3 - \tfrac{2}{4}t^4 + \tfrac{1}{5}t^5)\big|_0^1 \\ &= \tfrac{1}{3} - \tfrac{1}{2} + \tfrac{1}{5} - 0 = \tfrac{1}{30}. \end{aligned} \tag{6.58}$$

Again $W_C \neq 0$, although $q(t_0) = q(t_1)$, $v(t_0) = v(t_1)$. The same situation is true for the time-varying inductor.

In summary, a resistor, whether time-invariant or not, and whether linear or nonlinear, dissipates (or generates) energy. A time-invariant capacitor, which is either charge-controlled or voltage-controlled, and a time-invariant inductor, which is either flux-controlled or current-controlled, whether linear or nonlinear, only store energy in their electric and magnetic fields, respectively. A time-varying capacitor and a time-varying inductor may dissipate or generate energy in addition to storing energy in the electric or magnetic field. An inductor which is neither flux-controlled nor current-controlled may dissipate energy. A capacitor which is neither charge-controlled nor voltage-controlled may dissipate energy also.

EXERCISES

6.6.1 An initially uncharged capacitor $C = 1$ is connected to a current source whose waveform is shown in Figure 6.22(a). Compute the energy flow into the capacitor as a function of t. Repeat for the current waveform of Figure 6.22(b). For $t > 2$, compare the charges delivered by the two current sources. Compare the two energies for $t > 2$. What is the effect of reversing the terminals of the current sources?

6.6.2 A resistor $R = 2$ ohms is connected across a current source whose waveform is shown in Figure 6.22(a). Plot W_R vs. t. Repeat for the source in Figure 6.22(b). What is the effect of reversing the terminals of the current sources?

6.6.3 The v–i characteristic for a resistor is given by

$$v = \begin{cases} 2i, & \text{for} \quad i \geq 0, \\ 10i, & \text{for} \quad i < 0. \end{cases}$$

The resistor is then connected to a current source whose waveform is shown in Figure 6.22(a). Plot W_R vs. t. What is the effect of reversing the terminals of the current source?

6.6.4 Repeat Exercise 6.6.3 for the waveform shown in Figure 6.22(b).

6.6.5 An inductor $L = 3$ with zero initial flux is connected to a voltage source whose waveform is shown in Figure 6.22(a) where the ordinate i is replaced by v. Determine $W_L(t)$.

6.6.6 Repeat Exercise 6.6.5 for the waveform in Figure 6.22(a). What is the effect of reversing the terminals of the voltage source?

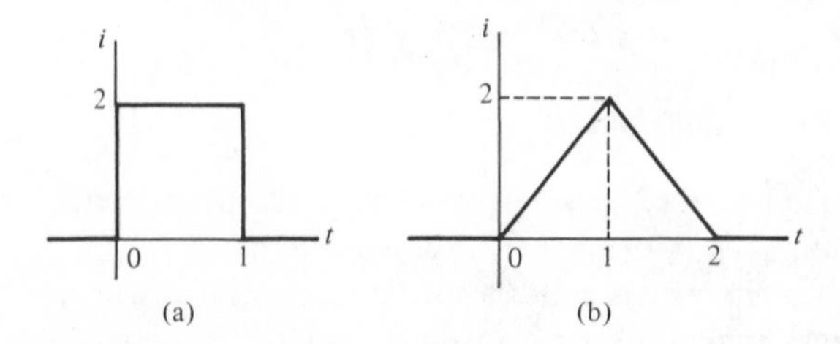

Figure 6.22

★6.6.7 A nonlinear inductor is represented by the equation

$$i = \psi^2 - \psi .$$

The flux at $t = 0$ is 0, and, at $t = t_1$, the flux is equal to 1. Compute the energy flow W_L into the inductor from $t = 0$ to $t = t_1$.

★6.6.8 Suppose that for the nonlinear inductor in Exercise 6.6.7 a voltage source as in Figure 6.22(b) is connected across L, where the ordinate i is replaced by v. Assuming that $\psi(0) = 0$, compute W_L into L from $t = 0$ to $t = 1$. What is the effect of reversing the terminals of the voltage source?

★6.6.9 Suppose that a nonlinear inductor is represented by $\psi = i^2 - i$. The inductor is connected across a current source with a waveform as in Figure 6.22(b). Suppose that at $t = 0$ there is no initial flux in the inductor and $i(0) = 0$. Compute the energy flow into L from $t = 0$ to $t = 1$.

Summary

1. Power is defined as $d(\text{energy})/dt = v(t)i(t)$ with sign conventions as shown in Figure 6.1. A positive sign for vi implies instantaneous time rate of energy flow *into* the network; a negative sign implies flow *out* of the network.
2. For a linear time-varying resistor,

$$p(t) = R(t)i^2(t) = G(t)v^2(t).$$

3. The average power of a periodic signal is

$$P_{av} = \frac{1}{T}\int_{t_1}^{t_1+T} p(t)\, dt$$

for all t_1 where T is the period. For a time-varying current-controlled resistor, $p(t)$ in this equation becomes

$$i(t)v[i(t), t]$$

and for a time-varying voltage-controlled resistor, $p(t)$ becomes

$$v(t)i[v(t), t].$$

For time-invariant charge- or voltage-controlled capacitors and for time-invariant flux-linkage-controlled or current-controlled inductors

$$P_{av} = 0.$$

Clearly, this is the case for linear time-invariant capacitors and inductors.

4. The energy in a network at time t, measured with respect to that at reference time t_1, is

$$W(t) = \sum_{k=1}^{n} \int_{t_1}^{t} v_k(\tau) i_k(\tau)\, d\tau .$$

For the linear time-invariant positive capacitor and inductor,

$$W_C(t) = \tfrac{1}{2}C[v^2(t) - v^2(t_1)], \qquad W_L(t) = \tfrac{1}{2}L[i^2(t) - i^2(t_1)].$$

5. A basic law which must be satisfied for all physical systems is the conservation of energy. It requires an energy balance considering energy stored, energy gained, and energy lost in a given system. It must be satisfied at each instant of time.
6. Tellegen's theorem has many applications in circuit analysis and design. It is based entirely on Kirchhoff's laws. It requires that

$$\sum_{k=1}^{N} v_k i_k = 0 .$$

The pair (v_k, i_k) need not satisfy the branch characteristics of the network. Any set of numbers $(v_1, v_2, \cdots, v_N)$ which satisfy KVL and any set of numbers $(i_1, i_2, \cdots, i_N)$ which satisfy KCL for the *same network structure* will satisfy Tellegen's theorem.
7. Energy considerations in elements which may be both time-varying and nonlinear are shown in Figures 6.23 and 6.24 and in the accompanying chart.

Element description	*Time-invariant*	*Time-varying*
Resistor	Dissipates or generates energy	Dissipates or generates energy
Capacitor (charge- or voltage-controlled) Inductor (flux- or current-controlled)	Stores energy. Lossless.	Stores energy. May dissipate or generate energy.
Capacitor or inductor (controlled by neither single variable)	Stores energy. May dissipate or generate energy.	Stores energy. May dissipate or generate energy.

PROBLEMS

6.1 Determine whether the gyrator defined in Chapter 4 is passive. Also determine whether it is lossless.

6.2 Repeat Problem 6.1 for the negative converter, voltage-inversion type, and current-inversion type.

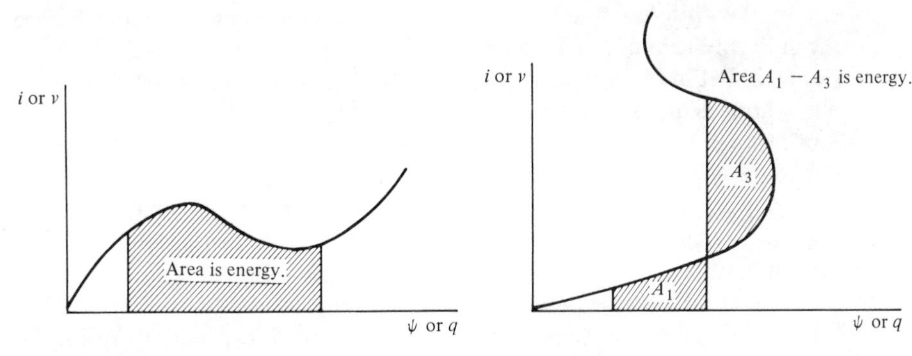

Figure 6.23

Figure 6.24

★6.3 Two linear time-invariant positive inductors are inductively coupled as in Figure 4.18(a) (p. 70) with $M \leq \sqrt{L_1 L_2}$. Determine whether the network is passive. Also determine whether the network is lossless.

6.4 In the network of Figure 6.13, suppose the elements are all resistors, $R_a = 1$, $R_b = 2$, $R_c = 3$, and $v_{12} = 2$ and $v_{23} = 4$. Compute the average power into each resistor. Now remove source v_{12} but add a source connected to terminals 1 and 3. What should the new source be so that the branch voltages and currents remain the same as before? Compute the power delivered from the new set of sources and compare to the power delivered from the former set of sources.

6.5 For the data in the first part of Problem 6.4, compute i_{21} and i_{23} in addition to the branch voltages, currents, and powers. Now replace v_{12} by a current source i_{21} with a value as computed previously. Likewise, replace v_{23} by a current source i_{23} with a value as computed before. Recompute the branch voltages, currents, and powers. Compute the powers from the current sources and compare with the powers into the R's.

6.6 Two networks are connected to each other by a cable containing three wires as in Figure 6.6. Certain measurements are made as follows: $I_{1'1} = 2$ amperes (constant), $I_{22'} = 1$ ampere, $v_{21} = 3$ volts, and $v_{13} = 4$ volts. Determine whether energy is flowing from B to A or A to B and how much.

6.7 For the network of Figure 6.13, suppose the data of the first part of Problem 6.4 are assumed. Now remove the sources and instead add three sources. Each new source is inserted in the lines, one to each line, and the three other terminals of the sources are connected together to form a new node. Determine the values with reference directions of each voltage source so that the branch voltages, currents, and powers remain the same.

6.8 Using a procedure analogous to the one used in Section 6.4, show that the energy flow into an n-terminal network such as the one in Figure 6.10 is given by the formula in Equation 6.31.

6.9 Suppose that in the network of Figure 6.12 element a is a resistor $R_a = -20$ ohms and element b is a resistor $R_b = 50$ ohms. Suppose that v_{13} is the input signal which is periodic and R_b is called the load or output resistor. Compute the average power delivered by the source and compute the average power delivered to the load. Is the output power less than, equal to, or greater than the input power? Explain the significance of your answer.

6.10 Repeat Problem 6.9 for $R_a = -20$ and $R_b = 10$.

6.11 Suppose that in the network of Figure 6.12 element b is a capacitor $C = 1$ microfarad and element a is a resistor $R = 10$ ohms. If $v_{21} = 5\sin(10^5 t + \theta)$, compute the average power from the source and the average power into R and C.

6.12 For a linear time-invariant capacitor C, show that the energy flow into C from t_1 to t_2 is

$$W_C = \int_{q(t_1)}^{q(t_2)} v\,dq = \int_{v(t_1)}^{v(t_2)} q\,dv\,.$$

6.13 Show that for a linear time-invariant inductor the energy flow into L from t_1 to t_2 is

$$W_L = \int_{\psi(t_1)}^{\psi(t_2)} i\,d\psi = \int_{i(t_1)}^{i(t_2)} \psi\,di\,.$$

★ 6.14 (a) Give an example of a nonlinear time-invariant capacitor where the energy is not equal to

$$\int_{v(t_1)}^{v(t_2)} q\,dv\,.$$

(b) Give an example of a nonlinear time-invariant inductor where the energy is not equal to

$$\int_{i(t_1)}^{i(t_2)} \psi\,di\,.$$

★ 6.15 In a mechanical system, momentum is a quantity which is associated with mass such that the time rate of momentum is the force accelerating the mass. Analogous to defining a capacitor by a q–v curve, a mass may be defined by a momentum–velocity curve. Similarly, a spring may be characterized by a force–displacement curve. Finally, a damper may be characterized by a force–velocity curve. Continuing the analogy, since force and velocity are analogous to voltage and current or current and voltage, the analogy of instantaneous electric power, voltage times current, is instantaneous mechanical power which is force times velocity, and mechanical energy is

$$W = \int_{t_1}^{t} f(\tau)v(\tau)\, d\tau,$$

where f is force and v is velocity.
(a) Show that the mechanical-energy flow into a mass is

$$W_M = \int_{t_1}^{t_2} fv\, dt = \int v\, d\psi,$$

where ψ is momentum. Show that for a linear time-invariant mass

$$W_M = \tfrac{1}{2}M[v^2(t_2) - v^2(t_1)].$$

(b) Show that the mechanical-energy flow into a spring is

$$W_K = \int_{t_1}^{t_2} fv\, dt = \int f\, dx,$$

where x is displacement. Show also that for a linear time-invariant spring,

$$W_K = \frac{1}{2K}[x^2(t_2) - x^2(t_1)],$$

where K is the compliance of the spring.
(c) Show that the instantaneous power into a linear damper is

$$P_D = Dv^2(t).$$

The time-invariant mass is said to store kinetic energy, the time-invariant spring is said to store potential energy, and the damper is said to dissipate energy. The above statement may be verified by a development analogous to that for R, L, and C. See Appendix A.

6.16 Suppose that in the network of Figure 6.12 branch b is a positive constant capacitor C and branch a is a positive constant resistor R. The source v_{13} is a battery of voltage $v = E$ connected at $t = 0$. The capacitor is initially uncharged. The current is

$$i_{12}(t) = \frac{E}{R} e^{-t/(RC)}.$$

Compute the energy delivered to the capacitor after an infinitely long period of time. Compute the total energy flow from the source. Note that the ratio of the energy delivered to the capacitor and the total energy delivered by the source is independent of the values of E, R, and C.

★ 6.17 Suppose that for the series RC network of Problem 6.16, C is a positive constant but R is a time-varying nonlinear resistor whose v–i characteristic is in the first and third quadrant. The capacitor is initially uncharged. The current is no longer an exponential but it is given that $i_{12} \to 0$ as $t \to \infty$. Compute the total energy delivered to C after an infinitely long period of time. Compute the total energy delivered by the battery after an infinitely long period of time. The ratio between the two energies is independent of the values of E and C and independent of the resistance characteristic: The efficiency of charging may not be improved upon by judicious choice of E, C, or R.

★6.18 Suppose that the constant capacitor in Problem 6.17 is replaced by a nonlinear time-invariant passive charge-controlled and voltage-controlled capacitor described by $v = f(q)$. If $i(t) \to 0$ as $t \to \infty$, determine the total energy delivered to the capacitor after an infinitely long period of time. Determine the total energy delivered by the battery after an infinitely long period of time. Note that these two energies do not depend on the resistor characteristic.

6.19 A four-terminal network like that in Figure 6.10 with $n = 4$ is such that

$$i_1(t) = 2\sin(\omega_0 t + \alpha), \qquad i_2(t) = 2\sin(\omega_0 t - 120° + \alpha),$$
$$i_3(t) = 2\sin(\omega_0 t - 240° + \alpha), \qquad i_4(t) \equiv 0,$$
$$v_{12}(t) = 100\sin(\omega_0 t + 30°), \qquad v_{23}(t) = 100\sin(\omega_0 t - 90°),$$
$$v_{34}(t) = (100/\sqrt{3})\sin(\omega_0 t - 240°).$$

Compute the average power into the network. What will happen if line 4 is disconnected?

★6.20 Give an example of a voltage-controlled, time-invariant, nonlinear active capacitor with a v–q characteristic in the first and third quadrants.

6.21 For the network in Figure 6.25, assign reference directions to the branch voltages and currents and verify Tellegen's theorem for any arbitrary distribution of voltages and currents which satisfy KVL and KCL.

6.22 (a) For the network in Figure 6.25, show that the energy flow from the voltage source v_s is equal to the energy flow into branch 1 plus the energy flow into branch 2 plus the energy flow into branch 3, using Tellegen's theorem.

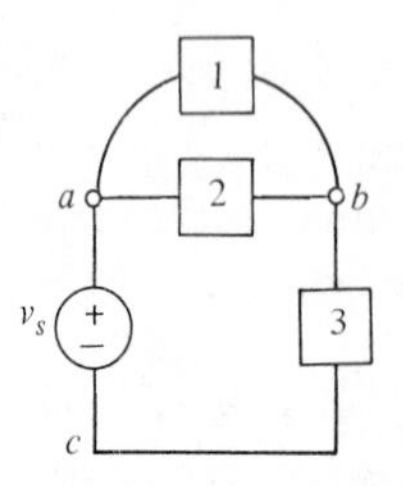

Figure 6.25

(b) Suppose that branch 3 of Figure 6.25 is a current source with a reference direction from b to c, through branch 3. Does the result of part (a) still hold?

(c) For the condition of part (b), show that the energy flow from the voltage source v_s plus the energy flow from the current source i_3 is equal to the energy flow into branch 1 plus the energy flow into branch 2, using Tellegen's theorem.

Chapter 7

Implications of Linearity in Networks

7.1 Characteristics of a linear system

In this chapter, we will consider systems with one designated input and one designated output. Such a system is shown in Figure 7.1(a) in which the voltage source $v_1(t)$ is the input and the voltage at the open-circuit terminals is the output, $v_2(t)$. Now in general the input can be a voltage or a current, and similarly the output can be some other voltage or current (or charge or flux, for that matter). A more general representation of the system we will study is shown in Figure 7.1(b) which is a block diagram with y labeled as output and x as input. Here x and y may be identified as voltages or currents depending on the particular problem under study. We will next discuss certain properties of linear systems having the representations of Figure 7.1. Our results will be easily extended to systems with more than one input or output.

First we define the *zero-input response* of the system as the output y when the input is identically zero. Such a response is not necessarily zero since there may be initial charges on capacitors or initial fluxes in inductors. The set of initial conditions will be called the initial *state* of the system. Next we define the *zero-state response* as the output y due to an arbitrary input when all initial conditions are zero, i.e., when the initial state is zero. A system or network is said to be *zero-state linear* if the following two conditions are satisfied:

(1a) If y_1 is the zero-state response due to an arbitrary input x_1, then cy_1 is the zero-state response due to an input cx_1, where c is an arbitrary constant. This is called the *homogeneity* condition.

(2a) If y_1 is the zero-state response due to an arbitrary input x_1, y_2, the zero-state response due to another arbitrary input x_2 not equal to x_1 identically, then the zero-state response due to $x_1 + x_2$ is $y_1 + y_2$. This is known as the *additivity* condition or *superposition* condition.

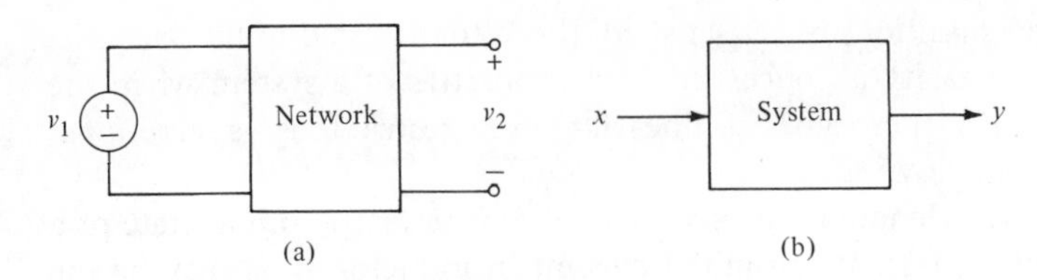

ure 7.1 A network with input v_1 and output v_2 identified and (b) the abstraction of (a) to a general system for which x is the single input and y is the single output.

AMPLE 7.1.1

Consider the network shown in Figure 7.1(a). Suppose that the zero-state response of the system when $v_1(t) = u(t)$ is

$$v_2(t) = (1 - e^{-t})u(t). \tag{7.1}$$

If the system satisfies the homogeneity condition (1a), then the zero-state response of the same system to the input $v_1(t) = 10u(t)$ is

$$v_2(t) = 10(1 - e^{-t})u(t). \tag{7.2}$$

Of course this result may be generalized by replacing 10 by K for both v_1 and v_2.

AMPLE 7.1.2

As an example of the superposition or additivity condition, suppose that the zero-state response due to a unit step is $(1 - e^{-t})u(t)$, and the zero-state response due to a delayed negative step $-u(t - 1)$ is $-(1 - e^{-(t-1)})u(t - 1)$. If the system obeys the superposition property (2a), then we may say the following: If

$$v_1(t) = u(t) - u(t - 1) \tag{7.3}$$

is the input (a rectangular pulse), the zero-state response is

$$v_2(t) = (1 - e^{-t})u(t) - (1 - e^{-(t-1)})u(t - 1). \tag{7.4}$$

AMPLE 7.1.3

Suppose that the systems of the two examples are identical. Then both systems would satisfy the superposition and homogeneity conditions. Suppose that the system is subjected to a pulse of amplitude 20 volts and duration 1 sec; this is written

$$v_1(t) = 20[u(t) - u(t - 1)]. \tag{7.5}$$

Then it follows that the zero-state response is

$$v_2(t) = 20[(1 - e^{-t})u(t) - (1 - e^{-(t-1)})u(t - 1)]. \tag{7.6}$$

If any of the above conditions is not satisfied, the system is said to be *zero-state nonlinear*. Zero-state linearity is concerned with properties of a system when the initial state is zero. Another type of linearity with restrictions is *zero-input linearity*, to be defined below.

Let $(\alpha_1, \alpha_2, \alpha_3, \ldots, \alpha_n)$ denote a set of initial conditions or the initial state of a system. The quantity α_1 may be the initial current in inductor 1, α_2 may be the initial voltage in capacitor 2, and so forth. A system is said to be *zero-input linear* if it satisfies the following two conditions:

(1b) If the zero-input response for any arbitrary initial state $(\alpha_1, \alpha_2, \ldots, \alpha_n)$ is y_0, then the zero-input response due to an initial state $(C\alpha_1, C\alpha_2, \ldots, C\alpha_n)$ is Cy_0, where C is any arbitrary constant. This is the homogeneity condition.

(2b) If the zero-input response for an arbitrary initial state $(\alpha_1, \alpha_2, \ldots, \alpha_n)$ is y_α and the zero-input response for another arbitrary initial state $(\beta_1, \beta_2, \ldots, \beta_n)$ is y_β, then the zero-input response for an initial state $(\alpha_1 + \beta_1, \alpha_2 + \beta_2, \ldots, \alpha_n + \beta_n)$ is $y_\alpha + y_\beta$. This is the additivity condition.

EXAMPLE 7.1.4

Suppose that in the network of Figure 7.2, there are no inputs, but initial conditions of $v_c(0) = 10$ volts and $i_L(0) = -3$ amperes are imposed. Let the zero-input response (due to initial conditions only) be given as

$$v(t) = (10e^{-t} \cos 2t)u(t). \tag{7.7}$$

If the network satisfies the homogeneity condition (1b), then if the initial conditions were changed to $v_c(0) = 40$ volts and $i_L(0) = -12$ amperes, the zero-input response would be

$$v(t) = (40e^{-t} \cos 2t)u(t). \tag{7.8}$$

This is because the new initial conditions are 4 times the old initial conditions, so that the new zero-input response is 4 times the old zero-input response.

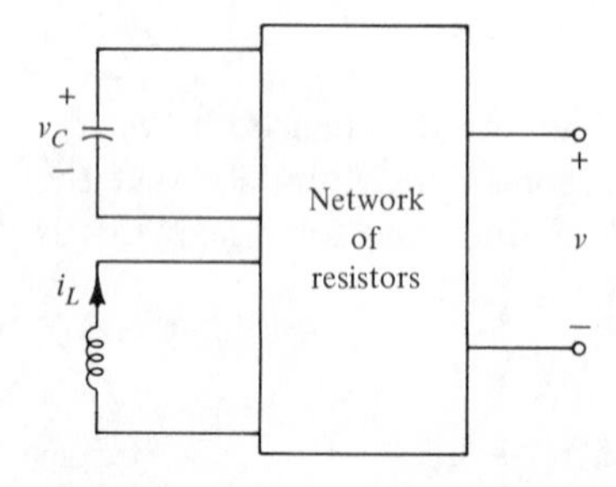

Figure 7.2 Network used to illustrate zero-input linearity.

AMPLE 7.1.5

For the network of Figure 7.2, suppose that the zero-input response due to $v_C(0) = 8$ volts and $i_L(0) = 2$ amperes is

$$v_1(t) = e^{-t}(7 \cos 2t + 4 \sin 2t)u(t), \tag{7.9}$$

and the zero-input response due to $v_C(0) = 2$ volts and $i_L(0) = -5$ amperes is

$$v_2(t) = e^{-t}(3 \cos 2t - 4 \sin 2t)u(t). \tag{7.10}$$

If the network satisfies the superposition property (2b), then the response due to the sum of initial conditions, namely $v_C(0) = 8 + 2 = 10$ volts and $i_L(0) = 2 - 5 = -3$ amperes is equal to the sum of the individual zero-input responses, namely

$$\begin{aligned} v(t) = v_1(t) + v_2(t) &= e^{-t}(7 \cos 2t + 4 \sin 2t)u(t) + e^{-t}(3 \cos 2t - 4 \sin 2t) \\ &= (10e^{-t} \cos 2t)u(t). \end{aligned} \tag{7.11}$$

AMPLE 7.1.6

Let us suppose that the network in Figure 7.2 is zero-input linear and the data in Examples 7.1.4 and 7.1.5 apply. The problem is to determine the zero-input response due to the initial conditions $v_C(0) = 6$ volts and $i_L(0) = 4$ amperes. Since the network is zero-input linear, conditions (1b) and (2b) are satisfied. From condition (1b), we know that if we multiply the initial conditions in Example 7.1.4 by a constant c_1, the zero-input response will be c_1 times the zero-input response given in Example 7.1.4. Similarly, if the initial conditions in Example 7.1.5 are multiplied by c_2, then the zero-input response will be c_2 times the zero-input response given in Example 7.1.5. Let us first determine c_1 and c_2 so that c_1 times the initial conditions in Example 7.1.4 added to c_2 times the initial conditions in Example 7.1.5 yields $v_C(0) = 6$ volts and $i_L(0) = 4$ amperes. Once c_1 and c_2 are found, then the required zero-input response may be computed by multiplying the zero-input response in Example 7.1.4 by c_1, that in Example 7.1.5 by c_2, and adding these two. Thus,

$$10c_1 + 8c_2 = 6 \tag{7.12}$$

and

$$-3c_1 + 2c_2 = 4. \tag{7.13}$$

Equation 7.12 is the condition for obtaining $v_C(0) = 6$ volts and Equation 7.13 is the condition for obtaining $i_L(0) = 4$ amperes. Solving for c_1 and c_2, we have

$$c_1 = \frac{\begin{vmatrix} 6 & 8 \\ 4 & 2 \end{vmatrix}}{\begin{vmatrix} 10 & 8 \\ -3 & 2 \end{vmatrix}} = \frac{12 - 32}{20 + 24} = -\frac{20}{44} = -\frac{5}{11} \tag{7.14}$$

and

$$c_2 = \frac{\begin{vmatrix} 10 & 6 \\ -3 & 4 \end{vmatrix}}{44} = \frac{40 + 18}{44} = \frac{58}{44} = \frac{29}{22}. \tag{7.15}$$

Thus the zero-input response due to $v_C(0) = -(5/11)(10)$ volts and $i_L(0) = -(5/11)(-3)$ amperes is $-(5/11)(10e^{-t} \cos 2t)u(t)$. The zero-input response due to $v_C(0) = (29/22)(8)$ volts and $i_L(0) = (29/22)(2)$ amperes is

$$\frac{29}{22} e^{-t}(7 \cos 2t + 4 \sin 2t)u(t).$$

Hence, by the additivity property, the zero-input response due to $v_C(0) = -(5/11)(10) + (29/22)(8) = 6$ volts and $i_L(0) = -(5/11)(-3) + (29/22)(2) = 4$ amperes is

$$\begin{aligned} \text{Zero-input response} &= -\frac{5}{11}(10e^{-t} \cos 2t)u(t) \\ &\quad + \frac{29}{22} e^{-t}(7 \cos 2t + 4 \sin 2t)u(t) \\ &= e^{-t}\left(\frac{103}{22} \cos 2t + \frac{58}{11} \sin 2t\right)u(t). \end{aligned} \tag{7.16}$$

This is the zero-input response due to $v_C(0) = 6$ volts and $i_L(0) = 4$ amperes.

A third notion relating those of zero-state response and zero-input response is that of *decomposition*. A system is said to have the *decomposition property* if it satisfies the following condition:

(3) If y_0 is the zero-input response to an arbitrary initial state $(\alpha_1, \alpha_2, \ldots, \alpha_n)$ and y_1 is the zero-state response for an arbitrary input x, then the total response for the same initial state and the same input is $y_0 + y_1$. This condition must hold for any arbitrary initial state and any arbitrary input.

EXAMPLE 7.1.7

Suppose that in the network of Figure 7.3 the zero-input response due to $v_C(0) = 10$ volts and $i_L(0) = -3$ amperes is $v(t) = (10e^{-t} \cos 2t)u(t)$, and the zero-state response due to $v_1(t) = u(t)$ is $v(t) = (1 - e^{-t} \cos 2t)u(t)$. If the network satisfies the decomposition property (3), then the response due to the input $v_1(t) = u(t)$ and the initial conditions $v_C(0) = 10$ volts and $i_L(0) = -3$ amperes acting simultaneously is

$$v(t) = (10e^{-t} \cos 2t)u(t) + (1 - e^{-t} \cos 2t)u(t). \tag{7.17}$$

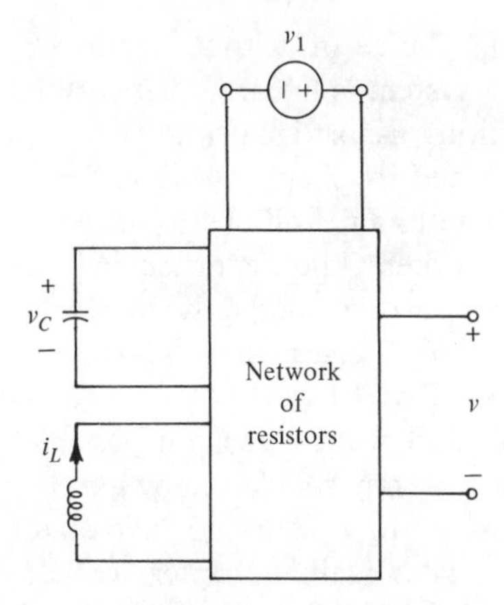

re 7.3 Network for illustrating linearity.

Finally, we define a system to be *linear* if it is zero-state linear, zero-input linear, and it has the decomposition property. In other words, a system is linear (with no other qualifiers) if and only if it satisfies (1a), (2a), (1b), (2b), and (3) above. We may refer to a system as zero-state homogeneous, zero-state additive, zero-input homogeneous, zero-input additive, or decomposable, if it satisfies (1a), (2a), (1b), (2b), or (3), respectively.

For a system with more than one input to be linear, the conditions (1a), (2a), (1b), and (2b) must be satisfied for each input with all other inputs set equal to zero. In addition it must satisfy the following:

(3′) If y_a is the zero-state response for an arbitrary input x_a with all other inputs set equal to zero, y_b is the zero-state response for an arbitrary input x_b with all other inputs set equal to zero, and so on up to the mth input if there are m inputs, and if y_0 is the zero-input response due to an arbitrary initial state, then the complete response due to the same initial state and all the same inputs acting simultaneously is $y_0 + y_a + y_b + \cdots + y_m$. This is the decomposition property. If there is more than one output port, then all the above conditions must hold for every output port for the system to be linear. Otherwise the system is nonlinear.

Note that in the definition of a linear system, the location of input and output signals must be identified. In a network, the signals are usually voltages and currents. A port is created by attaching wires to any pair of nodes or cutting a branch and exposing the pair of wires. For an n-port network, inputs must be restricted to port voltages and currents, and outputs must be restricted to port voltages and currents. Internal voltages and currents are not allowed as inputs and outputs. An n-port is said to be a linear n-port if for any choice of independent port voltages and port currents as inputs and for any choice of outputs from the remaining port currents and voltages, the resulting system is linear. A network is said to be linear if for any choice of ports, the resulting n-port is linear.

It is easy to verify that the network elements defined in Chapter 4 which have linear branch characteristics are linear networks. The two definitions of linearity for the network elements are consistent with each other. The definition in this section, however, applies to much more general networks, not only to single elements. For example, any arbitrary interconnection of linear elements can be shown to be a linear network.

The above defining properties of linear networks and systems make it possible to develop a general theory applicable to such systems. In particular, knowing the zero-state response due to a unit impulse allows us to obtain the zero-state response due to an arbitrary input as we shall see later in this chapter. Linear systems are simpler to study than nonlinear systems. In addition, the behavior of many nonlinear systems when the signals involved are "small" may be obtained by studying a linearized version of the nonlinear system. Finally, many systems, even with large signals, are approximately linear, and results obtained from a linear model are close enough to the actual case. For these reasons, the study of linear systems and networks is a major aspect of network or system theory.

EXERCISES

7.1.1 Show that the resistor model with a linear branch characteristic is a linear 1-port.

7.1.2 Repeat Exercise 7.1.1 for an inductor model with a linear ψ–i characteristic. Take i as output and v as input. Repeat for v as output and i as input.

7.1.3 A linear time-invariant R and a linear time-invariant L and a voltage source v are connected in a series. Taking v as the input and i as the output, derive the differential equation relating i to v. Show that the system is zero-state linear.

7.1.4 Show that the system in Exercise 7.1.3 is zero-input linear.

7.1.5 Show that the system in Exercise 7.1.4 satisfies the decomposition property.

7.1.6 Show that a capacitor with a nonlinear q–v characteristic is a nonlinear 1-port network.

7.1.7 A system is described by the equation $(dy/dt)^2 + y^2 = x^2$, where x is the input and y is the output. Is the system linear? Is the system zero-state homogeneous?

7.1.8 Show that if a system is zero-state additive, then if y_1 is the zero-state response for an input x_1, the zero-state response due to an input Cx_1 is

Cy_1, where C is a finite positive integer. (See Problem 7.3 for a more complicated version.)

7.1.9 Show that the zero-state response of a zero-state linear system to a zero input is zero.

7.1.10 For the network in Figure 7.2, suppose it is zero-input linear and the data in Example 7.1.5 apply. Determine the zero-input response due to $v_C(0) = 1$ volt and $i_L(0) = 0$.

7.1.11 Repeat Exercise 7.1.10 for $v_C(0) = 0$ and $i_L(0) = 1$ ampere.

7.1.12 Repeat Exercise 7.1.10 for $v_C(0) = 5$ volts and $i_L(0) = 8$ amperes.

7.1.13 For the network in Figure 7.3, assume that the data in Example 7.1.7 apply and assume that the network is linear. Determine the response due to an initial condition of $v_C(0) = -5$ volts and $i_L(0) = 1.5$ amperes, and an input of $3u(t)$.

2 The Thevenin and Norton theorems

Two theorems of importance in the analysis of networks were given independently by Helmholtz and later by Thevenin and by Norton. Because the theorems are widely known as Thevenin's theorem and Norton's theorem, we shall use these identifications in the discussion to follow. The Thevenin and Norton theorems provide a means for the simplification of network analysis by constructing equivalent networks. The original statements of these theorems were in terms of linear networks composed of linear, lumped, time-invariant elements. However, the theorems are valid for more general networks. In this section, we will state and prove the most general form of the theorems. In applications of the theorems in later chapters, various specializations will be made.

Let the 1-port network labeled A in Figure 7.4 be linear. It may contain sources as well as nonzero initial conditions (or initial state). Assume that the only connection between Network A and the arbitrary network is the pair of wires connected to terminals 1 and 2. There is no magnetic coupling between any coil in Network A and any coil in the arbitrary network. Furthermore, it is required that there be no

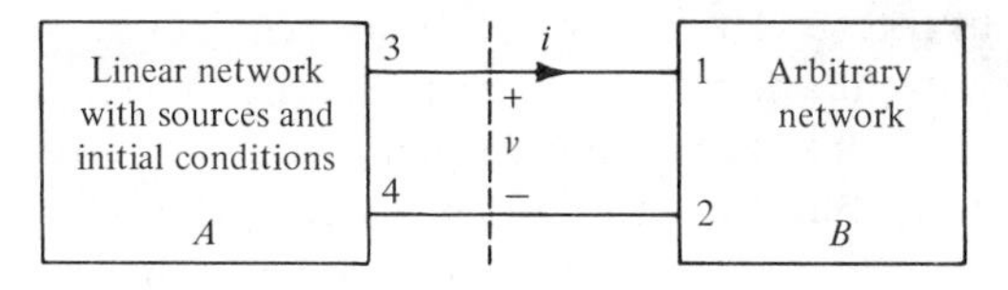

ure 7.4 A linear 1-port network connected to an arbitrary (possibly nonlinear) network.

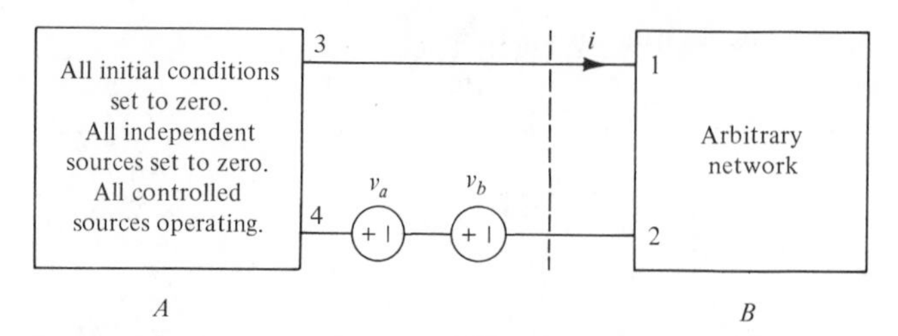

Figure 7.5 The linear network A of Figure 7.4 is replaced by the same linear network with the independent sources turned off and the initial conditions set equal to zero. Then v_a is made equal to v of Figure 7.4 when i is set to zero, and all initial conditions are set to zero. Also, v_b is made equal to v in Figure 7.4 when i is set to zero and all independent sources are turned off (but the dependent or controlled courses are left functioning).

controlled source in Network A which depends on any current or voltage in the arbitrary network except possibly the ones at the terminals, labeled v and i. Similarly, there is no controlled source in the arbitrary network which depends on any voltage or current in Network A except for v and i. The arbitrary network labeled Network B is not required to be linear.

To begin our discussion of the theorem, we construct the series network in Figure 7.5. The network is the same as Network A in Figure 7.4, except that all the independent sources are turned off; that is, the independent voltage sources and the independent current sources are all set to zero. This means that a voltage source becomes a short circuit, while a current source becomes an open circuit. This may eliminate some elements from the network being considered, since they are either shorted or left dangling! All the initial conditions are also set to zero. The voltage source $v_a(t)$ is made equal to the voltage $v(t)$ in Figure 7.4 when $i = 0$, and all the initial conditions in Figure 7.4 are equal to zero. The voltage source $v_b(t)$ is made equal to the voltage $v(t)$ in Figure 7.4 when i is zero, and all the independent sources in Network A of Figure 7.4 are set equal to zero. Thevenin's theorem states that for any arbitrary i, the voltage $v_{12}(t)$ in Figure 7.4 is identical to the voltage $v_{12}(t)$ in Figure 7.5, and the 1-port networks to the left of the dotted lines are said to be equivalent.

The proof follows as an immediate consequence of linearity. Since there are no connections between voltages and currents in Network A and those in the arbitrary network of Figure 7.4, the arbitrary current i may be treated as a source. We shall then determine the response v by adding the zero-input response and the zero-state response since Network A is linear. Since i is treated as a source, the zero-input response is the voltage v when all the independent sources are set equal to zero and when $i = 0$. Call this voltage v_b. Next determine the response v when the initial state is zero and $i = 0$. Call this voltage v_a. Finally, determine the response

v when the initial state is zero and all the independent sources inside Network A are set equal to zero. Call this voltage v_c. Thus the zero-state response is $v_a + v_c$ and the complete response is

$$v_{12} = v_a + v_b + v_c. \tag{7.18}$$

For the configuration of Figure 7.5, recall that the source v_a is made equal to the v_a above. Likewise, v_b is made equal to the v_b above. The voltage v_{34} in Figure 7.5 is equal to v_c by definition. Therefore, by KVL, the total response is

$$v_{12} = v_a + v_b + v_c, \tag{7.19}$$

which is identical to the v_{12} of Figure 7.4 as given in Equation 7.18. This proves Thevenin's theorem!

AMPLE 7.2.1

Consider the simple linear network in Figure 7.6(a) inside the dotted box. For $i = 0$ (arbitrary network is disconnected), v_{12} is

$$v_{12} = \frac{R_2}{R_1 + R_2} v. \tag{7.20}$$

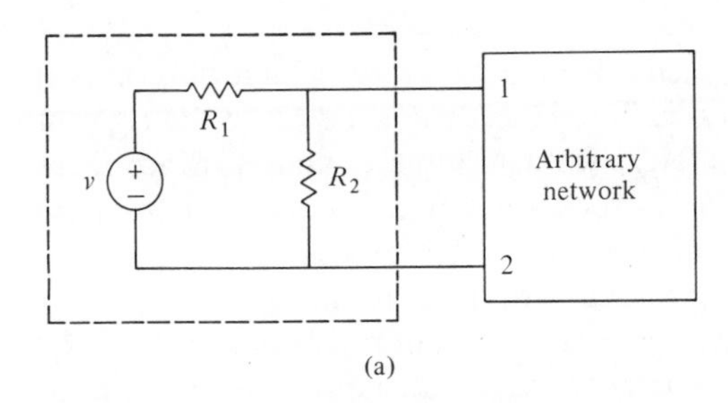

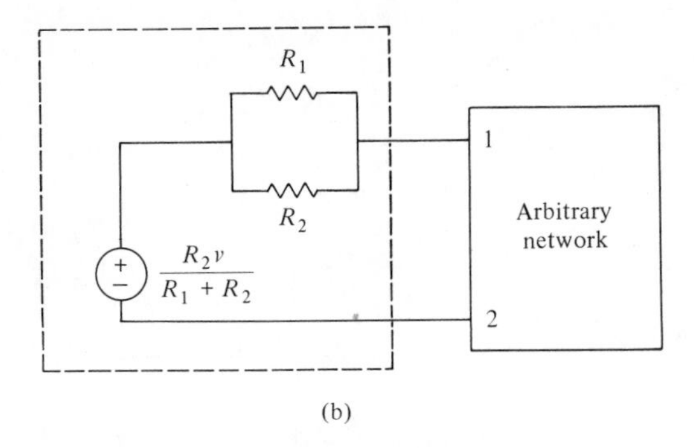

ure 7.6 Example for using Thevenin's theorem when R_1 and R_2 are linear. The two 1-port networks within the dotted lines are equivalent.

This is the voltage v_a in the network of Figure 7.5. When $v = 0$, R_1 is simply in parallel with R_2, and we obtain the equivalent network in Figure 7.6(b). In this simple example there is no initial condition, and $v_b = 0$.

EXAMPLE 7.2.2

Let us determine the Thevenin equivalent of the network in Figure 7.7(a). First we notice that there are no independent sources in the simple network so that $v_a = 0$. Next we determine the voltage v_b. This is the voltage that appears across terminals 1 and 2, when no external network is connected, due to initial conditions in the network. The capacitor is charged to Q coulombs so that the initial voltage across C is Q/C. Since there is no current flow under open-circuit conditions, the voltage v_b is equal to Q/C and it remains constant for all time. Finally, if we remove the charge Q, a simple uncharged capacitor remains. This is the Network A in Figure 7.5. The Thevenin equivalent is shown in Figure 7.7(b). Since the voltage v_b is a constant, it is drawn as a battery in Figure 7.7(b). Thus we conclude that insofar as the terminal behavior of a charged linear time-invariant capacitor is concerned, the charged capacitor may be replaced by an uncharged capacitor in series with a constant voltage source equal to Q/C.

EXAMPLE 7.2.3

Let us determine the Thevenin equivalent of the network to the left of terminals 1 and 2 in Figure 7.8. The first step is to disconnect the network to the right of terminals 1 and 2, that is, remove R_3. Next we determine v_a and v_b. Since there are no charged capacitors nor fluxed inductors, we expect a zero-input response of zero, that is, $v_b = 0$. The open-circuit voltage v_{12} due to V is the voltage v_a in the theorem and in Figure 7.5. Let us proceed with the determination of v_a. Note that the voltage across R_1 is V, and the current in R_1 from top to bottom is V/R_1. Since R_3 is disconnected, there is no current in R_3, and the current in R_2 from

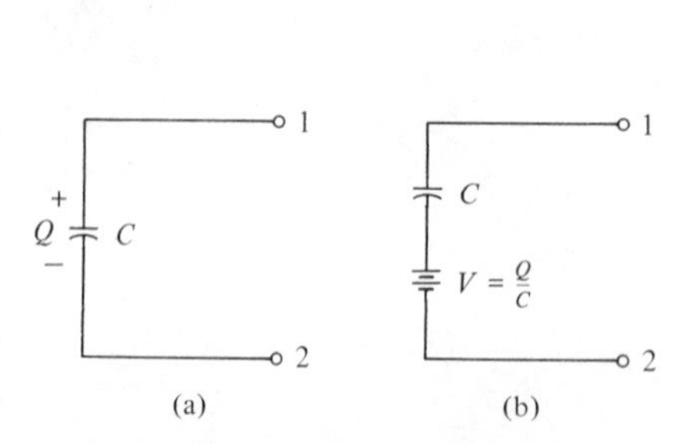

Figure 7.7 Thevenin's equivalent of a charged capacitor. See Example 7.2.2.

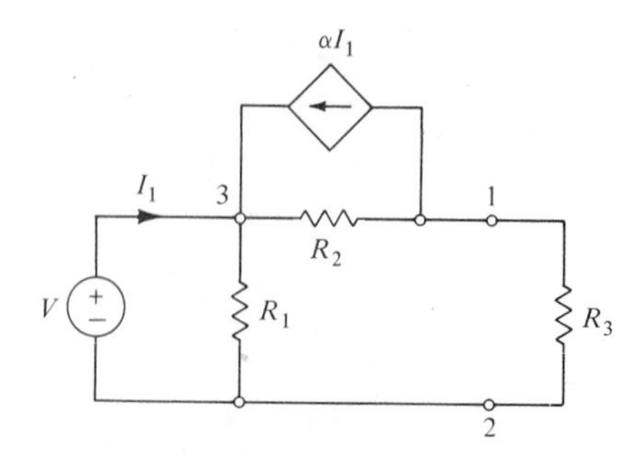

Figure 7.8 Network for Example 7.2.3.

left to right must be equal to αI_1 by KCL. By another application of KCL at node 3, we see that

$$I_1 + \alpha I_1 - \alpha I_1 - \frac{V}{R_1} = 0. \tag{7.21}$$

Hence

$$I_1 = \frac{V}{R_1}. \tag{7.22}$$

By KVL, the voltage v_{12} must be the sum of V and the voltage across R_2, which is v_{13}. Thus

$$v_{12}\big|_{\text{open-circuit}} = v_{13} + V = -\alpha I_1 R_2 + V = \left(1 - \frac{\alpha R_2}{R_1}\right)V. \tag{7.23}$$

The required v_a is v_{12} in Equation 7.23. Finally we obtain the network with no initial conditions and *no independent sources*. Since the current-controlled current source αI_1 is not an independent source (it depends on I_1) it must not be turned off. To turn V off, we replace it by a short circuit. Since R_1 is across the short circuit, there will be no voltage across nor current in R_1, and R_1 may be disconnected

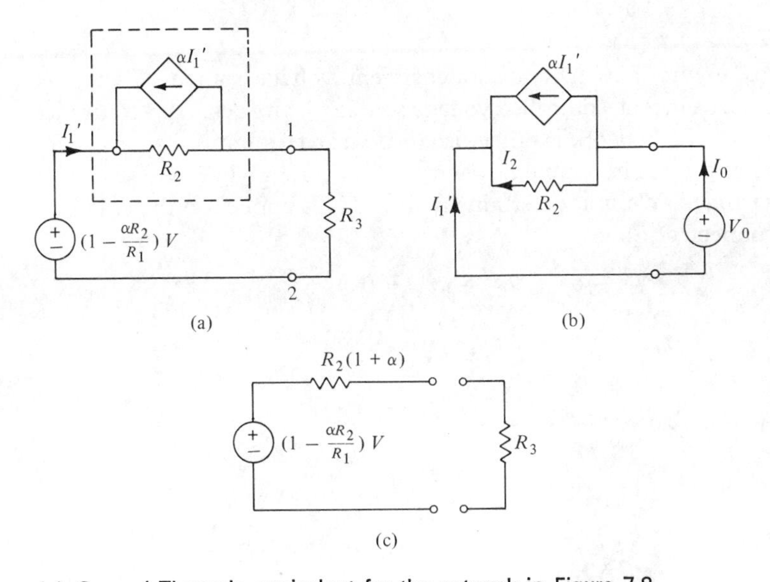

Figure 7.9 (a) General Thevenin equivalent for the network in Figure 7.8. (b) The application of voltage source V_0 makes it possible to simplify the network identified by dashed lines. (c) Simplified Thevenin equivalent network for the network of Figure 7.8.

without affecting the voltages and currents in the rest of the network. The complete Thevenin equivalent is shown in Figure 7.9. Observe that I_1' in Figure 7.9(a) is not the same as I_1 in Figure 7.8, since the Thevenin equivalent network applies only to external v–i characteristics at terminals 1 and 2. In the case of networks identified by the dashed lines of Figure 7.9(a), containing only resistive elements and controlled sources (and in other cases to be discussed in later chapters), it is possible to simplify the general Thevenin equivalent network. This is accomplished by applying a voltage source at terminals 1 and 2, as shown in Figure 7.9(b) as V_0, from which I_0 can be found. The analysis of the network of Figure 7.9(b) is carried out by noting that $I_0 = -I_1'$ and that $V_0 = R_2 I_2 = R_2 I_0(1 + \alpha)$. From this, we see that $V_0/I_0 = R_2(1 + \alpha)$ is an equivalent resistance of the network. This leads to the final form of the equivalent Thevenin network which is shown in Figure 7.9(c).

Whenever the network labeled A in Figure 7.5 contains only resistors and controlled sources, then Network A is equivalent to a single resistor whose value can be obtained by applying a voltage source and computing the current, as illustrated by the preceding example. The ratio of the voltage to current is the Thevenin equivalent resistance.

EXAMPLE 7.2.4

The network of Figure 7.10 has one independent voltage source v_1 and two controlled sources, a current-controlled voltage source bi_1 and a voltage-controlled voltage source av_1. The network is otherwise made up of 1-ohm resistors and R_L. It is required to find the Thevenin equivalent network from which the current in R_L may be determined. We first determine i_1 from KVL applied to the only loop after R_L is disconnected. Since

$$-v_1 + 2i_1 + av_1 = 0, \tag{7.24}$$

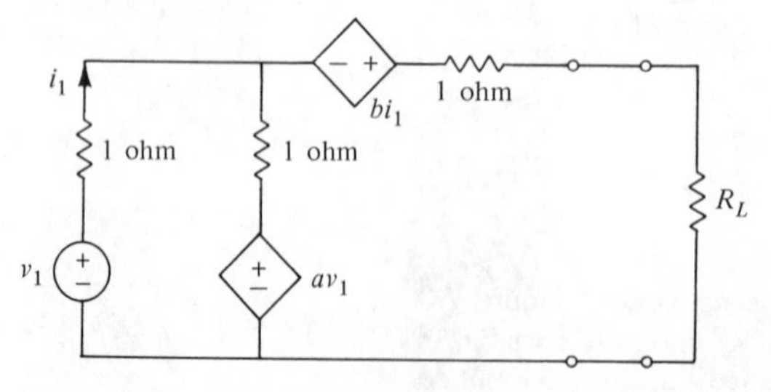

Figure 7.10 The network with two controlled sources used in Example 7.2.4.

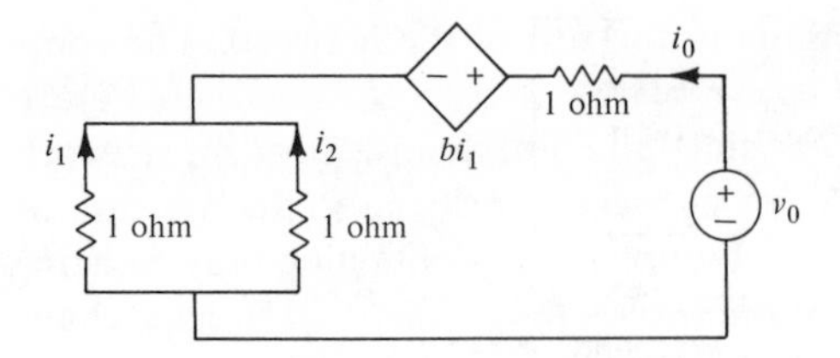

Figure 7.11 In determining the Thevenin equivalent network in Example 7.2.4, an auxiliary source v_0 is connected to the network, and $R_{eq} = v_0/i_0$ determined in Equation 7.29.

then

$$i_1 = v_1 \frac{1 - a}{2}. \tag{7.25}$$

With i_1 determined, we next compute the open-terminal voltage

$$v_a = v_1 \frac{1 - a}{2} \times 1 + av_1 + bv_1 \frac{1 - a}{2} \tag{7.26}$$

or

$$v_a = \frac{v_1}{2}(1 + a + b - ab). \tag{7.27}$$

To determine the remainder of the equivalent network, let us connect a voltage source to the open-circuit terminals of value v_0 as shown in Figure 7.11, and replace the independent voltage source by a short circuit. The resulting current i_0 will determine the equivalent resistor as $R_{eq} = v_0/i_0$. Since the parallel resistors have equal values, $i_1 = i_2 = -i_0/2$. Then

$$v_0 = i_0 \times 1 - \frac{bi_0}{2} + \frac{i_0}{2} \tag{7.28}$$

or

$$R_{eq} = \frac{v_0}{i_0} = \frac{3 - b}{2}. \tag{7.29}$$

The Thevenin equivalent network is the simple network shown in Figure 7.12.

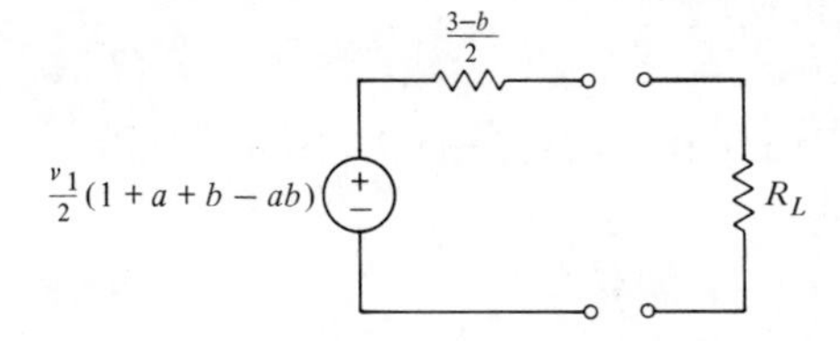

Figure 7.12 The Thevenin equivalent network with respect to R_L.

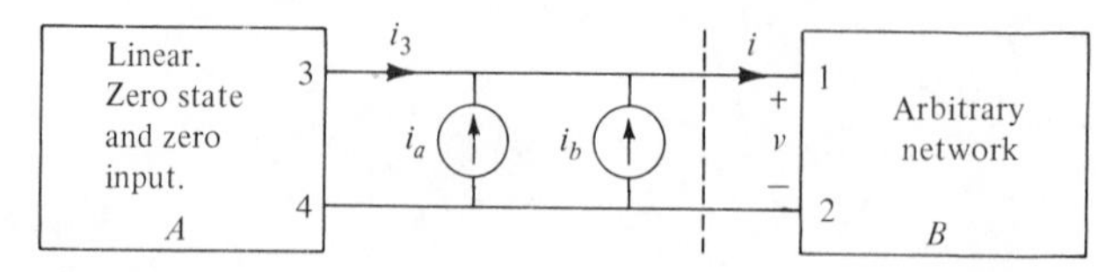

Figure 7.13 The linear network A of Figure 7.4 is replaced by the same network except that all initial conditions and all internal independent sources are set equal to zero. The current source i_a is equal to i in Figure 7.4 if $v = 0$ and the initial state is zero. The current source i_b is equal to the current i in Figure 7.4 if $v = 0$ and all internal independent sources are set equal to zero.

The situation in Figure 7.4 may be represented by another equivalent configuration as shown in Figure 7.13. This is the configuration for Norton's theorem. We assume the same set of hypotheses that we did for Thevenin's theorem: Network A is linear and there is no internal coupling between Network A and the arbitrary network, but Network B need not be linear.

To discuss Norton's theorem, we construct the network in Figure 7.13 as follows. Network A is the same as Network A of Figure 7.4, except that all the initial conditions are set equal to zero and all the inputs or independent sources are set equal to zero. By construction, the current source i_a is made equal to the current i that would flow in Figure 7.4 if all the initial conditions were set equal to zero and if v were set equal to zero. Likewise, i_b is made equal to the current i that would flow in Figure 7.4 if v were set equal to zero and if all the independent sources were set equal to zero. Norton's theorem states that for every arbitrary v the current in Figure 7.4 is identical to the current i in Figure 7.13, and the 1-ports to the left of the dotted lines are said to be equivalent.

The proof is similar to that for Thevenin's theorem. Since there are no internal couplings between the networks in Figure 7.4, we may treat the arbitrary voltage v as a voltage source. We shall then determine the response i by combining the effects of the inputs, the initial state, and v, since the network to the left of the dotted line is linear. Due to initial state above, with $v = 0$ and all inputs set to zero, the resulting response i is i_b. Due to the inputs above with initial state set to zero and v set to zero, the response i is i_a. Finally, denote the response i, when the initial state is set to zero and when the internal independent sources in A are set to zero, by i_c. Then, by linearity, the total response is

$$i = i_a + i_b + i_c . \tag{7.30}$$

Now for the network in Figure 7.13 we observe that the current i_3 is by definition the same as the current i_c. Hence, by KCL,

$$i = i_a + i_b + i_c , \tag{7.31}$$

which is identical to Equation 7.30, thus proving Norton's theorem.

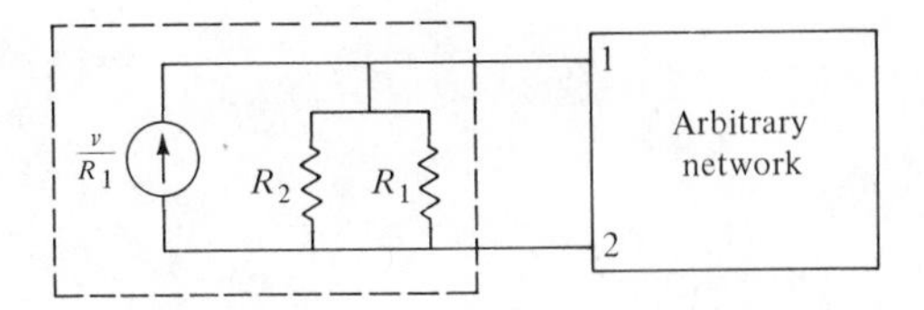

ure 7.14 Norton's equivalent for the 1-port of Figure 7.6.

AMPLE 7.2.5

Consider the simple network in Figure 7.6 again. The zero-state zero-input network is the same as in Thevenin's theorem, R_1 in parallel with R_2. The current source is the current that flows through 1–2 when v_{12} is zero. Hence, the current is

$$i = \frac{v}{R_1}. \tag{7.32}$$

This is i_a. The current i_b is zero since there is no initial condition. The complete Norton equivalent is shown in Figure 7.14.

AMPLE 7.2.6

Let us determine the Norton equivalent for the linear time-invariant fluxed inductor in Figure 7.15(a). Since there is no independent source to the left of terminals 1 and 2, $i_a = 0$. The current i_b is due to the initial current through the inductor. Following the procedure in the statement of Norton's theorem, we short terminals 1 and 2 and determine the resulting current. This is equal to I for all time since v_{12} is zero and $di/dt = 0$ so that i does not change. The zero-input zero-state network under consideration is simply an unfluxed inductor L. The Norton equivalent is shown in Figure 7.15(b).

In summary, a linear two-terminal network is equivalent to a series combination of a voltage source and the same network except that the initial state

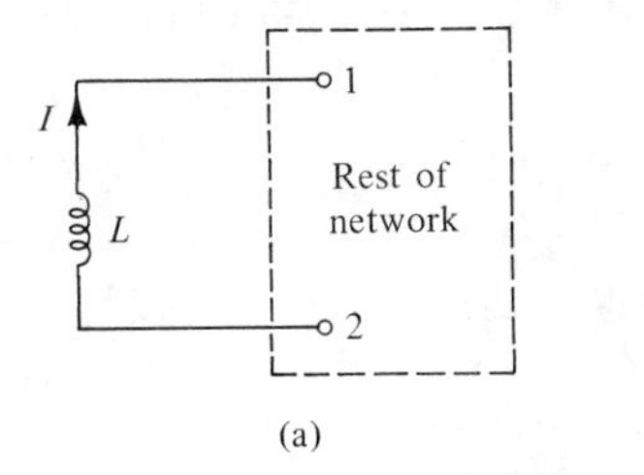

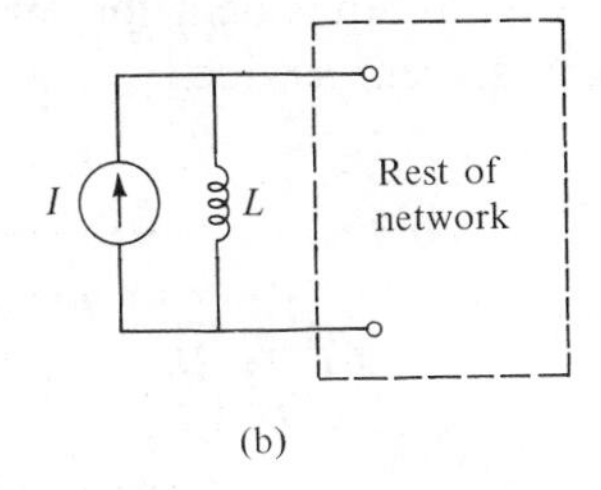

ure 7.15 Norton's equivalent of a fluxed constant inductor. See Example 7.2.6.

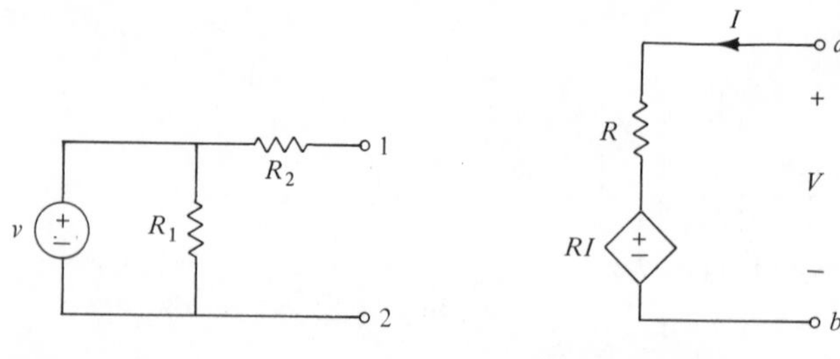

Figure 7.16

Figure 7.17

(initial conditions) and the inputs are set equal to zero. This is Thevenin's theorem. Similarly, the original linear 1-port is equivalent to a parallel combination of a current source and the same network except that the initial state and the inputs are set equal to zero. This is Norton's theorem.

When the network contains resistors and controlled sources only, it can be reduced to a single resistor which is determined with the aid of an auxiliary source. Other simplifications will be discussed later.

EXERCISES

7.2.1 Determine the Thevenin equivalent network for the linear network in Figure 7.16.

7.2.2 Determine the Norton equivalent network for the linear network in Figure 7.16.

7.2.3 Find the Thevenin equivalent for the network in Figure 7.17 containing a controlled source.

7.2.4 Find the Thevenin equivalent for the network in Figure 7.18 if $i(t)$ is a unit step $u(t)$ and C is initially uncharged and constant. Repeat for $i(t)$ equal to a unit-impulse function and compare with the equivalent of a charged capacitor as given in Example 7.2.2.

7.2.5 Find the Norton equivalent circuit for the network in Figure 7.19 where the voltage $v(t)$ is an arbitrary function of time.

7.2.6 Find the Thevenin equivalent for the network in Figure 7.20 where $i(t)$ is an arbitrary current source.

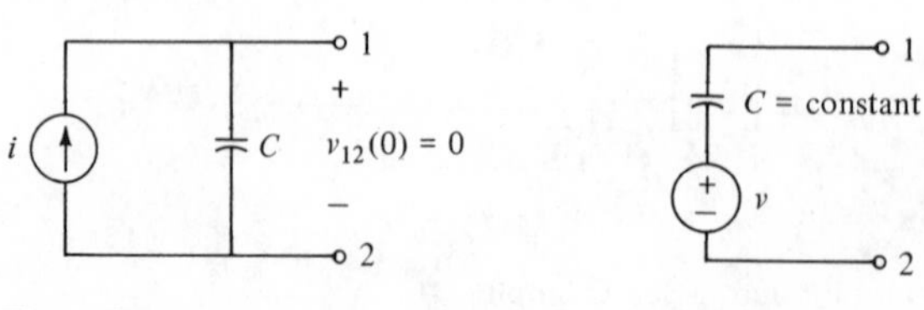

Figure 7.18

Figure 7.19

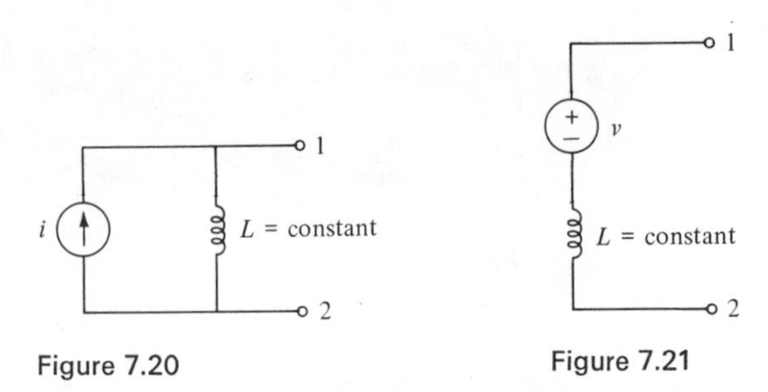

Figure 7.20

Figure 7.21

7.2.7 Find the Norton equivalent for the network in Figure 7.21 where $v(t)$ is an arbitrary voltage source. Assume that $i_L(0) = 0$.

3 Time-invariant systems

In Chapter 4, we described a time-invariant network element as one not varying with time. In terms of passive elements, R, L, and C as constants describe time-invariant elements, while $R(t)$, $L(t)$, and $C(t)$ do not. It seems intuitively reasonable that a network composed of time-invariant elements should itself be time-invariant. Similarly, let us consider a network in which a given input is applied at $t = 0$ and a given response results. Then if this same input should be applied to the identical network in the same initial state 3 sec later than $t = 0$, the response would be identical to that initially observed but delayed by 3 sec. While this seems like a simple concept to grasp, it is important and it is used frequently in analyzing networks.

XAMPLE 7.3.1

Consider a simple example. Suppose that a unit ramp is applied to a network and that this v_1, shown in Figure 7.22(a), results in a zero-state response

$$v_2(t) = (e^{-t} + t - 1)u(t). \tag{7.33}$$

Figure 7.22(b) shows a signal waveform known as a *truncated ramp* which is a unit ramp until $t = 1$ and then remains at unit value for all later time. If the truncated ramp is applied to the same network with zero initial conditions, what will be the response? We may answer this question by observing that the truncated ramp may be thought of as being made up of a ramp together with a delayed negative ramp, as illustrated in Figure 7.22(c). If the network is linear, then superposition applies and the response may be found for each part of the input

$$v_1(t) = tu(t) - (t - 1)u(t - 1). \tag{7.34}$$

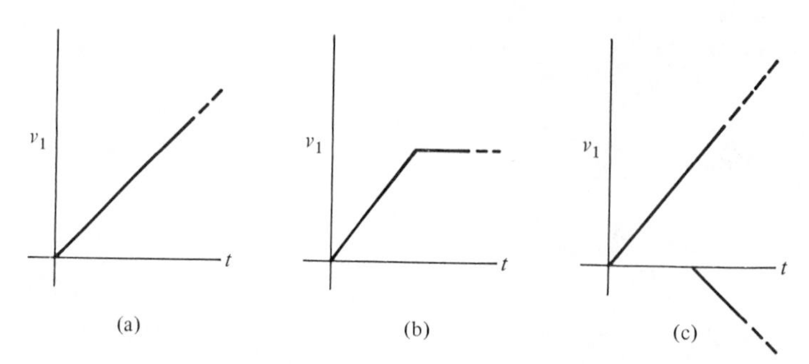

Figure 7.22 The truncated ramp waveform of (b) is represented by the ramp of (a) and a negative-going delayed ramp, as shown in (c).

The first is known and is given by Equation 7.33. If the network is also time-invariant, the second is determined from Equation 7.33 simply by replacing t by $t - 1$ and multiplying by -1. Hence

$$v_2(t) = (e^{-t} + t - 1)u(t) - [e^{-(t-1)} + (t - 1) - 1]u(t - 1). \quad (7.35)$$

This approach could not be used unless the network was known to be time-invariant.

This is a simple principle which may be stated in a very general form. Suppose that a system with an arbitrary input $x(t)$ and an arbitrary initial state $(\alpha_1, \alpha_2, \ldots, \alpha_n)$ gives the response $y(t)$. The system is said to be time-invariant if an input $x(t - \Delta)$ and an initial state $(\alpha_1, \alpha_2, \ldots, \alpha_n)$ causes the response $y(t - \Delta)$ for any $\Delta > 0$. No matter what the initial state is, an arbitrary delay on the input function causes the same delay on the output function for all possible inputs. The initial state for the delayed input case is applied Δ time units later.

EXAMPLE 7.3.2

Consider the network in Figure 7.3. Suppose that when $v_C(0) = 10$ volts, $i_L(0) = -3$ amperes, and $v_1(t) = u(t)$, the response $v(t)$ is as given in Equation 7.17. The initial state consists of the set of two quantities $v_C(0)$ and $i_L(0)$. Here α_1 is 10 volts and α_2 is -3 amperes. Suppose that the network is time-invariant and instead of an input $v_1(t)$, which is a unit step, we have a delayed unit step $v_1(t) = u(t - 3)$ which is delayed by 3 time units. For an initial condition of $\alpha_1 = v_C(3) = 10$ volts and $\alpha_2 = i_L(3) = -3$ amperes, the response $v(t)$ for $t \geq 3$ is given by

$$\begin{aligned} v(t) = &\{10e^{-(t-3)} \cos [2(t - 3)]\}u(t - 3) \\ &+ \{1 - e^{-(t-3)} \cos [2(t - 3)]\}u(t - 3). \end{aligned} \quad (7.36)$$

This is simply obtained from Equation 7.17 by replacing t by $t - 3$. The delay Δ is 3 time units. Notice that the initial condition for the delayed case is applied at $t = 3$, and not at $t = 0$.

It is easy to verify that the elements defined in Chapter 4 which have time-invariant branch characteristics are time-invariant systems in the sense of the new definition. For example, for a current-controlled time-invariant resistor, we have $v = f(i)$. If $v_1(t)$ is the response due to $i(t)$, so that

$$v_1(t) = f[i(t)], \tag{7.37}$$

then the response to $i(t - \Delta)$ is

$$v_2(t) = f[i(t - \Delta)] = v_1(t - \Delta). \tag{7.38}$$

The situation is similar for a time-invariant capacitor and inductor.

The distinction between a time-invariant n-port and a time-invariant network is analogous to the distinction between linear n-ports and linear networks. That is, in the n-port case only port variables are allowed as inputs and outputs. A trivial example of a time-invariant 1-port network with time-varying elements is a series connection of R_1 and R_2, where R_1 and R_2 are time-varying but such that $R_2 = K - R_1$, where K is a constant. However, the network is time-varying. Any network of time-invariant elements will be time-invariant in the sense of the definition of this section meaning that provided the initial states are the same, delaying the inputs causes only a delay in the output, giving exactly the same waveform.

ERCISES

7.3.1 A system is described by $dy(t)/dt + 3y(t) = 2tx(t)$, where $y(t)$ is the output and $x(t)$ is the input. Is the system time-invariant?

7.3.2 Demonstrate that a capacitor with a characteristic $q = f(v)$ is a time-invariant system, where i is the output and v is the input.

7.3.3 Demonstrate that an inductor with a characteristic $\psi = g(i)$ is a time-invariant system, where v is the output and i is the input.

.4 Linear time-invariant differential systems

A linear time-invariant differential system is a system which is linear in the sense of Section 7.1, time-invariant in the sense of Section 7.3, and describable by ordinary differential equations. If y is the output and x is the input, linearity and time-invariance require that the ordinary differential equation be of the form

$$a_n \frac{d^n y}{dt^n} + a_{n-1} \frac{d^{n-1} y}{dt^{n-1}} + \cdots + a_0 y = b_m \frac{d^m x}{dt^m} + \cdots + b_0 x, \tag{7.39}$$

where the a's and b's are constants.

An important part of circuit analysis is the solution of the class of differential equations represented by Equation 7.39, and so it is not unreasonable that we should devote the remainder of this chapter, most of Chapter 8, and also later chapters to this subject. Now any $y(t)$ satisfying Equation 7.39 is known as a *solution* of the differential equation. In general, we do not seek just any solution, but *all* solutions. The sum of all solutions is known as the *general solution.* If the general solution also satisfies additional conditions, known as *initial conditions,* then the general solution becomes the *particular solution* — particular to a specific set of constraints which are the initial conditions.

Before undertaking a study of techniques that will be useful in finding solutions, we digress to explain our plan of attack. It is convenient to subdivide the solution into more than one part. This is an arbitrary division, made for our convenience

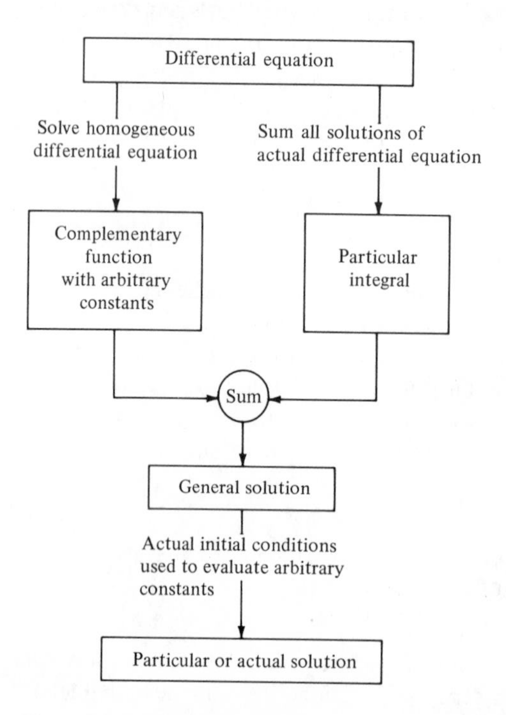

Figure 7.23 Classical method.

in the solution of the mathematical equations. Indeed, we will show that the division is not unique by displaying at least two different approaches.

Let us return to Equation 7.39 and ask what possible $y(t)$ might satisfy this differential equation for the specified $x(t)$. There will exist values for $y(t)$ which satisfy the homogeneous equation in the sense that when the required derivatives are found and multiplied by the specified constant coefficients, the left-hand side of Equation 7.39 will sum to zero. There will be another class of values for $y(t)$ which satisfy Equation 7.39. The first of these two is called the *complementary function* $y_c(t)$, which is the part of $y(t)$ satisfying the homogeneous differential equation, formed by setting the left-hand side of Equation 7.39 to zero. The remaining part is known as the particular integral $y_p(t)$ and it satisfies Equation 7.39. Then the total solution is

$$y(t) = y_c(t) + y_p(t) \tag{7.40}$$

and this $y(t)$ also satisfies Equation 7.39. As we shall see, $y_c(t)$ contains a number of

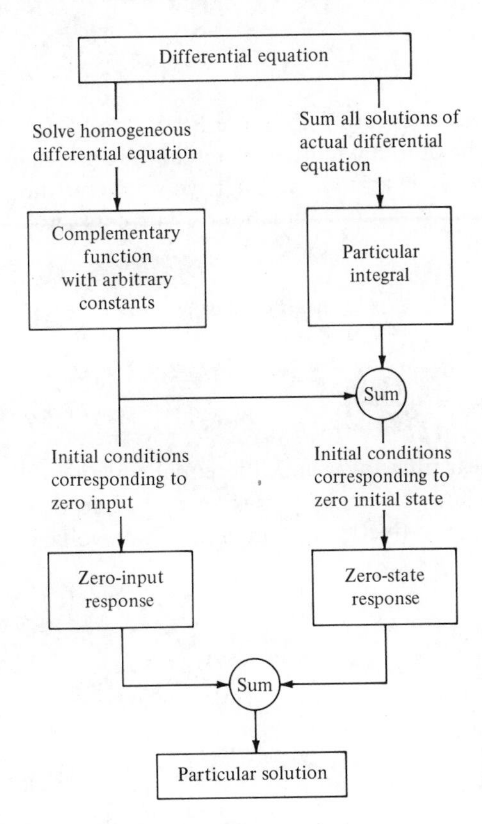

Figure 7.24 Alternative method.

undetermined constants equal in number to the order of the differential equation. These constants are determined by imposing initial conditions which $y(t)$ must satisfy. Although the particular integral satisfies Equation 7.39, it may not satisfy the initial conditions. In effect, the addition of y_c to y_p allows us to meet the initial conditions. The operations we have just described are illustrated by the flowchart in Figure 7.23. An alternative method of finding the solution which appears to be more complicated but offers certain advantages is shown in Figure 7.24.

In terms of Figures 7.23 and 7.24, we can describe our approach to finding particular solutions. In the remainder of this chapter, we will be concerned only with the complementary function and with the zero-input response. The determination of the particular integral, the zero-state response, and the particular solution will be postponed to Chapter 8 where a number of techniques will be studied, including the use of the convolution integral to determine the zero-state response.

To begin with, let us review the process by which equations like Equation 7.39 arise by considering Example 7.4.1.

EXAMPLE 7.4.1

Consider the linear time-invariant network in Figure 7.25 where R_1, R_2, and C are constants. Here v_1 is the input and v_2 is the output. In order to obtain the differential equation relating v_2 to v_1, we apply KCL at node 1. First we note that the voltage v_{21} across R_1 is equal to $v_2 - v_1$. Writing summation of currents away from node 2 equal to zero, we have

$$\frac{v_2 - v_1}{R_1} + \frac{v_2}{R_2} + C\frac{dv_2}{dt} = 0 \tag{7.41}$$

or

$$\frac{dv_2}{dt} + \frac{1}{RC}v_2 = \frac{1}{R_1C}v_1, \tag{7.42}$$

where $1/R_1 + 1/R_2 = 1/R$. This is a linear time-invariant differential equation of first order. The equation may be made homogeneous by making $v_1 = 0$ so that the $v_2(t)$ under this condition is the response due to the initial capacitor voltage only.

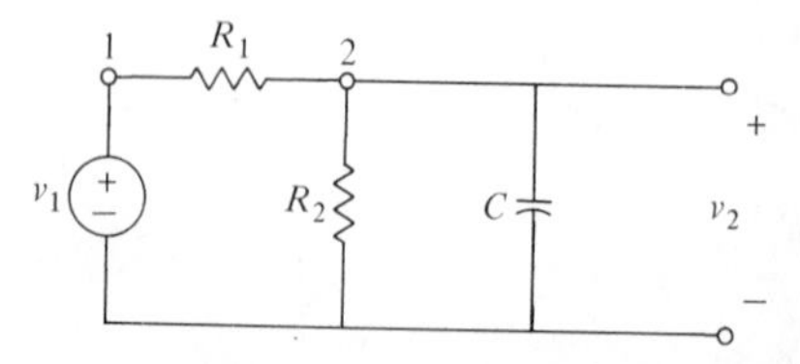

Figure 7.25 Three-element network described by a first-order differential equation.

The general form of the homogeneous differential equation of order n is

$$a_n \frac{d^n y}{dt^n} + a_{n-1} \frac{d^{n-1} y}{dt^{n-1}} + \cdots + a_1 \frac{dy}{dt} + a_0 = 0 \tag{7.43a}$$

or, more compactly,

$$\sum_{k=0}^{n} a_k \frac{d^k y(t)}{dt^k} = 0. \tag{7.43b}$$

The standard approach to the solution of Equation 7.43 is to assume a trial solution and test the assumption. We guess that the solution is $y = e^{st}$. If this is a solution, then it will satisfy Equation 7.43a. We compute the left-hand side expression in Equation 7.43a for $y = e^{st}$ obtaining

$$\sum_{k=0}^{n} a_k \frac{d^k}{dt^k}(e^{st}) = \sum_{k=0}^{n} a_k s^k e^{st}$$

$$= e^{st}(a_n s^n + a_{n-1} s^{n-1} + \cdots + a_0). \tag{7.44}$$

Since $e^{st} \neq 0$ for any finite s and t, the expression in Equation 7.44 can be zero only if

$$a_n s^n + \cdots + a_0 = 0. \tag{7.45}$$

This equation is known as the *characteristic equation*, and is important in the study of networks, or systems in general. From the fundamental theorem of algebra, an nth degree algebraic equation as in Equation 7.45 has n roots, that is, n values of s which make the left-hand side vanish. These n roots may or may not be distinct (or simple). For these values of s, e^{st} is a solution of Equation 7.43. Let s_1 be one of the roots. Then $e^{s_1 t}$ is a solution. Likewise $c_1 e^{s_1 t}$ is a solution, where c_1 is an arbitrary constant. This is easily verified by substituting in Equation 7.43.

Suppose first that all the roots of Equation 7.45 are distinct, i.e., different from each other. Then if $s_1, s_2, \ldots, s_n$ are the distinct roots of Equation 7.45, $e^{s_1 t}$, $e^{s_2 t}, \ldots, e^{s_n t}$ will be solutions of Equation 7.43. Likewise, any linear combination of the solutions will be a solution of Equation 7.43. The most general complementary solution is (for distinct roots)

$$y = C_1 e^{s_1 t} + C_2 e^{s_2 t} + \cdots + C_n e^{s_n t}, \tag{7.46}$$

where $C_1, C_2, \ldots, C_n$ are arbitrary constants. This general solution satisfies Equation 7.43. However, in order to satisfy the conditions at the initial instant, the C_k must be chosen properly. If there are n linearly independent initial conditions, then the C_k are uniquely determined.

EXAMPLE 7.4.2

Suppose that we have a series RLC circuit connected to a voltage source. Assuming that the elements are linear and time-invariant, we have

$$Ri + L\frac{di}{dt} + \frac{1}{C}\int_{t_1}^{t} i(\tau)\,d\tau + v_C(t_1) = v(t). \tag{7.47}$$

Differentiating once, we obtain

$$L\frac{d^2i(t)}{dt^2} + R\frac{di(t)}{dt} + \frac{1}{C}i(t) = \frac{dv(t)}{dt}, \tag{7.48}$$

where i may be considered the output or response and $v(t)$ the input. The differential equation specifies the relationship between $i(t)$ and $v(t)$ for $t \geq t_1$, where t_1 is the instant at which the input is applied. The initial state or initial conditions are specified at $t = t_1$. These conditions may consist of $v_C(t_1)$ and $i(t_1)$, the current through the inductor. It is convenient to convert $v_C(t_1)$ in terms of functions of i. At $t = t_1$, we have the KVL equation as in Equation 7.47 except that $v = 0$. Since the integral from t_1 to t_1 is zero, we have

$$Ri(t_1) + L\frac{di}{dt}\bigg|_{t=t_1} + 0 + v_C(t_1) = 0. \tag{7.49}$$

Therefore we have

$$\frac{di}{dt}\bigg|_{t=t_1} = \frac{-v_C(t_1) - Ri(t_1)}{L}, \tag{7.50}$$

and the initial conditions are expressed as conditions on $i(t_1)$ and $di(t_1)/dt$ at the initial instant. The zero-input response is a solution of

$$L\frac{d^2i}{dt^2} + R\frac{di}{dt} + \frac{1}{C}i = 0, \qquad t \geq t_1, \tag{7.51}$$

$$i(t_1) = \text{given number}, \tag{7.52}$$

$$i'(t_1) = \text{given number}. \tag{7.53}$$

That is, i is a zero-input response if it satisfies Equations 7.51, 7.52, and 7.53. As a numerical example, suppose $R/L = 3$ and $1/(LC) = 2$. Also, let $i(t_1) = 1$, $v_C(t_1) = -0.5$, and $L = 0.1$. Then Equation 7.51 becomes

$$\frac{d^2i}{dt^2} + 3\frac{di}{dt} + 2i = 0, \qquad t \geq t_1, \tag{7.54}$$

after dividing Equation 7.51 by L. Trying e^{st} as a solution, we have

$$\frac{d^2}{dt^2}e^{st} + 3\frac{d}{dt}e^{st} + 2e^{st} = e^{st}(s^2 + 3s + 2). \tag{7.55}$$

The expression in Equation 7.55 is zero if

$$s^2 + 3s + 2 = 0. \tag{7.56}$$

This is the case for $s = -1$ and $s = -2$. Hence, e^{-t} and e^{-2t} are solutions, and

$$i = C_1 e^{-t} + C_2 e^{-2t} \tag{7.57}$$

is a solution of Equation 7.54, where C_1 and C_2 are arbitrary constants. This is easily verified by direct substitution in Equation 7.54. Next we solve for C_1 and C_2 so that the i in Equation 7.57 satisfies Equations 7.52 and 7.53. For simplicity, suppose $t_1 = 0$. Then, from Equation 7.57 we have

$$C_1 + C_2 = i(0) = 1. \tag{7.58}$$

Differentiating Equation 7.57, we have

$$\frac{di(t)}{dt} = -C_1 e^{-t} - 2C_2 e^{-2t}. \tag{7.59}$$

At $t = 0$, using Equation 7.50, Equation 7.59 becomes

$$\left.\frac{di}{dt}\right|_{t=0} = -C_1 - 2C_2 = \frac{+0.5}{0.1} - 3(1) = 2. \tag{7.60}$$

From Equations 7.58 and 7.60 we may solve for C_1 and C_2 which turn out to be $C_1 = 4$ and $C_2 = -3$. So

$$i(t) = 4e^{-t} - 3e^{-2t} \tag{7.61}$$

satisfies Equation 7.54 for $t \geq 0$, and $i(0) = 1$ and $di(0)/dt = 2$ as specified. The $i(t)$ in Equation 7.51 is, then, the unique zero-input response.

Next consider the case in which the roots in Equation 7.55 are not all distinct. For example, suppose that a root s_1 is a double root so that $(s - s_1)^2$ is a factor of the nth degree polynomial in Equation 7.55. Then not only is $e^{s_1 t}$ a solution as discussed earlier, but so is $te^{s_1 t}$. This can be seen from the following discussion when we use the operator notation $D = d/dt$. Write Equation 7.53 as

$$\sum_{k=0}^{n} a_k p^k y = 0, \tag{7.62}$$

where

$$p^k y = \frac{d^k y}{dt^k}.$$

We may rewrite Equation 7.62 as

$$(a_n p^n + a_{n-1} p^{n-1} + \cdots + a_0)y = 0 \tag{7.63}$$

with the understanding that the expression on the left-hand side is defined as the left-hand side of Equation 7.62. From the n factors of the expression in the left-hand side of Equation 7.65, we may factor Equation 7.63 into

$$a_n(p - s_n)(p - s_{n-1})\cdots(p - s_1)y = 0. \tag{7.64}$$

Suppose now that s_2 and s_1 are equal so that s_1 is a double root of Equation 7.65. Then we want to check whether te^{s_1t} is a solution of the differential equation in Equation 7.64. We have

$$\begin{aligned}
&a_n(p - s_n)(p - s_{n-1})\cdots(p - s_1)(p - s_1)(te^{s_1t}) \\
&\quad = [a_n(p - s_n)(p - s_{n-1})\cdots(p - s_3)](p - s_1)(e^{s_1t} + s_1te^{s_1t} - s_1te^{s_1t}) \\
&\quad = [a_n(p - s_n)(p - s_{n-1})\cdots(p - s_3)](p - s_1)e^{s_1t} \qquad (7.65) \\
&\quad = a_n(p - s_n)(p - s_{n-1})\cdots(p - s_3)(s_1e^{s_1t} - s_1e^{s_1t}) \\
&\quad = a_n(p - s_n)(p - s_{n-1})\cdots(p - s_3)0 = 0.
\end{aligned}$$

Therefore te^{s_1t} is a solution if s_1 is a double root. By a similar procedure, it can be shown that if s_1 is a triple root of Equation 7.45, then not only are e^{s_1t} and te^{s_1t} solutions but also $t^2e^{s_1t}$. In general, if s_1 is a root of multiplicity r_1, then e^{s_1t}, te^{s_1t}, $t^2e^{s_1t}, \ldots, t^{r_1-1}e^{s_1t}$ are all solutions, so that

$$y = C_1e^{s_1t} + C_2te^{s_1t} + C_3t^2e^{s_1t} + \cdots + C_{r_1}t^{r_1-1}e^{s_1t} \tag{7.66}$$

is a solution also. This means that for the nth-order homogeneous differential equation of Equation 7.53, we may always write a general solution containing n linearly independent constants.

EXAMPLE 7.4.3

Suppose we have the differential equation

$$(p + 1)^3(p + 2)(p + 3)^2y = 0. \tag{7.67}$$

Then the solutions are e^{-3t}, te^{-3t}, e^{-2t}, e^{-t}, te^{-t}, and t^2e^{-t}, and the most general complementary solution is

$$Y = C_1e^{-3t} + C_2te^{-3t} + C_3e^{-2t} + C_4e^{-t} + C_5te^{-t} + C_6t^2e^{-t}, \tag{7.68}$$

which contains six arbitrary constants. The differential equation is, of course, sixth order, and six independent initial conditions are required.

EXERCISES

7.4.1 For the linear time-invariant network in Figure 7.26, derive the differential equation relating the output v_L to the input v.

7.4.2 For the linear time-invariant network in Figure 7.27, derive the differential equation relating the output v_{ab} to the input v.

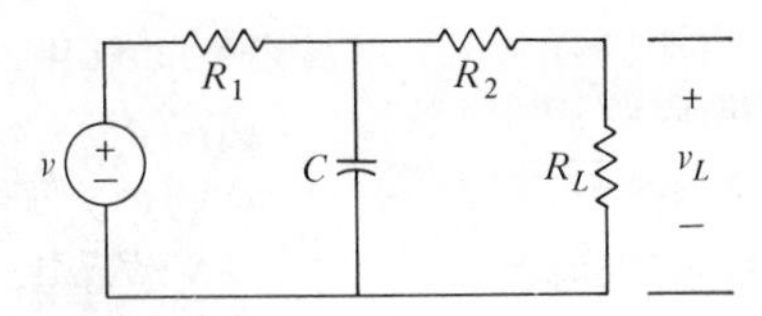

Figure 7.26

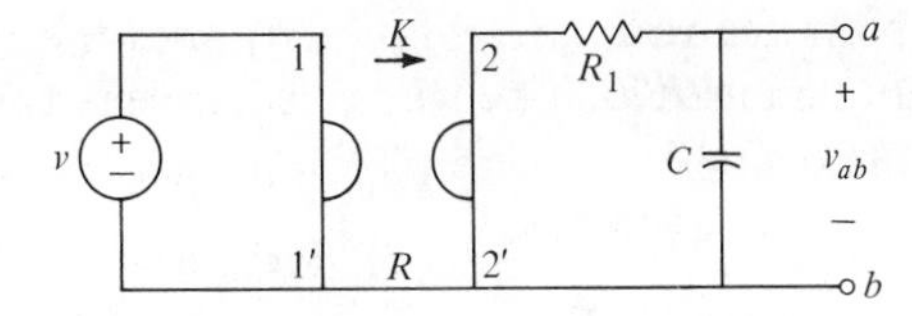

Figure 7.27

7.4.3 Find the general complementary solution for the differential equation

$$\frac{d^2y}{dt^2} + y = 0.$$

7.4.4 Find the general complementary solution for the differential equation

$$\frac{d^3y}{dt^3} + \frac{d^2y}{dt^2} + \frac{dy}{dt} + y = 0.$$

7.4.5 Find the general complementary solution for the differential equation

$$(p + 2)^4(p + 3)y = 0.$$

7.4.6 Find the zero-input response for the system described by the differential equation in Exercise 7.4.3 if $y(0) = 1$, $y'(0) = 1$.

5 Stability of linear networks

In this section, we will consider the possibility that a network composed of passive elements and controlled sources may be *unstable*. What do we mean by stable and unstable? To answer this question, we begin with the characteristic equation of the network given by Equation 7.45:

$$p(s) = a_n s^n + a_{n-1}s^{n-1} + \cdots + a_1 s + a_0 = 0. \tag{7.69}$$

Equation 7.69 may be presented in factored form like Equation 7.64,

$$p(s) = a_n(s - s_1)(s - s_2)\cdots(s - s_n) = 0. \tag{7.70}$$

The response of the network is then determined in terms of the roots of the characteristic equation as in Equation 7.46:

$$y(t) = c_1 e^{s_1 t} + c_2 e^{s_2 t} + \cdots + c_n e^{s_n t}, \tag{7.71}$$

where it is assumed only that the roots are distinct. Consider a typical term in this equation $c_k e^{s_k t}$ which is present because of the root

$$s_k = \sigma_k + j\omega_k. \tag{7.72}$$

Now the real part of s_k determines the form of each part of the response $y(t)$ in Equation 7.71. The way this happens is summarized as follows:

$$\begin{aligned} \sigma_k &< 0 \quad \text{is a damped response,} \\ \sigma_k &= 0 \quad \text{is an oscillatory response,} \\ \sigma_k &> 0 \quad \text{is an increasing response.} \end{aligned} \tag{7.73}$$

These waveforms were encountered in our study in Chapter 2. The damped response is illustrated in Figure 2.1, an oscillatory response in Figure 2.15, and an exponentially increasing response in Figure 2.1. These others are also shown in Figure 7.28 for $\omega \neq 0$ and for $\omega = 0$.

We define a stable network (or system) as one for which every component of the response is damped. An unstable system then includes an oscillatory response and an increasing response. You may want to later separate these last two possibilities since networks designed as oscillators are stable (as oscillators). For the present, we will exclude oscillators as stable networks.

An examination of the conditions of Equation 7.73 in light of the forms of response illustrated in Figure 7.28 leads us to a compact statement of the condition for stability: *A network is stable when the real parts of the roots of its characteristic equation are all negative.*

Since the mathematical study of systems began in the late 19th century, there have been numerous attempts to find a way to determine the stability of a system

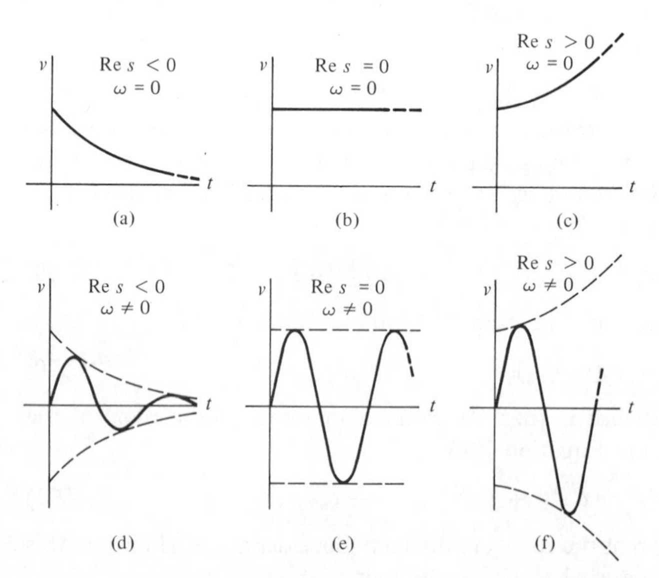

Figure 7.28 Signal waveforms corresponding to the possibilities with $\omega = 0$ and $\omega \neq 0$, and $\sigma > 0$, $\sigma = 0$, and $\sigma < 0$.

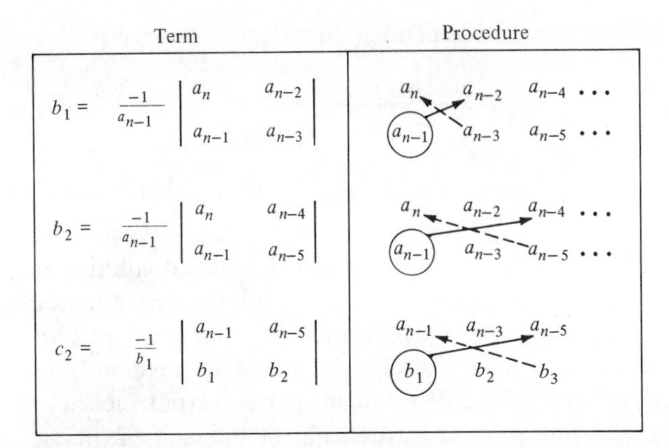

re 7.29 Chart illustrative of the algorithmic procedure for determining element values for the Routh array. Elements connected with a solid line have a positive sign, those connected by a dashed line a negative sign, and the denominator (pivotal element) is enclosed within a circle.

without the necessity of actually factoring the characteristic equation to determine its roots. One of the most successful of these is due to Routh, an English engineer, and was also described independently by Hurwitz, a German mathematician. The procedure and the associated criterion is known today as the *Routh-Hurwitz* criterion. We will present the procedure without proof, since it is complicated and usually treated only in advanced-level books. See Figure 7.29.

Step 1. Separate the polynomial $p(s)$ given by Equation 7.69 into even and odd parts. If n is even, these parts are

$$\text{Ev}\, p(s) = a_n s^n + a_{n-2} s^{n-2} + \cdots + a_0 \tag{7.74}$$

$$\text{Od}\, p(s) = a_{n-1} s^{n-1} + a_{n-3} s^{n-3} + \cdots + a_1 s. \tag{7.75}$$

Step 2. Arrange the coefficients of Ev $p(s)$ and Od $p(s)$ in the form of an *array* with the polynomial of higher order as the first row and the other polynomial as the second row. From Equations 7.74 and 7.75, the array is

$$\begin{matrix} a_n, a_{n-2}, a_{n-4}, \ldots, a_0 \\ a_{n-1}, a_{n-3}, a_{n-5}, \ldots, a_1. \end{matrix} \tag{7.76}$$

Step 3. This array is next extended downward by generating new elements. The scheme for generating the new elements is now described. Let the elements of the third row be designated as b_j. The first and second of these are

$$b_1 = \frac{-1}{a_{n-1}} \begin{vmatrix} a_n & a_{n-2} \\ a_{n-1} & a_{n-3} \end{vmatrix} = \frac{a_{n-1}a_{n-2} - a_n a_{n-3}}{a_{n-1}} \tag{7.77}$$

$$b_2 = \frac{-1}{a_{n-1}} \begin{vmatrix} a_n & a_{n-4} \\ a_{n-1} & a_{n-5} \end{vmatrix} = \frac{a_{n-1}a_{n-4} - a_n a_{n-5}}{a_{n-1}}. \tag{7.78}$$

The equation for b_3 will differ from these two only in that the second column of the determinant will have as elements a_{n-6} and a_{n-7}. In general, the denominator for the b's is always from the first column and is sometimes called the *pivotal element.* The first column of the determinant is always the pivotal element and the element above it. The second column of the determinant is taken from the same two rows as the first column and one column to the right of the element being determined.

Applying these rules to the two rows

$$\begin{aligned} &a_{n-1}, a_{n-3}, a_{n-5}, \ldots, a_1 \\ &b_1, b_2, b_3, \ldots, \end{aligned} \tag{7.79}$$

we see that the first two elements of the next row are

$$c_1 = \frac{-1}{b_1} \begin{vmatrix} a_{n-1} & a_{n-3} \\ b_1 & b_2 \end{vmatrix} = \frac{b_1 a_{n-3} - a_{n-1} b_2}{b_1} \tag{7.80}$$

$$c_2 = \frac{-1}{b_1} \begin{vmatrix} a_{n-1} & a_{n-5} \\ b_1 & b_2 \end{vmatrix} = \frac{b_1 a_{n-5} - a_{n-1} b_3}{b_1}. \tag{7.81}$$

This procedure is continued, moving downward one row each time and generating a new set of elements. Eventually, the procedure will terminate when we have $n + 1$ rows. The numbers generated by the procedure we have described are called the *Routh array.*

Step 4. Once the Routh array is formed, we may discard everything but the first column which contains the answer to our question concerning stability. Consider the first column to be a sequence of numbers, with the column arranged as a row

$$S = (a_n, a_{n-1}, b_1, c_1, d_1, \ldots, z_1). \tag{7.82}$$

If one element of this sequence is a number with a sign which is different from the sign of the number preceding it, then we say that there has been one sign change. The total number of sign changes is found by summing the individual sign changes. This number is the basis for the following statement: *The Routh-Hurwitz criterion states that the network described by the characteristic equation $p(s) = 0$ is stable if and only if there are no sign changes in the sequence S.* The elements of S are all positive or all negative if and only if $p(s)$ represents a stable system.

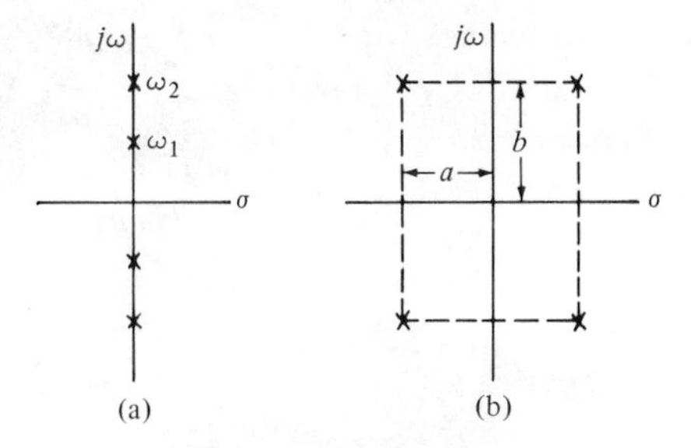

re 7.30 Root locations for even polynomials with positive coefficients: (a) $p_1(s) = (s^2 + \omega_1^2)(s^2 + \omega_2^2)$, and (b) $p_2(s) = s^4 + \alpha s^2 + \beta$.

The following additional information may sometimes be useful:

(1) If there are sign changes in S, then the total number of sign changes is equal to the number of roots with positive real parts.

(2) If an element in the first column vanishes, then the network is unstable. If you wish to determine the number of roots with positive real parts, replace the 0 by ϵ and continue noting that ϵ may be positive or negative.

(3) If an entire row vanishes — called *premature termination* — then $p(s)$ has roots on the imaginary axis or with positive real parts, and so the network is unstable. See Figure 7.30. Unless premature termination occurs, there are no j-axis roots.

(4) If $p(s)$ has any coefficient missing or any coefficients of unlike sign, then $p(s)$ has roots with nonnegative real parts and so will fail the test.

MPLE 7.5.1

Let us form our own polynomial

$$\begin{aligned} p(s) &= (s+1)(s+1)(s+2)(s+3) \\ &= s^4 + 7s^3 + 17s^2 + 16s + 3. \end{aligned} \tag{7.83}$$

There is no need to test this $p(s)$ since we know that the roots are $s = -1, -1, -2$, and -3. However, we are using this known $p(s)$ to illustrate the method. For this polynomial, the Routh array is

$$\begin{array}{ccc} 1 & 17 & 3 \\ 7 & 16 & \\ \frac{103}{7} & 3 & \\ \frac{1501}{103} & & \\ 3. & & \end{array} \tag{7.84}$$

From the sequence $S = (1, 7, 103/7, 1501/103, 3)$, we see that there are no sign changes, and therefore all the roots of $p(s)$ have negative real parts, agreeing with observation of the factored form of $p(s)$ in Equation 7.83.

EXAMPLE 7.5.2

Consider the polynomial

$$p(s) = 8s^3 + s^2 + s + 11.25. \tag{7.85}$$

The Routh array is

$$\begin{array}{cc} 8 & 1 \\ 1 & 11.25 \\ -89 & \\ 11.25 & \end{array} \tag{7.86}$$

The testing might have stopped as soon as the negative sign was found. However, if it is of interest, the sequence $S = (8, 1, -89, 11.25)$ indicates two changes of sign, meaning that $p(s)$ has two roots with positive real parts. This may be verified by factoring Equation 7.85, giving

$$p(s) = (s - \tfrac{1}{2} + j1)(s - \tfrac{1}{2} - j1)(s + \tfrac{9}{8}). \tag{7.87}$$

EXAMPLE 7.5.3

The characteristic equation for a network under study is

$$p(s) = s^4 + s^3 + 2s^2 + 2s + 1 = 0. \tag{7.88}$$

We construct the Routh array as follows:

$$\begin{array}{ccc} 1 & 2 & 1 \\ 1 & 2 & \\ \epsilon & 1 & \\ 2 - \dfrac{1}{\epsilon} & & \\ 1 & & \end{array} \tag{7.89}$$

For small ϵ, $-1/\epsilon$ makes the fourth number in the first column negative, indicating an unstable network and two roots of $p(s)$ with positive real parts.

EXERCISES

7.5.1 Determine the stability of the network represented by the characteristic equation

$$p(s) = s^4 + s^3 + 3s^2 + 2s + 1 = 0.$$

7.5.2 Determine the number of roots of the following characteristic equation which have positive real parts

$$p(s) = s^4 + 3s^3 + 8s^2 + s + 9.$$

7.5.3 Consider the following characteristic equation

$$p(s) = 0.5s^5 + 0.5s^4 + 4s^3 + 2s^2 + 5s + 1 = 0.$$

Determine the number of roots with positive real parts.

7.5.4 Repeat the analysis described in Exercise 7.5.3 for the characteristic equation

$$p(s) = s^5 + 2s^4 + 5s^3 + 2s^2 + 2.5s + 1 = 0.$$

7.5.5 Determine the stability of the network represented by the following characteristic equation:

$$p(s) = s^4 + 3s^3 + 4s^2 + 6s + 4 = 0.$$

6 Summary

1. For a linear network, the following properties apply relating response to input:
(a) If y_1 is the zero-state response due to an arbitrary input x_1, then cy_1 is the zero-state response due to an input cx_1, where c is an arbitrary constant. This is called the *homogeneity* condition.
(b) If y_1 is the zero-state response due to an arbitrary input x_1, y_2, the zero-state response due to another arbitrary input x_2 not equal to x_1 identically, then the zero-state response due to $x_1 + x_2$ is $y_1 + y_2$. This is known as the *additivity* condition or *superposition* condition.
(c) If the zero-input response for any arbitrary initial state $(\alpha_1, \alpha_2, \ldots, \alpha_n)$ is y_0, then the zero-input response due to an initial state $(C\alpha_1, C\alpha_2, \ldots, C\alpha_n)$ is Cy_0, where C is any arbitrary constant. This is the *homogeneity* condition.
(d) If the zero-input response for an arbitrary initial state $(\alpha_1, \alpha_2, \ldots, \alpha_n)$ is y_α and the zero-input response for another arbitrary initial state $(\beta_1, \beta_2, \ldots, \beta_n)$ is y_β, then the zero-input response for an initial state $(\alpha_1 + \beta_1, \alpha_2 + \beta_2, \ldots, \alpha_n + \beta_n)$ is $y_\alpha + y_\beta$. This is the *additivity* condition or superposition condition.
(e) If y_0 is the zero-input response to an arbitrary initial state $(\alpha_1, \alpha_2, \ldots, \alpha_n)$ and y_1 is the zero-state response for an arbitrary input x, then the total response for the same initial state and the same input is $y_0 + y_1$. This condition must hold for any arbitrary initial state and any arbitrary input. This is known as the *decomposition property*.

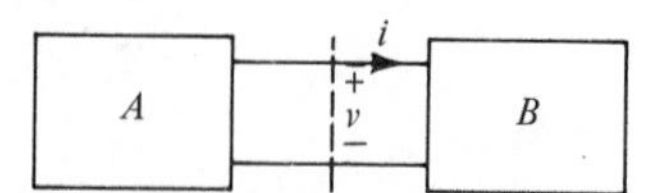

Figure 7.31 Network A: Linear elements, independent and dependent sources, internal coupling through mutual inductance or controlled sources, but no coupling to B. Network B: No dependent sources related to voltages or currents in Network A except i and v. Elements may be nonlinear or time-varying. (*Note:* In most applications, B will be a single passive element.)

2. Thevenin's and Norton's theorems apply to networks with the identifications shown in Figure 7.31 where A is a linear network.
3. *Thevenin's theorem:* Network A may be replaced by a voltage source equal to the open-circuit voltage at its terminals in series with a network in which all initial states are set to zero and all independent sources are set to zero, but in which controlled sources operate.
4. *Norton's theorem:* Network A may be replaced by a current source equal to the short-circuit current at its terminals in parallel with a network in which all initial states are set to zero and all independent sources are set to zero, but in which controlled sources operate.
5. A *time-invariant network* is one having the property that if input $x(t)$ and given initial conditions cause the response $y(t)$, then the input $x(t - \Delta)$ and the same initial conditions at $t = \Delta$ will cause the response $y(t - \Delta)$ for any $\Delta > 0$.
6. We consider an ordinary differential equation of order n with constant coefficients. The complementary function is the most general solution of the homogeneous differential equation. It is found by setting the left-hand side of the original equation to zero and assuming a solution of the form $c_k e^{s_k t}$ for distinct or nonrepeated s_k. Substitution of this solution into the homogeneous differential equation gives an algebraic equation called the *characteristic equation,* the roots of which are the values of s_k in the solution. The complementary function is the summation of the $c_k e^{s_k t}$ terms. If a root s_1 of the characteristic equation is repeated r_1 times, then the solution will contain $t^{r_1 - 1} e^{s_1 t}$ and similar terms for all lower powers of t. The zero-input response is obtained from the complementary function by choosing the constants so that the initial conditions associated with zero input are satisfied.
7. *The Routh-Hurwitz criterion* is employed to determine if all the roots of the characteristic equation

$$p(s) = a_n s_n + a_{n-1} s^{n-1} + \cdots + a_1 s + a_0 = 0$$

have negative real parts. Beginning with the coefficients of $p(s)$, we find that a Routh array with $n + 1$ rows is generated.

$$\begin{array}{ccccc} a_n & a_{n-2} & a_{n-4} & \cdots & a_0 \\ a_{n-1} & a_{n-3} & a_{n-5} & \cdots & a_1 \\ b_1 & b_2 & b_3 & \cdots & \\ c_1 & c_2 & c_3 & \cdots & \\ \vdots & & & & \end{array}$$

The sequence of numbers from the first column is examined,

$$S = (a_n, a_{n-1}, b_1, c_1, \ldots, z_1).$$

All the roots of $p(s)$ have negative real parts if and only if there are no sign changes in the sequence and no elements have zero value. The number of changes of signs equals the number of roots with positive real parts. If there are imaginary roots, then a whole row of the Routh array will contain only zeros. In terms of the summary just given, the following procedure is recommended:

(a) Examine $p(s)$ to see if any terms are negative (ignoring the trivial case when all are negative which is corrected by multiplying by -1) or zero. If so, further testing is unnecessary since the network is unstable.

(b) If (a) is passed, then complete the Routh array. If any element in the first column is negative or zero, the testing may terminate since the network is unstable.

(c) If (b) is passed and all elements in the first column are positive, then the network is stable.

(d) If the network is unstable, then additional information may be determined from the Routh array if desired (discussed in detail in sources cited in References):

(i) If there are negative numbers in S, the number of sign changes is equal to the number of roots with positive real parts.

(ii) If a zero appears in the first column, replace it by ϵ and continue, repeating with δ if a second zero is encountered, and continue to completion. The test of (i) may be applied to the completed sequence.

(iii) If an entire row vanishes, then this premature termination indicates that $p(s)$ contains an even polynomial as a factor which may represent either j-axis roots or a quad of four factors which are $s \pm a \pm jb$. This even polynomial may be formed from information in the row formed before premature termination.

PROBLEMS

7.1 Show that the ideal transformer model of Chapter 4 (p. 73) is a linear 2-port network.

7.2 Show that two linear, time-invariant, magnetically coupled coils as shown in Figure 4.18 (p. 70) form a linear network.

7.3 Show that if a system is zero-state additive, then, if y_1 is the zero-state response due to an input x_1, the output due to Cx_1 is Cy_1, where C is any finite integer (including negative integers).

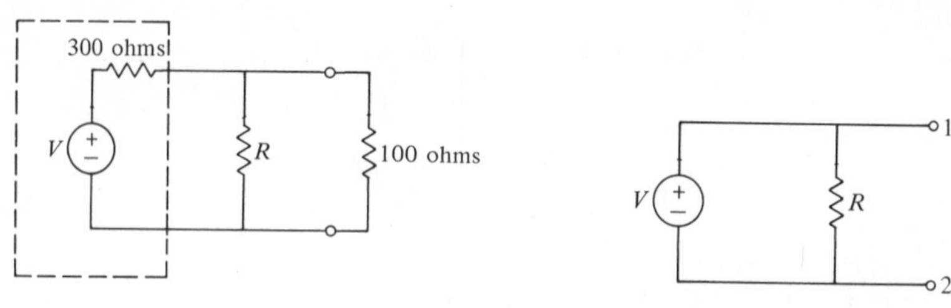

Figure 7.32

Figure 7.33

★ 7.4 Repeat Problem 7.3 for C equal to a finite rational number (finite ratio of two integers). [*Hint:* Do Problem 7.3 first, so that the result in Problem 7.3 may be assumed.] So if $x_1 \to y_1$, $nx_1 \to ny_1$. Let $x_2 = (n/m)x_1 \to y_2$. Then $(m/n)x_2 = x_1 \to y_1$ and $mx_2 \to ny_1 = my_2 \therefore x_2 = (n/m)x_1 \to (n/m)y_1$. The symbol $x \to y$ means that y is the zero-state response for an input x. Since any irrational number may be approximated arbitrarily closely by a rational number, this problem shows that a zero-state additive system is "practically" zero-state linear. This is why in practical cases, additivity or superposition is used as a definition of linearity. Similarly, using exactly the same proof, it can be shown that a zero-input additive system implies that for an arbitrary initial state $(\alpha_1, \alpha_2, \ldots, \alpha_n)$ yielding a zero-input response y, the arbitrary initial state $(C\alpha_1, C\alpha_2, \ldots, C\alpha_n)$ yields Cy, provided that C is any arbitrary finite *rational* number. Hence a system which is zero-input additive, zero-state additive, and decomposable, is "practically" linear.

7.5 Determine the Thevenin equivalent of a linear time-varying charged capacitor with a charge $q(t_1)$ at t_1.

7.6 Determine the Norton equivalent of a linear time-varying inductor with an initial current $i(t_1)$ at t_1.

7.7 A sinusoidal signal generator has a Thevenin equivalent resistance of 300 ohms and a Thevenin equivalent source voltage V volts rms. The signal

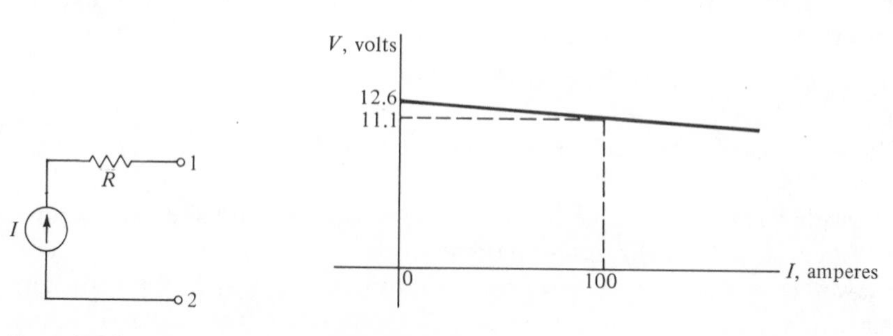

Figure 7.34

Figure 7.35

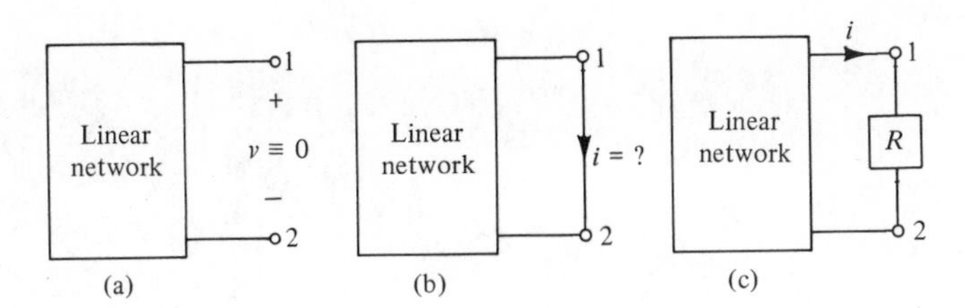

Figure 7.36

generator is to be connected to a 100-ohm load. For some reason, it is desired that the equivalent resistance to the left of the load be 100 ohms. This may be done by connecting a resistance R across the generator as shown in Figure 7.32. Determine the value for R that will do this.

7.8 Find the Thevenin equivalent for the network in Figure 7.33. Would your answer be affected if R were replaced by an arbitrary network?

7.9 Find the Norton equivalent for the network in Figure 7.34. Would your answer be affected if R were replaced by an arbitrary network?

7.10 The approximate terminal behavior of a practical battery is displayed as a straight line in Figure 7.35. Find its Thevenin equivalent.

7.11 The voltage v in the network of Figure 7.36(a) is identically zero. What will be the current i as in Figure 7.36(b) if the terminals in (a) are short circuited? How about the current in Figure 7.36(c)? Given any physical n-port network with sources as in (a), there is a *unique* set of values for currents and voltages throughout the network for any given set of port currents and voltages. In other words, the current through a branch cannot have two values at the same time. Likewise for voltages. From this fact, what can you say about the distribution of currents and voltages in (b) and (c) as compared to that of (a) of the figure?

7.12 The voltage v_{12} in the network of Figure 7.37(a) is equal to $v(t)$. Suppose a voltage source whose variation is identical to $v(t)$ is connected to the terminals as in (b). What will be the current $i(t)$? How about the current

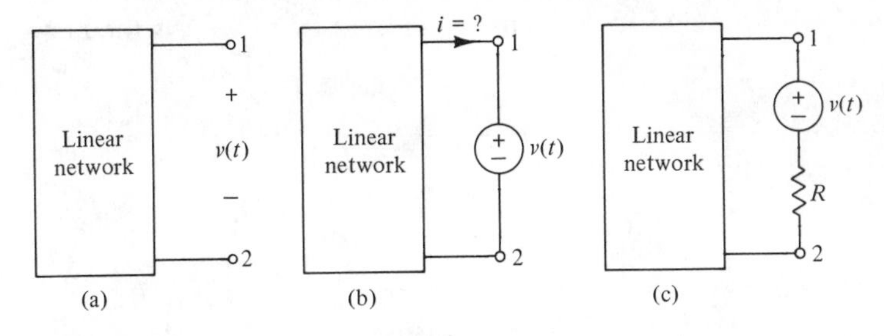

Figure 7.37

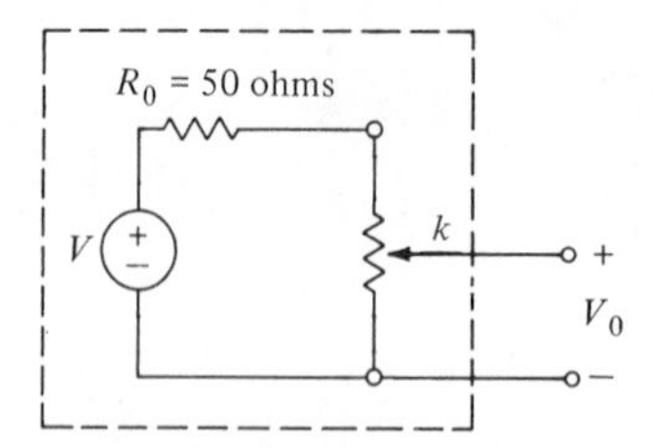

Figure 7.38

in (c)? See Problem 7.11 and answer the last question there for the networks in Figure 7.37.

7.13 A potentiometer, a resistance with an adjustable position tap, is commonly used to obtain an adjustable source of voltage from a fixed source of voltage, as shown in Figure 7.38. We have a sinusoidal signal generator of 10 volts rms and an internal resistance of 50 ohms. The tap on the potentiometer may be adjusted from lowest position (coincident with bottom terminal) to highest position (coincident with top terminal). Specify a value of potentiometer resistance so that the dissipation in the potentiometer is no more than 0.05 watt. If k denotes the fraction of potentiometer resistance included between the bottom terminal and the top, plot the magnitude of output voltage V_0 vs. k. If k is linearly proportional to the angular position of the potentiometer knob, is the variation of V_0 linear with respect to angular position of knob setting? Plot the Thevenin resistance as a function of k.

7.14 Calculate the voltage v_{ab} in Figure 7.39. [*Hint:* Use Norton's theorem repeatedly.]

7.15 For the network of Figure 7.39, calculate i_{ac}. It is desired to make the current i_{ac} equal to zero by inserting a battery in series with the resistor. What value of battery voltage must be used. Indicate polarity.

7.16 For the network of Figure 7.39, calculate i_1, the current through the branch with a battery of 2 volts in series with a 1-ohm resistor.

7.17 Suppose in Figure 7.39 we add a battery of 3 volts of zero internal resistance across ab with the positive terminal connected to a and similarly a battery of 1 volt across cb with the positive terminal connected to b. What is the new value for i_{ac}?

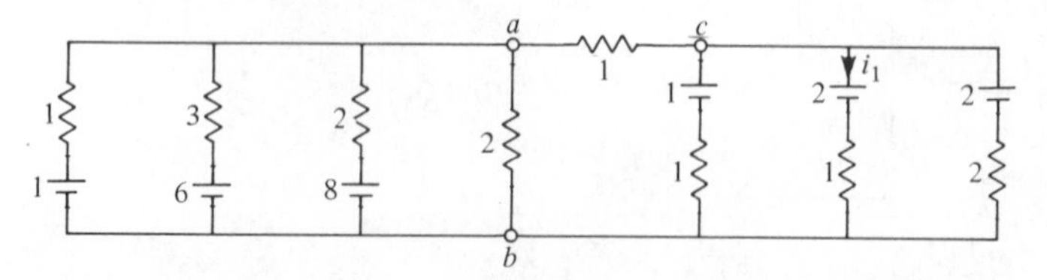

Figure 7.39

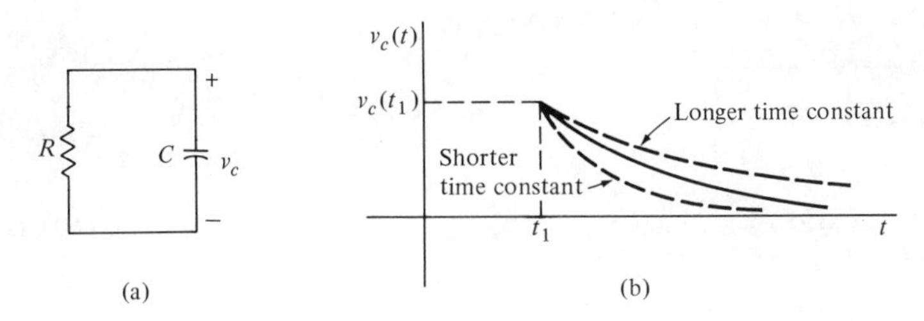

Figure 7.40

7.18 In the network of Figure 7.40(a), R and C are positive constants. Taking v_c as the output, show that the zero-input response is as shown in Figure 7.40(b) for $t \geq t_1$, the initial instant. Note the effect of a longer or shorter time constant on the shape of the waveform.

7.19 In the network shown in Figure 7.41, the switch K is in position a for a long period of time (say 24 hours) and then at $t = 0$ is switched from a to b. Find the current $i(t)$ that results after the switching occurs. Use numerical values wherever possible. The original voltage on the 1-farad capacitor is zero.

7.20 In the network of Figure 7.42, the switch K is closed at $t = 0$. Find the current $i(t)$ in the circuit. Evaluate any arbitrary constants and substitute numerical values wherever possible. The initial voltage on the right-hand side capacitor is zero.

7.21 In the linear time-invariant network shown in Figure 7.43, K is closed at $t = 0$. The capacitors have initial values of voltage on them marked V_a and V_b. (a) Find $v_{C_1}(t)$ and $v_{C_2}(t)$ for $t \geq 0$. (b) Check this result by verifying that $v_{C_1}(\infty) = v_{C_2}(\infty)$.

7.22 The series network shown in Figure 7.44 consists of two charged capacitors in series with a resistor. Before the switch is closed, the voltmeter marked VM 1 reads +10 volts and VM 2, +4 volts. At $t = 0$ the switch is closed. You are to find the two voltmeter readings when the system has again reached a steady state (i.e., at least 10 time constants after the closing of the switch).

7.23 Consider the network given for Problem 7.21. For that network, $R = 2$

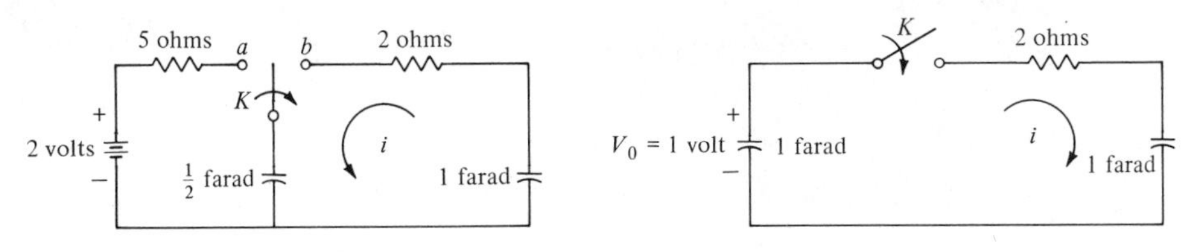

Figure 7.41

Figure 7.42

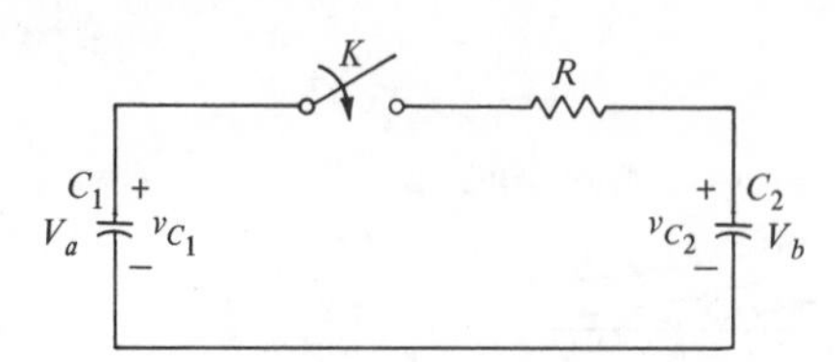

Figure 7.43

ohms, $C_1 = 1$ farad, and $C_2 = \frac{1}{2}$ farad. At the time switch K is closed, each capacitor carries a charge of 1 coulomb. Find the voltage $v_{C_2}(t)$ for $t \geq 0$.

★ 7.24 Show that the system represented by Equation 7.39 is zero-state linear.

★ 7.25 Assuming that initial conditions are given by $y(0), y'(0), \ldots, y^{(n-1)}(0)$, show that the system represented by Equation 7.39 is zero-input linear. Using the result of Problem 7.24, verify that the decomposition property holds. Thus the system represented by Equation 7.39 is linear.

7.26 Find the Thevenin equivalent network with respect to port *ab* for the linear time-invariant network of Figure 7.45. Note that I_1 is the current in R_1 and that this current controls the current source of the network. The network of this problem is a linear incremental equivalent network for a transistor. If possible, combine the resistors and controlled source into a single resistor.

7.27 Repeat Problem 7.26 for the network given in Figure 7.46. Observe that the voltage v_k is the voltage across R_k with respect to ground. This network is a linear incremental model for an electronic circuit known as a *cathode follower*.

7.28 For the linear time-invariant network in Figure 7.47 with v_{12} as output and v_1 as input, determine the differential equation relating v_1 to v_{12}.

7.29 For the network of Figure 7.13, suppose $R_1 = R_2 = 100$ ohms, $C = 0.01$ microfarad, and $v_2(0) = 1$ volt. Determine the zero-input response for $t \geq 0$.

7.30 Determine the Thevenin equivalent for a linear time-invariant fluxed inductor as shown in Figure 7.12(a). See Exercise 7.2.7 for v equal to a unit impulse.

7.31 Determine the Norton equivalent for a linear time-invariant charged

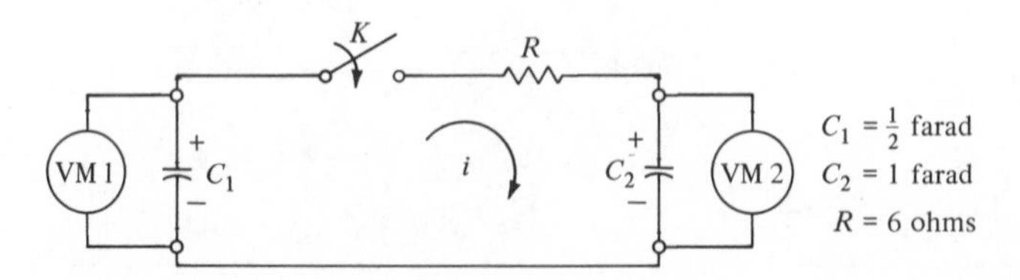

Figure 7.44

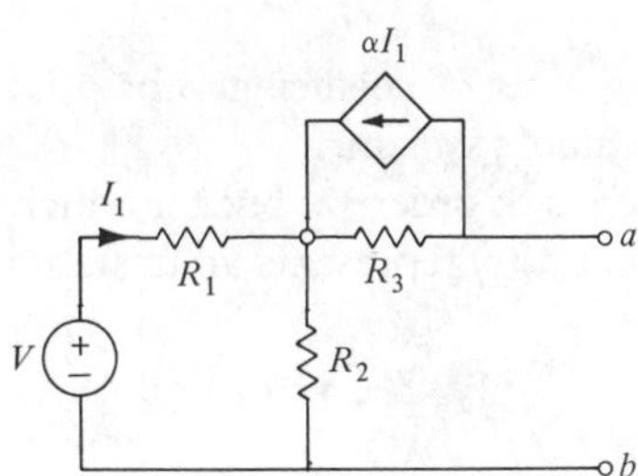

Figure 7.45

Figure 7.46

capacitor as shown in Figure 7.7(a). See Exercise 7.2.4 for i equal to a unit impulse.

7.32 Show that if $y(t)$ is the zero-state response for an input $x(t)$ of a linear time-invariant differential system, then $dy(t)/dt$ is the zero-state response for an input $dx(t)/dt$.

★ 7.33 Repeat Problem 7.32 for a linear time-invariant system which is not necessarily a differential system.

★ 7.34 Show that if $y(t)$ is the zero-state response for an input $x(t)$ of a linear time-invariant system, the zero-state response for

$$\int_{t_1}^{t} x(\tau)\, d\tau \qquad \text{is} \qquad \int_{t_1}^{t} y(\tau)\, d\tau,$$

where t_1 is the initial instant.

7.35 Given the characteristic equation

$$p(s) = s^2 + k_1 s + k_2 = 0,$$

for which values of k_1 and k_2 will the network represented by $p(s)$ be stable. Construct a k_1k_2-plane plot showing the regions in the plane for which the network is stable.

7.36 In the network of Figure 7.48, let $k = 2$, $C_1 = 1$ farad, and $R_2 = 1$ ohm. Determine the inequality that must exist involving R_1 and C_2 for the network to be stable.

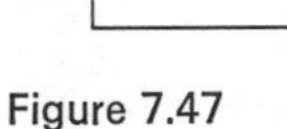
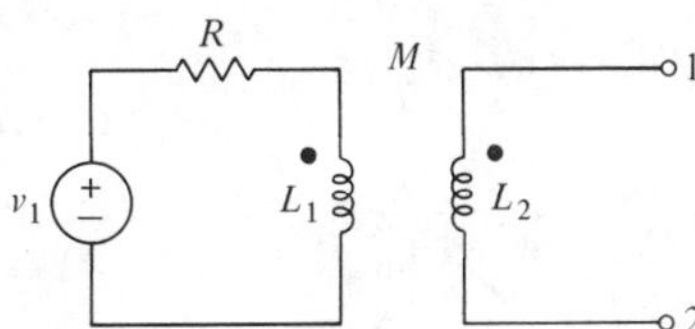

Figure 7.47

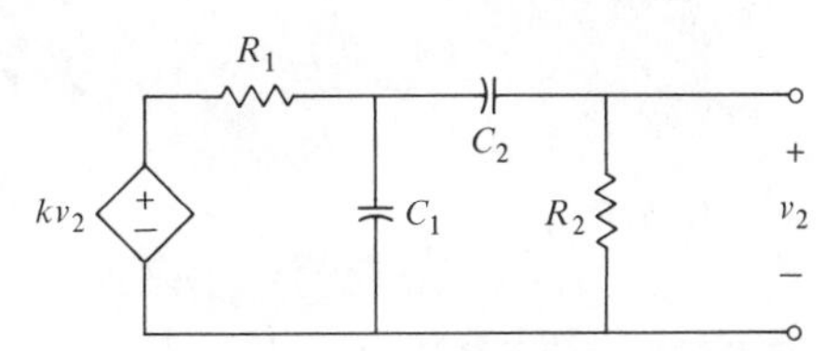

Figure 7.48

7.37 Repeat Problem 7.36 for $k = 1$.

7.38 Construct a polynomial $p_1(s)$ by determining a set of coefficients for $p_1(s)$ such that the order is six and it represents a stable system.

7.39 Repeat Problem 7.38 to determine $p_2(s)$ which is of order six but for which the Routh array has a row that vanishes so that $p_2(s)$ represents an unstable system.

Chapter 8

urther Implications of Linearity: onvolution and Impulse Response

Simple application of linearity and time-invariance

Suppose that the zero-state response of a linear time-invariant network to an input $x(t)$ that starts at $t = 0$ is $y(t)$. From the time-invariance principle studied in the last chapter, the response to $x(t - \Delta)$ is $y(t - \Delta)$ for any $\Delta > 0$. This means that if the input is delayed by Δ, then the output is also delayed by Δ. From linearity, we know that the combination of two inputs

$$x_1 = x(t) - x(t - \Delta) \tag{8.1}$$

will result in the output

$$y_1 = y(t) - y(t - \Delta). \tag{8.2}$$

Again making use of linearity, we know that when the input is multiplied by a constant $1/\Delta$, then the output is multiplied by that same constant $1/\Delta$; that is,

$$x_2 = \frac{1}{\Delta}[x(t) - x(t - \Delta)] \tag{8.3}$$

will result in the output

$$y_2 = \frac{1}{\Delta}[y(t) - y(t - \Delta)]. \tag{8.4}$$

In the limit, as Δ is taken to be arbitrarily small, x_2 given by Equation 8.3 will approach the time derivative of $x(t)$. Similarly, y_2 of Equation 8.4 will approach the time derivative of $y(t)$. If these derivatives exist, then we have the useful conclusion that if the zero-state response of a linear time-invariant network to an input $x(t)$ is $y(t)$, then the zero-state response of the same network to

$$\text{Input} = \frac{dx}{dt} \tag{8.5}$$

will be

$$\text{Response} = \frac{dy}{dt}. \tag{8.6}$$

Actually, it turns out that this conclusion still holds when x and y do not possess ordinary derivatives. For example, $x(t)$ may have a jump discontinuity at $t = t_1$ and in this case $dx(t)/dt$ contains an impulse at $t = t_1$. Hence we must exercise care in applying this principle to determine whether $x(t)$ has singularities at $t = 0$ to obtain the correct $dx(t)/dt$. We next illustrate this idea by means of a simple example.

EXAMPLE 8.1.1

In the linear time-invariant network of Figure 8.1(a), an input of $v_{in}(t) = u(t)$ with capacitor C initially uncharged results in the output $v_0(t) = e^{-t}u(t)$. We wish to determine the response to $dv_{in}/dt = \delta(t)$, which we do by differentiating $v_0(t)$. Now v_0 itself contains a jump discontinuity at $t = 0$ since it changes from zero to unit value at that time. We can avoid difficulty with functions of this type if we separate them into two parts. For a general function $f(t)$, for $t \geq 0$, let

$$f(t) = f(0)u(t) + [f(t) - f(0)]u(t), \tag{8.7}$$

where $f(t) - f(0) = f_1(t)$ has zero value at $t = 0$. Now if $f(0) = 0$, there is no difficulty since there is no jump discontinuity. However, when $f(0) \neq 0$, then the use of Equation 8.7 permits us to consider the discontinuity separately. Applying Equation 8.7 to $e^{-t}u(t)$, we have

$$v_0(t) = u(t) - (1 - e^{-t})u(t) \tag{8.8}$$

and the output resulting from an impulse input will be found by differentiating Equation 8.8 to give

$$\frac{dv_0}{dt} = \delta(t) - e^{-t}u(t). \tag{8.9}$$

The impulse response of the network which is Equation 8.9 is shown in Figure 8.1(e). This form of the response may be understood by examining the network shown in (a) of the figure. At $t = 0$, the capacitor is uncharged and so behaves as a short circuit ($v_C = 0$). The appearance of an impulse at the input causes the impulse to appear immediately as a voltage across the resistor R. The impulse

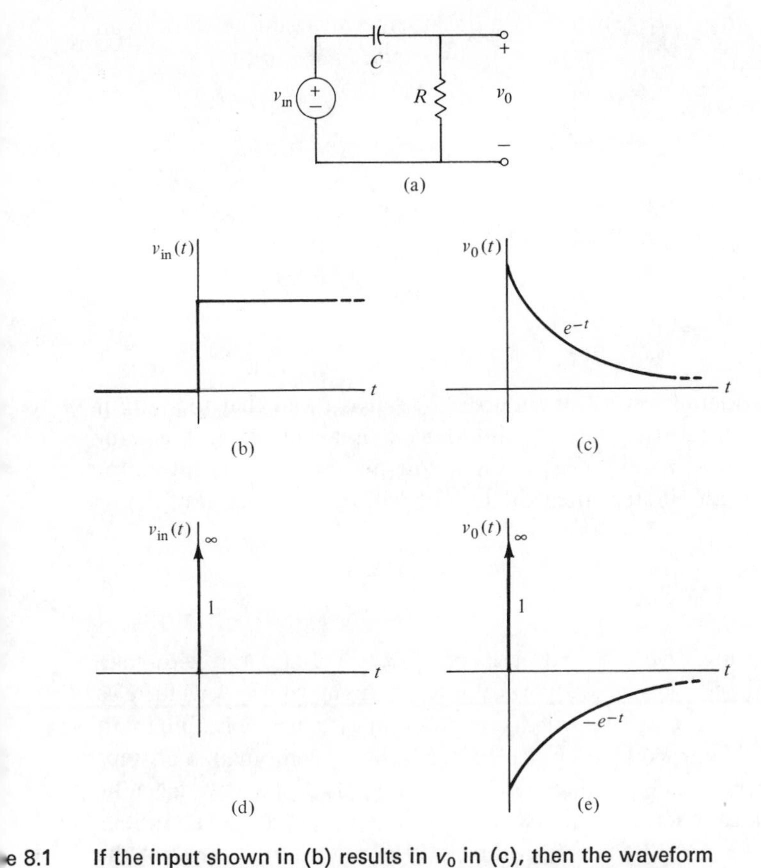

e 8.1 **If the input shown in (b) results in v_0 in (c), then the waveform of (d) which is the derivative of that in (b) will result in the output shown in (e) which is the derivative of that shown in (c).**

charges the capacitor, which then discharges for $t > 0$ as shown by the remainder of the response shown in (e).

By similar arguments as above, we see that if the zero-state response of a linear time-invariant network due to $i(t)$ is $v(t)$, where $i(t)$ starts at 0, then the zero-state response of the same network due to an input

$$\text{Input} = \int_0^t i(\tau)\, d\tau \tag{8.10}$$

is

$$\text{Response} = \int_0^t v(\tau)\, d\tau. \tag{8.11}$$

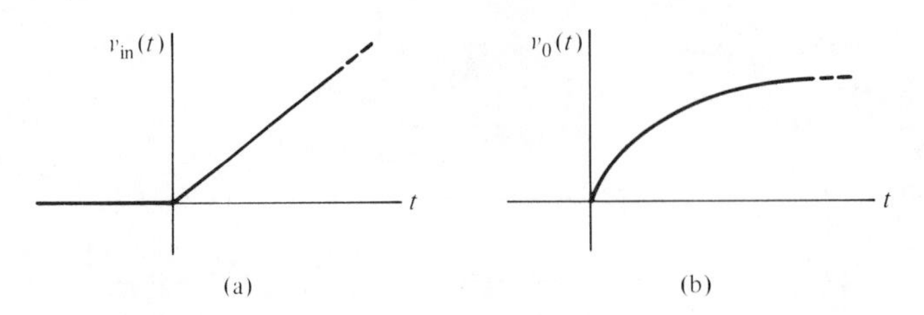

Figure 8.2 For the network of Figure 8.1(a), the v_{in} and v_0 found in Example 8.1.2 are shown.

EXAMPLE 8.1.2

In the linear time-invariant network of Figure 8.2(a), it is given that the output voltage $v_0(t)$, when C is initially uncharged and $v_{in}(t)$ is a unit step, is e^{-t} for $t \geq 0$ and zero for $t < 0$. If instead of v_{in} being a unit step, we have a unit ramp, which is the integral of a unit step, then the output will be the integral of e^{-t} or

$$v_{out} = \int_0^t e^{-t}\,dt = 1 - e^{-t}, \qquad t \geq 0. \tag{8.12}$$

From the above results, we see that it is possible to obtain the zero-state response of a linear time-invariant network due to an input whose waveform is piecewise linear in time from our knowledge of the unit-step response. The first step in the analysis is to express the input function as a linear combination of steps and ramps with appropriate delays. Since ramps are integrals of steps, we have the input expressed as a linear combination of steps and integrals of steps with various delays. The output is obtained by an application of superposition, in terms of the known response to a unit step.

EXAMPLE 8.1.3

Consider the triangular pulse of Figure 8.3. Suppose that for the network in Figure 8.1(a) the zero-state output due to a unit-step input $u(t)$ is $e^{-t}u(t)$. The problem is to determine the output due to the pulse input shown in Figure 8.3. The first step is to express the pulse $v_{in}(t)$ in terms of the step function. In this case, $v_{in}(t)$ is expressible as the linear combination of two delayed ramps and one delayed step:

$$v_{in}(t) = 2r(t-1) - 2r(t-2) - 2u(t-2). \tag{8.13}$$

From linear time-invariance, we obtain the output as

$$v_0(t) = 2\int_0^t e^{-(\tau-1)}u(\tau-1)\,d\tau - 2\int_0^t e^{-(\tau-2)}u(\tau-2)\,d\tau - 2e^{-(t-2)}u(t-2). \tag{8.14}$$

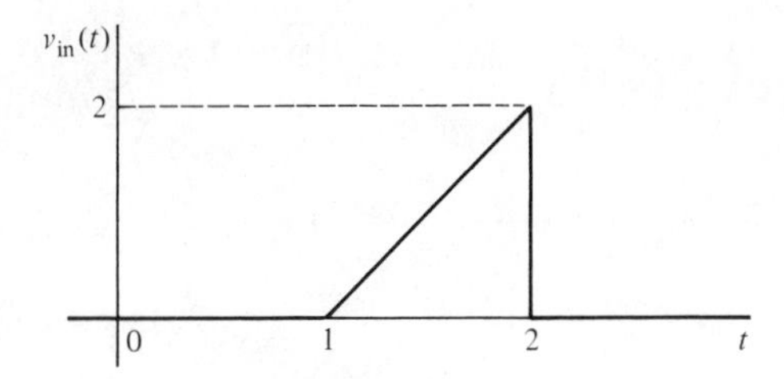

Figure 8.3 Pulse input for Example 8.1.3.

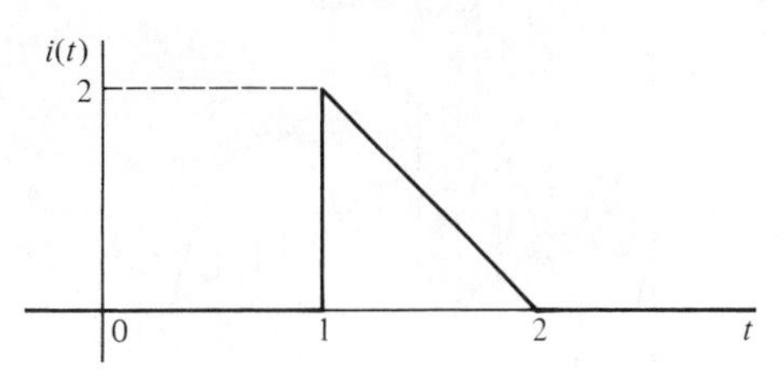

Figure 8.4

ERCISES

8.1.1 Sketch the ramps and step of Equation 8.13 and graphically verify that the sum is the pulse of Figure 8.3. Sketch the components of the total response due to the individual terms in Equation 8.13 of Example 8.1.3.

8.1.2 Express the pulse of Figure 8.4 in terms of steps and ramps. Let $v(t)u(t)$ be the zero-state response of a linear time-invariant network due to a unit-step-function input $u(t)$. Express the zero-state response of the same network due to the pulse input of Figure 8.4.

8.1.3 Repeat Exercise 8.1.2 if $v(t)u(t)$ is the zero-state response due to a unit impulse $\delta(t)$.

8.1.4 Determine the derivative of the pulse in Figure 8.3. You may have to use singularity functions.

8.1.5 Repeat Exercise 8.1.4 for the pulse in Figure 8.4.

2 The convolution integral

Consider a continuous function $x(t)$, which is 0 for $t < 0$, as the input to a linear time-invariant network. Let $s(t)$ denote the zero-state response due to a unit step $u(t)$ and let $h(t)$ denote the zero-state response due to a unit impulse $\delta(t)$. We note that the impulse response function $h(t)$ is zero for $t < 0$ because it is the response of a physical network and there can be no zero-state response before the input is applied. To ensure that this fact is not overlooked, it may be helpful to write $h(t)$ as $h(t)u(t)$.

Suppose that we approximate $x(t)$ by a staircase function which is a sum of rectangular pulses as in Figure 8.5. The approximate $x(t)$ is designated by an asterisk superscript as

$$x^*(t) = \sum_{k=0}^{[t/\Delta]} x(k\Delta)\,\frac{u(t-k\Delta) - u(t-k\Delta-\Delta)}{\Delta}\,\Delta, \tag{8.15}$$

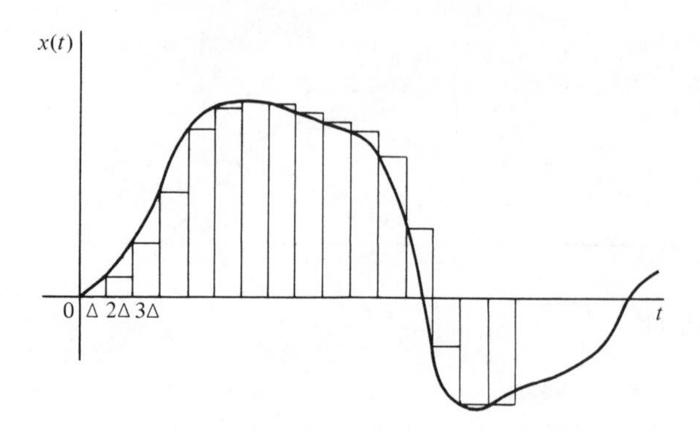

Figure 8.5 Staircase approximation of a continuous function.

where $[t/\Delta]$ denotes the largest integer closest in value to t/Δ. From the principle of linear time-invariance, the zero-state response to $x^*(t)$ is

$$y^*(t) = \sum_{k=0}^{[t/\Delta]} x(k\Delta)\,\frac{s(t-k\Delta) - s(t-k\Delta-\Delta)}{\Delta}\,\Delta. \tag{8.16}$$

As Δ becomes arbitrarily small, it is convenient to change notation and denote

$$k\Delta = \tau. \tag{8.17}$$

As we pass to the limit, $\Delta \to d\tau$, we see that

$$\frac{u(t-k\Delta) - u(t-k\Delta-\Delta)}{\Delta}\,\Delta \to \delta(t-\tau)\,d\tau, \tag{8.18}$$

$$\frac{s(t-k\Delta) - s(t-k\Delta-\Delta)}{\Delta} \to h(t-\tau), \tag{8.19}$$

$$x^*(t) \to \int_0^t x(\tau)\,\delta(t-\tau)\,d\tau = x(t), \tag{8.20}$$

and

$$y^*(t) \to \int_0^t x(\tau)h(t-\tau)\,d\tau = y(t). \tag{8.21}$$

That is, the zero-state response of a linear time-invariant network due to an input $x(t)$ which starts at $t = 0$ and which is continuous is

$$y(t) = \int_0^t x(\tau)h(t-\tau)\,d\tau. \tag{8.22}$$

Equation 8.22 is known as the *convolution integral*. Few equations are of more importance in electrical engineering!

If $x(t)$ is not continuous but if it is expressible as the sum of a continuous function $x_1(t)$ and a finite number of step functions, impulse functions, and possibly higher-order singularity functions, then we may write

$$x(t) = x_1(t) + \sum_{k=1}^{N_a} a_k u(t - \alpha_k) + \sum_{k=1}^{N_b} b_k \delta(t - \beta_k)$$
$$+ \sum_{k=1}^{N_c} c_k \delta'(t - \gamma_k) + \text{finite number of higher-order singularity functions.} \quad (8.23)$$

The singularity functions mentioned in the equation are the higher-order derivatives of the step, impulse, etc. The zero-state response may be obtained by considering the terms in Equation 8.23 one at a time, if the network is linear.

The response due to $x_1(t)$ is obtained by means of the convolution integral of Equation 8.22. The response due to the step functions may be obtained by integrating the impulse response with corresponding delay as in Section 8.1. Thus, the response due to $a_k u(t - \alpha_k) = x_2(t)$ is

$$a_k \int_0^t h(\tau - \alpha_k)\, d\tau = \begin{cases} 0, & \text{for } t < \alpha_k, \\ \displaystyle\int_{\alpha_k}^t a_k h(\tau - \alpha_k)\, d\tau, & \text{for } t \geq \alpha_k, \end{cases}$$
$$= \int_0^t a_k u(\tau - \alpha_k) h(\tau - \alpha_k)\, d\tau. \quad (8.24)$$

By a substitution of variables, replace $\tau - \alpha_k$ by $t - \xi$, $d\tau = -d\xi$. Then we have

$$\int_{\alpha_k}^t a_k h(\tau - \alpha_k)\, d\tau = \int_{\alpha_k}^t a_k h(t - \xi)\, d\xi = \int_0^t a_k u(\xi - \alpha_k) h(t - \xi)\, d\xi$$
$$= \int_0^t a_k u(\tau - \alpha_k) h(t - \tau)\, d\tau. \quad (8.25)$$

Since $a_k u(t - \alpha_k)$ is $x_2(t)$, Equation 8.25 is really a convolution integral as in Equation 8.22, where $a_k u(\tau - \alpha_k) = x_2(\tau)$. Similarly, for a typical impulse component $b_k \delta(t - \beta_k) = x_3(t)$, the zero-state response is

$$b_k h(t - \beta_k) u(t - \beta_k) = \int_0^t b_k \delta(\tau - \beta_k) h(t - \tau)\, d\tau. \quad (8.26)$$

Again, since $x_3(t) = b_k \delta(t - \beta_k)$, Equation 8.26 is really a convolution integral as in Equation 8.22.

In conclusion, if the input to a linear time-invariant network is a continuous function plus a finite number of higher-order singularity functions, the zero-state response may be obtained by a convolution integral as in Equation 8.22. In Equation 8.22, the lower limit 0 is replaced by t_1 if the signal input starts at t_1

instead of 0. The upper limit t may be replaced by ∞ without changing the value of the integral if $h(t)$ corresponds to a physical network, because $h(t - \tau) = 0$ for $\tau > t$, so that the integral from t to ∞ is zero and no contribution is made to the integral.

The convolution integral of Equation 8.22 may be written in a different form by a change of variables. Let $\tau = t - \xi$ so that $d\tau = -d\xi$, and when $\tau = 0$, $\xi = t$, and when $\tau = t$, $\xi = 0$. Then we have

$$\int_0^t x(\tau)h(t - \tau)\, d\tau = -\int_t^0 x(t - \xi)h(\xi)\, d\xi = \int_0^t x(t - \xi)h(\xi)\, d\xi. \tag{8.27}$$

Since ξ is a dummy variable in Equation 8.27, we may use τ, and so we obtain

$$\int_0^t x(\tau)h(t - \tau)\, d\tau = \int_0^t x(t - \tau)h(\tau)\, d\tau. \tag{8.28}$$

By the convolution of $x(t)$ and $h(t)$, where $x(t) = 0$ for $t < 0$ and $h(t) = 0$ for $t < 0$, we mean the integrations of Equation 8.28. Both forms yield the same answer, of course.

The convolution integral may be used to show that for a linear time-invariant system the zero-state response to an arbitrary input $x(t)$ may be obtained from a knowledge of the zero-state response due to a unit impulse $\delta(t)$ which is denoted by $h(t)$. By the time-invariance principle, the zero-state response to $\delta(t - \tau)$, where τ is a parameter, is $h(t - \tau)$. By linearity, we may scale the input by $x(\tau)\, d\tau$ and expect the response to be scaled by the same factor. Thus, an input $x(\tau)\delta(t - \tau)\, d\tau$ function of t will produce a zero-state response $x(\tau)h(t - \tau)\, d\tau$ which is a function of t. But the input function $x(t)$ may be resolved into a sum (or integral) of impulse functions as in Equation 8.20 or approximately as a sum of pulse functions as in Equation 8.15. Applying linearity once more, we then obtain the zero-state response by adding (integrating) the individual contributions $x(\tau)h(t - \tau)\, d\tau$ for various values of τ as in Equation 8.21. This is the essence of convolution.

EXAMPLE 8.2.1

The unit-impulse response of a linear time-invariant network is $h(t) = 1$ for $t \geq 0$ and $h(t) = 0$ for $t < 0$; that is, $h(t) = u(t)$. Let us compute the zero-state response of the network to an input pulse $x(t) = u(t) - u(t - 1)$ which is a pulse of unit height and unit duration. From the convolution integral in Equation 8.22, we have for $t \geq 0$

$$y(t) = \int_0^t x(\tau)h(t - \tau)\, d\tau = \int_0^t [u(\tau) - u(\tau - 1)]u(t - \tau)\, d\tau. \tag{8.29}$$

For $t < 1$, the integrand is simply equal to 1, so that the integral is equal to t.

For $t \geq 1$ the integrand is zero from $\tau = 1$ to $\tau = t$ and equal to 1 for $\tau < 1$. Hence, for $t \geq 1$,

$$y(t) = \int_0^1 1 \cdot d\tau + \int_1^t 0 \cdot d\tau = 1, \tag{8.30}$$

and the zero-state response due to the pulse is

$$y(t) = \begin{cases} t, & \text{for } 0 \leq t < 1, \\ 1, & \text{for } t \geq 1, \\ 0, & \text{for } t < 0. \end{cases} \tag{8.31}$$

AMPLE 8.2.2

The unit-impulse response of a linear time-invariant network is $e^{-t}u(t)$. To be determined is the zero-state response to an input

$$x(t) = \delta(t) + \delta(t - 1) + e^{-2t}u(t). \tag{8.32}$$

The zero-state response due to the individual terms in $x(t)$ may be determined first and then, by the principle of superposition, we may obtain the required zero-state response. The zero-state response due to an input $\delta(t)$ is $e^{-t}u(t)$ from the given data. By time-invariance, the zero-state response due to an input $\delta(t - 1)$ is $e^{-(t-1)}u(t - 1)$. The main problem that remains is the determination of the zero-state response due to $e^{-2t}u(t)$. From the convolution integral in Equation 8.22, we have

$$y(t) = \int_0^t e^{-2\tau}u(\tau)e^{-(t-\tau)}u(t - \tau)\, d\tau \tag{8.33}$$

for $t \geq 0$. Since $u(\tau)u(t - \tau) = 1$ for τ between 0 and t, we have

$$\begin{aligned} y(t) &= \int_0^t e^{-t}e^{-\tau}\, d\tau = e^{-t}\left[-e^{-\tau}\Big|_0^t\right] \\ &= e^{-t}(1 - e^{-t}) \qquad \text{for } t \geq 0. \end{aligned} \tag{8.34}$$

The zero-state response for $t < 0$ is, of course, zero so that we may multiply the response in Equation 8.34 by $u(t)$. The complete zero-state response due to the input $x(t)$ in Equation 8.32 is

$$y(t) = e^{-t}u(t) + e^{-(t-1)}u(t - 1) + e^{-t}(1 - e^{-t})u(t). \tag{8.35}$$

ERCISES

8.2.1 The unit-impulse response of a linear time-invariant network is $h(t) = e^{-t}$, $t \geq 0$, $h(t) = 0$, $t < 0$. Compute the zero-state response due to a unit-step input $x(t) = u(t)$ by the convolution of $h(t)$ and $x(t)$. Verify that the two forms of the convolution of $h(t)$ and $x(t)$ as given in Equation 8.28 yield the same answer.

8.2.2 Compute the convolution of $h(t)$ of Exercise 8.2.1 and $x(t) = u(t) - u(t - 1)$.

8.2.3 Repeat Exercise 8.2.2 for $x(t)$ equal to the pulse $v_{\text{in}}(t)$ in Figure 8.3.

8.2.4 From the convolution integral in Equation 8.22, verify that a linear network is zero-state homogeneous and zero-state additive. (Review definitions of Section 7.1.)

8.2.5 For the unit-impulse response of Exercise 8.2.1, compute the zero-state response for a sinusoidal input that is turned on at $t = 0$,

$$x(t) = (\sin t)u(t).$$

8.3 Graphical interpretation of convolution

The discussion of this section is centered on the convolution integral of Equation 8.22. From the given functions $x(t)$ and $h(t)$, which are assumed to be zero for $t < 0$, we form $x(\tau)$ and $h(t - \tau)$ as functions of τ for a specific value of t. Then we form the product $x(\tau)h(t - \tau)$. The area under the curve $x(\tau)h(t - \tau)$ plotted vs. τ from $\tau = 0$ or $\tau = t$ is the value of $y(t)$ for the chosen value of t. This may be carried out for every chosen value of t. Each of the steps we have mentioned above should be familiar to the reader. The only possible difficulty may be in the interpretation of $h(t - \tau)$ plotted vs. τ. From the conventional meaning of function notation, $h(t - \tau)$ is interpreted as the expression for $h(t)$, with t replaced by $t - \tau$. Let us investigate the graphical significance of this simple substitution of variables.

First recall that the plot of $h(\tau)$ vs. τ is exactly the same as the plot of $h(t)$ vs. t. For example, if $h(t)$ vs. t is as shown on Figure 8.6(a), then $h(\tau)$ vs. τ is as shown in Figure 8.6(b). Next form a function $g(\tau)$ which is obtained from $h(\tau)$ by replacing τ by $-\tau$ so that $g(\tau) = h(-\tau)$. The plot of $g(\tau)$ vs. τ would be the mirror image about the vertical axis of the plot of $h(\tau)$ vs. τ. This is illustrated by Figure 8.6(c). Finally, consider $g(\tau - t)$, where t is a positive constant. From Chapter 2 we recall that $g(\tau - t)$ vs. τ is a delayed version of $g(\tau)$ vs. τ, since t is a positive constant. This means that $g(\tau - t)$ is shifted to the right compared to $g(\tau)$. This is illustrated in Figure 8.6(d) for $t = 0.5$. Since $g(\tau) = h(-\tau)$, then $g(\tau - t) = h[-(\tau - t)] = h(t - \tau)$. Hence, given a plot of $h(t)$ vs. t, the plot of $h(t - \tau)$ vs. τ for a positive constant t is obtained by folding or reflecting the plot of $h(t)$ about the vertical axis, relabeling the abscissa by τ, and shifting to the right by an amount t, if t is positive.

From the given functions $x(t)$ and $h(t)$, convolution may be carried out graphi-

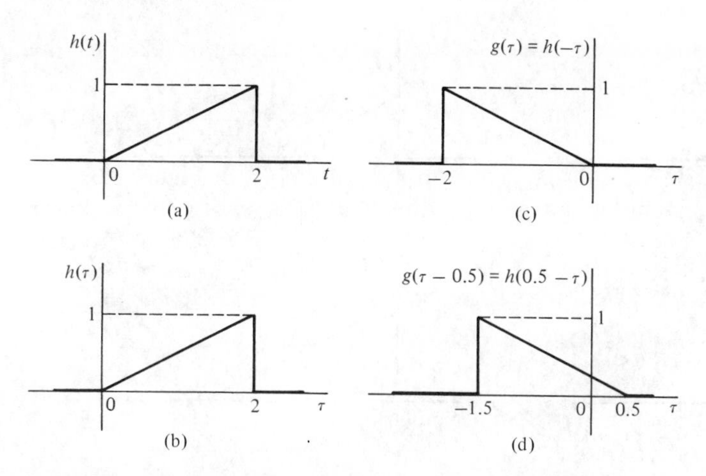

ure 8.6 Illustration of forming $h(t - \tau)$ vs. τ for $t = 0.5$ from $h(t)$ vs. t.

cally (or at least the mathematical operations may be visualized this way) for a specific value of t. Note that $x(\tau)h(t - \tau)$ vs. τ is the point-by-point product of $x(\tau)$ and $h(t - \tau)$ for various values of τ and a fixed value of t. Thus, if either $x(\tau)$ or $h(t - \tau)$ is zero for a value of τ, the product is zero for the same value of τ.

AMPLE 8.3.1

The convolution of $x(t)$ shown in Figure 8.7(a) and $h(t)$ shown in Figure 8.6(a) will be obtained. First note that for t negative, the product $x(\tau)h(t - \tau)$ is zero for all τ, so that the area under the curve is zero as expected. Next consider t to be some value between 0 and 1. Then, typically, $h(t - \tau)$ will be as shown in Figure 8.7(b). The product $x(\tau)h(t - \tau)$ is shown in Figure 8.7(c). The area under the curve between 0 and t is the value of the convolution integral for a specific value of t. This area is the cross-hatched portion of Figure 8.7(c). Since the slope of the triangle is $-\frac{1}{2}$, the ordinate of $x(\tau)h(t - \tau)$ for $\tau = 0$ is $\frac{1}{2}t$. Hence the area is $(\frac{1}{2})(\frac{1}{2})(t)t = \frac{1}{4}t^2$. This means that for t between 0 and 1, as the value of t increases, the value of the convolution integral increases as the square of t. Thus we have

$$y(t) = \int_0^t x(\tau)h(t - \tau)\, d\tau = \tfrac{1}{4}t^2, \qquad \text{for} \quad 0 \le t \le 1. \tag{8.36}$$

For t between 1 and 2, a typical plot of $h(t - \tau)$ is shown in Figure 8.7(d) and the product $x(\tau)h(t - \tau)$ is shown in Figure 8.7(e). The value of the integral is the area under the curve

$$y(t) = \tfrac{1}{2}[\tfrac{1}{2}t + (\tfrac{1}{2}t - \tfrac{1}{2})] \times 1 = \tfrac{1}{2}t - \tfrac{1}{4}, \qquad \text{for} \quad 1 \le t \le 2. \tag{8.37}$$

Thus for t between 1 and 2, $y(t)$ increases linearly with t. For $t > 2$, $h(t - \tau)$ is

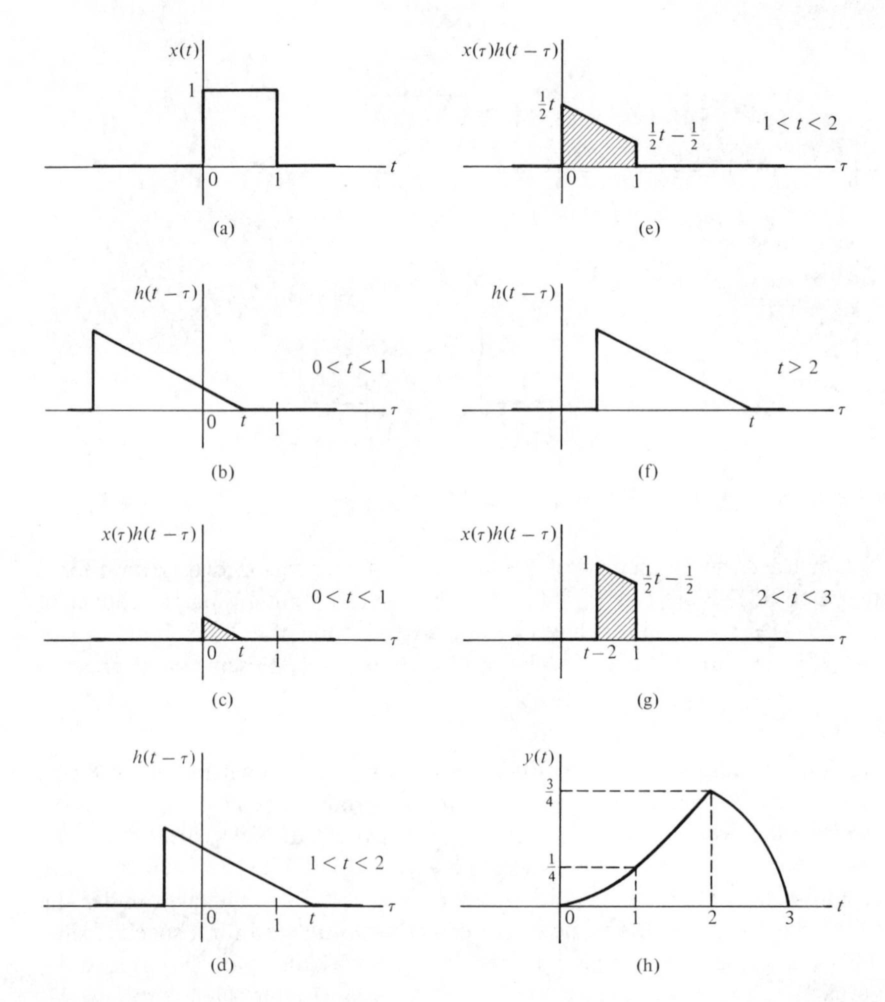

Figure 8.7 Illustration for graphical convolution.

shown in Figure 8.7(f). The product $x(\tau)h(t-\tau)$ for $2 < t < 3$ is shown in Figure 8.7(g). The area under the curve is

$$y(t) = \tfrac{1}{2}[1 + (\tfrac{1}{2}t - \tfrac{1}{2})] \times [1 - (t-2)] = \tfrac{1}{4}(t+1)(3-t), \qquad \text{for } 2 \le t \le 3, \tag{8.38}$$

which decreases parabolically with t until it reaches zero for $t = 3$. For $t > 3$, the product $x(\tau)h(t-\tau)$ is zero, so that the area under the curve is zero. The convolution $y(t)$ is plotted vs. t in Figure 8.7(h).

ERCISES

8.3.1 For $h(t)$ and $x(t)$ of Exercise 8.2.1, prepare a sequence of plots for the graphical determination of $y(1)$. Verify that the same value $y(1)$ is obtained regardless of whether $h(t)$ or $x(t)$ is folded and shifted.

8.3.2 Obtain the convolution of the pulse in Figure 8.4 and the rectangular pulse in Figure 8.7(a) with the aid of a sequence of appropriate plots.

8.3.3 Obtain the convolution of the pulse in Figure 8.7(a) with itself, and plot $y(t)$ vs. t.

8.3.4 Verify that for $t > 0$, $h(t - \tau)$ may be obtained from $h(t)$ by relabeling abscissa, shifting to the left by an amount t, and then folding with respect to the vertical axis. Give a series of equations to support your answer.

8.3.5 Repeat Exercise 8.3.3 for the pulse in Figure 8.3.

8.3.6 Prepare a sequence of sketches for the graphical determination of the convolution in Example 8.2.1.

4 Impulse response from differential equations

Suppose that in analyzing a linear time-invariant network the following differential equation is obtained:

$$a_n \frac{d^n y}{dt^n} + a_{n-1} \frac{d^{n-1}y}{dt^{n-1}} + \cdots + a_0 y = b_m \frac{d^m x}{dt^m} + b_{m-1} \frac{d^{m-1}x}{dt^{m-1}} + \cdots + b_0 x, \quad (8.39)$$

where y is the output or response-time function which may be voltage or current and $x(t)$ is the input or excitation-time function which is again either voltage or current. The zero-input response was treated in the previous chapter. In the previous section, it was demonstrated that the zero-state response can be obtained by convolution which involves the unit-impulse response $h(t)$. Then, by the decomposition property of linear systems, we can add the zero-state response to the zero-input response to obtain the complete response. The remaining question is how one obtains the unit-impulse response. This will be obtained by first considering the zero-state response for the modified equation

$$a_n \frac{d^n y}{dt^n} + a_{n-1} \frac{d^{n-1}y}{dt^{n-1}} + \cdots + a_0 y = x = u(t). \quad (8.40)$$

Denote the solution of Equation 8.40, for x equal to a unit-step input $u(t)$, by $y_0(t)$. Then the solution of Equation 8.40, with $u(t)$ replaced by a unit impulse $\delta(t)$, is

$$y_1(t) = \frac{dy_0(t)}{dt}.$$

This follows from the fact that $\delta(t)$ is the derivative of $u(t)$, and the fact that Equation 8.40 is linear time-invariant. By a repeated application of the results in Section 8.1, the solution of Equation 8.39 for x equal to a unit impulse $\delta(t)$ is

$$h(t) = b_m \frac{d^m}{dt^m}[y_1(t)] + b_{m-1}\frac{d^{m-1}}{dt^{m-1}}[y_1(t)] + \cdots + b_0 y_1(t). \qquad (8.41)$$

The crux of the problem then is the solution of Equation 8.40.

We proceed to obtain the zero-state response of a linear time-invariant network described by Equation 8.40. Now by zero-state response, we mean the output when initial values of all capacitor voltages and all inductor currents are set to zero. It can be shown that* for networks described by Equation 8.40 these conditions always lead to

$$y(0) = y'(0) = y''(0) = \cdots = y^{(n-1)}(0) = 0. \qquad (8.42)$$

We warn the reader that when derivative terms of the input are present, as in Equation 8.39, zero initial values of all capacitor voltages and all inductor currents do not necessarily imply that Equation 8.42 holds. For $t \geq 0$, we may replace $u(t)$ by 1. So we wish to solve

$$a_n \frac{d^n y_0}{dt^n} + \cdots + a_0 y_0 = 1, \qquad \text{for} \quad t \geq 0, \qquad (8.43)$$

with initial conditions

$$y_0(0) = 0, \qquad y_0'(0) = 0, \qquad \ldots, \qquad y_0^{(n-1)}(0) = 0. \qquad (8.44)$$

With $y_0(t)$ computed as described above, $y_1(t)$ is formed by

$$y_1(t) = \frac{dy_0(t)}{dt}. \qquad (8.45)$$

Finally, the unit-impulse response $h(t)$ for a network described by Equation 8.39 is obtained by substituting $y_1(t)$ in Equation 8.41.

* Equation 8.40 may be rewritten as a set of n differential equations

$$\frac{dx_i(t)}{dt} = x_{i+1}(t), \qquad \text{for } i = 1, 2, \ldots, n-1$$

and

$$\frac{dx_n(t)}{dt} = -a_0 x_1(t) - a_1 x_2(t) - \cdots - a_n x_n(t) + u(t),$$

where $x_1 = y$. The variables $x_1, x_2, \ldots, x_n$ are called state variables of the network (see Chapter 13 for further details on state variables). These state variables can always be written as linear combinations of capacitor voltages and inductor currents. Thus when these initial values of inductor currents and capacitor voltages are zero, $x_1(0) = 0$, $x_2(0) = 0, \ldots, x_n(0) = 0$, so that Equation 8.42 follows.

AMPLE 8.4.1

Consider the equation

$$\frac{d^2y}{dt^2} + 3\frac{dy}{dt} + 2y = 1. \tag{8.46}$$

Then $y = \frac{1}{2}$ satisfies Equation 8.46 and so does

$$y = \tfrac{1}{2} + c_1e^{-2t} + c_2e^{-t}. \tag{8.47}$$

We choose c_1 and c_2 so that $y(0) = 0$ and $y'(0) = 0$. Then we have

$$\left.\begin{aligned} y(0) &= \tfrac{1}{2} + c_1 + c_2 = 0 \\ y'(0) &= -2c_1 - c_2 = 0 \end{aligned}\right\}, \tag{8.48}$$

so that $c_1 = \frac{1}{2}$ and $c_2 = -1$. Hence

$$y(t) = \tfrac{1}{2} + \tfrac{1}{2}e^{-2t} - e^{-t} \tag{8.49}$$

satisfies Equation 8.47 and $y(0) = 0$ and $y'(0) = 0$.

Suppose that in Equation 8.43, $a_0 = 0$ but $a_1 \neq 0$. Then it is readily verified that $y = (1/a_1)t$ satisfies Equation 8.43 and so does $y = (1/a_1)t + y_c$. The constants in the y_c component are evaluated to satisfy the initial conditions as before.

AMPLE 8.4.2

For the equation

$$\frac{d^2y}{dt^2} + 3\frac{dy}{dt} = 1, \tag{8.50}$$

$y = \frac{1}{3}t$ is a solution. So is

$$y = \tfrac{1}{3}t + c_1 + c_2e^{-3t}. \tag{8.51}$$

Then c_1 and c_2 are chosen so that $y(0) = 0$, $y'(0) = 0$:

$$\left.\begin{aligned} y(0) &= c_1 + c_2 = 0 \\ y'(0) &= \tfrac{1}{3} - 3c_2 = 0 \end{aligned}\right\}. \tag{8.52}$$

Thus, $c_1 = -\frac{1}{9}$ and $c_2 = \frac{1}{9}$, and

$$y = \tfrac{1}{3}t - \tfrac{1}{9} + \tfrac{1}{9}e^{-3t} \tag{8.53}$$

satisfies Equation 8.43 as well as $y(0) = 0$ and $y'(0) = 0$.

In general, if the lowest-order derivative that is present in Equation 8.43 is the

kth, then the solution is of the form

$$y_0 = \frac{t^k}{k!\, a_k} + y_c, \tag{8.54}$$

and the constants in the complementary function y_c are evaluated so that the initial conditions $y_0(0), y_0'(0), \ldots, y_0^{(n-1)}(0)$ are all zero (see Exercise 8.4.5). Once the solution of Equation 8.43 (which satisfies the initial conditions) is obtained, then its derivative which is denoted by $y_1(t)$ is the solution of

$$a_n \frac{d^n y_1}{dt^n} + \cdots + a_0 y_1 = \delta(t), \qquad t \geq 0. \tag{8.55}$$

By the technique discussed in Section 8.1, the impulse response $h(t)$ for the network described by Equation 8.39 is obtained by means of Equation 8.41.

EXAMPLE 8.4.3

For the network in Figure 8.8 where R, L, and C are constants, v_1 is the input, and v_2 is the output, it is desired to find the unit-impulse response. Applying KVL around the loop, we have

$$\frac{1}{C}\int_0^t i(\tau)\, d\tau + Ri(t) + L\frac{di}{dt}(t) = v_1(t) \tag{8.56}$$

or

$$\frac{1}{C} i(t) + R\frac{di(t)}{dt} + L\frac{d^2 i(t)}{dt^2} = \frac{dv_1(t)}{dt}. \tag{8.57}$$

The output $v_2(t)$ is

$$v_2(t) = Ri(t) + L\frac{di(t)}{dt}. \tag{8.58}$$

We may eliminate i from Equations 8.57 and 8.58 to obtain a differential equation in v_2. Differentiate Equation 8.57 and multiply by L/R to obtain

$$\frac{L}{RC}\frac{di}{dt} + L\frac{d^2 i}{dt^2} + \frac{L^2}{R}\frac{d^3 i}{dt^3} = \frac{L}{R}\frac{d^2 v_1}{dt^2}. \tag{8.59}$$

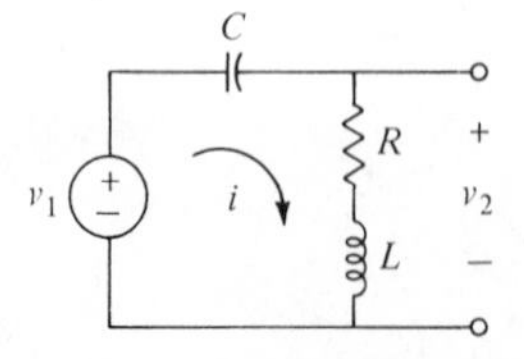

Figure 8.8 Network for Example 8.4.3.

Now add Equations 8.57 and 8.59. Thus we obtain

$$\frac{1}{RC}\left(Ri + L\frac{di}{dt}\right) + \frac{d}{dt}\left(Ri + L\frac{di}{dt}\right) + \frac{L}{R}\frac{d^2}{dt^2}\left(Ri + L\frac{di}{dt}\right) = \frac{L}{R}\frac{d^2v_1}{dt^2} + \frac{dv_1}{dt} \quad (8.60)$$

or

$$\frac{L}{R}\frac{d^2v_2}{dt^2} + \frac{dv_2}{dt} + \frac{1}{RC}v_2 = \frac{L}{R}\frac{d^2v_1}{dt^2} + \frac{dv_1}{dt}. \quad (8.61)$$

Now let us compute the impulse response for a given set of values for L/R and $1/(RC)$. Suppose $L/R = 1$ and $1/(RC) = 1$. Then the characteristic equation corresponding to the homogeneous differential equation is $m^2 + m + 1 = 0$, with roots $m_1 = -\frac{1}{2} + j\sqrt{3}/2$ and $m_2 = -\frac{1}{2} - j\sqrt{3}/2$. Then the solution of

$$\frac{d^2v_2}{dt^2} + \frac{dv_2}{dt} + v_2 = u(t), \quad (8.62)$$

$v_2(0) = 0$, and $v_1'(0) = 0$, is

$$\begin{aligned} v_2 &= 1 + c_1e^{(-1/2+j\sqrt{3}/2)t} + c_2e^{(-1/2-j\sqrt{3}/2)t}, \qquad t \geq 0, \\ v_2 &= 0, \qquad \text{for} \quad t < 0. \end{aligned} \quad (8.63)$$

Matching initial conditions, we have

$$v_2(0) = 0 = 1 + c_1 + c_2, \qquad v_2'(0) = 0 = \left(-\frac{1}{2} + j\frac{\sqrt{3}}{2}\right)c_1 + \left(-\frac{1}{2} - j\frac{\sqrt{3}}{2}\right)c_2.$$

So

$$c_1 = -\frac{1}{2} + j\frac{1}{2\sqrt{3}}, \qquad c_2 = -\frac{1}{2} - j\frac{1}{2\sqrt{3}}. \quad (8.64)$$

Hence, after some simplification, the solution for Equation 8.62 may be written as

$$v_2(t) = \left[1 + \frac{2e^{-(1/2)t}}{\sqrt{3}}\cos\left(\frac{\sqrt{3}}{2}t + \frac{5\pi}{6}\right)\right]u(t). \quad (8.65)$$

Note that $v_2(0+) = 0$. So the derivative of $v_2(t)$ in Equation 8.65 is

$$\begin{aligned} \frac{dv_2}{dt} &= \left[\frac{-1}{\sqrt{3}}e^{-(1/2)t}\cos\left(\frac{\sqrt{3}}{2}t + \frac{5\pi}{6}\right) - e^{-(1/2)t}\sin\left(\frac{\sqrt{3}}{2}t + \frac{5\pi}{6}\right)\right]u(t) \\ &= \left[\frac{2}{\sqrt{3}}e^{-(1/2)t}\sin\left(\frac{\sqrt{3}}{2}t\right)\right]u(t). \end{aligned} \quad (8.66)$$

Again, note that $dv_2/dt(0+) = 0$. The impulse response for the original network with $L/R = 1$, $1/(RC) = 1$ is obtained by applying the procedure of Equation 8.41. Thus we have

$$h(t) = \frac{d^2y_1}{dt^2} + \frac{dy_1}{dt}, \quad (8.67)$$

where

$$y_1(t) = \left(\frac{2}{\sqrt{3}} e^{-(1/2)t} \sin \frac{\sqrt{3}}{2} t\right) u(t). \tag{8.68}$$

Since $y_1(0+) = 0$, we have

$$\frac{dy_1}{dt} = \left(-\frac{1}{\sqrt{3}} e^{-(1/2)t} \sin \frac{\sqrt{3}}{2} t + e^{-(1/2)t} \cos \frac{\sqrt{3}}{2} t\right) u(t). \tag{8.69}$$

Since $y_1'(0+) = 1$ but $y_1'(0-) = 0$, $y_1'(t)$ has a jump discontinuity at $t = 0$ which is of height equal to one, and

$$\begin{aligned}\frac{d^2y_1}{dt^2} &= \delta(t) + \left(\frac{1}{2\sqrt{3}} e^{-(1/2)t} \sin \frac{\sqrt{3}}{2} t - \frac{1}{2} e^{-(1/2)t} \cos \frac{\sqrt{3}}{2} t\right. \\ &\qquad \left. - \frac{1}{2} e^{-(1/2)t} \cos \frac{\sqrt{3}}{2} t - \frac{\sqrt{3}}{2} e^{-(1/2)t} \sin \frac{\sqrt{3}}{2} t\right) u(t) \\ &= \left(-e^{-(1/2)t} \cos \frac{\sqrt{3}}{2} t - \frac{1}{\sqrt{3}} e^{-(1/2)t} \sin \frac{\sqrt{3}}{2} t\right) u(t) + \delta(t).\end{aligned} \tag{8.70}$$

Hence we obtain

$$h(t) = \delta(t) - \left(\frac{2}{\sqrt{3}} e^{-(1/2)t} \sin \frac{\sqrt{3}}{2} t\right) u(t). \tag{8.71}$$

The $\delta(t)$ component is due to a jump discontinuity of 1 at $t = 0$ for $dy_1(t)/dt$.

EXERCISES

8.4.1 For the linear time-invariant network in Figure 8.9, where v is the input, obtain the impulse response if v_{23} is the output.

8.4.2 Repeat Exercise 8.4.1 for v_{12} as the output.

8.4.3 A linear time-invariant system is represented by the differential equation

$$\frac{d^2y(t)}{dt^2} + 2\frac{dy(t)}{dt} + y(t) = \frac{dx(t)}{dt} + 3x(t),$$

where $y(t)$ is the output and $x(t)$ is the input. Determine the impulse response of the system.

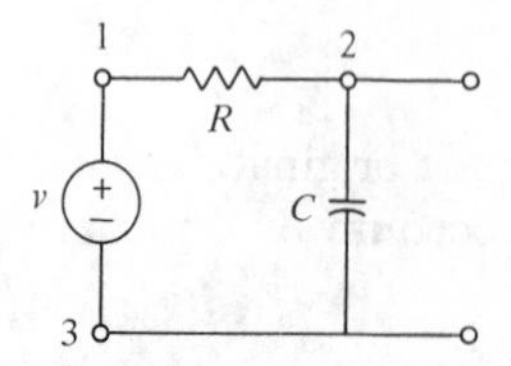

Figure 8.9

8.4.4 Repeat Exercise 8.4.3 if the differential equation is

$$\frac{d^3y(t)}{dt^3} + 4\frac{d^2y(t)}{dt^2} = \frac{dx(t)}{dt} + 2x(t).$$

8.4.5 Verify that the function y in Equation 8.54 satisfies the differential equation in Equation 8.43 if $a_k \neq 0$ and $a_{k-1} = a_{k-2} = \cdots = a_0 = 0$.

8.4.6 For the problem in Example 8.4.3, verify that if the values of c_1 and c_2 as obtained in Equation 8.64 are substituted in the expression for v in Equation 8.63, then the resulting expression for v may be simplified into that given in Equation 8.65.

5 Total solution by decomposition property

In this section, we will illustrate the calculation of the total response of a linear network by invoking the decomposition property

$$y_{\text{total}} = y_{zi}(t) + y_{zs}(t), \tag{8.72}$$

where $y_{zi}(t)$ is the zero-input response and $y_{zs}(t)$ is the zero-state response. In Chapter 7, we discussed a method for computing $y_{zi}(t)$ as the complementary solution $y_c(t)$, constrained to match the specified $y, y', y'', \ldots, y^{(n-1)}$ at the initial instant. A subsidiary network analysis may be necessary to determine these initial conditions $y(0), y'(0), y''(0), \ldots, y^{(n-1)}(0)$ from given initial values of capacitor voltages and inductor currents. Earlier in this chapter, we discussed one method for computing $y_{zs}(t)$ as the convolution of the input with the unit-impulse response. We also described a method for computing the unit-impulse response. Thus, we now have enough tools to compute the total response.

KAMPLE 8.5.1

The network of Figure 8.10 has an input $v_{\text{in}}(t)$ and an output $v_0(t)$. Let us compute the total response for $v_{\text{in}}(t) = u(t)$, $v_{C_1}(0) = 1$, and $v_{C_2}(0) = 2$ by the use of the decomposition property.

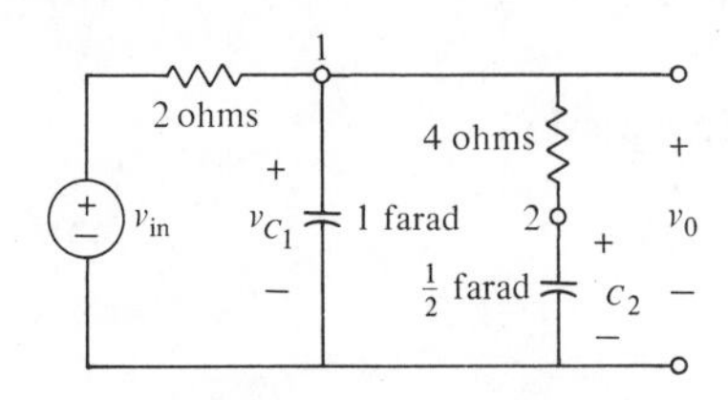

gure 8.10 Network for Example 8.5.1.

Applying KCL at node 1, we have

$$\frac{dv_{C_1}}{dt} = \frac{v_{\text{in}} - v_{C_1}}{2} + \frac{v_{C_2} - v_{C_1}}{4} \tag{8.73}$$

and applying KCL at node 2, we have

$$\frac{1}{2}\frac{dv_{C_2}}{dt} = \frac{v_{C_1} - v_{C_2}}{4}. \tag{8.74}$$

Solving for v_{C_2} from Equation 8.73, we obtain

$$v_{C_2} = 4\frac{dv_{C_1}}{dt} + 3v_{C_1} - 2v_{\text{in}}. \tag{8.75}$$

Differentiating this equation gives

$$\frac{dv_{C_2}}{dt} = 4\frac{d^2v_{C_1}}{dt^2} + 3\frac{dv_{C_1}}{dt} - 2\frac{dv_{\text{in}}}{dt}. \tag{8.76}$$

Substituting the last two results, Equations 8.75 and 8.76, into Equation 8.74, we have

$$2\frac{d^2v_{C_1}}{dt^2} + \frac{3}{2}\frac{dv_{C_1}}{dt} - \frac{dv_{\text{in}}}{dt} = \frac{v_{C_1}}{4} - \frac{1}{4}\left(4\frac{dv_{C_1}}{dt} + 3v_{C_1} - 2v_{\text{in}}\right). \tag{8.77}$$

Since $v_0 = v_{C_1}$, this equation simplifies to the form

$$\frac{d^2v_0}{dt^2} + \frac{5}{4}\frac{dv_0}{dt} + \tfrac{1}{4}v_0 = \frac{1}{2}\frac{dv_{\text{in}}}{dt} + \tfrac{1}{4}v_{\text{in}}. \tag{8.78}$$

Now this equation relates the output v_0 to the input v_{in} for the network under study, that of Figure 8.10, and we seek the solution of this equation.

To proceed with the zero-input response $y_{zi}(t)$, we obtain the complementary solution which is the solution of the homogeneous equation corresponding to Equation 8.78, $y_c(t)$. This equation is

$$\frac{d^2y_c}{dt^2} + \frac{5}{4}\frac{dy_c}{dt} + \tfrac{1}{4}y_c = 0. \tag{8.79}$$

For this equation, the corresponding characteristic equation is

$$s^2 + \tfrac{5}{4}s + \tfrac{1}{4} = (s + 1)(s + \tfrac{1}{4}) = 0. \tag{8.80}$$

Thus

$$y_c(t) = k_1e^{-t} + k_2e^{-t/4}. \tag{8.81}$$

To solve this equation by evaluating k_1 and k_2, we need the initial conditions $y_c(0)$ and $y_c'(0)$. From the network, we see that

$$y_c(0) = v_0(0) = v_{C_1}(0) = 1. \tag{8.82}$$

For the zero-input case, $v_{\text{in}} = 0$. Thus from Equation 8.73

$$y_c'(0) = v_{C_1}'(0) = \frac{-v_{C_1}(0)}{2} + \frac{v_{C_2}(0) - v_{C_1}(0)}{4} \tag{8.83}$$

$$= -\tfrac{1}{2} + \tfrac{1}{4} = -\tfrac{1}{4}.$$

Imposing the initial conditions of the last two equations on $y_c(t)$ of Equation 8.81 permits the evaluation of k_1 and k_2:

$$\left.\begin{aligned} 1 &= k_1 + k_2 \\ -\tfrac{1}{4} &= -k_1 - \frac{k_2}{4} \end{aligned}\right\}. \tag{8.84}$$

Solving these equations yields $k_1 = 0$ and $k_2 = 1$. Thus

$$y_{zi}(t) = e^{-t/4}. \tag{8.85}$$

Next, we investigate the unit-impulse response. Following the method of Section 8.4, we consider the equation

$$\frac{d^2y_0}{dt^2} + \frac{5}{4}\frac{dy_0}{dt} + \tfrac{1}{4}y = 1 \tag{8.86}$$

with

$$y_0(0) = y_0'(0) = 0. \tag{8.87}$$

The solution to Equation 8.86 is

$$y_0(t) = 4 + k_1e^{-t} + k_2e^{-t/4}. \tag{8.88}$$

Substituting the initial conditions of Equation 8.87 into this result gives us

$$0 = 4 + k_1 + k_2 \quad \text{and} \quad 0 = -k_1 - \frac{k_2}{4}. \tag{8.89}$$

From these two equations, we find that $k_1 = \frac{4}{3}$ and $k_2 = -\frac{16}{3}$. Thus the solution is

$$y_0(t) = 4 + \tfrac{4}{3}e^{-t} - \tfrac{16}{3}e^{-t/4}. \tag{8.90}$$

From this result, we determine $y_1(t)$:

$$y_1(t) = \frac{dy_0}{dt} = -\tfrac{4}{3}e^{-t} + \tfrac{4}{3}e^{-t/4}, \tag{8.91}$$

and applying Equation 8.41, we find

$$h(t) = \frac{1}{2}\frac{dy_1}{dt} + \tfrac{1}{4}y_1 = \tfrac{1}{3}e^{-t} + \tfrac{1}{6}e^{-t/4}. \tag{8.92}$$

Hence, since $v_{\text{in}}(t) = u(t)$,

$$\begin{aligned} y_{zs}(t) &= \int_0^t h(t-\tau)x(\tau)\,d\tau = \int_0^t \left(\tfrac{1}{3}e^{-(t-\tau)} + \tfrac{1}{6}e^{-(t-\tau)/4}\right) d\tau \\ &= 1 - \tfrac{1}{3}e^{-t} - \tfrac{2}{3}e^{-t/4}. \end{aligned} \tag{8.93}$$

Then the total response is

$$v_0(t) = y_{zi}(t) + y_{zs}(t) = 1 - \tfrac{1}{3}e^{-t} + \tfrac{1}{3}e^{-t/4}. \tag{8.94}$$

Observe that for this example the linearity property discussed in Section 8.1 allows us to obtain

$$\begin{aligned} y_{zs}(t) &= \frac{1}{2}\frac{dy_0}{dt} + \tfrac{1}{4}y_0 = -\tfrac{2}{3}e^{-t} + \tfrac{2}{3}e^{-t/4} + 1 + \tfrac{1}{3}e^{-t} - \tfrac{4}{3}e^{-t/4} \\ &= 1 - \tfrac{1}{3}e^{-t} - \tfrac{2}{3}e^{-t/4}, \end{aligned} \tag{8.95}$$

which is equal to that of Equation 8.93.

The method of expressing the total solution by means of the decomposition property displays the separate contributions of the input and the initial conditions on the capacitor voltages and inductor currents. Thus the solution may be investigated for various inputs without disturbing the contribution of the initial conditions on the capacitor voltages and currents in the inductors. Furthermore, various initial conditions may be investigated without disturbing the contribution of the input. This fact will prove to be of great advantage in the analysis of complicated systems, although its advantage in relatively simple networks is less evident.

EXERCISES

8.5.1 For the network in Figure 8.10, suppose $v_{C_1}(0) = 1$, $v_{C_2}(0) = 1$, and $v_{\text{in}}(t) = u(t)$. Find the total response $v_0(t)$ using the decomposition property.

8.5.2 Repeat Exercise 8.5.1 except that $v_{\text{in}}(t)$ is a rectangular pulse

$$v_{\text{in}}(t) = u(t) - u(t-1).$$

8.5.3 Suppose that for the network of Figure 8.10, v_{C_2} is the output. Derive the differential equation relating $v_{C_2}(t)$ to $v_{\text{in}}(t)$.

8.6 Total solution using the particular integral

When the input signal has a complicated waveform or when it is given in graphical or tabular form, the convolution integral (and its approximation, the convolution sum) are computationally attractive. However, if the network is to be analyzed only for one input and if the input waveform is simple (constant, exponential, or sinusoidal), the use of the particular integral may be more direct.

Suppose that a linear time-invariant network or system whose input is $x(t)$ and output is $y(t)$ is described by the equation

$$\frac{d^n y}{dt^n} + a_{n-1}\frac{d^{n-1}y}{dt^{n-1}} + \cdots + a_0 y = b_m \frac{d^m x}{dt^m} + \cdots + b_0 x. \qquad (8.96)$$

This is the same as Equation 7.19 except that we have taken $a_n = 1$ for simplicity (if it is not unity, divide both sides of the equation by a_n and redefine the coefficients). For a specified $x(t)$, a function $y_p(t)$ is called the *particular integral* of Equation 8.96 if it satisfies the equation.

EXAMPLE 8.6.1

For

$$\frac{d^2 y}{dt^2} + 3\frac{dy}{dt} + 2y = 2\frac{dx}{dt} + 5x, \qquad (8.97)$$

we consider the case when $x = 4$. For this case,

$$y_p = 10 \qquad (8.98)$$

satisfies Equation 8.97 so that it is a particular integral.

In general, whenever x is a constant K and $a_0 \neq 0$, then a particular integral of Equation 8.96 is

$$y_p(t) = \frac{b_0 K}{a_0}. \qquad (8.99)$$

This is readily verified by substitution into the general equation, Equation 8.96, since all derivatives are zero. If $a_0 = 0$ but $a_1 \neq 0$, then a particular integral for $x = K$ is

$$y_p(t) = \frac{b_0 K}{a_1} t. \qquad (8.100)$$

Similar answers are simply obtained when $a_1 = a_0 = 0$ but $a_2 \neq 0$ (see Exercise 8.6.1).

Another basic signal is

$$x(t) = e^{st}. \tag{8.101}$$

A particular integral for this input is

$$y_p(t) = He^{st}, \tag{8.102}$$

where H is an appropriate constant to be determined from the coefficients in Equation 8.96.

EXAMPLE 8.6.2

Let $x(t) = e^{2t}$ be the input to a network described by Equation 8.97. A suggested particular integral is

$$y_p(t) = He^{2t}. \tag{8.103}$$

Substituting this y_p into Equation 8.97, we obtain

$$[2^2 + 3(2) + 2]He^{2t} = (4 + 5)e^{2t}, \tag{8.104}$$

which requires that $H = \frac{9}{12}$. Hence we have determined the particular integral

$$y_p(t) = \tfrac{9}{12}e^{2t}. \tag{8.105}$$

Furthermore, if the input is

$$x(t) = e^{s_1 t}, \tag{8.106}$$

where s_1 is an arbitrary but fixed constant, then the particular integral is

$$y_p(t) = He^{s_1 t}, \tag{8.107}$$

where H is to be determined from the differential equation. For this particular example, routine substitution gives

$$H = \frac{2s_1 + 5}{s_1^2 + 3s_1 + 2}. \tag{8.108}$$

In general, the input of Equation 8.106 will yield for the general equation, Equation 8.96,

$$H = \frac{b_m s_1^m + b_{m-1}s_1^{m-1} + \cdots + b_0}{s_1^n + a_{n-1}s^{n-1} + \cdots + a_0}, \tag{8.109}$$

providing, of course, that the denominator of this equation is not zero. That is, provided that s_1 is not a root of the characteristic equation, the above procedure holds.

The third class of inputs we consider is the sinusoid,

$$x(t) = \cos \omega_0 t, \tag{8.110}$$

where ω_0 is frequency in radians/sec. The particular integral corresponding to this input is

$$y_p(t) = H \cos (\omega_0 t + \theta), \tag{8.111}$$

where H and θ are constants to be determined.

AMPLE 8.6.3

Applying the input of Equation 8.110 and the trial solution in Equation 8.111 to the system described by the equation of Example 8.6.1, Equation 8.97, we obtain the equation

$$\begin{aligned} H[-\omega_0^2 \cos (\omega_0 t + \theta) - 3\omega_0 \sin (\omega_0 t + \theta) + 2 \cos (\omega_0 t + \theta)] \\ = 5 \cos \omega_0 t - 2\omega_0 \sin \omega_0 t. \end{aligned} \tag{8.112}$$

To simplify this equation, we first expand the cosine and sine terms, and then collect terms and equate corresponding coefficients of $\cos \omega_0 t$ and $\sin \omega_0 t$. The result of these algebraic operations is two equations:

$$H(-\omega_0^2 \cos \theta - 3\omega_0 \sin \theta + 2 \cos \theta) = 5$$

and (8.113)

$$H(\omega_0^2 \sin \theta - 3\omega_0 \cos \theta - \sin \theta) = -2\omega_0.$$

The two unknowns in these two equations are H and θ. They may be found by dividing the two equations, followed by algebraic reduction, and adding with simplification. The result is

$$\tan \theta = \frac{-\omega_0(2\omega_0^2 + 11)}{\omega_0^2 + 10} \tag{8.114}$$

and

$$H = \frac{5}{(2 - \omega_0^2) \cos \theta - 3\omega_0 \sin \theta}. \tag{8.115}$$

It is assumed that the denominator of this equation does not vanish which will be the case provided $j\omega_0$ is not a root of the characteristic equation. The calculation of H and θ is drastically simplified with the aid of phasors described in the next chapter.

To find the total response, the $y_p(t)$ is added to the complementary solution $y_c(t)$ and the constants in $y_c(t)$ are determined by forcing $y_p(t) + y_c(t)$ to satisfy the specified initial conditions. A preliminary network analysis may be needed to obtain these initial conditions from the specified capacitor voltages and inductor currents, and the value of the input at the initial instant.

EXAMPLE 8.6.4

Consider the network in Figure 8.8 which was discussed in Example 8.5.1 and is described by Equation 8.78. By inspection of this equation, the particular integral is $y_p(t) = 1$, for $x(t) = u(t)$, and so the total solution is of the form

$$y(t) = 1 + k_1e^{-t} + k_2e^{-t/4}. \tag{8.116}$$

It is now necessary to compute $y(0)$ and $y'(0)$ from the given $v_{C_1}(0) = 1$ and $v_{C_2}(0) = 2$ and $v_{\text{in}}(0) = 1$ of Example 8.5.1. Examination of the network reveals that

$$y(0) = v_{C_1}(0) = 1 \tag{8.117}$$

and

$$y'(0) = v'_{C_1}(0) = \frac{1-1}{2} + \frac{2-1}{4} = \frac{1}{4}. \tag{8.118}$$

Imposing these conditions on $y(t)$ leads to

$$\left.\begin{aligned} 1 &= 1 + k_1 + k_2 \\ \tfrac{1}{4} &= -k_1 - \frac{k_2}{4} \end{aligned}\right\} \tag{8.119}$$

or $k_1 = -\frac{1}{3}$ and $k_2 = \frac{1}{3}$. Thus

$$y_{\text{total}}(t) = 1 - \tfrac{1}{3}e^{-t} + \tfrac{1}{3}e^{-t/4}, \tag{8.120}$$

which agrees with Equation 8.94.

The particular integral may be used for computing the zero-state response. Compared to the discussion above, the only modification needed is that the initial values on all capacitor voltages and inductor currents must be set to zero. This makes it necessary to recompute $y(0)$, $y'(0)$, etc.

EXAMPLE 8.6.5

For the network of Figure 8.10, let us compute the zero-state response using the particular-integral approach. The particular integral is the same as in Example 8.6.4. The difference is in the calculation of the initial conditions corresponding to $v_{C_1}(0) = 0$ and $v_{C_2}(0) = 0$, but with $v_{\text{in}}(0) = 1$. The new values are $y(0) =$

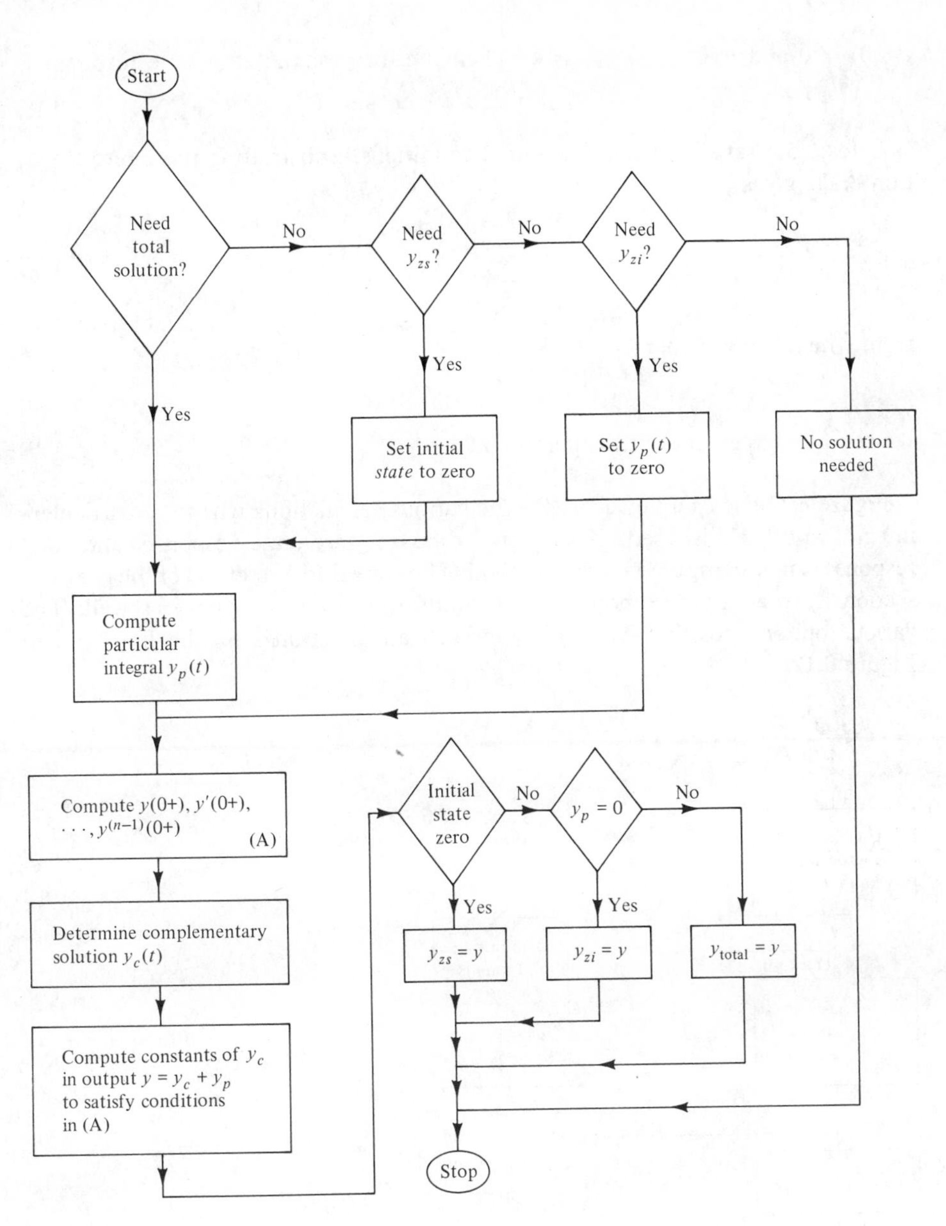

gure 8.11 Flowchart illustrating the steps and decisions to be made in solving a differential equation using the classical approach.

$v_{C_1}(0) = 0$ and $y'(0) = v'_{C_1}(0) = \frac{1}{2}$. Then the form of the zero-state response is

$$y_{zs}(t) = 1 + k_1 e^{-t} + k_2 e^{-t/4} \tag{8.121}$$

as before, but satisfying the new initial conditions. Substituting these into Equation 8.121 gives

$$\left.\begin{aligned} 0 &= 1 + k_1 + k_2 \\ \tfrac{1}{2} &= -k_1 - \frac{k_2}{4} \end{aligned}\right\} \tag{8.122}$$

from which we find that $k_1 = -\frac{1}{3}$ and $k_2 = -\frac{2}{3}$, so that

$$y_{zs}(t) = 1 - \tfrac{1}{3}e^{-t} - \tfrac{2}{3}e^{-t/4}, \tag{8.123}$$

which is in agreement with Equation 8.93.

Figure 8.11 shows a flowchart for the various calculations when the particular-integral approach is used. Zero-input response, zero-state response, and total response can be computed by this method. The convolution method of the previous section is an alternative method for computing the total response as well. The various options possible with this approach are illustrated by the flowchart in Figure 8.12.

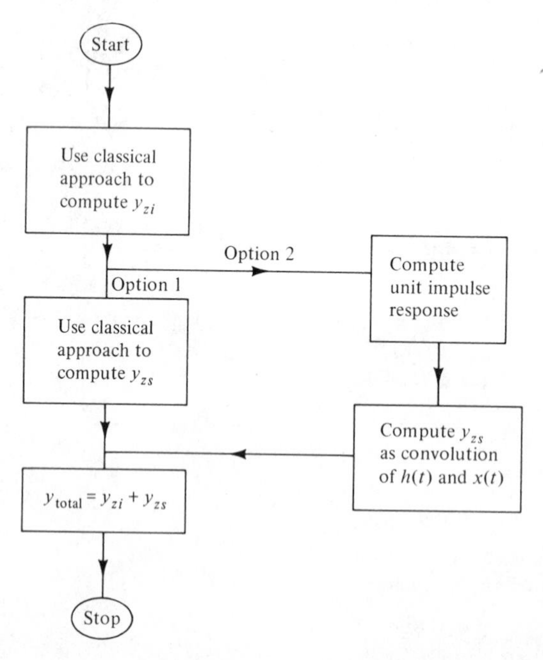

Figure 8.12 Flowchart illustrating options for computing total response.

ERCISES

8.6.1 Suppose that in Equation 8.96, $a_0 = a_1 = 0$ but $a_2 \neq 0$. Determine a particular-integral solution for $x = K$ (a constant).
8.6.2 Repeat Exercise 8.6.1 for $a_0 = a_1 = \cdots = a_k = 0$ but $a_{k+1} \neq 0$.
8.6.3 Determine the particular-integral solution for $x = e^{j\omega t}$ in Equation 8.96.
8.6.4 Suppose that the network of Figure 8.10 is such that $v_{C_2}(t)$ is the output. For $v_{\text{in}}(t) = u(t)$, $v_{C_1}(0) = 1$, and $v_{C_2}(0) = 2$, determine the total response by the particular-integral approach.

7 Summary

1. If a linear time-invariant network has a zero-state response $y_1(t)$ due to an input $x_1(t)$, then its zero-state response to

$$x_2(t) = \frac{dx_1(t)}{dt}$$

is

$$y_2(t) = \frac{dy_1(t)}{dt}$$

and its zero-state response to

$$x_3(t) = \int_0^t x_1(\tau)\, d\tau$$

is

$$y_3(t) = \int_0^t y_1(\tau)\, d\tau.$$

2. The unit-impulse response $h(t)$ of a linear time-invariant network is defined as the zero-state response of the network to a unit-impulse input $\delta(t)$. For physical networks, $h(t) = 0$ for $t < 0$. In terms of $h(t)$, the zero-state response due to any other input $x(t)$ is

$$y(t) = \int_0^t h(t - \tau)x(\tau)\, d\tau.$$

By a simple change of variables, it is shown that

$$y(t) = \int_0^t x(t - \tau)h(\tau)\, d\tau.$$

These integrals are known as *convolution integrals.* The zero-state response $y(t)$ is said to be the convolution of the unit-impulse response $h(t)$ and the input $x(t)$.

3. The convolution of two functions $f_1(t)$ and $f_2(t)$ can be interpreted graphically as follows: Plot $f_2(\tau)$ and $f_1(t-\tau)$. This latter function is interpreted as turning $f_1(\tau)$ around the vertical axis and translating the result to the right by t units. Determine the area under the product curve $f_1(t-\tau)f_2(\tau)$ from 0 to t. Repeat for several values of t.
4. To find $h(t)$ for an nth-order linear time-invariant network equation: (a) set the left-hand side of the differential equation equal to 1 for $t \geq 0$ with zero initial conditions and find y_0, (b) set y_0 to zero for $t < 0$, (c) find $y_1 = dy_0/dt$, and (d) determine $h(t)$ by substituting y_1 into the right-hand side of the differential equation (see Equations 8.39, 8.41, 8.43, 8.44, and 8.45).
5. Denoting the zero-input response of a linear time-invariant network by $y_{zi}(t)$ and the zero-state response by $y_{zs}(t)$, the total or complete response is

$$y_{\text{total}}(t) = y_{zi}(t) + y_{zs}(t)$$

from the decomposition property for linear networks.
6. The zero-input response $y_{zi}(t)$ may be obtained by setting it equal to the complementary function $y_c(t)$ and evaluating the constants in y_c so that the initial conditions are satisfied. A preliminary network analysis may be needed to compute the initial conditions on the output from the capacitor voltages and inductor currents. See Chapter 7.
7. Two methods are given for finding the zero-state response. One is the convolution integral of 2. The other involves the determination of the particular integral. Specific operations are shown in the flowcharts of Figures 8.11 and 8.12. The particular-integral method is simple for constant, exponential, or sinusoidal inputs.
8. When the decomposition property is invoked in determining the total response of 5, the separate contributions of the input and the initial conditions of the capacitor voltages and inductor currents are prominently displayed. This is useful in studying a class of inputs or a class of initial conditions.

PROBLEMS

8.1 The signal $v_{\text{in}}(t)$ is applied to the RC network of Figure 8.1(a). The time constant of the network is 1 millisec. Using superposition, determine the

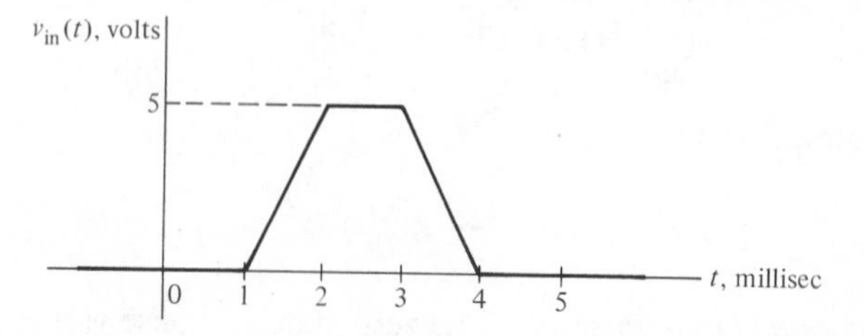

Figure 8.13

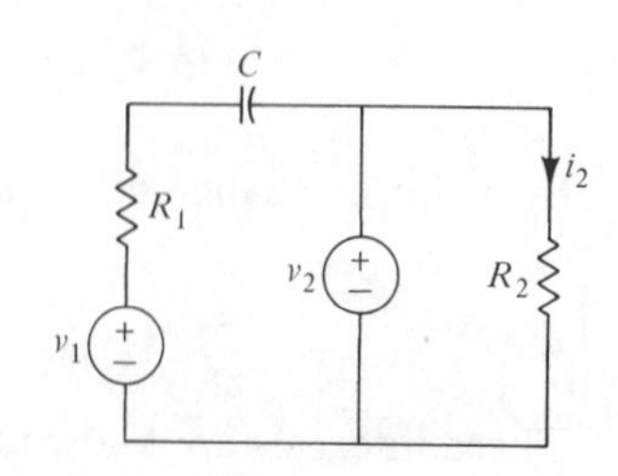

Figure 8.14

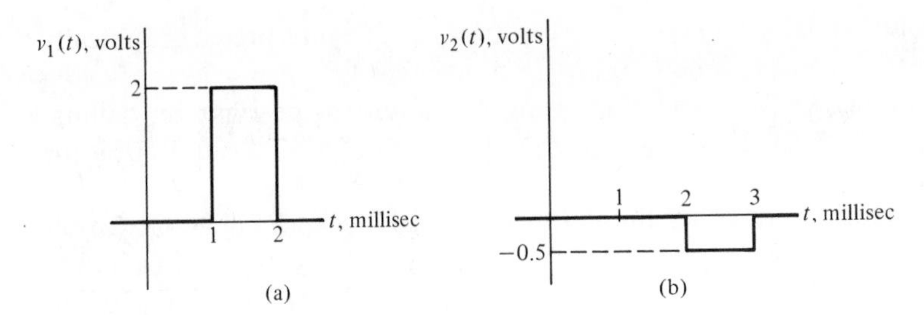

Figure 8.15

output $v_0(t)$. [*Hint:* Express $v_{\text{in}}(t)$ as a combination of step and ramp functions.] The capacitor is initially uncharged. See Figure 8.13 for $v_{\text{in}}(t)$.

8.2 For the network in Figure 8.14, if $R_1 = 100$ ohms, $R_2 = 300$ ohms, and $C = 10$ microfarads, determine $i_2(t)$ for $v_1(t) = v_2(t) = u(t)$, a unit-step voltage. Assume that $v_c(0) = 0$.

8.3 Repeat Problem 8.2 for $v_1(t)$ and $v_2(t)$ as shown in Figure 8.15.

8.4 What is the power delivered by each of the batteries in Figure 8.16? How does the sum of these powers compare with the sum of the powers delivered by each battery when the other one is turned off? Explain.

8.5 Suppose that in the network of Figure 8.14, v_2 is a controlled source $v_2 = Ri_2$, where R is a constant. Assuming $v_1(t) = e^{st}$, determine the current through R_1. Assume that s is not a characteristic root.

8.6 In Figure 8.17, suppose v_1 is a controlled source $v_1 = Av_2$, where A is a constant. Assuming $i_2(t) = e^{st}$, find the voltage across the current source that satisfies the differential equations of the network.

8.7 In the network of Figure 8.14, suppose $v_1(t) = \sin t$ and $i_2(t) = 2 \sin t$. Determine $v_2(t)$ with numerical values for A and θ in $v_2(t) = A \sin(t + \theta)$, which satisfies the network differential equations.

8.8 Using superposition, find the voltage v_2 across the 1-ohm resistor in the network given in Figure 8.18.

8.9 In the network of Figure 8.19, it is given that $v_1(t) = 10 \sin t$ and $i_2(t) = \sin(t + \pi/2)$. We wish to connect a sinusoidal current source across ab such

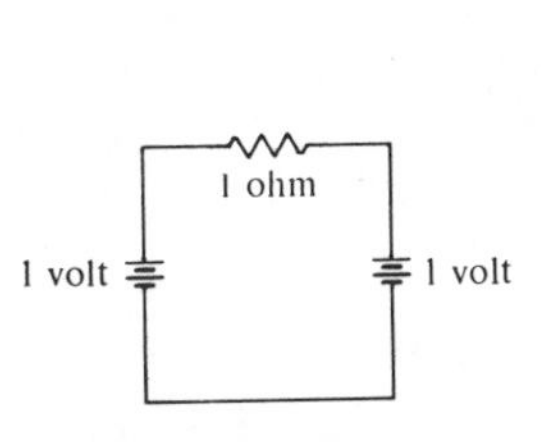

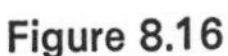

Figure 8.16

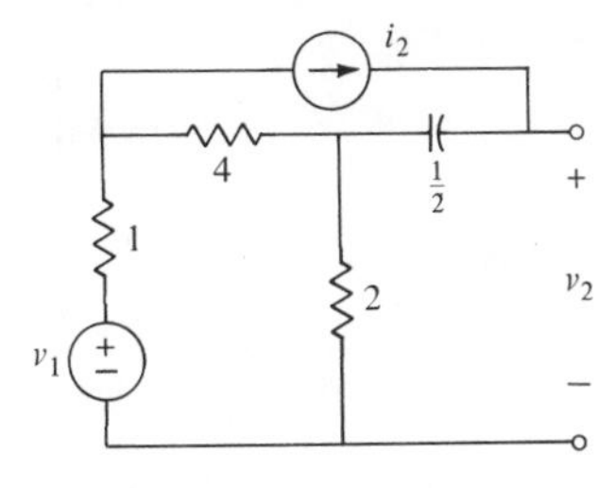

Figure 8.17

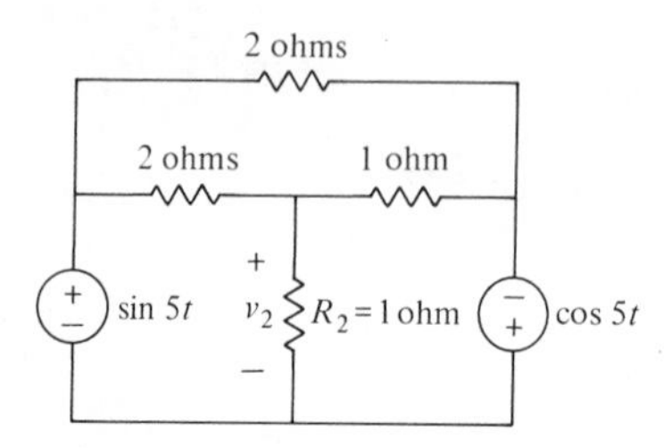

Figure 8.18

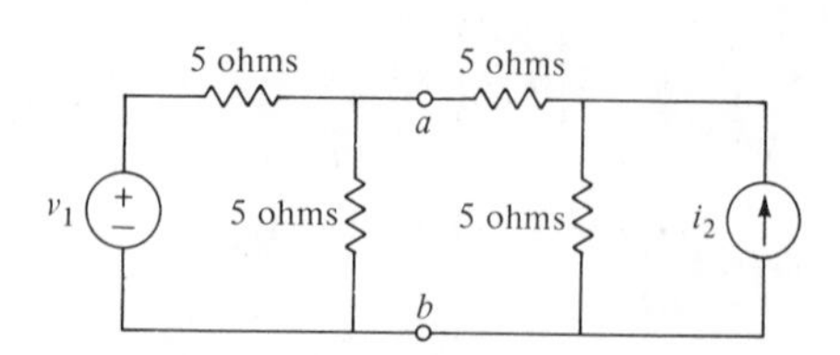

Figure 8.19

that the voltage there is zero. If this is possible, find the equation of the current source and indicate the reference direction you assumed. If it is not possible, explain why not.

8.10 In Figure 8.19, suppose that v_1 and i_2 are related by $v_1 = Ri_2$, where R is a real constant. For what constant R is $v_{ab} \equiv 0$? The voltage v_1 is arbitrary but not equal to zero.

8.11 In the network of Figure 8.20, $R_1 = R_2 = 1000$ ohms, $C = 100$ microfarads, $v(t)$ is a step function $100u(t)$, and the initial voltage V_0 on the capacitor is 50 volts. Determine the voltage across the capacitor as a function of time.

8.12 Repeat Problem 8.11 with $v(t)$ equal to a pulse as shown in Figure 8.21.

8.13 For a unit-step input of current to the network of Figure 8.22(a), the response voltage is $v(t) = (1 - e^{-t})u(t)$. Find the response voltage for the same network but with the triangular pulse input shown in Figure 8.22(b).

8.14 The network shown in Figure 8.23 is excited by a voltage source having a waveform described by the equation $v_s(t) = V_1[\sin(\omega t + \phi)]u(t)$. The current in the network, $i(t)$, contains a term of the general form $i_t(t) = K_1 e^{-\alpha t}$ which will be identified as the transient current term. Assume that C is initially uncharged and that there is no initial current in L. (a) With the switch in position a, determine the value of ϕ that will cause the transient current term to vanish. (b) Repeat part (a) but with the switch in position b.

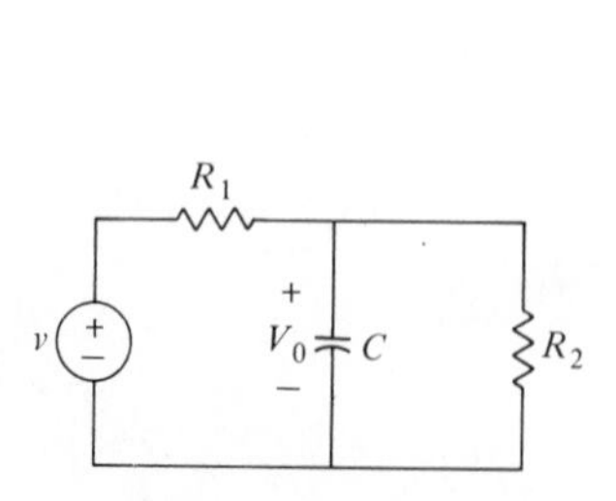

Figure 8.20

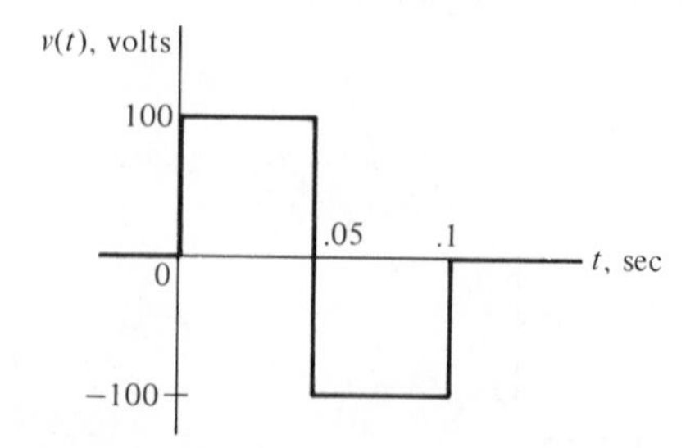

Figure 8.21

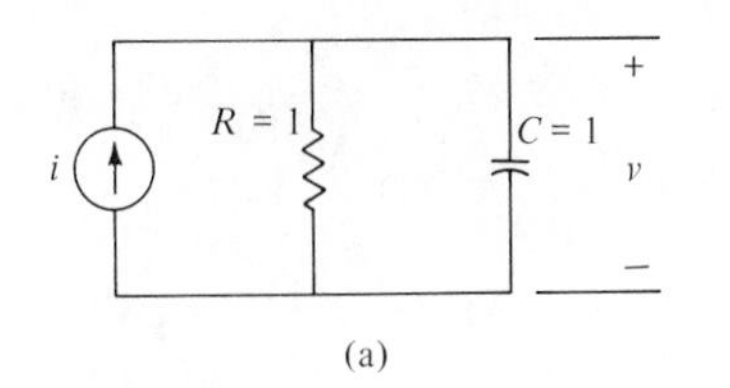

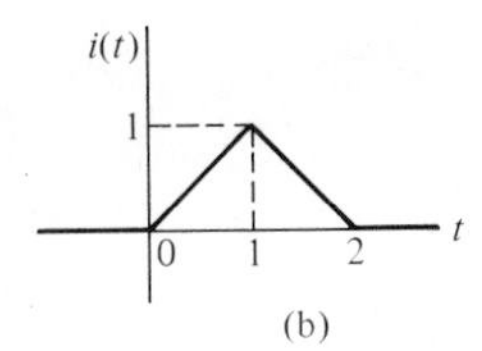

Figure 8.22

8.15 The network in Figure 8.24 contains a switch K which is moved from a to b instantaneously at $t = 0$. Find $v_2(t)$ that results using numerical values where possible. Assume that the switch was in position a for a long time and that the initial current in the 2-henry inductor is zero.

8.16 The voltage source in the network of Figure 8.25 is $v_s(t) = e^{-\alpha t}u(t)$. Find $v_2(t)$ if $v_2(0) = 0$.

8.17 Repeat Problem 8.16 if $v_s(t)$ is as given, but $v_2(0) = 1$ volt.

8.18 For the network of Figure 8.17, (a) determine the impulse response v_2 due to v_1 and (b) determine the impulse response v_2 due to i_2.

8.19 (a) Determine the impulse response of the network in Figure 8.22. (b) Determine the zero-state response due to the triangular pulse, by convolution.

8.20 Determine and sketch the convolution of the two pulses in Figure 8.15.

8.21 Determine the impulse response of the network in Figure 8.25.

8.22 (a) Determine the impulse response of the network in Figure 8.22. (b) From (a), determine the zero-state response due to $i(t) = e^{-2t}u(t)$, by convolution. (c) Verify the answer in (b) by finding a function $v = Ae^{-2t}$ which satisfies the network differential equation for $i = e^{-2t}$ and then adding the complementary function. The constants must be determined to satisfy initial conditions.

★ 8.23 In Equation 8.31, suppose $x = e^{s_1 t}$, where s_1 is a simple (not multiple) characteristic root. Let

$$Q(s) = \sum_{i=0}^{n} a_i s^i \quad \text{and} \quad P(s) = \sum_{i=0}^{m} b_i s^i .$$

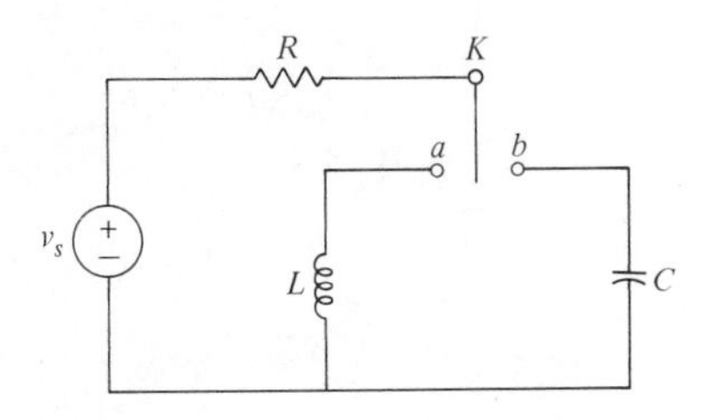

Figure 8.23

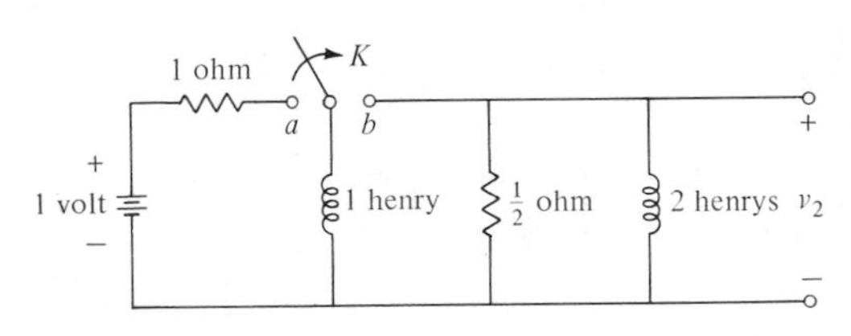

Figure 8.24

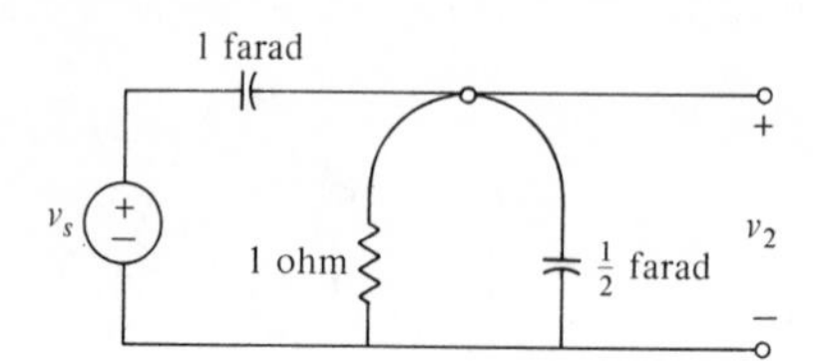

Figure 8.25

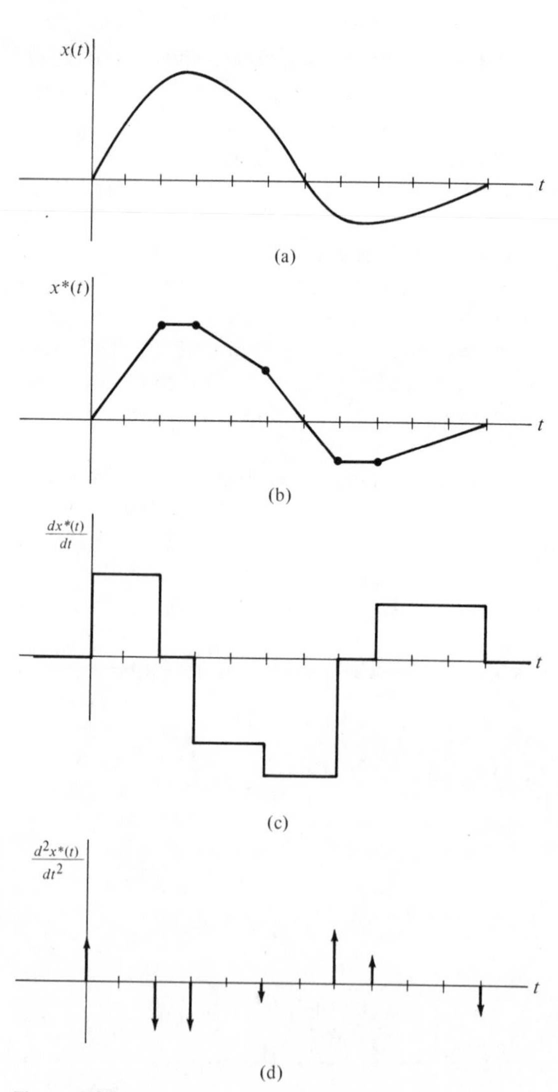

Figure 8.26

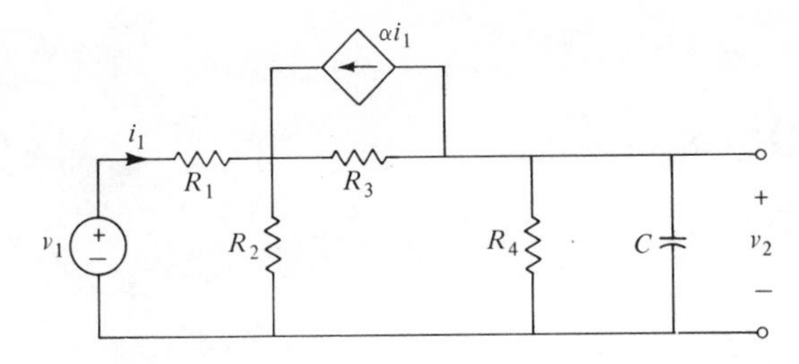

Figure 8.27

Show that

$$y = \frac{P(s_1)}{Q'(s_1)} te^{s_1 t}$$

satisfies Equation 8.31 where $Q'(s)$ is the derivative of $Q(s)$, for $n = 2$, $m = 2$.

★ 8.24 Repeat Problem 8.23 for a general n and a general m.

★ 8.25 The following approximation for convolution was suggested by E. A. Guillemin:* To find the convolution of $x(t)$ and $h(t)$, approximate $x(t)$ by $x^*(t)$ which consists of straight-line segments joined together (piecewise linear approximation). The derivative of $x^*(t)$ consists of a staircase curve, and the second derivative of $x^*(t)$ is a train of impulses of various delays and various areas. The response to $x''^*(t)$ is simply obtained. Then by linear time-invariance, the response due to $x^*(t)$ is obtained by a double integration of the response to $x''^*(t)$. Suppose we choose n points on the curve of $x(t)$, $t_1, t_2, \ldots, t_n$ and suppose $x^*(t)$ is obtained by joining the points $x(t_1), x(t_2), \ldots, x(t_n)$ by straight lines. Find a formula for the response to $x^*(t)$ if the impulse response is denoted by $h(t)$. See Figure 8.26.

8.26 For the network in Figure 8.14, use the classical approach with particular integrals, to determine the voltage across C, $v_C(t)$. Assume a plus reference mark toward v_1, $v_C(0) = 1$, $v_1(t) = u(t)$, and $v_2(t) = 3u(t)$.

8.27 For the network in Figure 8.27, the input $v_1(t)$ is a unit-step function $u(t)$. With $v_2(t)$ as output, and $v_2(0) = 2$, compute the response $v_2(t)$ by the particular-integral method. The elements have normalized values $R_1 = R_2 = R_3 = R_4 = 1$, $C = 1$, and $\alpha = 0.9$.

8.28 Determine the unit-impulse response for the network of Problem 8.27.

* E. A. Guillemin, *Theory of Linear Physical Systems* (New York: John Wiley and Sons, Inc., 1963), p. 390.

Chapter **9**

Phasors for Sinusoidal Steady-State Analysis

For the simple networks we studied in the previous chapter, the assumption of linear time-invariant models for the network elements leads to ordinary linear differential equations with constant coefficients. This type of equation will also be obtained for more complicated networks if the elements are assumed to be linear and time-invariant. The choice of linear time-invariant models is a simplifying one. The resulting mathematical problem is a few orders of magnitude simpler than that for time-varying or nonlinear models in most cases. Fortunately, for many purposes, the simple models yield sufficiently accurate results. Most of the methods of analysis in this book are slanted toward solving linear network problems.

We have already encountered the sinusoidal signal to some extent in Chapter 2. We now study useful methods of analyzing the behavior of linear time-invariant networks with sinusoidal excitations or input signals. In particular, only the *steady-state* behavior will be considered. By sinusoidal steady-state, we shall mean that we are considering network behavior *after* the sinusoidal excitations have been applied for an arbitrarily long time in the past. We consider only the class of networks for which the complementary solution y_c eventually approaches zero. We shall also restrict the frequencies of the sinusoidal sources to be identical for simplicity. At this point, we reiterate the reasons, of which there are several, for devoting considerable time to the study of very special input signals under very special conditions. First, many signal sources are sinusoidal and, in most cases, a steady-state analysis is sufficient. Giant generators that supply electricity to industry are almost all sinusoidal; many electronic oscillators have sinusoidal waveforms. Second and more significant, as we mentioned in the previous chapter, many nonsinusoidal signals may be represented by a sum (or integral) of sinusoids of various frequencies. Linearity, then, allows us to express the response as a sum of responses to the various sinusoids. Linearity also makes it possible to break up the analysis into two simpler problems: analysis of the network excited by an elementary signal; and superposition (addition or integration) of responses due to elementary signals.

Addition of sinusoids of the same frequency

We shall demonstrate the important fact that the sum of any number of sinusoids of the same frequency, but each with arbitrary amplitude and phase, is itself a sinusoid with the same frequency. Let us start by considering

$$v(t) = V_1 \sin \omega t + V_2 \cos \omega t. \tag{9.1}$$

It is claimed that this is of the form

$$v(t) = V \sin (\omega t + \theta). \tag{9.2}$$

Expanding the right-hand side of Equation 9.2, we have

$$v(t) = V \cos \theta \sin \omega t + V \sin \theta \cos \omega t. \tag{9.3}$$

If the $v(t)$'s in Equations 9.1 and 9.2 are to be equal, then we must have

$$V_1 = V \cos \theta, \qquad V_2 = V \sin \theta \tag{9.4}$$

or

$$V = \sqrt{V_1^2 + V_2^2} \qquad \text{and} \qquad \theta = \tan^{-1} \frac{V_2}{V_1}. \tag{9.5}$$

Thus if V and θ are chosen as specified in Equation 9.5, the two terms in Equation 9.1 may be combined into a single term as in Equation 9.2.

Suppose that we now add an arbitrary number of sinusoids of the same frequency but with arbitrary amplitudes and phase, that is,

$$v(t) = \sum_{n=1}^{N} V_n \sin (\omega t + \theta_n). \tag{9.6}$$

Expanding each term, we obtain

$$\begin{aligned} v(t) &= \sum_{n=1}^{N} (V_n \cos \theta_n \sin \omega t + V_n \sin \theta_n \cos \omega t) \\ &= \left(\sum_{n=1}^{N} V_n \cos \theta_n \right) \sin \omega t + \left(\sum_{n=1}^{N} V_n \sin \theta_n \right) \cos \omega t. \end{aligned} \tag{9.7}$$

We see that this is of the same form as Equation 9.1. The following conclusion follows immediately. The amplitude of the resulting sinusoid is equal to the square root of the sum of the squares of the coefficient of $\sin \omega t$ and the coefficient of $\cos \omega t$ in Equation 9.7. The phase angle θ is the arctangent of the ratio of the coefficient of $\cos \omega t$ to the coefficient of $\sin \omega t$. In equation form, the result is expressed by Equation 9.5.

EXAMPLE 9.1.1

Let it be required to express the signal

$$\begin{aligned} v(t) &= \sin [2\pi(60)t] + 3\sqrt{2} \sin [2\pi(60)t + \pi/4] \\ &= \sin 377t + 3\sqrt{2} \sin (377t + \pi/4) \end{aligned}$$

as a single sinusoid. Expanding the second term and collecting, we have

$$v(t) = 4 \sin 377t + 3 \cos 377t = 5 \sin (377t + \tan^{-1} \tfrac{3}{4}).$$

EXERCISES

9.1.1 Express the following summations of sinusoids in the general form $K \sin (\omega t + \theta)$:
(a) $i_1(t) = 3 \sin t + 4 \cos t$,
(b) $v_2(t) = 15 \sin 3t + 7 \cos 3t$,
(c) $i_3(t) = 3 \sin t + 4 \cos (t + \pi/4)$,
(d) $q_1(t) = 13 \sin (t - \pi/4) - 5 \cos (t + \pi/4)$,
(e) $v_1(t) = 2 \sin (t - \pi/2) + 5 \cos t$.

9.1.2 Repeat Exercise 9.1.1 but expressing the summations of sinusoids in the general form $K' \cos (\omega t + \phi)$.

9.2 Complex-number arithmetic

In Chapter 2, we noted that the sinusoidal function is related to the exponential function $e^{j\omega t}$. It turns out that analysis using $e^{j\omega t}$ is much simpler and that sinusoidal steady-state response is readily obtainable from the response due to $e^{j\omega t}$. Notice, however, that $e^{j\omega t}$ is a complex function. In this section we summarize the basic complex-number arithmetical operations that we shall often use.

First we consider the representation of complex numbers. A complex number C is said to be in rectangular or Cartesian form if it is written as

$$C = a + jb, \tag{9.8}$$

where a and b are real numbers and $j = \sqrt{-1}$. A convenient geometric interpretation of a complex number is that of a point in a plane called the *complex plane*.

In Equation 9.8, a is called the *real part* of C and b is called the *imaginary part* of C. In symbols, we have

$$a = \text{Re}(C) \qquad \text{and} \qquad b = \text{Im}(C). \tag{9.9}$$

Notice that both the real part and imaginary part are real numbers. In the complex plane, the real and imaginary parts are considered as perpendicular components. As a convention, the real part is taken as the horizontal component and the imaginary part as the vertical component. For example, the number $-3 - j2$ is plotted as point P in Figure 9.1. The location of a point in the complex plane may alternatively be specified by giving its radial distance from the origin and the angle made between the radial line passing through the point and the positive real axis. Angle is measured positively in the counterclockwise direction. In symbolic form, we may write $C = |C|\underline{/\theta}$, where the complex number C is specified by the *magnitude* or radial distance $|C|$ and the *phase angle* θ. This representation is called the *polar form*. The real quantities $|C|$ and θ are illustrated in Figure 9.1. Since the two representations refer to the same complex number, a, b, $|C|$, and θ must be interrelated. Referring to Figure 9.1, we see that

$$a = |C| \cos \theta \tag{9.10}$$

and

$$b = |C| \sin \theta, \tag{9.11}$$

or

$$|C| = \sqrt{a^2 + b^2} \tag{9.12}$$

and

$$\theta = \tan^{-1} \frac{b}{a}. \tag{9.13}$$

Equations 9.10 and 9.11 may be used for converting from polar to rectangular

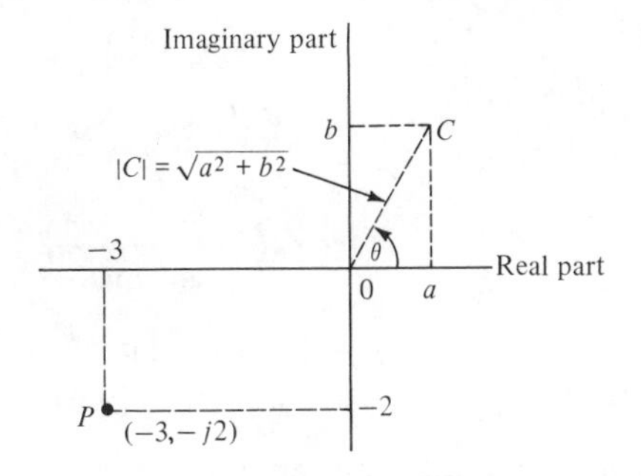

Figure 9.1 The complex plane.

form and Equations 9.12 and 9.13 for converting from rectangular to polar form. In actual numerical problems, Equations 9.12 and 9.13 may not be the most efficient to use. For instance, in using a slide rule it may be more convenient to use

$$|C| = a \sec \theta \qquad \text{or} \qquad |C| = b \operatorname{cosec} \theta, \tag{9.14}$$

where θ is calculated from Equation 9.13 or

$$\theta = \cot^{-1} \frac{a}{b}. \tag{9.15}$$

The student should familiarize himself with the solution of these equations with his own slide rule. *In using Equation 9.13 or Equation 9.15, care should be taken in determining the correct quadrant.*

EXAMPLE 9.2.1

Suppose we wish to convert $2 - j2$ to polar form. The angle θ is

$$\theta = \tan^{-1} \frac{-2}{2} = \frac{-\pi}{4}, \tag{9.16}$$

so that the number is $2\sqrt{2}\,\underline{/-\pi/4}$.

EXAMPLE 9.2.2

The angle θ for $-2 + j2$ is

$$\theta = \tan^{-1} \frac{2}{-2} = \frac{3\pi}{4}, \tag{9.17}$$

so that

$$-2 + j2 = 2\sqrt{2}\,\underline{/3\pi/4}. \tag{9.18}$$

Referring to Figure 9.1 again, we see that

$$C = a + jb = |C| \cos \theta + j|C| \sin \theta = |C|(\cos \theta + j \sin \theta). \tag{9.19}$$

But from Euler's formula we have

$$\cos \theta + j \sin \theta = e^{j\theta}, \tag{9.20}$$

so that we may write

$$C = |C|e^{j\theta}. \tag{9.21}$$

Equation 9.21 is known as the *exponential form*. It contains the same information as the polar form.

The *conjugate* of a complex number C is defined as one whose real part is equal to that of C and whose imaginary part is equal to the negative of that of C. It is denoted by C^*. Thus if $C = a + jb$, we have

$$C^* = a - jb. \tag{9.22}$$

The *sum* of two complex numbers C_1 and C_2 is defined as

$$C = C_1 + C_2 = C_2 + C_1 = (a_1 + a_2) + j(b_1 + b_2), \tag{9.23}$$

where a_1 and a_2 are the real parts and b_1 and b_2 the imaginary parts of C_1 and C_2, respectively. Graphically, the complex numbers may be added by the parallelogram law as shown in Figure 9.2. Likewise, for the *difference* $C_1 - C_2$, we have

$$C = C_1 - C_2 = (a_1 - a_2) + j(b_1 - b_2). \tag{9.24}$$

The *product* $C = C_1C_2$ is defined as

$$C_1C_2 = (a_1 + jb_1)(a_2 + jb_2) = a_1a_2 + j(b_1a_2 + a_1b_2) + j^2b_1b_2. \tag{9.25}$$

Since $j^2 = -1$, we have

$$C_1C_2 = (a_1a_2 - b_1b_2) + j(a_1b_2 + a_2b_1). \tag{9.26}$$

If C_1 and C_2 are in exponential or polar form, the answer is very easily obtained as

$$C_1C_2 = (|C_1|e^{j\theta_1})(|C_2|e^{j\theta_2}) = |C_1||C_2|e^{j(\theta_1+\theta_2)}. \tag{9.27}$$

Thus the magnitude of the product of two complex numbers is equal to the product of the magnitudes of the individual complex numbers, and the phase angle of the product is equal to the sum of the individual phase angles. For the *quotient* or ratio of two complex numbers, we have

$$\frac{C_1}{C_2} = \frac{|C_1|e^{j\theta_1}}{|C_2|e^{j\theta_2}} = |C_1|e^{j\theta_1}\left(\frac{1}{|C_2|}e^{-j\theta_2}\right) = \frac{|C_1|}{|C_2|}e^{j(\theta_1-\theta_2)}. \tag{9.28}$$

The magnitude of the ratio of two complex numbers is equal to the ratio of the

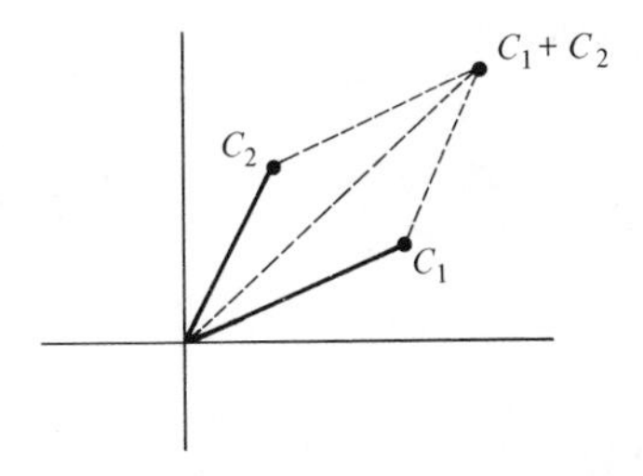

Figure 9.2 The sum of two complex numbers.

magnitudes of the individual complex numbers. If the numbers are in rectangular form, they may be converted to exponential form and then Equation 9.28 can be applied. Alternatively, we may proceed to make the calculation in rectangular form. If

$$\frac{C_1}{C_2} = \frac{a_1 + jb_1}{a_2 + jb_2} \tag{9.29}$$

is multiplied by C_2^*/C_2^*, we have

$$\frac{(a_1 + jb_1)(a_2 - jb_2)}{(a_2 + jb_2)(a_2 - jb_2)} = \frac{(a_1a_2 + b_1b_2) + j(b_1a_2 - a_1b_2)}{a_2^2 + b_2^2}. \tag{9.30}$$

Note that a complex number multiplied by its conjugate yields a real number equal to the square of the magnitude of the complex number, that is

$$CC^* = a^2 + b^2 = |C|^2. \tag{9.31}$$

EXERCISES

9.2.1 Convert the following to polar form:
(a) $3 - j4$, (b) $-3 - j4$, (c) $-3 + j4$, (d) $6.5 + j2.0$, (e) $1.5 + j13$, (f) $-12 + j5$, (g) $-5 + j0$, (h) $0 + j8$, (i) $0 - j8$, (j) $15 - j0.5$.

9.2.2 Convert the following to rectangular form:
(a) $5e^{j\pi/2}$, (b) $2e^{j\pi/4}$, (c) $10e^{-j\pi/2}$, (d) $8e^{j5\pi/6}$, (e) $3e^{j\pi}$, (f) $6\underline{/30^\circ}$, (g) $6\underline{/-30^\circ}$, (h) $12\underline{/105^\circ}$, (i) $2\underline{/135^\circ}$, (j) $4.5\underline{/360^\circ}$.

9.2.3 Evaluate the following sums:
(a) $(6.5 + j2.0) + (8 + j13)$, (b) $12\underline{/105^\circ} + 6\underline{/30^\circ}$, (c) $8e^{j5\pi/6} + 6\underline{/-30^\circ}$.

9.2.4 Evaluate the following products:
(a) $(3 - j4)(-3 - j4)$, (b) $(-12 + j5)(1.5 + j13)$, (c) $3e^{j\pi}(15 - j0.5)$.
Answers should be in exponential form.

9.2.5 Verify that the conjugate of $e^{j\theta}$ is $e^{-j\theta}$.

9.2.6 Verify that the conjugate of the product of two complex numbers is equal to the product of the conjugates of the two complex numbers.

Phasors and rotating phasors

In the preceding section, we introduced the complex number as a point in a plane called the complex plane. Geometrically, the addition of two complex numbers may be carried out by drawing a parallelogram where the rays passing through the origin and the two points form adjacent sides. The diagonal between the two sides then represents the sum. Alternatively, we may use the triangle as shown in Figure 9.3(a) and (b). Note that we have attached arrows to the lines. These directed lines will be called *phasors*. Insofar as addition is concerned, these phasors behave exactly like two-dimensional vectors. However, in situations where both complex numbers and space vectors are involved (representation of time-varying electromagnetic fields, for example), some confusion may arise if the same name is used for two different concepts. The use of the word phasor to represent the complex number is widely accepted and is now conventional in electrical engineering.

To add phasors, we simply arrange them in head-to-tail fashion as shown in Figure 9.3(b). The sum is obtained as a phasor whose head is at the unconnected head of one of the component phasors and whose tail is at the unconnected tail of another component phasor. Thus, in Figure 9.3(b), the head of B is unconnected and the tail of A is unconnected. The order in which the component phasors are joined end-to-end is immaterial. Note that the resultant phasor in Figure 9.3(c) is the same as that of Figure 9.3(b).

We might wonder what the difference is between an ordinary complex number and a phasor. Recall that a complex number is considered a point in the complex plane whose location or coordinate is specified in several ways, one of which is by its distance from the origin and the angle θ the ray makes with the positive real axis. Another is by its Cartesian or rectangular coordinates. To visualize the geometrical addition of complex numbers, however, it is convenient to detach the rays from the origin and move them around. When this is done, it is important to keep track of which end of the ray was formerly connected to the origin. For the phasor, it is the tail end that corresponds to the origin. The length of the phasor

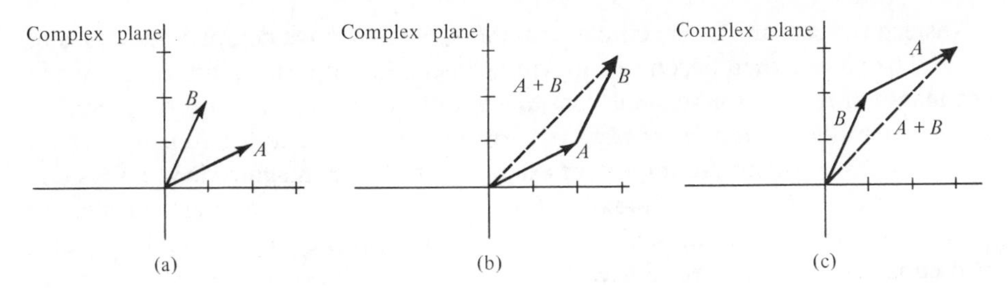

Figure 9.3 Addition of complex numbers by the use of directed quantities called phasors.

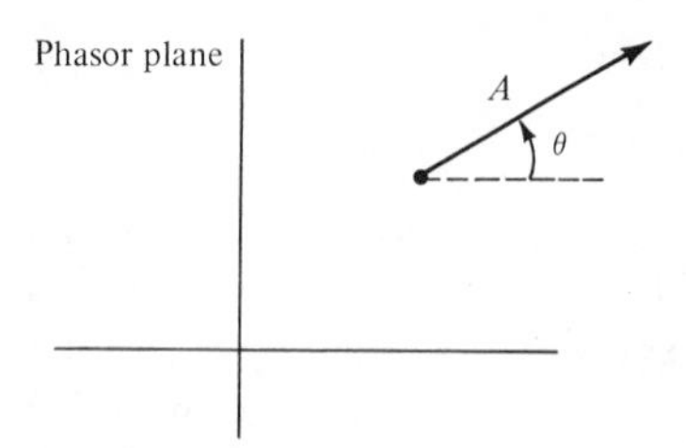

Figure 9.4 Phasor specified by length and direction.

corresponds to the magnitude of the complex number, and the angle θ of the complex number corresponds to the angle the phasor makes with the positive real axis as shown in Figure 9.4. Note that the angle is measured positively in the counterclockwise direction from the positive real axis. The plane in which the phasor is drawn is really not the same complex plane where numbers are represented as points whose distances from the origin are important. Here distance from the origin is not important. We must know just the direction of the positive real axis. Conventionally, this is horizontal and to the right. We shall call this plane the *phasor plane*. Thus the difference between the complex number and the phasor is geometric. Algebraically, they are the same.

Phasors may be used to actually add several complex quantities graphically. However, its more important use is as an aid in visualizing the various steps in an analysis. Usually, a simple sketch of the phasors together with the additions and other required operations roughly carried out serve as a rough guide in detecting arithmetical mistakes in the calculations made later on.

EXAMPLE 9.3.1

Suppose we wish to add $3\underline{/0}$, $4\underline{/30^\circ}$, and $6\underline{/-45^\circ}$. A graphical addition is shown in Figure 9.5. The phasor in dotted lines is the resultant.

Instead of ordinary fixed complex numbers, we may have complex functions of time. The phasor then becomes a moving phasor. An important case occurs when the length of the phasor remains constant but its angle changes with time. Such a phasor is called a *rotating phasor*. It corresponds to a complex function whose magnitude is constant. An important example is $e^{j\omega t}$. The magnitude is always one and the angle at time t is ωt radians. The phasor $e^{j\omega t}$ rotates counterclockwise for positive ω and positive t and, if ω is constant, the frequency of rotation is constant and equal to ω radians/sec. Since

$$e^{j\omega t} = \cos \omega t + j \sin \omega t,$$

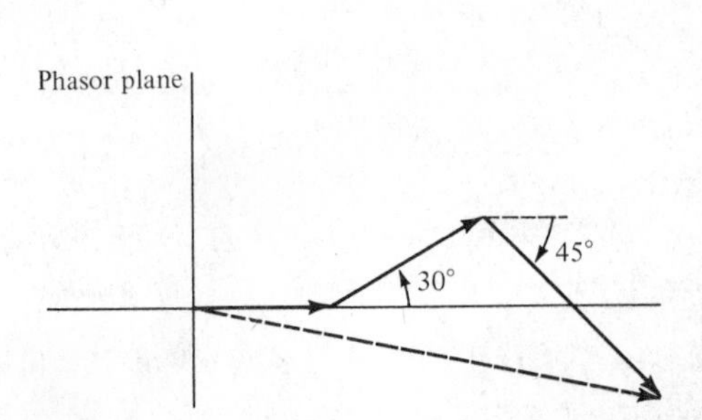

Figure 9.5 Example of phasor addition.

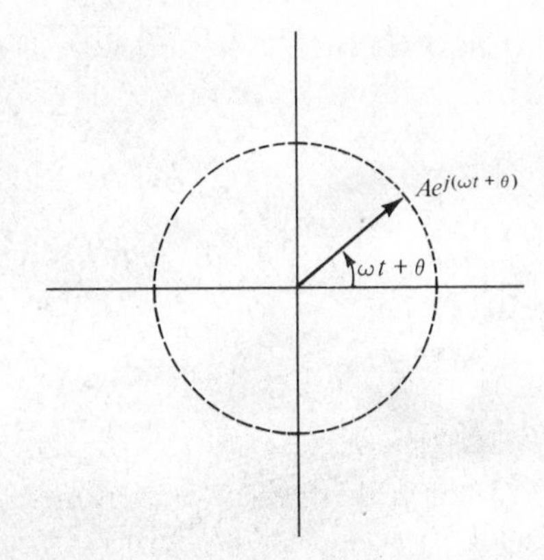

Figure 9.6 A rotating phasor.

the real part and imaginary part are both sinusoidal. Another example of a rotating phasor is $(1 + j1)e^{jt} = \sqrt{2}\, e^{j\pi/4}e^{jt} = \sqrt{2}\, e^{j(t+\pi/4)}$. The real part of this is $\sqrt{2} \cos(t + \pi/4)$. Figure 9.6 shows the rotating phasor $Ae^{j(\omega t+\theta)}$.

If two rotating phasors have the same speed of rotation, their sum is a rotating phasor of the same speed of rotation. For example, if we have

$$V_1(t) = (1 + j2)e^{j10t}, \tag{9.32}$$

$$V_2(t) = (3 + j1)e^{j10t}, \tag{9.33}$$

then

$$V_1(t) + V_2(t) = (4 + j3)e^{j10t} = 5e^{j[10t+\tan^{-1}(3/4)]}. \tag{9.34}$$

The frequency ω is the rate of change or time derivative of the total phase angle of the various rotating phasors. In each case, ω is 10 radians/sec if t is in seconds.

If the real part of $V_1(t)$ above is added to the real part of $V_2(t)$, the result is equal to the real part of $[V_1(t) + V_2(t)]$. Likewise, the sum of the imaginary parts of two complex numbers or functions is equal to the imaginary part of the sum of the complex numbers or functions. This commutative property of the real-part operation (or the imaginary-part operation) and addition is a key to the use of rotating phasors for sinusoidal steady-state analysis. Thus, to add several sinusoidal functions of the same frequency, we write the individual functions as real parts (or imaginary parts) of appropriate rotating phasors. Then we perform the addition on the rotating phasors and take the real part (imaginary part) of the sum. For example, if we have

$$v_1(t) = 2\sqrt{2} \sin(10t + \pi/4), \tag{9.35}$$

$$v_2(t) = 2 \sin 10t + \cos 10t, \tag{9.36}$$

then if we write $v_1(t)$ and $v_2(t)$ as imaginary parts of $V_1(t)$ and $V_2(t)$, we have

$$V_1(t) = 2\sqrt{2}\, e^{j(10t + \pi/4)} = 2e^{j10t} + j2e^{j10t}, \tag{9.37}$$

$$V_2(t) = 2e^{j10t} + je^{j10t}, \tag{9.38}$$

$$V_1(t) + V_2(t) = (4 + j3)e^{j10t} = 5e^{j[10t + \tan^{-1}(3/4)]}, \tag{9.39}$$

and

$$\begin{aligned} v_1(t) + v_2(t) &= \text{Im}\,[V_1(t) + V_2(t)] = 5 \sin(10t + \tan^{-1}\tfrac{3}{4}) \\ &= 5 \sin(10t + 36.9^\circ). \end{aligned} \tag{9.40}$$

We may alternatively represent $v_1(t)$ and $v_2(t)$ as real parts of a new set of $V_1(t)$ and $V_2(t)$, so that we have

$$V_1(t) = 2\sqrt{2}\, e^{j(10t - \pi/4)} = -j2e^{j10t} + 2e^{j10t}, \tag{9.41}$$

$$V_2(t) = -2je^{j10t} + e^{j10t}, \tag{9.42}$$

$$V_1(t) + V_2(t) = (3 - j4)e^{j10t}, \tag{9.43}$$

$$\begin{aligned} v_1(t) + v_2(t) &= \text{Re}\,[V_1(t) + V_2(t)] = \text{Re}\,(5e^{j[10t - \tan^{-1}(4/3)]}) \\ &= 5 \cos(10t - 53.1^\circ) = 5 \sin(10t + 36.9^\circ), \end{aligned} \tag{9.44}$$

which is, of course, the same answer as before.

The procedure is summarized in Figure 9.7. Here we have used $V_k \cos(\omega t + \theta_k) = \text{Re}\,(V_k e^{j(\omega t + \theta_k)})$. Alternatively, we may use $V_k \sin(\omega t + \alpha_k) = \text{Im}\,(V_k e^{j(\omega t + \alpha_k)})$.

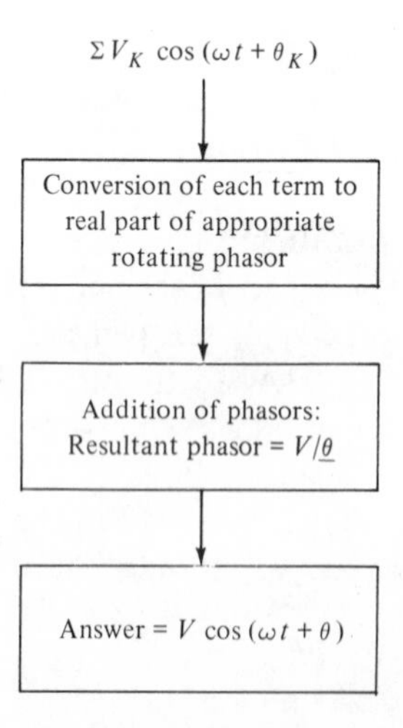

Figure 9.7 Steps in the use of phasors for adding sinusoidal functions.

RCISES

9.3.1 With the aid of a protractor and scale, graphically determine the sum

$$A = 5\underline{/20^\circ} + 3\underline{/60^\circ} + 8\underline{/180^\circ} + 2\underline{/-240^\circ}.$$

9.3.2 If A is an arbitrary phasor, draw
(a) jA relative to A,
(b) $-2jA$ relative to A,
(c) $(1 - 2j)A$ relative to A,
(d) A^* relative to A.

9.3.3 Add $10 \sin (377t + 30^\circ)$ and $5 \cos (377t + 45^\circ)$ using phasors.

9.3.4 Suppose in Exercise 9.3.3 we add 10° to the angles of both terms, what will the new sum be? Suppose, instead, we subtract 30° from the angles of both terms, what will the new sum be? In view of the above consideration, is it possible to factor out a complex number so as to make one term in the phasor additions always a real number?

9.3.5 The current through a series RL circuit is a sinusoid

$$i(t) = I_0 \sin (\omega t + \theta_0).$$

Determine the magnitude and phase angle of the voltage across the series connection in terms of R, L, I_0, ω, and θ_0. Specify reference directions in your circuit diagram.

Steady-state response to $e^{j\omega t}$

We have already seen that the sum of rotating phasors of the same frequency ω results in a rotating phasor of the same frequency. We shall now demonstrate that if we apply a hypothetical rotating-phasor input signal to a linear time-invariant network or system, the steady-state output is also a rotating phasor of the same frequency.

MPLE 9.4.1

Let us first consider as an example a network or system described by

$$2\frac{dy(t)}{dt} + 8y(t) = x(t), \tag{9.45}$$

where $y(t)$ is the output signal and $x(t)$ is the input signal. This equation may describe, for example, a series RL circuit excited by a voltage source, where $R = 8$ ohms, $L = 2$ henrys, and current is the output variable. Let us apply a hypothetical input $x(t) = e^{j\omega t}$ in Equation 9.45. The fact that no such physical signal exists does not preclude its usefulness in analysis. We shall show how it relates to a physical sinusoidal response. We shall verify that the steady-state output is of the form

$$y(t) = He^{j\omega t}, \tag{9.46}$$

where H may be complex and may depend on ω. More specifically, we shall investigate if the *particular solution* of Equation 9.45 is of the form as given in Equation 9.46. That is, we shall substitute Equation 9.46 and $x(t) = e^{j\omega t}$ in Equation 9.45 and see if the equality holds. Thus we have

$$\begin{aligned} 2j\omega He^{j\omega t} + 8He^{j\omega t} &\overset{?}{=} e^{j\omega t} \\ H(8 + j2\omega)e^{j\omega t} &\overset{?}{=} e^{j\omega t}. \end{aligned} \tag{9.47}$$

Clearly, if H is chosen as

$$H = \frac{1}{8 + j2\omega}, \tag{9.48}$$

then Equation 9.47 holds. So the particular solution for an input $e^{j\omega t}$ is

$$y(t) = \frac{1}{8 + j2\omega} e^{j\omega t}. \tag{9.49}$$

Note that this is a complex time function, i.e., it has real and imaginary parts. This is not surprising since our hypothetical input is also complex. However, for our physical system, real input signals necessarily produce only real output signals. Thus, if the input is $\cos \omega t$, the output is a real signal $y_r(t)$, and, if the input signal is $\sin \omega t$, the output signal is a real signal $y_i(t)$. Furthermore, since our system is linear, if we scale (multiply) the input by a constant A, the output will also be scaled by the same constant. The constant could be anything including a complex number. Thus, if the input is $j \sin \omega t$, the output is $jy_i(t)$, where $y_i(t)$ is the output due to $\sin \omega t$. Since $e^{j\omega t}$ is equal to $\cos \omega t + j \sin \omega t$ and since our system is linear, superposition tells us that the output due to $e^{j\omega t}$ is

$$y(t) = y_r(t) + jy_i(t). \tag{9.50}$$

But we have calculated the output due to $e^{j\omega t}$ previously, and it is given in Equation 9.49. That is, we have

$$y(t) = \text{Re}\left(\frac{1}{8 + j2\omega} e^{j\omega t}\right) + j\,\text{Im}\left(\frac{1}{8 + j2\omega} e^{j\omega t}\right). \tag{9.51}$$

We conclude that in this case an input $x(t) = \cos \omega t$ will produce

$$y_r(t) = \text{Re}\left(\frac{1}{8 + 2j\omega} e^{j\omega t}\right) = \frac{1}{\sqrt{64 + 4\omega^2}} \cos\left(\omega t - \tan^{-1} \frac{2\omega}{8}\right). \tag{9.52}$$

Likewise, for an input $x(t) = \sin \omega t$, the output will be

$$y_i(t) = \text{Im}\left(\frac{1}{8 + 2j\omega}\, e^{j\omega t},\right) = \frac{1}{\sqrt{64 + 4\omega^2}} \sin\left(\omega t - \tan^{-1}\frac{2\omega}{8}\right). \tag{9.53}$$

If instead of $e^{j\omega t}$ we have $Ae^{j\theta}e^{j\omega t}$ as input, then because of linearity the rotating-phasor output is

$$y(t) = \frac{Ae^{j\theta}}{8 + j2\omega}\, e^{j\omega t}, \tag{9.54}$$

and for the real input $A \cos(\omega t + \theta)$, the output would be

$$y_r(t) = \frac{A}{\sqrt{64 + 4\omega^2}} \cos\left(\omega t + \theta - \tan^{-1}\frac{2\omega}{8}\right). \tag{9.55}$$

Thus we see that sinusoidal inputs produce sinusoidal responses as particular solutions.

For more general situations, our linear time-invariant network or system will be described by a differential equation:

$$a_n \frac{d^n y(t)}{dt^n} + a_{n-1}\frac{d^{n-1}y(t)}{dt^{n-1}} + \cdots + a_0 y(t) = b_m \frac{d^m x(t)}{dt^m} + b_{m-1}\frac{d^{m-1}x(t)}{dt^{m-1}} + \cdots + b_0 x(t). \tag{9.56}$$

Just as before, a hypothetical rotating-phasor input $e^{j\omega t}$ will produce a rotating-phasor output. We again assume the form in Equation 9.46 and substitute in Equation 9.56. We obtain

$$[a_n(j\omega)^n + a_{n-1}(j\omega)^{n-1} + \cdots + a_0]He^{j\omega t} = [b_m(j\omega)^m + b_{m-1}(j\omega)^{m-1} + \cdots + b_0]e^{j\omega t}. \tag{9.57}$$

So that for

$$H = \frac{b_m(j\omega)^m + b_{m-1}(j\omega)^{m-1} + \cdots + b_0}{a_n(j\omega)^n + a_{n-1}(j\omega)^{n-1} + \cdots + a_0} = |H|\underline{/\theta}, \tag{9.58}$$

Equation 9.57 becomes an identity provided that $j\omega$ is not a characteristic root. Again, the response due to $\cos \omega t$ is

$$\text{Re}\,(|H|e^{j\theta}e^{j\omega t}) = |H| \cos(\omega t + \theta). \tag{9.59}$$

Also, if the input is $A \cos(\omega t + \alpha)$, the response will be

$$\text{Re}\,(Ae^{j\alpha}|H|e^{j\theta}e^{j\omega t}) = A|H| \cos(\omega t + \theta + \alpha). \tag{9.60}$$

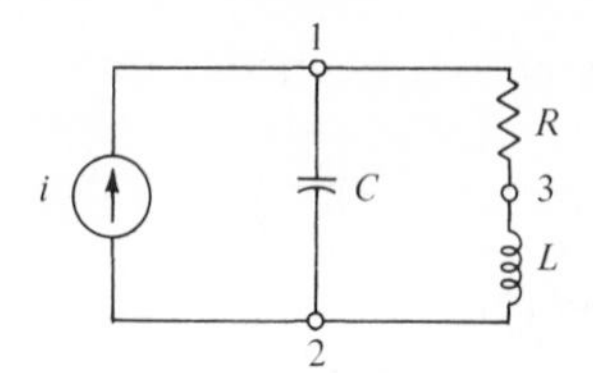

Figure 9.8

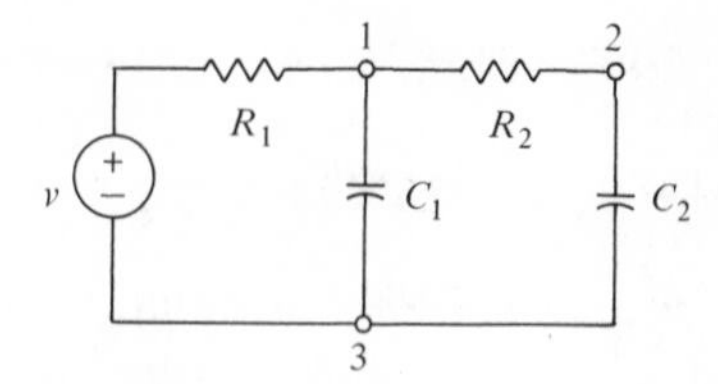

Figure 9.9

The use of phasors has something to do with the determination of $|H|$ and θ in Equation 9.58. The phasor method of this section is a special case of the analysis of the previous chapter using exponential signals e^{st}, where $s = j\omega$.

EXERCISES

9.4.1 For the network of Figure 9.8, obtain the differential equation relating the input $i(t)$ to the output $v_{12}(t)$, where R, L, and C are constants. For $C = 0.05$ microfarads, $R = 10$ ohms, and $L = 100$ millihenrys, and $i(t) = 10^{-3} \sin(10^4 t + \pi/6)$, determine the steady-state component of v_{12}.

9.4.2 Using the circuit parameters of Exercise 9.4.1 for the network in Figure 9.8, determine the input that will cause the steady-state output $v_{12}(t)$ to be $\sin 10^4 t$.

9.4.3 For simplicity, assume that all the elements in the network of Figure 9.9 have value 1. Determine the differential equation relating $v_{13}(t)$ to $v(t)$ and express $v_{13}(t)$ in the form $v_{13}(t) = V \cos(t + \alpha)$ if $v(t) = \sin t + 2 \cos t$.

9.4.4 Repeat Exercise 9.4.3 for $v_{23}(t)$ as output instead of $v_{13}(t)$.

9.5 Summary

1. We can combine two sinusoids as follows:

$$V_1 \sin \omega t + V_2 \cos \omega t = V \sin(\omega t + \theta),$$

$$V = \sqrt{V_1^2 + V_2^2}, \qquad \theta = \tan^{-1} \frac{V_2}{V_1}.$$

Repeated application of this principle together with the following trigonometric identities

$$\sin(\omega t + \theta) = \sin \omega t \cos \theta + \cos \omega t \sin \theta$$
$$\cos(\omega t + \theta) = \cos \omega t \cos \theta - \sin \omega t \sin \theta$$

makes it possible to combine any number of sinusoids of one frequency into a single sinusoid.

2. Conversion from polar to rectangular form or vice versa:

$$\text{rectangular form} \qquad C = a + jb$$
$$\text{polar form} \qquad C = |C|\underline{/\theta}$$
$$\text{exponential form} \qquad C = |C|e^{j\theta}$$
$$a = |C| \cos \theta \qquad |C| = \sqrt{a^2 + b^2}$$
$$b = |C| \sin \theta \qquad \theta = \tan^{-1}\frac{b}{a}.$$

3. Operations with complex numbers where $C = a + jb$:

$$C_1 + C_2 = (a_1 + a_2) + j(b_1 + b_2);$$
$$C_1 - C_2 = (a_1 - a_2) + j(b_1 - b_2);$$
$$C_1C_2 = a_1a_2 - b_1b_2 + j(a_1b_2 + a_2b_1);$$
$$C_1C_2 = |C_1||C_2|e^{j(\theta_1+\theta_2)};$$
$$\frac{C_1}{C_2} = \left|\frac{C_1}{C_2}\right|e^{j(\theta_1-\theta_2)};$$
$$CC^* = |C|^2, \qquad C^* = a - jb \text{ (conjugate)}.$$

4. A rotating phasor has the representation

$$V(t) = Ve^{j(\omega t+\theta)},$$

where V is the magnitude of the phasor, θ is its position in the complex plane at $t = 0$, and ω is the frequency of rotation in hertz. Note that

$$\text{Re}\,[V(t)] = V\cos(\omega t + \theta), \qquad \text{Im}\,[V(t)] = V\sin(\omega t + \theta).$$

5. The exponential input $x(t) = e^{j(\omega t+\phi)}$ may be used to determine the steady-state response of a network in the form

$$y(t) = He^{j(\omega t+\phi)}, \qquad H = |H|e^{j\theta}.$$

With H determined, then the input

$$x(t) = \text{Re}\,(e^{j(\omega t+\phi)}) = \cos(\omega t + \phi)$$

produces the response

$$y(t) = |H|\cos(\omega t + \phi + \theta).$$

Similarly, the input

$$y(t) = \text{Im}\,(e^{j(\omega t+\phi)}) = \sin(\omega t + \phi)$$

produces the response

$$y(t) = |H| \sin (\omega t + \phi + \theta).$$

PROBLEMS

9.1 The signal voltage for a given element is known to be given by the equation $v_a(t) = 10 \sin 3t - 20 \sin (3t + \pi/4)$ volts. When an rms reading voltmeter is connected across this element, what value will it read? Show the method.

9.2 Given the phasor in rectangular coordinates $V_1 = 5\sqrt{3} + j5$. Rotate this phasor $-75°$. What is the rectangular representation of the phasor in its rotated position?

9.3 Given the phasor $5 + j5\sqrt{3}$. Rotate this phasor $-15°$ and express the result in rectangular form, $C = A + jB$.

9.4 In the equation $V_0 \cos (\omega t + \phi_0) = V_1 \cos (\omega t + \phi_1) + V_2 \cos (\omega t + \phi_2)$, find an expression for V_0 and ϕ_0 in terms of the real, positive quantities, V_1, V_2, ϕ_1, and ϕ_2.

9.5 Three wires in a circuits laboratory are connected to a common point. Ammeters are used to establish that the effective (or rms) current in one wire is 5 amperes and in another 10 amperes. A cathode-ray oscillograph is used to establish the fact that the currents are substantially sinusoidal. A frequency meter reads 60 Hz (which is 377 radians/sec). The oscillograph shows that the current read as 5 amperes reaches its maximum 1/200 sec before the 10-ampere current. Write equations for the three currents, determining numerical values whenever possible. Positive reference directions are towards the junction.

9.6 Two components connected in series are labeled A and B. An rms-reading voltmeter records 10 volts across A, 25 volts across B, and 15 volts across the combination of A and B. All these voltages are sinusoidal. Assuming that the voltage across A is the reference (it has the form $V \sin \omega t$), write expressions for the three voltages, determining numerical values for constants whenever possible.

9.7 Using Euler's relationship, show that

$$\sin \omega t = \frac{1}{2j} (e^{j\omega t} - e^{-j\omega t})$$

and

$$\cos \omega t = \tfrac{1}{2}(e^{j\omega t} + e^{-j\omega t}).$$

9.8 The results given in the last problem show than $\sin \omega t$, which is a real function, is expressed in terms of complex rotating phasors. Draw a phasor diagram to show that this particular combination of rotating phasors is

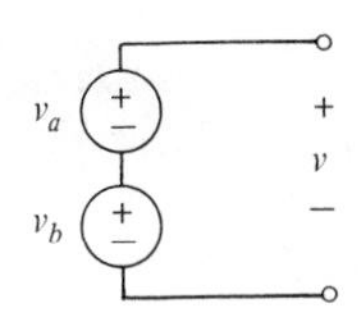

Figure 9.10

indeed real (having values only on the real axis of the complex plane) for all values of ωt.

9.9 Repeat Problem 9.8 for the cosine function of Problem 9.7.

9.10 In the equation

$$V_0 \sin(\omega t + \phi_0) = \sum_{k=1}^{N} V_k \cos(\omega t + \phi_k),$$

find expressions for V_0 and ϕ_0 in terms of the real, positive quantities $V_1, V_2, V_3, \ldots, V_n$ and $\phi_1, \phi_2, \ldots, \phi_n$.

9.11 In Figure 9.10, the voltage source v_b is sinusoidal and has the same frequency as v_a which is $v_a(t) = 2 \sin \omega_0 t$. The voltage $v_a + v_b$ has the value $v_a + v_b = 10 \sin(\omega_0 t + \theta_t)$.
(a) Find v_b if $\theta_t = 0°$; if $\theta_t = 90°$; if $\theta_t = 180°$; if $\theta_t = -90°$.
(b) Draw the appropriate phasors for V_a, V_b, and $V_a + V_b$ for the values of θ_t in (a).

9.12 Verify that the conjugate of the ratio of two complex numbers is equal to the ratio of their conjugates.

9.13 Determine the polar form of A:

$$A = \frac{(8 + j6)(3.6 - j5)}{10\underline{/120°} - (-8.5 - j2)}.$$

9.14 Is the product of the real parts of two complex numbers equal to the real part of the product of the complex numbers? Answer the same question for the ratio.

9.15 Show that the real part of the derivative of a rotating phasor is equal to the derivative of the real part of the rotating phasor. Likewise for the imaginary part.

9.16 Given a linear system described by a linear ordinary differential equation with real constant coefficients, show that the response due to $e^{-j\omega t}$ is equal to the conjugate of the response due to $e^{j\omega t}$. Thus, since

$$\cos \omega t = \frac{e^{j\omega t} + e^{-j\omega t}}{2},$$

demonstrate that if $e^{j\omega t}$ produces $He^{j\omega t}$, $\cos \omega t$ produces Re $(He^{j\omega t})$.

9.17 The waveforms of the currents in six of seven wires connected at a junction are sinusoidal and have the same frequency. What can you say about the waveform of the current in the seventh wire? Explain.

9.18 If the steady-state voltage across a certain portion of a linear time-invariant network is $v(t) = 45 \sin(377t + 20°) + 10 \cos(377t - 10°)$ due to a source $i(t) = \cos 377t$ somewhere in the network, what would the voltage response be at the same terminal pair if the current were $3 \sin(377t - 30°)$?

9.19 The sum of three phasors, A, B, and C, is equal to zero. Phasor A has length 5 and zero angle. Phasors B and C have angles $-105°$ and $150°$, respectively. Compute the magnitudes of B and C both graphically and analytically.

9.20 Expand the functions e^{jx}, $\cos x$, and $\sin x$ in Maclaurin's series and demonstrate that $e^{jx} = \cos x + j \sin x$.

9.21 If $A(t)$ is a rotating phasor, what can you say of the product $A(t)A^*(t)$? How about the ratio $A(t)/A^*(t)$? The sum $A(t) + A^*(t)$? The difference $A(t) - A^*(t)$?

9.22 (a) If we use the phasor $1\underline{/0°}$ corresponding to $\cos \omega t = \text{Re}(e^{j\omega t})$, what should be the phasors corresponding to

(1) $v_1(t) = \sin \omega t$,

(2) $v_2(t) = 5 \sin(\omega t + 30°)$,

(3) $v_3(t) = 8 \cos(\omega t - 30°) + 6 \cos \omega t$.

(b) Repeat (a) if we use the phasor $1\underline{/0°}$ to correspond to $\sin(\omega t + 30°) = \text{Im}(e^{j(\omega t + 30°)})$.

(c) Add the sinusoidal functions $v_1(t)$, $v_2(t)$, and $v_3(t)$ in (a) by using the phasors in (a). Repeat, using the phasors in (b).

9.23 The signal $v(t)$ is equal to the sum

$$v(t) = 8.5 \sin(1000t - 30°) + 5 \sin(1000t - 90°) + 4 \cos(1000t - 45°) + 10 \sin(1000t + \theta).$$

For what value of θ is the peak value of $v(t)$ the largest?

9.24 Repeat Problem 9.23 for the value of θ that makes the peak value $v(t)$ the smallest.

9.25 The input and output voltages, $x(t)$ and $y(t)$, respectively, in a linear time-invariant network are related by

$$\frac{d^3y(t)}{dt^3} + 10\frac{d^2y(t)}{dt^2} + 8\frac{dy(t)}{dt} + 5y(t) = 13\frac{dx(t)}{dt} + 7x(t).$$

For what input will the steady-state output be $y(t) = \sin t$?

9.26 Restate Kirchhoff's current law and Kirchhoff's voltage law for a linear time-invariant network in the sinusoidal steady-state where the sources are all sinusoidal with the same frequency. Use phasors.

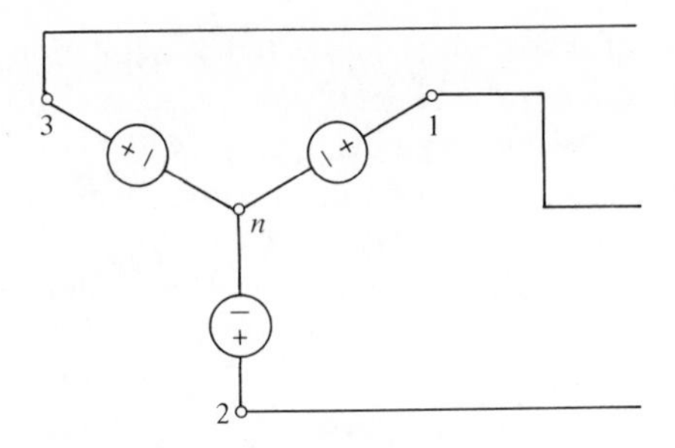

Figure 9.11

9.27 Large amounts of electrical energy are usually transmitted by three-phase power networks. The energy is usually converted from other forms into electrical energy where the final stage of the conversion system is a rotating machine called a three-phase generator. Three sinusoidal voltages with the same frequency and the same peak amplitude are produced by the generator. The three voltages are 120° out of phase with respect to each other. The voltage sources are usually internally connected in a Y-configuration as in Figure 9.11. In the figure, if v_{2n} lags v_{1n} and if v_{3n} lags v_{2n}, we say that the phase sequence is 1–2–3. Otherwise, the phase sequence is said to be 1–3–2. Suppose that $v_{1n}(t) = \sqrt{2}\, V \cos [2\pi(60)t]$, where V is a positive constant, and the phase sequence is 1–2–3. Using the phasor $\sqrt{2}\, V\underline{/0°}$ for representing $v_{1n}(t)$, draw the phasors V_{1n}, V_{2n}, and V_{3n}. On the same phasor diagram, draw V_{12}, V_{23}, and V_{31}. From the phasor diagram, determine $v_{12}(t)$, $v_{23}(t)$, and $v_{31}(t)$.

Chapter **10**

The Impedance Concept and Network Functions

10.1 Impedance for sinusoidal signals

In the previous chapter, it was demonstrated that for a linear time-invariant network, the steady-state response to a sinusoid is another sinusoid of the same frequency, provided that the complementary solution eventually reduces to zero. The most convenient procedure for carrying out analysis is to obtain the response to a hypothetical input $e^{j\omega t}$. The response to a sinusoid is then easily obtained from the real or imaginary part.

Consider a 1-port linear time-invariant network which may be a portion of a larger network as shown in Figure 10.1. Applying Kirchhoff's laws and the branch constraints characterizing the linear time-invariant network, a single differential equation relating v and i may be obtained. The differential equation would be linear with constant coefficients of the type considered in the previous chapter. It is assumed that the linear 1-port network contains no independent sources. The quantities i and v are not identically zero because of sources outside the 1-port network. Suppose that i or v is sinusoidal. Then in the steady-state, v or i will also be sinusoidal with the same frequency. Suppose that

$$i = \text{Re}\,(Ie^{j\omega t}), \tag{10.1}$$

where I is a complex number $|I|e^{j\theta}$ and

$$v = \text{Re}\,(Ve^{j\omega t}), \tag{10.2}$$

where V is also a complex number $|V|e^{j\phi}$. If $i = Ie^{j\omega t}$ and $v = Ve^{j\omega t}$ satisfy a differential equation with *real* constant coefficients, then $i = \text{Re}\,(Ie^{j\omega t})$ and $v = \text{Re}\,(Ve^{j\omega t})$ also satisfy the same differential equation. The same statements may be made for the imaginary parts of the complex time functions $Ie^{j\omega t}$ and $Ve^{j\omega t}$. These facts are important in much of the discussion of this chapter.

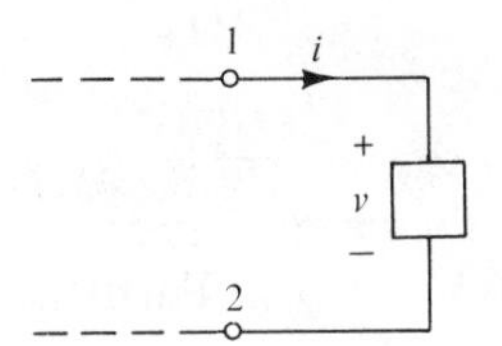

ure 10.1 Reference conventions for i and v used in defining impedance for a linear time-invariant network.

The ratio of the voltage-rotating phasor $Ve^{j\omega t}$ to the current-rotating phasor $Ie^{j\omega t}$ is defined as *impedance* Z:

$$Z = \frac{Ve^{j\omega t}}{Ie^{j\omega t}} = \frac{V}{I}. \tag{10.3}$$

Note that because the network is linear and time-invariant, scaling I by any amount will result in a scale of V by the same amount, and time shifting i will result in a time shift of v by the same amount. This means that Z does not depend on the magnitude of I nor the angle of I. It does depend on the *ratio* of the magnitudes of V and I and the *difference* in the angles between V and I. The reciprocal of impedance also finds frequent use, and it is known as *admittance* Y:

$$Y = \frac{1}{Z} = \frac{I}{V}. \tag{10.4}$$

Equations 10.3 and 10.4 are frequently written in the form of Ohm's law as

$$V = ZI \tag{10.5}$$

and

$$I = YV. \tag{10.6}$$

Steinmetz first used the words "apparent resistance" in describing what we now call impedance.* The motivation in introducing this and like terms just before the turn of the century was to make the analysis of networks with sinusoidal signals similar to the analysis of resistive networks with time-invariant (dc) signals. Impedance is intended to be the means of relating a voltage to a current when both are sinusoidal. Note that for the generalized Ohm's law of Equation 10.5, the current-reference direction through the 1-port network is from the positive terminal to the negative terminal of the voltage reference. Otherwise, the equation would be $V = -ZI$. The following example will illustrate how impedance may be used in relating sinusoidal voltage and current.

* A. E. Kennelly, "Impedance," *Trans. AIEE*, **10** (1893), 175–216; C. P. Steinmetz and F. Bedell, "Reactance," *Trans. AIEE*, **11** (1894), 640–648.

EXAMPLE 10.1.1

The impedance of a given element in a network is $10e^{j30°}$. If the current is $i(t) = 10 \sin(\omega t + 45°)$, find the voltage related to this current by the given impedance. We first represent the current by the phasor $I = 10e^{j45°}$ (or $10\underline{/45°}$). Then the voltage phasor is

$$V = ZI = (10e^{j30°})(10e^{j45°}) = 100e^{j75°}. \tag{10.7}$$

The voltage $v(t)$ corresponding to this phasor is

$$v(t) = 100 \sin(\omega t + 75°). \tag{10.8}$$

The impedance and admittance in Equations 10.5 and 10.6 may be written in rectangular form. These complex numbers have real and imaginary parts which are identified by specific names. Thus, in the equation

$$Z = R + jX \text{ ohms}, \tag{10.9}$$

R is the *resistance* and X is the *reactance*, while in

$$Y = G + jB \text{ mhos}, \tag{10.10}$$

G is the *conductance* and B the *susceptance*, where R, X, G, and B are real numbers. Since $Z = 1/Y$, we see that R and X may be expressed in terms of G and B as follows:

$$R + jX = \frac{1}{G + jB} = \frac{G - jB}{G^2 + B^2}. \tag{10.11}$$

Equating real and imaginary parts, we have

$$R = \frac{G}{G^2 + B^2} \quad \text{and} \quad X = \frac{-B}{G^2 + B^2}. \tag{10.12}$$

EXERCISES

10.1.1 Given a voltage and current related by the admittance function $Y = 1 + j1$. If $v(t) = 5 \sin(3t - 30°)$, determine the associated $i(t)$. Assume that v and i have the reference directions as in Figure 10.1.

10.1.2 Write expressions similar to Equations 10.12, expressing G and B in terms of R and X.

10.1.3 Write expressions for G and B in terms of $|Z|$ and θ.

Impedance of R, L, and C elements

We next make use of the definition of impedance given in the last section to determine expressions for Z and Y for the three linear time-invariant passive elements: R, L, and C. For the resistor, the voltage and current are related by Ohm's law

$$v_R(t) = Ri_R(t), \tag{10.13}$$

where v_R and i_R are identified in Table 10.1. To find impedance, we let the current be $i_R(t) = Ie^{j\omega t}$, so that $v_R(t) = RIe^{j\omega t}$. To use the phasor definition of impedance, we see that $I_R = Ie^{j0}$ and $V_R = RIe^{j0}$. Thus we have

$$Z_R = \frac{V_R}{I_R} = R \text{ ohms}. \tag{10.14}$$

Table 10.1

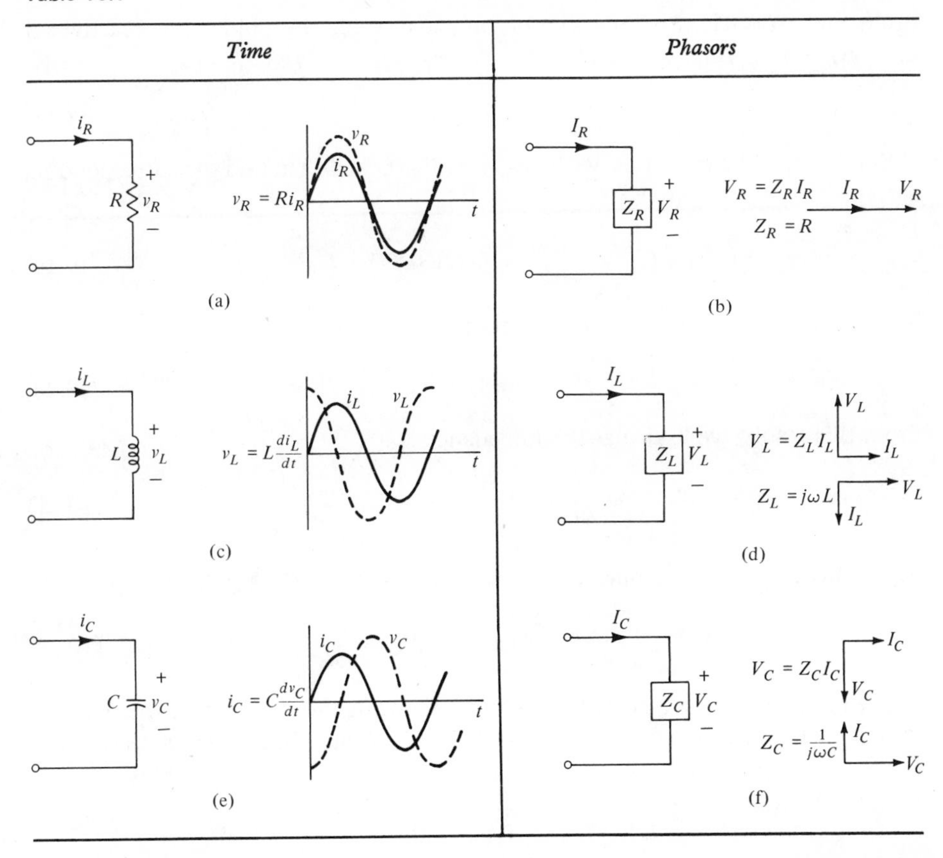

While the use of the phasor definition for impedance is not necessary for this particular derivation, its use in analyzing the resistor sets the pattern we will follow in finding the impedance expression for L and C.

For the inductor, we again write the current as $i = I_L e^{j\omega t}$ and the voltage as $V_L e^{j\omega t}$. Then the voltage is

$$v_L(t) = L\frac{di_L}{dt} = j\omega L I_L e^{j\omega t} = V_L e^{j\omega t}. \tag{10.15}$$

The corresponding current and voltage phasors are

$$\mathbf{I}_L = I_L e^{j0} \qquad \text{and} \qquad \mathbf{V}_L = \omega L I_L e^{j\pi/2}. \tag{10.16}$$

Then the impedance for the inductor is

$$Z_L = \frac{\mathbf{V}_L}{\mathbf{I}_L} = \omega L e^{j\pi/2} = j\omega L. \tag{10.17}$$

From this expression, we see that the impedance for the inductor is reactive; if $Z_L = R_L + jX_L$, then we have

$$R_L = 0 \qquad \text{and} \qquad X_L = \omega L. \tag{10.18}$$

Similarly for the capacitor, we let $v_C(t) = V_C e^{j\omega t}$ and then find the corresponding current:

$$i_C(t) = C\frac{dv_C}{dt} = j\omega C V_C e^{j\omega t} = I_C e^{j\omega t}. \tag{10.19}$$

The phasors for this particular current and voltage are

$$\mathbf{V}_C = V_C e^{j0} \qquad \text{and} \qquad \mathbf{I}_C = \omega C V_C e^{j\pi/2}. \tag{10.20}$$

From this result, we find that the impedance is

$$Z_C = \frac{1}{\omega C} e^{-j\pi/2} = -j\frac{1}{\omega C} = \frac{1}{j\omega C}. \tag{10.21}$$

Again this impedance is purely reactive since, if $Z_C = R_C + jX_C$, then we have

$$R_C = 0 \qquad \text{and} \qquad X_C = \frac{-1}{\omega C}. \tag{10.22}$$

The admittance expressions for the three elements are found from the relationship $Y = 1/Z$. Then, from Equations 10.14, 10.17, and 10.21, we see that

$$Y_R = \frac{1}{R} = G, \qquad Y_L = \frac{-j}{\omega L}, \qquad Y_C = j\omega C. \tag{10.23}$$

These impedance and admittance relationships for R, L, and C are summarized in Table 10.2.

Table 10.2

Element	*Impedance*	*Admittance*
R	R	$1/R$
L	$j\omega L$	$-j/(\omega L)$
C	$-j/(\omega C)$	$j\omega C$

From the relationship $V = ZI$, we see that if the current is the reference quantity, then the phase of the voltage with respect to this current is the phase of the impedance. Similarly, the expression $I = YV$ indicates that with the voltage as the reference quantity, the phase of the current with respect to the voltage is the phase of the admittance. This information is summarized in Table 10.3* and is illustrated in the right-hand column of Table 10.1.

Table 10.3

Element	*Phase of $v(t)$ in relationship to $i(t)$*	*Phase of $i(t)$ in relationship to $v(t)$*
R	0°	0°
L	90°	−90°
C	−90°	90°

When considered singly, the capacitor and inductor elements may have nonzero arbitrary constants for complementary functions associated with the responses. For instance, if there is an initial charge on a capacitor and if the excitation is current, even if i is sinusoidal, the voltage will be a sinusoid plus a constant. We are assuming that the initial condition is such that the response is simply a sinusoid. Alternatively, we take the point of view that even if the storage elements have non-zero initial conditions, when the element is combined with others to form a network, the complementary function for the network response eventually decays to zero for t approaching infinity.

* Table 10.3 is summarized by the mnemonic, "ELI, the ICE man," where E represents voltage and I current. Thus we read ELI as "Voltage leads current in an inductor," and ICE as "Current leads voltage in a capacitor." While this crutch may seem hardly necessary working at your desk, it will be helpful under the duress of an examination.

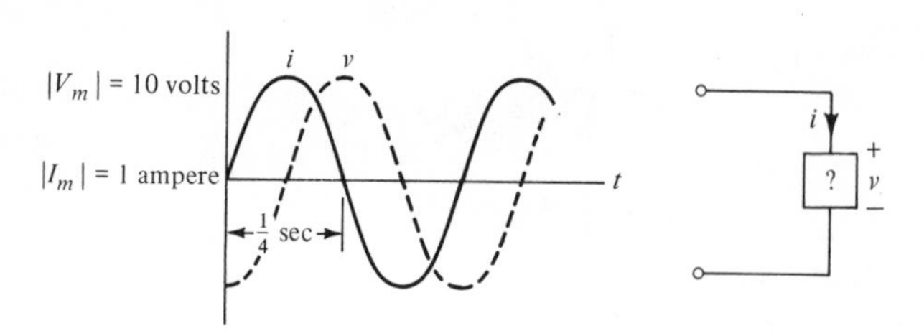

Figure 10.2

EXERCISES

10.2.1 Figure 10.2 shows the sinusoidal voltage and current for a constant network element. What is the element? What is its value in ohms, henrys, or farads?

10.2.2 With the references for voltage and current indicated in Figure 10.3 together with the waveforms shown, find the kind of element and its value in ohms, henrys, or farads.

10.2.3 Repeat Exercise 10.2.1 for the waveforms of Figure 10.4.

10.3 The impedance of series-parallel networks

Figure 10.5 shows a series *RLC* network connected to a sinusoidal source of voltage. For this network, we will consider two problems: (a) given $i(t)$, find $v(t)$; and (b) given $v(t)$, find $i(t)$. For the first problem, we will assume that we are at liberty to select $i(t)$ as the reference quantity by which we mean that we will select its phase angle to be zero. The Kirchhoff voltage law tells us that (see Exercise 10.3.1)

$$V_R + V_L + V_C = V \tag{10.24}$$

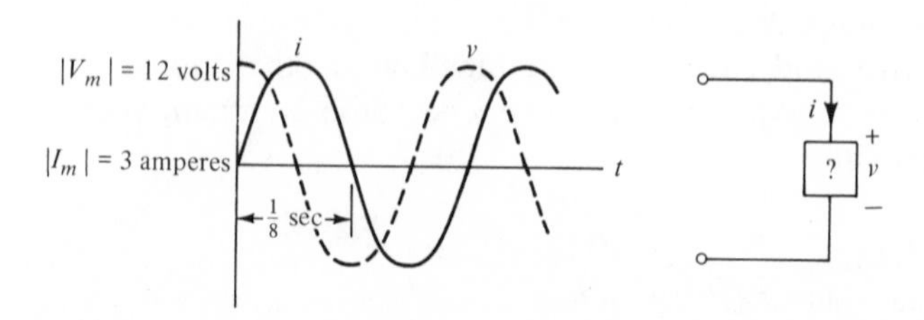

Figure 10.3

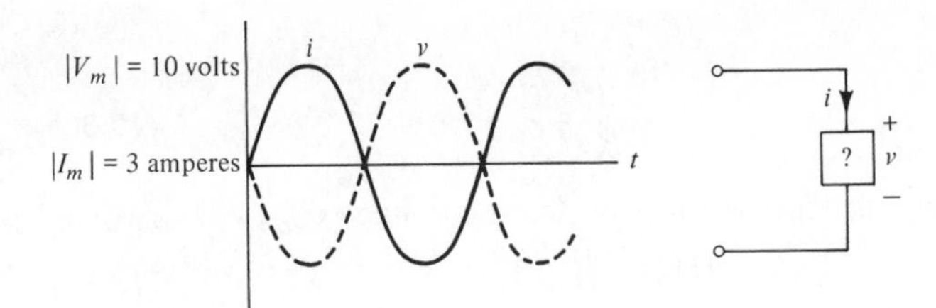

Figure 10.4

or

$$Z_R I + Z_L I + Z_C I = V. \tag{10.25}$$

If we next divide through the equation by the phasor I and define the phasor ratio V/I to be the total impedance of the network, Z_t, then we have

$$Z_t = Z_R + Z_L + Z_C. \tag{10.26}$$

This expression may be generalized for any number of elements in series as

$$Z_t = \sum_{i=1}^{n} Z_i. \tag{10.27}$$

If we substitute appropriate quantities from Table 10.2 into Equation 10.26, we find that Z_t in Equation 10.27 has the form

$$Z_t = R + j\left(\omega L - \frac{1}{\omega C}\right). \tag{10.28}$$

The magnitude and phase of the total impedance Z_t are

$$|Z_t| = \sqrt{R^2 + [\omega L - 1/(\omega C)]^2} \tag{10.29}$$

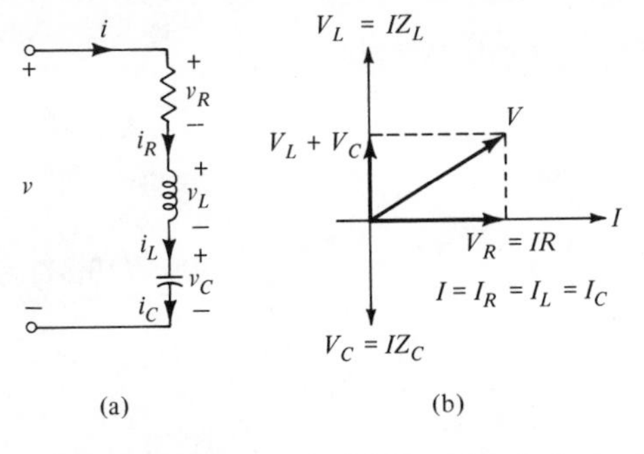

Figure 10.5 (a) Series *RLC* network. (b) The related phasor diagram showing the relationship of all voltages and currents.

and

$$\theta = \tan^{-1} \frac{\omega L - 1/(\omega C)}{R}. \tag{10.30}$$

From these values, we may write an expression for the voltage $v(t)$; if

$$i(t) = |I| \sin \omega t, \tag{10.31}$$

then the voltage $v(t)$ is

$$v(t) = \sqrt{R^2 + [\omega L - 1/(\omega C)]^2}\, |I| \sin \left[\omega t + \tan^{-1} \frac{\omega L - 1/(\omega C)}{R}\right]. \tag{10.32}$$

These operations are illustrated by the phasor diagram of Figure 10.5. This diagram can be constructed by observing that $I = I_R = I_L = I_C$ because of the series connection. These quantities are represented on the diagram as the reference phasor. The phasors V_R, V_L, and V_C are located on this diagram by superimposing the individual phasor diagrams shown in Table 10.1. Once this is accomplished, the phasor V is found from

$$V = V_R + V_L + V_C. \tag{10.33}$$

Directly from the figure, we see that the phase of V with respect to I is that angle described by Equation 10.30. We also see that V has a magnitude equal to the magnitude of Z_t times the magnitude of I. Observe that

$$|V_R| + |V_L| + |V_C| \neq |V|. \tag{10.34}$$

For this particular figure, both $|V_L|$ and $|V_C|$ are larger than $|V|$.

For the second problem, given $v(t)$ and required to find $i(t)$, we assume that $v(t)$ is the reference quantity. This does not necessarily imply that the phase of $v(t)$ is zero, but rather that other quantities will be expressed in terms of the $v(t)$ reference. Let the voltage $v(t)$ be given by

$$v(t) = |V| \sin (\omega t + \phi) = \text{Im}\,(|V| e^{j(\omega t + \phi)}). \tag{10.35}$$

For convenience, choose the real quantity $|V|$ for the V phasor:

$$V = |V|. \tag{10.36}$$

The phasor current is related to V by

$$I = Y_t V = \frac{1}{Z_t} V. \tag{10.37}$$

Now the phase and magnitude of Z_t are given by Equations 10.29 and 10.30, and we see that

$$I = \frac{V}{\sqrt{R^2 + [\omega L - 1/(\omega C)]^2}} \exp \left\{j\left[-\tan^{-1} \frac{\omega L - 1/(\omega C)}{R}\right]\right\}. \tag{10.38}$$

But from the choice of the reference phasor V in Equation 10.36 and from the relationship of $v(t)$ to V in Equation 10.35, we see that

$$i(t) = \text{Im}\,(Ie^{j(\omega t+\phi)}). \tag{10.39}$$

Hence we have

$$i(t) = \frac{|V|}{|Z_t|}\sin\left[\omega t + \phi - \tan^{-1}\frac{\omega L - 1/(\omega C)}{R}\right], \tag{10.40}$$

which is the desired current.

In Equation 10.27, it is shown that the impedance of a number of elements connected in series is the sum of the impedances of the individual elements. The analysis of the representation of a number of elements connected in parallel shown in Figure 10.6 requires that we recognize that the same voltage is applied to all the parallel elements (or combinations of elements within any one box in the figure). The total current is the summation of the individual currents to the elements by Kirchhoff's current law, so that (see Exercise 10.3.2) we have

$$I = I_1 + I_2 + I_3 + \cdots + I_n. \tag{10.41}$$

Now each of the phasor currents is related to the phasor voltage V by the appropriate admittance, so that we obtain

$$I = Y_1V + Y_2V + Y_3V + \cdots + Y_nV. \tag{10.42}$$

Again dividing through by V and defining the ratio of I to V to be the total admittance of the network, we have

$$Y_t = Y_1 + Y_2 + Y_3 + \cdots + Y_n. \tag{10.43}$$

The impedance of the network of Figure 10.6 is

$$Z_t = \frac{1}{Y_t} = \frac{1}{Y_1 + Y_2 + \cdots + Y_n}. \tag{10.44}$$

A special case of this equation for $n = 2$ is

$$Z_t = \frac{Z_1Z_2}{Z_1 + Z_2}, \tag{10.45}$$

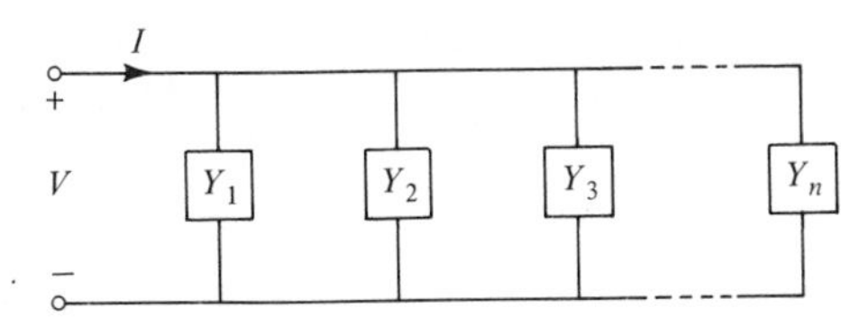

Figure 10.6 Representation of elements connected in parallel with each one characterized by its admittance.

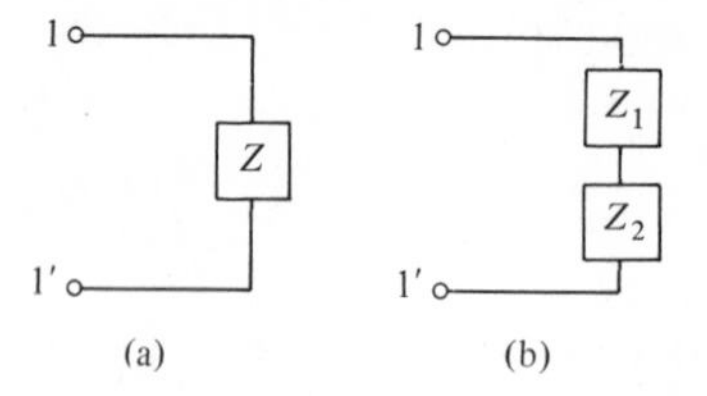

Figure 10.7 The two networks are equivalent for sinusoidal signals (in the steady-state) for measurements made at terminals $1-1'$ when $Z = Z_1 + Z_2$.

for which the pattern "product over sum" of impedances is easily remembered. In this equation, $Z_1 = 1/Y_1$ and $Z_2 = 1/Y_2$.

Two 1-port networks are said to be *equivalent* if they have the same impedance at the port terminals called the *driving-point* terminals. The two networks of Figure 10.7 are equivalent *for sinusoidal signals* when

$$Z = Z_1 + Z_2. \tag{10.46}$$

Note that this definition for equivalence of 1-port networks is for sinusoidal steady-state only. More general cases will be given later. The same conclusions apply to the parallel connection shown in Figure 10.8 when

$$Y = \frac{1}{Z} = Y_1 + Y_2. \tag{10.47}$$

The pattern suggested by these two simple examples may be used successively to find the equivalent impedance for series-parallel networks of the form shown in Figure 10.9(a). (Non-series-parallel networks formed by the bridging of nodes will be considered in a later chapter.) Reduction for this network starts at the "far end" of the network where we recognize that $Z_a = Z_3 + Z_4$ as in (b). The impedance Z_b is found from the parallel combination of Z_2 and Z_a from Equation 10.45. Thus we have

$$Z_b = \frac{Z_2 Z_a}{Z_2 + Z_a}. \tag{10.48}$$

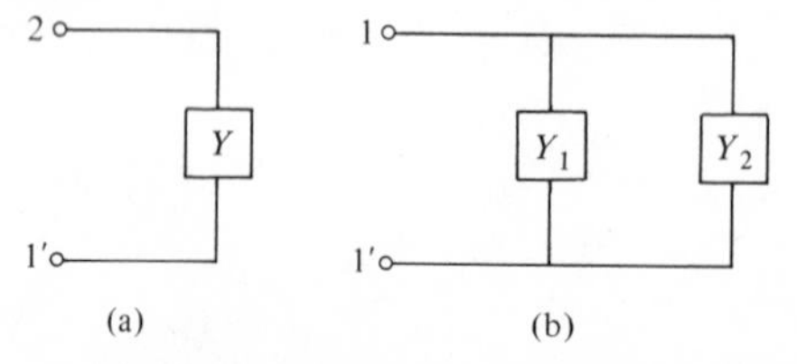

Figure 10.8 Two networks which are equivalent for sinusoidal signals for measurements at terminals 1–1′ when $Y = 1/Z = Y_1 + Y_2$.

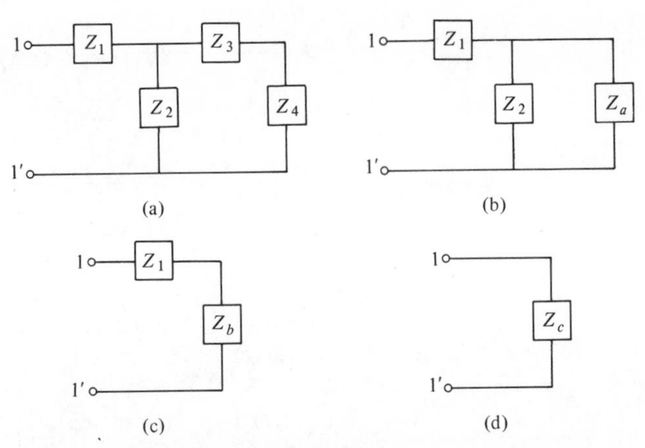

ure 10.9 Steps in finding an equivalent network by successive series and parallel combinations of impedances for sinusoidal signals.

Finally, we see that the total impedance of the network at terminals 1 and 1′ is the sum of Z_1 and Z_b. Thus we have

$$Z_t = Z_c = Z_1 + \frac{Z_2(Z_3 + Z_4)}{Z_2 + Z_3 + Z_4}. \tag{10.49}$$

ERCISES

10.3.1 Verify Equation 10.24, starting from

$$v_R(t) + v_L(t) + v_C(t) = v(t)$$

as given by Kirchhoff's voltage law and writing each of the sinusoidal voltages as the imaginary part of appropriate rotating phasors.

10.3.2 Verify Equation 10.41. Use a hint similar to that in Exercise 10.3.1.

10.3.3 For the network in Figure 10.10, an rms-reading voltmeter records the following values:

Voltmeter terminals at	*Voltage (volts)*
a and b	1
b and c	3
c and d	2
d and e	2
e and f	1
f and g	2
g and h	3
h and i	2

What will the voltmeter read when connected from a to i? From c to g?

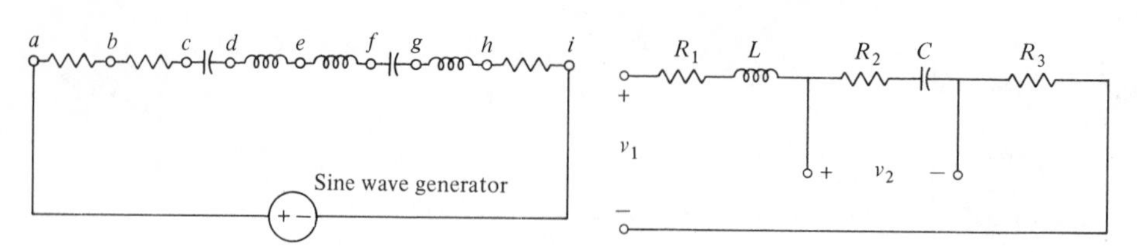

Figure 10.10

Figure 10.11

10.3.4 In the network of Figure 10.11, $R_1 = 4$ ohms, $R_2 = 6$ ohms, and $R_3 = 2$ ohms. The reactance of L has a magnitude of 3 ohms and that of C a magnitude of 8 ohms. If $|V_1| = 100$ volts, find $|V_2|$.

10.3.5 In the series RLC network of Figure 10.12, $i(t)$ lags $v(t)$ by 45°. If $|V| = 1$ volt, $|I| = 1$ ampere, $L = 1$ henry, and $\omega = 2$ radians/sec, find the magnitude of the voltage across each of the elements: $|V_L|$, $|V_C|$, and $|V_R|$.

10.3.6 The following voltmeter readings are made on the series RLC network given in Figure 10.13 where the subscripts refer to the voltmeter connections:

$$|V_{ac}| = 20 \text{ volts}, \qquad |V_{bd}| = 9 \text{ volts}, \qquad |V_{ad}| = 15 \text{ volts}.$$

Find the voltmeter readings across each of the three elements under these conditions.

10.3.7 Find the impedance of the network shown in Figure 10.14(a) when $\omega = 1$ radian/sec. Repeat for $\omega = 2$ radians/sec.

10.3.8 Determine the impedance of the network of Figure 10.14(b) at terminals 1 and 1′ for $\omega = 2$ radians/sec.

10.3.9 In Exercise 10.3.4, if $|V_2| = 100$ volts, find $|V_1|$.

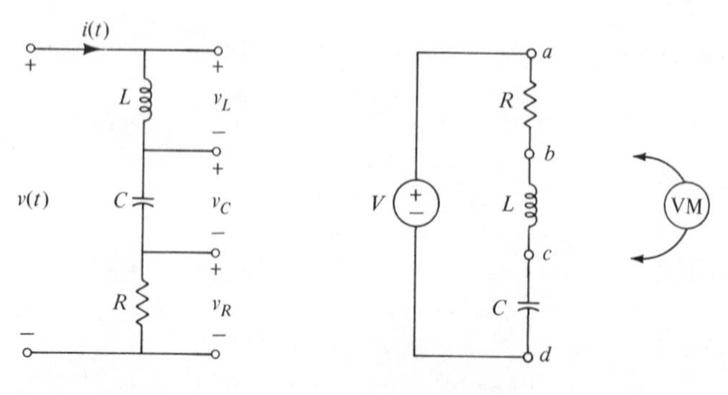

Figure 10.12

Figure 10.13

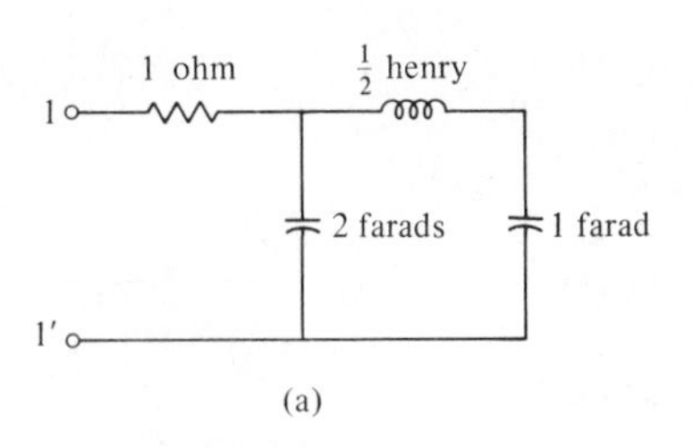

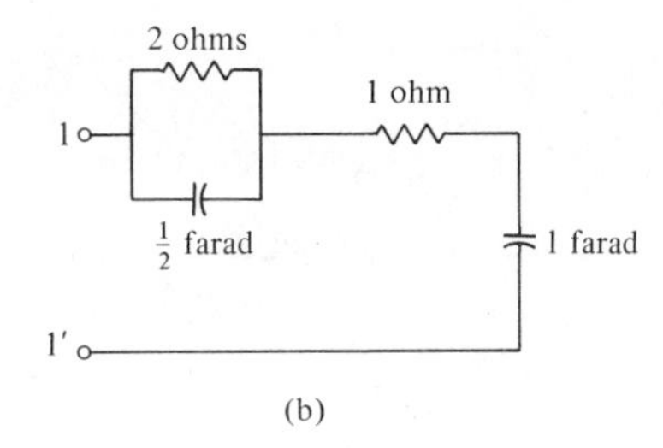

Figure 10.14

.4 Analysis with phasor diagrams

The *phasor diagram* of Figure 10.5(b) shows important phase and magnitude relationships for the voltages and currents in the series *RLC* network of Figure 10.5(a). It shows at a glance that the phasor voltage V_L leads the phasor current I by 90°, that the phasor sum of $V_L + V_C$ and V_R is equal to the phasor voltage V. We also see at a glance that the current lags the applied voltage by an angle between 0° and 90°. If we wish to find the voltage across the resistor and inductor, it may be done routinely by adding the phasors V_L and V_R. Phasor diagrams of this type find frequent use for the analysis of networks with sinusoidal signals, not often for detailed computation but instead for a "rough check" to guide computations and analysis.

The *reference angle* or *reference* of a phasor diagram is the line $\theta = 0$. It is not really necessary that any voltage or current coincide with the reference, although analysis is often more convenient when a given voltage or current phasor is the reference. Different phasors of Figure 10.5(b) may be made to be the reference simply by rotating the entire phasor diagram in either the clockwise or counterclockwise direction. For example, when all phasors are rotated in the clockwise direction, the phasor V may be made to coincide with the reference angle $\theta = 0$.

The selection of the reference quantity to be used in the construction of a phasor diagram turns out to be an important matter. In the series *RLC* network of Figure 10.5(a), the current common to the three elements turned out to be a good choice as a reference. Try working the problem with V as the reference to prove that it is possible to make a poor choice. When several elements are in parallel, then the voltage common to these elements is a good choice as a reference. No firm rule applies to all situations. A number of examples will be given to indicate the choice of the reference quantity.

The phasor diagram may be drawn in a number of ways such as in the form of a *polar* (or ray) diagram with all phasors originating at the origin, or in the form of a *polygon* with one phasor located at the end of another. These two forms of phasor diagrams are illustrated by the examples of Figure 10.15.

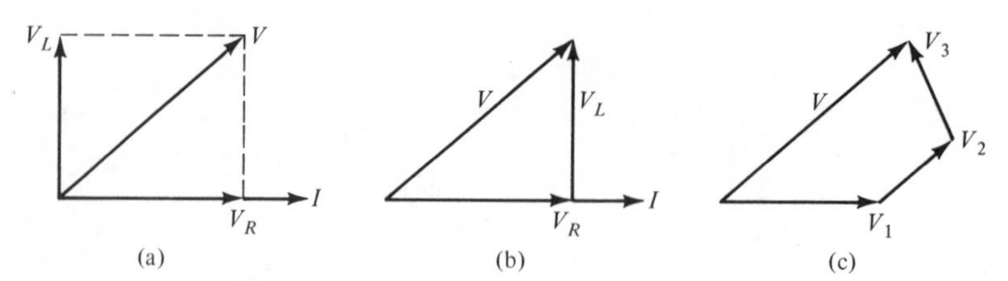

Figure 10.15 (a) A polar phasor diagram. (b) A phasor diagram in polygon form. (c) Another polygon phasor diagram.

EXAMPLE 10.4.1

For the network in Figure 10.16, it is given that $v_L(t) = \sin t$ and it is required that we find $v(t)$. The phasor diagram is constructed with $V_L = 1$ as the reference. We know that the current in the inductor lags the voltage by 90°. Since this current must also be the current of the resistor, we see that V_R will be in phase with I_L. Finally, as shown in Figure 10.16(b), $V = V_L + V_R$, and from this phasor voltage $v(t)$ may be written. With the numerical values given: $V_L = 1$, $I_L = V_L/Z_L = -j2$, $V_R = RI_L = -j4$, and $V = 1 - j4$ or $V = \sqrt{17}\, \underline{/-\tan^{-1} 4}$. Finally, we have

$$v(t) = \sqrt{17} \sin (t - \tan^{-1} 4) \text{ volts}. \tag{10.50}$$

EXAMPLE 10.4.2

We are required to draw a phasor diagram for the voltages and currents of the network in Figure 10.17(a). We select the two voltage phasors V_R and V_C as the reference as shown in Figure 10.17(b). We know that I_R is in phase with V_R and that I_C leads V_C by 90°. By KCL, $I_L = I_C + I_R$, and the construction of this phasor is shown by the broken lines. Now we also know that V_L leads I_L by 90°, which fixes the position of V_L. Finally, from KVL, $V_L + V_C$ and again broken lines indicate the construction that determines the position of V. This phasor

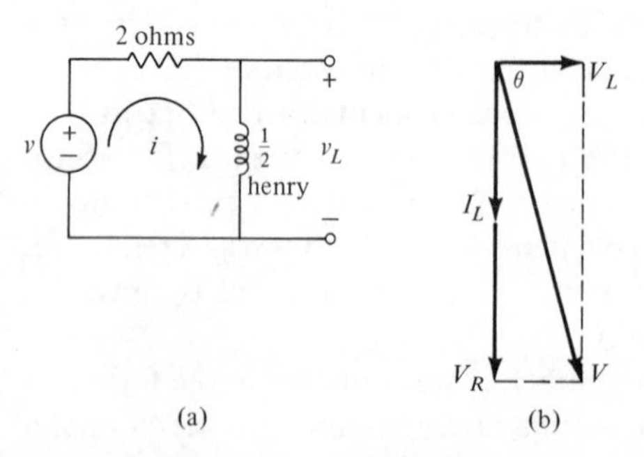

Figure 10.16 The network and corresponding phasor diagram used in Example 10.4.1.

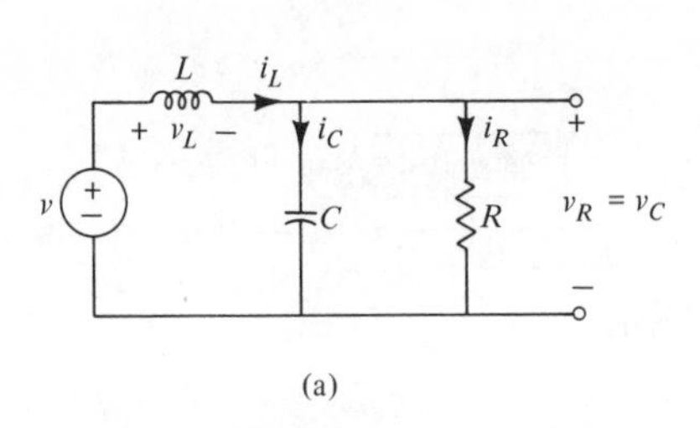

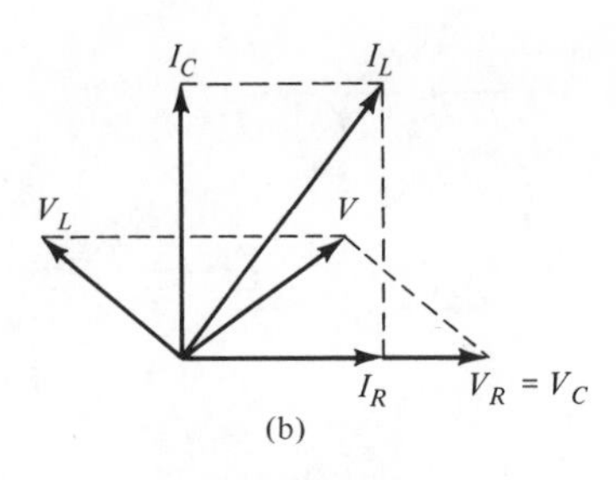

ure 10.17 A series-parallel network and a complete phasor diagram showing all voltages and currents.

diagram was constructed without knowledge of element values or frequency of the sinusoidal signal, and it is clear that the actual phasor diagram for a specific network may differ in detail (magnitudes and angles) from that shown. If $v_R(t) = |V_R| \cos(\omega t + \phi)$, then $v(t) = \text{Re}(Ve^{j(\omega t + \phi)})$, where V is the phasor obtained from the phasor diagram.

‹ERCISES

10.4.1 In the network of Figure 10.18, it is given that $v_L(t) = 2\sqrt{2} \sin(t - 30°)$. Find $v(t)$, using phasors and show each step on a phasor diagram.

10.4.2 In the network of Figure 10.19, $v_2(t) = 2 \sin(2t - \pi/4)$ for all t. (a) Find $v_1(t)$, using phasors. (b) Draw a phasor diagram showing all voltages and currents.

10.4.3 For the RC network of Figure 10.20, $v_2(t) = \sqrt{2} \sin t$. (a) Draw a phasor diagram (approximately to scale) showing all voltages in the network. (b) From the phasor diagram, show that $v_1(t) = 3\sqrt{2} \sin(t - 90°)$ volts. (c) What value will an rms meter record for the voltage $v_1(t)$?

10.4.4 This exercise is intended to provide practice in drawing and interpreting phasor diagrams. In Table 10.4, the network of Figure 10.21 is described as well as the voltage $v_2(t) = V_m \sin(\omega t + \phi)$. For each row, you are to determine $v_1(t)$ and draw a phasor diagram used in finding this voltage. (A double entry in the Network 1 and Network 2 columns implies a series connection of two elements.)

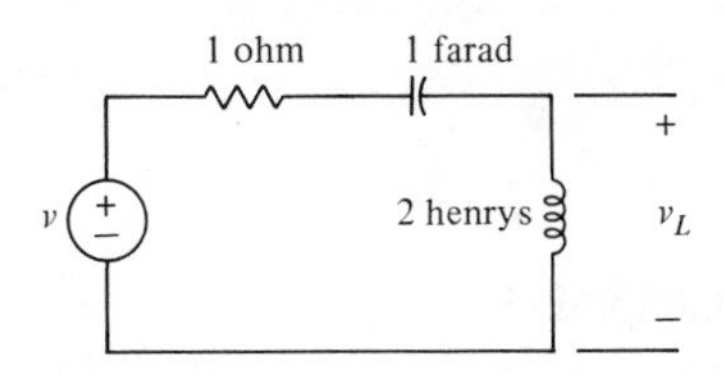

Figure 10.18

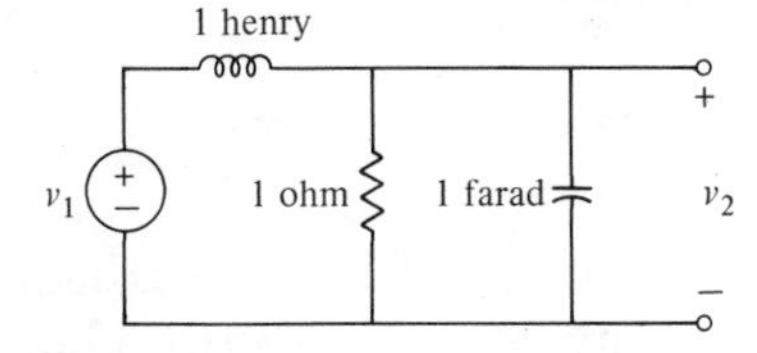

Figure 10.19

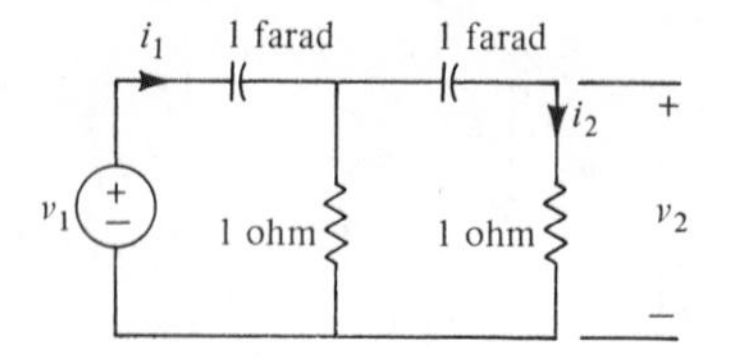

Figure 10.20

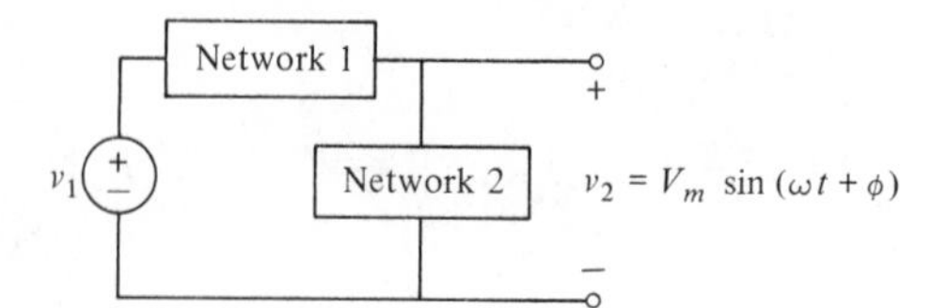

Figure 10.21

Table 10.4

Number	*Network* 1	*Network* 2	V_m	ω	ϕ
a	$R = 1$	$C = 2$	2	$\frac{1}{2}$	$-30°$
b	$R = 2$	$C = 1$	$2\sqrt{2}$	2	$45°$
c	$R = 20$	$C = \frac{1}{2}$	1	0.1	$0°$
d	$R = 2$	$L = 2$	2	$\frac{1}{2}$	$30°$
e	$L = \frac{1}{2}$	$R = 1$	$2\sqrt{2}$	$\frac{1}{2}$	$0°$
f	$C = 2$	$R = 2$	3	1	$45°$
g	$L = 3$	$C = 1$	$5\sqrt{2}$	$\frac{1}{2}$	$-45°$
h	$C = 1$	$L = \frac{1}{2}$	1	2	$0°$
i	$R = 2$	$L = \frac{1}{2}$	$2\sqrt{2}$	4	$37.5°$
j	$L = 3$	$R = 4$	$120\sqrt{2}$	1	$-37.5°$
k	$C = 4$	$R = 3$	1	$\frac{1}{16}$	$20°$
l	$L = 1$	$C = 2$	2	$\frac{1}{2}$	$120°$
m	$C = 3$	$L = 4$	1	1	$240°$
n	$R = 3$	$R = 4$	$2\sqrt{2}$	27	$713°$
o	$R = 5$	$C = 1$	$2\sqrt{3}$	$\frac{1}{4}$	$0°$
p	$R = 1,\ C = 1$	$L = 2$	1	$\frac{1}{2}$	$-45°$
q	$R = 2,\ C = 2$	$L = 1$	3	1	$30°$
r	$R = 1,\ L = 2$	$C = \frac{1}{2}$	$3\sqrt{2}$	1	$45°$
s	$R = \frac{1}{2},\ L = 1$	$C = 2$	1	2	$120°$
t	$L = 1,\ C = 2$	$R = 1$	$3\sqrt{2}$	$\frac{1}{2}$	$25°$
u	$R = 1,\ C = 1$	$R = 1,\ L = 1$	$4\sqrt{2}$	1	$0°$
v	$R = 2,\ C = 2$	$R = 2,\ L = 1$	1	$\frac{1}{2}$	$90°$
w	$L = \frac{1}{2},\ C = 1$	$R = 1,\ C = 1$	120	1	$-90°$
x	$R = 3,\ L = 2$	$R = 1,\ C = \frac{1}{2}$	1	$\frac{1}{2}$	$0°$
y	$L = 1,\ C = 2$	$L = 2,\ C = 1$	$5\sqrt{2}$	1	$-90°$

10.5 Complex power

Our study of power and energy was initiated in Chapter 3 by introducing the root-mean-square value of a signal and continued in Chapter 6 for time-domain voltages and currents. In this section, we will discuss power relationships expressed in terms of phasor voltages and phasor currents. The average power was defined by Equation 6.9 as

$$P_{av} = \frac{1}{T} \int_{t_1}^{t_1+T} p(t)\, dt \tag{10.51}$$

for a periodic $p(t)$ of period T. Since $p(t) = v(t)i(t)$, we let $v = |V_m| \cos \omega t$ and $i = |I_m| \cos (\omega t - \theta)$. Then

$$\begin{aligned} p(t) &= vi = |V_m| \cos \omega t |I_m| \cos (\omega t - \theta) \\ &= \frac{|V_m||I_m|}{2} [\cos \theta + \cos (2\omega t - \theta)] \end{aligned} \tag{10.52}$$

through the use of a trigonometric identity. Now the average value of the cosine of frequency 2ω over a complete period T is 0, so that the second term of this equation contributes nothing to P_{av}. Thus we conclude that

$$P_{av} = \frac{|V_m||I_m|}{2} \cos \theta = |V_{rms}||I_{rms}| \cos \theta\,. \tag{10.53}$$

In this equation, V_{rms} and I_{rms} are phasor voltage and phasor current, respectively, scaled down from V_m and I_m by a factor of $1/\sqrt{2} \simeq 0.707$. The calculation of P_{av} involves only the magnitude of these phasors as the equation shows. The constant $\cos \theta$ is known as the *power factor* (p.f.), and θ is the *power-factor angle*, the angle of the phasor V_{rms} measured with I_{rms} as the reference. Observe that θ is also the angle of $Z(j\omega) = V(j\omega)/I(j\omega) = |Z|\underline{/\theta}$. It is conventional to speak of the power factor as lagging when θ is positive and as leading when θ is negative; the origins of these two terms will be evident later when we study the power relationships for the individual elements.

The measurement of the quantities in Equation 10.53 is routine since P_{av} is measured by a standard wattmeter, $|V_{rms}|$ by a voltmeter, and $|I_{rms}|$ by an ammeter. From the three measurements, $\cos \theta$ is determined and thus $|\theta|$ is fixed.

Now if we are dealing with time-invariant sources (batteries, dc generators, etc.) and resistive networks, then $\theta = 0$ and the average power is the product of $|V_{rms}|$ and $|I_{rms}|$. When the sources are sinusoidal rather than time-invariant, then this product is no longer P_{av}. In order to speak of it, let it be called apparent power, a real positive number which will be identified as

$$|S| = |V_{rms}||I_{rms}|\,. \tag{10.54}$$

In earlier sections of this chapter and in Chapter 9, we have made extensive use of the complex-plane and right-triangle relationships, and these concepts turn out to be useful in discussing power. We now have two equations,

$$P_{av} = |V_{rms}||I_{rms}| \cos \theta \tag{10.55}$$

and

$$|S| = |V_{\text{rms}}||I_{\text{rms}}|, \tag{10.56}$$

so that

$$P_{\text{av}} = |S| \cos \theta. \tag{10.57}$$

With the right triangle in mind, we define a new quantity

$$Q = |S| \sin \theta, \tag{10.58}$$

and add P_{av} and Q as complex numbers:

$$P_{\text{av}} + jQ = |S| \cos \theta + j|S| \sin \theta = |S|(\cos \theta + j \sin \theta). \tag{10.59}$$

We can now see that this complex number is, in polar coordinates, $|S|\underline{/\theta}$, or simply complex S. Thus we have developed the idea that

$$S = P_{\text{av}} + jQ \tag{10.60}$$

is *complex power*. Has anything been gained by this definition?

In answer to this question, we note that the specification of Q with a sign will permit $|\theta|$ to be resolved as either $+\theta$ or $-\theta$. Of greater importance, it is more convenient to speak of P_{av} and Q as rectangular coordinates than to speak of $|S|$ and θ as polar coordinates, primarily because it is routine to measure Q with a meter which is similar to the wattmeter. In fact, the use of complex power and of Q is standard in the electrical energy industry.

To distinguish the three quantities involved in complex power, different units are used for each. The normal unit for power is the watt. The other two units are the *var* for volt-ampere-reactive, and the *va* for volt-ampere. They are used as follows:

$$P_{\text{av}} = \text{average power in watts,}$$
$$Q = \text{reactive power in vars,}$$
$$S = \text{apparent or complex power in va.}$$

These units are more familiar in power applications when the prefix *kilo-* is attached: kilowatts (kW), kilovars (kvar), and kilovolt-ampere (kva).

Certain other conventions are associated with the concept of complex power. Earlier in the chapter, we discussed the relationship of the phasor voltage and phasor current in the three passive elements R, L, and C. In the resistor, the voltage and current are in phase so that $\theta = 0$ and

$$S_R = P_{\text{av}} + j0. \tag{10.61}$$

In the inductor, current lags the voltage and since θ was defined as the angle of the voltage phasor measured with respect to the current phasor, $\theta = 90°$ and

$$S_L = 0 + jQ_L. \tag{10.62}$$

Similarly, for the capacitor, current leads the voltage; hence $\theta = -90°$ and

$$S_C = 0 - jQ_C. \tag{10.63}$$

From these results, we see the origin of the words leading and lagging, defined earlier. When a network as measured at a port is primarily inductive, then S will have a positive imaginary part, and θ is positive. In this case $\cos\theta$ is spoken of as a *lagging power factor* (current lags voltage). For the other case, when S has a negative imaginary part, the network as measured at a port is primarily capacitive, and $\cos\theta$ is spoken of as a *leading power factor* (current leads voltage). In effect, we use the words *lagging* and *leading* to replace the sign associated with $-\theta$.

Observe also from the last two equations, Equations 10.62 and 10.63, that Q is associated with the inductor and the capacitor, the two so-called *reactive* elements. For this reason, Q is known as *reactive power*. In the time domain, energy is supplied to the capacitor and the inductor, and it is stored by these elements and then returned to the source or to other elements. In a sense, Q represents lossless power. It must be supplied to the network, but it is not dissipated, only stored and returned. Good practice in a power system requires that the value of Q be kept at some minimum level, and this is accomplished by *power-factor correction*. Since most industrial loads are inductive, correction is accomplished by the addition of a capacitor (or equivalent) to the system.

We next return to a consideration of complex power S with the objective of expressing it in terms of phasor voltage and phasor current, the generalization of Equation 10.56. Now the power factor angle θ is the angle of the voltage phasor measured with respect to the current phasor. If we let

$$V_{rms} = |V_{rms}|\underline{/\phi_v} \qquad \text{and} \qquad I_{rms} = |I_{rms}|\underline{/\phi_i}, \tag{10.64}$$

then the power-factor angle θ is

$$\theta = \phi_v - \phi_i. \tag{10.65}$$

However, when we multiply V_{rms} and I_{rms}, we get

$$|V_{rms}|\underline{/\phi_v}|I_{rms}|\underline{/\phi_1} = |V_{rms}||I_{rms}|\underline{/\phi_v + \phi_i}. \tag{10.66}$$

To obtain the difference we require rather than the sum, it is necessary to use the conjugate of the current

$$I^*_{rms} = |I_{rms}|\underline{/-\phi_i} \tag{10.67}$$

to give θ as in Equation 10.65. Thus we see that

$$S = V_{rms}I^*_{rms} \tag{10.68}$$

and that

$$P_{av} = \text{Re}\,(V_{rms}I^*_{rms}) = |V_{rms}||I_{rms}|\cos\theta \tag{10.69}$$

and

$$Q = \mathrm{Im}\,(V_{\mathrm{rms}} I^*_{\mathrm{rms}}) = |V_{\mathrm{rms}}||I_{\mathrm{rms}}| \sin\theta\,. \tag{10.70}$$

These are the basic equations that we use in the study of complex power in networks under sinusoidal steady-state conditions. Several other forms of the equations are also useful and are based on the relationship of voltage and current in phasor form, $V_{\mathrm{rms}} = ZI_{\mathrm{rms}}$ and $I_{\mathrm{rms}} = YV_{\mathrm{rms}}$. Substituting these into Equation 10.68, we see that

$$S = V_{\mathrm{rms}}(YV_{\mathrm{rms}})^* = V_{\mathrm{rms}} V^*_{\mathrm{rms}} Y^* = |V_{\mathrm{rms}}|^2 Y^* \tag{10.71}$$

and

$$S = (ZI_{\mathrm{rms}})I^*_{\mathrm{rms}} = I_{\mathrm{rms}} I^*_{\mathrm{rms}} Z = |I_{\mathrm{rms}}|^2 Z \tag{10.72}$$

since the product of a phasor and its conjugate is the phasor magnitude squared. It should be noted from the same equations that the angle of S is the angle of Z or the angle of Y^* which is θ. Identifying the real and imaginary parts of impedance and admittance in the usual form

$$Z = R + jX = |Z|\underline{/\theta^\circ} \quad \text{and} \quad Y = G + jB = \frac{1}{Z} = \frac{1}{|Z|}\underline{/-\theta^\circ}, \tag{10.73}$$

we see that

$$P_{\mathrm{av}} = |V_{\mathrm{rms}}|^2 G = |V_{\mathrm{rms}}|^2 |Y| \cos\theta = |I_{\mathrm{rms}}|^2 R = I^2_{\mathrm{rms}} |Z| \cos\theta \tag{10.74}$$

and

$$Q = -|V_{\mathrm{rms}}|^2 B = |V_{\mathrm{rms}}|^2 |Y| \sin\theta = |I_{\mathrm{rms}}|^2 X = |I_{\mathrm{rms}}|^2 |Z| \sin\theta\,. \tag{10.75}$$

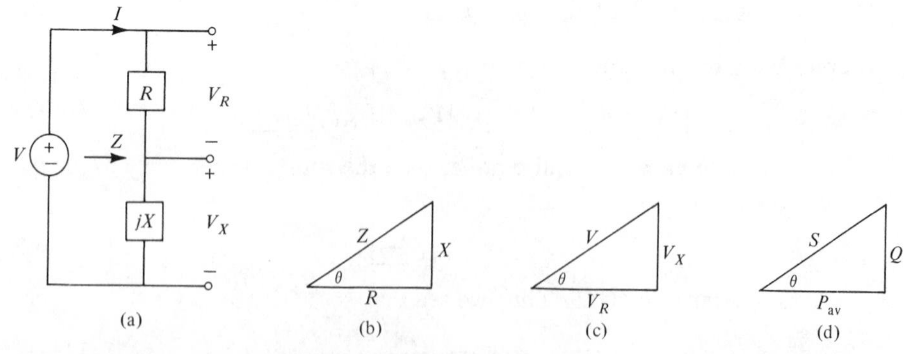

Figure 10.22 For the network shown in (a), there are three related right-triangle relationships: (b) that for impedance, (c) that for the voltages of the network, and (d) that for complex power $S = P_{\mathrm{av}} + jQ$.

From this last equation, it is seen again that a positive sign of Q implies inductive reactive power, whereas a negative sign implies capacitive reactive power.

The phasor diagrams studied in the last section have useful counterparts for complex power, suggested by Equation 10.68. For the network of Figure 10.22(a), part of the figure shows the familiar impedance triangle with Z, R, and X identified. If each of these quantities is multiplied by I_{rms}, then the voltage diagram of (c) results. Finally, multiplication of each quantity in (b) of the figure by $|I_{rms}|^2$ gives the relationships that exist for complex power, $S = P_{av} + jQ$. From the figures, we see that the power factor is

$$\text{p.f.} = \cos\theta = \frac{P_{av}}{|S|} = \frac{R}{|Z|}. \tag{10.76}$$

EXAMPLE 10.5.1

Consider the network shown in Figure 10.23(a) with $V_{rms} = 1\underline{/0°}$ volt, $G = 1$ mho, and $B = -\frac{1}{2}$ mho. The current is

$$\begin{aligned} I_{rms} = V_{rms}Y &= (1 + j0)(1 - j\tfrac{1}{2}) \\ &= 1 - j\tfrac{1}{2} \text{ amperes}. \end{aligned} \tag{10.77}$$

Complex power is

$$\begin{aligned} S = V_{rms}I_{rms}^* &= (1 + j0)(1 + j\tfrac{1}{2}) \\ &= 1 + j\tfrac{1}{2} \text{ va} \end{aligned} \tag{10.78}$$

so that

$$P_{av} = 1 \text{ watt}, \tag{10.79}$$

$$Q = \tfrac{1}{2} \text{ var}, \tag{10.80}$$

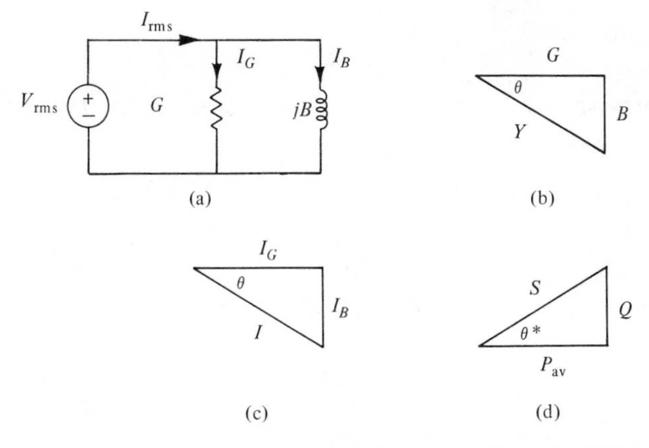

Figure 10.23 The network and right-triangle relationships of Example 10.5.1.

and

$$\text{p.f.} = \cos\theta = \frac{P}{|S|} = \frac{1}{\frac{5}{4}} = \frac{2}{5}\ \text{lagging}. \tag{10.81}$$

The relationships for Y, the currents, and the components of phasor power are shown in the figure.

In Chapter 6, we introduced Tellegen's theorem in Equation 6.32. We shall now show that this has important implications with respect to phasor power in a network. We will make use of Tellegen's theorem with v_k and i_k replaced by V_k and I_k^* and with k applying to the same network rather than two different networks with the same topology. Stated in these terms, Tellegen's theorem is based on the equation

$$\sum_{k=1}^{N} V_k I_k^* = 0. \tag{10.82}$$

Now the product $V_k I_k^*$ is simply complex power S_k. Clearly the real and imaginary parts of S_k must separately sum to zero if this equation is to be satisfied. Hence we may separate Equation 10.82 into three equations:

$$\sum_{k=1}^{N} S_k = 0, \tag{10.83}$$

$$\sum_{k=1}^{N} P_{\text{av}\,k} = 0, \tag{10.84}$$

and

$$\sum_{k=1}^{N} Q_k = 0. \tag{10.85}$$

EXAMPLE 10.5.2

Figure 10.24 shows a 2-port network in which the output is $V_2 = 1$ volt. Using the element impedances given, we find by routine analysis that

$$I_4 = 1, \qquad I_3 = 1, \qquad I_2 = -1 + j1, \qquad I_1 = j1, \qquad V_1 = 1 + j2$$

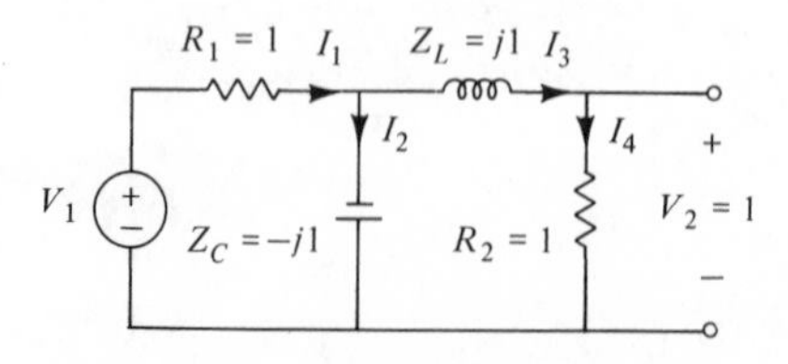

Figure 10.24 The network of Example 10.5.2.

in amperes and volts. The complex power supplied by the source must be

$$S = V_1 I_1^* = (1 + j2)(-j1) = 2 - j1. \tag{10.86}$$

This must be equal to the phasor power of the four elements. These values are:

$$\begin{aligned} P_4 &= I_4^2 R_4 = 1 \text{ watt}, \\ Q_3 &= I_3^2 X_{L_3} = 1 \text{ var}, \\ Q_2 &= I_2^2 X_{C_2} = 2(-1) = -2 \text{ vars}, \\ P_1 &= I_1^2 R_1 = 1 \text{ watt}. \end{aligned}$$

The complex power supplied by the source must be identical to that of the individual branches of the network. In applying Tellegen's theorem, it is necessary that a sign be given to all values of $P_{\text{av}\,k}$ and jQ_k depending on the reference voltage and current. If we employ the convention that with the + sign at the tail of the arrow power is positive, then Tellegen's theorem requires that

$$P_4 + jQ_3 + jQ_2 + P_1 - V_1 I_1^* = 0 \tag{10.87}$$

or

$$1 + j1 - j2 + 1 - (2 - j1) = 0. \tag{10.88}$$

ERCISES

10.5.1 If $V_{\text{rms}} = 1\underline{/0^\circ}$ and $Z_R = 1$ ohm in the given network (Figure 10.25), determine the values for I_{rms}, P_{av}, S, and θ.

10.5.2 Repeat Exercise 10.5.1 for the same V_{rms} but $Z_R = \frac{1}{2}$ ohm.

10.5.3 In the network of Figure 10.26, it is given that $V_2 = 1\underline{/0^\circ}$ volt and $Z_R = 1$ ohm. Determine all branch currents in the network and V_1. From this compute all values of complex power in the network, showing the balance between the power supplied by the source and that of the elements.

10.5.4 Repeat Exercise 10.5.3 for the given V_2, but with $Z_R = \frac{1}{2}$ ohm.

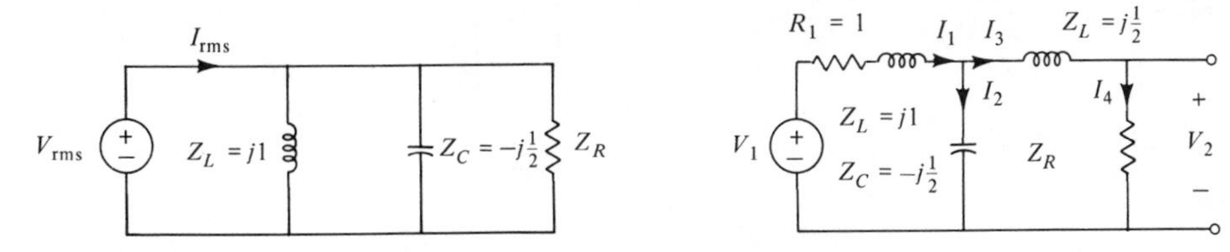

Figure 10.25

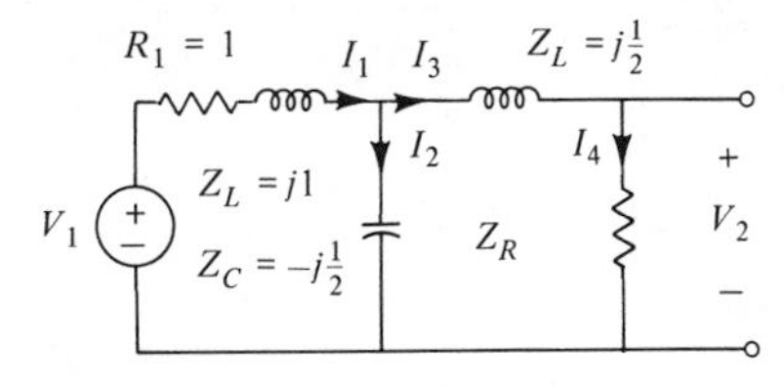

Figure 10.26

10.6 Generalization of impedance: network functions

In Example 10.4.1, we let $V_L = 1$ and found that $V = 1 - j4$ or $V = \sqrt{17}\, e^{-j \tan^{-1} 4}$. The quotient of these two phasors is

$$\frac{V_L}{V} = \frac{1}{\sqrt{17}}\, e^{j \tan^{-1} 4}. \tag{10.89}$$

By this example, we see that the quotient of two voltage phasors is itself a phasor similar to impedance. Furthermore, this ratio may be used in exactly the same way as impedance. Thus, if $v(t) = 2 \sin t$ and the associated phasor is $V = 2$, then we have

$$V_L = \frac{1}{\sqrt{17}}\, e^{j \tan^{-1} 4}\, V. \tag{10.90}$$

From this, we see that

$$v_L(t) = \frac{2}{\sqrt{17}} \sin\,(t + \tan^{-1} 4). \tag{10.91}$$

Now the ratio of V_L and V is not impedance according to our definition. Phasor ratios of this kind are given the name *network functions*. Impedance is one member of the family of network functions. Before introducing you to the other members of this family, we will distinguish between driving-point and transfer functions. When the voltage and current are identified with one pair of terminals, as in Figure 10.27(a), then the network function which is the quotient of the phasors V and I or its reciprocal is known as a *driving-point function*. The network is being driven by a voltage source or current source at these terminals; in this sense, these terminals are the driving points. In the situation depicted in Figure 10.27(b), other pairs of terminals are involved. When the phasors of a network function are identified with two different pairs of terminals, the network function is said to be *transfer* in nature. Knowing the transfer network function, and one phasor for one

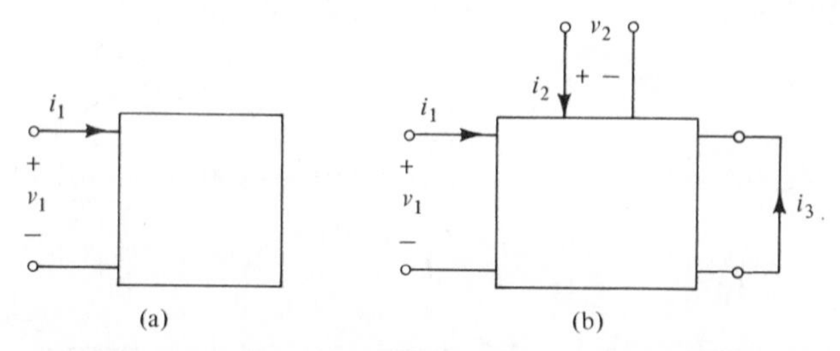

Figure 10.27 Network described by (a) driving-point functions, and (b) both driving-point and transfer functions.

pair of terminals, we may compute a phasor for another pair of terminals and in this sense transfer from one terminal pair to another.

Driving-point functions may be written without subscript identification if there is only one port. Thus

$$Z = \frac{V_1}{I_1} \quad \text{and} \quad Y = \frac{I_1}{V_1} \tag{10.92}$$

are the *driving-point impedance* and *admittance* functions at terminal pair 1. Transfer functions, on the other hand, are written with double subscripts identifying the pairs of terminals involved. If the two terminal pairs are identified as 1 and 2 (the terminals themselves are ordinarily marked 1 and 1′, 2 and 2′), then

$$Z_{21} = \frac{V_2}{I_1} \tag{10.93}$$

is the transfer impedance function, while

$$Y_{21} = \frac{I_2}{V_1} \tag{10.94}$$

is the *transfer admittance.* Voltage ratios are denoted by G (from gain of similar amplifying devices), so that

$$G_{21} = \frac{V_2}{V_1} \tag{10.95}$$

is the *voltage-ratio transfer function.* The current ratio

$$\alpha_{21} = \frac{I_2}{I_1} \tag{10.96}$$

is similarly transfer in nature and is known as the *current-ratio transfer function.* Transfer functions are conventionally defined as ratios of output phasors to input phasors.

AMPLE 10.6.1

In Exercise 10.4.3, it was shown that when $v_2(t) = \sqrt{2}\sin t$, then $v_1(t) = 3\sqrt{2}\sin(t - 90°)$. This information may be summarized by a voltage-ratio transfer function; thus, from $V_2 = \sqrt{2}\,e^{j0}$ and $V_1 = 3\sqrt{2}\,e^{-j\pi/2}$, we have

$$G_{21} = \frac{V_2}{V_1} = \frac{1}{3}e^{j\pi/2} \tag{10.97}$$

in polar form, or in rectangular form we have

$$G_{21} = 0 + j\tfrac{1}{3}. \tag{10.98}$$

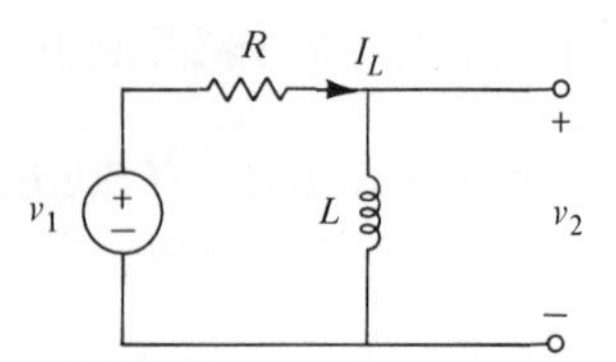

Figure 10.28 Network with voltage ratio transfer function given by Equation 10.103.

Now suppose that we change the magnitude and phase of $v_1(t)$, not changing ω, of course, since G_{21} is a function of ω. Then we may find the corresponding $v_2(t)$ from the transfer function we have computed. For example, if $v_1(t) = 10 \sin (t - 30°)$, then using the corresponding phasor $V_1 = 10e^{-j\pi/6}$, we have

$$V_2 = G_{21}V_1 = \tfrac{1}{3}e^{j\pi/2}10e^{-j\pi/6} = 3.33e^{j\pi/3}. \tag{10.99}$$

Finally, from this phasor we construct the time function $v_2(t)$ as

$$v_2(t) = 3.33 \sin (t + 60°). \tag{10.100}$$

The network functions of these examples have been complex numbers. In many cases, we need the network functions expressed in terms of ω. In the network of Figure 10.28, for example, we see that

$$I_L = V_2 Y_L = \frac{V_2}{j\omega L}. \tag{10.101}$$

The driving-voltage phasor is

$$V_1 = I_L R + V_2 = \left(\frac{R}{j\omega L} + 1\right)V_2. \tag{10.102}$$

Finally, we see that

$$\frac{V_2}{V_1} = \frac{j\omega L}{R + j\omega L}. \tag{10.103}$$

From this general expression, the voltage-ratio transfer function may be found for given values of R, L, and ω.

EXERCISES

10.6.1 (a) For the network of Figure 10.11 and the element values of Exercise 10.3.4, find the voltage-ratio transfer function, $G_{21} = V_2/V_1$. Assume that for the given reactance data, the frequency is 1000 radians/sec. (b) If $v_1(t) = 10 \sin (\omega t + 45°)$, use the transfer function of (a) to find $v_2(t)$.

10.6.2 Using the network of Figure 10.19, determine $G_{21} = V_2/V_1$ as a function of frequency, ω. Express your results in polar form.

10.6.3 For the network of Figure 10.20 with $\omega = 1$ radian/sec, find: (a) $Z_{21} = V_2/I_1$; (b) $Y_{21} = I_2/V_1$. (c) Does $Y_{21} = 1/Z_{21}$? Discuss your conclusion.

10.6.4 For the network of Figure 10.20 with $\omega = 1$, find the current-ratio transfer function, $\alpha_{21} = I_2/I_1$.

10.6.5 For the network of Figure 10.20, determine an expression for $G_{21} = V_2/V_1$ as a function of frequency, ω. To simplify notation, let $j\omega = s$, and express your solution in the form of a quotient of polynomials in s.

10.6.6 Repeat Exercise 10.6.5 for $I_2/I_1 = \alpha_{21}$.

10.7 Network functions for signals of the form Ke^{st}

The same philosophy of impedance and other network functions just developed for sinusoidal signals may be applied to signals of the form Ke^{st}. For this type of input signals, the particular solution to the differential equation relating the input and output also has the form K_1e^{st}, where K_1 is not a function of time but may depend on s. For example, consider the series RLC network in Figure 10.5. Applying Kirchhoff's voltage law, we have

$$v(t) = L\frac{di(t)}{dt} + Ri(t) + \frac{1}{C}\int i(t)\,dt, \tag{10.104}$$

and differentiating once, we obtain the second-order differential equation

$$L\frac{d^2i(t)}{dt^2} + R\frac{di(t)}{dt} + \frac{1}{C}i(t) = \frac{dv(t)}{dt}. \tag{10.105}$$

Suppose the hypothetical input signal (not necessarily physically realizable) is $v(t) = Ve^{st}$. Then we wish to verify that $i(t) = Ie^{st}$ satisfies Equation 10.105 for an appropriate choice of I, provided that s is not equal to a characteristic root. For $v(t) = Ve^{st}$, the right-hand side of Equation 10.105 is sVe^{st}, and for $i(t) = Ie^{st}$, the left-hand side is equal to $[Ls^2e^{st} + Rse^{st} + (1/C)e^{st}]I$. In order for the two sides to be equal, the complex quantity I must be

$$I = \frac{sVe^{st}}{Ls^2e^{st} + Rse^{st} + (1/C)e^{st}} = \frac{sV}{Ls^2 + Rs + 1/C}. \tag{10.106}$$

We define the ratio V/I as the *impedance function* $Z(s)$. In this example, it is

$$Z(s) = \frac{Ls^2 + Rs + 1/C}{s} = Ls + R + \frac{1}{Cs}. \tag{10.107}$$

Note that this impedance expression yields the same answer for the rotating phasor $e^{j\omega t}$ if we simply replace s by $j\omega$. Thus, for an ideal constant inductor, the impedance expression is Ls; for a constant resistor, it is R; and for a constant capacitor, it is $1/(Cs)$.

The procedures developed for finding the impedance of series-parallel combinations of elements for the signal $e^{j\omega t}$ also apply when the signal is e^{st}. The apparent difference is that the term $j\omega$ is replaced by s. Thus the network functions $Z(j\omega)$, $Y(j\omega)$, $G(j\omega)$, $\alpha(j\omega)$ which are functions of $j\omega$ become $Z(s)$, $Y(s)$, $G(s)$, and $\alpha(s)$ which are functions of s. The parameter s is a generalized frequency which is conventionally called *complex frequency*. The complex frequency s may have real and imaginary parts,

$$s = \sigma + j\omega, \tag{10.108}$$

where σ and ω are real. The signal e^{st} may be written

$$e^{st} = e^{\sigma t}e^{j\omega t}. \tag{10.109}$$

This form may be considered as a *damped* or *growing rotating phasor* depending on whether σ is negative or positive as shown in Figure 10.29(a) and Figure 10.29(b), respectively. Aside from the minor algebraic simplification that results when $j\omega$ is replaced by s, the major significance in the use of a complex s is in the representation of general signals as sums or integrals of complex exponentials, as mentioned in Chapter 8.

From Equation 10.107, it is apparent that there will be some values of s such that $Z(s)$ becomes zero. When

$$Ls^2 + Rs + \frac{1}{C} = 0, \tag{10.110}$$

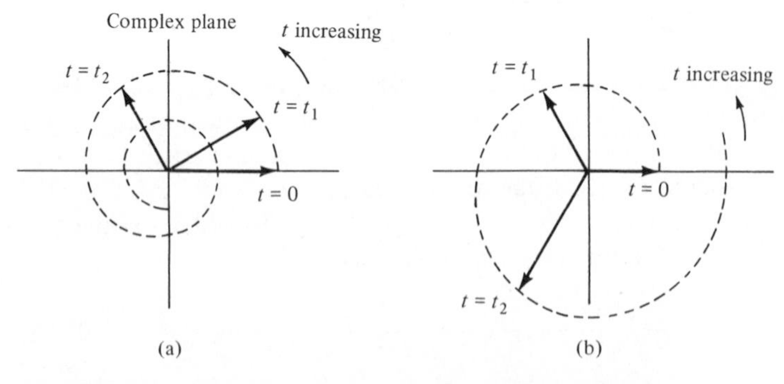

Figure 10.29 (a) A damped or shrinking rotating phasor. (b) A growing or expanding rotating phasor. The magnitude of the rotating phasor $e^{j\omega t}$ is scaled by $e^{\sigma t}$. In (a), σ is negative, and in (b) σ is positive.

which is the case when

$$\left.\begin{matrix} s_1 \\ s_2 \end{matrix}\right\} = \frac{-R}{2L} \pm \sqrt{[R/(2L)]^2 - 1/(LC)}, \tag{10.111}$$

where s_1 and s_2 are roots of Equation 10.110, then $Z(s_1) = Z(s_2) = 0$. These values of s for which $Z(s)$ becomes zero are called *zeros* of the function $Z(s)$. Likewise, in the expression for $Z(s)$ which is a ratio of polynomials in s, there are values of s, for which $1/Z(s)$ becomes zero. These are the values of s for which the denominator of $Z(s)$ becomes zero. For example, in Equation 10.107, the denominator is zero when $s = 0$. Those values of s which make $1/Z(s)$ take zero value are called *poles* of $Z(s)$. Remember that we are talking about linear time-invariant lumped networks which give rise to ordinary linear differential equations with constant coefficients. For such networks, the impedance $Z(s)$ is always a ratio of polynomials. A ratio of two polynomials is called a *rational function*. That is, in general, we have

$$Z(s) = \frac{b_m s^m + b_{m-1}s^{m-1} + \cdots + b_0}{a_n s^n + a_{n-1}s^{n-1} + \cdots + a_0}. \tag{10.112}$$

The polynomials may be factored so that

$$Z(s) = \frac{b_m}{a_n} \frac{(s - z_1)(s - z_2)\cdots(s - z_m)}{(s - p_1)(s - p_2)\cdots(s - p_n)}, \tag{10.113}$$

where the z_j are the zeros and the p_j are the poles. Thus $Z(s)$ may be specified by determining the locations of the zeros, the locations of the poles, and the value of a constant scale factor (in this case, b_m/a_n). In the complex s-plane, the locations of zeros are marked by o and the locations of poles by **x**. An example of a *pole-zero* configuration is shown in Figure 10.30. Other network functions such as $Y(s)$, $G(s)$, and $\alpha(s)$ are treated in a similar way; that is, they have zeros and poles too.

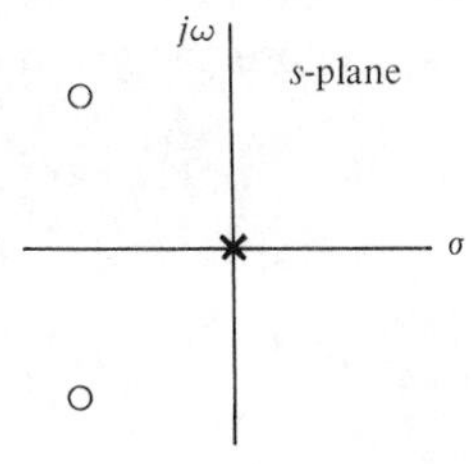

ure 10.30 Example of a pole-zero plot in the s-plane.

EXAMPLE 10.7.1

Consider the network in Figure 10.14(b). Let us determine the driving-point impedance $Z(s)$ at terminals 1 and 1′. The impedance of the parallel combination is the reciprocal of $\frac{1}{2}s + \frac{1}{2}$. Hence

$$Z(s) = \frac{1}{\frac{1}{2}s + \frac{1}{2}} + 1 + \frac{1}{s} = \frac{(s+1)^2 + 2s}{s(s+1)}. \tag{10.114}$$

This function has poles at $s + 1 = 0$ and $s = 0$, or

$$s = -1 \quad \text{and} \quad s = 0. \tag{10.115}$$

The zeros are the roots of

$$(s+1)^2 + 2s = 0. \tag{10.116}$$

Since

$$(s+1)^2 + 2s = s^2 + 4s + 1 = 0, \tag{10.117}$$

then

$$(s+2)^2 - 3 = 0 \tag{10.118}$$

and

$$s = -2 \pm \sqrt{3}. \tag{10.119}$$

Thus $s_1 = -2 + \sqrt{3}$ and $s_2 = -2 - \sqrt{3}$ are the two zero locations of $Z(s)$.

EXAMPLE 10.7.2

The voltage transfer function for a linear time-invariant network has a double zero at the origin and poles at

$$s = -\frac{1}{2} \pm j\frac{\sqrt{3}}{2}. \tag{10.120}$$

The constant scale factor for the transfer function is 5. The problem is to determine the steady-state response when the input voltage is 10 cos t. From the given data,

$$G_{21}(s) = \frac{5s^2}{\left(s + \frac{1}{2} - j\frac{\sqrt{3}}{2}\right)\left(s + \frac{1}{2} + j\frac{\sqrt{3}}{2}\right)} \tag{10.121}$$

$$= \frac{5s^2}{(s + \frac{1}{2})^2 + \frac{3}{4}} = \frac{5s^2}{s^2 + s + 1}.$$

The response due to e^{st} is $G_{21}(s)e^{st}$. In particular, the response due to e^{jt} is $G_{21}(j1)e^{jt}$. Hence the response due to 10 cos t which is 10 Re (e^{jt}) is

$$\begin{aligned} y(t) &= 10 \text{ Re } [G_{21}(j1)e^{jt}] \\ &= 10 \text{ Re} \left[\frac{5(-1)}{-1 + j1 + 1} e^{jt} \right] \\ &= 10 \text{ Re } (j5e^{jt}) \\ &= 50 \text{ Re } (e^{jt + \pi/2}) = 50 \cos (t + \pi/2). \end{aligned} \tag{10.122}$$

ERCISES

10.7.1 A network is composed of constant R, L, and C connected in parallel and is driven by a current source $i(t)$. For this network, show that

$$Y(s) = \frac{1}{R} + Cs + \frac{1}{Ls}.$$

10.7.2 For the network of Figure 10.20, find: (a) $Z_{21}(s)$, (b) $Y_{21}(s)$, (c) $G_{21}(s)$, and (d) $\alpha_{21}(s)$. Find numerical values for each of these functions with $s = -1$.

10.7.3 Find $G_{21}(s) = V_L(s)/V(s)$ for the network of Figure 10.18.

10.7.4 Find $G_{21}(s)$ for the network of Figure 10.19. If $v_1(t) = 10e^{3t}$, find $v_2(t)$ using this transfer function.

.8 Thevenin's and Norton's theorems again

Thevenin's and Norton's theorems, as given in Section 7.2, refer to the network configuration shown in Figure 10.31 where separate subnetworks, Network A and Network B, are identified. Network A is assumed to be linear, but permitted to contain both dependent and independent sources. Coupling between A and B is not permitted, through either mutual inductance or controlled sources, except through the two connecting wires. Network B may be nonlinear. As studied in Section 7.2, the results were in general form. When the network is operating in the sinusoidal steady-state, then there are simplifications in the application of the theorems which we describe.

First, if the network is operating in the sinusoidal steady-state, then initial conditions play no role and need not be considered. Second, we can substitute the determination of the impedance or admittance of the network in place of finding the network itself.

The theorems may be stated in terms of the networks shown in (b) and (c) of

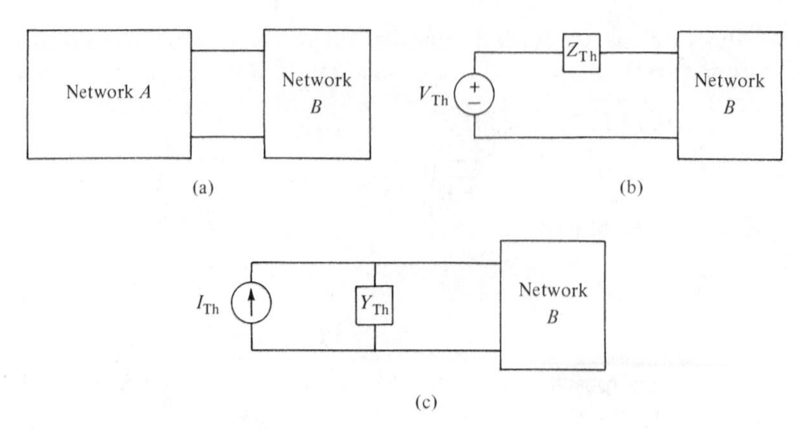

Figure 10.31 For the network of (a), that shown in (b) is the Thevenin equivalent for the sinusoidal steady-state, and that in (c) is the Norton equivalent for the sinusoidal steady-state.

Figure 10.31. For networks A and B satisfying the requirements that have been given, Network A may be replaced by a phasor representation of a voltage source V_{Th} in series with impedance Z_{Th}. The voltage V_{Th} is that appearing at the open-circuit terminals of Network A with all sources operating. The impedance Z_{Th} is that calculated at the open-circuit terminals with all independent sources turned off (voltage sources replaced by a short-circuit current sources replaced by open circuits). Norton's theorem is the dual of Thevenin's theorem, and can be stated in terms of the configuration of (c) in Figure 10.31. Here I_{Th} is a phasor representation of a current source having the value of current that will appear when the terminals of Network A are shorted, and Y_{Th} is the reciprocal of Z_{Th} or the admittance of the network at the output terminals under the same conditions given for Z_{Th}.

EXAMPLE 10.8.1

Figure 10.32(a) shows a network for which we wish to determine the current in the load resistor R_L. The independent source is $v_1(t) = \sqrt{2}\, V_1 \cos 2t$, where V_1 is the

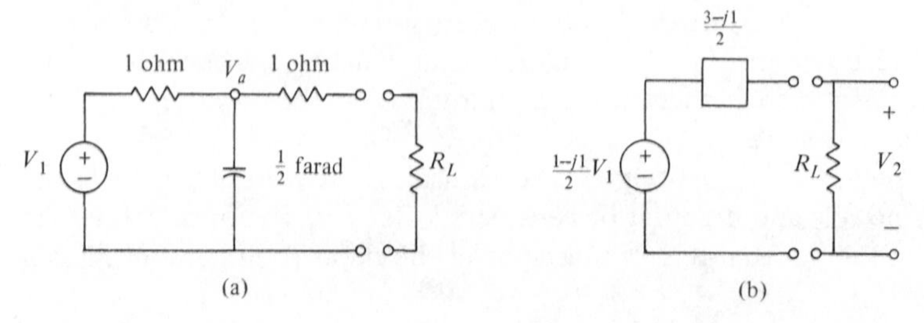

Figure 10.32

rms or effective value of voltage. With R_L removed, there is no current in the 1-ohm resistor in series with R_L and $V_{Th} = V_a$. We may find V_a from the product of the current in the RC series current and the impedance of the capacitor; thus

$$V_{Th} = V_1 \frac{-j1}{1 - j1} = \frac{1 - j1}{2} V_1 . \qquad (10.123)$$

The impedance Z_{Th} is found with V_1 replaced by a short circuit and is

$$Z_{Th} = 1 + \frac{1}{1 + j1} = \frac{3 - j1}{2} . \qquad (10.124)$$

Then the Thevenin equivalent network is that shown in (b) of Figure 10.32. From this figure, the current in R_L is

$$I_2 = \frac{V_1[(1 - j1)/2]}{(3 - j1)/2 + \frac{1}{2}} \qquad (10.125)$$

$$= \frac{3 - j5}{17} V_1 , \qquad (10.126)$$

with $R_L = \frac{1}{2}$ ohm. We can also determine the output voltage V_2 by multiplying I_2 by $\frac{1}{2}$. In the time domain, the output is, from the last equation multiplied by $\frac{1}{2}$,

$$v_2(t) = \sqrt{2} \frac{V_1}{\sqrt{34}} \cos\left(2t - \tan^{-1} \tfrac{5}{3}\right) . \qquad (10.127)$$

EXAMPLE 10.8.2

The network shown in Figure 10.33(a) is similar to that of the last example, except that a controlled source of value $3I_1$ has been inserted. Using all the numerical values of the last example, but with $V_1 = 1$ volt, we will analyze the network, making use of Thevenin's theorem. From the network shown in Figure 10.33(b), we determine V_{Th}. The open-circuit terminals permit us to neglect the 1-ohm resistor connected to the controlled source. The current I_1 passes through R and C and produces the voltage V_a. Since the current through the RC combination is $I_1 = 1/(1 - j1) = (1 + j1)/2$, then

$$V_a = Z_C I_1 = \frac{(-j1)(1 + j1)}{2} = \frac{1 - j1}{2} . \qquad (10.128)$$

Then

$$V_{Th} = V_a + 3I_1 = 2 + j1 \text{ volts} . \qquad (10.129)$$

The impedance of the network Z_{Th} is found through the use of an auxiliary source V, from which the current I is found, and then V/I calculated to give the required impedance. From the network of Figure 10.33(c), KVL tells us that

$$V - I \times 1 - 3I_1 + I_1 = 0 . \qquad (10.130)$$

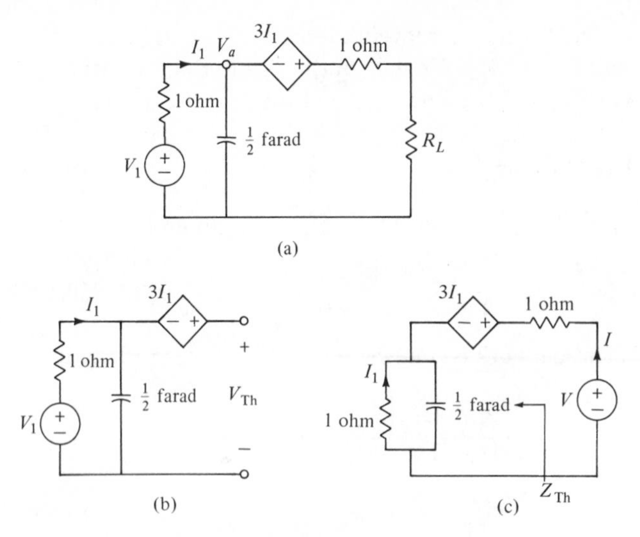

Figure 10.33 For the network of (a), the Thevenin equivalent network is found for V_{Th} by the network of (b), and for Z_{Th} from the network of (c).

We eliminate I_1 from this expression by considering the RC parallel network from which we compute I_1 in terms of I as

$$I_1 = \frac{j}{1-j} I. \tag{10.131}$$

Substituting this result into the last equation, we find the required impedance

$$\frac{V}{I} = Z_{Th} = \frac{1+j}{1-j} = j \text{ ohms}. \tag{10.132}$$

The voltage across the resistor R_L is found from the network of Figure 10.34. We see that

$$V_{R_L} = R_L \frac{2 + j1}{R_L + j}. \tag{10.133}$$

An especially simple result is obtained when $R_L = 2$ ohms, which is

$$V_{RL} = 2 \text{ volts}, \tag{10.134}$$

from which we see that the output has twice the magnitude of the input and is in phase with it.

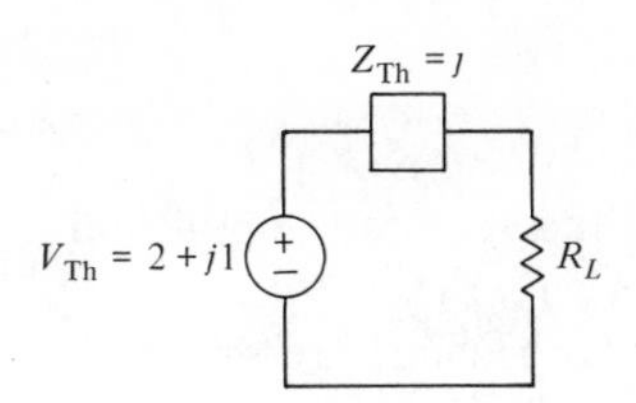

Figure 10.34

EXERCISES

10.8.1 Repeat the problem of Example 10.8.2 except that the controlled source has the value $-2I_1$ rather than $3I_1$.

10.8.2 Consider the network of Figure 10.33(a) in which the two resistors are marked R_1 and R_2 (left to right), the capacitor is C_1, and the controlled source has the value kI_1. Determine a general expression for V_{Th} and Z_{Th}.

10.8.3 Repeat Exercise 10.8.2 for a network in which the controlled source has the value $k_1 V_a$ where k_1 is real, but may be positive or negative.

10.9 Summary

1. Impedance Z is defined initially for the sinusoidal steady-state and is later generalized for exponential signals (complex frequency). The definition is

$$Z = \frac{Ve^{j\omega t}}{Ie^{j\omega t}} = \frac{V}{I} = \frac{1}{Y},$$

where V and I are phasors (complex numbers) and Y is the admittance.

2. The real and imaginary parts of Z and Y are

$$Z = R + jX, \qquad Y = G + jB,$$

where R is resistance in ohms, X is reactance in ohms, G is conductance in mhos, and B is susceptance in mhos.

3. Impedance and admittance expressions for the three passive elements are given in the following table:

	Resistor	*Inductor*	*Capacitor*
Impedance	R	$j\omega L$	$\frac{1}{j\omega C}$
Admittance	$G = \frac{1}{R}$	$\frac{1}{j\omega L}$	$j\omega C$

4. For parallel networks,

$$Y_t = Y_1 + Y_2 + \cdots + Y_n;$$

and for series networks,

$$Z_t = Z_1 + Z_2 + \cdots + Z_m.$$

Series-parallel networks may sometimes be reduced by the successive application of these two equations.

5. One-port networks having the same driving-point impedance are said to be equivalent networks. The concept may be generalized to 2-port networks where networks are equivalent if they have the same network functions to describe them.

6. There are two forms of phasor diagrams — polar and polygon. Phasor diagrams are useful in showing all voltages and currents in a network in phasor form in such a way that the relationships of all voltages and currents in magnitude and phase are shown clearly.

7. Two-port networks with specified terminations (other networks and/or sources) are typically described by six network functions:

$$Z_{11} = \frac{V_1}{I_1}, \qquad Z_{22} = \frac{V_2}{I_2},$$

$$Z_{21} = \frac{V_2}{I_1}, \qquad Y_{21} = \frac{I_2}{V_1} \neq \frac{1}{Z_{21}},$$

$$G_{21} = \frac{V_2}{V_1}, \qquad \alpha_{21} = \frac{I_2}{I_1}.$$

It is conventional to express the four transfer functions as the ratio of output to input. (This will be discussed further in Chapter 15.)

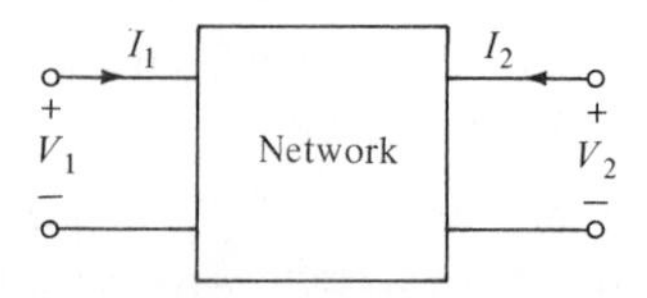

8. Complex-power relationships

$$S = V_{\text{rms}} I^*_{\text{rms}} = P_{\text{av}} + jQ,$$

$$P_{\text{av}} = |V_{\text{rms}}||I_{\text{rms}}| \cos \theta = \text{Re}\,(V_{\text{rms}} I^*_{\text{rms}}),$$

$$Q = |V_{\text{rms}}||I_{\text{rms}}| \sin \theta = \text{Im}\,(V_{\text{rms}} I^*_{\text{rms}}),$$

where $\cos \theta$ is the power factor (p.f.) and θ is the power-factor angle. Other equations:

$$S = |V_{\text{rms}}|^2 Y^* = |I_{\text{rms}}|^2 Z,$$

$$P_{\text{av}} = |V_{\text{rms}}|^2 G = |I_{\text{rms}}|^2 R,$$

$$Q = -|V_{\text{rms}}|^2 B = |I_{\text{rms}}|^2 X,$$

$$\text{p.f.} = \frac{P_{\text{av}}}{|S|} = \frac{R}{|Z|}.$$

$$\text{Leading power factor implies that}\begin{cases}\theta \text{ is negative,}\\ \text{current leads voltage,}\\ \text{network is primarily capacitive.}\end{cases}$$

$$\text{Lagging power factor implies that}\begin{cases}\theta \text{ is positive,}\\ \text{current lags voltage,}\\ \text{network is primarily inductive.}\end{cases}$$

9. Tellegen's theorem equations:

$$\sum V_k I_k^* = 0, \qquad \sum P_{\text{av}\,k} = 0, \qquad \sum Q_k = 0.$$

10. (a) Thevenin's theorem for the sinusoidal steady-state: In terms of the network specified in Figure 10.31(a), Network A may be replaced by a voltage source V_{Th} which is the open-circuit voltage of Network A with all sources acting, in series with a network of impedance Z_{Th}, which is the impedance of Network A with all independent sources set to zero but dependent (or controlled) sources acting. The equivalent network applied only for calculations in Network B.
(b) Norton's theorem for the sinusoidal steady-state: In terms of the network specified in Figure 10.31(a), Network A may be replaced by a current source I_{Th} which is the short-circuit current of Network A at the output terminals, in parallel with a network of admittance Y_{Th}, which is the admittance of Network A with all independent sources set to zero but dependent (or controlled) sources acting. The equivalent network applies only for calculations in Network B.

ROBLEMS

10.1 In the series RLC network shown in Figure 10.35, the current is known to be $i(t) = 2 \sin (10t - 30°)$. It is given that $R = 3$ ohms, $C = 0.05$ farad, and $L = 0.6$ henry. Determine the following quantities: (Let $\sin 10t$ correspond to $1\underline{/0°}$.) (a) I, (b) V_R, V_L, and V_C, (c) V, (d) the impedance $Z = V/I$, (e) the admittance, $Y = 1/Z$, (f) R and X in $Z = R + jX$, (g) G and B in $Y = G + jB$, (h) $v(t)$, $v_r(t)$, $v_L(t)$, and $v_C(t)$, and (i) the rms value of $v(t)$.

10.2 For the two-terminal network shown in Figure 10.36, it is given that

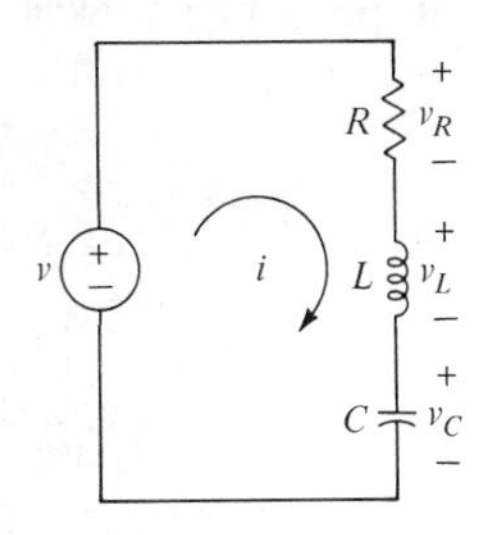

Figure 10.35

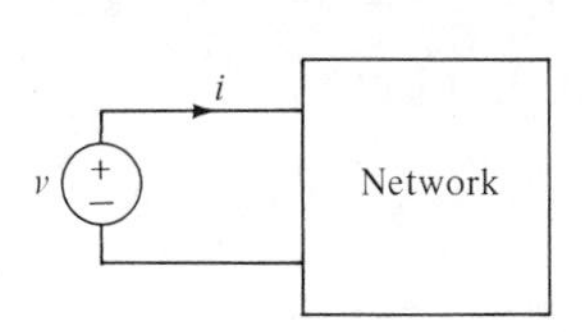

Figure 10.36

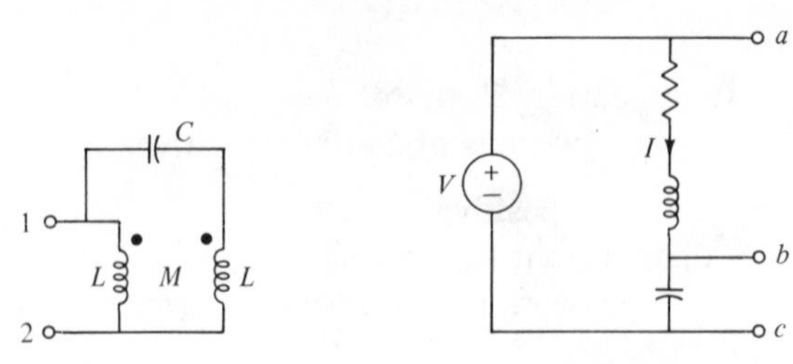

Figure 10.37 Figure 10.38

$v(t) = 100 \sin(100t + 25°)$ volts and $i(t) = 2 \sin(100t + 65°)$ amperes. (a) Determine the equivalent series resistance of the impedance at the input terminals of the network. Is the network capacitive or inductive? (b) Compute the average power into the network by taking the average of the instantaneous power.
(c) Compute the average power into the network by taking the average of the instantaneous i^2R_{eq}, and compare with the answer in (b). Is the average of i^2R_{eq} equal to $I^2_{rms}R_{eq}$, where I_{rms} is the rms value of i?
(d) Show that for this example, average power $P_{av} = V_{rms}I_{rms} \cos \theta$, where V_{rms} and I_{rms} are rms values of the instantaneous v and i, respectively, and θ is the difference in the phase angles of v and i.
(e) Let V_{rms} be $1/\sqrt{2}$ times the phasor V of v, and let I_{rms} be $1/\sqrt{2}$ times the phasor I of i. Form $V_{rms}I^*_{rms}$, where I^*_{rms} is the complex conjugate of I_{rms}. Show that for this example, the average power into the network is equal to the real part of $V_{rms}I^*_{rms}$.

10.3 In the given network of Figure 10.37, $C = 1$ farad, $L = 1$ henry, and $M = \frac{1}{2}$ henry. Find the impedance of the network at terminals 1–2 when $\omega = 1$ radian/sec.

10.4 The source in the network of Figure 10.38 is sinusoidal. An rms-reading voltmeter is used to measure the following: $V_{ab} = 10$ volts and $V_{bc} = 15$ volts. Using I as the reference, draw a phasor diagram showing V_{bc}, V_{ab}, and V. Is your phasor diagram unique?

10.5 The network of Figure 10.39 is excited by a sinusoidal source and is in the steady-state. Using V_2 as the reference, draw a neat phasor diagram showing I_C, I_L, I_R, I, and V_1.

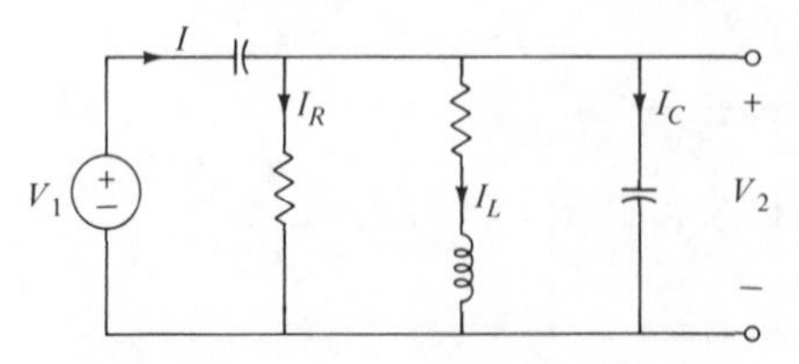

Figure 10.39

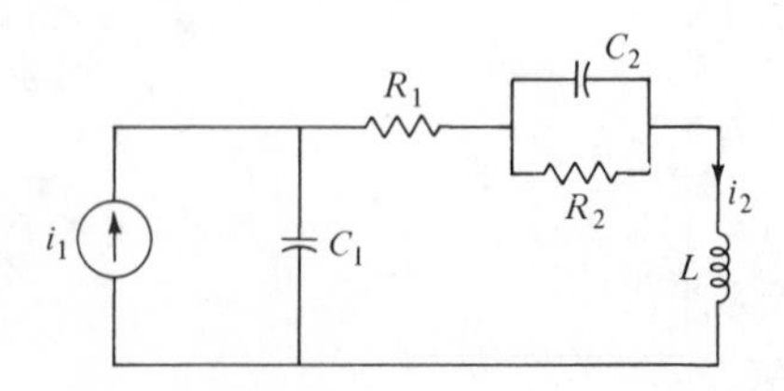

Figure 10.40

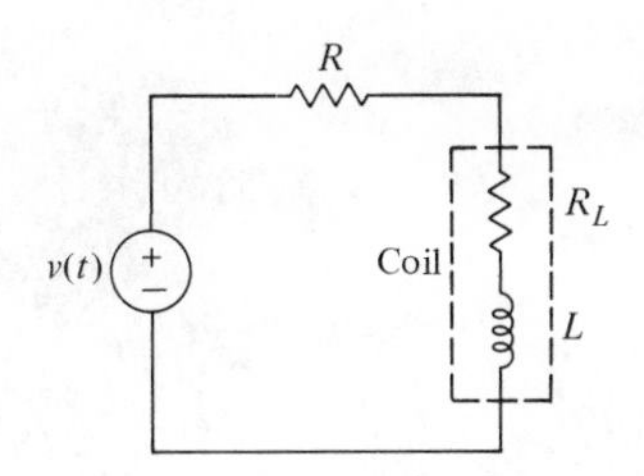

Figure 10.41

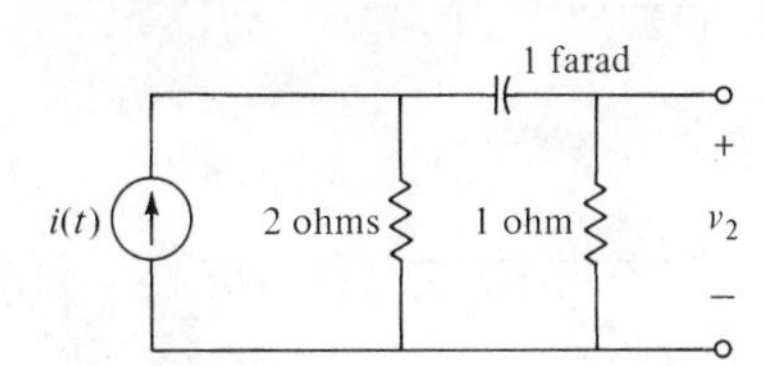

Figure 10.42

10.6 The network of Figure 10.40 is excited by a sinusoidal source and is in the steady-state. Use the phasor I_2 as the reference and draw a complete phasor diagram showing all voltage and current phasors.

10.7 The network shown in Figure 10.41 is a model for a resistor in series with a lossy coil. With the system operating in the sinusoidal steady-state, the rms voltage of the source is 15 volts, that of the coil is 13 volts, and that of the resistor, 5 volts. If R is known to be 10 ohms, and $\omega = 1$ radian/sec, find R_L and L.

10.8 In the network of Figure 10.42, $i(t) = \sqrt{2} \sin t$. Find $v_2(t)$.

10.9 For the given network in Figure 10.43, find $v_a(t)$ if $v_1(t) = \sqrt{2} \cos 2t$ and the system is in the steady-state. Draw a complete phasor diagram showing all voltage and current phasors.

10.10 Determine the transfer function $V_2(s)/I(s)$ for the network in Figure 10.42.

10.11 Determine the transfer function $V_a(s)/V_1(s)$ for the network in Figure 10.43.

10.12 In the network of Figure 10.44, $v_2(t) = 2 \sin 2t$. Determine $v_1(t)$ in the form $K_1 \sin (2t + \theta)$.

10.13 It is determined that the current in the capacitor, $i_a(t)$, is $i_a(t) = \frac{1}{2} \sin t$. For this network (Figure 10.45), find the phasor ratio $Z_{21} = V_2/I_1 = a + jb$.

10.14 In the network of Figure 10.46, it is known from an oscillograph that $v_3(t) = 2 \sin 2t$. Determine the transfer function V_2/V_1 in polar form.

10.15 For the network in Figure 10.44, find the transfer function, $G_{21} = V_2/V_1$.

10.16 For the network of Figure 10.46, determine the following voltage-ratio transfer functions: (a) V_3/V_1, (b) V_2/V_1.

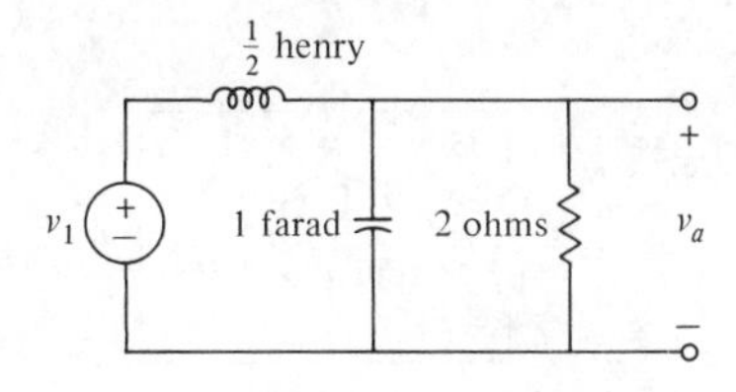

Figure 10.43

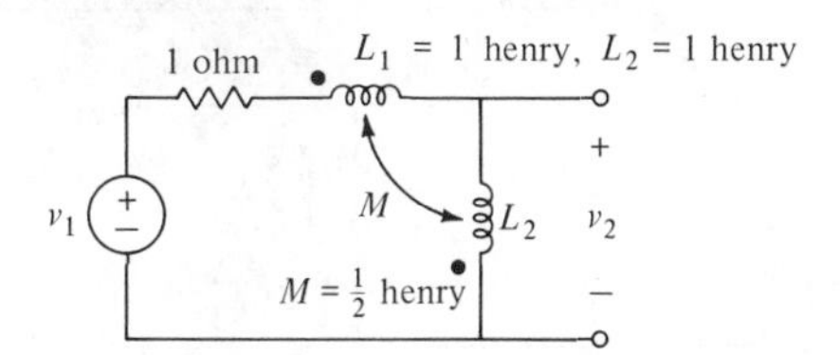

Figure 10.44

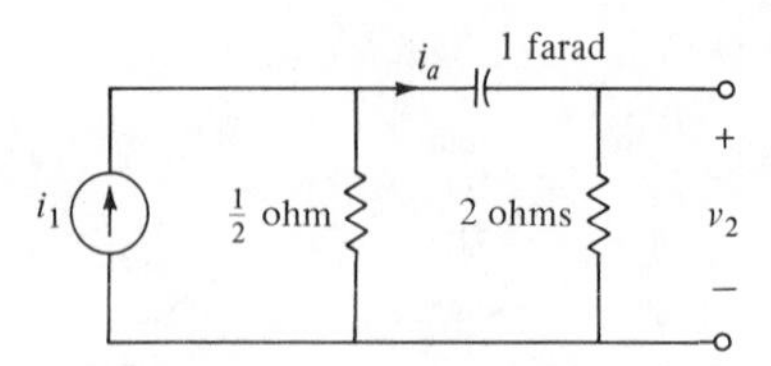

Figure 10.45

10.17 In the network of Figure 10.47, it is given that $R = 10$ ohms, $1/(\omega_0 C) = 10$ ohms, and $i(t) = 10 \sin \omega_0 t$. We are required to find the source voltage, $v(t)$. The following solution is proposed:

$$
\begin{aligned}
v(t) &= \left(R + \frac{1}{j\omega_0 C}\right) 10 \sin \omega_0 t \\
&= (10 - j10)10 \sin \omega_0 t \\
&= 100 \sin \omega_0 t - j100 \sin \omega_0 t = 100\sqrt{2} \sin (\omega_0 t - 45^\circ).
\end{aligned}
$$

Is this solution correct? Comment on the method of solution and the justification of each of the steps.

10.18 Repeat Problem 10.1 but let cos $10t$ correspond to $1\angle 0^\circ$. Are the answers to (d), (e), (f), (g), (h), and (i) affected by the choice of reference phasor?

10.19 Repeat Problem 10.18 but let $\sin (10t - 30^\circ)$ correspond to $1\angle 0^\circ$.

10.20 Repeat Problem 10.1 but for $i(t) = 5 \cos (10t - 45^\circ)$. Are the answers to (d), (e), (f), and (g) affected by this change? Why or why not?

10.21 Repeat Problem 10.1 but for $i(t) = 2 \sin (20t - 30^\circ)$. Are the answers to (d), (e), (f), and (g) affected by this change? Why or why not?

10.22 For the network in Figure 10.46, $v_1(t) = 2 \sin 2t$. Use phasors to compute $v_2(t)$. [*Hint:* Make use of linearity.]

10.23 Repeat Problem 10.13 but for $i_a(t) = 10 \cos (t + \pi/6)$. Is the answer the same as that of Problem 10.13? Why or why not?

10.24 For the network in Figure 10.14(a), does the resistive component of Z (real part of Z) depend on the value of ω? How about the real part of Y?

10.25 Determine the driving point impedance $Z(s)$ for the networks in Figure 10.14 and plot the pole-zero configuration for $Z(s)$.

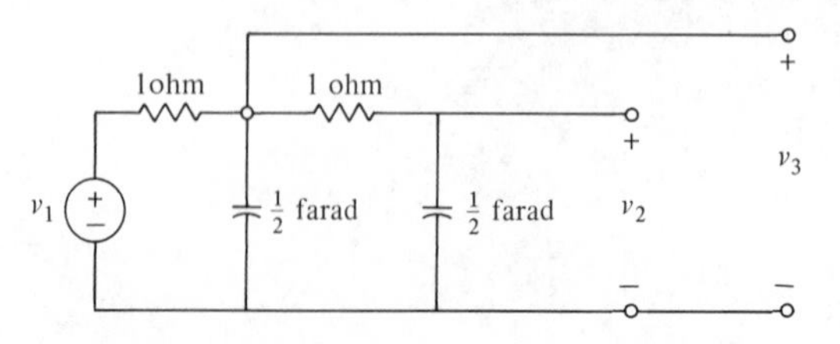

Figure 10.46

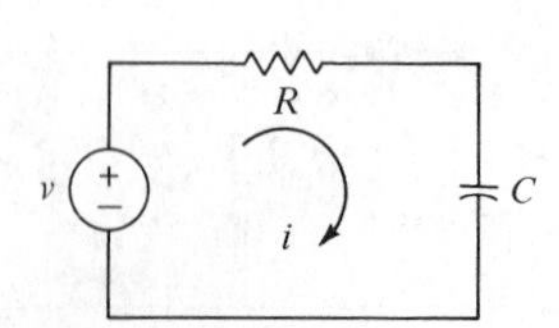

Figure 10.47

10.26 Plot the pole-zero configuration for $Y(s)$ for the same networks in Problem 10.25.

10.27 Determine the voltage transfer function $G(s) = V_L(s)/V(s)$ for the network in Figure 10.18.

10.28 Determine the voltage transfer function $G(s) = V_2(s)/V_1(s)$ for the network in Figure 10.19 and plot the pole-zero configuration of $G(s)$.

10.29 Repeat Problem 10.28 for the network of Figure 10.20.

10.30 Determine the transfer impedance $Z_{21}(s)$ for the network in Figure 10.42.

10.31 Repeat Problem 10.30 for Figure 10.45.

10.32 Restate Thevenin's theorem for linear time-invariant networks with sinusoidal sources of the same frequency, if only the steady-state is of interest. Use phasors.

10.33 Restate Norton's theorem for linear time-invariant networks in the sinusoidal steady-state where the sources are all sinusoidal with the same frequency. Use phasors.

10.34 Restate the principle of superposition for linear time-invariant networks in the sinusoidal steady-state where the sources are all sinusoidal with the same frequency. Use phasors.

10.35 A sinusoidal voltage source is in series with a linear time-invariant network, forming a composite two-terminal network. Assuming sinusoidal steady-state conditions, determine an equivalent composite two-terminal network consisting of a sinusoidal current source in parallel with a linear time-invariant network. Denote the driving-point impedance of the original network by Z, and use phasors.

10.36 Find the Thevenin equivalent for the linear time-invariant network in Figure 10.48, where v is Ve^{st}.

10.37 The network consists of a voltage-controlled voltage source connected to an arbitrary linear time-invariant network with driving-point impedance $Z(j\omega)$. Determine the Thevenin equivalent from terminals 1 and 2. How does the controlled source compare with an NIC? (Find relations between the phasors V_1 and V_2, and between the phasors I_1 and I_2.) See Figure 10.49.

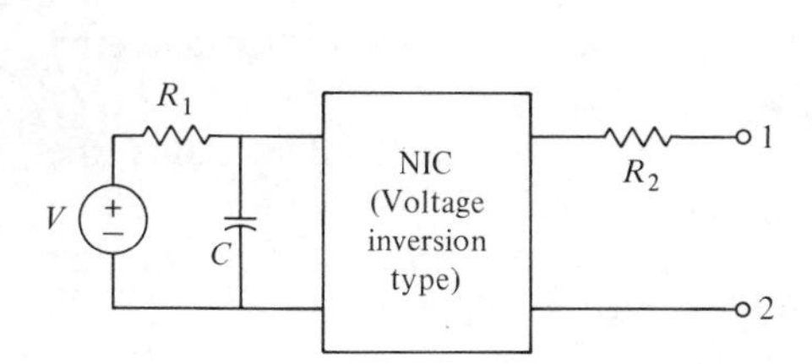

Figure 10.48

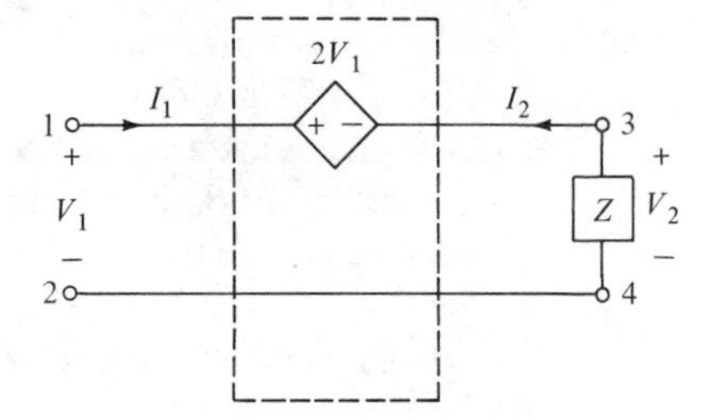

Figure 10.49

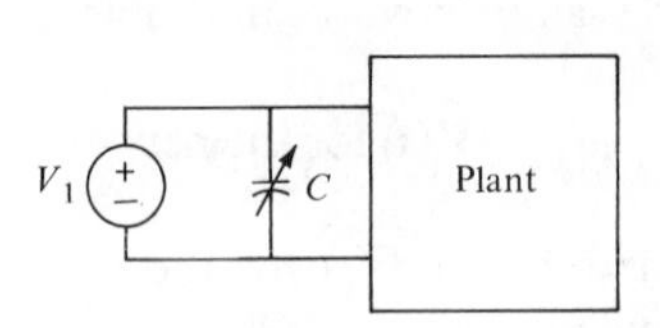

Figure 10.50

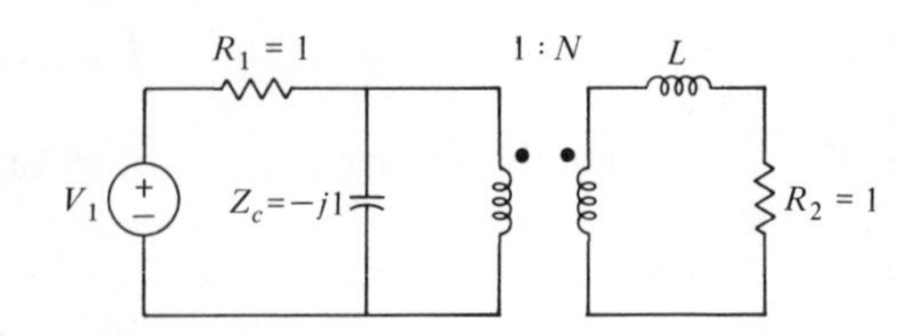

Figure 10.51

10.38 The average power P_{av} for the plant shown in the network represented by Figure 10.50 has a rating of 1000 kilowatts, and the power factor is 0.8 lagging. The voltage source has an effective value of 2300 volts and the frequency is $\omega = 377$ radians/sec (60 Hz). A capacitor C is to be placed in parallel with the plant as shown by the figure to correct the power factor. Determine the value of C such that the following values of power factor are obtained: (a) p.f. = 0.9 lagging; (b) p.f. = 1.0; (c) p.f. = 0.9 leading.

10.39 Refer to the network represented in Figure 10.50. Let the rating of the plant be 2000 kva and the power factor be 0.8 lagging with the voltage of the source being 1000 volts. The system is operating in the sinusoidal steady-state with $\omega = 377$ radians/sec (60 Hz). Determine the values of C such that (a) p.f. = 0.95 lagging, (b) p.f. = 0.95 leading.

10.40 The network shown in Figure 10.51 contains an ideal transformer of turns-ratio 1:N. Let $L = 0$. With the element impedances as given, determine the value of N that will cause maximum power to be delivered to the load resistor R_2.

10.41 Repeat Problem 10.40 if L has a value such that its impedance is $Z_L = j1$ ohm.

10.42 A model for a sinusoidal signal generator consists of sinusoidal voltage source in series with an impedance $Z_s = R_s + jX_s$. The source is connected to a load $Z_L = R_L + jX_L$. If R_L and X_L are independently adjustable, determine values of R_L and X_L such that the power delivered to Z_L is maximized. It is given that $R_s > 0$.

10.43 Repeat Problem 10.42 if $|Z_L|$ is adjustable but X_L/R_L is fixed.

10.44 Repeat Problem 10.42 if X_L is fixed but R_L is adjustable.

10.45 Suppose that for the network in Problem 10.42, R_s and X_s are adjustable, but R_L and X_L are fixed. Determine R_s and X_s so that the power delivered to Z_L is maximized. It is also required that $R_s \geq 0$.

10.46 The maximum power delivered to Z_L in Problem 10.42 is called the *available power* of the source. Determine the available power for the source in Problem 10.42. Note that the available power is the most power that can be extracted from a given practical source whose model includes an impedance.

10.47 Restate Tellegen's theorem for linear time-invariant networks in the

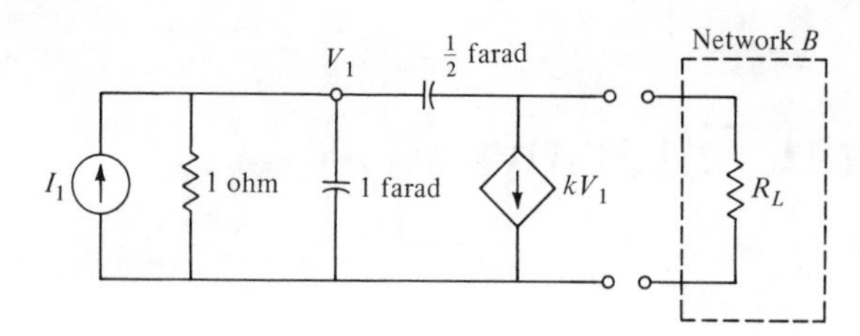

Figure 10.52

sinusoidal steady-state where the sources are all sinusoidal with the same frequency. Use phasors.

10.48 Restate the formula for the average power flow into an n-terminal linear time-invariant network in the sinusoidal steady-state where the sources are all sinusoidal with the same frequency. Use phasors. See Problem 6.8 (p. 131) and Figure 6.10 (p. 117).

10.49 The network of Figure 10.52 contains one controlled source. Assuming that the network is operating in the sinusoidal steady-state, determine the equivalent Thevenin network giving numerical values for V_{Th} and Z_{Th} when possible. Is there any value of k (a real number) for which the current in R_L will be zero?

10.50 Repeat Problem 10.49 but determine the Norton equivalent network instead of the Thevenin.

Chapter **11**

Resonance and Signal Filtering

11.1 Filtering

The word *filter* usually connotes something that separates one thing from a mixture of things. For instance, a sieve may let liquid and small particles pass through but obstruct solid particles larger than the mesh size of the sieve from passing through. Likewise, an electric signal filter seeks to extract one electric signal from a mixture of electric signals. We may have a signal composed of the summation of various sinusoids (or exponentials) of different frequencies. This signal may be the input to a filter which yields as an output a single sinusoid. This is of course an idealization. In an actual practical case, the undesired components are not completely suppressed. Let us examine a simple example involving the network in Figure 11.1.

Suppose the input to the network is a sinusoid of frequency ω. Then, if $v_{\text{in}}(t) = V_{\text{in}} \sin \omega t$ is represented by the phasor $V_{\text{in}}e^{j0}$, then the output phasor is

$$V_0 = \left[\frac{1/(j\omega C)}{R + 1/(j\omega C)}\right] V_{\text{in}} = \frac{V_{\text{in}}}{1 + j\omega RC}. \tag{11.1}$$

This means that the function $v_0(t)$ is

$$v_0(t) = \frac{V_{\text{in}}}{\sqrt{1 + (\omega RC)^2}} \sin\,(\omega t - \tan^{-1} \omega RC). \tag{11.2}$$

If $R = 1000$ ohms, $C = 0.001$ microfarad, and $\omega = 1000$ radians/sec, then we have

$$\begin{aligned} v_0(t) &= \frac{V_{\text{in}}}{\sqrt{1 + [(1000)(1000)(10^{-3})(10^{-6})]^2}} \sin\,(1000t - \tan^{-1} 10^{-3}) \\ &\approx V_{\text{in}} \sin 1000t. \end{aligned} \tag{11.3}$$

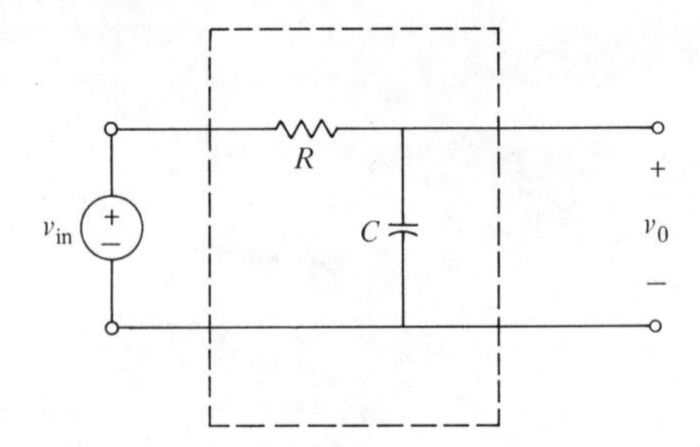

e 11.1 A simple low-pass filter.

If instead of $\omega = 1000$ we have $\omega = 10^7$ radians/sec with the same R and C, the output voltage becomes

$$\begin{aligned} v_0(t) &= \frac{V_{in}}{\sqrt{1 + [(10^7)(10^3)(10^{-9})]^2}} \sin(10^7 t - \tan^{-1} 10) \\ &= 0.099 V_{in} \sin(10^7 t - 88.5°). \end{aligned} \tag{11.4}$$

Suppose that the input voltage now consists of two sinusoids

$$v_{in}(t) = V_1 \sin 1000t + V_2 \sin 10^7 t. \tag{11.5}$$

Since the network is linear, we may use the principle of superposition. From the previous calculations of Equations 11.3 and 11.4, the output due to the input of Equation 11.5 is

$$v_0(t) = V_1 \sin 10^3 t + 0.099 V_2 \sin(10^7 t - 88.5°). \tag{11.6}$$

Notice that peak values of the two sinusoids comprising the input are V_1 and V_2, whereas the output of the network has the peak value of the higher-frequency sinusoid relatively suppressed. By examining Equation 11.2, we see that the higher the frequency ω, the less the magnitude of the output. The ratio of the magnitudes of V_0 to V_{in} is

$$\left| \frac{V_0}{V_{in}} \right| = \frac{1}{\sqrt{1 + (\omega RC)^2}}. \tag{11.7}$$

The variation with respect to ω is shown in Figure 11.2. Because of the shape of the curve, this network is called a *low-pass filter*. It is low-pass in the sense that high-frequency signals are attenuated much more than those of low frequency. By convention, the range of frequencies for which the relative frequency response or characteristic is at least 0.707 is called the *passband*. In Figure 11.2, the passband is the band of frequencies from $\omega = 0$ to $\omega = \omega_c$. The passband is also called the *bandwidth*. The band of frequencies for which the relative frequency characteristic

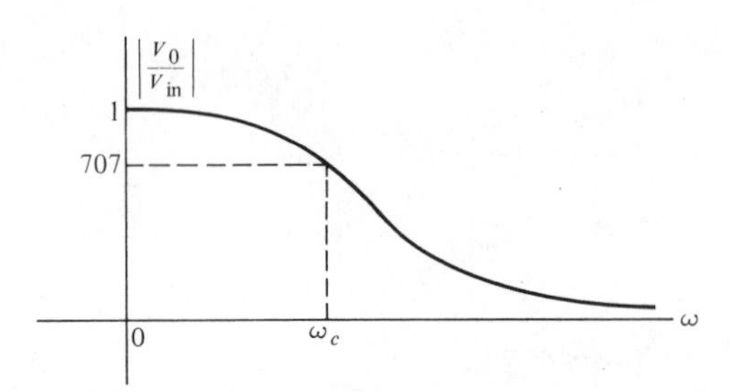

Figure 11.2 Frequency characteristic of the low-pass filter of Figure 11.1.

or response is less than $1/\sqrt{2}$ is referred to as the *stopband.** In Figure 11.2, the stopband is the range of frequencies from $\omega = \omega_c$ up to infinity. The *ideal* low-pass response is a constant characteristic of 1 from $\omega = 0$ to $\omega = \omega_c$ and zero characteristic or response for $\omega > \omega_c$. Such a characteristic is not physically realizable (meaning that it cannot be built), but it is still a useful reference for comparing actual low-pass responses. With more network elements, it is possible to design more elaborate filters which have a response nearly equal to one in the passband, and closer to zero in the stopband than the one for the simple RC network we have considered.

Another simple filter is shown in Figure 11.3. It is called a *high-pass filter* because it attenuates low-frequency signals much more than it does the high-frequency ones. An examination of the network reveals immediately that if $\omega = 0$, then there is no output voltage because there is no current through the capacitor and hence no current through R. For low frequencies, the impedance of C, which is $1/(j\omega C)$, is relatively high. The impedance of the resistor is R, a constant. The voltage-divider effect of these two impedances is such that the output voltage is

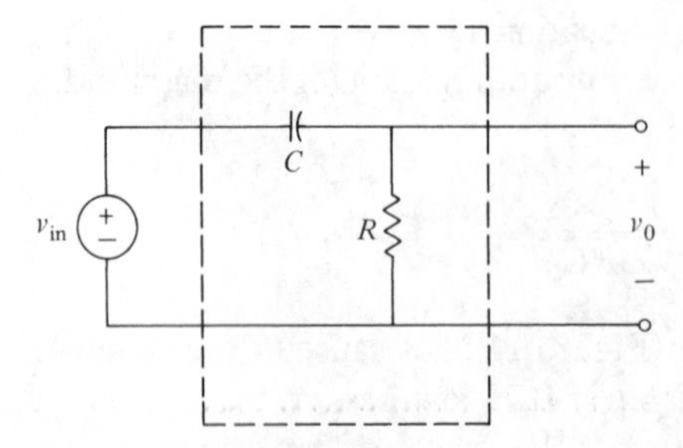

Figure 11.3 A simple high-pass filter.

* If the voltage is $1/\sqrt{2}$ of the largest possible voltage, then assuming that we have a resistive load, the power is proportional to V^2. Hence, the power is one-half of the largest possible power. Although a figure other than $\frac{1}{2}$ could be arbitrarily chosen, $\frac{1}{2}$ is frequently used.

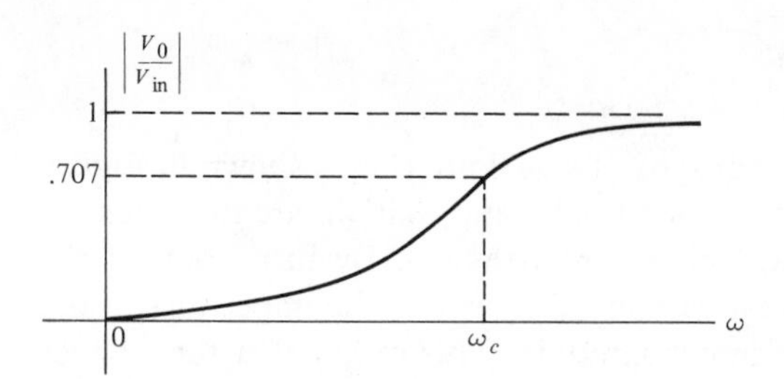

Figure 11.4 Frequency characteristic of the high-pass filter of Figure 11.3.

relatively low. However, when ω is very large, then the magnitude of the impedance $1/(j\omega C)$ is small and the voltage drop across C is also very small. Consequently, for large ω, the magnitude of the voltage at the output is almost equal to that of the input. The voltage transfer function is

$$\frac{V_0}{V_{in}} = \frac{R}{R + 1/(j\omega C)} = \frac{j\omega RC}{1 + j\omega RC}. \tag{11.8}$$

Hence the magnitude is

$$\left|\frac{V_0}{V_{in}}\right| = \frac{\omega RC}{\sqrt{1 + (\omega RC)^2}}, \tag{11.9}$$

which is plotted in Figure 11.4. From our definition of the passband, this high-pass filter has a passband corresponding to the range of frequencies from ω_c to infinity. Again ω_c, the *cut-off frequency*, corresponds to the point where the magnitude of the transfer function is 0.707 times maximum possible magnitude. The stopband is the frequency range from 0 to ω_c. For more elaborate networks, the response can be made more nearly constant in the passband and more nearly zero in the stopband.

EXERCISES

11.1.1 For the network of Figure 11.1, what is the cut-off frequency ω_c in terms of R and C? What is the *time constant* of the network? How is the time constant related to the bandwidth?

11.1.2 Suppose the network of Figure 11.1 has $R = 1000$ ohms and $C = 100$ microfarads. The input voltage is $v_{in}(t) = 100 - 20 \cos [2\pi(60)t]$. Determine $v_0(t)$. Sketch $v_{in}(t)$ and $v_0(t)$ for t from 0 to 0.1 sec.

11.1.3 Repeat Exercise 11.1.2 for Figure 11.3.

11.1.4 Determine an expression for the phase angle of the voltage transfer function as a function of ω for the low-pass filter of Figure 11.1. What is the phase angle at the cut-off frequency?

11.1.5 Repeat Exercise 11.1.4 for the high-pass filter of Figure 11.3.

11.2 Series resonance

A simple circuit which exhibits another type of signal filtering is shown in Figure 11.5. Suppose that the source voltage is sinusoidal and that we are interested in how the current magnitude depends on the frequency ω. By inspection of the series RLC circuit, we observe that for very low frequencies the impedance of the capacitor, $1/(j\omega C)$, has a relatively large magnitude. This means that the current magnitude is low. In fact, for $\omega = 0$, there is zero current. Likewise, for very large frequency, the impedance of the inductor, $j\omega L$, has a large magnitude resulting in a small magnitude of current. It turns out that for some intermediate value of ω, the current magnitude attains its largest value. Let us determine this critical frequency.

Let the voltage be

$$v(t) = V_m \sin(\omega t + \theta). \tag{11.10}$$

Since the impedance is

$$Z = R + j\omega L + \frac{1}{j\omega C} = R + j\left(\omega L - \frac{1}{\omega C}\right), \tag{11.11}$$

then the current is

$$i(t) = I_m \sin(\omega t + \theta - \phi), \tag{11.12}$$

where the phase angle ϕ is

$$\phi = \tan^{-1} \frac{\omega L - 1/(\omega C)}{R}, \tag{11.13}$$

and I_m, the peak value of the sinusoidal current, is

$$I_m = \frac{V_m}{\sqrt{R^2 + [\omega L - 1/(\omega C)]^2}}. \tag{11.14}$$

For a given set of R, L, C, and V_m which are constants, I_m and ϕ depend only on the frequency ω. We see that I_m is largest if the value of ω is such that the denomi-

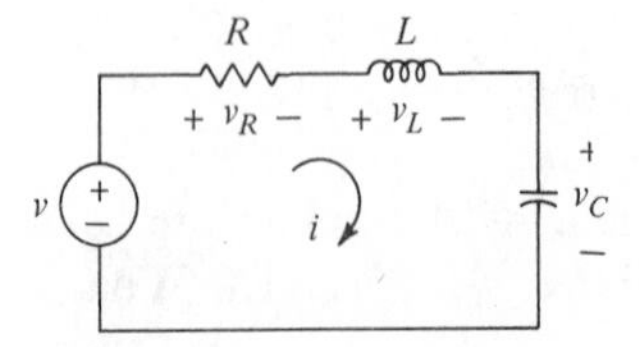

Figure 11.5 A series RLC circuit.

nator of the right-hand side of Equation 11.14 is the smallest. This occurs when $[\omega L - 1/(\omega C)]^2$ is smallest. Clearly, this occurs when

$$\omega L - \frac{1}{\omega C} = 0 \tag{11.15}$$

or

$$\omega = \frac{1}{\sqrt{LC}}. \tag{11.16}$$

From Equation 11.13, we have $\phi = 0$ if Equation 11.15 is satisfied. For a series RLC circuit, the condition which obtains when $\phi = 0$ is called *resonance*. The corresponding value of ω is called *resonant frequency*, usually denoted by ω_r. Figure 11.6 shows a typical plot of I_m vs. ω. From Equation 11.14, we see that the largest value of I_m is V_m/R. The bandwidth is $\Delta\omega = \omega_2 - \omega_1$, which is the range of frequencies for which I_m is at least $1/\sqrt{2}$ of its largest value.

Let us derive the bandwidth $\Delta\omega$ in terms of the parameters of the circuit. To calculate ω_1 and ω_2, we solve for the frequencies which satisfy

$$I_m = \frac{V_m}{\sqrt{R^2 + [\omega L - 1/(\omega C)]^2}} = \frac{1}{\sqrt{2}}\frac{V_m}{R}. \tag{11.17}$$

These points are indicated on Figure 11.6.

Simplifying Equation 11.17 we have

$$\left(\omega L - \frac{1}{\omega C}\right)^2 = R^2 \quad \text{or} \quad \omega L - \frac{1}{\omega C} = \pm R. \tag{11.18}$$

Taking the positive sign first and simplifying, we have

$$LC\omega^2 - RC\omega - 1 = 0 \tag{11.19}$$

or

$$\omega = \frac{RC \pm \sqrt{(RC)^2 + 4LC}}{2LC}. \tag{11.20}$$

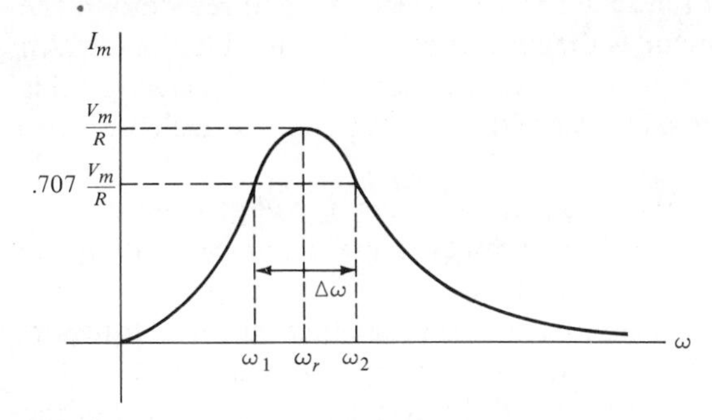

Figure 11.6 Plot of I_m vs. ω for the RLC circuit of Figure 11.5.

Since we are considering positive ω's only, we discard the negative solution; and the positive solution in Equation 11.20 is either ω_1 or ω_2. Taking the negative sign in Equation 11.18 and simplifying, we have

$$LC\omega^2 + RC\omega - 1 = 0 \tag{11.21}$$

or

$$\omega = \frac{-RC \pm \sqrt{(RC)^2 + 4LC}}{2LC}. \tag{11.22}$$

Again, the positive solution in Equation 11.22 is either ω_1 or ω_2, and since we are assuming ω_2 to be larger than ω_1, then we have

$$\omega_1 = \frac{\sqrt{(RC)^2 + 4LC} - RC}{2LC} \tag{11.23}$$

and

$$\omega_2 = \frac{\sqrt{(RC)^2 + 4LC} + RC}{2LC}. \tag{11.24}$$

Then from Equations 11.23 and 11.24, it follows that

$$\Delta\omega \equiv \omega_2 - \omega_1 = \frac{R}{L} \tag{11.25}$$

and

$$\omega_1\omega_2 = \frac{1}{LC} = \omega_r^2, \tag{11.26}$$

where ω_r is the resonant frequency.

Equation 11.25 indicates that in order to have a narrow bandwidth the ratio of R to L must be small. Now R is the total series resistance of the circuit. In practical circuits, the components are not ideal and have loss. The coil component is represented by the model with an inductor in series with a small resistor. In the series circuit of Figure 11.5, R includes the resistance of the coil. The bandwidth as given in Equation 11.25 cannot be made arbitrarily small because of this resistance. If we make L larger by using more turns in the coil, we use more wire and the resistance of the coil also increases.

Equation 11.26 also shows that the resonant frequency is the geometric mean of the cut-off frequencies ω_1 and ω_2. The formula is useful for calculating the resonant frequency from the cut-off frequencies and also for calculating one cut-off frequency in terms of the resonant frequency and the other cut-off frequency.

In an ideal resonant circuit, there is no dissipation. Energy is simply stored in the associated magnetic field of the inductor and the associated electric field of the capacitor. In a practical circuit with nonzero resistance, part of the energy is dissipated. A useful figure of merit for resonant circuits is based on the ratio of total stored energy at the resonant frequency to energy dissipated per cycle. Let us first find expressions for energy stored and energy dissipated per cycle before we define this figure of merit precisely. Let the current through the circuit be

$$i(t) = I_m \sin(\omega t + \alpha). \tag{11.27}$$

Then the energy stored in the inductor at any time t is

$$W_L(t) = \tfrac{1}{2}Li^2 = \tfrac{1}{2}LI_m^2 \sin^2(\omega t + \alpha). \tag{11.28}$$

The voltage $v_C(t)$ with polarity reference as shown in Figure 11.5 lags the current by 90°, and it is

$$v_C(t) = \frac{I_m}{\omega C}\sin(\omega t + \alpha - 90°) = -\frac{I_m}{\omega C}\cos(\omega t + \alpha). \tag{11.29}$$

The energy stored in the capacitor at any time t is

$$W_C(t) = \tfrac{1}{2}Cv_C^2 = \tfrac{1}{2}C\left(\frac{I_m^2}{\omega^2 C^2}\right)\cos^2(\omega t + \alpha) = \frac{I_m^2}{2\omega^2 C}\cos^2(\omega t + \alpha). \tag{11.30}$$

The total stored energy at any instant of time is then

$$W_T(t) = W_L(t) + W_C(t) = \tfrac{1}{2}|I_m|^2\left[L\sin^2(\omega t + \alpha) + \frac{1}{\omega^2 C}\cos^2(\omega t + \alpha)\right]\cdot \tag{11.31}$$

At resonance, $\omega^2 = 1/(LC)$ so that $1/(\omega^2 C) = L$ and Equation 11.31 becomes

$$W_T(t) = \tfrac{1}{2}|I_m|^2 L[\sin^2(\omega t + \alpha) + \cos^2(\omega t + \alpha)] = \tfrac{1}{2}L|I_m|^2 = \text{constant}. \tag{11.32}$$

We note that at resonance, although the current and voltages in the series *RLC* circuit are sinusoidal, the total *instantaneous* stored energy is a *constant*. For frequencies other than ω_r, the instantaneous stored energy as given by Equation 11.31 fluctuates with time. The energy dissipated at resonant frequency in one cycle is

$$W_d = \text{average power} \times \text{period} = I_{\text{rms}}^2 R \times T = \frac{I_m^2 R}{2} \times T. \tag{11.33}$$

We now *define* a figure of merit or quality factor Q as

$$Q = 2\pi\,\frac{\text{total energy stored at resonant frequency}}{\text{energy dissipated per cycle at resonant frequency}}\cdot \tag{11.34a}$$

Since one cycle corresponds to 2π radians, then the energy dissipated per cycle at resonant frequency divided by 2π is the average energy dissipated per radian at resonant frequency. Hence, the definition of Q in Equation 11.34a may be written

$$Q = \frac{\text{total energy stored at resonant frequency}}{\text{average energy dissipated per radian at resonant frequency}}\cdot \tag{11.34b}$$

For the series circuit of Figure 11.5, Q is

$$Q = 2\pi \frac{\frac{1}{2}LI_m^2}{\frac{1}{2}RI_m^2 T} = \frac{2\pi}{T}\frac{L}{R} = \frac{\omega_r L}{R}. \tag{11.35}$$

The energy definition of Q of Equation 11.34a is quite universal and applies to parallel resonance to be described later as well as to more complicated resonant networks. In fact, it is used in describing resonant mechanical networks, resonant acoustical devices, and resonant microwave cavities. For series RLC circuits, it reduces to the expression in Equation 11.35. Another figure of merit used to characterize resonant systems is *selectivity* S which is defined as

$$S = \frac{\text{resonant frequency}}{\text{bandwidth}} = \frac{\omega_r}{\Delta\omega}. \tag{11.36}$$

We can verify that for the series circuit, since $\Delta\omega = R/L$, $S = \omega_r L/R$, which is equal to the Q factor as in Equation 11.35. Note that R in Equation 11.35 is the total resistance in the circuit. In general, Q and S are different. A quality factor for the practical inductor alone is defined as

$$Q_{\text{coil}} = \frac{\omega_r L}{R_L}, \tag{11.37}$$

which is the same as the expression in Equation 11.35 except that total resistance R is replaced by R_L, the resistance of the practical inductor.

Figure 11.7 shows the relative responses for different values of S. For the *same resonant frequency* ω_r, the higher the selectivity, the narrower the bandwidth. Similarly, for the *same* S, the higher the resonant frequency ω_r, the larger the bandwidth in accordance with Equation 11.36.

Thus far, we have considered the current as the output variable for the network in Figure 11.5. Other possible quantities we may consider are the voltages across each of the elements. The variation of voltage magnitude as a function of ω is

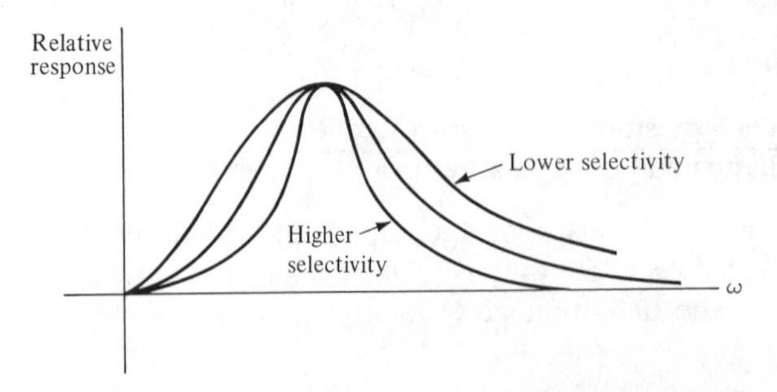

Figure 11.7 Effect of selectivity on resonance curve for *RLC* series network of Figure 11.5.

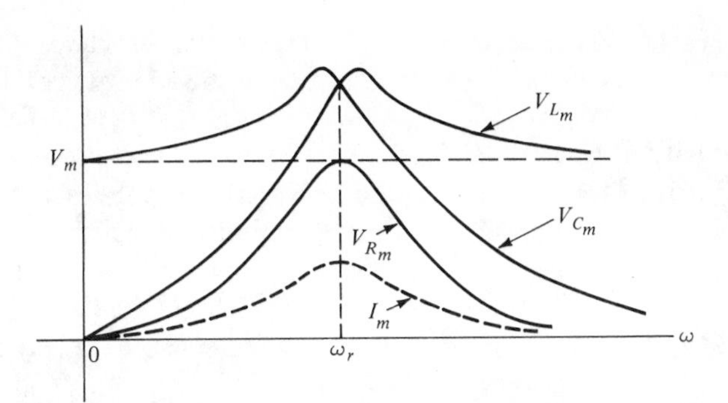

re 11.8 Plots of V_{Rm}, V_{Lm}, and V_{Cm} vs. ω for the series *RLC* circuit.

found by multiplying the magnitude of the current by the magnitude of the impedance. That is, we have

$$V_{Rm} = R|I_m| = \frac{|V_m|R}{\sqrt{R^2 + [\omega L - 1/(\omega C)]^2}}, \tag{11.38}$$

$$|V_{Cm}| = \frac{1}{\omega C}|I_m| = \frac{|V_m|}{\omega C\sqrt{R^2 + [\omega L - 1/(\omega C)]^2}}, \tag{11.39}$$

and

$$|V_{Lm}| = \omega L|I_m| = \frac{\omega L|V_m|}{\sqrt{R^2 + [\omega L - 1/(\omega C)]^2}}. \tag{11.40}$$

These expressions give the peak values of the corresponding sinusoidal voltages. Typical plots of these magnitudes vs. ω are shown in Figure 11.8. Observe that V_{Rm}, the peak value of v_R, and I_m have the same variation with frequency, but that the peak values V_{Cm} and V_{Lm} plotted as a function of ω each have their own characteristic shape. The largest value of V_{Cm} and V_{Lm} occur at frequencies slightly different from resonant frequency (see Problem 11.5).

ERCISES

11.2.1 Show that the Q of a series *RLC* circuit is equal to (a) $Q = 1/(\omega_r RC)$, (b) $Q = (1/R)\sqrt{L/C}$.

11.2.2 Show that $V_{Cm} = V_{Lm}$ when $\omega = \omega_r$ as shown in Figure 11.8 for the series *RLC* circuit.

11.2.3 Show that V_L approaches V_m asymptotically as ω approaches infinite value, as shown in Figure 11.8 for the *RLC* circuit.

11.2.4 Determine an expression for the magnitude of the admittance of the series *RLC* circuit of Figure 11.5. Compare your expression to that for I_m

in Equation 11.14. From this expression, we see that series resonance corresponds to the frequency for which the magnitude of the admittance is a maximum.

11.2.5 Show that the selectivity of a series *RLC* circuit is equal to the expressions in Exercise 11.2.1.

11.3 Parallel resonance

In the preceding section, we have considered a series *RLC* circuit where a voltage source is the excitation and current is the response of interest. We will now examine another simple and commonly used network where the usual response is the voltage across the network as shown in Figure 11.9 and the excitation is a current source.

As before, we are interested in the dependence of the response on the frequency of the source, ω. Suppose the input current is

$$i(t) = I_m \sin(\omega t + \theta) \tag{11.41}$$

with a corresponding phasor

$$I = I_m e^{j\theta}. \tag{11.42}$$

Then the phasor voltage is

$$V = IZ, \tag{11.43}$$

where Z is the total complex impedance

$$Z = \frac{1}{j\omega C + \dfrac{1}{R + j\omega L}} = \frac{1}{j\omega C + \dfrac{R}{R^2 + (\omega L)^2} + \dfrac{-j\omega L}{R^2 + (\omega L)^2}}. \tag{11.44}$$

At resonance, $v(t)$ and $i(t)$ must be in phase by definition. This requires that ω

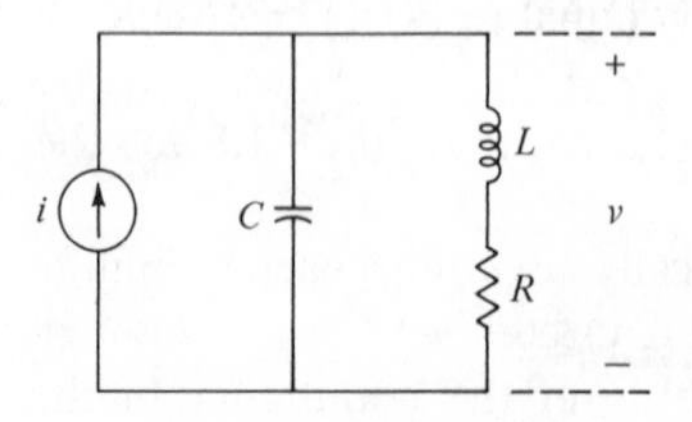

Figure 11.9 A parallel network with a current source input.

have a value such that Z is real at that frequency. The impedance Z is real provided the denominator in Equation 11.44 is real; that is,

$$\omega C - \frac{\omega L}{R^2 + (\omega L)^2} = 0. \tag{11.45}$$

This is satisfied when either $\omega = 0$ or

$$\omega = \frac{1}{\sqrt{LC}} \sqrt{1 - R^2C/L}. \tag{11.46}$$

Equation 11.46 assumes that

$$\frac{R^2C}{L} \leq 1. \tag{11.47}$$

If $R^2C/L > 1$, then there is no nonzero frequency for which $v(t)$ and $i(t)$ are in phase.

Examining Equation 11.46, it is clear that if $R = 0$, then the formula for resonant frequency is the same as that of Equation 11.16 for series resonance. However, for practical coil components, a pure inductance model is not satisfactory for many purposes. Hence, the R in Figure 11.8 is meant to represent the coil resistance.

For a constant current source, the magnitude of the voltage response is proportional to the magnitude of the complex Z as given in Equation 11.44. For a given set of R, L, and C, we may plot $|Z|$ as a function of ω and graphically determine the value of ω for which $|Z|$ is largest. We may also determine the range of frequencies for which the response is at least $1/\sqrt{2}$ of the largest $|Z|$. If an exact analytical expression is sought for the bandwidth $\Delta\omega$ and the peak frequency, the resulting equations are quite unwieldy. For this reason, let us consider a special case for which an approximation can be made. Suppose that in the frequency range of interest, such as in the bandwidth range $\Delta\omega$, R is small compared to ωL. Then in Equation 11.44, $R^2 + (\omega L)^2 \approx (\omega L)^2$, and Equation 11.44 can be approximated by

$$Z \approx \frac{1}{j\omega C + \dfrac{R}{(\omega L)^2} + \dfrac{-j\omega L}{(\omega L)^2}} = \frac{1}{\dfrac{R}{(\omega L)^2} + j\left(\omega C - \dfrac{1}{\omega L}\right)}. \tag{11.48}$$

Hence we have

$$|Z| \approx \frac{1}{\sqrt{[R/(\omega L)^2]^2 + [\omega C - 1/(\omega L)]^2}}. \tag{11.49}$$

Assuming that the selectivity of the circuit is high, then for ω inside the range of $\Delta\omega$ the value of ω is approximately equal to the value of ω for which $|Z|$ is maximum. Then $R/(\omega L)^2$ is approximately constant in the bandwidth $\Delta\omega$. If $R/(\omega L)^2$

is considered constant in the range of $\Delta\omega$ in Equation 11.49, then the problem of determining $\Delta\omega$ approximately is simplified. First, we notice from Equation 11.49 that $|Z|$ is approximately maximum when

$$\omega C - \frac{1}{\omega L} = 0$$

or

$$\omega = \frac{1}{\sqrt{LC}}. \tag{11.50}$$

That is, the peak in the response curve occurs in the vicinity of $\omega = 1/\sqrt{LC}$, which is also in the vicinity of resonance as given by Equation 11.46. Next, we notice that the dependence of $|Z|$ on ω, for the range of ω in $\Delta\omega$, is similar to that of I_m for series resonance as given in Equation 11.14. Comparing Equation 11.14 and 11.49, we see that R in Equation 11.14 corresponds to $R/(\omega L)^2$ in Equation 11.49, L and C in Equation 11.14 correspond to C and L, respectively, in Equation 11.49. The mathematical problem of determining $\Delta\omega$ is the same for both. Since we have done it already for Equation 11.14, we need not repeat. We simply look at the result in Equation 11.26 and make appropriate substitutions in symbols. Thus in Equation 11.26, we replace R by $R/(\omega L)^2$ and L by C. We note also that $\omega \approx \omega_r$. The approximate bandwidth for the parallel circuit of Figure 11.9 is then

$$\Delta\omega \approx \frac{R/(\omega L)^2}{C} \approx \frac{R}{\omega_r^2 L^2 C}. \tag{11.51}$$

Since $\omega_r^2 = 1/(LC)$, this reduces further to

$$\Delta\omega \approx \frac{R}{L}, \tag{11.52}$$

which is the same for the series case. The approximate selectivity S is

$$S = \frac{\omega_r}{\Delta\omega} \approx \frac{\omega_r L}{R}. \tag{11.53}$$

To arrive at Equations 11.52 and 11.53, we assumed that $R \ll \omega L$ and that S is high. The two conditions turn out to be equivalent. In order for our approximations to be justified, the resulting value of S should be high. They should not be used for S less than 10. To determine the general shape of the response curve, we note, by examining Figure 11.9, that the higher the frequency, the lower the impedance of the capacitor. For extremely large ω, the capacitor is essentially a short circuit. Hence, as ω approaches infinity, $|Z|$ approaches zero. Likewise for very low frequencies, the capacitor becomes almost an open circuit and the inductor acts like a short circuit. The net impedance is low but not zero. As ω

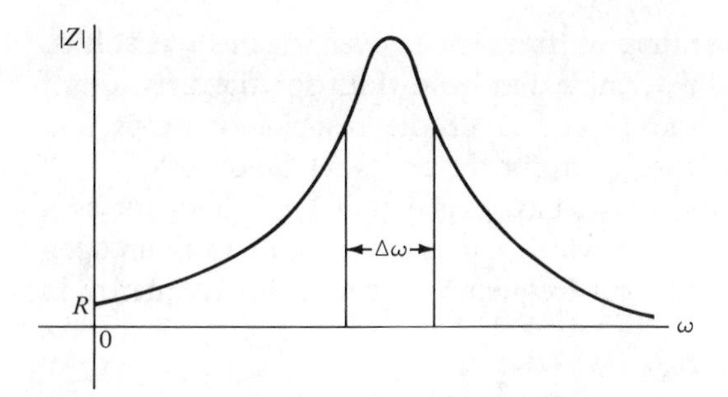

Figure 11.10 A typical response for the network of Figure 11.9 when R is small.

approaches zero, $|Z|$ approaches R. A typical response curve is shown in Figure 11.10. Thus for high values of S, the parallel circuit of Figure 11.9 has approximately the same equations for bandwidth, Q, and peak frequency.

ERCISES

11.3.1 Suppose for the network in Figure 11.9, $C = 100$ picofarads (1 picofarad $= 10^{-12}$ farad), $L = 0.01$ henry, $R = 1000$ ohms. What is the approximate resonant frequency? What is the selectivity S? What is the bandwidth? What is the Q?

11.3.2 For the network in Exercise 11.3.1, what is the approximate magnitude of the impedance at resonance? Calculate $|Z|$ at the following additional values of ω: $\omega = 0$, $\omega = \frac{1}{2}\omega_r$, $\omega = \omega_r - \frac{1}{4}\Delta\omega$, $\omega = \omega_r + \frac{1}{4}\Delta\omega$, $\omega = 2\omega_r$, $\omega = \infty$. Sketch the variation of $|Z|$ as a function of ω.

11.3.3 Suppose the current $i(t)$ in Figure 11.9 is $i(t) = 0.1 \cos \omega_r t + 0.1 \cos 2\omega_r t$, where $\omega_r = 1/\sqrt{LC}$. Determine $v(t)$ if the network parameters are as given in Exercise 11.3.1. Is there a filtering effect?

11.3.4 The parallel RLC network of Figure 11.11 is connected to a sinusoidal current source of fixed maximum value but adjustable frequency. All measurements are made when the system is in the steady-state. We define resonance as occurring when $v(t)$ and $i(t)$ are in phase. Determine the frequency of resonance, ω_r, in terms of the R, L, and C of the network.

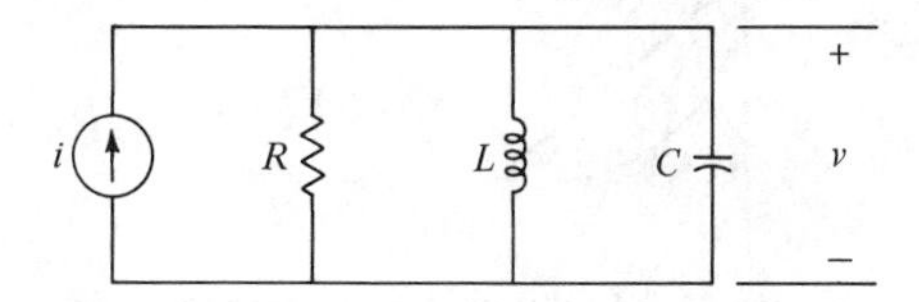

Figure 11.11

11.3.5 Defining bandwidth as the range of frequencies over which V_m is at least 0.707 of the largest V_m, determine the bandwidth for the network of Figure 11.11. Calculate the Q and S of the resonant network and compare the answer with that given for the series RLC network.

11.3.6 Plot the magnitude of the impedance of a parallel LC network as a function of ω. The frequency at which the circuit appears to be an open circuit is defined as the resonant frequency. What is this frequency in terms of L and C?

11.4 Locus diagrams

Suppose that we plot the current phasor I in the series RLC circuit for various values of ω. Then we will obtain a set of phasors as in Figure 11.12. If we connect the tips of the arrows of the phasors, we obtain a curve known as the *locus* of I with ω as a parameter. For the RLC series circuit, the locus turns out to be a circle as shown in Figure 11.13. That this is true may be seen by the following consideration. The phasor voltage V_R is always 90° out of phase with the phasor voltage V_x across the LC combination. The phasor sum, however, is equal to V, the phasor voltage corresponding to the source. No matter what the current is

$$V = V_R + V_x = \text{constant}. \tag{11.54}$$

A typical phasor diagram is shown in Figure 11.14. Since the V_R and V_x phasors are always perpendicular and the sum is constant, the locus of the right-angle corner will be a circle. Hence, the locus of V_R is a circle. Since $V_R = IR$ and since R is constant, the locus of I is also a circle. It is clear that V_R is largest when $V_x = 0$ and the corresponding I is $|V|/R$, which is the diameter of the circle of

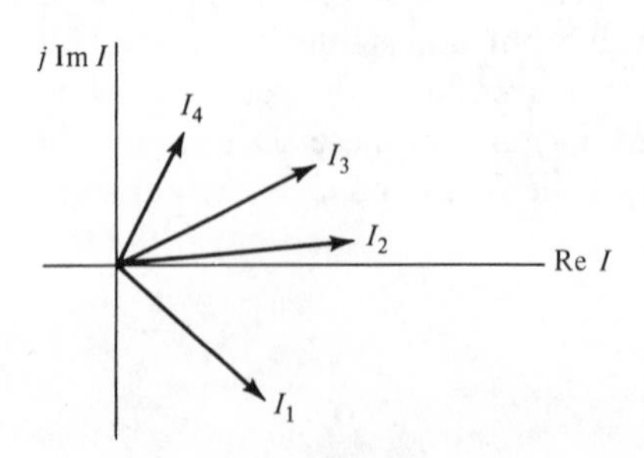

Figure 11.12 Current phasors for various values of ω.

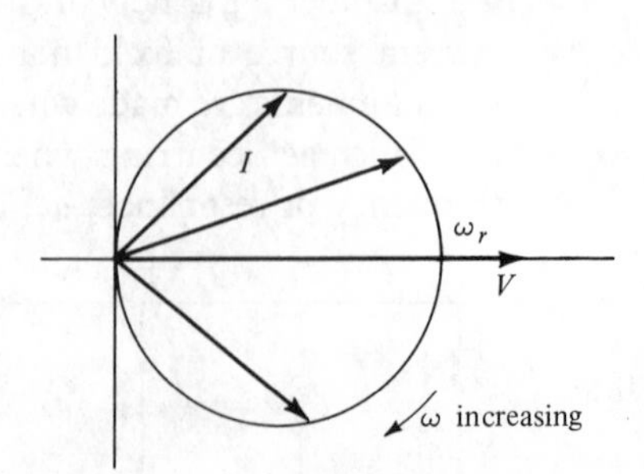

Figure 11.13 Locus for the I phasor with ω as the parameter.

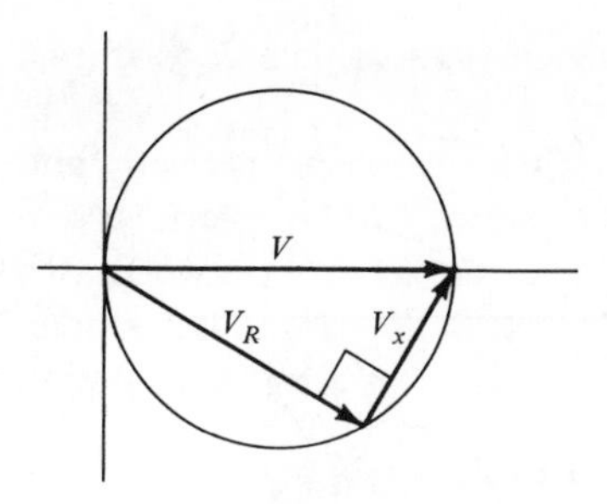

Figure 11.14 Typical phasor diagram for a series *RLC* network.

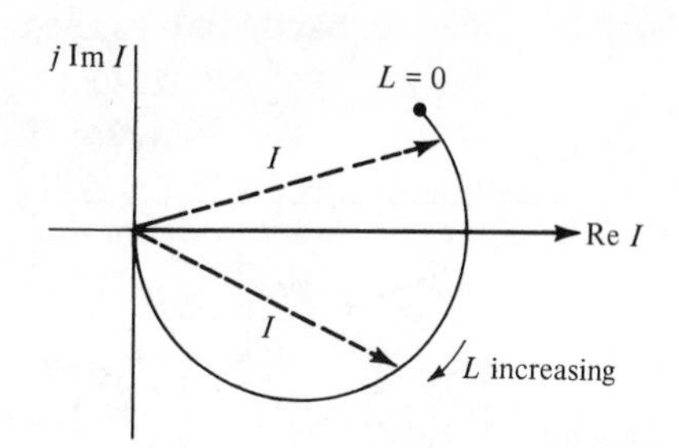

Figure 11.15 Locus of *I* in the *RLC* series network with *L* as a parameter

the I locus. Note that we are using V as a reference phasor (on the horizontal axis) and that I lagging corresponds to

$$\omega L - \frac{1}{\omega C} > 0$$

or

$$\omega > \frac{1}{\sqrt{LC}}, \tag{11.55}$$

and I leading corresponds to

$$\omega L - \frac{1}{\omega C} < 0,$$

or

$$\omega < \frac{1}{\sqrt{LC}}. \tag{11.56}$$

At the resonant frequency ω_r, I and V are in phase. From the locus it is seen that $|I|$ is largest at $\omega = \omega_r$.

Instead of plotting I for various values of ω, we may investigate the dependence of I on L or C at a fixed value of ω. For example, we may allow L to be adjustable from zero to very large values. For the sake of drawing locus diagrams, let L be adjustable from 0 to ∞. The reactance $\omega L - 1/(\omega C)$ will vary between $-1/(\omega C)$ and ∞. Thus, when $L = 0$, I leads V. As L increases, the angle of lead decreases and eventually I becomes a lagging current. Finally, as L approaches ∞, the lag angle will approach $-90°$, but the magnitude of I approaches 0 at the same time. If the tips of the I phasor for various values of L are connected, we obtain the *locus* of I with L as a parameter, and we obtain a *part* of a circle as shown in

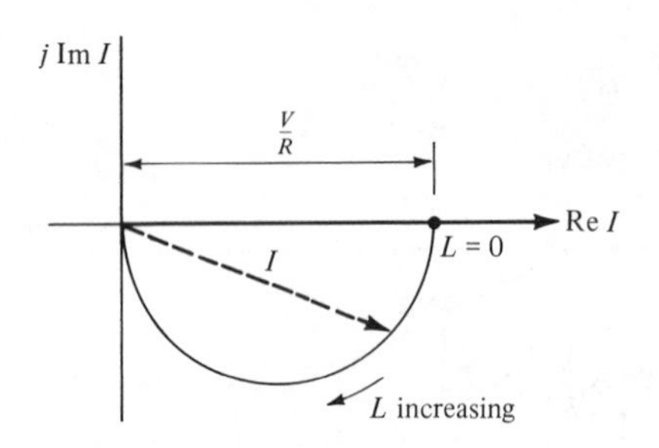

Figure 11.16 Locus of I in an RL series network with L as a parameter.

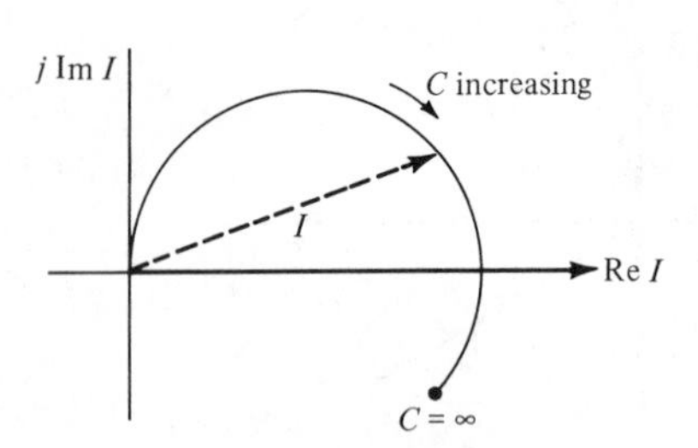

Figure 11.17 Locus of I in an RLC series network with C as a parameter.

Figure 11.15. For I to be in phase with V, L must be such that $\omega L - 1/(\omega C) = 0$ or

$$L_r = \frac{1}{\omega^2 C}. \tag{11.57}$$

If C is short-circuited in the RLC series circuit, i.e., if we have an RL circuit only, then the locus of I as L is adjusted will be a semicircle as shown in Figure 11.16. The reason for the exclusion of portions of the circle in Figures 11.15 and 11.16 is that we are not allowing L to be negative. Recall that the complete circle in Figure 11.13 corresponds to the variation of the reactance $\omega L - 1/(\omega C)$ from $-\infty$ to $+\infty$. In Figure 11.15, the reactance varies from $-1/(\omega C)$ to $+\infty$, and in Figure 11.16, from 0 to $+\infty$. Similarly, if we have an adjustable C in an RLC series circuit, the reactance varies between $-\infty$ to ωL, and the locus of I with C as a parameter is shown in Figure 11.17. For an RC circuit with an adjustable C, the reactance varies between $-\infty$ and 0, and the locus of I is shown in Figure 11.18.

The locus of I with R as a parameter in an RC series circuit is as shown in Figure 11.19. For an RL series circuit with an adjustable R, the locus of I is also a semicircle as shown in Figure 11.20.

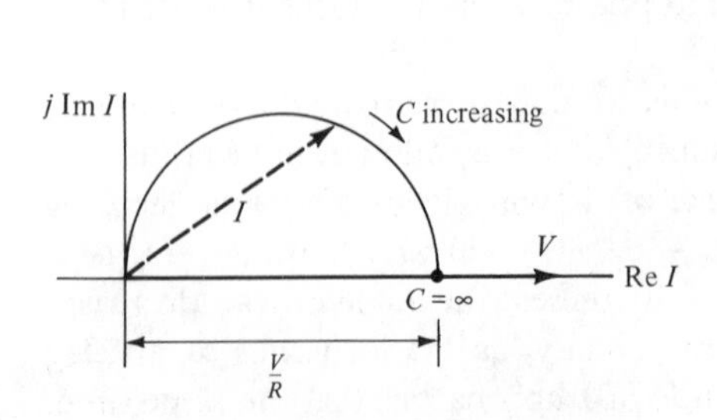

Figure 11.18 Locus of I in an RC circuit with C as the parameter.

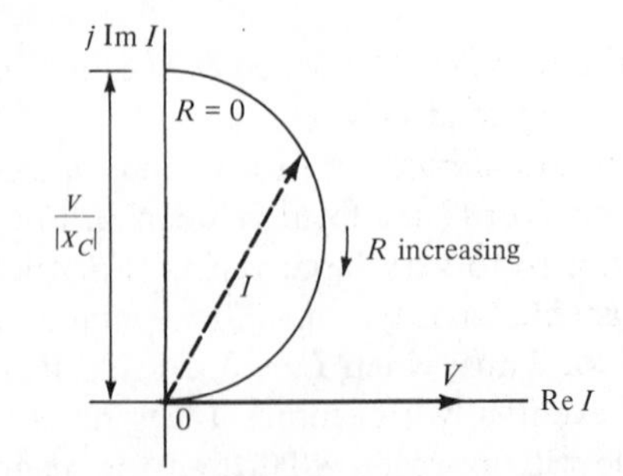

Figure 11.19 Locus of I in an RC circuit with R as the parameter.

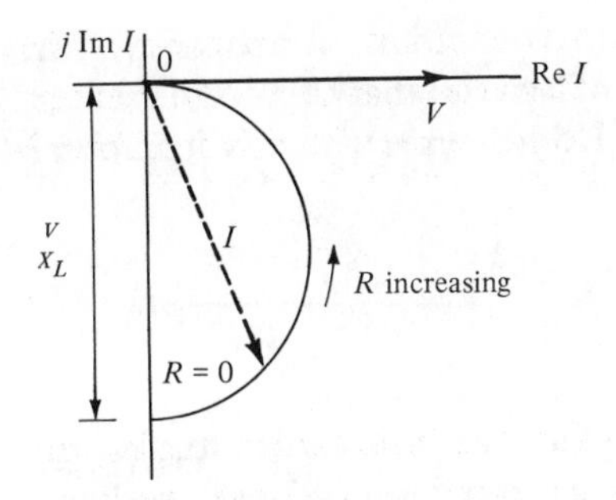

Figure 11.20 Locus of I in an RL circuit with R as the parameter.

EXERCISES

11.4.1 From the locus diagram of I in an RLC series circuit, with ω as a parameter, show that when $|I|$ is $1/\sqrt{2}$ of its maximum value, V and I are out of phase by 45°.

11.4.2 For a series RLC circuit, verify the locus of I with C as a parameter by calculating I for several values of C. Assume $V_{\text{rms}} = 10$ volts, $R = 1$ ohm, $L = 1$ henry, $\omega = 1$ radian/sec, and take C to be 0, 1, 10, and 100.

11.4.3 For a series RLC circuit with adjustable C, draw the locus of I if $v(t) = 10 \sin 100t$, $R = 10$ ohms, and $L = 0.1$ henry. Draw the I phasor corresponding to $C = 10^3$ microfarads.

11.4.4 The voltage source in Figure 11.21 has a given peak value V_m and a given ω and all the network parameters are given except C. Assuming that V is the reference phasor, complete the following: (a) draw I_L; (b) draw the locus of I_C with C as a parameter; (c) draw the locus of I with C as a parameter.

11.4.5 Suppose that in the network of Figure 11.21, L instead of C is adjustable. Draw a locus of I with L as a parameter.

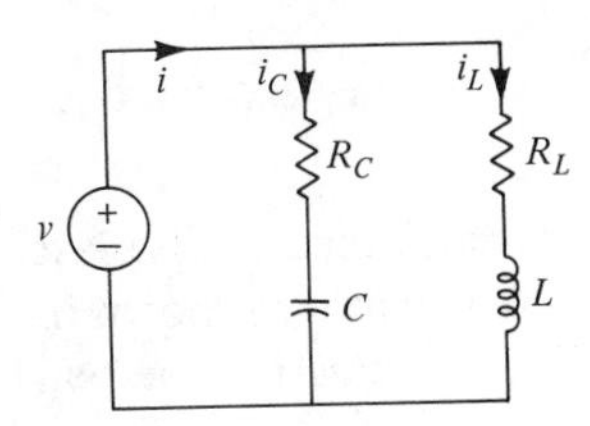

Figure 11.21

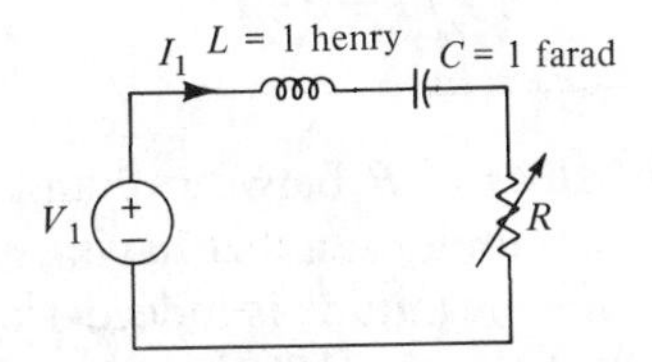

Figure 11.22 An RLC series circuit in which the resistor R is allowed to vary from 0 to ∞.

11.4.6 A popular dictionary gives the following definition of resonance: "In electricity, the condition of adjustment of a circuit that allows the greatest flow of current of a certain frequency." Do you agree with this definition? Discuss why or why not.

11.5 Locus of roots

The series *RLC* circuit shown in Figure 11.22 is familiar from earlier studies, the most recent of which was in Section 11.2. For an exponential signal, we know from Equation 10.68 that the current is

$$I_1 = \frac{V_1}{L} \frac{s}{s^2 + Rs/L + 1/(LC)}. \tag{11.58}$$

For simplicity, let $L = 1$ henry, $C = 1$ farad, so that the denominator of Equation 11.58 has a particularly simple form

$$s^2 + Rs + 1 = 0. \tag{11.59}$$

We do this so that we can study the change of location of the roots, the so-called *locus of roots*, of the characteristic equation with variations of one element's values — R in this case. The two roots of Equation 11.59 are

$$s_1, s_2 = -\frac{R}{2} \pm \sqrt{\left(\frac{R}{2}\right)^2 - 1}, \tag{11.60}$$

showing that the roots are determined by the value of R. We will study the locus of roots as R takes values from 0 to ∞. A few of the significant values of root location are shown in the table that follows:

Value of R	*Root location*
large R	$0, -R$
2	$-1, -1$
0	$\pm j1$
0 to 2	$-\frac{R}{2} \pm j\sqrt{1 - \left(\frac{R}{2}\right)^2}$

Root locations for values of R between 0 and 2 are of special interest. The last entry in the table is a complex number having rectangular coordinates as shown in Figure 11.23. That the magnitude is indeed 1 can be verified by squaring the real and imaginary parts of the root location

$$\frac{R^2}{4} + 1 - \frac{R^2}{4} = 1. \tag{11.61}$$

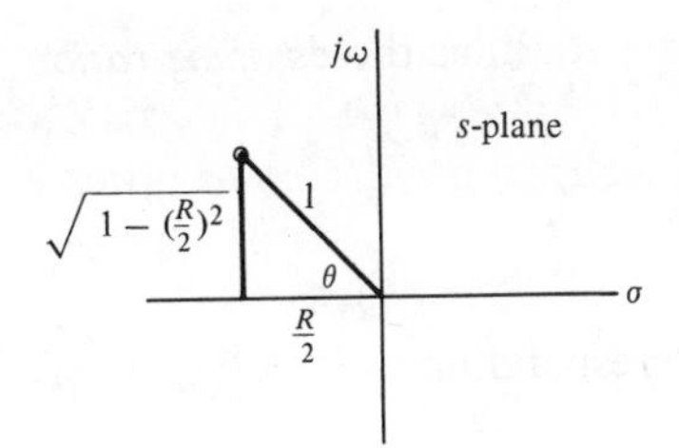

Figure 11.23 The s-plane location of the root of the characteristic equation given by Equation 11.59.

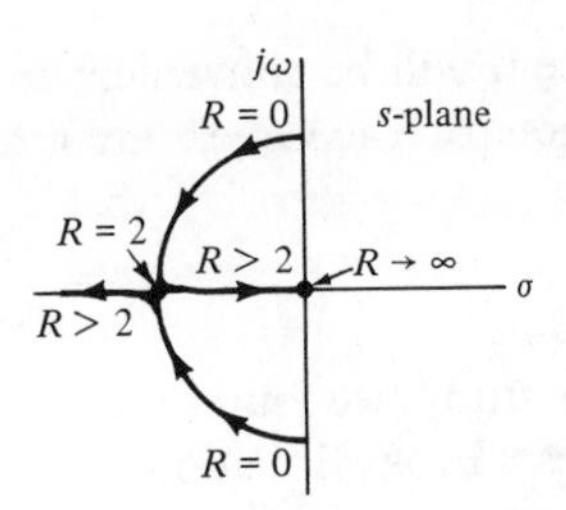

Figure 11.24 The locus of the roots of Equation 11.59 as R in the network of Figure 11.22 takes values from 0 to ∞.

And as shown by Figure 11.23, the angle θ can be specified by any two of the sides of the right triangle, the most convenient choice being

$$\theta = \cos^{-1}\frac{R}{2} \tag{11.62}$$

for $R \leq 2$.

Combining the information provided by the table with the polar form of the location of roots, $1\underline{/\theta}$ where θ is the angle of Equation 11.62, we see the total locus of roots as shown in Figure 11.24. When $R = 0$, the roots are on the imaginary axis. As R increases, the root locus is a circle of radius 1, the two roots being a conjugate pair. When R reaches the value 2, the two roots coincide at $\sigma = -1$. For larger values of R, the two roots remain negative and real, with one root approaching $-\infty$ and the other approaching 0. The path of all root locations for all possible values of R is the root locus, shown in Figure 11.24.

While this appears to be a very simple example, it is worth exploring further, for the general case will illustrate interesting aspects of second-order systems. The root locations for the denominator of Equation 11.58 are, for general R, L, and C:

$$s_1, s_2 = -\frac{R}{2L} \pm \sqrt{\left(\frac{R}{2L}\right)^2 - \frac{1}{LC}}. \tag{11.63}$$

Let R_{cr} be the value of R causing the term inside the radical to vanish. It is given by the equation

$$\left(\frac{R_{cr}}{2L}\right)^2 = \frac{1}{LC} \tag{11.64}$$

or

$$R_{cr} = 2\sqrt{\frac{L}{C}}. \tag{11.65}$$

Now we are not as interested in R_{cr} as we are in the relationship of the value of R

to R_{cr}. Hence it will be convenient to define the ratio R/R_{cr} as the *damping ratio*. This we denote by the Greek letter zeta,

$$\zeta = \frac{R}{R_{cr}} = \frac{R}{2}\sqrt{\frac{C}{L}}. \tag{11.66}$$

In our earlier study, we found the values of roots corresponding to $R = 0$ to be of special interest. From Equation 11.63, these are

$$s_1, s_2 = \pm j\sqrt{\frac{1}{LC}}, \tag{11.67}$$

which we have encountered before in Equation 11.16 as the frequency of resonance. The square of this value is

$$\omega_r^2 = \frac{1}{LC}, \tag{11.68}$$

which is the last term of the characteristic equation. The coefficient of s in the characteristic equation can also be expressed in terms of ζ and ω_r. It is

$$\frac{R}{L} = 2\left(\frac{R}{2}\sqrt{\frac{C}{L}}\right)\left(\frac{1}{\sqrt{LC}}\right) = 2\zeta\omega_r, \tag{11.69}$$

so that the generalized form of the characteristic equation becomes in terms of ζ and ω_r,

$$s^2 + 2\zeta\omega_r s + \omega_r^2 = 0. \tag{11.70}$$

We may now repeat the operations carried out previously for the special case. For this general case, the root locations are

$$s_1, s_2 = -\zeta\omega_r \pm \omega_r\sqrt{\zeta^2 - 1}. \tag{11.71}$$

Again, a few of the significant root locations are summarized in the following table:

Values of ζ	*Root locations*
large	$0, -2\omega_r$
1	$-\omega_r, -\omega_r$
0	$\pm j\omega_r$
0 to 1	$-\zeta\omega_r \pm j\omega_r\sqrt{1 - \zeta^2}$

From the last entry in the table, the complex-root location is displayed in terms of real and imaginary components in Figure 11.25. The distance from the origin to

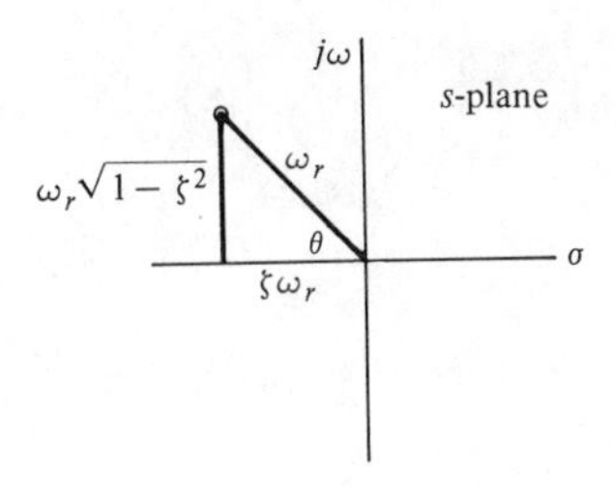

Figure 11.25 The location of one root of the characteristic equation of Equation 11.70.

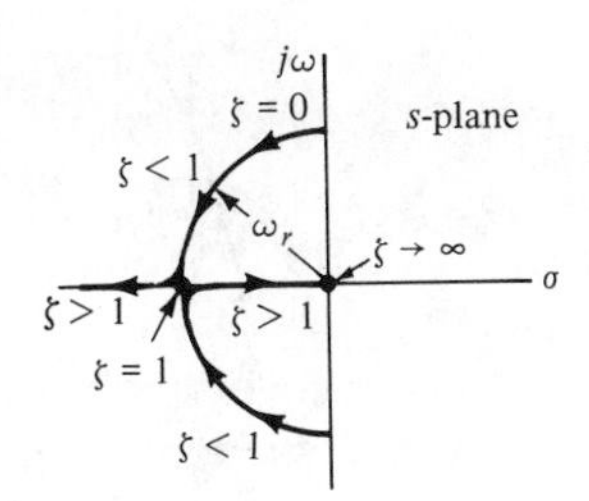

Figure 11.26 The locus of roots of Equation 11.70 as ζ varies from 0 to ∞.

the root is seen to be ω_r, and the angle is defined by

$$\theta = \cos^{-1}\zeta \tag{11.72}$$

provided $|\zeta| \leq 1$.

The root locus plot is shown in Figure 11.26 showing the same regions identified in the discussion of Figure 11.24, except that the identifications are now made in terms of ζ rather than R.

While the results we have obtained thus far may appear to be specialized, pertaining only to a second-order system similar to a simple *RLC* circuit that was the original motivation, they turn out to be much more important. First, higher-order networks will have characteristic equations made up of roots which are either real or occur in conjugate complex pairs. For stable systems the roots must have negative real parts. Each pair of these roots can be described in terms of ζ, and ω_r will be those dominating the response of the system. This will be the case whether the response is due to a step input or related signal or due to the application of a sinusoidal signal.

The concept of locus of roots applies to networks described by characteristic equations of higher order than 2. In general, one element or parameter variation is considered, and the migration of the roots in the complex plane is investigated. We are interested in the locus of roots since it will tell us the behavior of the system for a particular value of the element or parameter. Suppose that the characteristic equation of a network under study is

$$s^3 + 6s^2 + 5s + k = 0, \tag{11.73}$$

where k represents the value of a varying parameter. The most direct method of determining the locus of roots is to let k have a given value and then factor Equation 11.73. This is laborious, of course, unless accomplished with the aid of a computer. A set of rules and useful plotting aids* are available when a large

* William R. Perkins and Jose B. Cruz, Jr., *Engineering of Dynamic Systems* (New York: John Wiley & Sons, Inc., 1969), Chapter 8.

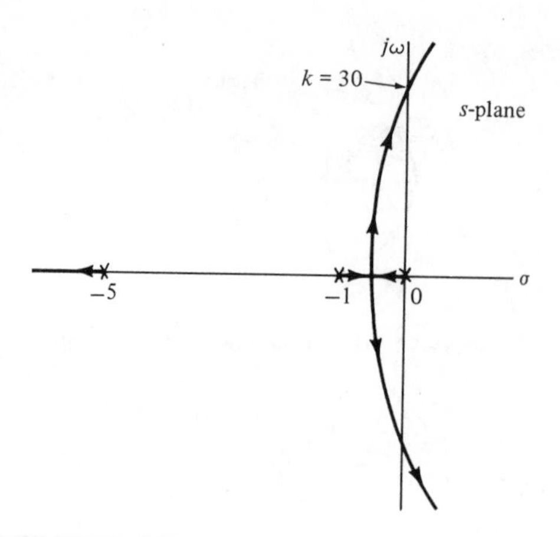

Figure 11.27 The locus of roots of Equation 11.73 as k varies from 0 to ∞.

amount of plotting must be done. The root locus for Equation 11.73 is shown in Figure 11.27 for values of k ranging from 0 to a large value. For small k, the roots are all real and negative. For larger k, two roots break away from the real axis and become a complex conjugate pair. When $k = 30$, these roots move to the right half of the s-plane and then represent an unstable system.

EXERCISES

11.5.1 Plot the locus of roots of the characteristic equation for $V_2(s)$ in the network of Figure 11.28 as R changes from 0 to ∞.

11.5.2 The characteristic equation of a network is

$$s^2 + s + k = 0.$$

Plot the locus of roots of this equation for all positive values of k.

11.5.3 Repeat Exercise 11.5.2 for the characteristic equation

$$s^2 + 100s + 500k = 0.$$

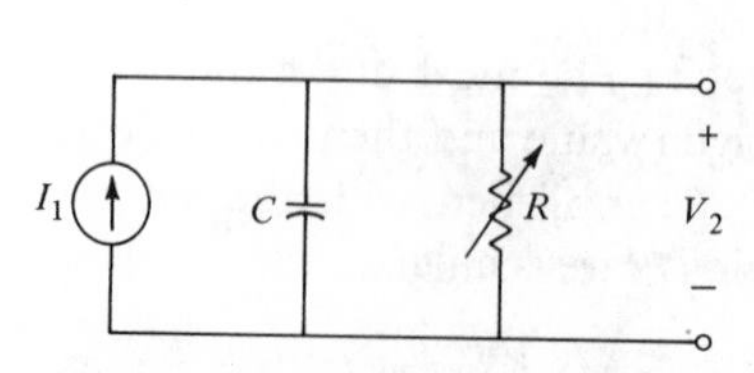

Figure 11.28

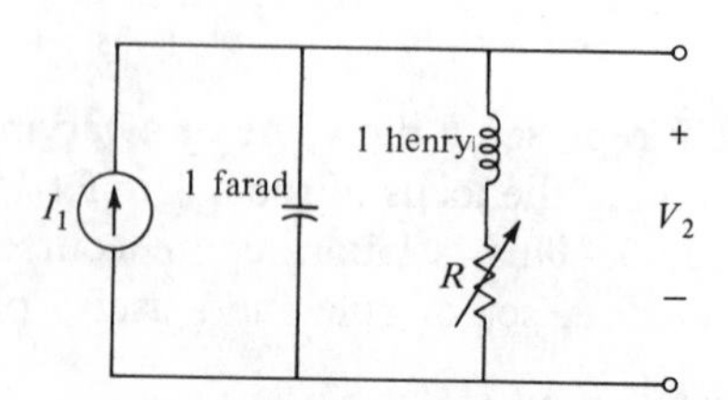

Figure 11.29

11.5.4 The RLC network shown in Figure 11.29 has fixed values of L and C, but R is permitted to vary from 0 to ∞. Plot the locus of roots of the characteristic for $V_2(s)$ in this network.

.6 Magnitude and phase functions from pole-zero diagrams

From previous chapters, we have seen that if we apply an exponential signal e^{st} as input to a linear time-invariant system or network, the particular integral is of the form $H(s)e^{st}$, where $H(s)$ is the system or network transfer function. That is, if $x(t)$ is the input and $y(t)$ the particular-integral component of the zero-state output, we have

$$H(s) = \left.\frac{y(t)}{x(t)}\right|_{x=e^{st}}. \tag{11.74}$$

For example, for the series RLC circuit, if $i(t) = e^{st}$, then

$$v(t) = \left(R + Ls + \frac{1}{Cs}\right)e^{st},$$

and the impedance is

$$Z(s) = R + Ls + \frac{1}{Cs} = \frac{LCs^2 + RCs + 1}{Cs} = L\,\frac{s^2 + (R/L)s + 1/(LC)}{s}. \tag{11.75}$$

This function $Z(s)$ has a pole at $s = 0$ and two zeros at the roots of the equation

$$s^2 + \frac{R}{L}s + \frac{1}{LC} = 0. \tag{11.76}$$

Suppose $L = 0.1$ henry, $C = 10$ microfarads, $R = 100$ ohms. Then the zeros are

$$s_1, s_2 = -500 \pm j500\sqrt{3}. \tag{11.77}$$

The pole-zero configuration for the impedance is shown in Figure 11.30. Note that s_1 and s_2 are conjugates of each other. We may write this impedance function as

$$\begin{aligned} Z(s) &= 0.1\,\frac{(s - s_1)(s - s_2)}{s} \\ &= \frac{0.1(s + 500 - j500\sqrt{3})(s + 500 + j500\sqrt{3})}{s}. \end{aligned} \tag{11.78}$$

If we wish to evaluate $Z(s)$ at some value of s, we simply substitute that value of s in Equation 11.78. In particular, we are usually interested in $s = j\omega$. It is

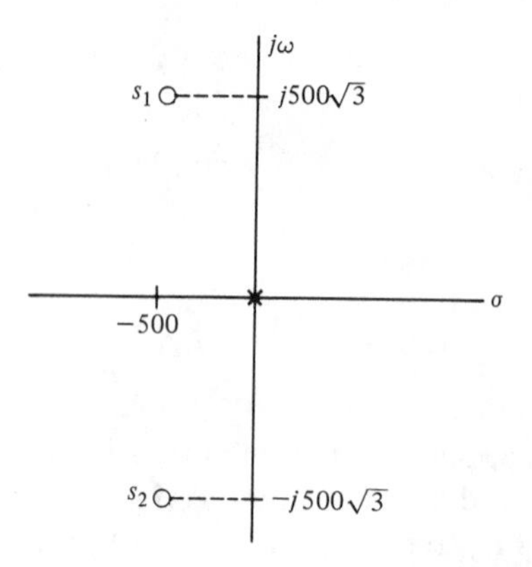

Figure 11.30 Pole-zero configuration for the impedance of an *RLC* series circuit with $R = 100$ ohms, $L = 0.1$ henry, and $C = 10$ microfarads.

expedient in many cases to do (or at least visualize) the calculation graphically. First we note that the complex number s minus the complex number s_1 is equal to another complex number $s - s_1$. The length and angle of the complex number $s - s_1$ are indicated in Figure 11.31. That is, l is the distance between the points s and s_1, and θ is the angle made by the ray directed from s_1 to s and the positive real axis. Applying this idea to the calculation of $Z(j\omega_1)$, we see from Figure 11.32 that

$$\begin{aligned} Z(j\omega_1) &= 0.1\,\frac{(j\omega_1 - s_1)(j\omega_1 - s_2)}{j\omega_1} = 0.1\,\frac{l_1\underline{/\theta_1}\; l_2\underline{/\theta_2}}{l_3\underline{/\theta_3}} \\ &= 0.1\,\frac{l_1 l_2}{l_3}\,\underline{/\theta_1 + \theta_2 - \theta_3}\,. \end{aligned} \tag{11.79}$$

The quantities l_1, l_2, l_3, θ_1, θ_2, and θ_3 can be measured by a ruler and protractor if the pole and zero locations are accurately plotted to scale. To compute $Z(j\omega)$ for several values of ω, the procedure in Figure 11.32 is repeated for various values of ω. This technique may be applied to any rational function

$$H(s) = K\frac{(s - z_1)(s - z_2)\cdots(s - z_m)}{(s - p_1)(s - p_2)\cdots(s - p_n)}. \tag{11.80}$$

To calculate $H(s)$ at some value $s = j\omega_1$, we plot the pole-zero configuration and draw phasors from the zero locations and the pole locations to the point $s = j\omega_1$. Then the magnitude of $H(j\omega_1)$ is K times the product of the lengths of the phasors from the zeros to $j\omega_1$ divided by the product of the lengths of the phasors from the poles to $j\omega_1$. If some zeros and poles are multiple, i.e., two or more are located at the same point, then care should be taken to consider the multiplicity. For ex-

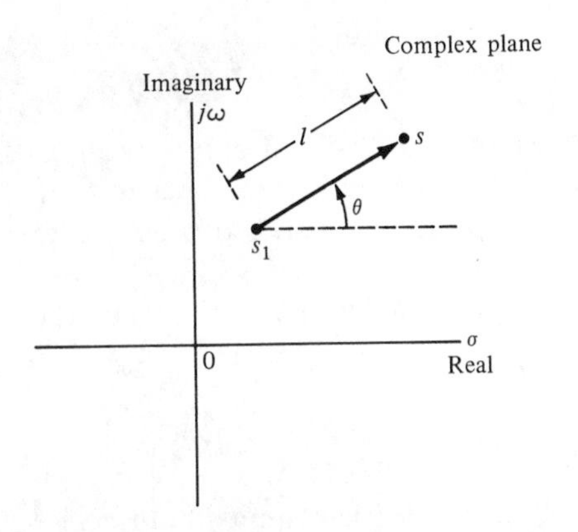

Figure 11.31 Graphical representation of the phasor $s - s_1$.

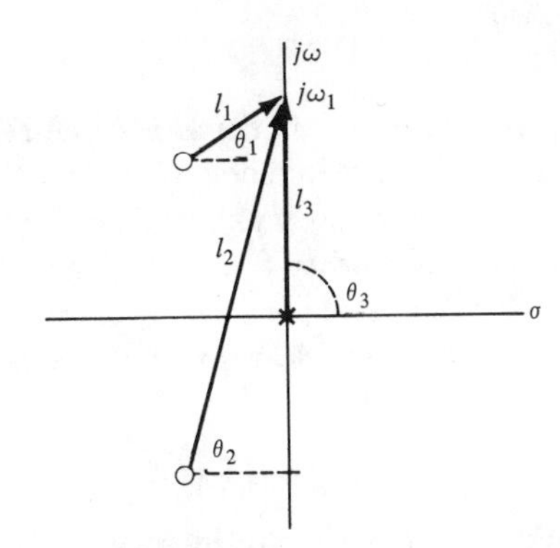

Figure 11.32 Phasors which enter into the calculation of $Z(j\omega)$, shown at $s = j\omega_1$.

ample, if there is a double zero at a point Z_1, then we imagine that there are two phasors drawn from Z_1 to $j\omega_1$. Thus we square the length and double the angle. After some practice, the graphical technique described above enables one to sketch the magnitude and phase response vs. frequency rapidly. Furthermore, certain properties of the response may be determined by simply looking at the pole-zero plot. For instance, in Figure 11.32 we note that the magnitude curve vs. frequency becomes very large for small ω because the length of the phasors from the pole at the origin is very small. Likewise, for extremely large ω, we have very large l_1, l_2, and l_3. Since they are about the same order of magnitude, the magnitude of Z is of the order of $0.1l$. We also guess that somewhere in the neighborhood of the resonant frequency near the largest zero, l_1 will be very short and $|Z(j\omega)|$ will be smallest.

EXERCISES

11.6.1 Plot the pole-zero diagram for the admittance $Y(s)$ of an RCL series network whose impedance is described in Figure 11.32. Use the graphical procedure described in this section to evaluate $|Y(j\omega)|$ and $\angle Y(j\omega)$ for several values of ω and plot these functions of ω.

11.6.2 Do Exercise 11.3.2 using a graphical procedure.

11.6.3 Plot the pole-zero configurations for the voltage transfer functions of the networks in Figures 11.1 and 11.3.

11.6.4 Plot the pole-zero configuration of the impedance of the network in Figure 11.11. There are three cases, depending on the relative values of R, L, and C.

11.7 Summary

1. For *RLC* 2-port networks, the response which is the ratio of output to input varies with the frequency of the input. An important application of such a network is as a *filter* which makes the response small for prescribed bands of frequencies and larger for other frequency bands.
2. Resonance phenomena are important in explaining filtering action. For the series *RLC* network, resonance occurs at

$$\omega_r = \frac{1}{\sqrt{LC}},$$

where $v(t)$ and $i(t)$ are in phase, $Z_{in} = R$, and $I = V/R$ has a maximum value. If ω_1 and ω_2 are frequencies at which current is the maximum divided by $\sqrt{2}$, then

$$\omega_2 - \omega_1 = \frac{R}{L} \text{ is bandwidth}.$$

3. For a coil with inductance and resistance, Q is a figure of merit defined as

$$Q = \frac{\text{maximum energy stored} \times 2\pi}{\text{energy dissipated per cycle}}$$

with both quantities measured at resonant frequency. For the series *RLC* circuit

$$Q = \frac{\omega_r L}{R}.$$

4. Networks other than a series *RLC* network exhibit resonance effects. Such a network is said to be in resonance when, in the sinusoidal steady-state, $v(t)$ is in phase with $i(t)$.
5. *Locus diagrams.* Select one phasor as a reference and plot the path of the tip of another phasor resulting from the variation of ω or some network parameter. The resulting plot is known as a *locus diagram.*
6. The *root locus* is the plot of the path of roots in the s-plane as some parameter in the network changes values from one limit to another (usually 0 to ∞). For a second-order system, the location of the roots may be expressed in terms of general parameters ζ and ω_r as

$$s_1, s_1^* = -\zeta\omega_r \pm j\omega_r\sqrt{1 - \zeta^2}$$

for a root having the polar coordinates $\omega_r \underline{/\cos^{-1}\zeta}$ with the angle measured from the negative real axis, for $|\zeta| \leq 1$.
7. If

$$Z = \frac{(s - z_1)(s - z_2)\cdots(s - z_m)}{(s - p_1)(s - p_2)\cdots(s - p_n)},$$

then when $s = s_1$ (any complex frequency), $Z(s_1)$ may be written in polar form by letting

$$s_1 - z_j = l_{z_j}\underline{/\theta_{z_j}},$$
$$s_1 - p_j = l_{p_j}\underline{/\theta_{p_j}}.$$

Then the magnitude of $Z(s_1)$ is

$$M_t = \frac{l_{z_1} l_{z_2} \cdots l_{z_m}}{l_{p_1} l_{p_2} \cdots l_{p_n}}$$

and the phase is

$$\theta_t = \theta_{z_1} + \theta_{z_2} + \cdots - \theta_{p_1} - \theta_{p_2} - \cdots.$$

Plots of M_t and θ_t may be made graphically by measuring the l's and θ's for different values of s_1. This technique is especially useful and important in the sinusoidal steady-state when $s_1 = j\omega_1$ and all measurements are made to the imaginary axis of the s-plane.

PROBLEMS

11.1 The series RLC network of Figure 11.33 is in resonance. The voltage V_1 has an rms value of 1 volt, and V_2 across R and C is 5 volts (rms). If $\omega = 1$ radian/sec, what is the value of L? [*Answer:* $\sqrt{24}$ henrys.]

11.2 The Q of a series RLC network at resonance is 10. The magnitude of the current at resonance is 1 ampere when the applied voltage is 10 volts. If $L = 0.1$ henry, find the value of C in microfarads. Assume that all voltage and currents are sinusoidal.

11.3 A 5-volt (rms) sinusoidal voltage source is connected to a series RLC network. When $C = \frac{1}{5}$ farad, $|I| = 1$ ampere (rms) and the average power is $P = |I|^2R = 3$ watts. With the same voltage source connected to the network, but with the capacitor changed such that $C = \frac{1}{45}$ farad, the magnitude of the current and the power are the same. Find the value of L in henrys.

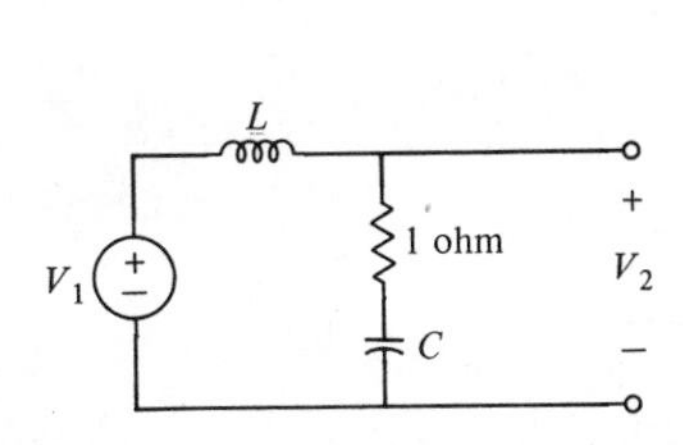

Figure 11.33

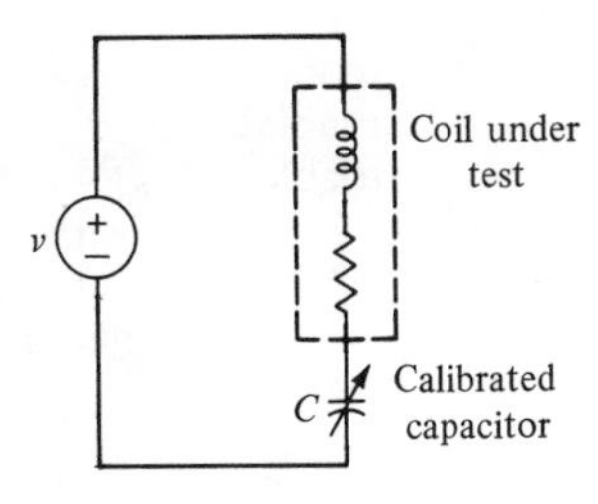

Figure 11.34

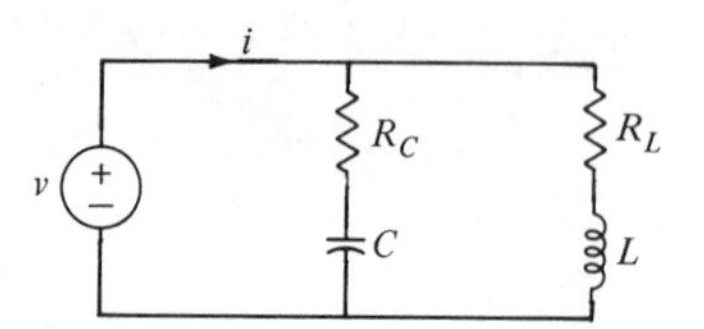

Figure 11.35

Figure 11.36

11.4 A coil under test is connected in series with a calibrated capacitor as shown in Figure 11.34. A sine wave generator of 10 volts rms at a frequency $\omega = 1000$ radians/sec is connected to the network. The capacitor is adjusted, and it is found that the current is a maximum when $C = 10.0$ microfarads. Further, when $C = 12.5$ microfarads, the current is 0.707 of the maximum value. Find the Q of the coil at $\omega = 1000$ radians/sec.

★11.5 As frequency is varied, the voltages $|V_L|$ and $|V_C|$ of a series RLC network vary as shown in Figure 11.8. Show that

$$|V_{Cm}|_{\max} = |V_{Lm}|_{\max}.$$

11.6 The capacitor C is varied from zero to infinity in the series-parallel network of Figure 11.35. If $R_C = 5$ ohms, $R_L = 10$ ohms, and $X_L = 20$ ohms, how many conditions of resonance will be observed as C is adjusted?

The following problems refer to the network of Figure 11.35. In each case, draw a locus diagram showing the variation of I as the parameter is varied; determine the number of values of the parameter that will cause resonance; and plot $|I|$ and the phase angle of I as functions of the element value. For each problem, $\omega = 1$ radian/sec.

Problem	R_C *(ohms)*	C *(farads)*	R_L *(ohms)*	L *(henrys)*
11.7	variable	1	1	1
11.8	variable	2	1	1
11.9	variable	1	$\frac{1}{2}$	$\frac{1}{2}$
11.10	0.06	variable	$\frac{1}{4}$	$\frac{1}{3}$
11.11	1	variable	2	2
11.12	0	variable	1	1
11.13	$\frac{1}{3}$	4	variable	$\frac{25}{4}$
11.14	$\frac{1}{2}$	2	variable	1
11.15	$\frac{1}{3}$	4	variable	0.8
11.16	$\frac{1}{3}$	4	1	variable
11.17	$\frac{1}{2}$	2	$\frac{1}{2}$	variable
11.18	1	1	$\frac{1}{4}$	variable

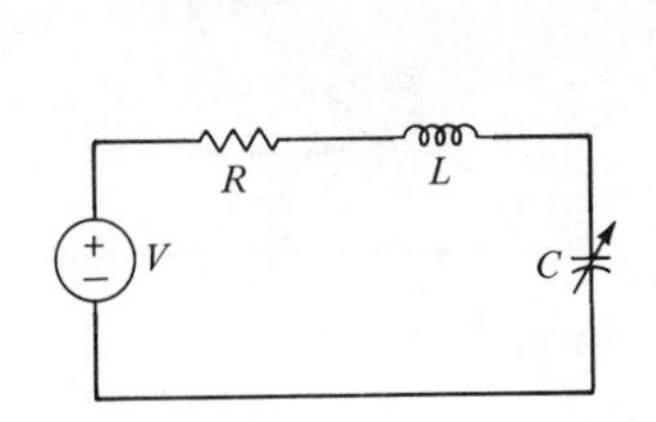

Figure 11.37

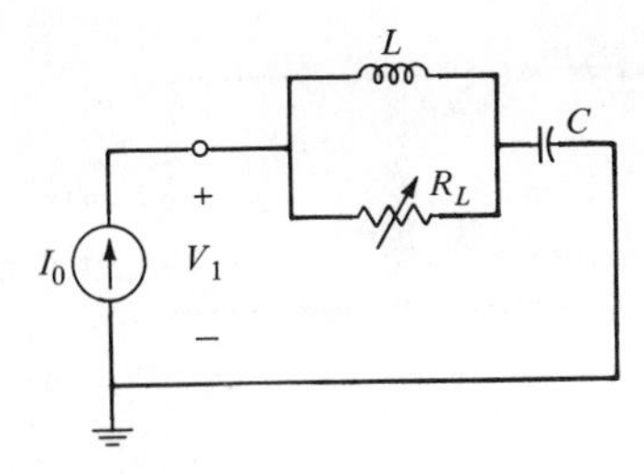

Figure 11.38

11.19 The network of Figure 11.36 is driven by a current source which is sinusoidal, and the frequency is such that $|X_C| = |X_L|$. Determine the locus of the voltage phasor V_R as R is varied from zero to infinity.

11.20 In the network of Figure 11.37, the generator voltage is sinusoidal and has an rms value of 10 volts. Also $R = 2$ ohms, L and ω are constant, but C is varied. (a) Draw a locus diagram of I with V as the reference. Label important points on the locus. (b) If I leads V by 60°, what must be the value of $|I|$?

11.21 For the network shown in Figure 11.38, draw a locus diagram for the voltage V_1 as R_L varies from zero to infinite value. Use I_0 as the reference.

11.22 In the network given in Figure 11.39, the value of C_1 is adjusted until $|I_1|$ is minimum. Then C_2 is adjusted until $|I_2|$ is minimum. Calculate the value of $|I_2|$ (rms) for these settings of C_1 and C_2.

★11.23 The derivation of Section 11.2 found the bandwidth $\Delta\omega$ defined by the current being at least $1/\sqrt{2}$ of the maximum value. Repeat the derivation for ω_1 and ω_2 at which the current is the fraction $1/k$ of the maximum value (at resonance) and show that the bandwidth using this definition is

$$\Delta\omega = \frac{R\sqrt{k^2 - 1}}{L}.$$

★11.24 The capacitor in the network of Figure 11.35 is varied from zero to infinity and the following observations are made: resonance occurs at only

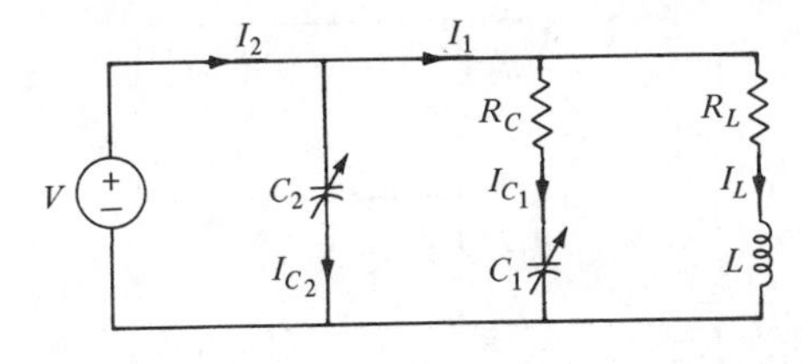

Figure 11.39

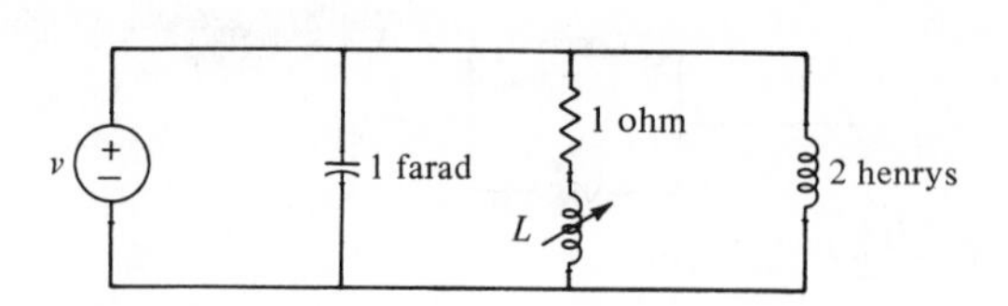

Figure 11.40

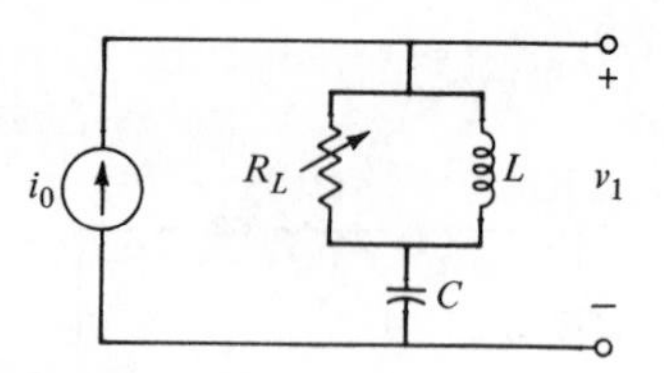

Figure 11.41

one value of C, and the value of capacitance at resonance is C_r. Show that L is given by the expression

$$L = \frac{1 \pm \sqrt{1 - C_r^2 R_L^2 \omega^2}}{\omega^2 C_r}.$$

11.25 For the network of Figure 11.40, how many values of L will cause resonance and what are these values of L (a) for $\omega = 1$, (b) for $\omega = \frac{1}{2}$?

11.26 For the network shown in Figure 11.41, draw a locus diagram for the voltage V_1 as R_L varies from zero to infinity, using I_0 as the reference.

11.27 Draw a locus diagram for V_1 as R_C varies from zero to infinity with I_0 as a reference for the network in Figure 11.42. Repeat for L varying from zero to infinity.

11.28 The voltage source in the network of Figure 11.43 is sinusoidal, and the network is operating in the steady-state. Show that if the elements are adjusted such that

$$R_L = R_C = \sqrt{\frac{L}{C}} \equiv R,$$

then the impedance of the network at the driving-point terminals is

$$Z = \frac{V_1}{I_1} = R$$

for all values of ω. In other words, show that the impedance of this network is independent of ω.

11.29 For the network of Figure 11.43 under the conditions of Problem 11.28,

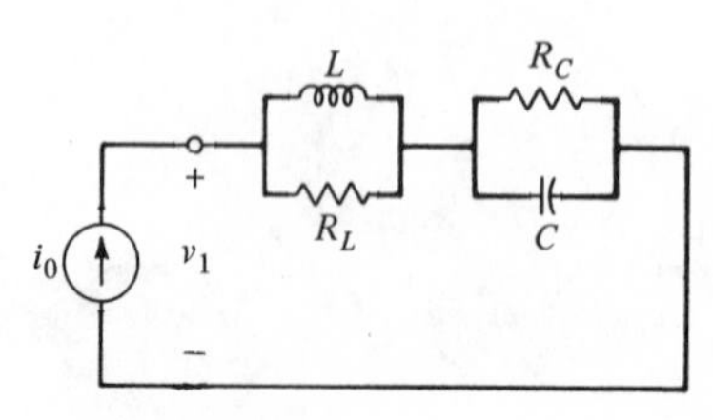

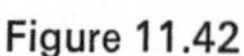

Figure 11.42

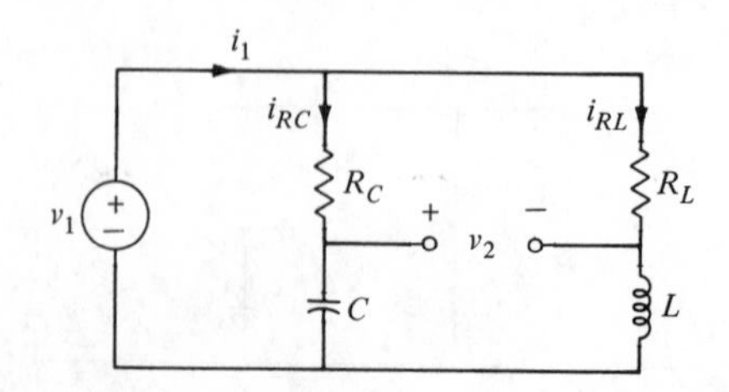

Figure 11.43

show that the current through the RC branch and the voltage across the inductor are in phase for all frequencies.

11.30 For the conditions of Problem 11.28, it is clear that the voltage V_1 must be in phase with the current I_1 for all values of ω. It is interesting to study how the various voltages and currents in the network adjust so that this is always the case. For the given network complete the following: (a) draw a complete phasor diagram for the condition $|V_L| = |V_{R_L}|$; (b) let the frequency for the condition of part (a) be ω_1. Draw a phasor diagram to approximately the same scale as in (a) for a frequency ω_2 larger than ω_1. (c) Repeat part (b) for a frequency ω_3 smaller than ω_1.

11.31 For the network of Figure 11.43, let frequency ω_y be the frequency at which $v_1(t)$ and $v_2(t)$ are 90° out of phase. Determine an expression for ω_y and simplify. Does v_2 lead or lag v_1?

11.32 For the network of Figure 11.43, let ω_z be the frequency at which $|V_1| = |V_2|$. Determine the value of ω_z and simplify your expression. Is there only one value?

11.33 In the network of Figure 11.44, $v_1(t)$ is sinusoidal and the network is in the steady-state. The capacitor C_1 is varied from 0 to ∞. (a) Show that the phase of $v_2(t)$ with respect to $v_1(t)$ may be varied from 0 to 180°. It is suggested that this be done by a phasor diagram. (b) For this part of the problem, the limits of the capacitor are such that it varies from the value for which $1/(\omega C_1) = 0.1R_1$ to that for which $1/(\omega C_1) = 10R_1$. Determine the range of the phase angle between $v_1(t)$ and $v_2(t)$ which is possible with these limits on C_1.

11.34 In the network of Figure 11.44, the capacitor C_1 is replaced by an inductor L_1. Repeat parts (a) and (b) of Problem 11.33 with the quantity ωL_1 replacing $1/(\omega C_1)$ in part (b).

11.35 It is desired to design a simple low-pass filter as in Figure 11.1, such that signal components with frequencies of 60 Hz or higher are attenuated by at least 100 to 1 compared to dc (or $\omega = 0$) signals. Select practical and satisfactory values for R and C. Is your answer unique?

11.36 It is usually convenient to normalize the resonance curve so that the maximum value is unity. For the series RLC circuit, we define a normalized

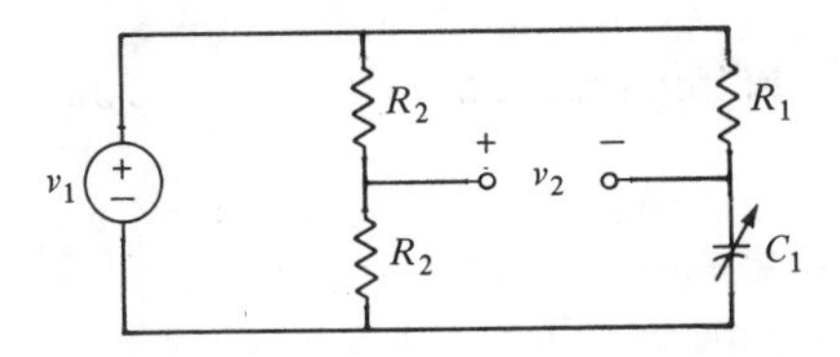

Figure 11.44

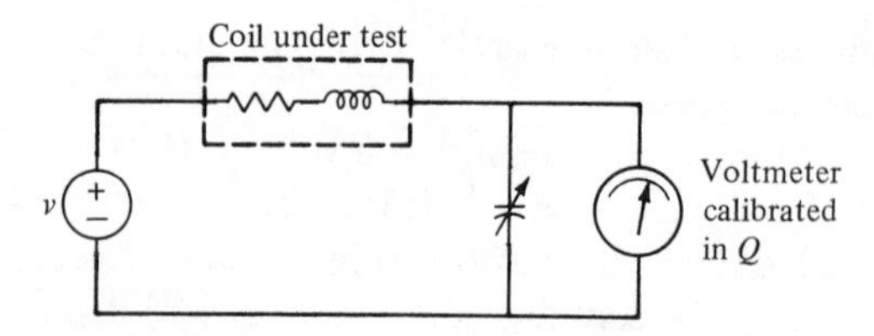

Figure 11.45

response as $I/I_{\max}$, where I is the phasor current and $I_{\max}$ is the largest value for the magnitude of I. Show that

$$\frac{I}{I_{\max}} = \frac{1}{1 + j\left(\frac{\omega L}{R} - \frac{1}{\omega CR}\right)} = \frac{1}{1 + jQ\left(\frac{\omega}{\omega_r} - \frac{\omega_r}{\omega}\right)}.$$

11.37 The network in Figure 11.45 is commonly used to measure Q of coils with relatively high Q's. The procedure is to use a voltage signal generator with an adjustable frequency and provision for making the peak value remain the same. The signal generator is adjusted at the desired frequency, and the capacitor is adjusted until the voltmeter reads maximum. For a high Q circuit, this value of C is approximately the value to set the circuit in resonance. Assuming that the circuit is thus adjusted to resonance, show that $V_C = QV$, where V_C is the rms voltage across C, and V is the rms voltage across the source. Thus, if $V = 1$, V_C reads directly in Q.

★11.38 For the circuit of Figure 11.45, derive an exact formula for the value of C which causes the voltmeter to read the largest. Compare this C with C_r, the value needed for resonance.

11.39 Derive an expression for the voltage transfer function of the network in Figure 11.46. Assume that $R_L \ll \omega_r L$ and $R \approx R_L$. Plot the pole-zero configuration.

11.40 For the parallel network of Figure 11.9 and assuming $R \ll \omega_r L$ at the resonant frequency ω_r, apply the energy definition for Q to derive an approximate expression for Q. [*Hint:* Assume that the voltage across L is approximately equal to the voltage across the RL combination.]

11.41 For the network in Figure 11.5, let V_x be the phasor corresponding to the voltage across the LC combination. Assume that R, L, and C are given constants and v is a sinusoidal source with the peak value given. Using the frequency ω as a parameter, show that the locus of the V_x phasor is a circle.

11.42 For the parallel network in Figure 11.21, the following parameters are given: $V_{\text{rms}} = 100$ volts, $R_C = R_L = 5$ ohms, $1/(\omega C) = 12$ ohms, $\omega = 1000$ radians/sec, and L is a free parameter. Draw an accurate locus diagram for I with L as a parameter. What is the angle of lead of I with

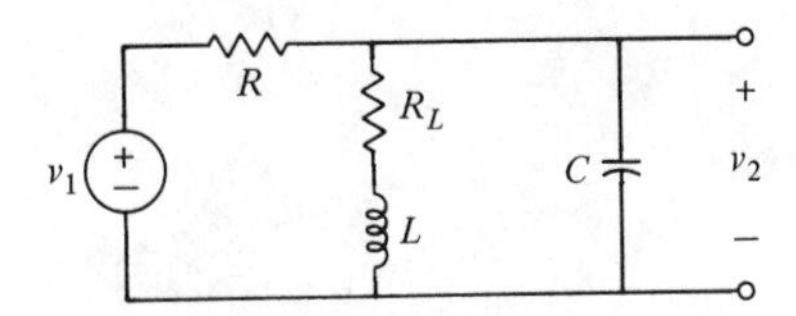

Figure 11.46

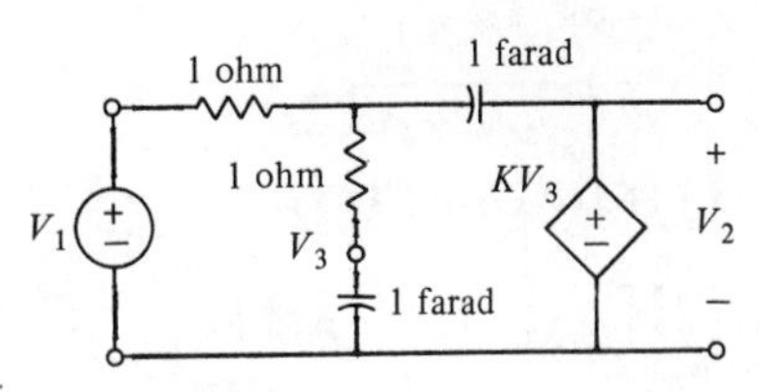

Figure 11.47

respect to V when $|I|$ is minimum? Using a graphical calculation, find values for L which cause resonance.

11.43 Suppose that in the network of Figure 11.21 $R_C = 0$ and C is adjustable. (a) Calculate the maximum $|Z|$. (b) Calculate the nonzero value of C such that $|I| = |I_L|$. It is given that $R_L = 5$ ohms, $\omega L = 12$ ohms, and $\omega = 1000$ radians/sec.

11.44 The locus diagrams studied in the chapter are diagrams of I with a parameter adjustable and V fixed. Since V is usually assumed to be the reference (on the positive real axis), I/V, which is the admittance, will have the same form as I. The condition for resonance, for example, corresponds to the case when the admittance is purely conductive. From the locus diagram in Figure 11.21, find the admittance Y at several points of the locus and calculate the corresponding complex Z. Plot the complex values of Z. Do the points lie on a circle? Note that minimum $|Y|$ corresponds to maximum $|Z|$. Also, note that the resonant points will be zero reactance.

11.45 The network of Figure 11.47 contains one voltage-controlled voltage source. (a) Show that the characteristic equation associated with V_2 expressed in terms of V_1 is

$$s^2 + (3 - K)s + 1 = 0.$$

(b) Sketch the root locus for all positive values of K. For what value of K does the network become unstable?

11.46 The network shown in Figure 11.48 contains a voltage-controlled voltage source of value KV_2. (a) Show that the characteristic equation associated with V_2 in terms of I_1 is

$$s^2 + (3 - K)s + 2 = 0.$$

(b) Sketch the root locus for all positive values of K. For what value of K does the network become unstable?

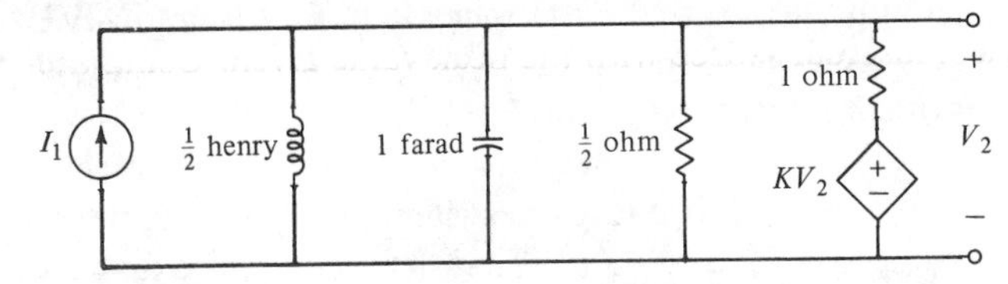

Figure 11.48

Chapter **12**

Linear Independence of Topological Equations

12.1 Linearly independent variables and equations

In our development of network analysis, there are two sets of fundamental postulates. One set is the assumption of voltage-current relationships in the various branches of the network. These branch characteristics or branch constraints may be considered as the set of generalized Ohm's-law equations governing the terminal behavior of the elements. These equations, as discussed in Chapter 4, may be nonlinear differential equations. The other set of fundamental postulates is the assumption of Kirchhoff's current law and Kirchhoff's voltage law. These laws may be reduced to simple linear algebraic equations.

The approach we have used thus far is to consider only very simple networks with few variables. We have also developed analysis concepts which were applicable only if the branch constraints are linear and time-invariant. To facilitate the understanding of these concepts, only simple networks were treated. In this chapter, we finally face the question of what to do when we have a network with a large number of variables or when we have a complex topological configuration.

In analyzing a network with b branches, we have b branch voltages and b branch currents. The branch constraints provide us with b equations. On the other hand, there are more than b loops and nodes, so that the application of KCL and KVL to all nodes and loops will give us more than b equations. Since there are only $2b$ variables, some of these equations must be redundant. In order to develop a general method for writing a set of as many nonredundant equations as there are variables or unknowns, we must have a procedure for systematically choosing the loops and nodes to which we apply KVL and KCL. First, let us illustrate that the application of KVL and KCL may yield redundant equations.

MPLE 12.1.1

Consider the network in Figure 12.1(a). Applying KVL to loops *abcfa*, *bcdeb*, and *abcdefa*, we have

$$v_{ab} + v_{be} + v_{ef} + v_{fa} = 0, \tag{12.1}$$

$$v_{bc} + v_{cd} + v_{de} + v_{eb} = 0, \tag{12.2}$$

and

$$v_{ab} + v_{bc} + v_{cd} + v_{de} + v_{ef} + v_{fa} = 0. \tag{12.3}$$

We note that since *f*, *e*, and *d* are the same node, we have $v_{de} = v_{ef} = 0$. We also note that if we add Equations 12.1 and 12.2, we obtain Equation 12.3. Hence the application of KVL to loop *abcdefa* is redundant if KVL has been applied already to loops *abefa* and *bcdeb*. Thus, in the network of Figure 12.1(a) where there are six branch voltages, although it is possible to write six equations by applying KVL to six different loops, some of the equations are redundant. Similarly, some variables may be redundant. For example, in Figure 12.1(a) again, consider the variables v_{ab}, v_{bc}, and v_{ac}. If we are given the three values for v_{ab}, v_{bc}, and v_{ac}, one specification is really redundant since it is always true that

$$v_{ac} = v_{ab} + v_{bc}. \tag{12.4}$$

Thus, knowing two voltages automatically specifies the third. For the same network, the set of voltages v_{ae}, v_{be}, and v_{ab} also constitute a redundant set because, from KVL, we have

$$v_{ab} + v_{be} - v_{ae} = 0, \tag{12.5}$$

and knowing two of the voltages automatically fixes the value of the third.

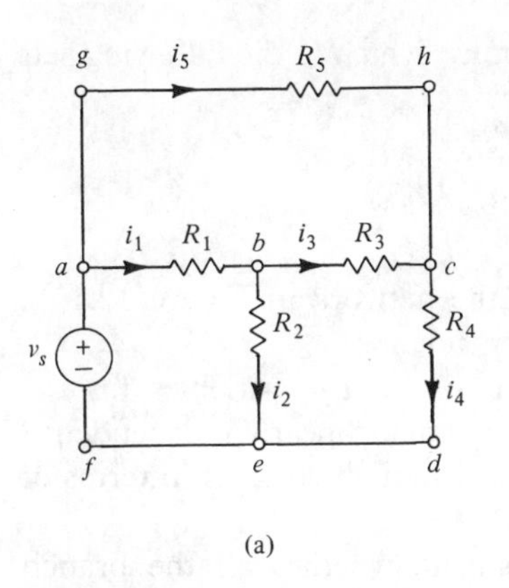

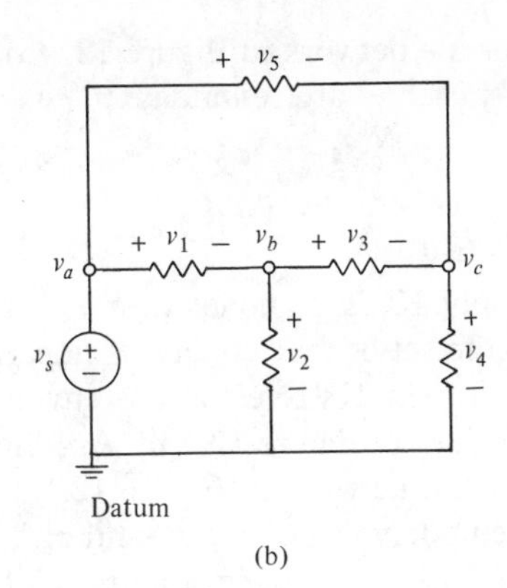

ure 12.1 Networks for Example 12.1.1.

Redundancy in a set of equations is now defined. Consider a set of p algebraic equations in n variables. Hence we have

$$\begin{aligned} f_1(x_1, x_2, \ldots, x_n) &= 0, \\ f_2(x_1, x_2, \ldots, x_n) &= 0, \\ &\vdots \\ f_p(x_1, x_2, \ldots, x_n) &= 0, \end{aligned} \tag{12.6}$$

where the f_k functions may be linear or nonlinear in the arguments $x_1, x_2, \ldots, x_n$. The *set* of p equations in Equation 12.6 is said to be *linearly dependent* if there exist constants $c_1, c_2, \ldots, c_p$, at least one of which is not zero, such that

$$c_1 f_1(x_1, x_2, \ldots, x_n) + c_2 f_2(x_1, x_2, \ldots, x_n) + \cdots + c_p f_p(x_1, x_2, \ldots, x_n) \equiv 0 \tag{12.7}$$

for all arbitrary values of $x_1, x_2, \ldots, x_n$. If the identity in Equation 12.7 which is for all arbitrary values of $x_1, x_2, \ldots, x_n$ cannot be satisfied except for $c_1 = c_2 = \cdots = c_p = 0$, then the set of equations is said to be *linearly independent*. For instance, the set of Equations 12.1, 12.2, and 12.3 constitutes a linearly dependent set because the left-hand side of Equation 12.1 plus the left-hand side of Equation 12.2 minus the left-hand side of Equation 12.3 is identically zero for all arbitrary values of the voltage variables.

The remainder of the chapter is devoted to the determination of a set of linearly independent network equations whose solutions completely characterize the network.

EXERCISES

12.1.1 For the network in Figure 12.1(a), determine which of the following sets of variables are redundant:

(a) v_{ab}, v_{ac}, v_{be};
(b) $v_{be}, v_{cd}, v_{ab}, v_{gh}$;
(c) i_1, i_2, i_4;
(d) i_1, i_2, i_3;
(e) i_2, i_3, i_4.

12.1.2 Apply KCL to nodes a, b, c, and e of the network in Figure 12.1(a). Is the set of four equations linearly independent?

12.1.3 Write six KVL equations involving the six voltage variables for the network in Figure 12.1(b). Are these six equations linearly independent?

12.1.4 For the network in Figure 12.1, express the branch voltages in terms of the node voltages v_a, v_b, and v_c.

12.1.5 For the network in Figure 12.1, investigate whether all the branch currents can be expressed in terms of i_1, i_3, and i_5.

2 Network topology

In Chapter 5, we introduced the notions of *branches, nodes, paths,* and *loops.* The structural relationship among the elements or branches, or the manner in which they are interconnected, has nothing to do with the type of elements in the branches; it does not matter whether the branch is a resistor, capacitor, inductor, voltage source, or current source. The skeleton that remains when all elements are replaced by lines is known as a *linear graph* of the network. The graph portrays the topological relationship of the elements and nodes in the network. Thus two graphs may be topologically identical even though they contain different kinds of elements in corresponding branches.

AMPLE 12.2.1

Consider the network in Figure 12.2. Replacing the element symbols by lines or arcs, we get the graph in Figure 12.3. The graph in Figure 12.4 also represents the network in Figure 12.2. A little checking of the connecting lines in the graph of Figure 12.5 shows that it is no different from those in Figures 12.3 and 12.4. It also represents the network in Figure 12.2. The example shows that identical graphs may have different geometric shapes. Also different networks may be represented by the same graph. The network in Figure 12.6, for example, also has a graph shown in Figure 12.3. Again, the graphs of Figures 12.4 and 12.5 are equivalent representations.

The definitions of *branches* and *nodes* for networks have counterparts in linear graphs. A network *branch* corresponds to an *edge* of a graph. Thus in Figure 12.5, *a, b, c, d, e,* and *f* are edges. On the other hand, *ab* is not an edge. A network node corresponds to a *vertex* of a graph. Thus in Figure 12.5, 1, 2, 3, and 4 are vertices. While all these vertices are junctions of three branches, in general a junction of two branches is also called a vertex.

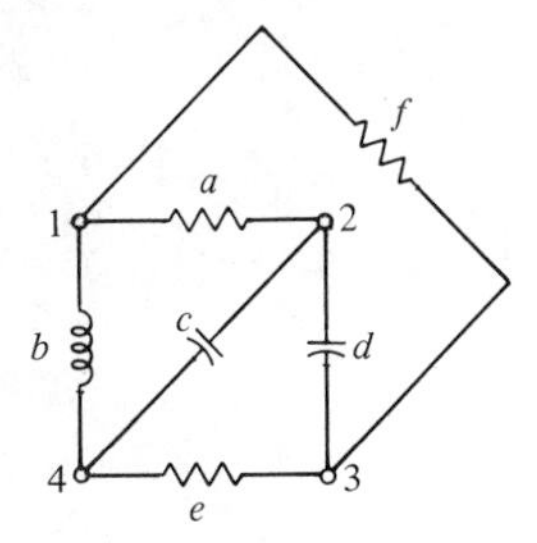

Figure 12.2 Network for Example 12.2.1.

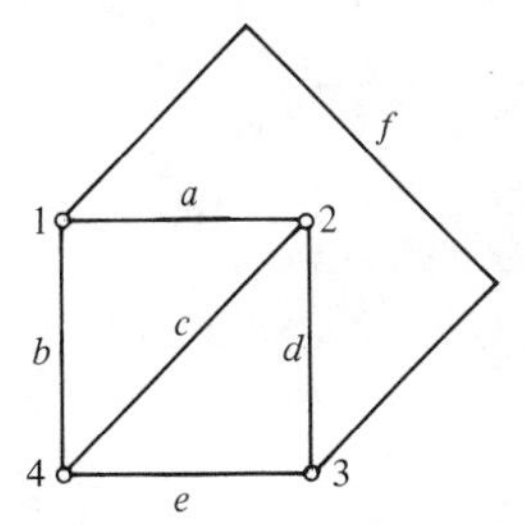

Figure 12.3 Graph corresponding to the network of Figure 12.2.

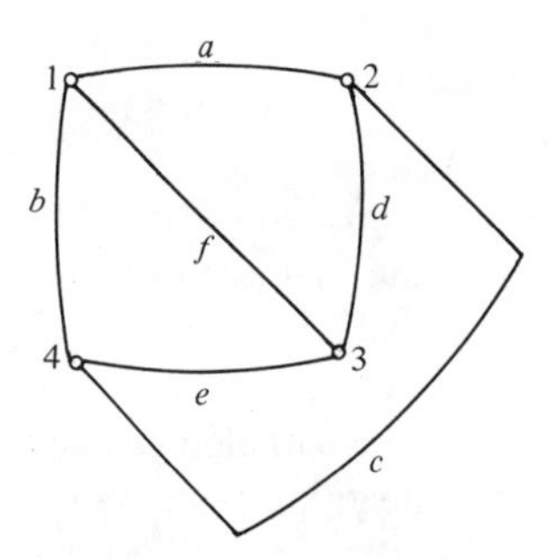

Figure 12.4 A different version of the graph for the network of Figure 12.2.

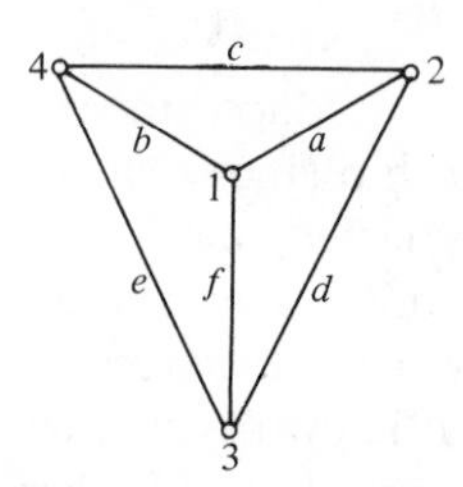

Figure 12.5 A third version of the grapn for the network of Figure 12.2.

In a graph, a *path* is a specified train of edges which becomes a *closed path* or a *loop* when the starting vertex and ending vertex are the same. Thus in Figure 12.5 *cd* is a path, *ade* is also a path, *bf* is a path, and *e* is a path. These paths are shown in Figure 12.7. Note that the intermediate vertices in the paths connect exactly two edges. On the other hand, the end or terminal vertices in the paths connect only one edge. A path is defined so that it may consist of only one edge.

Figure 12.8 shows more examples of paths in a graph. These are all the paths between vertices 4 and 1 in the graph of Figure 12.5.

Another entity which we encountered in Chapter 5 which applies here is the *closed loop* or simply the *loop*. Examples for some loops in the graph of Figure 12.5 are shown in Figure 12.9.

The entities mentioned above — edges, vertices, path, and loops — are portions of a graph. They may be classified under the heading *subgraphs*. Another kind of subgraph which is very fundamental in the study of graphs is the *tree*, a name first used by Kirchhoff in 1846. A tree is a subgraph containing all the vertices of the graph, but not containing any loop. Furthermore, the tree must be a *connected* subgraph in the sense that a path should exist between any two vertices.

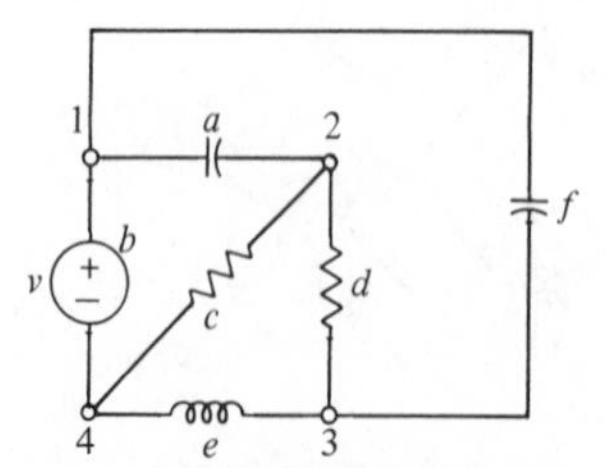

Figure 12.6 Another network for Example 12.2.1.

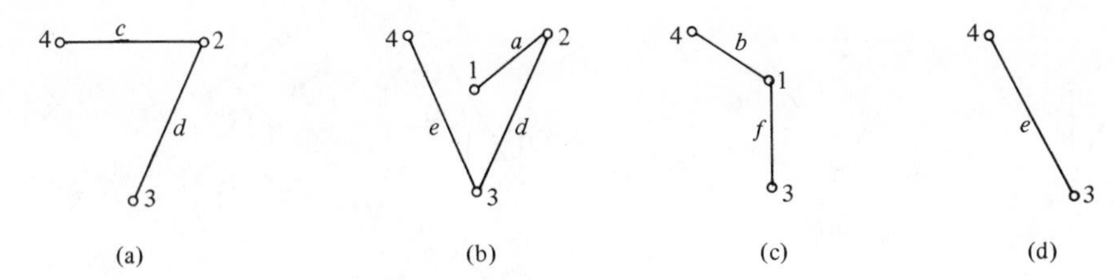

12.7 Paths for the graph given in Figure 12.5.

MPLE 12.2.2

Figure 12.10 shows some of the trees for the graph in Figure 12.5. We observe that in a tree there must be only one path between any two vertices. If there were two or more, then a loop would be present and this would violate our definition for a tree. In Figure 12.10(a), edges *a*, *b*, and *f* together constitute a tree. Similarly, from Figures 12.10(b) and 12.10(c), branches *b*, *c*, and *e* constitute another tree, and *c*, *d*, and *f* constitute still a third tree. Figure 12.11 shows some subgraphs which, although containing all the nodes, are *not* trees. Figure 12.11(a) and 12.11(b) violate the condition that there must be a path between any two nodes. For example, in Figure 12.11(a) there is no path between 4 and 2. Similarly, in Figure 12.11(b), there is no path between 2 and 1. In Figure 12.11(c), the condition violated is that there must be no loops in the subgraph, because *adf* constitutes a loop.

The edges of a graph belonging to a particular tree are called *tree edges*. For instance, edges *a*, *b*, and *f* are tree edges for the tree in Figure 12.10(a). After a tree is extracted from a graph, the edges that remain are called *chords* or *links*.

MPLE 12.2.3

The chords for the graph in Figure 12.5 corresponding to the tree in Figure 12.10(a) are *c*, *e*, and *d*. The chords for the tree in Figure 12.10(b) are *a*, *f*, and *d*. For the tree in Figure 12.10(c), the chords are *b*, *a*, and *e*.

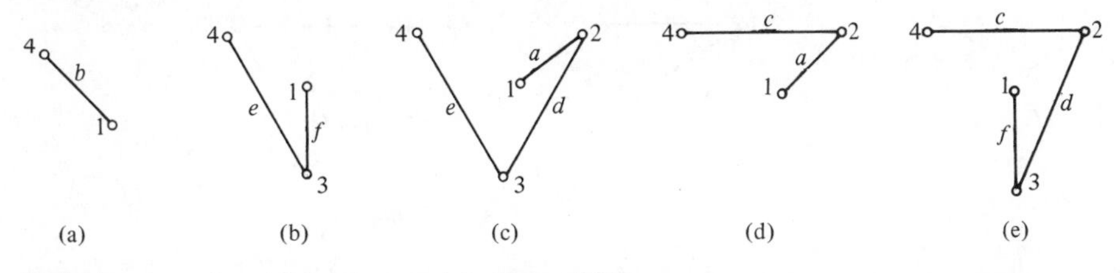

e 12.8 Paths between 4 and 1 for the graph in Figure 12.5.

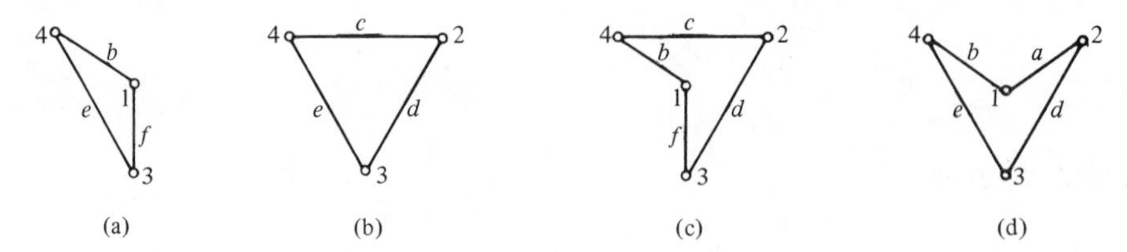

Figure 12.9 Some loops in the graph in Figure 12.5.

Observe next that if one chord is restored in a tree, the chord closes a loop in the graph. For instance, if chord c is added to the tree in Figure 12.10(a), loop abc is produced. Similarly, the chord d added to the tree in Figure 12.10(a) produces loop afd. For this reason, chords are sometimes called *links* because they provide the "missing link" to produce loops.

Let b be the number of edges in a connected graph, and let n be the number of vertices. Then the number of tree edges in a given tree is clearly $n - 1$. If there are two vertices in a graph, only one edge is needed to connect them. If there are three vertices, two tree edges are involved. For example, in the graph of Figure 12.5 there are six edges, four vertices, and the trees shown in Figure 12.10 have three edges each. In general, for a graph with n nodes, $n - 1$ tree edges are needed to connect them. To get the number of chords corresponding to a tree, we can simply subtract the number of tree edges from the total number of edges. Hence, the number of chords is

$$l = b - (n - 1) = b - n + 1 \tag{12.8}$$

for connected graphs. If a graph has s separate parts, then the 1 of Equation 12.8 is replaced by s. The loops generated by the chords of a tree are called chord-set loops.

EXERCISES

12.2.1 Find trees other than those given in Figure 12.10 for the graph in Figure 12.5.

12.2.2 Draw chord-set loops for the trees in Exercise 12.2.1.

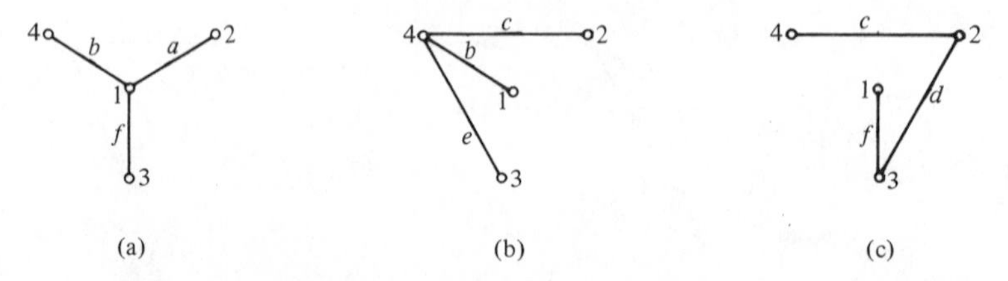

Figure 12.10 Some trees for the graph in Figure 12.5.

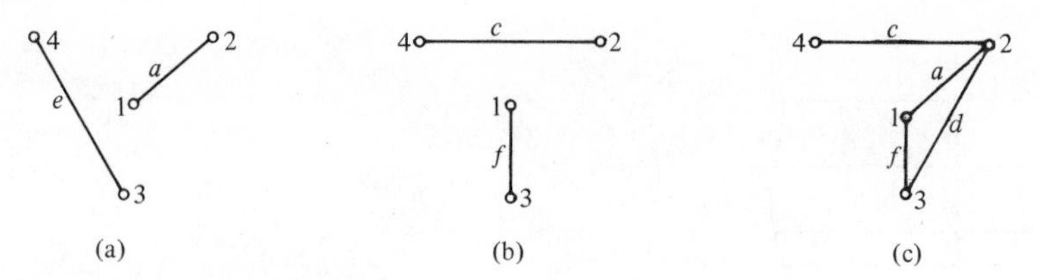

re 12.11 Some subgraphs which are not trees (although all the nodes are included) for the graph in Figure 12.5.

12.2.3 Find all the paths connecting nodes 2 and 4 in the graph of Figure 12.3.

12.2.4 For the graph in Figure 12.12, which of the following are trees of the graph: 1234, 1576, 1673, 5678, 5724, 1683, 1237, 1238, 1235, 2347, 2456, and 1268?

12.2.5 Draw chord-set loops for the trees in Exercise 12.2.4.

3 Number of independent topological constraints

We shall now establish the maxima for the number of linear independent topological equations based on KCL and KVL. First let us recapitulate Kirchhoff's laws:

(a) The voltage law (KVL) states that if any loop of a network is traced in one direction (say clockwise), the sum of voltages with like reference direction minus the sum of voltages with the opposite reference location is zero.

(b) The current law (KCL) states that the sum of the currents entering any vertex in a network minus the sum of currents leaving the same vertex is zero.

KCL may be generalized by adding appropriate sets of KCL equations. Thus, if we stretch a network so as to expose n wires as in Figure 12.13, the net current flowing through a hypothetical plane across the wires is zero. This may be derived by simply adding the KCL equations for all the nodes in N_1. The set of edges cut by the plane is called a *cut set*.

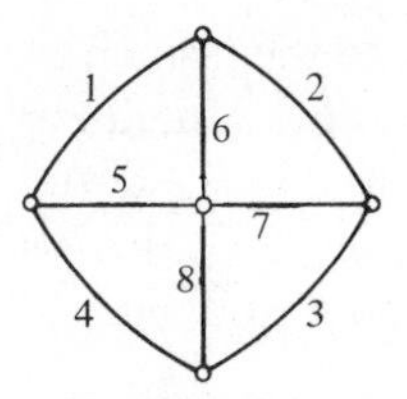

Figure 12.12

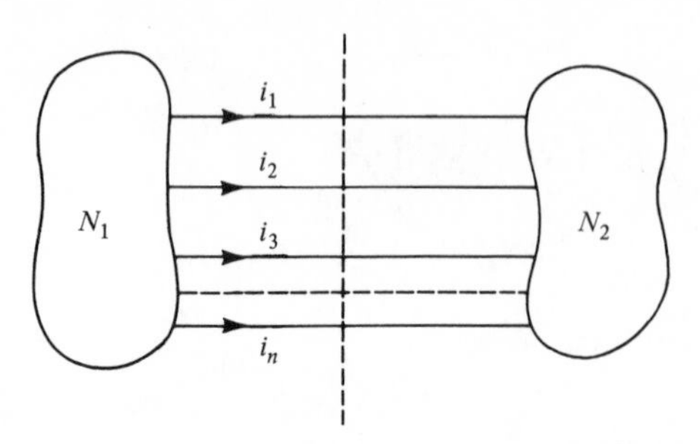

Figure 12.13 The net current flowing from N_1 to N_2 is zero.

From the previous section, given a connected network with b branches and n nodes, we see that any tree of the network has exactly $n - 1$ tree branches and $b - n + 1$ chords. We also recall that for a given tree, every time we replace a chord, one and only one loop is formed. Hence, we can find $b - n + 1$ loops, each one of which contains an edge not contained in any other loop. Hence, if we apply KVL to these chosen loops, each equation would contain a voltage variable not contained in any other KVL equation. We conclude that there must be *at least* $b - n + 1$ linearly independent KVL equations.

We *assume* that a unique nontrivial solution exists. Since there are $2b$ unknowns, there must be exactly $2b$ linearly independent equations. This means that we are assuming that it is possible to obtain answers (values for v and i), at least one of which is not zero. These answers satisfy KCL, KVL, and the differential equations defining the various network elements.

The equations which relate the branch currents to the branch voltages must be linearly independent. This is the case because each differential equation contains at least one variable not contained in any other. Since there are b branches, there are b such equations.* Hence, there must be exactly $2b - b$ linearly independent topological constraint equations. But we have established that there are at least $b - n + 1$ linearly independent KVL equations. Therefore, there must be *at most* $b - (b - n + 1) = n - 1$ linearly independent KCL equations.

Let us now turn our attention to the particular tree from which we obtained the $b - n + 1$ chords above. Consider a typical tree as shown in solid lines in Figure 12.14 and a typical tree edge ab as labeled in Figure 12.14. We group the subnetwork containing all the tree edges connected to node a (imagine the branch ab to be temporarily cut) and call it N_1. Likewise, we group the subnetwork containing all tree branches connected to node b and call it N_2. The dotted lines in the figure are the chords corresponding to the tree being used. Now we apply KCL across the hypothetical plane separating N_1 and N_2. The cut-set equation we obtain will

* If a branch is a voltage source (or current source), the branch equation becomes degenerate in that it simply specifies the branch voltage (or current).

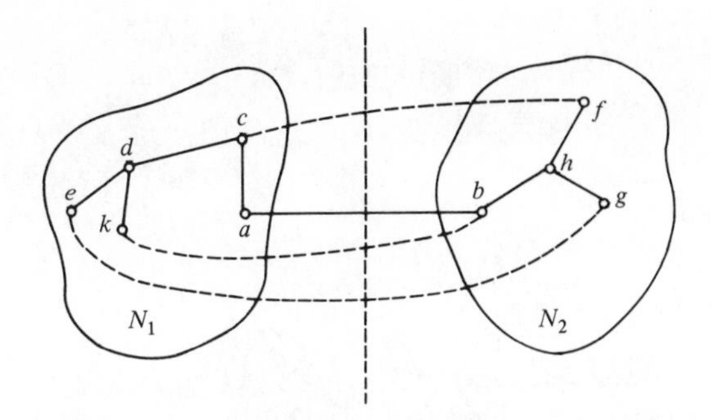

e 12.14 Illustrating the application of KCL across a plane intersecting exactly one tree branch.

contain exactly one tree-branch current and some chord currents. We repeat this procedure for all other tree branches. Since we have $n - 1$ tree branches, we obtain $n - 1$ KCL cut-set equations. Since each equation contains a tree-branch current variable not appearing in any of the other equations, these $n - 1$ equations must be linearly independent. Since we do not know whether there may be more, we conclude that there must be *at least* $n - 1$ linearly independent KCL equations.

Recall that from a previous argument we also concluded that there must be *at most* $n - 1$ linearly independent equations. The only way we can have both conditions satisfied is to have exactly $n - 1$ linearly independent KCL equations. Since the number of linearly independent KCL equations plus the number of linearly independent KVL equations equals b, there must be exactly $b - (n - 1) = b - n + 1$ linearly independent KVL equations. Thus, we have shown that if we assume that for any given network a unique nontrivial solution exists, then there are exactly

$$N_c = n - 1 \tag{12.9}$$

linearly independent KCL equations and exactly

$$N_v = b - n + 1 \tag{12.10}$$

linearly independent KVL equations.*

In the above development, we have assumed that the graph of the network is *connected*. In case the graph is disconnected, then the above equations apply to the separate subgraphs. For instance, suppose that a graph has two separate parts with b_1 branches and n_1 nodes in one part and b_2 branches and n_2 nodes in another part. Then $N_{c_1} = n_1 - 1$, $N_{c_2} = n_2 - 1$, $N_{v_1} = b_1 - n_1 + 1$, and $N_{v_2} = b_2 - n_2 + 1$.

* In advanced treatments, it is possible to prove the same results without assuming the existence of nontrivial solutions. However, the proof is not as simple as we have presented here. See S. Seshu and M. B. Reed, *Linear Graphs and Electrical Networks* (Reading, Mass.: Addison-Wesley Publishing Co., Inc., 1961) or W. Mayeda, *Linear Graphs* (New York: John Wiley & Sons, Inc., 1972).

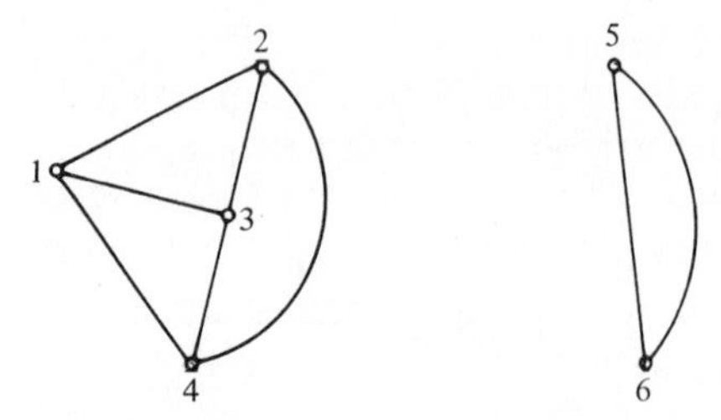

Figure 12.15

In other words, the total number of N_c is $N_c = N_{c_1} + N_{c_2} = (n_1 + n_2) - 2$ and $N_v = N_{v_1} + N_{v_2} = (b_1 + b_2) - (n_1 + n_2) + 2$. In general, for a network with s separate or disconnected parts, we have

$$N_c = n - s \tag{12.11}$$

and

$$N_v = b - n + s, \tag{12.12}$$

where n is the *total* number of nodes and b is the *total* number of branches.

EXERCISES

12.3.1 For the graph of Figure 12.5, what is the maximum number of linearly independent KCL equations that can be written? How about KVL?

12.3.2 Repeat Exercise 12.3.1 for Figure 12.6.

12.3.3 Repeat Exercise 12.3.1 for Figure 12.12.

12.3.4 Repeat Exercise 12.3.1 for Figure 12.15.

12.3.5 For the graph of Figure 12.14 and the tree indicated, apply KCL in such a way that each equation contains a tree-edge current not contained in any of the other equations.

12.3.6 For the graph of Figure 12.14, apply KVL such that each equation contains a voltage variable not contained in any of the other equations.

12.4 Choosing independent equilibrium equations

In the preceding section, we proved that there are $n - s$ linearly independent KCL equations and $b - n + s$ linearly independent KVL equations. In arriving at these results, we demonstrated a method for obtaining $n - s$ linearly independent KCL equations and another method for obtaining $b - n + s$ linearly independent KVL equations. We may pick a tree, and since each link forms exactly one loop with the tree edges, we use the $b - n + s$ loops as the loops to which we apply KVL. The equations are guaranteed to be independent. For KCL, we also pick a tree and apply KCL to cut sets in such a way that each equation contains a tree-edge current not contained in any other equation, as illustrated in Figure 12.14. Again, the $n - s$ KCL equations are guaranteed to be independent. Of course, for a graph with s separate parts, we pick a tree in each separate part. Since a graph does not have a unique tree, we see that there are several ways of obtaining the maximum number of equilibrium equations. As a matter of fact, there are other ways of systematically choosing linearly independent sets of equations without using trees. We shall now describe two such commonly used procedures. In most cases, especially simple ones, these choices are easier to apply.

Node Method. In any connected graph with n nodes, the application of KCL to any $n - 1$ of the n nodes results in $n - 1$ linearly independent equations. If the graph has s separate parts, then if n_K is the number of nodes in the Kth part, we apply KCL to any $n_K - 1$ nodes of the Kth part to obtain $n_K - 1$ linearly independent equations. Since there are s separate parts, we have $(n_1 - 1) + (n_2 - 1) + \cdots + (n_s - 1) = n - s$ linearly independent KCL equations. To prove the validity of this method, it is sufficient to consider only one separate part, that is, a connected graph. We shall carry out the proof by contradiction. Thus, we shall suppose that the theorem is not true. We shall then show that this assumption always leads to a contradiction. This implies that the theorem must be true because if it is not, we always get a contradiction.

So suppose the theorem is not true; that is, suppose that after applying KCL to any $n - 1$ of the n nodes of a network, we find that one of the equations can be expressed as a linear combination of the remaining $n - 2$ equations. Since we may label the nodes arbitrarily, there is no loss of generality if we assume that the first equation obtained from node 1 is a linear combination of equation number 2 from node 2, equation number 3 from node 3, and so on up to equation $n - 1$ from node $n - 1$. Let $y_1 = 0$ denote equation 1 and $y_2 = 0$ denote equation 2, and so on. Note that y_K is a sum of currents at the Kth node. Then we have

$$y_1 = c_2y_2 + c_3y_3 + \cdots + c_{n-1}y_{n-1}, \tag{12.13}$$

where the c's are constants, not all of which are zero. Suppose r of these constants are not zero where $1 \leq r \leq n - 2$. Again, we may relabel the nodes and their corresponding equations for convenience and without loss of generality say that

$$\begin{aligned} c_K &\neq 0, \qquad 2 \leq K \leq r + 1, \\ c_K &= 0, \qquad r + 1 < K \leq n - 1. \end{aligned}$$

That is, we have

$$y_1 = c_2y_2 + c_3y_3 + \cdots + c_{r+1}y_{r+1}, \tag{12.14}$$

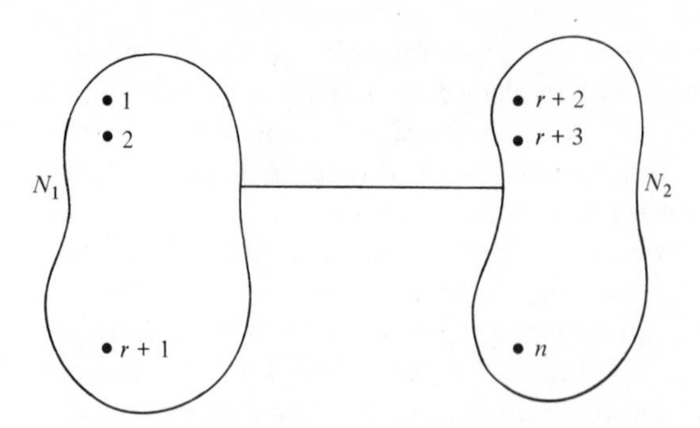

Figure 12.16 Figure relating to the establishment of the node method.

so that it is assumed that the first equation depends on r of the other $n - 2$ equations. Draw the network in such a way that nodes 1 to $r + 1$ are on one side and nodes $r + 2$ to n are on another, as shown in Figure 12.16.

The portion N_2 in Figure 12.16 contains *at least* one node, namely node n. If some of the c's are zero, then N_2 contains two or more nodes, and if all the c's are nonzero, then N_2 contains just node n. Since we are considering a connected network (or a connected part of a network with s separate parts), then there must be at least one branch joining N_1 to N_2, carrying the current i_x, for instance. If i_x goes to node 1, then equation 1 will contain i_x, but it will be absent from equations $2, 3, \ldots, r + 1$. Therefore, Equation 12.14 cannot be satisfied. Likewise, if i_x goes to either node $2, 3, \ldots$ or $r + 1$, say node K, then it will be contained in equation K, but not in any other. Since $c_K \neq 0$, then Equation 12.14 cannot be satisfied either. Since these are all the possibilities which all imply that Equation 12.14 cannot be satisfied, and since we assumed at the start of the proof that such an equation is satisfied, we arrive at a contradiction. Hence the theorem is true.

In summary, given any connected network with n nodes, we may choose any $n - 1$ of the n nodes at which we apply KCL. The resulting $n - 1$ equations are guaranteed to be linearly independent. Furthermore, from the results of the previous section, this is the maximum number of linearly independent KCL equations that may be written.

Mesh Method. For writing linearly independent KVL equations, the mesh method which we will describe shortly is useful for a class of networks called *planar networks*. A planar network is one which has a planar graph. A *planar graph* is a type of graph which can be laid flat on a plane surface without any of its edges crossing each other. If it is not possible to rearrange the positions of the vertices and edges such that there is no crossing of edges, then the graph is called *nonplanar*.

EXAMPLE 12.4.1

The graph in Figure 12.5 is planar because no lines cross. The graph shown in Figure 12.17(a) is also planar because it can be redrawn as shown in Figure 12.17(b)

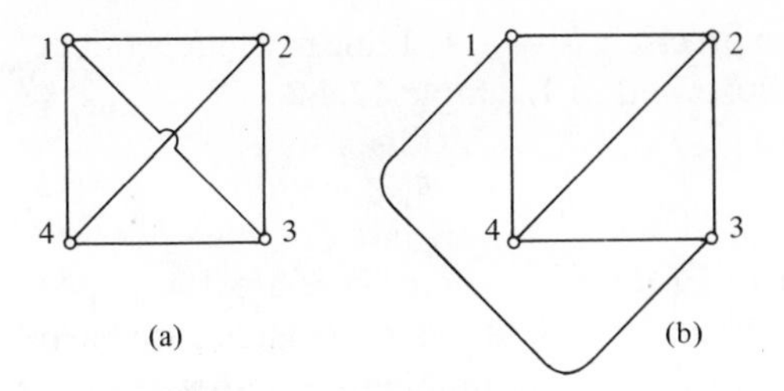

re 12.17 Example of a planar graph.

where no edges cross. Figure 12.18 shows an example of a nonplanar graph. A special kind of loop defined for planar graphs is the *mesh* or *window*. If a planar graph is drawn in planar form as in Figure 12.17(b), then a *mesh* of the graph is a loop such that no other loop of the network is inside it. In Figure 12.17(b), the loop 2342 is an example of a mesh. It is analogous to the mesh of a fishing net. If we draw a planar graph in planar form on a piece of paper and cut along the edges, then the boundaries of the resulting small pieces of paper represent the meshes. Note that there is no unique way of drawing a planar network in planar form. As we have mentioned previously, the graphs in Figures 12.3, 12.4, and 12.5 represent the same network. Hence, for a given planar network, the set of meshes is not unique.

The mesh method depends on two key theorems. These theorems pertain to connected planar graphs drawn in planar form and are as follows:

(1) For every planar form, there are exactly $b - n + 1$ meshes.

(2) The set of equations resulting from the application of KVL around each of the $b - n + 1$ meshes is linearly independent.

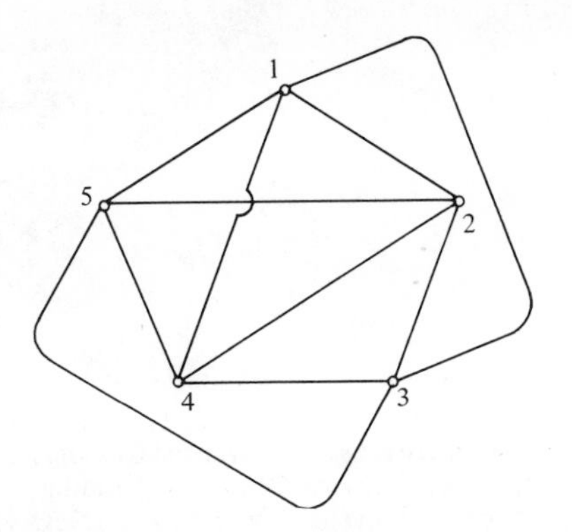

ure 12.18 An example of a nonplanar graph.

These two theorems enable us to write precisely $b - n + 1$ linearly independent KVL equations. The first theorem is illustrated in Example 12.4.2.

EXAMPLE 12.4.2

Consider the planar network in Figure 12.19. By inspection, there are 15 meshes or windows. We also note that there are 24 nodes and 38 branches. Hence we have $b - n + 1 = 38 - 24 + 1 = 15$, which is the correct number of meshes.

Let us prove the first theorem. Suppose we have a connected planar graph with b edges and n vertices. We shall show that if we draw any planar graph in planar form, we always have $b - n + 1$ meshes.

Consider retracing the graph. First, trace a tree. Since the planar graph is assumed to be drawn in planar form, the tree is also in planar form, containing exactly $n - 1$ edges and all the n vertices. We now have $b - (n - 1) = b - n + 1$ untraced edges left. Any edge retraced at this stage will produce one loop. Geometrically, this produces a polygon. Each additional edge retraced will increase the number of polygons that are not contained in other polygons by exactly one. Since there are $b - n + 1$ such edges, $b - n + 1$ such polygons will be generated. These polygons are the meshes for the planar form.*

In terms of our paper-cutting model, we first cut along the tree edges. Then the first additional edge cut will generate a polygon. The second additional edge cut will either generate an entirely separate polygon or else split the first polygon into two polygons. In either case, we now have two polygons. Similarly, the third additional edge cut will either generate a separate polygon or split one of the previous two polygons, giving a total of three polygons. The end result for making

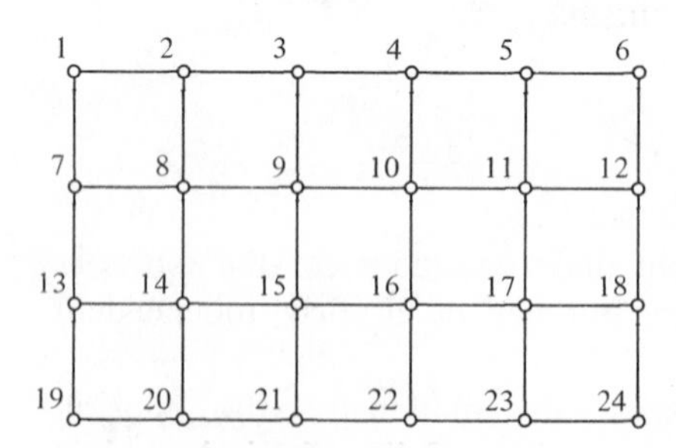

Figure 12.19 Graph for illustrating that the number of meshes is $b - n + 1$.

* A more elegant proof using a mapping of the graph onto a sphere is used in advanced treatises. See, for example, S. Seshu and M. B. Reed, *Linear Graphs and Electrical Networks* (Reading, Mass.: Addison-Wesley Publishing Co., Inc., 1961), 39–45, or W. Mayeda, *Linear Graphs* (New York: John Wiley & Sons, Inc., 1972).

the additional $b - n + 1$ cuts is to partition the planar form into $b - n + 1$ polygons which are identified with $b - n + 1$ meshes.

To prove the second theorem, we will use the method of contradiction. That is, we suppose that the theorem is false. Then the mesh equations are linearly dependent, so that we have

$$c_1 y_1 + c_2 y_2 + \cdots + c_{b-n+1} y_{b-n+1} = 0, \tag{12.15}$$

where not all the c's are zero, and y_K is the left-hand side of the Kth mesh equation. Suppose there are r nonzero coefficients in Equation 12.15. Without loss of generality, we may assume that these are the first r coefficients, since the mesh labeling is arbitrary. Then we have

$$c_1 y_1 + c_2 y_2 + \cdots + c_r y_r = 0. \tag{12.16}$$

That is, mesh equation 1 is dependent on mesh equations 2, 3, . . ., r. Suppose we remove the edges of the planar form not involved in the first r meshes. Then in the resulting graph there will be at least one *boundary* edge. A boundary edge is an edge which belongs to only one mesh. Hence, one of the y's contains a term not present in other terms of Equation 12.16. This means that these r equations cannot be linearly dependent which contradicts Equation 12.16. We conclude that the $b - n + 1$ mesh equations are linearly independent.*

We note that from Section 12.3 the maximum number of linearly independent KVL equations that can be written for a connected graph with b edges and n vertices is $b - n + 1$. Hence the set of $b - n + 1$ mesh equations represents one such set of a maximum number of linearly independent KVL equations.

RCISES

12.4.1 Determine the meshes in Figure 12.3. The graph of Figure 12.4 represents the same network as that of the graph of Figure 12.3. Trace the meshes of Figure 12.3 on the graph of Figure 12.4. Likewise, trace the meshes of Figure 12.3 on the graph of Figure 12.5.

12.4.2 Repeat Exercise 12.4.1 for the meshes of the graph on Figure 12.4 and trace on the graphs of Figures 12.3 and 12.5.

12.4.3 Write a set of linearly independent node equations for the graph in Figure 12.3, using the maximum number of linearly independent equations.

12.4.4 Write a set of linearly independent mesh equations for the graph in Figure 12.3, using the maximum number of linearly independent equations.

* Consider our paper model again. We draw our planar graph in planar form on a piece of paper and color the first r meshes in question. Then we cut out the colored portion of the paper. Obviously, the colored cutout will have at least one edge or boundary. This edge corresponds to an edge in a mesh not found in any of the other $r - 1$ meshes.

12.4.5 Repeat Exercise 12.4.3 for the graph in Figure 12.18.

12.4.6 Is the mesh method applicable to the graph in Figure 12.18? How about that in Figure 12.17(a)? Explain.

12.5 Choosing network variables

From what has been described in this chapter so far, given any network with b branches, we can always write $2b$ linearly independent equations in b voltage variables and b current variables. However, these equations may be readily reduced to $n - 1$ linearly independent equations in $n - 1$ voltage variables. Once these voltage variables are found, the other $b - n + 1$ voltage variables can be obtained very easily. Furthermore, once the branch voltages are known, the branch currents are known. We will also show that it is always possible to write $b - n + 1$ linearly independent equations in $b - n + 1$ current variables. Then, once these variables are determined, all the other variables can be determined readily.

Voltage Variables. We have seen from the previous section that there are $b - n + 1$ linearly independent KVL equations. This means that we have $b - n + 1$ independent equations relating the b branch voltage variables. From these equations it is possible to solve for each of $b - n + 1$ branch voltage variables in terms of the remaining $n - 1$ branch voltage variables. If we can determine the values of these $n - 1$ voltage variables from some other considerations, then the other $b - n + 1$ voltage variables follow. Thus of the b branch voltage variables, it is sufficient to first solve for $n - 1$ of them. The set of $n - 1$ voltages which completely specifies all the voltages in a network is not unique. There may be several such sets which will do. But not any set of $n - 1$ voltages may work.

The problem is to determine a method which will guarantee that our choice of $n - 1$ voltages is sufficient to subsequently determine the other $b - n + 1$ voltages.

One way is to pick a tree and *choose the $n - 1$ tree-edge voltages* as the set to use. The remaining edges are links of the tree. We may apply KVL around each loop formed by a chord. This enables us to solve for each link voltage in terms of the tree-edge voltages. Hence, if we know the tree-edge voltages, we know all the branch voltages of the network.

Another way, which is the one most commonly used, is to pick a node (any node) and call it the datum or reference node. Then we define the voltage between any node, say node K, and the reference node as the node voltage at node K. We then have $n - 1$ node voltages. Note that every node voltage can be expressed as a linear combination of some tree-edge voltages. Likewise every tree-edge voltage for any tree can be expressed as the difference between the node voltages for the two nodes of the branch. That is, the set of node voltages uniquely determines a set of tree-edge voltages and vice versa. Hence, if we know the node voltages, we automatically know all the branch voltages.

AMPLE 12.5.1

For the graph of Figure 12.5, some trees are shown on Figure 12.10. The tree-edge voltages in Figure 12.10(a) are v_{14}, v_{12}, and v_{13}. The three other edge voltages which are the chord voltages are determined from these three tree-edge voltages. We have

$$\begin{aligned} v_{42} &= v_{41} + v_{12} = -v_{14} + v_{12}, \\ v_{23} &= v_{21} + v_{13} = -v_{12} + v_{13}, \\ v_{34} &= v_{31} + v_{14} = -v_{13} + v_{14}. \end{aligned} \tag{12.17}$$

Hence the tree-branch voltages v_{14}, v_{12}, and v_{13} are sufficient to determine all branch voltages.

AMPLE 12.5.2

In Figure 12.5 again, suppose we assign vertex 3 as the reference vertex. Let v_1, v_2, and v_4 be the voltages with respect to vertex 3 of vertices 1, 2, and 4, respectively. Then the edge voltages are expressible in terms of the vertex node voltages. Thus we have

$$\begin{aligned} v_{42} &= v_4 - v_2, \qquad & v_{43} &= v_4, \qquad & v_{23} &= v_2, \\ v_{41} &= v_4 - v_1, \qquad & v_{21} &= v_2 - v_1, \qquad & v_{13} &= v_1. \end{aligned} \tag{12.18}$$

We conclude that v_1, v_2, and v_4 are sufficient to determine all the branch voltages of the corresponding network.

We now have a method for choosing $n - 1$ voltage variables which completely determine all other network voltages. Recall from the previous section that there are $n - 1$ linearly independent KCL equations. We therefore write these KCL equations in terms of these $n - 1$ voltage variables. Note that the current through any branch is related to the difference between the node voltages of the nodes touching the branch, by means of the branch constraint equations. Similarly, tree-branch voltage variables may be employed. The full development of the details in writing these equations is carried out in the next chapter.

Current Variables. For a connected network with b branches and n nodes, we have b branch currents. Recall that we have $n - 1$ linearly independent KCL equations relating these b currents. Analogous to our discussion of voltage variables, this means that from these equations, we may solve for each of $n - 1$ currents in terms of the other $b - (n - 1) = b - n + 1$ currents. If the latter $b - n + 1$ currents are known, then automatically we also have the other $n - 1$ currents. What we are looking for then is a method for choosing $b - n + 1$ current variables such that knowing these, all other currents are determined.

One way is to pick a tree and *choose the* $b - n + 1$ *link currents* as an appropriate set of current variables. To see this we recall the discussion in Section 12.3 of passing hypothetical planes through one tree edge and a set of chords as in Figure 12.14, thus dividing the network into two (*a cut set*). This enables us to solve for the tree-edge current in terms of the chord currents. Since we can do this for all the $n - 1$ tree-branch currents, all currents are determined if we know the values of the chord currents. To mechanize the procedure more efficiently, it is advantageous to imagine the link currents as causing *loop currents*. Thus, given a tree of a network, every time we put a link back (only one at a time) we generate exactly one loop. We imagine a current flowing through this loop. For instance, for the tree of Figure 12.10(b) we have the three loops in Figure 12.20. The three chord currents are equal to these loop currents and the tree currents may be obtained by inspection of the graph in Figure 12.20 with the loop currents superimposed.

The above rule, requiring that we find a tree, restore the chords one at a time, and locate an appropriate set of loops, is guaranteed to work for the most general conditions. No matter how complicated a network graph is, the resulting set of loop-current variables will always be sufficient for solving the network problem.

Many network graphs, however, do not come under the heading of "very general" nor "very complicated." Although the chord-set procedure just described works all the time, it is usually much simpler to use special procedures for a special type of graphs. The procedure we will next describe is based on choosing loop-current variables around meshes. It works for *planar graphs*.

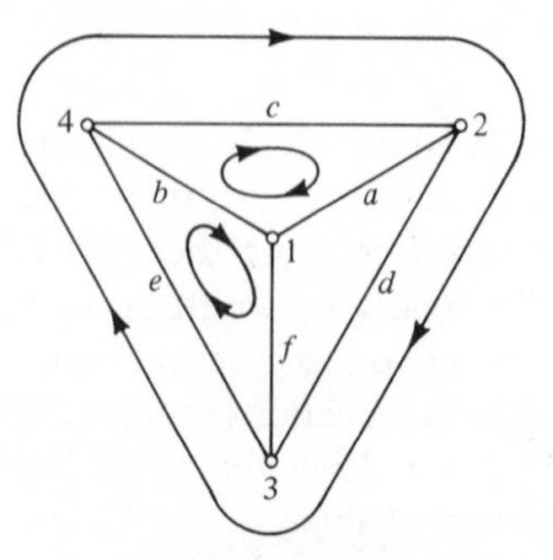

Figure 12.20 Loop-current variable assignment corresponding to the tree in Figure 12.10(b).

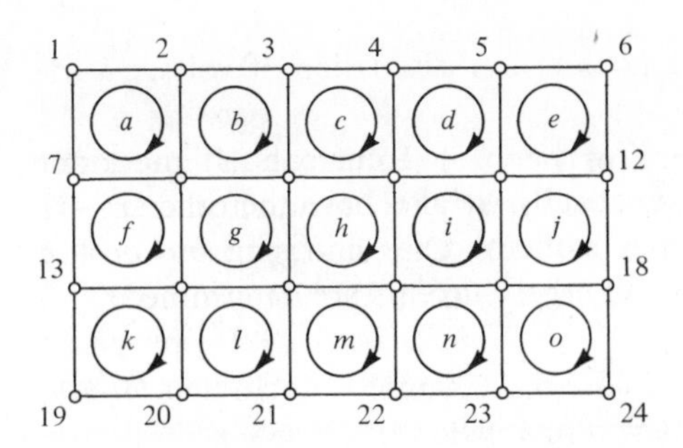

ure 12.21 Mesh-current assignments for the graph in Figure 12.19.

This method assigns $b - n + 1$ loop currents or *mesh currents* around the meshes of the network as shown in Figure 12.21 for the graph in Figure 12.19. By inspection of the graph in Figure 12.21, we see that every branch current is obtainable from knowledge of the mesh currents.

We have shown two methods of choosing $b - n + 1$ currents which determine all branch currents. We also recall that there are exactly $b - n + 1$ linearly independent KVL equations. So if we could write these KVL equations using only these $b - n + 1$ current variables, we would have $b - n + 1$ linearly independent equations in $b - n + 1$ unknowns. Every branch voltage is related to the corresponding branch current by the branch constraint equation. We have already seen that every branch current is expressible in terms of link currents or mesh currents. Hence the branch voltages may be expressed in terms of the chosen $b - n + 1$ current variables. The details for obtaining these equations are described in the next chapter.

ERCISES

12.5.1 By inspection of Figure 12.20, determine the tree-edge currents for the tree in Figure 12.10(b) in terms of the loop currents. Verify this by the technique of passing planes through the graph as illustrated in Figure 12.14.

12.5.2 In Figure 12.21, which branch currents are equal to mesh currents?

12.5.3 Choose an appropriate set of loop-current variables for the graph of Figure 12.18.

12.5.4 Can mesh-current variables be assigned to the graph in Figure 12.18? Explain.

12.5.5 Choose an appropriate set of voltage variables for the graph in Figure 12.18, using node voltages.

12.5.6 Repeat Exercise 12.5.5 using tree-branch voltages.

12.6 Summary

1. The concept of linear independence of network equations is important because the sets of equations we write must be linearly independent if there are to exist unique solutions that describe the network. Our strategy is to ensure linear independence by proper choice of variables and a correct procedure to follow for formulating equations.

2. Some terms for linear graphs are familiar from Chapter 5, including *branch*, *node*, *path*, and *loop* (*closed path*). Some new terms include:
Tree — a connected subgraph containing all nodes with no closed paths.
Chords or *links* — branches remaining when a tree is removed from a graph; there are $b - n + 1$ of them.
Connected graph has one part. Graphs associated with networks with mutual inductance or controlled sources may have more than one part.

3. Number of variables required: To describe a connected network on the node basis with KCL requires $n - 1$ voltage variables and equations. To describe a connected network on the loop basis with KVL requires $b - n + 1$ current variables and equations.

4. A connected network with n nodes may be described by applying KCL at any $n - 1$ of the n nodes; the resulting equations will be linearly independent. A connected network with n nodes and b branches may be described by applying KVL around each of $b - n + 1$ loops defined by the link currents; the resulting equations will be linearly independent. It may also be described by applying KVL around each of $b - n + 1$ meshes when the network is planar; the resulting equations will be linearly independent.

5. Preferred variables:
(a) In applying KCL, use the $n - 1$ node-to-datum voltage variables.
(b) In applying KVL to planar networks, use $b - n + 1$ mesh currents.
(c) In applying KVL to nonplanar networks, use $b - n + 1$ link currents.

PROBLEMS

12.1 Verify that the number of meshes equals the number of branches minus the number of nodes plus 1, for the network in Figure 12.22.

12.2 (a) Draw a set of nonredundant mesh currents for the graph in Figure 12.12. (b) Draw a set of nonredundant mesh currents different from those in (a) for the same network. [*Hint:* Rearrange the geometrical location of the nodes to change the shape of the figure, but not the topology of the graph. Then find the meshes.]

12.3 For the network in Figure 12.14, show that the sum of the branch currents intersecting the hypothetical plane is zero by applying KCL for the nodes in N_2 and adding the equations. Show that this result is true for any arbitrary network, as shown in Figure 12.13.

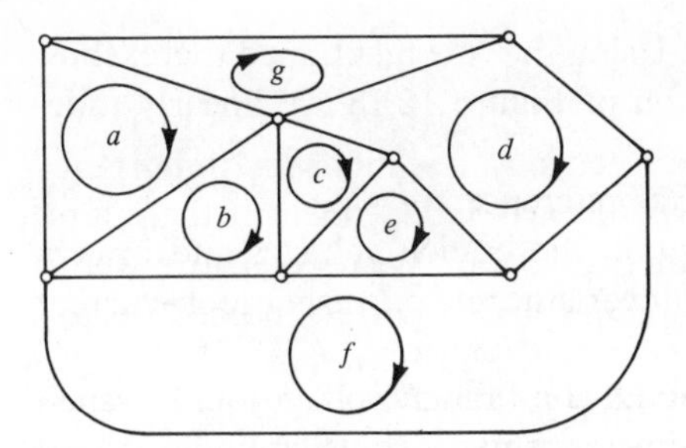

Figure 12.22

12.4 If a branch is added to a connected network N_1 without introducing any new nodes, what is the relationship between the number of current variable assignments in N_1 compared to the new network. Show that for the modified network, a suitable current variable assignment consists of a loop involving the added branch and the loops (based on chords) of N_1.

12.5 Suppose that there are two paths from node a to node b in a network. Is the sum of the branch voltages along one path from a to b equal to the sum of the branch voltages along the other path from a to b? Prove your answer.

12.6 Describe two methods of choosing current variables for a graph with s separate parts.

12.7 Describe two methods of choosing voltage variables for a graph with s separate parts.

12.8 Given a set of m linear equations in n unknowns, $m \leq n$. If it is possible to renumber (rearrange) the equations in such a way that the second equation contains a variable not found in the first equation, then the first two equations are linearly independent. Likewise, if we can find a third equation which contains a variable not found in the first two, then the three equations are linearly independent. If all the equations are so arranged that the rth equation contains a variable not contained in the first $r - 1$ equations, for all r up to $m \leq n$, then we say that the m equations are linearly independent. This argument is more general than the one we have often used in this chapter for proving several of the theorems. The earlier argument is that if *each* equation contains a variable not found in *any other* equation, then the equations are linearly independent. Apply the more general rule to show that the mesh equations for Figure 12.17 are linearly independent.

12.9 Read the argument in Problem 12.8 and use it to obtain a set of linearly independent KVL equations for the graph of Figure 12.19. Don't use mesh equations.

12.10 Read the argument in Problem 12.8. Using the rule mentioned there, show that the node equations for the graph of Figure 12.18 are linearly independent. Use node 5 as a reference.

12.11 Repeat Problem 12.10 using node 1 as reference.

★12.12 Outline a procedure for determining all the branch voltages and branch currents in a connected linear time-invariant network using loop-current variables.

★12.13 Outline a procedure for determining all the branch voltages and branch currents in a connected linear time-invariant network using node-voltage variables.

12.14 If in Figure 12.18 edge 34 is removed, is the resulting graph planar? How about removing 42 instead of 34?

12.15 Write KVL equations for the two loops of Figure 12.23. Are these equations independent? What is the maximum number of linearly independent KVL equations that can be written for this network? For the two indicated loop variables, verify that KCL is satisfied at all nodes. Suppose that all resistances are constant and equal to 1 ohm and $v_s = 3$ volts; solve for i_1 and i_2 from your two equations. Verify whether KVL is satisfied in loops *abefa, bcdeb, aghcba*. Is there any loop for which KVL is satisfied? This problem demonstrates that although the equations are linearly independent and there are as many unknowns as equations, the equation solutions may not be network solutions. This is because the number of loop variables used is less than the maximum number of linearly independent loops. Thus the two loop variables in this problem do not completely determine the branch currents for this network.

12.16 Repeat Problem 12.15 for loops as shown in Figure 12.24.

12.17 For the graph of Figure 12.18, show that the sum of the KCL equations applied at all vertices is zero, and thus the number of linearly independent node equations is less than five.

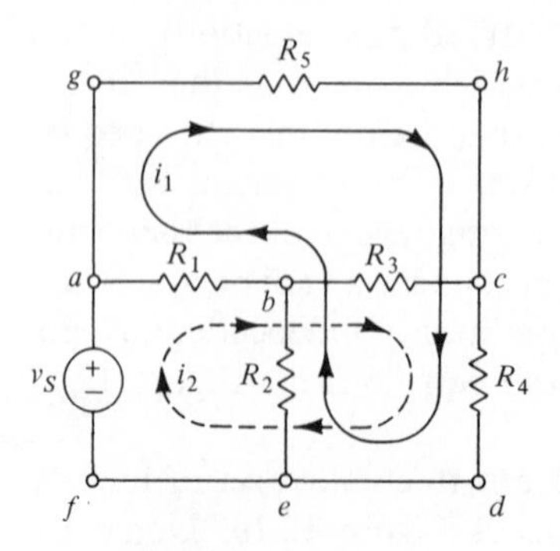

Figure 12.23

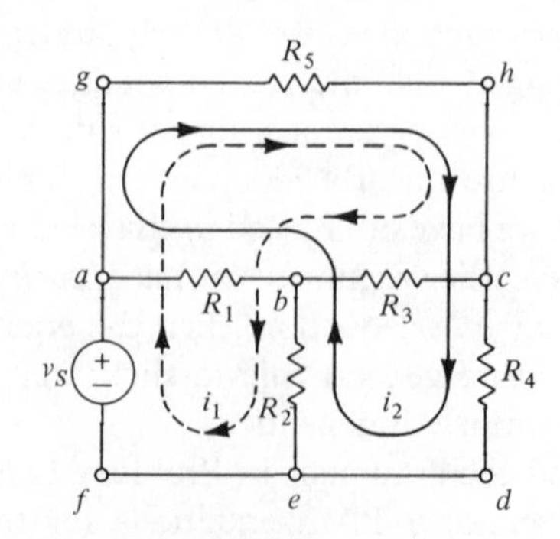

Figure 12.24

★12.18 Generalize Problem 12.17 to a connected graph with n vertices. Show that the sum of the KCL equations applied at all vertices is zero, and hence that the number of linearly independent node equations is less than n.

12.19 Add a loop-current variable to the set given in Figure 12.23. Write an appropriate set of three linearly independent KVL equations.

12.20 Repeat Problem 12.19 for a different third loop-current variable.

12.21 Suppose a set of p equations is linearly independent. Take any subset $q (q < p)$ of the equations. Is the set of q equations linearly independent? Explain.

Chapter **13**

Loop, Node, and State Equations

In this chapter, we shall consider in more detail how we formulate network equations using the fundamentals developed in the last chapter. In all cases, we attempt to write as many linearly independent equations as there are unknowns. The general procedure for solving these equations will be given in the last chapter. However, it should be emphasized that although the techniques described here are applicable to general networks, the amount of algebra and arithmetic involved in solving a specific problem can usually be reduced drastically if one keeps in mind the powerful tools in Chapters 7 through 10 for linear networks. It is a common temptation on the part of a student to apply loop analysis or nodal analysis to every network problem encountered. The formulation of the equations is relatively simple, but the amount of actual computation involved is by no means trivial. A short reflection on how Thevenin's theorem can be utilized almost always pays off greatly. This is not to say that we can forget about general loop and node methods. They are useful in general theoretical studies, but they should not be blindly used for every network problem.

13.1 Formulation of linear time-invariant network equations on the node basis

In this section, our variables will be the node-to-datum voltages. We will apply KCL at $n - 1$ nodes (assuming a connected graph) to arrive at a set of $n - 1$ linearly independent equations in the $n - 1$ voltage variables. We will explain the method by considering a number of examples first.

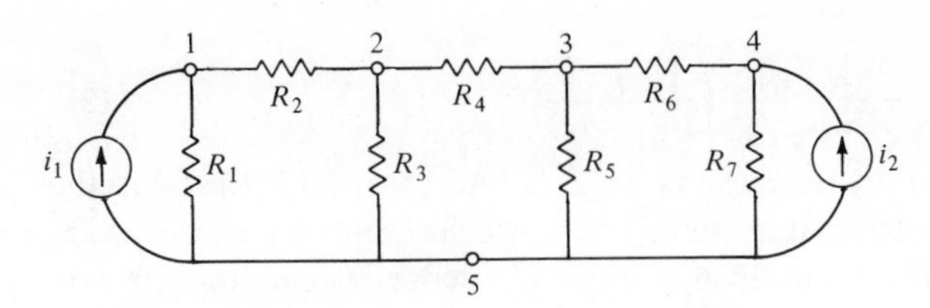

ure 13.1 Linear time-invariant network for Example 13.1.1.

AMPLE 13.1.1

Consider the linear time-invariant network shown in Figure 13.1. For this network, let us select node 5 as the datum node so that the voltage variables are v_{15}, v_{25}, v_{35}, and v_{45}. For simplification, we will drop the second number in the double-subscript notation. Applying KCL at node 1, and making use of the linear time-invariant branch characteristics, we have

$$\frac{v_1}{R_1} + \frac{v_1 - v_2}{R_2} - i_1 = 0 \tag{13.1}$$

or

$$\left(\frac{1}{R_1} + \frac{1}{R_2}\right)v_1 - \frac{1}{R_2}v_2 + 0 + 0 = i_1 . \tag{13.2}$$

For node 2, KCL yields

$$\frac{v_2 - v_1}{R_2} + \frac{v_2}{R_3} + \frac{v_2 - v_3}{R_4} = 0 \tag{13.3}$$

or

$$-\frac{1}{R_2}v_1 + \left(\frac{1}{R_2} + \frac{1}{R_3} + \frac{1}{R_4}\right)v_2 - \frac{1}{R_4}v_3 + 0 = 0 . \tag{13.4}$$

For node 3, the equation is

$$\frac{v_3 - v_2}{R_4} + \frac{v_3}{R_5} + \frac{v_3 - v_4}{R_6} = 0 \tag{13.5}$$

or

$$0 - \frac{1}{R_4}v_2 + \left(\frac{1}{R_4} + \frac{1}{R_5} + \frac{1}{R_6}\right)v_3 - \frac{1}{R_6}v_4 = 0 . \tag{13.6}$$

Finally, applying KCL at node 4, we find that

$$\frac{(v_4 - v_3)}{R_6} + \frac{v_4}{R_7} - i_2 = 0 \tag{13.7}$$

or

$$0 + 0 - \frac{1}{R_6} v_3 + \left(\frac{1}{R_6} + \frac{1}{R_7}\right) v_4 = i_2 . \tag{13.8}$$

The set of equations given by Equations 13.2, 13.4, 13.6, and 13.8 specifies sufficient and independent relationships to solve for the variables v_1, v_2, v_3, and v_4, so that the remaining problem is purely algebraic. We may eliminate the variables one at a time and end up with one equation in one unknown, or we may apply Cramer's rule* to obtain the solution in terms of determinants. We will postpone studying general methods of solution of simultaneous sets of network equations until Chapter 14. Just now, we are interested only in *formulating a correct set of network equations.*

Consider the following array of numbers:

$$\mathscr{Y} = \begin{bmatrix} \left(\frac{1}{R_1} + \frac{1}{R_2}\right) & -\frac{1}{R_2} & 0 & 0 \\ -\frac{1}{R_2} & \left(\frac{1}{R_2} + \frac{1}{R_3} + \frac{1}{R_4}\right) & -\frac{1}{R_4} & 0 \\ 0 & -\frac{1}{R_4} & \left(\frac{1}{R_4} + \frac{1}{R_5} + \frac{1}{R_6}\right) & -\frac{1}{R_6} \\ 0 & 0 & -\frac{1}{R_6} & \left(\frac{1}{R_6} + \frac{1}{R_7}\right) \end{bmatrix} . \tag{13.9}$$

This array of numbers is called the *node-admittance matrix* for this network with node 5 as reference. Note that each entry in the array has the dimension of admittance. Similarly, the array of quantities on the right side of Equations 13.2, 13.4, 13.6, and 13.8 may be symbolized by

$$\mathscr{I} = \begin{bmatrix} i_1 \\ 0 \\ 0 \\ i_2 \end{bmatrix} . \tag{13.10}$$

This array is called a *column matrix.* Denote the array of voltage variables by

$$\mathscr{V} = \begin{bmatrix} v_1 \\ v_2 \\ v_3 \\ v_4 \end{bmatrix} , \tag{13.11}$$

* See Chapter 14.

so that the four network equations are compactly expressed in the matrix form

$$\mathscr{Y}\mathscr{V} = \mathscr{I}. \tag{13.12}$$

The theory of matrices is well developed in mathematics, and certain operations may be performed on Equation 13.12. Our sole intention in using matrices in this chapter is to simplify the writing of large arrays of coefficients. For the present purpose we may just as well call the array a *table of coefficients* or a *coefficient filing cabinet* instead of a matrix. The formulation of the set of node equations is to a large extent a matter of constructing a table which we call the *node-admittance matrix*.

AMPLE 13.1.2

Let us consider the linear time-invariant network in Figure 13.2 and select node 4 as the datum. Applying KCL at nodes 1, 2, and 3, and expressing each branch current in terms of the node voltages and the appropriate branch characteristics, we have

$$\frac{1}{R_1} v_1 + C_1 \frac{d}{dt}(v_1 - v_2) - i_1 = 0, \tag{13.13}$$

$$C_1 \frac{d}{dt}(v_2 - v_1) + \frac{1}{R_2} v_2 + \frac{1}{L_1}\int_0^t (v_2 - v_3)\, d\tau + i_{23}(0) = 0, \tag{13.14}$$

$$\frac{1}{L_1}\int_0^t (v_3 - v_2)\, d\tau + i_{32}(0) + \frac{1}{R_3} v_3 - i_2 = 0. \tag{13.15}$$

Using operational notation,* we have

$$p = \frac{d}{dt} \tag{13.16}$$

and

$$\frac{1}{p} = \int_0^t d\tau, \tag{13.17}$$

we see that Equations 13.13, 13.14, and 13.15 become

$$\left(\frac{1}{R_1} + C_1 p\right) v_1 - C_1 p v_2 + 0 = i_1, \tag{13.18}$$

$$-C_1 p v_1 + \left(C_1 p + \frac{1}{R_2} + \frac{1}{L_1 p}\right) v_2 - \frac{1}{L_1 p} v_3 = -i_{23}(0), \tag{13.19}$$

$$0 - \frac{1}{L_1 p} v_2 + \left(\frac{1}{L_1 p} + \frac{1}{R_3}\right) v_3 = i_2 + i_{23}(0). \tag{13.20}$$

* See any book on elementary differential equations.

The array of coefficients in the node-admittance matrix is

$$\mathscr{Y} = \begin{bmatrix} \left(\frac{1}{R_1} + C_1 p\right) & -C_1 p & 0 \\ -C_1 p & \left(C_1 p + \frac{1}{R_2} + \frac{1}{L_1 p}\right) & -\frac{1}{L_1 p} \\ 0 & -\frac{1}{L_1 p} & \left(\frac{1}{L_1 p} + \frac{1}{R_3}\right) \end{bmatrix}, \quad (13.21)$$

and the current matrix is

$$\mathscr{I} = \begin{bmatrix} i_1 \\ 0 \\ i_2 \end{bmatrix}. \quad (13.22)$$

If the signals are of the form e^{st}, and if we disregard complementary solutions, then the node-admittance matrix is

$$\mathscr{Y} = \begin{bmatrix} \left(\frac{1}{R_1} + C_1 s\right) & -C_1 s & 0 \\ -C_1 s & \left(C_1 s + \frac{1}{R_2} + \frac{1}{L_1 s}\right) & -\frac{1}{L_1 s} \\ 0 & -\frac{1}{L_1 s} & \left(\frac{1}{L_1 s} + \frac{1}{R_3}\right) \end{bmatrix}. \quad (13.23)$$

If the sources are all sinusoidal and of the same frequency, and considering only the steady-state, then phasor network equations may be written and the resulting node-admittance matrix is

$$\mathscr{Y} = \begin{bmatrix} \left(\frac{1}{R_1} + j\omega C_1\right) & -j\omega C_1 & 0 \\ -j\omega C_1 & \left(j\omega C_1 + \frac{1}{R_2} + \frac{1}{j\omega L_1}\right) & -\frac{1}{j\omega L_1} \\ 0 & -\frac{1}{j\omega L_1} & \left(\frac{1}{j\omega L_1} + \frac{1}{R_3}\right) \end{bmatrix}, \quad (13.24)$$

and the current-source matrix is

$$\mathscr{I} = \begin{bmatrix} I_1 \\ 0 \\ I_2 \end{bmatrix}. \quad (13.25)$$

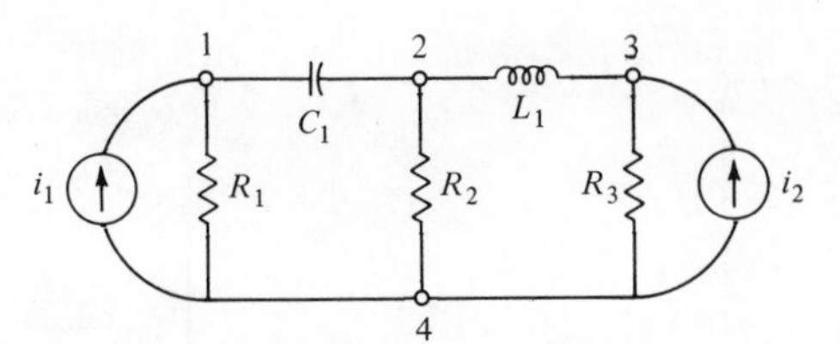

Figure 13.2 Linear time-invariant network for Example 13.1.2.

In writing phasor equations for the sinusoidal steady-state components, we disregard the complementary solution as well as the initial conditions. This is also the case when we write equations for the particular solution when the input signals are of the form e^{st}.

EXAMPLE 13.1.3

Let us now consider a linear time-invariant network with a controlled source. Suppose that in the network of Figure 13.2, i_2 is a controlled source $i_2 = Gv_{24}$ and $i_1(t) = I_1e^{st}$. Choosing node 4 as reference, we have

$$C_1s(V_1 - V_2) + \frac{1}{R_1} V_1 = I_1, \tag{13.26}$$

$$C_1s(V_2 - V_1) + \frac{1}{R_2} V_2 + \frac{1}{L_1s}(V_2 - V_3) = 0, \tag{13.27}$$

and

$$\frac{1}{L_1s}(V_3 - V_2) + \frac{V_3}{R_3} = GV_2. \tag{13.28}$$

Simplifying the equations, we have

$$\begin{aligned} \left(C_1s + \frac{1}{R_1}\right)V_1 - C_1sV_2 + 0 &= I_1, \\ -C_1sV_1 + \left(C_1s + \frac{1}{R_2} + \frac{1}{L_1s}\right)V_2 - \frac{1}{L_1s} V_3 &= 0, \\ 0 - \left(\frac{1}{L_1s} + G\right)V_2 + \left(\frac{1}{L_1s} + \frac{1}{R_3}\right)V_3 &= 0. \end{aligned} \tag{13.29}$$

The general idea in the nodal method for linear time-invariant networks is to write the network equations in the form

$$\begin{aligned} y_{11}V_1 + y_{12}V_2 + y_{13}V_3 + \cdots + y_{1m}V_m &= I_1, \\ y_{21}V_1 + y_{22}V_2 + y_{23}V_3 + \cdots + y_{2m}V_m &= I_2, \\ &\vdots \\ y_{m1}V_1 + y_{m2}V_2 + y_{m3}V_3 + \cdots + y_{mm}V_m &= I_m, \end{aligned} \tag{13.30}$$

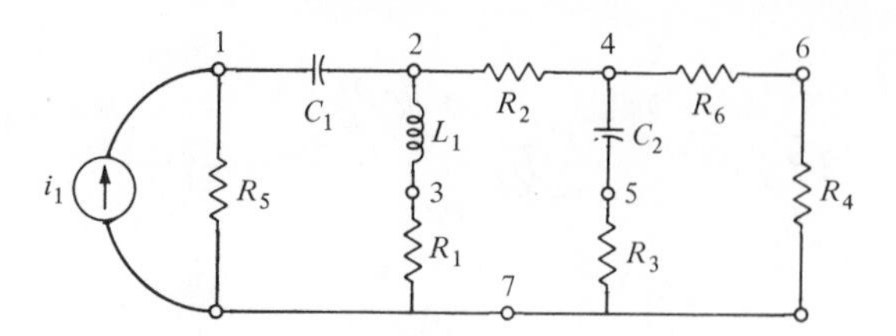

Figure 13.3 The network used for Example 13.1.4.

for a network with $m + 1$ nodes (considering only a connected network). The basic procedure involved is to assign a reference node and to assign a node voltage variable to all the other $m = n - 1$ nodes. Then Kirchhoff's current law is applied at each of the $n - 1$ nodes. After some algebraic simplification, Equation 13.30 is obtained. For networks which contain no controlled sources as in Examples 13.1.1 and 13.1.2, there is some symmetry in the coefficients of the above formulation. Specifically, $y_{jk} = y_{kj}$ (check Examples 13.1.1 and 13.1.2). For networks with controlled sources, this may not be so (check Example 13.1.3). For the special case where no controlled sources are present, the y_{jk}'s may be obtained with hardly any effort at all. The rules for linear time-invariant networks containing only R's, L's, and C's are as follows:

(1) If $i = k$, then $y_{ik} = y_{kk}$ is obtained by adding all the admittances connected from node k to *all other* nodes.

(2) If $i \neq k$, then y_{ik} is the negative of the admittance connected directly between nodes i and k.

If we are writing the equations in differential-equation form, we use p instead of $j\omega$, where p is the differential operator. If we are writing the equation for e^{st}-type signals, then we use s to replace $j\omega$.

EXAMPLE 13.1.4

For the linear time-invariant network in Figure 13.3, we choose node 7 as the datum and our variables are $v_1, v_2, v_3, v_4, v_5, v_6$. By inspection of the network, we get

$$\mathscr{Y} = \begin{bmatrix} \left(\frac{1}{R_5}+C_1p\right) & -C_1p & 0 & 0 & 0 & 0 \\ -C_1p & \left(C_1p+\frac{1}{R_2}+\frac{1}{L_1p}\right) & -\frac{1}{L_1p} & -\frac{1}{R_2} & 0 & 0 \\ 0 & -\frac{1}{L_1p} & \left(\frac{1}{R_1}+\frac{1}{L_1p}\right) & 0 & 0 & 0 \\ 0 & -\frac{1}{R_2} & 0 & \left(\frac{1}{R_2}+\frac{1}{R_6}+C_2p\right) & -C_2p & -\frac{1}{R_6} \\ 0 & 0 & 0 & -C_2p & \left(C_2p+\frac{1}{R_3}\right) & 0 \\ 0 & 0 & 0 & -\frac{1}{R_6} & 0 & \left(\frac{1}{R_6}+\frac{1}{R_4}\right) \end{bmatrix} \tag{13.31}$$

For the column matrix $\mathscr{I}$, the kth row i_k is the sum of the current sources with reference direction into node k minus the sum of the current sources with reference direction away from node k. For the network in Equation 13.31, the $\mathscr{I}$ matrix is

$$\mathscr{I} = \begin{bmatrix} i_1 \\ 0 \\ 0 \\ 0 \\ 0 \\ 0 \end{bmatrix}. \tag{13.32}$$

AMPLE 13.1.5

The linear time-invariant network shown in Figure 13.4 is known as a twin-T. Using node 5 as the reference node, the network differential equations are

$$\left(C_1 \frac{d}{dt} + \frac{1}{R_1}\right) v_1 - \frac{1}{R_1} v_2 - C_1 \frac{dv_3}{dt} + 0 = i_1, \tag{13.33}$$

$$-\frac{1}{R_1} v_1 + \left(\frac{1}{R_1} + \frac{1}{R_2} + C_3 \frac{d}{dt}\right) v_2 + 0 - \frac{1}{R_2} v_4 = 0, \tag{13.34}$$

$$-C_1 \frac{dv_1}{dt} + 0 + \left(C_1 \frac{d}{dt} + C_2 \frac{d}{dt} + \frac{1}{R_3}\right) v_3 - C_2 \frac{dv_4}{dt} = 0, \tag{13.35}$$

$$0 - \frac{1}{R_2} v_2 - C_2 \frac{dv_3}{dt} + \left(\frac{1}{R_2} + \frac{1}{R_4} + C_2 \frac{d}{dt}\right) v_4 = 0. \tag{13.36}$$

ERCISES

13.1.1 Write the node equations in differential-equation form for the network in Figure 13.4, using node 3 as the reference.

13.1.2 Repeat Example 13.1.1, using node 4 as the reference.

13.1.3 Write the node equations in phasor form for the network in Figure 13.3, using node 3 as reference.

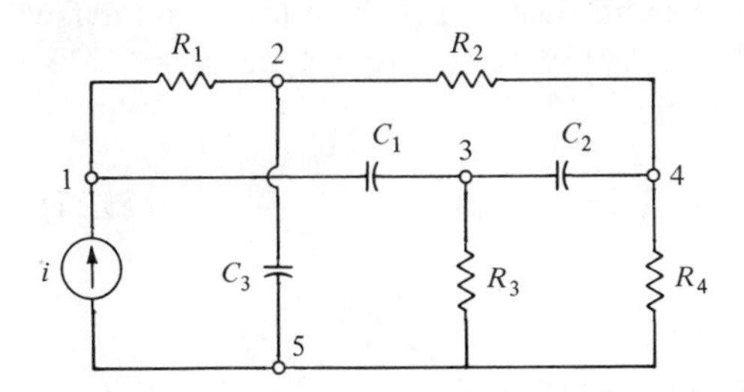

re 13.4 A twin-T network used in Example 13.1.5.

13.1.4 Repeat Exercise 13.1.3, using node 5 as reference.

13.1.5 Repeat Exercise 13.1.3, using node 2 as reference.

13.1.6 Write the node equations, assuming input signals of the form e^{st}, for the network in Figure 13.3, using node 4 as a reference.

13.1.7 Repeat Exercise 13.1.6, using node 6 as reference.

13.1.8 Repeat Exercise 13.1.6, using node 1 as reference.

13.1.9 Write the node equations in differential-equation form for the network in Figure 13.2, using node 2 as the reference.

13.1.10 Repeat the Exercise 13.1.9, using node 3 as reference.

13.1.11 Suppose in the network of Figure 13.1 that i_2 is a voltage-controlled current source $i_2 = Gv_{35}$, where G is a constant with dimension mhos. Using node 5 as reference, write the node equations. Is y symmetric?

13.1.12 For Figure 13.3, add a current source i_2 connected from node 7 to node 5, with the reference arrow toward node 5. Suppose this i_2 is a controlled source $i_2 = Ki_{23}$. Write the node equations using node 7 as reference and $i_1(t) = I_1 e^{st}$.

13.2 Formulation of linear time-invariant network equations on the loop basis

The next topic in this chapter is the formulation of network equations using loop-current variables. As we have discussed in the previous chapter, $b - n + 1$ loop-current variables when properly chosen are sufficient to describe the behavior of a connected network. Furthermore, we can write $b - n + 1$ linearly independent KVL equations. This gives $b - n + 1$ equations in $b - n + 1$ unknowns if we express each branch voltage in terms of the loop-current variables and the appropriate branch characteristics. Again, we illustrate by examples.

EXAMPLE 13.2.1

Consider the linear time-invariant network shown in Figure 13.5 with the three loop currents indicated. We apply KVL around these same three loops and make use of the branch constraints. First, let us apply KVL around the loop 1236. Thus we have

$$R_1(i_a - i_c) + L_1 \frac{d}{dt}(i_a - i_c) + \frac{1}{C}\int_0^t (i_a - i_b)\,d\tau + v_{36}(0) - v_1 = 0. \tag{13.37}$$

For the loop 3456, KVL yields

$$\frac{1}{C}\int_0^t (i_b - i_a)\,d\tau - v_{36}(0) + R_2(i_b - i_c) + R_3 i_b + L_2 \frac{di_b}{dt} = 0, \tag{13.38}$$

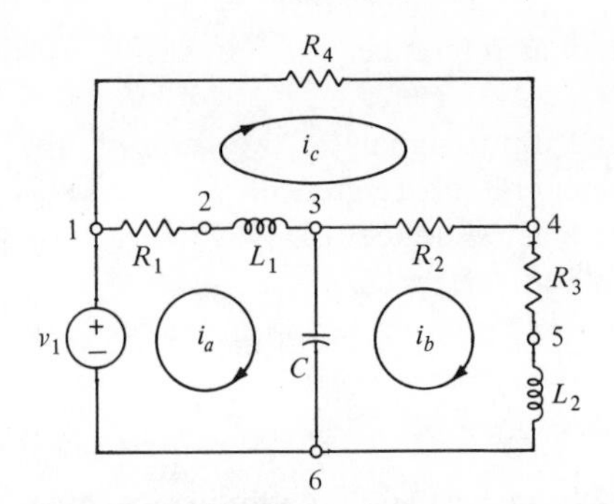

ure 13.5 The network for Example 13.2.1 with loops assigned.

and for the loop 1432, the equation is

$$R_4 i_c - R_2(i_b - i_c) + L_1 \frac{d}{dt}(i_c - i_a) + R_1(i_c - i_a) = 0. \tag{13.39}$$

Collecting terms for i_a, i_b, and i_c, assuming e^{st} signals, and neglecting the initial conditions, we get the set of equations

$$\begin{aligned} \left(L_1 s + R_1 + \frac{1}{Cs}\right) I_a - \frac{1}{Cs} I_b - (L_1 s + R_1) I_c &= V_1, \\ -\frac{1}{Cs} I_a + \left(L_2 s + R_2 + R_3 + \frac{1}{Cs}\right) I_b - R_2 I_c &= 0, \\ -(L_1 s + R_1) I_a - R_2 I_b + (R_1 + R_2 + R_4 + L_1 s) I_c &= 0. \end{aligned} \tag{13.40}$$

The array of coefficients in the loop equations is called the *loop-impedance matrix* and for the example it is as follows:

$$\mathscr{L} = \begin{bmatrix} \left(L_1 s + R_1 + \frac{1}{Cs}\right) & -\frac{1}{Cs} & -(L_1 s + R_1) \\ -\frac{1}{Cs} & \left(L_2 s + R_2 + R_3 + \frac{1}{Cs}\right) & -R_2 \\ -(L_1 s + R_1) & -R_2 & (R_1 + R_2 + R_4 + L_1 s) \end{bmatrix}. \tag{13.41}$$

Like the node-admittance matrix, we may replace s by $j\omega$ if the sources are sinusoidal and the network is in the steady-state.

XAMPLE 13.2.2

The problem in this example is to choose an appropriate set of loop-current variables for the linear time-invariant network in Figure 13.6, and to write the corresponding loop equations. We will use the chord-set method of choosing loops.

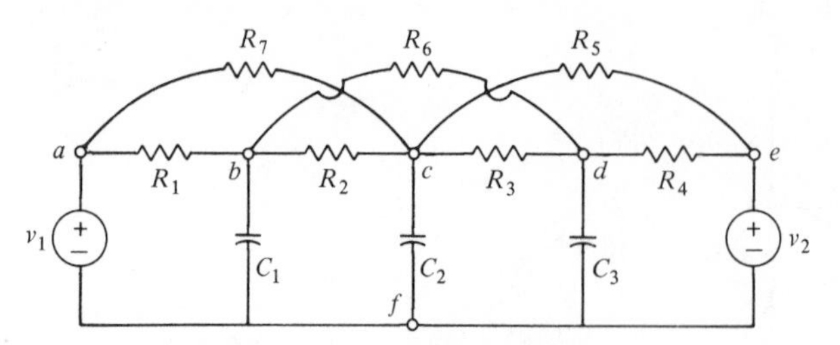

Figure 13.6 The nonplanar network for Example 13.2.2.

The network or its corresponding graph has 12 branches, and 6 nodes, so we must use 12 − 6 + 1 or 7 loop variables. A tree of the graph corresponding to the network is shown in Figure 13.7. The choice of loop variables is also shown. As discussed in the last chapter, these chord-set loops are generated by replacing the chords one at a time. The 7 loops are $R_1C_1V_1$, $C_1R_2C_2$, $C_2R_3C_3$, $C_3R_4V_2$, $R_5V_2C_2$, $R_6C_3C_1$, and $R_7C_2V_1$. Applying KVL around these loops, and using the linear branch constraint equations, the following set of equations is obtained:

$$\left(R_1 + \frac{1}{C_1s}\right)I_1 - \frac{1}{C_1s}I_2 + 0 + \cdots + 0 + 0 - \frac{1}{C_1s}I_6 + 0 = V_1, \tag{13.42}$$

$$-\frac{1}{C_1s}I_1 + \left(\frac{1}{C_1s} + \frac{1}{C_2s} + R_2\right)I_2 - \frac{1}{C_2s}I_3 + 0 - \frac{1}{C_2s}I_5 + \frac{1}{C_1s}I_6 + \frac{1}{C_2s}I_7 = 0, \tag{13.43}$$

$$0 - \frac{1}{C_2s}I_2 + \left(\frac{1}{C_2s} + \frac{1}{C_3s} + R_3\right)I_3 - \frac{1}{C_3s}I_4 + \frac{1}{C_2s}I_5 + \frac{1}{C_3s}I_6 - \frac{1}{C_2s}I_7 = 0, \tag{13.44}$$

$$0 + 0 - \frac{1}{C_3s}I_3 + \left(\frac{1}{C_3s} + R_4\right)I_4 + 0 - \frac{1}{C_3s}I_6 + 0 = -V_2, \tag{13.45}$$

$$0 - \frac{1}{C_2s}I_2 + \frac{1}{C_2s}I_3 + 0 + \left(\frac{1}{C_2s} + R_5\right)I_5 + 0 - \frac{1}{C_2s}I_7 = -V_2, \tag{13.46}$$

$$-\frac{1}{C_1s}I_1 + \frac{1}{C_1s}I_2 + \frac{1}{C_3s}I_3 - \frac{1}{C_3s}I_4 + 0 + \left(\frac{1}{C_1s} + \frac{1}{C_3s} + R_6\right)I_6 + 0 = 0, \tag{13.47}$$

$$0 + \frac{1}{C_2s}I_2 - \frac{1}{C_2s}I_3 + 0 - \frac{1}{C_2s}I_5 + 0 + \left(\frac{1}{C_2s} + R_7\right)I_7 = V_1. \tag{13.48}$$

The purpose of this example is to demonstrate how to choose an appropriate set of loop-current variables and to write the corresponding loop equations. There is no implication that the loop basis is preferable to the node basis for this example. In fact, on the basis of number of unknowns and number of equations, this example is best treated by the node method. An examination of Figure 13.6 reveals that if node f is chosen as reference, then only three node voltages need to be determined, namely, v_b, v_c, and v_d. The node voltages v_a and v_e are known, since

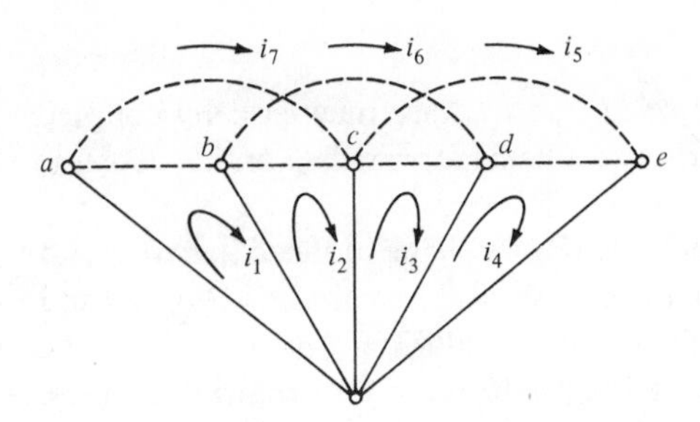

re 13.7 A tree selected for the graph of the network shown in Figure 13.6.

they are equal to the source voltages v_1 and v_2, respectively. The method of Section 13.1 may be used to write the three node equations for nodes b, c, and d.

AMPLE 13.2.3

The linear time-invariant network in Figure 13.8 contains a controlled voltage source

$$v_2(t) = [i_a(t) - i_b(t)]R,$$

where R is a constant. Applying KVL and the usual branch constraint equations, and assuming $v_1(t) = V_1 e^{st}$, we have

$$(R_1 + R_2)I_a - R_2 I_b + 0 = V_1, \tag{13.49}$$

$$-R_2 I_a + \left(R_2 + R_3 + \frac{1}{Cs}\right)I_b - \frac{1}{Cs}I_c = 0, \tag{13.50}$$

and

$$-\frac{1}{Cs}I_b + \left(Ls + R_4 + \frac{1}{Cs}\right)I_c = -R(I_a - I_b)$$

or

$$RI_a - \left(R + \frac{1}{Cs}\right)I_b + \left(Ls + R_4 + \frac{1}{Cs}\right)I_c = 0. \tag{13.51}$$

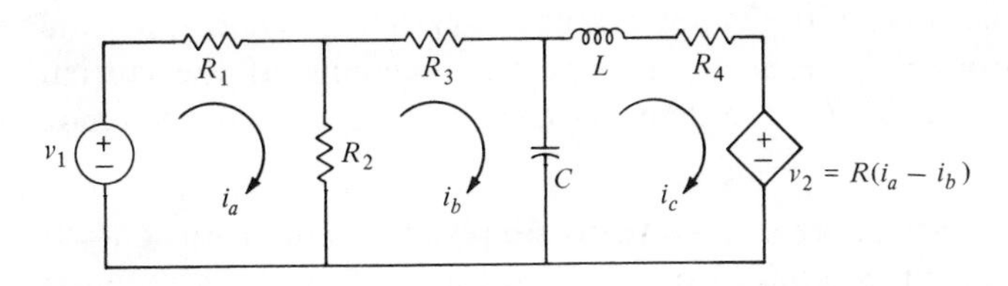

ure 13.8 The network used in Example 13.2.3.

EXAMPLE 13.2.4

Suppose that in the network of Figure 13.5, L_1 and L_2 are magnetically coupled with mutual inductance M. Let the polarity dots be at terminals 2 and 5. Assume $v_1(t) = V_1 e^{st}$.

Because of the change of i_b with time, there will be a voltage induced in coil 1 with a + polarity on terminal 2. Hence for loop a we have to add a voltage $M(di_b/dt)$ and for loop c we add $-M(di_b/dt)$. Likewise for loop b we add $M[(d/dt)(i_a - i_c)]$. The other terms are the same as those in Equation 13.40 for no mutual inductance. The final equations are as follows:

$$\begin{aligned}\left(L_1 s + R_1 + \frac{1}{Cs}\right)I_a + \left(Ms - \frac{1}{Cs}\right)I_b - (L_1 s + R_1)I_c &= V_1, \\ \left(Ms - \frac{1}{Cs}\right)I_a + \left(L_2 s + R_2 + R_3 + \frac{1}{Cs}\right)I_b - (R_2 + Ms)I_c &= 0, \\ -(L_1 s + R_1)I_a - (R_2 + Ms)I_b + (R_1 + R_2 + R_4 + L_1 s)I_c &= 0.\end{aligned} \tag{13.52}$$

In all the above examples, our goal is to write KVL in the form

$$\begin{aligned} z_{11}I_1 + z_{12}I_2 + z_{13}I_3 + \cdots + z_{1m}I_m &= V_1, \\ z_{21}I_1 + z_{22}I_2 + z_{23}I_3 + \cdots + z_{2m}I_m &= V_2, \\ &\vdots \\ z_{m1}I_1 + z_{m2}I_2 + z_{m3}I_3 + \cdots + z_{mm}I_m &= V_m. \end{aligned} \tag{13.53}$$

The basic procedure involved is the writing of KVL equations around the same loops traversed by the assigned loop-current variables. As in the node method, z_{ik} is simply determined if the network contains only R, L, and C elements, and independent voltage sources. For this special case, $z_{ik} = z_{ki}$, and the impedance coefficients z_{ik} are obtained using the following rules:

(1) If $k = i(z_{ik} = z_{kk})$, z_{kk} may be calculated by adding up all the impedances in loop k.

(2) If $k \neq i$, z_{ik} is the negative of the impedance common to loops i and k, if i_i and i_k are directed through the element in opposite directions. If the current directions are the same through the elements, then z_{ik} is equal to the common impedance.

The entries in the $\mathscr{V}$ matrix are also obtained simply. The kth element V_K is equal to the sum of the source-voltage rises minus the sum of the source-voltage drops in the kth loop, following the reference orientation of the kth loop current.

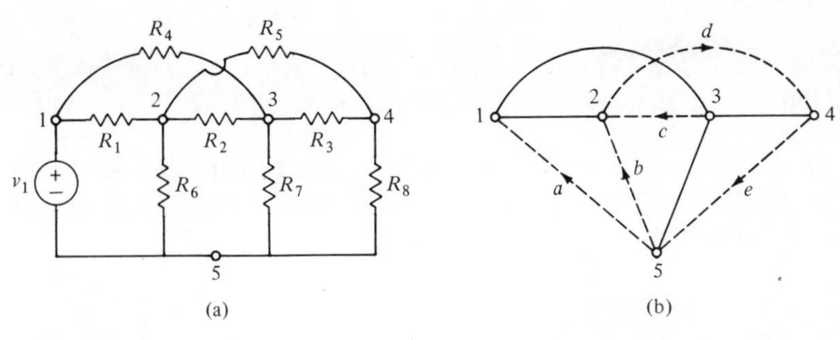

Figure 13.9

ERCISES

13.2.1 If the reference direction of i_b in Figure 13.5 is reversed, write the loop equations in phasor form. Form the loop-impedance matrix.

13.2.2 Repeat Exercise 13.2.1 if i_c is reversed instead of i_b.

13.2.3 A tree for the network in Figure 13.9(a) is shown in Figure 13.9(b). Chords are shown dotted and arrows and letters are also indicated. Using the variables i_a, i_b, i_c, i_d, and i_e, write an appropriate set of loop equations.

13.2.4 Repeat Exercise 13.2.3 for the chord set in Figure 13.10.

13.2.5 A tree for the network in Figure 13.6 is shown in Figure 13.11. The chords are shown dotted with arrows and numbers also are indicated. Using the variables i_0 to i_6, write an appropriate set of loop equations in differential-equation form.

13.2.6 Suppose the source v_2 in the network of Figure 13.8 is a current-controlled voltage source $v_2 = Ri_b$, where R is a constant of dimension ohms, instead of the one indicated. Assuming that the signals are of the form e^{st}, find the loop-impedance matrix corresponding to the choice of loop variables as shown. Is the matrix symmetric?

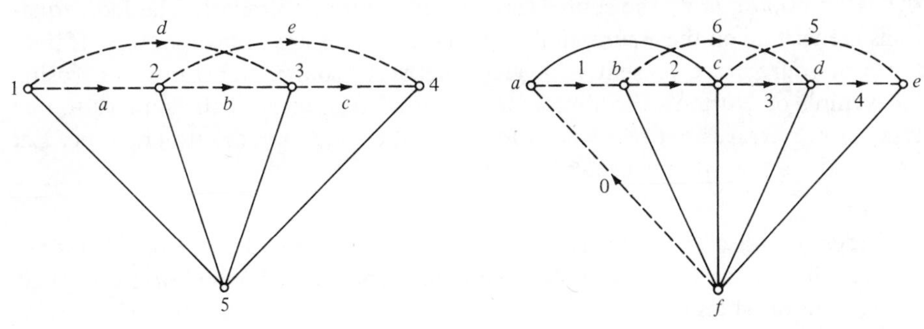

Figure 13.10

Figure 13.11

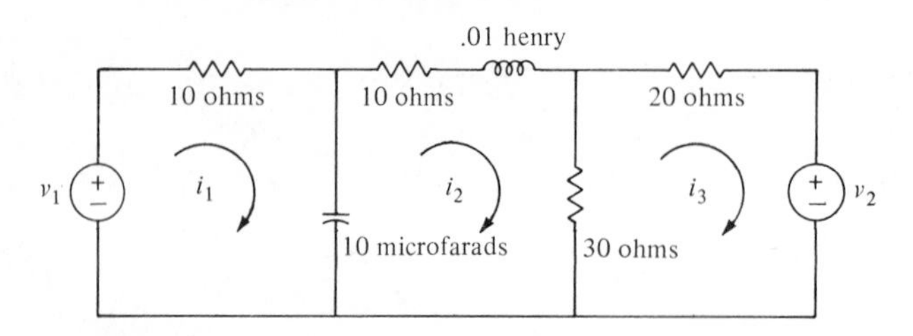

Figure 13.12

13.2.7 For the linear time-invariant network in Figure 13.12, we have

$$v_1(t) = 10 \sin 1000t \qquad \text{and} \qquad v_2(t) = 5 \sin (1000t - 30°).$$

Using the loop variables indicated, show that the loop equations in phasor form are as follows:

$$\begin{aligned} (10 - j100)I_1 + j100I_2 + 0I_3 &= 10, \\ j100I_1 + (10 - j90)I_2 - 30I_3 &= 0, \\ 0I_1 - 30I_2 + 50I_3 &= -5e^{-j30°}. \end{aligned}$$

13.2.8 Suppose that in the linear time-invariant network of Figure 13.5 we insert a controlled voltage source v_2 in series with R_4, such that $v_2 = R(i_a - i_b)$ with a + polarity towards node 4. Assume also that L_1 and L_2 are magnetically coupled with dots at terminals 3 and 5. Write loop equations for $v_1(t) = V_1 e^{st}$.

13.3 Linear time-invariant networks with mixed sources

In the two preceding sections, the formulation of node equations is carried out on the assumption that all the sources are current sources. Similarly, the loop equations are written on the assumption that all the sources are voltage sources. If both types of sources are present in a given network, source transformation (using Thevenin's or Norton's theorems) may be applied to convert the sources to one type. However, it is not necessary to transform all the sources into one type. Let us consider a few specific examples again.

EXAMPLE 13.3.1

Consider the linear time-invariant network of Figure 13.13. We note that since there are five nodes, four node voltages are independent. Using node e as reference, the voltage variables are $v_{ae}, v_{be}, v_{ce}, v_{de}$. However, $v_{ae} = v_1$ is a known quantity, so that there are only three unknowns. Hence three independent equations should

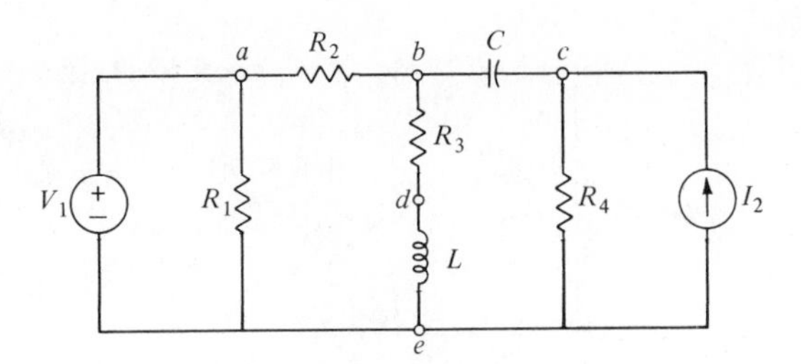

ure 13.13 Figure showing the network used in Example 13.3.1.

suffice. We may apply KCL at nodes b, c, and d, and in phasor form we obtain

$$\left(\frac{1}{R_2} + \frac{1}{R_3} + j\omega C\right) V_{be} - j\omega C V_{ce} - \frac{1}{R_3} V_{de} - \frac{1}{R_2} V_1 = 0, \qquad (13.54)$$

$$-j\omega C V_{be} + \left(j\omega C + \frac{1}{R_4}\right) V_{ce} + 0 = I_2, \qquad (13.55)$$

$$-\frac{1}{R_3} V_{be} + 0 + \left(\frac{1}{R_3} + \frac{1}{j\omega L}\right) V_{de} = 0. \qquad (13.56)$$

Since V_1 is known, we may transpose V_1/R_2 in Equation 13.54 to the right, and the equations will be in the same form as in Section 13.1. If voltage sources appear in series with elements, there is no difficulty since they can be transformed into current sources in parallel with the impedances.

AMPLE 13.3.2

Consider the network in Figure 13.14 which is the same as the one in Figure 13.13 except that a voltage source is inserted in branch bc. First we transform the branch bc into a capacitor in parallel with a current source. Then using node e as reference, KCL applied at nodes b, c, and d yields

$$\left(\frac{1}{R_2} + \frac{1}{R_3} + j\omega C\right) V_{be} - j\omega C V_{ce} - \frac{1}{R_3} V_{de} = \frac{1}{R_2} V_1 + j\omega C V_2, \qquad (13.57)$$

$$-j\omega C V_{be} + \left(\frac{1}{R_4} + j\omega C\right) V_{ce} + 0 = I_2 - j\omega C V_2, \qquad (13.58)$$

$$-\frac{1}{R_3} V_{be} + 0 + \left(\frac{1}{R_3} + \frac{1}{j\omega L}\right) V_{de} = 0. \qquad (13.59)$$

Once V_{be}, V_{ce}, and V_{de} are found, then V_{fe} is obtained simply as $V_{fe} = V_{ce} + V_2$.

Similarly, in the application of KVL to obtain loop equations, source transformation may be used to convert all the sources into voltage sources. However,

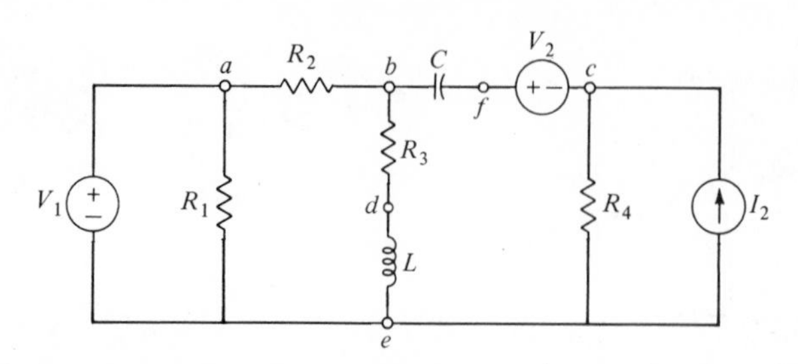

Figure 13.14 The network of Example 13.3.2.

the current sources which are not directly in parallel with any impedance are not directly transformable. In this case, it is best to leave these sources alone.

EXAMPLE 13.3.3

Consider the network in Figure 13.15. Suppose we wish to write loop equations. We note that I_1 cannot be transformed directly. By inspection there are three meshes or windows, but since I_1 is known, only I_a and I_b are unknown. Hence we need to write only two independent equations. Applying KVL to loops a and b, we have

$$\left(R_1 + \frac{1}{j\omega C}\right)I_a - \frac{1}{j\omega C}I_b - R_1 I_1 = V_1, \tag{13.60}$$

$$-\frac{1}{j\omega C}I_a + \left(j\omega L + R_2 + \frac{1}{j\omega C}\right)I_b - j\omega L I_1 = 0. \tag{13.61}$$

Since I_1 is known, the last terms on the left-hand side of these equations may be transposed to the right to reduce them to standard form.

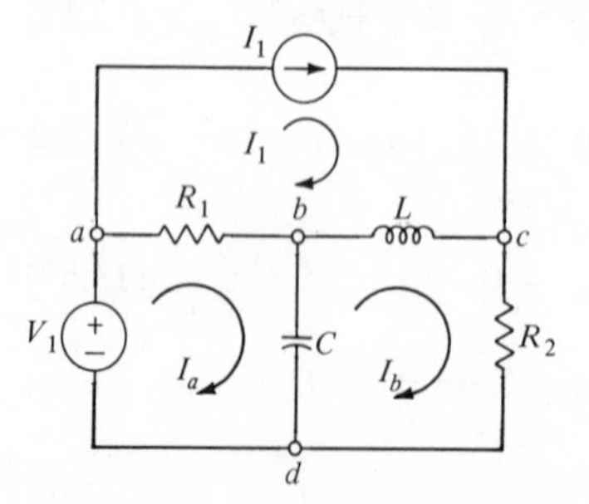

Figure 13.15 The network for Example 13.3.3.

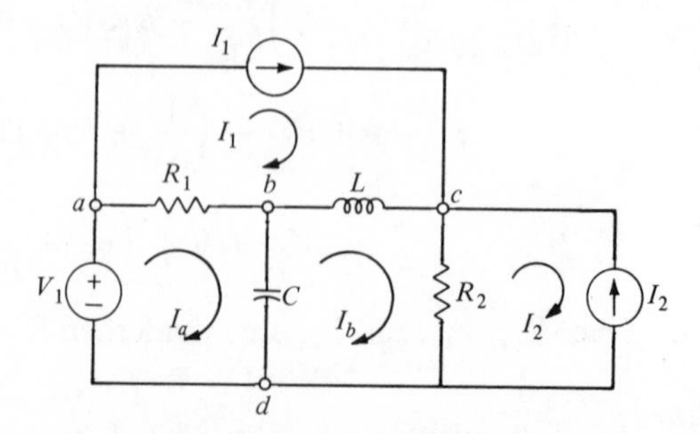

Figure 13.16 Network for Example 13.3.4.

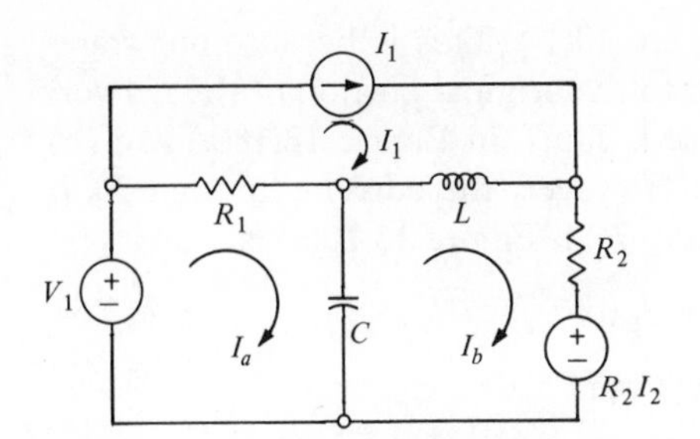

e 13.17 The network of Figure 13.16 with I_2 and R_2 transformed into $R_2 I_2$ and R_2.

MPLE 13.3.4

The network in Figure 13.16 is the same as the one in Figure 13.15 except that a current source I_2 is added in parallel with R_2. Again, since I_1 and I_2 are known, we need to write only two loop equations. They are as follows:

$$\left(R_1 + \frac{1}{j\omega C}\right) I_a - \frac{1}{j\omega C} I_b = V_1 + R_1 I_1, \tag{13.62}$$

$$-\frac{1}{j\omega C} I_a + \left(j\omega L + R_2 + \frac{1}{j\omega C}\right) I_b = j\omega L I_1 - R_2 I_2. \tag{13.63}$$

The $R_1 I_1$ and $R_2 I_2$ terms have already been transposed to the right in the above equations, and I_a and I_b may be determined from the two simultaneous equations. Alternatively, we may apply a source transformation on I_2 first and obtain the network in Figure 13.17. Writing two loop equations for loop currents I_a and I_b, we again obtain Equations 13.62 and 13.63. Suppose that after we have obtained I_a and I_b for Figure 13.17 we wish to determine the current through R_2 in Figure 13.16. A point of caution here is that the current through R_2 of Figure 13.16 is not

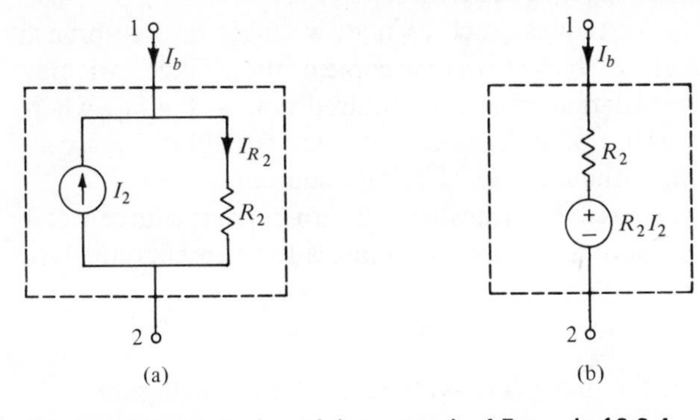

re 13.18 The transformed portion of the network of Example 13.3.4.

the same as the current through R_2 of Figure 13.17. This is because our transformed portion of the network is equivalent to the original portion of the network only insofar as external behavior is concerned. Thus, in Figure 13.18(a) and (b), the V_{12}'s are equal and the I_b's are equal. However, the current through R_2 in Figure 13.18(b) is I_b, but the current through R_2 in Figure 13.18(a) is

$$I_{R_2} = I_b + I_2. \tag{13.64}$$

Likewise, from Figure 13.18(a), the voltage V_{12} is

$$V_{12} = R_2(I_b + I_2) \tag{13.65}$$

and not simply $R_2 I_b$. Equation 13.65 is also obtained by examining Figure 13.18(b), which shows the voltage V_{12} to be

$$V_{12} = R_2 I_b + R_2 I_2, \tag{13.66}$$

which is the same as Equation 13.65. Similar caution should be observed when dealing with voltage-to-current source transformation as in Example 13.3.2.

The following rules summarize the important steps in analyzing connected networks with mixed sources.

(A) If a formulation on the loop basis is desired:
  (1) Choose a set of mesh-current variables or chord-set variables so that the current sources become some of the mesh variables or chord-set variables.
  (2) Apply KVL to the loops which do not correspond to the current sources. The number of KVL equations that are required is $(b - n + 1) - n_i$, where $b - n + 1$ is the usual number of meshes or chords and n_i is the number of current sources.
  (3) If some of the current sources are transformed into voltage sources before applying 1, care should be taken in computing currents or voltages for elements which are involved in the transformation (see Example 13.3.4).

(B) If a formulation on the node basis is desired:
  (1) Choose a set of voltage variables (such as node voltages or tree-branch voltages) so that the voltage sources become some of the voltage variables.
  (2) The number of KVL equations that are required is $n - 1 - n_v$, where $n - 1$ is the usual number of independent voltage variables for a connected network, and n_v is the number of voltage sources.
  (3) If some of the voltage sources are transformed into current sources, care should be taken in computing currents or voltages for elements involved in the transformation.

EXAMPLE 13.3.5

As a final example, we consider the linear time-invariant network of Figure 13.19 with two voltage sources and a current source. Suppose we wish to write node

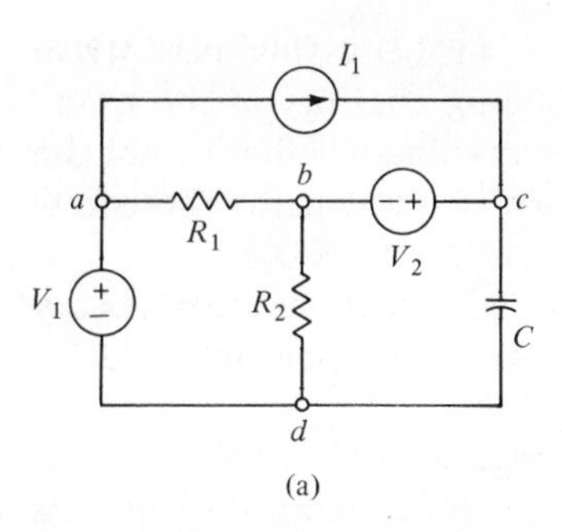

(a)

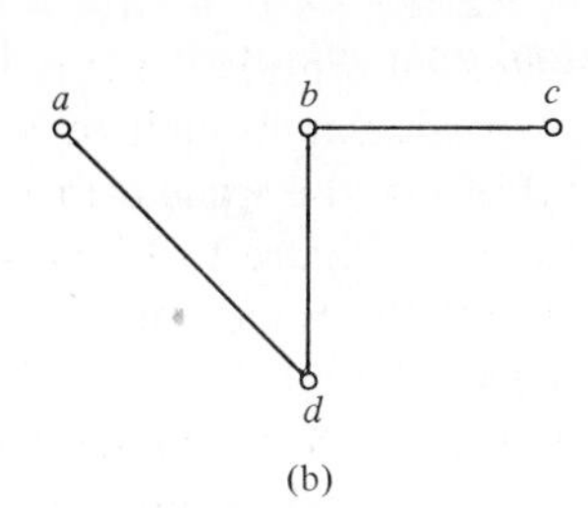

(b)

re 13.19 The network of Example 13.3.5.

equations. Since there are four nodes, three voltages are required to characterize the network. Since two voltage sources are present, only one voltage remains to be computed. Figure 13.19(b) shows a tree for the network. If we choose tree-branch voltages as variables, we notice that v_{ad} and v_{bc} correspond to the two voltage sources. So we pick v_{bd} as the unknown voltage variable.

Writing KCL at node b, we have

$$\frac{V_{bd}}{R_2} + I_{ba} + I_{bc} = 0. \tag{13.67}$$

But we see that

$$I_{ba} = \frac{V_{bd} - V_1}{R_1} \tag{13.68}$$

and

$$I_{bc} = -I_1 + I_{cd} = -I_1 + j\omega C V_{cd} = -I_1 + j\omega C(V_{bd} + V_2). \tag{13.69}$$

Hence we have

$$\frac{V_{bd}}{R_2} + \frac{V_{bd} - V_1}{R_1} - I_1 - j\omega C(V_{bd} + V_2) = 0$$

or

$$\left(\frac{1}{R_2} + \frac{1}{R_2} + j\omega C\right) V_{bd} = \frac{V_1}{R_1} + I_1 - j\omega C V_2. \tag{13.70}$$

RCISES

13.3.1 Suppose that in Example 13.3.2 with the network transformed, and the variables V_{be}, V_{ce}, and V_{de} determined from Equations 13.57, 13.58, and 13.59, we wish to determine the current through C in the *original* network. Find an expression for this current. Is this the same as the current through C in the transformed network?

13.3.2 Are the currents through R_1, R_2, R_3, and R_4 in the original network in Example 13.3.2 the same as the corresponding currents in the transformed network? Formulate a rule for determining whether or not the current in a particular branch of a network is the same as the corresponding current in the transformed network.

13.3.3 For the network in Figure 13.15, how many node equations are necessary to determine all the node-pair voltages? Write an appropriate set of node equations.

13.3.4 Repeat Exercise 13.3.3 for the network in Figure 13.16.

13.3.5 How many loop equations are necessary for determining the currents in the network of Figure 13.13? Write an appropriate set of loop equations.

13.4 State-variable method

In the nodal method, we use node-voltage variables in writing KCL equations, while in the loop method we use loop-current variables in writing KVL equations. In either case, both derivative terms and integral terms appear in general, resulting in a set of integro-differential equations. If the steady-state response for sinusoidal inputs is desired, the phasor approach may be used in conjunction with either the nodal method or loop method, thus converting the integro-differential equations into algebraic phasor equations. For such steady-state applications, the nodal method enjoys widespread preference because in most of the networks considered the number of nodes is less than the number of loops. Digital computer programs for sinusoidal steady-state analysis based on the nodal method are widely available.

The integro-differential equations in either the loop method or the node method may be converted to a set of second-order differential equations by simply differentiating all equations containing integrals. If a particular loop or node variable is desired, the other variables may be eliminated, and an nth-order differential equation is obtained. If either the zero-input response or the total network response is desired, it is necessary to know the initial conditions, $y(0), y'(0), \ldots, y^{(n-1)}(0)$, where y is the loop or node variable being studied. As discussed in Chapters 7 and 8, the initial state of the network is usually known in terms of initial voltages across capacitors and initial currents through inductors. A preliminary network analysis is necessary to determine the initial conditions from the known capacitor voltages and inductor currents. This preliminary analysis must be repeated for each different loop variable or node variable for which a solution is sought.

When the effects of initial conditions are to be taken into account, it is advantageous to use capacitor voltages and inductor currents as the network variables, with no other variable appearing except for the input. Such variables are called *state variables*. The identification of state variables as the i_L's and v_C's for a network is a specific case of a more general concept; state variables are widely used to study mechanical and other kinds of systems. The equations involving the state variables are written by applying KVL and KCL in a systematic way that will be described. The standard form of the *state equations*, as we will see, is a set of first-order differential equations in which the derivative of each state variable is expressed as a function of the state variables and the inputs and possibly input derivatives. For linear networks, the functions are linear.

AMPLE 13.4.1

Consider the simple series RLC network of Figure 11.5 (see page 282). The state variables that will be used to describe the network are v_C and i_L. Applying KVL around the loop, we obtain

$$-v + Ri_L + L\frac{di_L}{dt} + v_C = 0. \tag{13.71}$$

This equation may be rearranged in the following form:

$$\frac{di_L}{dt} = -\frac{R}{L}i_L - \frac{1}{L}v_C + \frac{1}{L}v. \tag{13.72}$$

Next, we apply KCL to the junction between L and C to obtain

$$C\frac{dv_C}{dt} - i_L = 0 \tag{13.73}$$

or

$$\frac{dv_C}{dt} = \frac{1}{C}i_L. \tag{13.74}$$

The two equations we have obtained, Equations 13.72 and 13.74, constitute the state equations of the network in standard form, where v is the input and i_L and v_C are the state variables. The initial conditions that will be required are $i_L(0)$ and $v_C(0)$. Digital computer programs are widely available for the solution of sets of first-order differential equations such as those we have found in Equations 13.72 and 13.74.

AMPLE 13.4.2

For the network of Figure 11.9 (see page 288), let us assign a downward reference direction to the current i_L of the inductor. The voltage across the capacitor is v as shown in the figure. We choose i_L and v as the two state variables for the network. Applying KVL around the RLC loop, we obtain

$$-v + L\frac{di_L}{dt} + Ri_L = 0 \tag{13.75}$$

and applying KCL to the top node yields

$$-i + C\frac{dv}{dt} + i_L = 0. \tag{13.76}$$

The two equations may be rearranged to the standard form

$$\frac{di_L}{dt} = -\frac{R}{L}i_L + \frac{1}{L}v \tag{13.77}$$

and

$$\frac{dv}{dt} = -\frac{1}{C}i_L + \frac{1}{C}i. \tag{13.78}$$

It is not always possible to write the state equations by inspection, where the only variables that appear are the state variables and input variables. Sometimes it is necessary to eliminate intermediate variables. We will shortly state a general guide for a systematic procedure for writing state equations for a class of networks. We consider only linear time-invariant networks whose graphs are connected and whose elements are resistors, inductors, capacitors, and independent sources. We exclude networks which have loops of capacitors and voltage sources or cut sets of inductors and current sources. We also assume that each independent voltage source is in series with a resistor, capacitor, or inductor, and that each independent current source is in parallel with a resistor, capacitor, or inductor. Thevenin's theorem or Norton's theorem may be used to advantage in modifying the networks to meet these specifications, if necessary. In defining the branches, the sources are assumed to be part of the elements that are connected in series or parallel with them. For example, a voltage source and a resistor in series will be considered as one branch. However, a resistor and a capacitor in series will be considered as separate branches.

With those qualifications, we are now prepared to state a *procedure for writing network state equations.*

Step 1. Choose a tree containing all capacitors and none of the inductors. There may be several such trees which qualify. Each is called a *proper tree*. Pick only one. Observe that all inductors are always links of a proper tree. Sources in the tree must be current sources and sources in the links must be voltage sources. Apply Thevenin's theorem if necessary to modify the sources.

Step 2. Assign reference directions for all branch variables. Choose capacitor voltages and inductor currents as the state variables.

Step 3. Each tree branch of a given tree uniquely defines a cut set which contains that tree branch and some links but no other tree branches. This cut set is called the *fundamental cut set* associated with the tree branch of that tree. Identify the fundamental cut sets of the proper tree chosen in step 1.

Step 4. Apply KCL to the fundamental cut sets associated with each capacitor. If convenient, express each branch current of the fundamental cut set in terms of the state variables chosen in step 2. Each equation contains exactly one time derivative of a capacitor voltage. The other terms of each equation constitute a linear combination of state variables and possibly link resistor currents.

Step 5. Each link of a given tree uniquely defines a loop containing that link and some tree branches but no other links. This loop is called the *fundamental loop* associated with the link of that tree. Identify the fundamental loops of the proper tree chosen in step 1.

Step 6. Apply KVL to the fundamental loop associated with each inductor. If convenient, express each branch voltage of the fundamental loop in terms of the state variables. Each equation contains exactly one time derivative of an inductor current. The other terms of each equation constitute a linear combination of state variables, inputs, and possibly tree-branch resistor voltages.

Step 7. If the equations in steps 4 and 6 contain no link resistor currents and no tree-branch resistor voltages, rewrite the equations in standard form. Otherwise proceed to step 8.

Step 8. Apply KCL to the fundamental cut set associated with each tree-branch resistor. Each equation expresses a tree-branch resistor current as a linear combination of state variables, inputs, and link resistor voltages.

Step 9. Apply KVL to the fundamental loop associated with each link resistor. Each equation expresses a link resistor voltage as a linear combination of state variables, inputs, and tree-branch resistor currents.

Step 10. Solve for tree-branch resistor currents and link resistor voltages in terms of state variables and inputs, from the equations of steps 8 and 9, and substitute into the equations of steps 4 and 6. This leaves state variables and inputs as the only variables, with each equation containing only one derivative term. Rewrite the equations in standard form.

MPLE 13.4.3

Consider the network in Figure 13.2. The object in analyzing this network is to write state equations in standard form. First, we pick a proper tree. We include C_1 in the tree and exclude L_1. Next we add resistors to complete the tree. Choose R_2, R_3, and i_2. The proper tree that has been selected is drawn in heavy lines in Figure 13.20. Observe that R_3 and i_2 are treated as a single branch. Step 1 in the

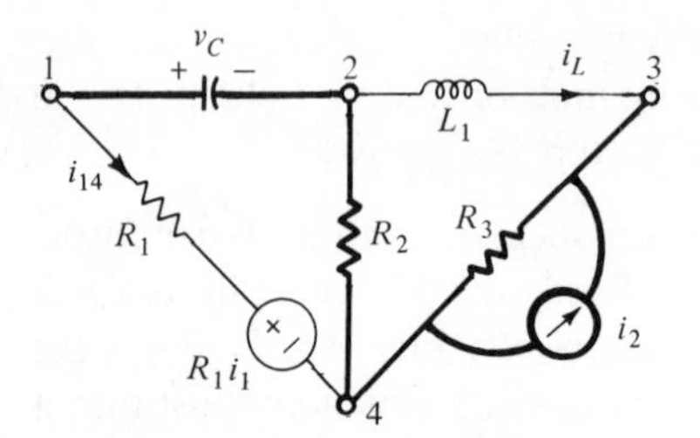

Figure 13.20 Network for Example 13.4.3.

procedure we have given suggests converting sources in the links to voltage sources. The branch containing i_1 and R_1 which is a link is modified accordingly as shown in Figure 13.20. Choose v_C and i_L as state variables. Applying KCL to the fundamental cut set associated with C_1, and noting that the cut-set branches are 12 and 14, we have

$$C_1 \frac{dv_C}{dt} + \frac{v_{14} - R_1 i_1}{R_1} = 0. \tag{13.79}$$

This equation contains v_{14}, a link resistor voltage which we will eliminate later. Applying KVL to the fundamental loop associated with L_1 and noting that the loop branches are 23, 34, and 42, we have

$$L_1 \frac{di_L}{dt} + (i_L + i_2)R_3 - R_2 i_{24} = 0. \tag{13.80}$$

This equation contains i_{24}, a tree-branch resistor current which we will eliminate later. Next apply KCL to the cut set associated with R_2 (the cut-set branches are 24, 14, and 23),

$$i_{24} + i_L + \frac{v_{14} - R_1 i_1}{R_1} = 0. \tag{13.81}$$

Furthermore, application of KVL to the fundamental loop associated with branch 14 yields

$$v_{14} - R_2 i_{24} - v_C = 0. \tag{13.82}$$

Solving for v_{14} and i_{24} from Equations 13.81 and 13.82, we obtain

$$i_{24} = \frac{-R_1}{R_1 + R_2} i_L - \frac{1}{R_1 + R_2} v_C + \frac{R_1}{R_1 + R_2} i_1 \tag{13.83}$$

and

$$v_{14} = \frac{-R_1 R_2}{R_1 + R_2} i_L + \frac{R_1}{R_1 + R_2} v_C + \frac{R_1 R_2}{R_1 + R_2} i_1 . \tag{13.84}$$

Substituting these in Equations 13.79 and 13.80, we obtain

$$C_1 \frac{dv_C}{dt} - i_1 - \frac{R_2}{R_1 + R_2} i_L + \frac{1}{R_1 + R_2} v_C + \frac{R_2}{R_1 + R_2} i_1 = 0 \qquad (13.85)$$

and

$$L_1 \frac{di_L}{dt} + (i_L + i_2)R_3 + \frac{R_1 R_2}{R_1 + R_2} i_L + \frac{R_2}{R_1 + R_2} v_C - \frac{R_1 R_2}{R_1 + R_2} i_1 = 0\,. \qquad (13.86)$$

Finally, rewriting the equations in standard form, we have

$$\frac{dv_C}{dt} = \frac{-1}{C_1(R_1 + R_2)} v_C + \frac{R_2}{C_1(R_1 + R_2)} i_L + \frac{R_1}{C_1(R_1 + R_2)} i_1 \qquad (13.87)$$

and

$$\frac{di_L}{dt} = \frac{-R_2}{L_1(R_1 + R_2)} v_C - \frac{1}{L_1}\left(R_3 + \frac{R_1 R_2}{R_1 + R_2}\right) i_L + \frac{R_1 R_2}{L_1(R_1 + R_2)} i_1 - \frac{R_3}{L_1} i_2\,. \qquad (13.88)$$

Given the initial conditions $v_C(0)$ and $i_L(0)$ together with inputs i_1 and i_2 for $t \geq 0$, these equations are ready to be solved. Analytical techniques are available, as well as standard digital computer subroutines.

MPLE 13.4.4

Consider the network shown in Figure 13.13, where the sources are replaced by time-varying signals v_1 and i_2. Since v_1 is in parallel with R_1, this network does not meet the specifications for the class of networks that can be analyzed by the procedure we have just explained. However, it is clear that the current through R_1, with a downward reference direction chosen, is always equal to v_1/R_1 regardless of what happens to the rest of the network. The Thevenin's equivalent network for the parallel combination consists of only v_1, and so we may replace the combination by v_1 alone, thus eliminating R_1. Next we choose a tree containing C, but not containing L. Figure 13.21 shows such a choice indicated by heavy lines.

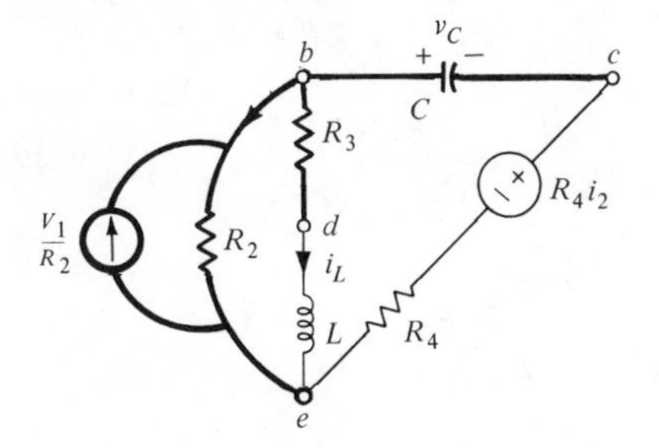

Figure 13.21 Network for Example 13.4.4.

Since the series connection of R_2 and v_1 is in the tree, it is converted to an equivalent current source in parallel with R_2 as shown in Figure 13.21. Furthermore, since the combination of i_2 and R_4 is a link branch, i_2 is transformed into an equivalent voltage source as shown. The state variables chosen are v_C and i_L.

Applying KCL to the fundamental cut set associated with C,

$$-C\frac{dv_C}{dt} + \frac{v_{ce} - R_4 i_2}{R_4} = 0. \tag{13.89}$$

Applying KVL to the fundamental loop associated with L,

$$L\frac{di_L}{dt} - R_2\left(i_{be} + \frac{v_1}{R_2}\right) + R_3 i_L = 0. \tag{13.90}$$

The variables v_{ce} and i_{be} are not state variables and so must be eliminated. Applying KCL to the fundamental cut set associated with branch *be*, we obtain

$$i_{be} + i_L + \frac{v_{ce} - R_4 i_2}{R_4} = 0. \tag{13.91}$$

Applying KVL to the fundamental loop associated with branch *ce*, we have

$$v_{ce} - R_2\left(i_{be} + \frac{v_1}{R_2}\right) + v_C = 0. \tag{13.92}$$

Solving for i_{be} and v_{ce} from Equations 13.91 and 13.92, we have

$$i_{be} = -\frac{R_4}{R_2 + R_4} i_L + \frac{1}{R_2 + R_4} v_C + \frac{R_4}{R_2 + R_4} i_2 - \frac{1}{R_2 + R_4} v_1 \tag{13.93}$$

and

$$v_{ce} = -\frac{R_2 R_4}{R_2 + R_4} i_L - \frac{R_4}{R_2 + R_4} v_C + \frac{R_2 R_4}{R_2 + R_4} i_2 + \frac{R_4}{R_2 + R_4} v_1. \tag{13.94}$$

Substituting these results into Equations 13.89 and 13.90 and then rearranging the equations in standard form, we have the final results,

$$\frac{dv_C}{dt} = \frac{-1}{C(R_2 + R_4)} v_C - \frac{R_2}{C(R_2 + R_4)} i_L + \frac{1}{C(R_2 + R_4)} v_1 - \frac{R_4}{C(R_2 + R_4)} i_2 \tag{13.95}$$

and

$$\frac{di_L}{dt} = \frac{R_2}{L(R_2 + R_4)} v_C - \frac{1}{L}\left(R_3 + \frac{R_2 R_4}{R_2 + R_4}\right) i_L + \frac{R_4}{L(R_2 + R_4)} v_1 + \frac{R_2 R_4}{L(R_2 + R_4)} i_2. \tag{13.96}$$

Any network variable can be expressed as a function of the state variables and the inputs. For example, if v_{ce} is the output, Equation 13.94 may be used to express it in terms of i_L, v_C, i_2, and v_1. Thus, after the state variables are found by integrating the last two equations, Equations 13.95 and 13.96, either v_{ce} or i_{be} can be obtained from Equations 13.92 or 13.91. If v_{be} is considered to be the output, it is seen from Figure 13.13 that

$$v_{be} = v_C + v_{ce}. \tag{13.97}$$

By algebraic manipulation of the equations already written, the output v_{be} may be obtained as a linear combination of state variables and inputs. Similarly, if v_{de} is selected as the output, it is computed from

$$v_{de} = v_{be} - R_3 i_L = v_{ce} + v_C - R_3 i_L. \tag{13.98}$$

By routine algebraic manipulation of the results already given, v_{de} may be expressed as a linear combination of state variables and inputs.

In addition to the convenience of having the initial conditions directly available in terms of specifications for the inductor and capacitor at the initial instant, the writing of state equations in standard form is important for several significant reasons:

(a) The state equations are in convenient form for analog or digital computer programming.

(b) The extension of the concepts of state equations to time-varying and/or nonlinear networks is straightforward.

(c) Modern system theory makes extensive use of models represented in standard-state form. Understanding that state concepts apply equally well to network theory and system theory provides unity, and makes the tools developed for one field available for application in the other.

ERCISES

13.4.1 For the network of Figure 13.2, draw a proper tree different from that in Figure 13.20. How many proper trees does this network have?

13.4.2 For the network of Figure 13.13, draw a proper tree different from that in Figure 13.21. How many proper trees does the network have?

13.4.3 Write the state equations in a standard form for the network of Figure 10.17(a). (See page 249.)

13.4.4 Repeat Exercise 13.4.3 for the network of Problem 10.14.

13.5 Summary

1. The variables used in the nodal method are node-to-datum voltages, where the datum or reference node is any one of the n nodes of a connected network. In an n-node connected network, there are $n - 1$ node-to-datum voltage variables. The node equations are obtained by applying KCL to each of the $n - 1$ nodes. The result is of the form

$$\mathscr{Y}\mathscr{V} = \mathscr{I},$$

where $\mathscr{Y}$ is the node-admittance matrix, $\mathscr{V}$ is the column matrix of node voltages, and $\mathscr{I}$ is the column matrix of equivalent input currents at each node.
2. The variables used in the loop method are loop-current variables, which are linked to branch-current variables. If there are b branches and n nodes in a connected network, there are $b - n + 1$ loop-current variables. The loop equations are obtained by applying KVL to the $b - n + 1$ loops. The result is of the form

$$\mathscr{Z}\mathscr{I} = \mathscr{V},$$

where $\mathscr{Z}$ is the loop-impedance matrix, $\mathscr{I}$ is the column matrix of loop-current variables, and $\mathscr{V}$ is a column matrix of equivalent input voltages in each loop. If the network is planar, mesh currents associated with the windows of the graph may be used. The number of mesh currents is also $b - n + 1$.
3. If the inputs are sinusoidal and of the same frequency, the phasor approach may be used in conjunction with either the loop method or nodal method. The resulting system of linear equations with complex-number coefficients are readily solved using standard digital computer programs.
4. In the state-variable approach, the number of state variables required is equal to the number of independent initial conditions. For connected networks with no loops containing only capacitors and voltage sources and with no cut sets containing only inductors and current sources, all the capacitor voltages and inductor currents constitute an acceptable choice of state variables. The standard form for the state equation is that of a set of first-order differential equations in which the time-derivative of each state variable is expressed as a function of the state variables and the inputs. If the network is linear and time-invariant, each function is a linear combination (with constant coefficients) of the state variables and inputs. Each branch voltage or current of the network is expressible as a linear combination of the state variables and the inputs. In some cases, derivatives of inputs may also appear.

PROBLEMS

13.1 For the linear time-invariant network of Figure 13.22, write the node

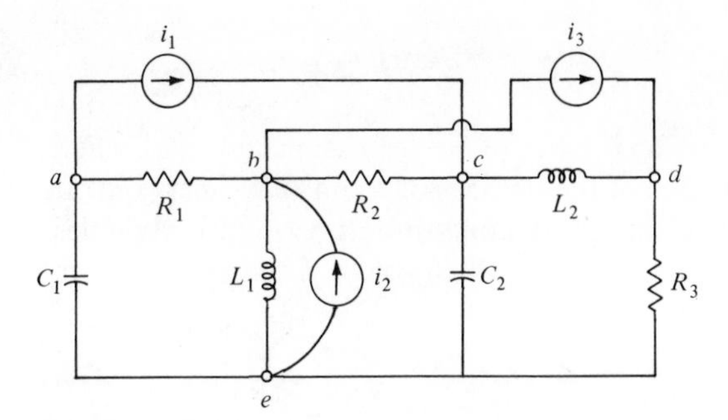

Figure 13.22

equations in matrix form

$$\mathscr{Y}\begin{bmatrix} v_1 \\ v_2 \\ v_3 \\ v_4 \end{bmatrix} = \mathscr{I},$$

where the variables are node-to-datum voltages, using node b as the reference and with $v_1 = v_{cb}$, $v_2 = v_{db}$, $v_3 = v_{ab}$, and $v_4 = v_{eb}$. Use the differential operator p in your formulation.

13.2 For the linear time-invariant network of Figure 13.23, use node 7 as the datum and make the voltage identification $v_{j7} = v_j$ for $j = 1, \ldots, 6$. Write the node equations for this network in a matrix form similar to that of Problem 13.1 with six voltage variables.

13.3 For the linear time-invariant network given in Figure 13.24, determine the differential equation for $v_1(t)$, the voltage of node 1 with respect to the datum in terms of the voltage sources v_a, v_b, and v_c. Write the equation with the coefficient of the highest-order derivative of $v_1(t)$ normalized to unity.

13.4 Repeat Problem 13.3 with $v_2(t)$ replacing $v_1(t)$. The voltage $v_2(t)$ is that of node 2 with respect to the datum.

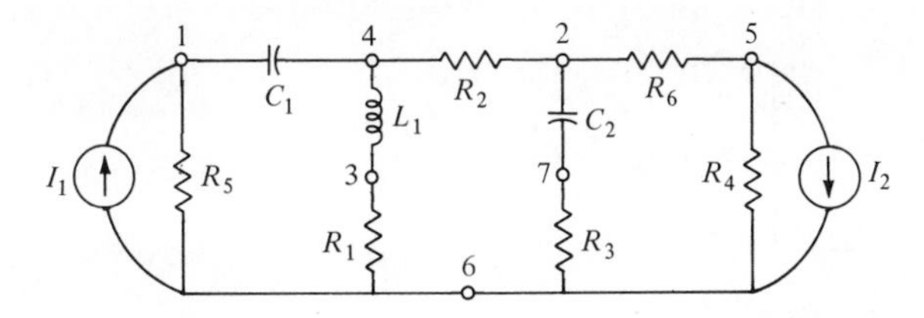

Figure 13.23

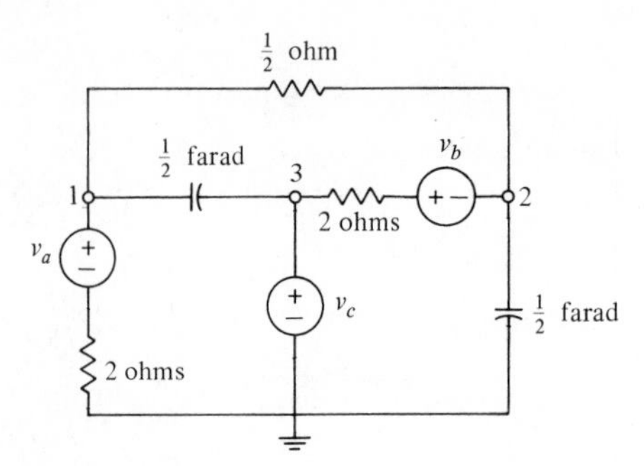

Figure 13.24

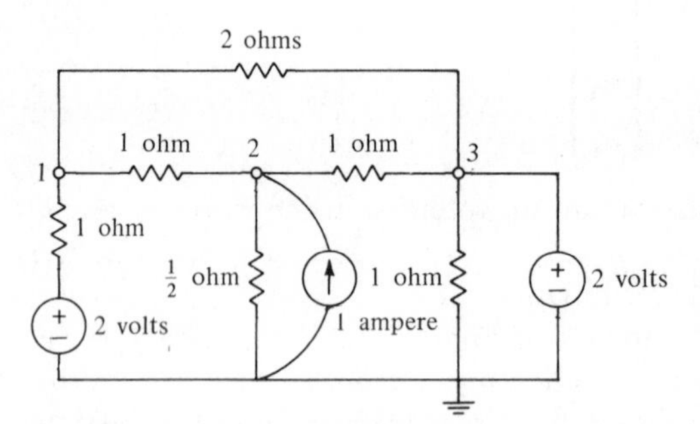

Figure 13.25

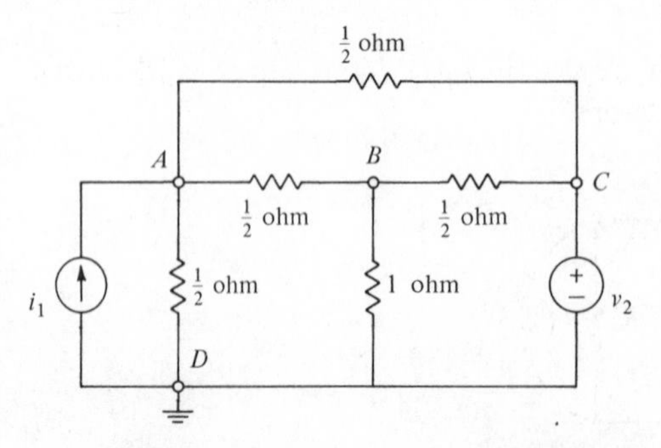

Figure 13.26

13.5 Determine the voltage at each node with respect to the datum for the linear time-invariant network of Figure 13.25. The sources are time-invariant.

13.6 For the linear time-invariant network given in Figure 13.26, determine the node-to-datum voltages in terms of $i_1(t)$ and $v_2(t)$.

13.7 The network of Figure 13.27 is operating in the steady-state. Find the rms value of the voltage at node B with respect to the datum.

13.8 Both sinusoidal sources in the network in Figure 13.28 operate at the frequency $\omega = 1$ radian/sec, and both are in phase with the reference. The values indicated are rms values of voltage. Find the steady-state component of voltage at node B with respect to the datum node.

13.9 For the conditions of Problem 13.8, determine the steady-state voltage at node B with respect to node A.

13.10 The network shown in Figure 13.29 is operating in the sinusoidal steady-state. If $v_1(t) = \cos t$, and $v_2(t) = \sin t$, find the voltage at node B with respect to that at node A.

13.11 The network of Figure 13.30 is in the steady-state and the source is sinusoidal, described by the equation, $v(t) = 2 \sin (t + 90°)$ volts. If $L_1 = L_2 = 1$ henry, $M = \frac{1}{4}$ henry, and $C = 1$ farad, find the voltage at node A with respect to the datum.

13.12 In the given network (Figure 13.31) the capacitors are charged to Q_0 each at time $t = 0^-$. What will be the time variation of the voltage at the two nodes with respect to the datum for $t > 0$? Use numerical values where possible. Assume the node voltages initially positive at $t = 0^-$.

13.13 For the network of Figure 13.32, determine the steady-state voltage at node B with respect to the datum.

13.14 Repeat Problem 13.13, solving for the steady-state voltage at node C with respect to the datum.

13.15 Figure 13.33 shows a network containing two voltage sources each turned

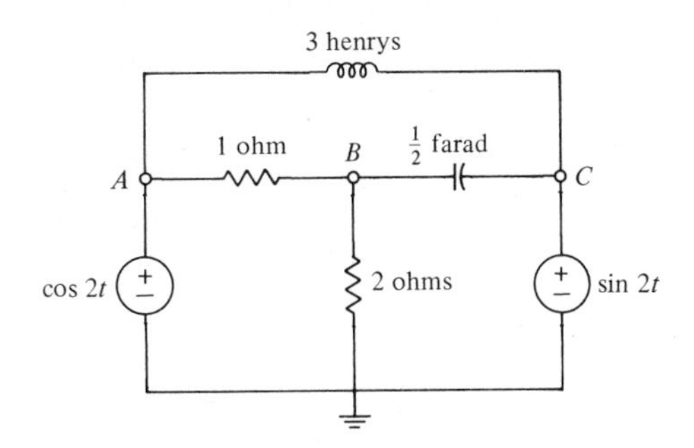

Figure 13.27

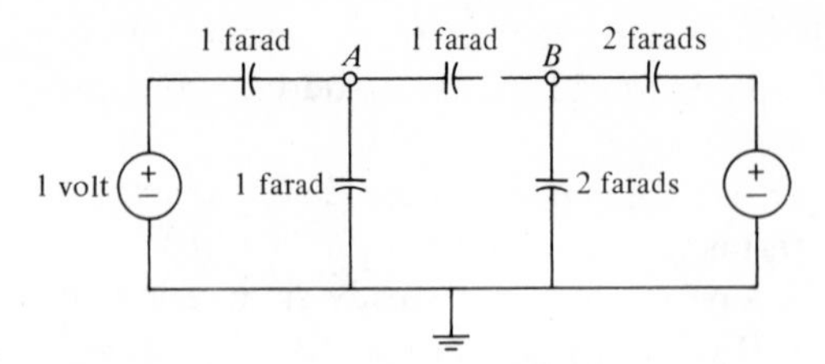

Figure 13.28

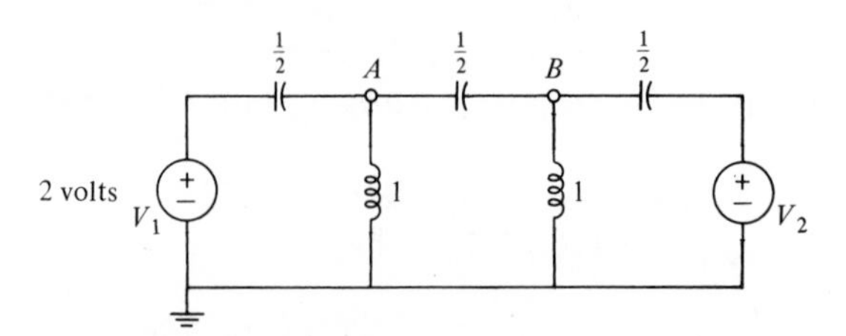

Figure 13.29

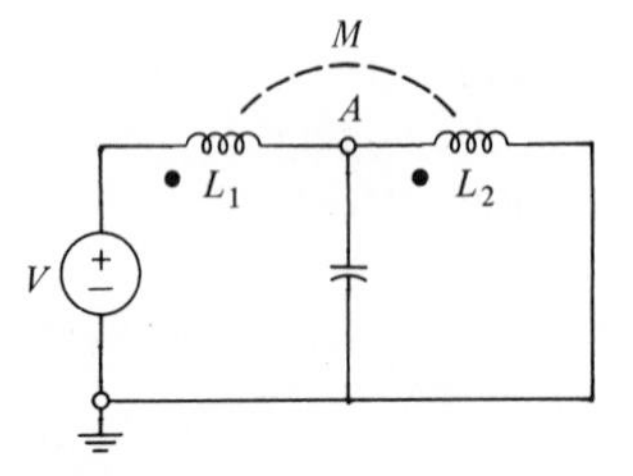

Figure 13.30

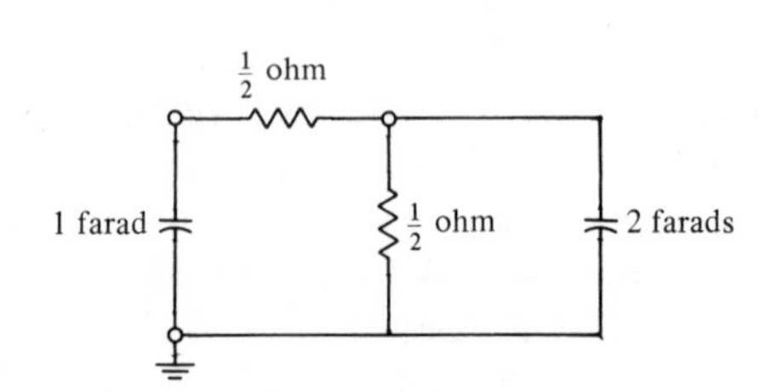

Figure 13.31

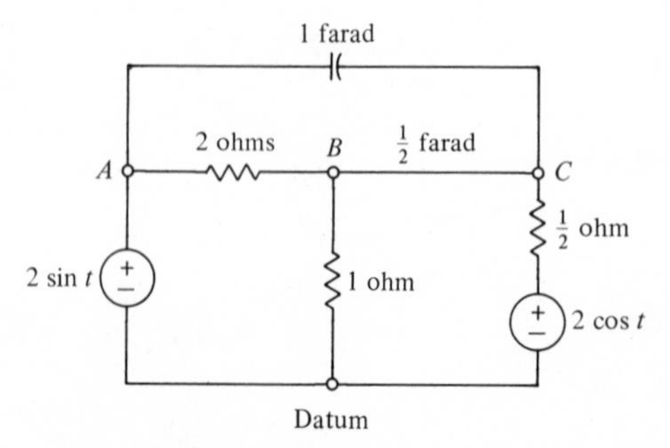

Figure 13.32

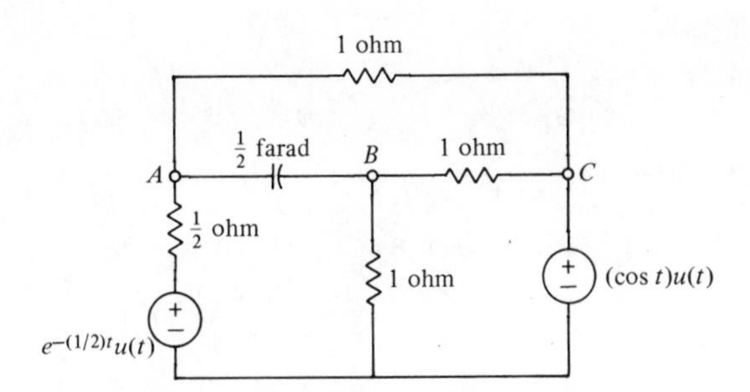

Figure 13.33

on at $t = 0$. For the element values given, find the time variation of the voltage at node B with respect to the datum. Assume that the capacitor is uncharged prior to $t = 0$.

13.16 Repeat Problem 13.15 for the voltage of node A with respect to the datum.

13.17 The current source of a linear time-invariant network in Figure 13.34(a) is shifted as in (b) and transformed as in (c). The solutions for i_a and i_b for

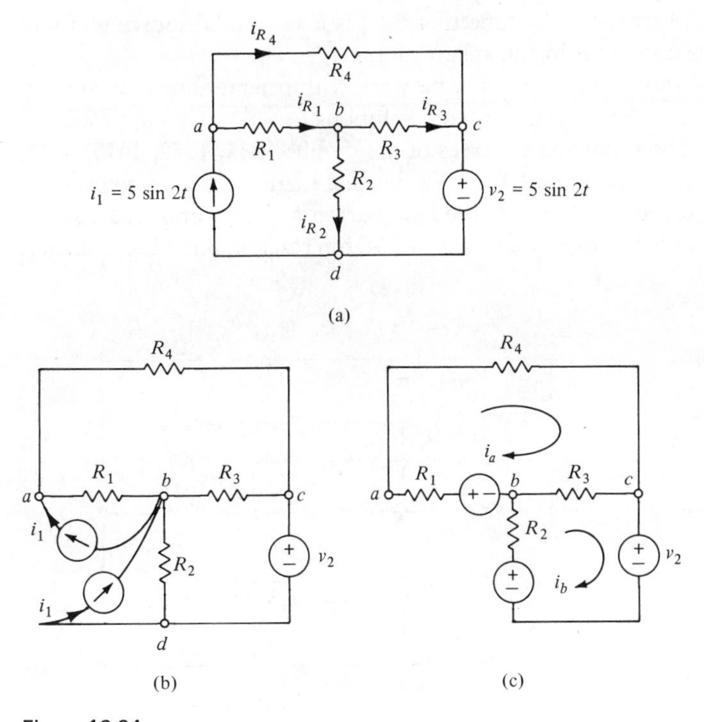

Figure 13.34

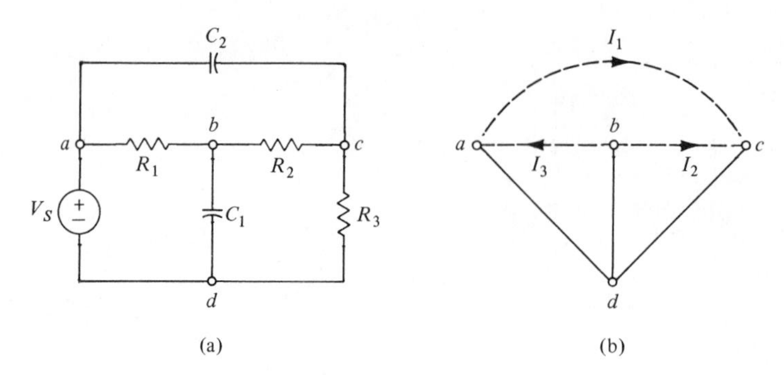

Figure 13.35

the network of (c) are $i_a(t) = i_b(t) = 2.5 \sin 2t$. Find $i_{R_1}(t)$, $i_{R_2}(t)$, $i_{R_3}(t)$, and $i_{R_4}(t)$ in the network of (a).

13.18 A tree for the linear time-invariant network in Figure 13.35(a) is shown in (b) by solid lines. The corresponding chord set is shown by dashed lines. Choosing the directions indicated on the chords for the three current variables, I_1, I_2, and I_3, write three equations which are sufficient for solving for the currents. Assume that $v_S(t)$ is a sinusoidal source and that the system is operating in the steady-state.

13.19 Figure 13.36 shows the graph of a network with orientations indicated by arrow directions. For this graph, find solutions to the following problems. (a) Which of the following are trees of the graph: 1234, 1572, 1673, 5678, 5724, 1683, 1237, 1238, and 1235; (b) write a matrix equation expressing the branch currents in terms of the loop currents; (c) let branch 1 contain a 1-volt sinusoidal source in series with a 1-ohm resistor. All other branches

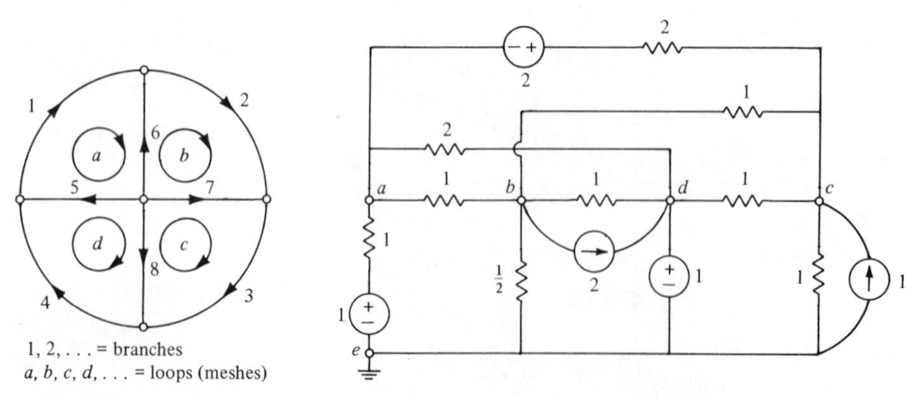

Figure 13.36

Figure 13.37

contain only a 1-ohm resistor. Write a matrix equation of the form

$$\mathscr{R}\mathscr{I}_b = \mathscr{V},$$

where $\mathscr{I}_b$ is the branch-current matrix.

13.20 In the network of Figure 13.37, all resistor values are in ohms and all sources are sinusoidal, having an rms voltage or current as given. Write a matrix equation from which all unknown voltages in the network can be determined, in the form

$$\mathscr{Y}\mathscr{V} = \mathscr{I},$$

where $\mathscr{V}$ is the node-to-datum voltage matrix.

13.21 For the network of Problem 13.20, write a matrix equation, from which all unknown currents in the network can be determined, in the form

$$\mathscr{Z}\mathscr{I}_l = \mathscr{V},$$

where $\mathscr{I}_l$ is the loop-current matrix. Also write a matrix equation which relates all branch currents to the loop currents you have used in working this problem.

13.22 Suppose that for the linear time-invariant network of Figure 13.5 L_1 and L_2 are magnetically coupled with dots at nodes 3 and 6. A controlled current source i is connected from 6 to 4 with the reference arrow pointing toward 4. If $i = K(i_a - i_c)$ and $v_1(t) = V_1e^{st}$, write the loop equations using I_a, I_b, and I_c as variables. Is the impedance matrix symmetric?

13.23 Write loop equations for the network in Figure 13.2.

13.24 Draw the network of Figure 13.4 in planar form, choose mesh variables, and write a corresponding set of loop equations.

13.25 For the network of Figure 13.3, replace R_4 by L_2, assign polarity dots on terminals 3 and 6, and write loop equations.

13.26 For the network of Figure 13.3, replace R_1 by L_2, assign polarity marks at terminals 2 and 7, and write loop equations. Is your $\mathscr{L}$ matrix symmetric?

13.27 Show that the effect of mutual inductance can be represented by controlled sources by comparing the loop equations in Figure 13.38(a) and (b). The elements are linear and time-invariant.

13.28 Suppose the polarity dot for L_2 in Figure 13.38(a) is at the bottom terminal. What is its equivalent circuit using controlled sources?

13.29 Using the inductor current and capacitor voltage as state variables x_1 and x_2, respectively, write two first-order differential equations for the linear time-invariant network in Figure 13.39.

13.30 Draw a block diagram based on the equations of Problem 13.29, using two integrators, adders, and amplifiers.

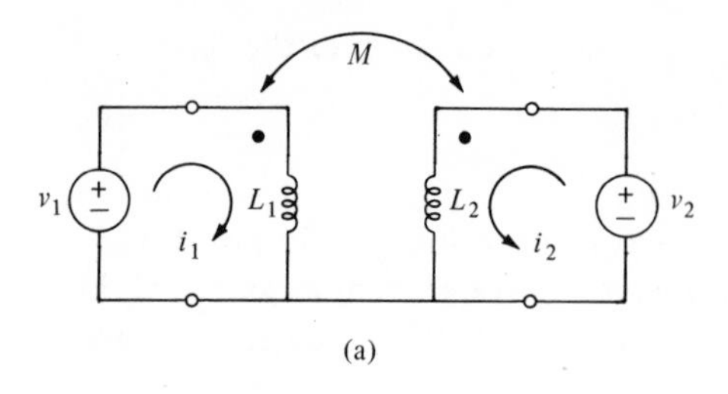

(a)

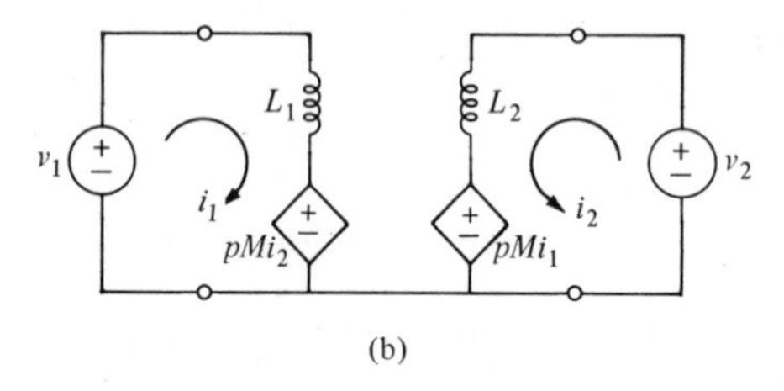

(b)

Figure 13.38

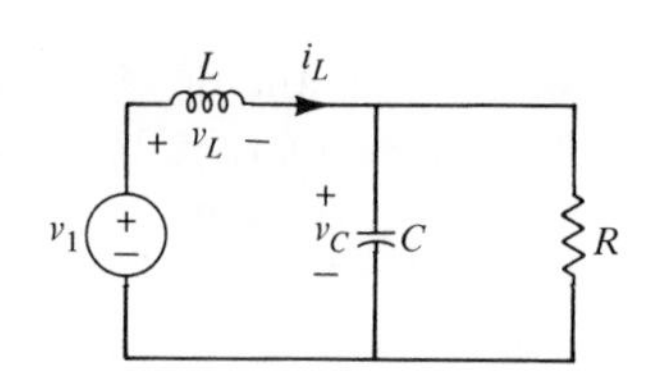

Figure 13.39

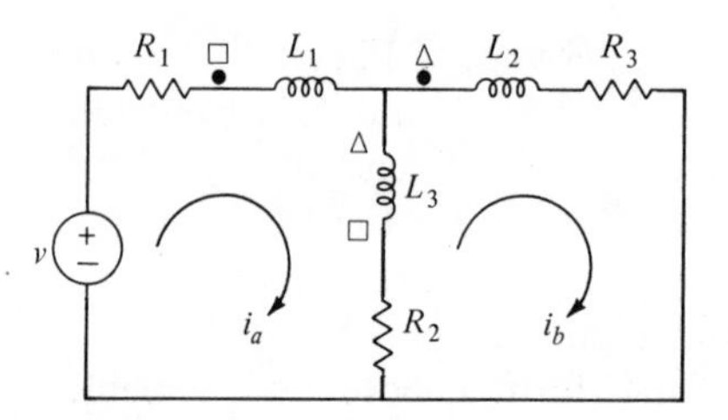

Figure 13.40

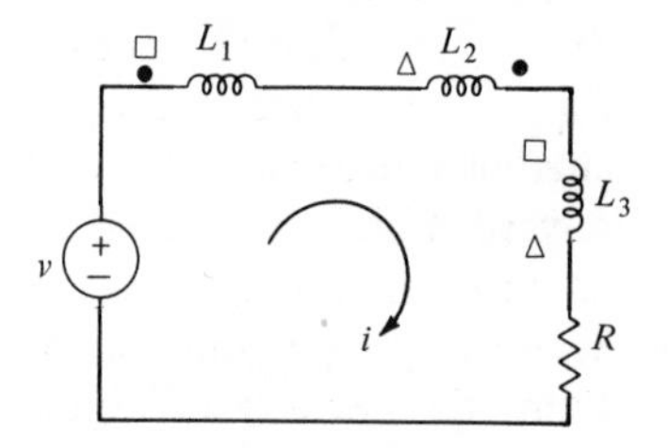

Figure 13.41

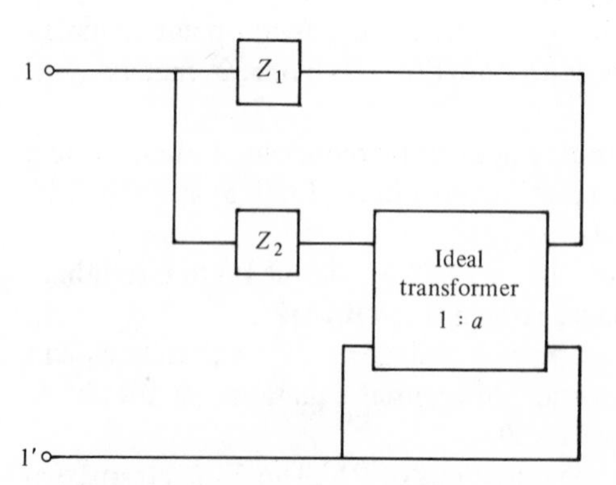

Figure 13.42

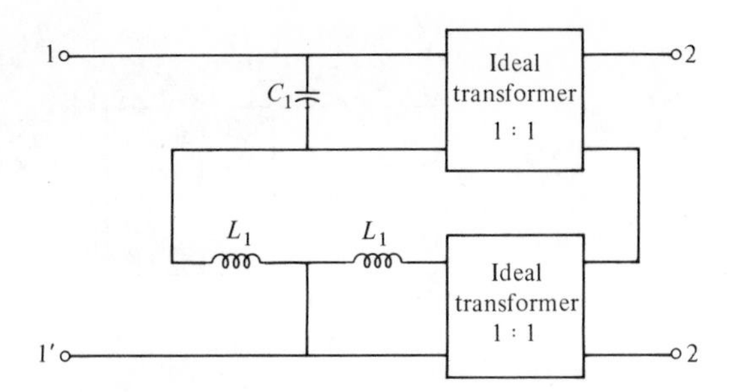

Figure 13.43

13.31 For the network in Figure 13.39, use the inductor flux linkage ψ and capacitor charge q as state variables and write the state equations in standard form.

13.32 The three inductors in the linear time-invariant network of Figure 13.40 are all mutually coupled. The polarity marks are given by three different sets of symbols for the three pairings of coils. Assuming $v = Ve^{st}$, write the loop equations, using i_a and i_b as loop variables.

13.33 Write the loop equation for the network (Figure 13.41) for $v = Ve^{j\omega t}$, assuming the network to be linear and time-invariant.

13.34 For the linear time-invariant network of Figure 13.42, find the driving-point impedance at terminals 1–1′ in terms of Z_1, Z_2, and a, the transformation ratio of the ideal transformer.

13.35 The linear time-invariant network shown in Figure 13.43 contains two 1:1 ideal transformers, one capacitor and two inductors. (a) Find the impedance at terminals 1–1′ with terminals 2–2′ open. (b) Find the impedance at terminals 2–2′ with 1–1′ open.

13.36 Find the driving-point impedance of the network of Figure 13.44. Assume that the system is operating in the sinusoidal steady-state.

13.37 In the network of Figure 13.45, $v_2(t) = \sin 2t$ and the network is in the steady-state. Determine $v_1(t)$, using numerical values where given.

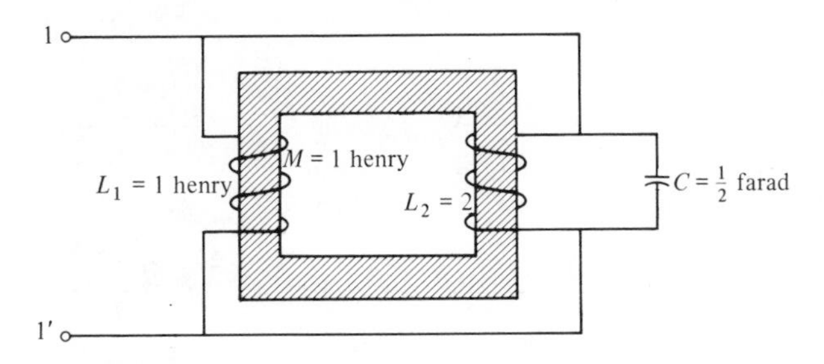

Figure 13.44

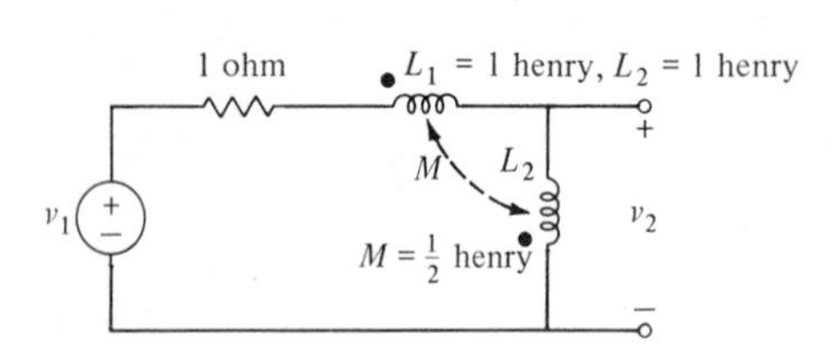

Figure 13.45

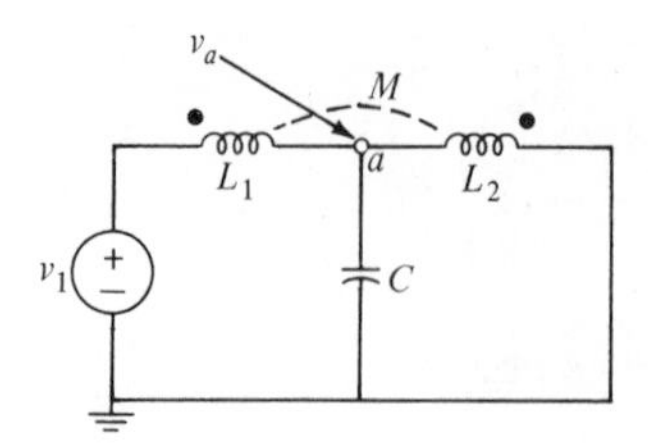

Figure 13.46

13.38 In the linear time-invariant network of Figure 13.46, determine an expression for $v_a(t)$ if $v_1(t) = V \sin \omega t$ and the system is in the steady-state.

13.39 In the network of Figure 13.47, it is given that $L_1 = L_2 = 1$ henry, $L_3 = 2$ henrys, $M_{12} = \frac{1}{2}$ henry, $M_{23} = 1$ henry, and $M_{31} = 1$ henry. If $v_1(t) = 2 \sin t$, solve for the two loop currents, $i_1(t)$ and $i_2(t)$.

13.40 In the network of Figure 13.48, all self-inductance values are 1 henry, and all mutual inductance values are $\frac{1}{2}$ henry. If the network is in the steady-state and $v(t) = 2 \sin t$, find an expression for $i(t)$. What is the driving-point impedance of the network at $\omega = 1$ radian/sec?

13.41 Redraw the network shown in Figure 13.49 without the magnetic core and with appropriate polarity marks on the coils (like that shown in Figure 13.40). Assuming that all inductors are linear and time-invariant, write an appropriate set of loop equations in phasor form.

13.42 The current source i_1 in the linear time-invariant network of Figure 13.50(a) is shifted as shown in (b). Show that the node equations are not affected and hence that the node voltages should remain the same. The current sources are further transformed as in (d). Show that the loop equations for (a) and (d) of the figure are the same.

13.43 Show that the node equations for the linear time-invariant networks in Figure 13.51 are identical.

13.44 Examine how the voltage sources are shifted out of the branches of the linear time-invariant network. Thus v_1 in branch 14 is pushed out of

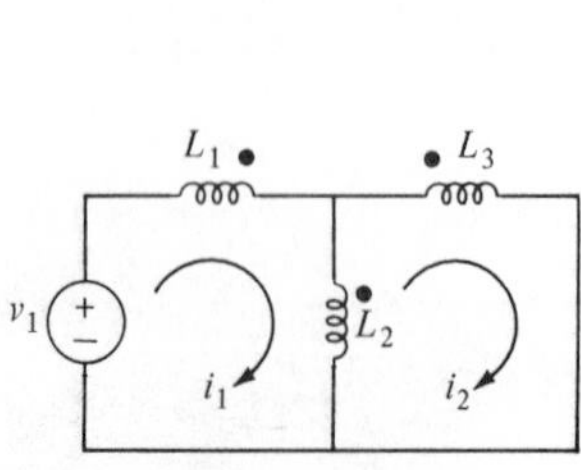

Figure 13.47

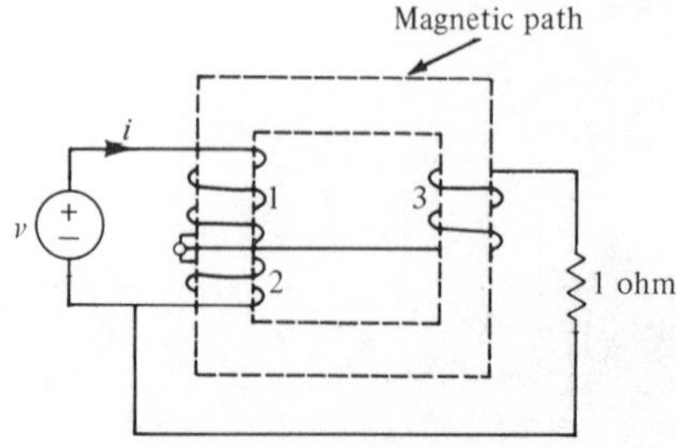

Figure 13.48

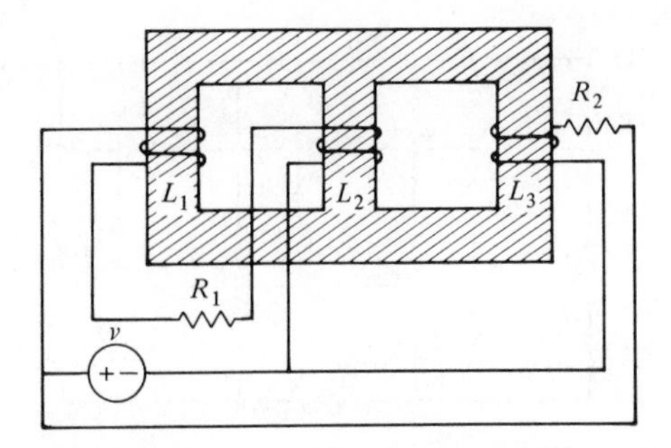

Figure 13.49

node 1. As many replicas of v_1 emerge as there are branches connected to node 1 aside from branch 14. The polarity positions are preserved (no twisting allowed as v_1 goes through the node). The result is Figure 13.52(b). Similarly, if v_1 is pushed through node 4 and replicas of v_1 are manufactured as it emerges from node 4, Figure 13.52(c) results. Show that the loop equations for the networks in the first three figures are identical.

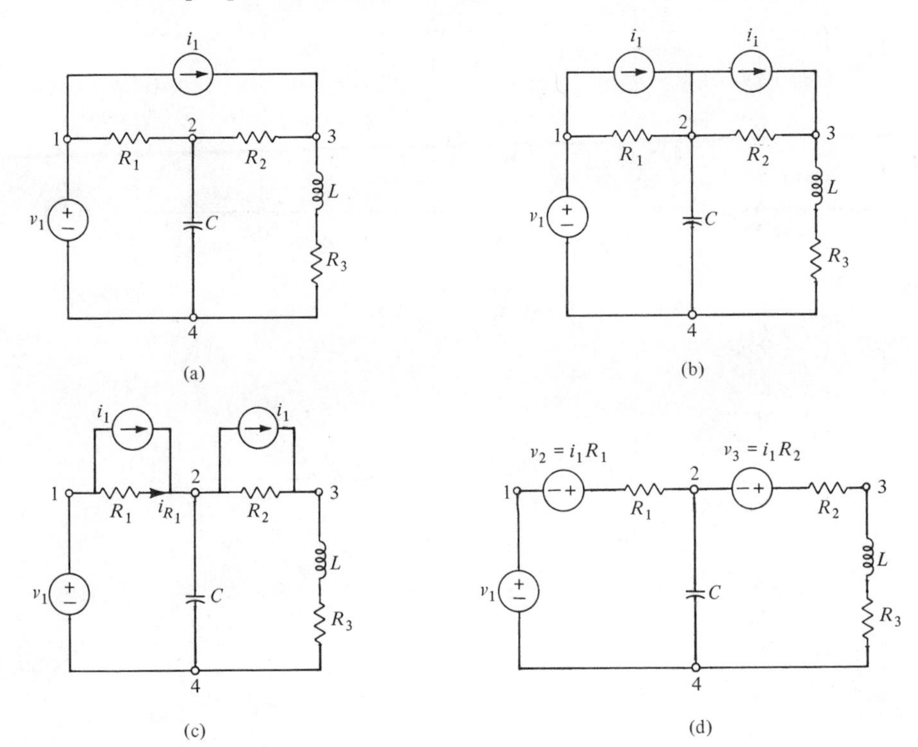

Figure 13.50

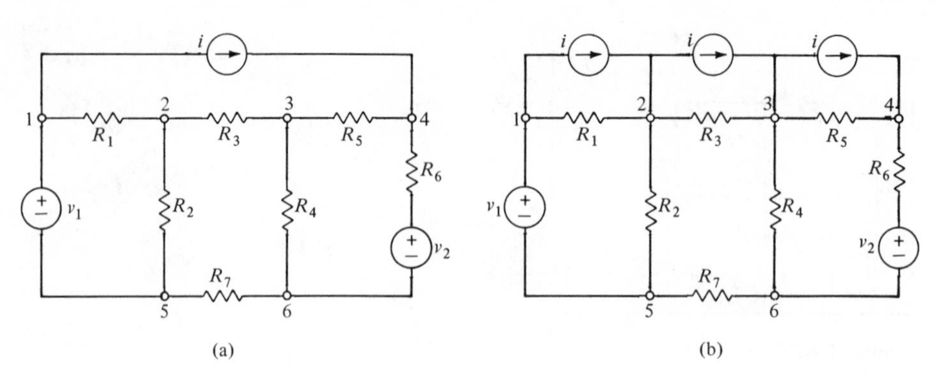

Figure 13.51

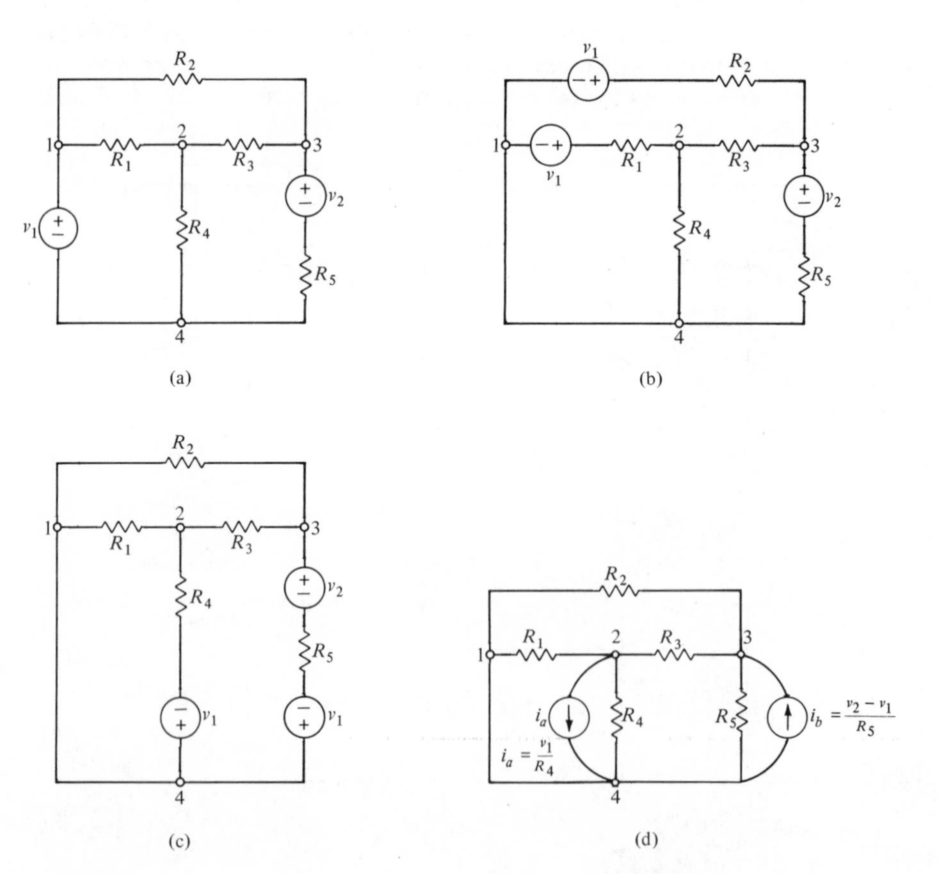

Figure 13.52

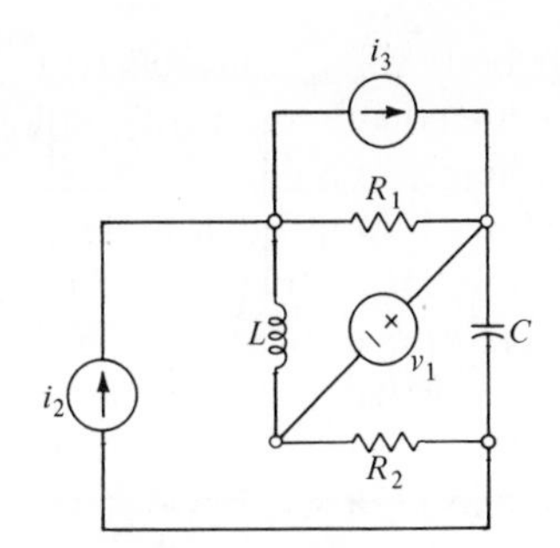

Figure 13.53

13.45 Write loop equations for the linear time-invariant network of Figure 13.53. Assume sinusoidal steady-state.

13.46 Write node equations for the linear time-invariant network in Figure 13.53.

13.47 Write node equations for the network in Figure 13.54. Assume that the sources are of the form $v_1 = V_1 e^{st}$, $i_2 = I_2 e^{st}$, $v_3 = V_3 e^{st}$.

13.48 Write loop equations for the network in Figure 13.54 with sources as specified in Problem 13.47.

★13.49 How many state variables are there for the network in Figure 13.41? How about for the network in Figure 13.40? Write the state-variable equations for the latter.

★13.50 Assume that the elements in the network of Figure 13.22 are linear time-varying and write an appropriate set of node equations.

★13.51 Write an appropriate set of loop equations for the linear time-varying network of Problem 13.50.

13.52 For the network specified in Problem 13.15, it is desired to determine $v_{ab}(t)$ for $t \geq 0$. Find the Thevenin equivalent network as seen by the capacitor and then write the state equation for the resulting network.

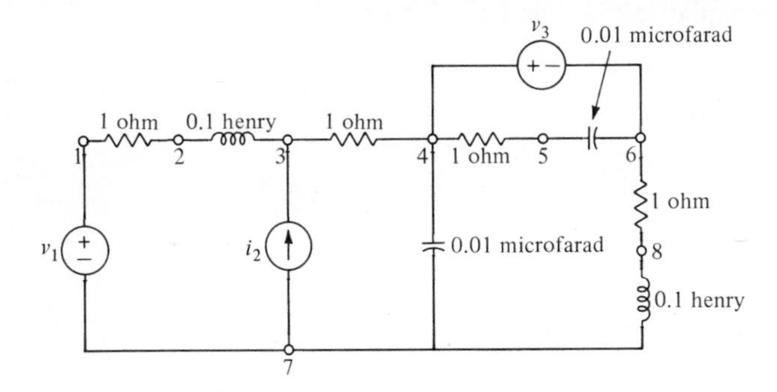

Figure 13.54

13.53 Consider a linear time-invariant network with both voltage and current sources. Suppose it is desired to write a set of node equations. Justify and briefly discuss the following pointers. (a) If a voltage source is connected from a node k to the reference node, KCL need not be applied at node k. (b) If a voltage source is connected from node k to node m, then v_k and v_m are simply related. Moreover, if the node equations applied at nodes k and m are added together, the equation that is obtained is the cut-set equation that separates nodes k and m from the rest of the network. For such situations, KCL need not be applied to nodes k and m singly, but instead KCL may be applied to the cut set separating nodes k and m from the rest of the network.

13.54 For the network in Figure 13.24, apply KCL to nodes 1 and 2. Eliminate all variables except v_{13}, v_2, v_a, v_b, and v_c, and thus reduce the equations to standard-state form with v_{13} and v_2 as state variables.

13.55 For the network given in Figure 13.27, apply KVL around the outer loop and KCL at node B. Express the branch currents at node B in terms of v_A, v_C, and v_{BC}. Reduce the equation to standard-state form with i_L (from A to B) and v_{BC} as state variables.

13.56 For the network given in Figure 13.32, express v_B and v_C in terms of the capacitor voltages v_{BC} and v_{AC}, and the two input voltages. Express the currents through the 1-ohm resistor and the $\frac{1}{2}$-ohm resistor in terms of the capacitor voltages and input voltages. The sum of these currents equals the current through the voltage source $2 \sin t$ with a positive reference toward node A. Using the above expressions, apply KCL at nodes A and B and obtain the state equations in standard form.

13.57 For the network given in Figure 13.46, suppose that $M = 0$. Assign appropriate state variables and write the state equations in standard form.

13.58 Write state equations in standard form for the network given in Figure 13.50(d).

13.59 For the network in Figure 13.3, choose v_{12}, v_{45}, and i_{23} as state variables and write the state equations in standard form.

Chapter **14**

ətermination of Solutions for ɿear Time-invariant Networks

1 Gauss's elimination method

From an engineering standpoint, writing a set of equilibrium equations which completely characterizes an arbitrary network is only part of an over-all goal. Obtaining numerical solutions to specific sets of equations is usually at least as important. In this section, we review a well-known method, Gauss's elimination method, for solving one of the simplest sets of equations, a set of linear algebraic equations. A linear time-invariant resistive network, for instance, has loop or node equations which are linear algebraic equations with real constant coefficients. A linear time-invariant network with sinusoidal sources (of the same frequency) in the steady-state has steady-state solutions which are obtainable through the use of phasors. The phasor variables (say, loop currents or node voltages) satisfy a set of linear algebraic equations with complex constant coefficients.

The Gauss algorithm is a systematic way for organizing the elimination of variables. Clearly, both sides of any equation may be multiplied by the same constant without disturbing the equality. Also, the validity of an equation is not changed by adding another equation to it. The general goal in the Gauss elimination procedure is to end up with a set of equations such that one equation contains only one unknown, say x_n, another equation contains only two unknowns, x_n and x_{n-1}, a third equation contains only three unknowns, x_n, x_{n-1}, and x_{n-2}, and so on. Clearly, the solution for x_n may be readily obtained. With x_n computed, x_{n-1} may be readily obtained from the second equation. The algorithm is continued until all solutions are obtained. Before describing the general procedure, let us consider a simple specific case.

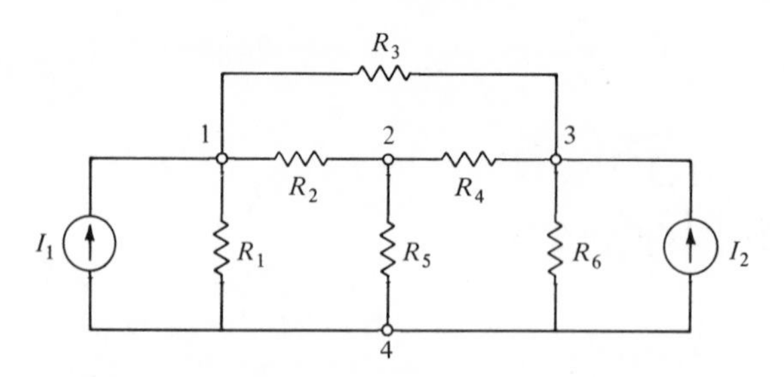

Figure 14.1 Network for Example 14.1.1.

EXAMPLE 14.1.1

Suppose that the resistor values for the network in Figure 14.1 are $\frac{1}{2}$ ohm each, and suppose $I_1 = 1$ ampere and $I_2 = 2$ amperes. Using node 4 as the reference, the node equations are as follows:

$$\begin{aligned} 6v_1 - 2v_2 - 2v_3 &= 1, && \text{(a)} \\ -2v_1 + 6v_2 - 2v_3 &= 0, && \text{(b)} \\ -2v_1 - 2v_2 + 6v_3 &= 2. && \text{(c)} \end{aligned} \qquad (14.1)$$

We may divide Equation (a) by 3 and add the result to Equation (b), with the result that we have

$$0 + \tfrac{16}{3}v_2 - \tfrac{8}{3}v_3 = \tfrac{1}{3}. \qquad \text{(b')}$$

Similarly, we may divide Equation (a) by 3 and add the result to (c), with the result that we have

$$0 - \tfrac{8}{3}v_2 + \tfrac{16}{3}v_3 = \tfrac{7}{3}. \qquad \text{(c')}$$

Equations (a), (b), and (c) are now equivalent to Equations (a), (b′), and (c′). Note that v_1 was eliminated from Equations (b) and (c), yielding (b′) and (c′). Now we eliminate v_2 from (c′). This is accomplished by adding $\frac{1}{2}$ of (b′) to (c′). Thus we have

$$0 \cdot v_1 + 0 \cdot v_2 + 4v_3 = \tfrac{15}{6}. \qquad \text{(c'')}$$

By judicious elimination of variables, the set of equations in 14.1 generated Equations (b′) and (c″). Note that from (c″), v_3 can be obtained as $\frac{5}{8}$. Then (b′) readily gives the value of v_2 as

$$v_2 = \tfrac{3}{16}[\tfrac{1}{3} + (\tfrac{8}{3})(\tfrac{5}{8})] = \tfrac{3}{8}.$$

Finally, from (a) we have

$$v_1 = \tfrac{1}{6}[2(\tfrac{3}{8}) + 2(\tfrac{5}{8}) + 1] = \tfrac{1}{2}.$$

To make the procedure more systematic, observe that the set of equations (a), (b′), and (c″) is in triangular form as follows:

$$\begin{aligned} 6v_1 - 2v_2 - 2v_3 &= 1, \\ 0 + \tfrac{16}{3}v_2 - \tfrac{8}{3}v_3 &= \tfrac{1}{3}, \\ 0 + 0 + 4v_3 &= \tfrac{15}{6}. \end{aligned} \tag{14.2}$$

Notice the triangular array of zeros. Solution for v_1, v_2, and v_3 was accomplished by steps of back substitution: v_3 from the third equation back into the second equation to solve for v_2, and finally both v_3 and v_2 back into the first equation to determine v_1. This pattern is a general one, as we will see.

We now discuss the general case. Suppose we have a set of n linear algebraic equations in n unknowns, $x_1, x_2, \ldots, x_n$. We assume that the set of equations has a unique solution. Among the n equations, at least one will contain a term in x_1, otherwise there is no unique solution. Denote this equation by

$$a_{11}x_1 + a_{12}x_2 + a_{13}x_3 + \cdots + a_{1n}x_n = b_1, \tag{14.3}$$

where $a_{11} \neq 0$. The other equations are of the form

$$a_{i1}x_1 + a_{i2}x_2 + \cdots + a_{in}x_n = b_i,$$

where $i = 2, 3, \ldots, n$. If Equation 14.3 is multiplied by $-a_{i1}/a_{11}$ and the result is added to the equation whose coefficient for x_1 is a_{i1}, then we have

$$0 + \left(a_{i2} - \frac{a_{12}a_{i1}}{a_{11}}\right)x_2 + \cdots + \left(a_{in} - \frac{a_{1n}a_{i1}}{a_{11}}\right)x_n = b_i - \frac{a_{i1}}{a_{11}}b_1,$$

thus eliminating x_1. This is done for all the $n - 1$ equations of $i = 2, \ldots, n$. At this stage, we have one equation in n unknowns, and $n - 1$ equations in the $n - 1$ unknowns, $x_2, x_3, \ldots, x_n$. From the set of $n - 1$ equations, there is at least one equation with a term in x_2; otherwise there is no unique solution. Let this equation be

$$a_{22}^{(1)}x_2 + a_{23}^{(1)}x_3 + \cdots + a_{2n}^{(1)}x_n = b_2^{(1)}, \tag{14.4}$$

where $a_{22}^{(1)} \neq 0$. The other equations are of the form

$$a_{i2}^{(1)}x_2 + a_{i3}^{(1)}x_3 + \cdots + a_{in}^{(1)}x_n = b_i^{(1)},$$

for $i = 3, \ldots, n$. Now we eliminate x_2 from the $n - 2$ equations, $i = 3, \ldots, n$, using the same algorithm for eliminating x_1. That is, multiply Equation 14.4 by $-a_{i2}^{(1)}/a_{22}^{(1)}$ and add the resulting equation to the equation whose coefficient for x_2 is $a_{i2}^{(1)}$, for $i = 3, 4, \ldots, n$. After this is done $n - 2$ times, we have one equation in n unknowns, Equation 14.3, one equation in $n - 1$ unknowns, $x_2, x_3, \ldots, x_n$,

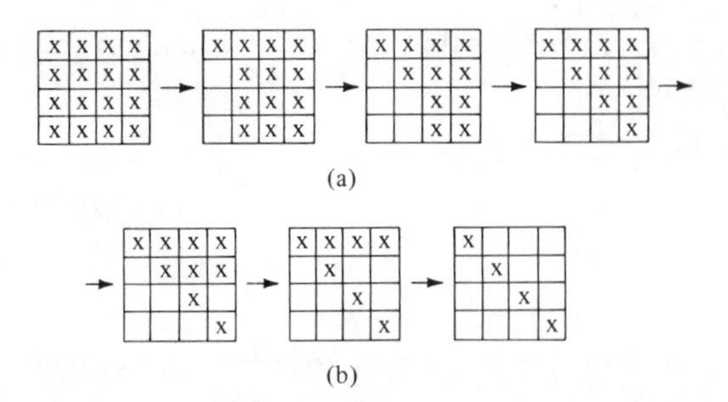

Figure 14.2 Pattern of the Gauss elimination scheme: (a) steps in the reduction to triangular form; (b) steps in back substitution.

which is labeled Equation 14.4, and $n - 2$ equations in the $n - 2$ unknowns, $x_3, x_4, \ldots, x_n$. The algorithm is continued until we obtain a triangular form of the system of equations of the following form:

$$\begin{aligned} a_{11}x_1 + a_{12}x_2 + a_{13}x_3 + \cdots + a_{1n}x_n &= b_1, \\ 0 + a_{22}^{(1)}x_2 + a_{23}^{(1)}x_3 + \cdots + a_{2n}^{(1)}x_n &= b_2^{(1)}, \\ 0 + 0 + a_{33}^{(2)}x_3 + a_{34}^{(2)}x_4 + \cdots + a_{3n}^{(2)}x_n &= b_3^{(2)}, \\ 0 + 0 + 0 + a_{44}^{(3)}x_4 + \cdots + a_{4n}^{(3)}x_n &= b_4^{(3)}, \\ &\vdots \\ 0 + 0 + 0 + \cdots + a_{nn}^{(n-1)}x_n &= b_n^{(n-1)}. \end{aligned} \tag{14.5}$$

The next steps involved in the solution for $x_1, x_2, \ldots, x_n$ are known as *back substitution.* From the last equation in the triangularized form, we solve for x_n and substitute that result into the next-to-last equation. Having made the substitution, we are in a position to solve for x_{n-1}. This back-substitution process is then continued, equation by equation, until we finally solve the first equation for x_1, and the solution is complete. The pattern of operations we have described comprise the basic Gauss elimination method which is illustrated in Figure 14.2 with x's substituting for the actual numbers. In (a) of the figure are shown steps that result in the reduction of the equations (or the matrix of the coefficients) to triangular form. The steps in (b) illustrate the operation known as back substitution by which we finally solve for the unknowns.

The triangularization method and variants of it are widely used for machine computation. The procedure is usually economical in terms of arithmetical calculations. Any method for hand solution of equations becomes tedious when the number of unknowns becomes large, of course, and the Gauss method is no exception. It does, however, provide a systematic approach to the solution of equations, whether simple or difficult.

MPLE 14.1.2

Consider the following set of four equations:

$$x_1 - 2x_2 + 2.5x_3 + 3x_4 = 16.5, \tag{14.6}$$

$$-8x_1 + 20x_2 + 14x_3 - 16x_4 = 10, \tag{14.7}$$

$$5x_1 + 7x_2 + 2x_3 - 10x_4 = -15, \tag{14.8}$$

$$6x_1 - 8x_2 - 10x_3 + 12x_4 = 8. \tag{14.9}$$

Multiplying Equation 14.6 by 8 and adding this to Equation 14.7, we obtain

$$0 + 4x_2 + 34x_3 + 8x_4 = 142. \tag{14.7a}$$

Similarly, multiplying Equation 14.6 by 5 and subtracting this from Equation 14.8, we obtain

$$0 + 17x_2 - 10.5x_3 - 25x_4 = -97.5. \tag{14.8a}$$

Multiplying Equation 14.6 by 6 and subtracting the result from Equation 14.9 give us

$$0 + 4x_2 - 25x_3 - 6x_4 = -91. \tag{14.9a}$$

Our system of equations is now 14.6, 14.7a, 14.8a, and 14.9a, where x_1 appears in Equation 14.6 only. Next we eliminate x_2 from Equations 14.8a and 14.9a. This is obtained by adding $-\frac{17}{4}$ of Equation 14.7a and subtracting Equation 14.7a from Equations 14.8a and 14.9a, respectively. Thus we have

$$0 + 0 - 155x_3 - 59x_4 = -701, \tag{14.8b}$$

$$0 + 0 - 59x_3 - 14x_4 = -233. \tag{14.9b}$$

Finally, we eliminate x_3 from Equation 14.9b by adding $-\frac{59}{155}$ of Equation 14.8b to Equation 14.9b to obtain

$$0 + 0 + 0 + 8.458x_4 = 33.832. \tag{14.9c}$$

The triangularized set of equations is then

$$\begin{aligned} x_1 - 2x_2 + 2.5x_3 + 3x_4 &= 16.5, \\ 0 + 4x_3 + 34x_3 + 8x_4 &= 142, \\ 0 + 0 - 155x_3 - 59x_4 &= -701, \\ 0 + 0 + 0 + 8.458x_4 &= 33.832. \end{aligned} \tag{14.10}$$

The solution is found by back substitution and is

$$x_4 = \frac{33.832}{8.458} = 4.00, \qquad x_3 = \frac{-701 + 59(4.00)}{-155} = 3.00,$$

$$x_2 = \frac{142 - 8(4.00) - 34(3.00)}{4} = 2.00,$$

$$x_1 = \frac{16.5 - 3(4.00) - 2.5(3.00) + 2(2.00)}{1} = 1.00.$$

The same algorithm applies to equations with complex constant coefficients and complex-valued solutions, although the arithmetic is approximately doubled as compared to that for a corresponding set of equations with real coefficients but the same number of variables.

EXERCISES

14.1.1 (a) Suppose that in the set of Equations (a), (b), and (c) of Example 14.1.1, we multiply (a) by $\frac{1}{2}$ to get a new equation (a′), multiply (b) by -10 to get a new equation (b′), and divide (c) by -5 to get (c′), what can you say of the solution of the new system of Equations (a′), (b′), and (c′) as compared to the solution for (a), (b), and (c)?
(b) Give equations (a), (b), and (c) of Example 14.1.1, form new equations (a′) = (a) + (b), (b′) = (a) − (b), and (c′) = (b) − (c). What is the relationship between the solution of (a), (b), and (c) as compared to the solution of (a′), (b′), and (c′).

14.1.2 For the network of Figure 14.1, suppose that $R_1 = R_2 = R_3 = R_4 = R_5 = R_6 = 1$ ohm, $I_1 = 1$ ampere, and $I_2 = 2$ amperes. Using node 4 as reference, solve for v_1, v_2, and v_3, using Gauss's elimination procedure. Compare this network to the network of Example 14.1.1. Compare the two solutions.

14.1.3 (a) For the network of Figure 14.1, suppose that $R_1 = 1$, $R_2 = \frac{1}{2}$, $R_3 = \frac{1}{3}$, $R_4 = \frac{1}{4}$, $R_5 = \frac{1}{5}$, and $R_6 = \frac{1}{6}$ ohm. Let $I_1 = 1$ ampere and $I_2 = 2$ amperes. Solve for v_1, v_2, and v_3, using node 4 as reference.
(b) Suppose that instead of $I_1 = 1$ ampere and $I_2 = 2$ amperes in part (a), we have $I_1 = \sin 1000t$ amperes and $I_2 = 2\sin 1000t$ amperes. Determine the solution.
(c) Suppose that instead of the current sources in part (b) we have $I_1 = \sin 500t$ amperes and $I_2 = 2\sin 3000t$ amperes. Determine v_1, v_2, and v_3.

14.1.4 In Example 14.1.1, note that we also get a triangular set of equations by considering Equations (b), (c′), and (c″). Solve for v_1, v_2, and v_3 from this new set of equations and compare with the answer in Example 14.1.1.

From the given equations (a), (b), (c), (b′), (c′), and (c″) of Example 14.1.1, how many other triangular sets of equations are there? Should they all yield the same solutions?

14.1.5 Suppose that for the network of Figure 14.1, R_2 is replaced by a capacitor C_2. Let all the other resistors have a value of $\frac{1}{2}$ ohm and let I_1 and I_2 be sinusoidal currents of the same frequency. Thus we have $I_1 = \sin \omega t$ amperes, $I_2 = 2 \sin (\omega t + 30°)$ amperes, and $\omega C_2 = 2$ mhos. Use the Gauss elimination procedure for computing v_1, v_2, and v_3.

2 Cramer's rule

Instead of the Gauss elimination procedure or any of its variants, the solutions to linear algebraic equations may be obtained as ratios of determinants. However, if the number of equations is more than four, the arithmetic involved in evaluating the determinants is much more than that in the Gauss procedure. In theoretical work where specific numerical solutions are not the primary goal, determinants are often useful tools. For instance, it may be desirable to express the solutions in terms of literal coefficients a_{ij} rather than in terms of specific values of the coefficients. This form might be sought because the sensitivity of a solution to changes in a_{ij} might be under investigation, and thus it is desirable to obtain the functional dependence of the solution on a_{ij}. For such studies, the exploitation of properties of determinants yields useful results. In this section, we shall state without proof some results from determinant theory.

A determinant is a function of a square array of elements (which are either real or complex numbers) or a matrix usually written in the form

$$\det A = \begin{vmatrix} a_{11} & a_{12} & \cdots & a_{1n} \\ a_{21} & a_{22} & \cdots & a_{2n} \\ \cdots & \cdots & \cdots & \cdots \\ a_{n1} & a_{n2} & \cdots & a_{nn} \end{vmatrix} \tag{14.11}$$

where a_{ij}'s are the elements of the array or matrix A. Notice that in the double subscript the first symbol denotes the row location and the second denotes the column location. Thus a_{ij} is an element in the ith row and jth column. The determinant, det A, has the following defining properties:

(a) The value of det A is not changed if the elements of any row (or any column) are added to the respective elements of another row (or column).

(b) The value of det A is multiplied by K if all the elements of any row (or column) of A are multiplied by K.

(c) The value of a determinant is unity if the elements on the principal diagonal (elements a_{ij}, where $i = j$) are unity and all other elements are zero. (An element is said to be on the principal diagonal if its row and column locations are identical. Thus if it is on the third row it must also be on the third column in order for the element to be on the principal diagonal.) A degeneracy occurs when the matrix has only one element. In this case when the element is unity, the determinant is unity.

From the above defining properties, it can be shown that for a two-by-two matrix, the determinant is

$$\det A = \begin{vmatrix} a_{11} & a_{12} \\ a_{21} & a_{22} \end{vmatrix} = a_{11}a_{22} - a_{12}a_{21}. \tag{14.12}$$

Also, for the degenerate one-by-one matrix, we have

$$\det a_{11} = a_{11}. \tag{14.13}$$

The *minor* M_{ij} of a matrix A is defined as the determinant of a matrix which is obtained by deleting the ith row and jth column from A. The cofactor A_{ij} of the element a_{ij} of a matrix A is defined as $A_{ij} = (-1)^{i+j}M_{ij}$.

We state without proof the *Laplace expansion* of a determinant as follows:

$$\det A = \sum_{k=1}^{n} a_{ik}A_{ik}, \qquad \text{for} \quad i = 1, 2, \ldots, n, \tag{14.14}$$

and

$$\det A = \sum_{k=1}^{n} a_{kj}A_{kj}, \qquad \text{for} \quad j = 1, 2, \ldots, n. \tag{14.15}$$

Equation 14.14 gives the expansion along the ith row where i may be any integer between 1 and n, inclusive, and Equation 14.15 gives the Laplace expansion along the jth column where j is an integer, $1 \leq j \leq n$. We also state without proof that

$$\sum_{k=1}^{n} a_{ik}A_{rk} = 0, \qquad i \neq r, \tag{14.16}$$

and

$$\sum_{k=1}^{n} a_{kj}A_{kr} = 0, \qquad j \neq r. \tag{14.17}$$

The expression in Equation 14.16 is like a Laplace expansion along the ith row, except that cofactors for the rth row are used. Similarly, the expression in Equation 14.17 is like a Laplace expansion along the jth column, except that cofactors for the rth column are used. All the above statements can be proved using the defining properties of determinants.

Now consider the following set of equations:

$$\begin{aligned} a_{11}x_1 + a_{12}x_2 + \cdots + a_{1n}x_n &= b_1, \\ a_{21}x_1 + a_{22}x_2 + \cdots + a_{2n}x_n &= b_2, \\ &\vdots \\ a_{n1}x_1 + a_{n2}x_2 + \cdots + a_{nn}x_n &= b_n. \end{aligned} \tag{14.18}$$

Suppose that we multiply the first equation by A_{1k}, the second equation by $A_{2k}, \ldots,$ and the nth equation by A_{nk} and then add all the resulting equations. We obtain

$$\sum_{i=1}^{n} A_{ik} \sum_{j=1}^{n} a_{ij}x_j = \sum_{i=1}^{n} A_{ik}b_i. \tag{14.19}$$

Interchanging the order of summation, we have

$$\sum_{j=1}^{n} x_j \sum_{i=1}^{n} a_{ij}A_{ik} = \sum_{i=1}^{n} A_{ik}b_i. \tag{14.20}$$

But from Equations 14.15 and 14.17, we have

$$\sum_{i=1}^{n} a_{ij}A_{ik} = \begin{cases} \det A, & j = k, \\ 0, & j \neq k. \end{cases} \tag{14.21}$$

Hence, provided $\det A \neq 0$, Equation 14.20 reduces to

$$x_k = \frac{\sum_{i=1}^{n} A_{ik}b_i}{\det A}, \qquad k = 1, 2, \ldots, n. \tag{14.22}$$

Note that

$$\sum_{i=1}^{n} A_{ik}b_i = \begin{vmatrix} a_{11} & a_{12} & \cdots & a_{1k-1} & b_1 & a_{1k+1} & \cdots & a_{1n} \\ a_{21} & a_{22} & \cdots & a_{2k-1} & b_2 & a_{2k+1} & \cdots & a_{2n} \\ \cdots & \cdots & \cdots & \cdots & \cdots & \cdots & \cdots & \cdots \\ a_{n1} & a_{n2} & \cdots & a_{nk-1} & b_n & a_{nk+1} & \cdots & a_{nn} \end{vmatrix} \tag{14.23}$$

by expanding along the kth column. Let D_k denote the determinant in Equation 14.23 which is the determinant of an array obtained by replacing the kth column of A by the column of b_i's. Denote $\det A$ by D. Then equation 14.22 may be written as

$$x_k = \frac{D_k}{D}, \qquad k = 1, 2, \ldots, n. \tag{14.24}$$

Equation 14.24 is *Cramer's rule* for solving linear equations. If $D = 0$ and if the b_i's are not all zero, the system has no solution.

The Laplace expansion may be applied to the minors of a determinant, and the process may be repeated until an nth-order determinant is expressed in terms of second-order determinants. Although, in principle, this process is simple, it involves $n!$ multiplications and if n is large, the required arithmetic is astronomical. Nevertheless, for theoretical developments, Cramer's rule may be useful. For example, from the Laplace expansion it is clear that D is a linear function of any of the elements a_{ij}. By expanding along the jth column, none of the cofactors involves a_{ij}, and the only term that depends on a_{ij} is $a_{ij}A_{ij}$. Hence D is linear in a_{ij}. Similarly, D_K is linear in any a_{ij}. Hence the solution x_k from Cramer's rule in Equation 14.24 is bilinear in a_{ij} (that is, a ratio of linear functions in a_{ij}). This important theoretical result is obtained without having to evaluate specific determinants.

EXAMPLE 14.2.1

Consider the following system of equations:

$$a_{11}x_1 + a_{12}x_2 + a_{13}x_3 = b_1, \tag{14.25}$$

$$a_{21}x_1 + a_{22}x_2 + a_{23}x_3 = b_2, \tag{14.26}$$

$$a_{31}x_1 + a_{32}x_2 + a_{33}x_3 = b_3, \tag{14.27}$$

To solve for x_1, assuming det $A \neq 0$, we have

$$x_1 = \frac{\begin{vmatrix} b_1 & a_{12} & a_{13} \\ b_2 & a_{22} & a_{23} \\ b_3 & a_{32} & a_{33} \end{vmatrix}}{\begin{vmatrix} a_{11} & a_{12} & a_{13} \\ a_{21} & a_{22} & a_{23} \\ a_{31} & a_{32} & a_{33} \end{vmatrix}}. \tag{14.28}$$

The determinant det A may be expanded along the first column to obtain

$$\begin{aligned} \det A &= a_{11}A_{11} + a_{21}A_{21} + a_{31}A_{31} \\ &= a_{11}\begin{vmatrix} a_{22} & a_{23} \\ a_{32} & a_{33} \end{vmatrix} - a_{21}\begin{vmatrix} a_{12} & a_{13} \\ a_{32} & a_{33} \end{vmatrix} + a_{31}\begin{vmatrix} a_{12} & a_{13} \\ a_{22} & a_{23} \end{vmatrix} \\ &= a_{11}(a_{22}a_{33} - a_{23}a_{32}) - a_{21}(a_{12}a_{33} - a_{13}a_{32}) \\ &\quad + a_{31}(a_{12}a_{23} - a_{13}a_{22}). \end{aligned} \tag{14.29}$$

The numerator of Equation 14.28 may be expanded along the first column to yield

$$x_1 = [b_1(a_{22}a_{33} - a_{23}a_{32}) - b_2(a_{12}a_{33} - a_{13}a_{32}) + b_3(a_{12}a_{23} - a_{13}a_{22})]/\det A. \tag{14.30}$$

Note that all of the elements appear linearly in det A.

ERCISES

14.2.1 Expand the determinant in Example 14.2.1 along the second column and compare with Equation 14.29. Repeat for an expansion along the third column.

14.2.2 (a) Expand the determinant in Example 14.2.1 along the first row and compare with Equation 14.29. (b) Repeat for an expansion along the second row. (c) Repeat for an expansion along the third row.

14.2.3 (a) Solve for x_2 in Example 14.2.1. (b) Solve for x_3 in Example 14.2.1.

14.2.4 Starting from a second-order determinant, use the defining properties of determinants to transform the array of coefficients so that off-diagonal elements are zero. By factoring out certain quantities, reduce the determinant to

$$\begin{vmatrix} a_{11} & a_{12} \\ a_{21} & a_{22} \end{vmatrix} = K \begin{vmatrix} 1 & 0 \\ 0 & 1 \end{vmatrix} = K$$

and verify that $K = a_{11}a_{22} - a_{12}a_{21}$ as given by Equation 14.12.

14.2.5 Count the number of multiplications, divisions, and additions involved in applying Cramer's rule for solving for x_1 and x_2 from a system of two equations. Do the same counting for Gauss's elimination procedure and compare.

.3 Simultaneous differential equations

The methods of the preceding sections may be applied to the solution of systems of linear differential equations. We confine our attention to the simple but common situation when the coefficients are constants. The first step is to use operational notation by replacing d/dt by the differential operator p. The coefficient matrix A will now contain elements which are polynomials in p. Thus, in Equation 14.18, the coefficients a_{ij} may be regarded as polynomials in p. The quantities b_i may be constants or functions of time. In applying Cramer's rule, we note that det A is now a polynomial in p and, likewise, the various cofactors are now polynomials in p. Instead of Equation 14.22, we have

$$\det A[x_k] = \sum_{i=1}^{n} A_{ik}[b_i], \tag{14.31}$$

which is an ordinary linear differential equation with constant coefficients.

EXAMPLE 14.3.1

Suppose we are given

$$\frac{dx_1}{dt} + 3x_1 - x_2 = u_1, \tag{14.32}$$

$$-x_1 + \frac{2dx_2}{dt} + 2x_2 = 3u_2. \tag{14.33}$$

We wish to obtain a single differential equation containing the dependent variable x_1 but not x_2. Using Cramer's rule, we have

$$\begin{vmatrix} p+3 & -1 \\ -1 & 2p+2 \end{vmatrix} x_1 = \begin{vmatrix} u_1 & -1 \\ 3u_2 & 2p+2 \end{vmatrix},$$

$$[(p+3)(2p+2) - 1]x_1 = (2p+2)u_1 + 3u_2, \tag{14.34}$$

$$\frac{2d^2x_1}{dt^2} + \frac{8dx_1}{dt} + 5x_1 = \frac{2du_1}{dt} + 2u_1 + 3u_2.$$

Similarly, for x_2 we have

$$(2p^2 + 8p + 5)x_2 = \begin{vmatrix} p+3 & u_1 \\ -1 & 3u_2 \end{vmatrix} = (p+3)3u_2 + u_1. \tag{14.35}$$

The techniques of Chapters 7 and 8 may be applied to the solutions of Equations 14.34 and 14.35 for specified input functions $u_1(t)$ and $u_2(t)$. The order of the system is given by the degree of the polynomial in p of det A. In this example, det A is second-degree in p so the system is second-order. This means that we require two linearly independent initial conditions to determine the solution uniquely. The complementary solution for Equation 14.34 involves two constants. Likewise, the complementary solution for Equation 14.35 involves two constants. However, these four constants are linearly dependent since there are only two linearly independent initial conditions.

EXAMPLE 14.3.2

For the network of Figure 14.3, if we assume that the elements are constant, applying KVL to loop 1 and 2 yields

$$\frac{1}{C_1}\int i_1\,dt + R_1 i_1 - R_1 i = 0 \tag{14.36}$$

and

$$\frac{1}{C_2}\int i_2\,dt + R_2 i_2 - R_2 i = 0. \tag{14.37}$$

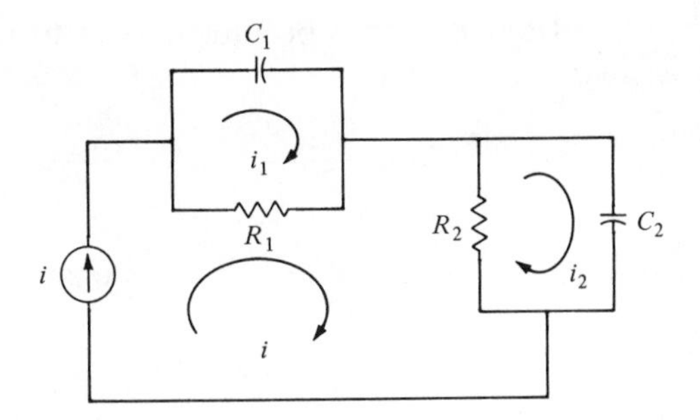

Figure 14.3 Network for Example 14.3.2.

In terms of charges q_1 and q_2, we have

$$\left(R_1 p + \frac{1}{C_1}\right) q_1 + 0 = R_1 i, \tag{14.38}$$

$$0 + \left(R_2 p + \frac{1}{C_2}\right) q_2 = R_2 i. \tag{14.39}$$

For this example, since the A matrix is diagonal, the equations are not coupled so that the equations may be solved separately. However, if we insist on applying Cramer's rule, we have

$$\left(R_1 p + \frac{1}{C_1}\right)\left(R_2 p + \frac{1}{C_2}\right) q_1 = \left(R_2 p + \frac{1}{C_2}\right) R_1 i \tag{14.40}$$

and

$$\left(R_1 p + \frac{1}{C_1}\right)\left(R_2 p + \frac{1}{C_2}\right) q_2 = \left(R_1 p + \frac{1}{C_1}\right) R_2 i. \tag{14.41}$$

Clearly, no matter what R_1, R_2, C_1, and C_2 are, $R_2 p + 1/C_2$ is a common factor in Equation 14.40, so that it may be reduced to Equation 14.38. Similarly, $(R_1 p + 1/C_1)$ is a common factor in Equation 14.41 so that after cancellation we obtain Equation 14.39. Although it is not necessary to cancel the common factors, unnecessary computation is avoided if cancelable factors are canceled!

Suppose that for this example, we wish to solve for the zero-input response ($i = 0$) for $q_1(0) = Q_1$ and $q_2(0) = Q_2$. Then, from Equation 14.38, we have

$$q_1 = Q_1 e^{-t/(R_1 C_1)}, \qquad t \geq 0, \tag{14.42}$$

and from Equation 14.39 we have

$$q_2 = Q_2 e^{-t/(R_2 C_2)}, \qquad t \geq 0. \tag{14.43}$$

But suppose that we did not notice the cancellation of factors in Equations 14.40 and 14.41. Then our tentative solutions would be

$$q_1 = K_1 e^{-t/(R_1C_1)} + K_2 e^{-t/(R_2C_2)} \tag{14.44}$$

and

$$q_2 = K_3 e^{-t/(R_1C_1)} + K_4 e^{-t/(R_2C_2)}. \tag{14.45}$$

From the initial-condition specification, we have

$$Q_1 = K_1 + K_2, \tag{14.46}$$

$$Q_2 = K_3 + K_4. \tag{14.47}$$

But from Equation 14.36, with $i \equiv 0$, and setting t equal to zero, we have

$$\frac{1}{C_1} q_1(0) + R_1 q_1'(0) = 0$$

or

$$q_1'(0) = -\frac{1}{R_1C_1} Q_1. \tag{14.48}$$

Likewise, from Equation 14.37, with $i \equiv 0$ and $t = 0$, we have

$$\frac{1}{C_2} q_2(0) + R_2 q_2'(0) = 0$$

or

$$q_2'(0) = -\frac{1}{R_2C_2} Q_2. \tag{14.49}$$

Differentiating $q_1(t)$ in Equation 14.44 and setting $t = 0$, we have

$$\frac{-K_1}{R_1C_1} - \frac{K_2}{R_2C_2} = q_1'(0) = \frac{-1}{R_1C_1} Q_1. \tag{14.50}$$

Similarly, differentiating $q_2(t)$ in Equation 14.45 and setting $t = 0$, we have

$$\frac{-K_3}{R_1C_1} - \frac{K_4}{R_2C_2} = q_2'(0) = \frac{-1}{R_2C_2} Q_2. \tag{14.51}$$

From Equations 14.46, 14.47, 14.50, and 14.51 we obtain

$$K_1 = Q_1, \qquad K_2 = 0, \qquad K_3 = 0, \qquad K_4 = Q_2,$$

which checks with our earlier solution. As expected, we have more computation although the answer comes out the same. In general, it is desirable to reduce each equation to minimum order by canceling the common factors.

EXERCISES

14.3.1 (a) $(2p + 3)x_1 - px_2 - 2x_3 = 2u_1$.
(b) $-px_1 + (3p + 5)x_2 - (p + 1)x_3 = u_2$.
(c) $-2x_1 - (p + 1)x_2 + (p + 2)x_3 = 4u_3$.
Obtain a single differential equation in x_1 whose solution satisfies the given system of differential equations. How many linearly independent initial conditions are needed?

14.3.2 Repeat Exercise 14.3.1 for x_2.

14.3.3 Repeat Exercise 14.3.1 for x_3.

14.3.4 Triangularize the system of equation in Exercise 14.3. [*Suggestion:* (1) Label (c) as Equation 1 and eliminate x_1 from (a) and (b). (2) Eliminate x_2 from one of the remaining equations. This elimination step may require several multiplications and additions. Avoid unnecessary increase in the order of the differential equations.] Compute the product of the main diagonal elements of the triangularized system and compare with the determinant of the original array of coefficients in Exercise 14.3.1.

14.3.5 (a) For the linear time-invariant network of Figure 14.4, where v_1 is the source and v_2 is the output, write node equations in differential-equation form from which v_2 and v_3 may be determined.
(b) Determine a differential equation involving v_2 but not v_3 whose solution satisfies the node equations.
(c) Assume that $R_1 = R_2 = R_3 = 100$ ohms, $L = 10^{-3}$ henry, $C = 10^{-8}$ farad, $k = 0.01$, $v_{12}(0+) = 1$ volt, $i_{34}(0+) = 10^{-3}$ ampere, and $v_1 \equiv 0$. Solve for $v_2(t)$.

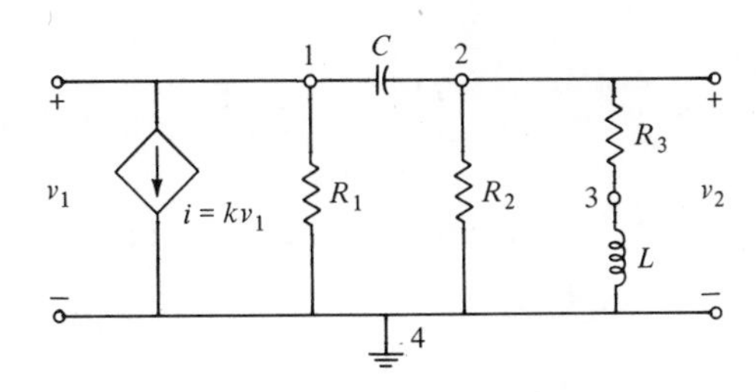

Figure 14.4

14.4 Summary

1. We are concerned with the solution of the simultaneous equations

$$\begin{aligned} a_{11}x_1 + a_{12}x_2 + \cdots + a_{1n}x_n &= b_1, \\ a_{21}x_1 + a_{22}x_2 + \cdots + a_{2n}x_n &= b_2, \\ &\vdots \\ a_{n1}x_1 + a_{n2}x_2 + \cdots + a_{nn}x_n &= b_n. \end{aligned}$$

If we let A be the array of a_{ij} coefficients or the matrix of a_{ij} coefficients, then det A is the determinant symbolized as

$$\det A = D = \begin{vmatrix} a_{11} & a_{12} & \cdots & a_{1n} \\ a_{21} & a_{22} & \cdots & a_{2n} \\ \cdots & \cdots & \cdots & \cdots \\ a_{n1} & a_{n2} & \cdots & a_{nn} \end{vmatrix}.$$

2. One efficient method for the solution of the equations is the Gauss elimination method. It is described as a series of steps which first triangularizes the equations and then follows with a procedure known as back substitution, that produces the solution. The algebraic steps are illustrated by the following sequence for a third-order set of equations:

Triangularization → Back substitution

3. Cramer's rule requires several definitions:

M_{ij} = a determinant known as a *minor* formed from A by deleting the ith row and jth column.

$A_{ij} = (-1)^{i+j}M_{ij}$ is known as a *cofactor*.

The following operation

$$\det A = \sum_{k=1}^{n} a_{ik}A_{ik}, \qquad i = 1, 2, \ldots, n,$$

is known as the *Laplace expansion*.

Providing det $A \neq 0$, Cramer's rule provides the solution of the simultaneous equations as

$$x_k = \frac{\sum_{i=1}^{n} b_i A_{ik}}{\det A}, \qquad k = 1, 2, \ldots, n.$$

4. For differential equations, the operators p and $1/p$ for differentiation and integration, respectively, give a_{ij} coefficients like

$$a_{ij} = L_{ij}p + R_{ij} + \frac{1}{C_{ij}p}$$

and algebraic equations in the variable p result from the solution for any x_j.

PROBLEMS

14.1 Determine x_1, x_2, x_3, and x_4 which satisfy the equations

$$\begin{aligned} x_1 - 2x_2 + 2.5x_3 + 3x_4 &= 8, \\ -8x_1 + 20x_2 + 14x_3 - 16x_4 &= 5, \\ 5x_1 + 7x_2 + 2x_3 - 10x_4 &= 4, \\ 6x_1 - 8x_2 - 10x_3 + 12x_4 &= 6. \end{aligned}$$

14.2 Rewrite the equations of Problem 14.1 by solving for x_1 from the first equation, x_2 from the second equation, x_3 from the third equation, and x_4 from the fourth equation, in terms of the other variables. Assume that x_1, x_2, x_3, and x_4 are dc signals that may be measured from an analog-computer set-up involving amplifiers, adders, and voltage sources. Using your rewritten equations as a guide, draw a suitable block diagram and label the points where you would measure x_1, x_2, x_3, and x_4.

14.3 Compute the determinant of the A matrix of Problem 14.1, using a Laplace expansion along the first column.

14.4 For the system of equations in Problem 14.1, write x_2 as a ratio of determinants.

14.5 Consider the matrix

$$\begin{bmatrix} 20 & 14 & -16 \\ 7 & 2 & -10 \\ -8 & -10 & 12 \end{bmatrix}.$$

(a) Compute A_{11}, A_{21}, and A_{31}.
(b) Verify that $a_{12}A_{11} + a_{22}A_{21} + a_{32}A_{31} = 0$ and $a_{13}A_{11} + a_{23}A_{21} + a_{33}A_{31} = 0$.

14.6 Given a square matrix $A(n \times n)$ whose elements are a_{ij}, form a matrix B whose elements b_{ij} are obtained from $b_{ij} = a_{ji}$. That is, the ith-row jth-column element of B is equal to the jth row ith column of A. Show that $\det A = \det B$. Here B is said to be the transpose of A. [*Hint:* Use Laplace expansion.]

14.7 For the linear time-invariant network of Figure 14.5, where $R_1 = R_2 = L_1 = L_2 = C = 1$ (normalized values) and $i_1(0^-) = 0$, $i_2(0^-) = 0$, $v_c(0^-) = 1$, $v_1 \equiv 0$, solve for $i_1(t)$ and $i_2(t)$ for $t \geq 0$.

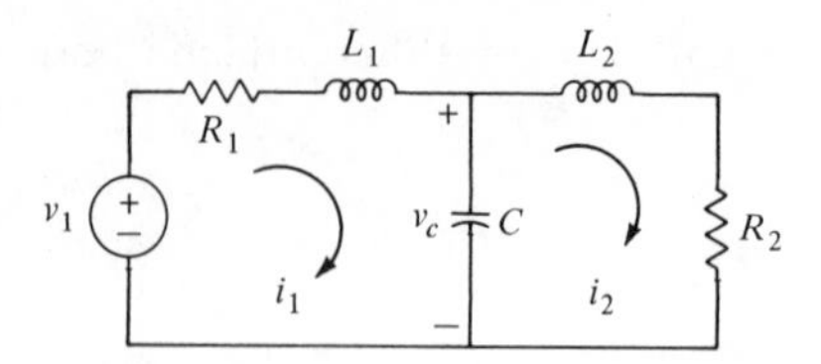

Figure 14.5

14.8 Repeat Problem 14.7 for v_1 equal to a unit step function $u(t)$ instead of zero.

14.9 The sources in the linear time-invariant network of Figure 14.6 are all sinusoidal, and they have the same frequency. It is desired to examine the steady-state behavior of the network. At the frequency of the sources, the impedances are as follows: $Z_1 = Z_2 = Z_3 = -j1$, $Z_5 = Z_6 = 1 + j1$, and $Z_7 = j1$. The sources are $i_a(t) = 2 \sin \omega t$, $i_b(t) = 2 \sin \omega t$, $v_4(t) = \sin \omega t$, and $v_7(t) = \sin \omega t$. Write an appropriate set of node equations for solving for the steady-state components of $v_{ae}(t)$, $v_{be}(t)$, $v_{ce}(t)$, and $v_{de}(t)$.

14.10 Write an appropriate set of loop currents for the network in Problem 14.9 for solving for the steady-state components of the branch currents $i_{ae}(t)$, $i_{be}(t)$, $i_{ce}(t)$, and $i_{de}(t)$.

14.11 In the three-phase power system of Figure 14.7, the impedances at 377 radians/sec (60 Hz) are as follows: $Z_{01} = Z_{02} = Z_{03} = 0.1 + j1$, $Z_{11'} = Z_{22'} = Z_{33'} = 0.5 + j0.8$, $Z_{1'n} = 5 + j0$, $Z_{2'n} = 0 + j5$, and $Z_{3'n} = 0 - j5$. Also v_a, v_b, v_c form a balanced set of voltages where the phase sequence is *acb*, that is, v_c lags v_a by 120°, and v_b lags v_c by 120°. The voltage $v_a = 141.4 \sin (377t + 30°)$. Calculate the steady-state components of the line currents.

14.12 (a) Repeat Problem 14.11 if the load impedances are balanced as follows: $Z_{1'n} = Z_{2'n} = Z_{3'n} = 5 + j0$. (b) Repeat (a) if, in addition, nodes n and 0 are joined by a wire of impedance $0.1 + j1$.

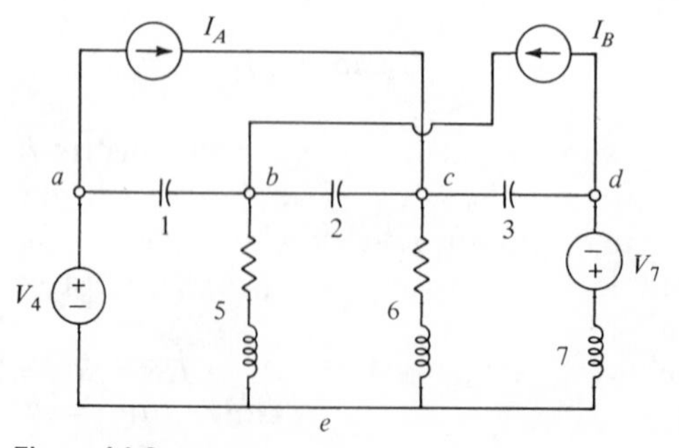

Figure 14.6

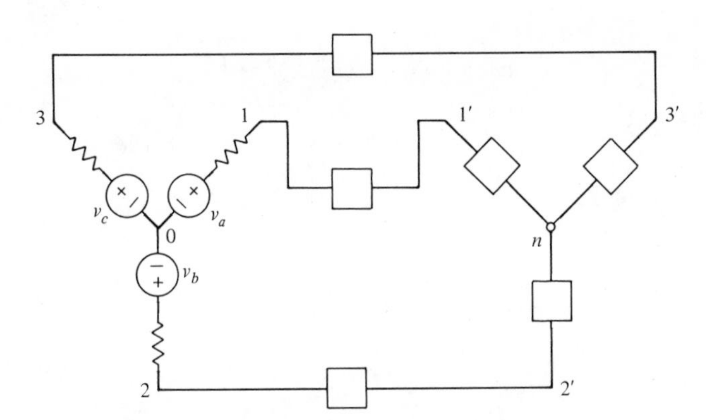

Figure 14.7

★14.13 Suppose that we have a set of linear equations with complex coefficients and possibly complex solutions of the form as shown in Equation 14.18, except that the *a*'s and *b*'s are complex numbers, and the *x*'s may be complex. By writing the complex numbers in rectangular form, derive an expanded set of linear equations with real coefficients involving $2n$ real unknowns.

14.14 Suppose the current source in Figure 14.3 is replaced by a voltage source. Assume that the network elements have normalized values of $R_1 = 1$, $C_1 = 1$, $R_2 = 2$, $C_2 = 3$. Let $q_1(0^-) = 1$ and $q_2(0^-) = 3$. (a) Assuming $v \equiv 0$, solve for $i_1(t)$ and $i_2(t)$. (b) Assuming $v = r(t)$, a unit ramp function, solve for $i_1(t)$ and $i_2(t)$.

14.15 Using $q_1(t)$ and $q_2(t)$ as state variables, write an appropriate set of state equations for the network of Figure 14.3. Draw an elementary analog-computer diagram for solving the network problem.

14.16 Repeat Problem 14.15 replacing the source $i(t)$ by a voltage source $v(t)$.

14.17 (a) For the three-phase network in Problem 14.12(a), compute the average power delivered to the three-phase load.
(b) Compute the average power delivered to the three-phase load for the three-phase network in Problem 14.12(b).

14.18 For the network in Problem 14.11, (a) Compute the average power delivered to the individual phase loads $Z_{1'n}$, $Z_{2'n}$, and $Z_{3'n}$. (b) Compute the total power into the three-phase load using the result of Problem 10.48 (see p. 277). Check your answer against the sum of the load phase powers in part (a).

Chapter **15**

Terminals and Ports

15.1 Classes of networks: breadboards and black boxes

In our discussions in past chapters, we have most often assumed that we had access to all elements and all nodes in the network under study. In other discussions, we have dealt with networks to which access was possible only at the network ports. Electrical engineers deal with both classes of networks. In the early stages of design, it is common to "breadboard" the network — meaning to temporarily arrange the components such that easy access may be had to any node or any component for measurement or adjustment. When the preliminary design is complete, the network is transferred to a chassis or perhaps it is encased in plastic. The network then becomes a "black box" in the sense that we can no longer measure voltages or currents except through the connections to the outside world, the terminals or the ports. The black box of which we speak may be created by encasing it in metal or plastic, as we have mentioned, or it may be created by spatial separation as in the case of the transmission line, or it may be the result of the network being fabricated by integrated-circuit technology so that only input, output, and power-supply terminals are accessible.

Clearly, the black-box network is as important in electrical engineering as is the breadboard-type network; we must know how to deal with either situation. Breadboard-type networks are analyzed by the application of the Kirchhoff laws, as studied in Chapter 13. For the black-box type of network, it is convenient to work in terms of one of the set of port parameters which describe the network.

We find it useful to talk about networks in terms of individual terminals and associated pairs of terminals which constitute the ports. In Figure 15.1, a network is illustrated in which the boundaries of the black box are shown by the dashed lines. The terminals are formed by soldering on lead wires; in this example, there are three such wires to form a three-terminal network. This idea may be general-

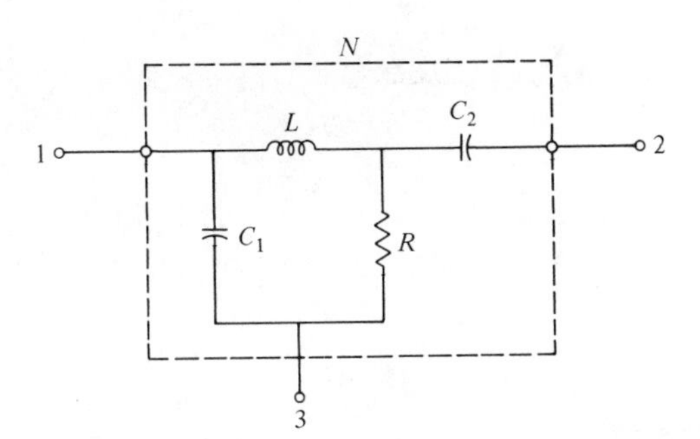

Figure 15.1 An example of a three-terminal network.

ized as shown in Figure 15.2 by identifying n such terminals and so forming an n-terminal network. These terminals may be used in pairs to connect energy sources. The terminals may be used in various combinations of two for the measurement of voltages. Or there may be other parts of the network connected to these terminals. In this sense, the n-terminal network may be embedded in another network.

Examples of networks that are frequently embedded in more complex networks are the three-terminal networks shown in Figure 15.3. That of (a) of the figure is known as a *T-network*; that of (b) as a *π-network*. If the elements of these networks are all of the same kind, then it is always possible to determine appropriate element values such that one network may always replace the other in the sense that one may be substituted for the other without disturbing voltages and currents of the network within which the T- or π-network is embedded.

The n-port network of Figure 15.4 is similar in appearance to the n-terminal network of Figure 15.2. The important difference is that the terminals are identified in associated pairs; we will not be interested in terminal-to-terminal voltages other than port voltages. Sources of energy will be connected to one or more of the ports.

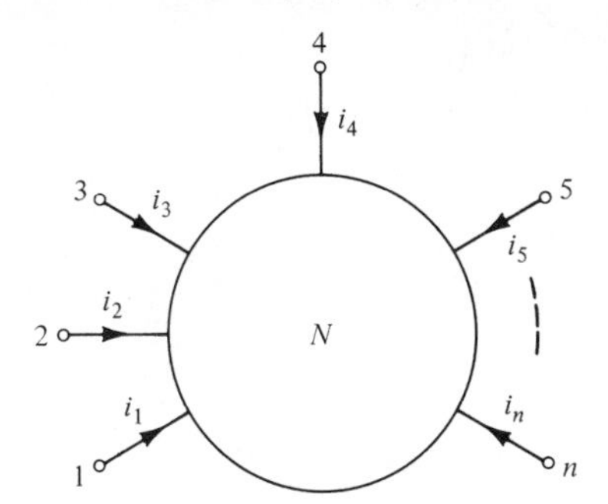

Figure 15.2 A network N in which n terminals are identified. Note the reference direction assigned for the n currents.

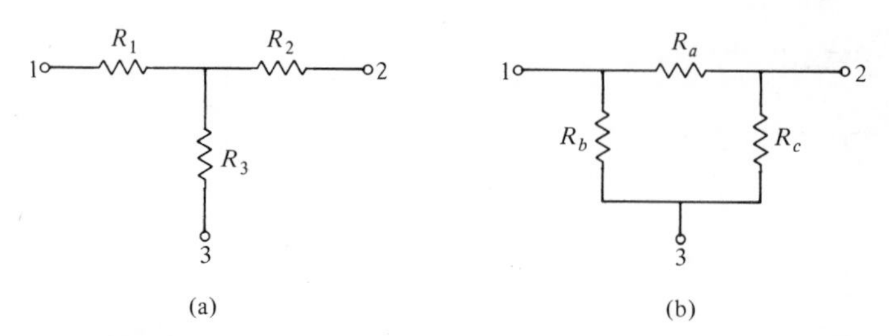

Figure 15.3 Two common three-terminal networks:
(a) the resistive T-section; (b) the resistive π-section.

These ports of energy entry will be distinguished from ports of energy exit which are connected to the loads. A port may not involve energy transfer but may be provided for the recording of a voltage of interest.

Applications of the port concept are shown in Figure 15.5. In some cases, we are interested in separating the lossless L and C elements from the R elements in accomplishing design. Thus in (a) of the figure, all ports of the lossless network are terminated in resistors, while the opposite situation takes place in (b) of the figure. This type of display may be used simply to give prominence to a given part of a network.

While we find the generality of the n-port descriptions useful in advanced applications, most problems of the electrical engineer may be formulated in terms of the familiar 1-port network, the 2-port network of Figure 15.6, or, at most, the 3-port network of Figure 15.7. The 3-port network finds application in the analysis of tunnel-diode amplifiers or oscillators.

Reference directions for the two kinds of network representations are illustrated by the figures. In the n-terminal network, the current reference direction is taken to be directed into the network. In the n-port network, the reference direction for

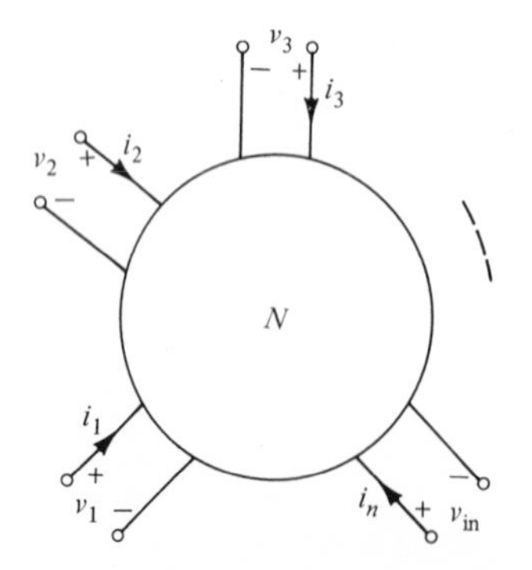

Figure 15.4 An n-terminal pair or n-port network with the assigned reference directions for the n currents and n voltages.

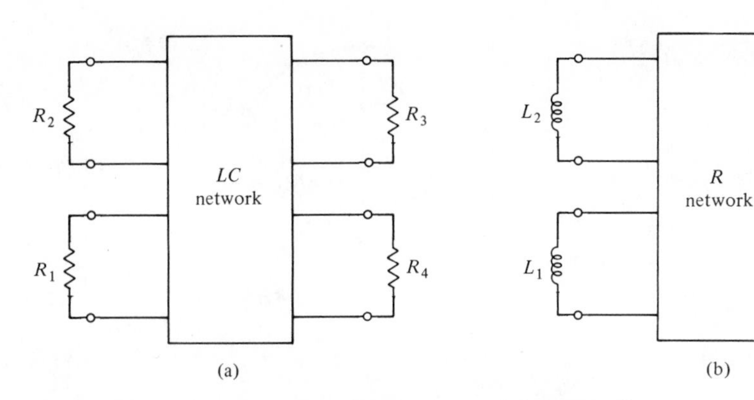

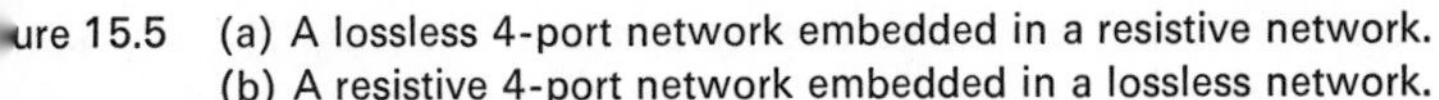
ure 15.5 (a) A lossless 4-port network embedded in a resistive network.
(b) A resistive 4-port network embedded in a lossless network.

the current is into the network at the positive terminal (and, of course, out for the negative terminal).

ERCISES

15.1.1 (a) Rearrange the network shown in Figure 15.8 in the form shown in Figure 15.5(a), showing the connections in the LC network in detail. (b) Rearrange the network of Figure 15.8 in the form of Figure 15.5(b), showing the connections in the R network in detail.

15.1.2 Repeat parts (a) and (b) of Exercise 15.1.1 for the network shown in Figure 15.9.

15.1.3 For the two networks of Figure 15.3, determine R_1, R_2, and R_3 as functions of R_a, R_b, and R_c in order that the two networks be equivalent for embedding within a more complex network.

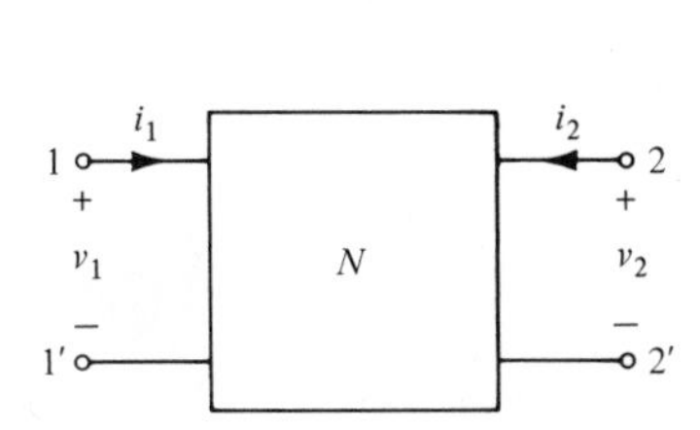

Figure 15.6 Standard representation of the 2-port network, including voltage and current reference directions for the two ports.

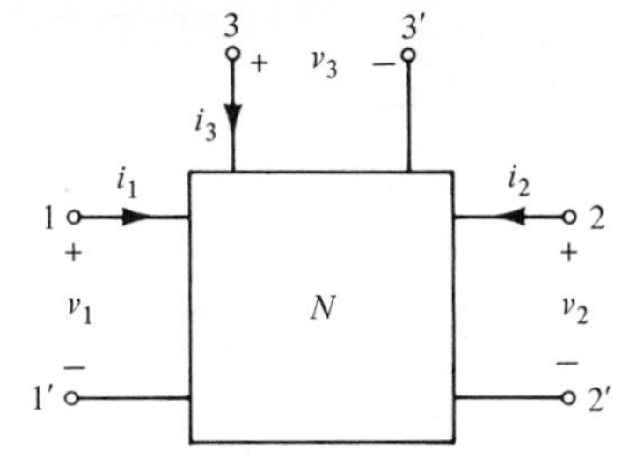

Figure 15.7 A 3-port network with reference directions.

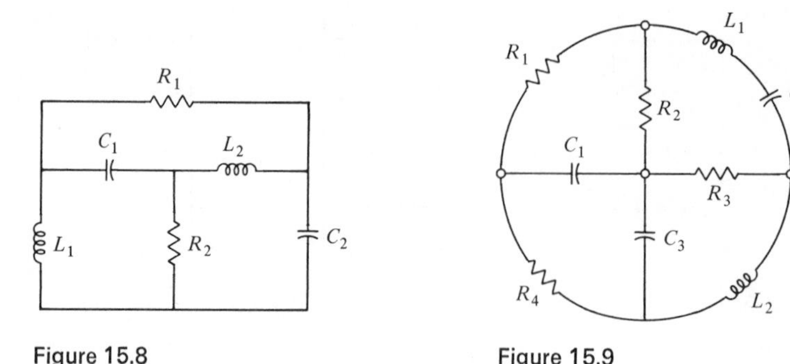

Figure 15.8

Figure 15.9

15.2 Two-port parameters

In this discussion, we will consider only the time-invariant linear 2-port network, containing no independent sources, with the assurance that the 2-port approach can be generalized for the 3-port, 4-port, or the n-port case. We recall from Section 10.2 that a 1-port network is characterized by a single network function, the driving-point impedance, or the reciprocal of this quantity which is the driving-point admittance. For the 2-port network, there are four variables, as shown in Figure 15.6, and thus more possible functions for describing the network.

As in Chapter 10, we will use $I(s)$ and $V(s)$ as the network variables for exponential signals of the form Ke^{st}. The four variables in Figure 15.6 are the quantities

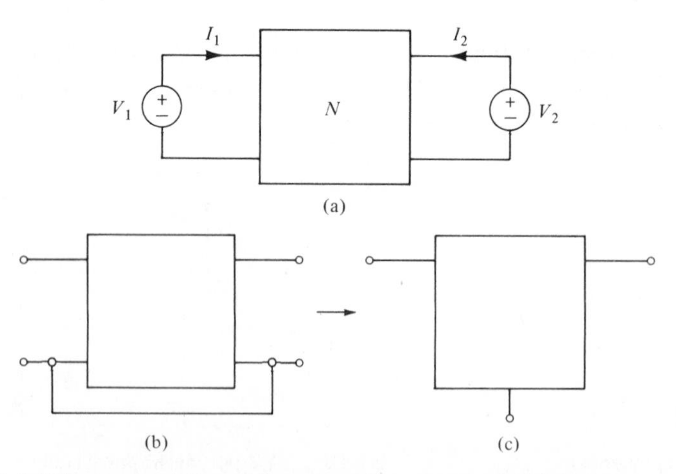

Figure 15.10 (a) The representation of the general 2-port network. (b) In the special case that two of the terminals are connected (and sometimes grounded), the 2-port network becomes a three-terminal network as shown in (c).

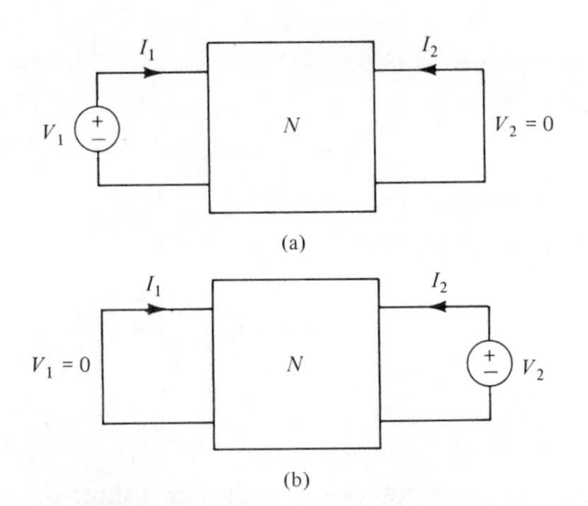

ure 15.11 The network of Figure 15.10, with one port shorted such that the short-circuit admittance functions are defined.

V_1 and I_1 at port 1 and the quantities V_2 and I_2 at port 2. Our objective is to express any two of these variables in terms of the remaining two; clearly there are a number of possibilities.

Consider first a linear time-invariant system for which there are two input signals, X_1 and X_2, and for which two responses, Y_1 and Y_2, are to be found. The outputs are

$$Y_1 = Q_{11}X_1 + Q_{12}X_2, \tag{15.1}$$

$$Y_2 = Q_{21}X_1 + Q_{22}X_2. \tag{15.2}$$

The quantities Q_{11}, Q_{12}, Q_{21}, and Q_{22} are network functions, of course, with their nature depending on the dimensions of the Y's and X's.

For the 2-port network of Figure 15.10, the inputs are V_1 and V_2; and the responses are I_1 and I_2. Then the general equations of Equations 15.1 and 15.2 specialize to

$$I_1 = y_{11}V_1 + y_{12}V_2, \tag{15.3}$$

$$I_2 = y_{21}V_1 + y_{22}V_2. \tag{15.4}$$

Here we have used y_{jk} in place of Q_{jk} because the dimensions of these network functions are clearly admittances.

The y-functions of Equation 15.3 and 15.4 have a meaningful interpretation in terms of the 2-port network with one of the two ports shorted as shown in Figure 15.11. The shorting of port 2 in (a) of the figure causes $V_2 = 0$ and makes possible the following definitions from Equations 15.3 and 15.4:

$$y_{11} = \left.\frac{I_1}{V_1}\right|_{V_2=0} \tag{15.5}$$

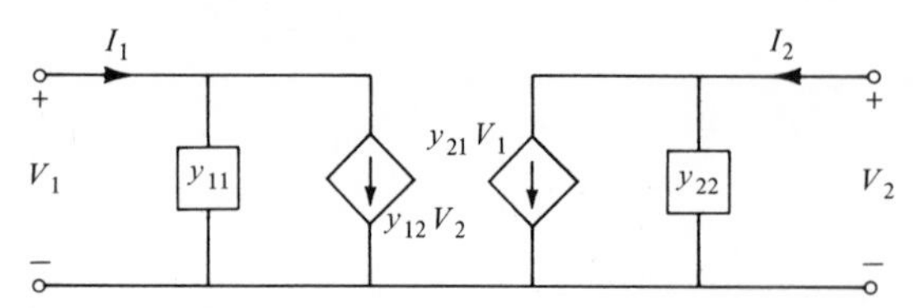

Figure 15.12 A 2-port representation using the short-circuit admittance y-parameters.

and

$$y_{21} = \left.\frac{I_2}{V_1}\right|_{V_2=0}. \tag{15.6}$$

In other words, y_{11} is the driving-point admittance at port 1 with port 2 shorted, and y_{21} is the transfer admittance relating port-2 current to port-1 voltage with port 2 shorted.

Similarly, the shorting of port 1 in Figure 15.11(b) results in the simplification of Equations 15.3 and 15.4, and we find that

$$y_{22} = \left.\frac{I_2}{V_2}\right|_{V_1=0} \tag{15.7}$$

and

$$y_{12} = \left.\frac{I_1}{V_2}\right|_{V_1=0}. \tag{15.8}$$

Thus we see that y_{22} is the driving-point admittance at port 2 with port 1 shorted, and y_{12} is the transfer admittance relating port-1 current and port-2 voltage with port 1 shorted. Because of these interpretations and the dimension of the y-functions, they are known as the *short-circuit admittance functions*. They play an important role in the description of black-box networks.

Figure 15.12 shows a 2-port with two admittances and two voltage-controlled current sources. Application of KCL immediately reveals that this 2-port is described by Equations 15.3 and 15.4. Thus the network is a 2-port representation using the short-circuit admittance parameters.

EXAMPLE 15.2.1

The π-network of Figure 15.13 is an example of a network for which the y-functions can be found by inspection. Thus we see that

$$y_{11} = Y_A + Y_C \tag{15.9}$$

and

$$y_{22} = Y_B + Y_C. \tag{15.10}$$

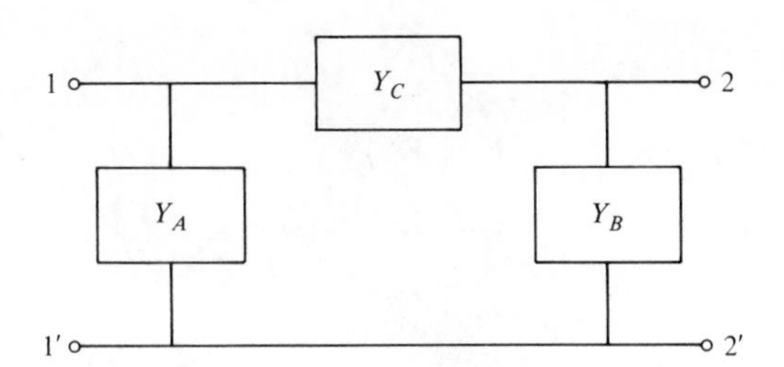

Figure 15.13 A general π-network in which the three arm admittances are Y_A, Y_B, and Y_C. This network is used in Example 15.2.1.

Shorting port 1, we have $I_1 = -Y_C V_2$, and shorting port 2 gives $I_2 = -Y_C V_1$, the minus sign arising because of the reference directions chosen for the currents. Then we have

$$y_{12} = y_{21} = -Y_C. \tag{15.11}$$

If the arm admittances of the network of Figure 15.14 satisfy the condition $Y_A = Y_B$, then we see that

$$y_{11} = y_{22}, \tag{15.12}$$

and the network is said to be *symmetrical*. As we shall see in the next section, the condition

$$y_{12} = y_{21} \tag{15.13}$$

implies that the network is *reciprocal*.

For the 2-port network of Figure 15.14, the inputs are now I_1 and I_2, and the responses are V_1 and V_2. For this case, Equations 15.1 and 15.2 become

$$V_1 = z_{11}I_1 + z_{12}I_2 \tag{15.14}$$

and

$$V_2 = z_{21}I_1 + z_{22}I_2. \tag{15.15}$$

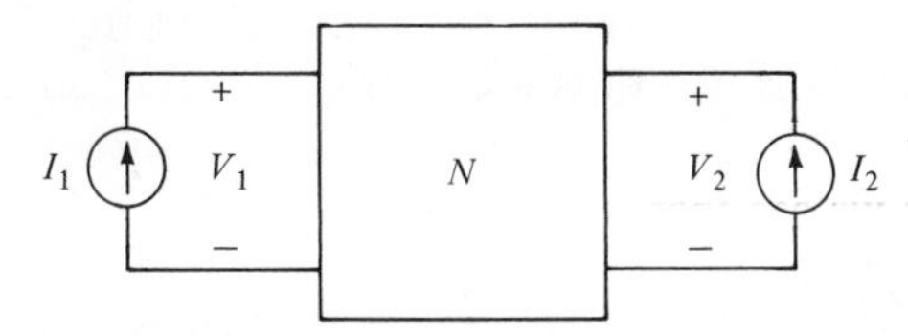

Figure 15.14 The standard 2-port network in which current sources provide the excitation. The network is used in deriving the open-circuit impedance functions.

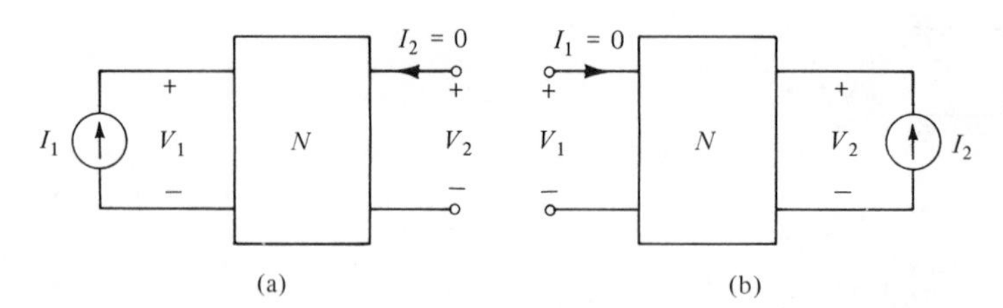

Figure 15.15 The 2-port network of Figure 15.14 modified such that the open-circuit impedance functions are defined.

We have selected z_{ij} notation for the network functions because the dimensions are impedance. The functions are also called *open-circuit functions* for reasons that parallel those given for the y-functions. This time we open-circuit the ports as shown in Figure 15.15. When either $I_1 = 0$ or $I_2 = 0$, Equations 15.14 and 15.15 may be manipulated such that

$$z_{11} = \left.\frac{V_1}{I_1}\right|_{I_2=0}, \tag{15.16}$$

$$z_{21} = \left.\frac{V_2}{I_1}\right|_{I_2=0}, \tag{15.17}$$

$$z_{12} = \left.\frac{V_1}{I_2}\right|_{I_1=0}, \tag{15.18}$$

and

$$z_{22} = \left.\frac{V_2}{I_2}\right|_{I_1=0}. \tag{15.19}$$

Thus we see that z_{11} is the driving-point impedance at port 1 with port 2 open, and z_{22} is the driving-point impedance of port 2 with port 1 open. Further, z_{12} and z_{21} are transfer impedances computed under open-circuit conditions. The four functions constitute the *open-circuit impedance functions* which describe a 2-port network. Figure 15.16 shows a 2-port with two impedances and two current-controlled voltage sources. Writing KVL around the two loops immediately yields Equations 15.14 and 15.15, and so the network is a 2-port representation using open-circuit impedance parameters.

EXAMPLE 15.2.2

For some networks such as the T-network or a ladder network, the defining equations, Equations 15.16 through 15.19, together with their interpretation as shown in Figure 15.16, may be used for routine determination of the z-functions. In

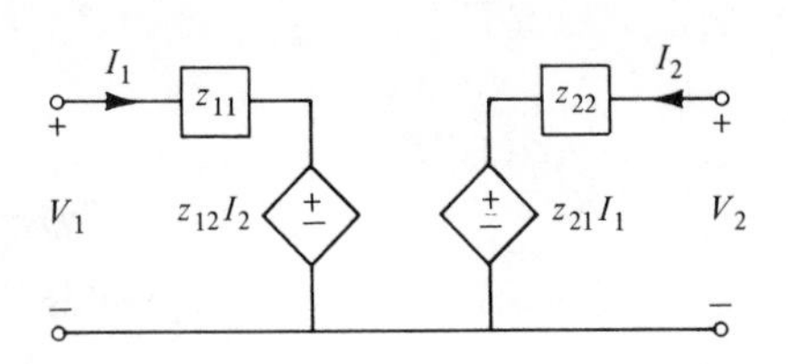

Figure 15.16 A representation using the open-circuit impedance parameters.

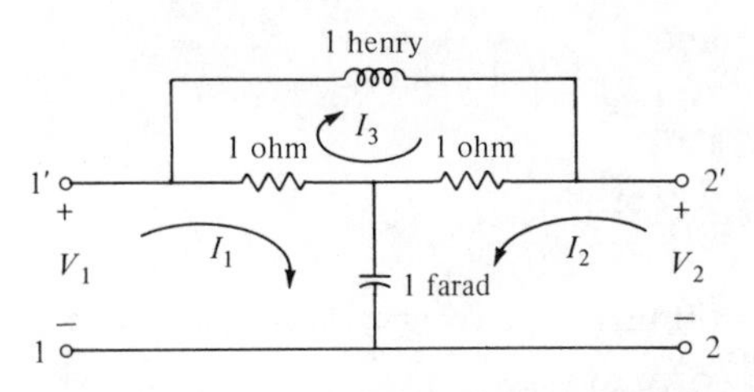

Figure 15.17 A bridged-*T* network which is analyzed in Example 15.2.2.

some cases, it may be necessary to write the network equations. For the network of Figure 15.17, from KVL we may write

$$\left(1+\frac{1}{s}\right)I_1 + \frac{1}{s}I_2 - I_3 = V_1, \tag{15.20}$$

$$\frac{1}{s}I_1 + \left(1+\frac{1}{s}\right)I_2 + I_3 = V_2, \tag{15.21}$$

$$-I_1 + I_2 + (2+s)I_3 = 0. \tag{15.22}$$

Eliminating I_3 from the first two equations, we obtain

$$V_1 = \frac{s^2+2s+1}{s^2+2s}I_1 + \frac{2s+2}{s^2+2s}I_2, \tag{15.23}$$

$$V_2 = \frac{2s+2}{s^2+2s}I_1 + \frac{s^2+2s+1}{s^2+2s}I_2. \tag{15.24}$$

Thus we have

$$z_{11} = z_{22} = \frac{s^2+2s+1}{s^2+2s} \tag{15.25}$$

and

$$z_{12} = z_{21} = \frac{2s+2}{s^2+2s}. \tag{15.26}$$

The network is both symmetrical and reciprocal.

Since Equations 15.3 and 15.4 and Equations 15.14 and 15.15 describe the same 2-port network, it is clear that the z-functions can be expressed in terms of the y-functions and vice versa. If we start with Equations 15.3 and 15.4 and solve for V_1 and V_2, we obtain

$$V_1 = \frac{y_{22}}{\Delta_y}I_1 + \frac{-y_{12}}{\Delta_y}I_2 \tag{15.27}$$

and

$$V_2 = \frac{-y_{21}}{\Delta_y} I_1 + \frac{y_{11}}{\Delta_y} I_2, \tag{15.28}$$

where

$$\Delta_y = y_{11}y_{22} - y_{12}y_{21}. \tag{15.29}$$

Similarly, starting with Equations 15.14 and 15.15 and solving for I_1 and I_2 yields

$$I_1 = \frac{z_{22}}{\Delta_z} V_1 + \frac{-z_{12}}{\Delta_z} V_2 \tag{15.30}$$

and

$$I_2 = \frac{-z_{21}}{\Delta_z} V_1 + \frac{z_{11}}{\Delta_z} V_2, \tag{15.31}$$

where

$$\Delta_z = z_{11}z_{22} - z_{12}z_{21}. \tag{15.32}$$

The relationships we have found between the y- and z-functions are summarized compactly in Table 15.1, together with other relationships important to studies of this section.

Table 15.1 Relationships of 2-port parameters; $\Delta_x = x_{11}x_{22} - x_{12}x_{21}$

	z		y		h		g	
z	z_{11}	z_{12}	$\frac{y_{22}}{\Delta_y}$	$\frac{-y_{12}}{\Delta_y}$	$\frac{\Delta_h}{h_{22}}$	$\frac{h_{12}}{h_{22}}$	$\frac{1}{g_{11}}$	$\frac{-g_{12}}{g_{11}}$
	z_{21}	z_{22}	$\frac{-y_{21}}{\Delta_y}$	$\frac{y_{11}}{\Delta_y}$	$\frac{-h_{21}}{h_{22}}$	$\frac{1}{h_{22}}$	$\frac{g_{21}}{g_{11}}$	$\frac{\Delta_g}{g_{11}}$
y	$\frac{z_{22}}{\Delta_z}$	$\frac{-z_{12}}{\Delta_z}$	y_{11}	y_{12}	$\frac{1}{h_{11}}$	$\frac{-h_{12}}{h_{11}}$	$\frac{\Delta_g}{g_{22}}$	$\frac{g_{12}}{g_{22}}$
	$\frac{-z_{21}}{\Delta_z}$	$\frac{z_{11}}{\Delta_z}$	y_{21}	y_{22}	$\frac{h_{21}}{h_{11}}$	$\frac{\Delta_h}{h_{11}}$	$\frac{-g_{21}}{g_{22}}$	$\frac{1}{g_{22}}$
h	$\frac{\Delta_z}{z_{22}}$	$\frac{z_{12}}{z_{22}}$	$\frac{1}{y_{11}}$	$\frac{-y_{12}}{y_{11}}$	h_{11}	h_{12}	$\frac{g_{22}}{\Delta_g}$	$\frac{-g_{12}}{\Delta_g}$
	$\frac{-z_{21}}{z_{22}}$	$\frac{1}{z_{22}}$	$\frac{y_{21}}{y_{11}}$	$\frac{\Delta_y}{y_{11}}$	h_{21}	h_{22}	$\frac{-g_{21}}{\Delta_g}$	$\frac{g_{11}}{\Delta_g}$
g	$\frac{1}{z_{11}}$	$\frac{-z_{12}}{z_{11}}$	$\frac{\Delta_y}{y_{22}}$	$\frac{y_{12}}{y_{22}}$	$\frac{h_{22}}{\Delta_h}$	$\frac{-h_{12}}{\Delta_h}$	g_{11}	g_{12}
	$\frac{z_{21}}{z_{11}}$	$\frac{\Delta_z}{z_{11}}$	$\frac{-y_{21}}{y_{22}}$	$\frac{1}{y_{22}}$	$\frac{-h_{21}}{\Delta_h}$	$\frac{h_{11}}{\Delta_h}$	g_{21}	g_{22}

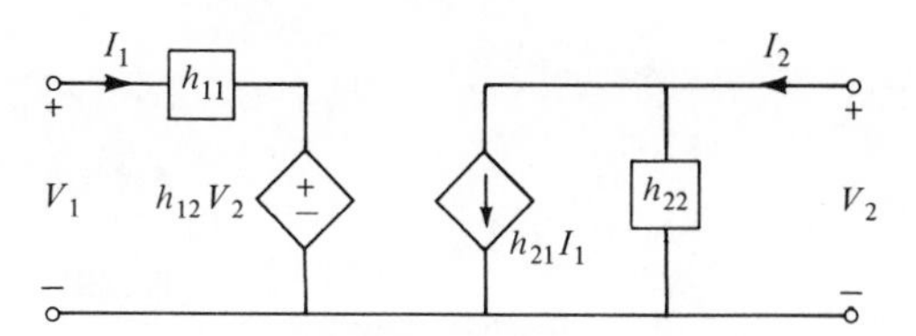

re 15.18 A 2-port representation using the hybrid h-parameters.

Another set of parameters are defined by the equations

$$V_1 = h_{11}I_1 + h_{12}V_2 \tag{15.33}$$

and

$$I_2 = h_{21}I_1 + h_{22}V_2 \tag{15.34}$$

and are known as the *hybrid parameters*. These are particularly important in the representation of transistor networks where it is inconvenient to measure all the y- or z-parameters. Following the pattern of our previous studies, we note that

$$h_{11} = \left.\frac{V_1}{I_1}\right|_{V_2=0}, \tag{15.35}$$

$$h_{21} = \left.\frac{I_2}{I_1}\right|_{V_2=0}, \tag{15.36}$$

$$h_{12} = \left.\frac{V_1}{V_2}\right|_{I_1=0}, \tag{15.37}$$

and

$$h_{22} = \left.\frac{I_2}{V_2}\right|_{I_1=0}. \tag{15.38}$$

We may now see the justification for calling the parameters hybrid. Note that they are dimensionally inhomogeneous, being an impedance, current ratio, voltage ratio, and an admittance, respectively. Note also that h_{11} and h_{21} are short-circuit parameters, while h_{12} and h_{22} are open-circuit parameters. In fact, a simple relationship exists for h_{11} and h_{22} in terms of other parameters:

$$h_{11} = \frac{1}{y_{11}} \quad \text{and} \quad h_{22} = \frac{1}{z_{22}}. \tag{15.39}$$

Figure 15.18 shows a 2-port representation using the hybrid h-parameters.

Still another set of parameters are defined by the equations

$$I_1 = g_{11}V_1 + g_{12}I_2 \tag{15.40}$$

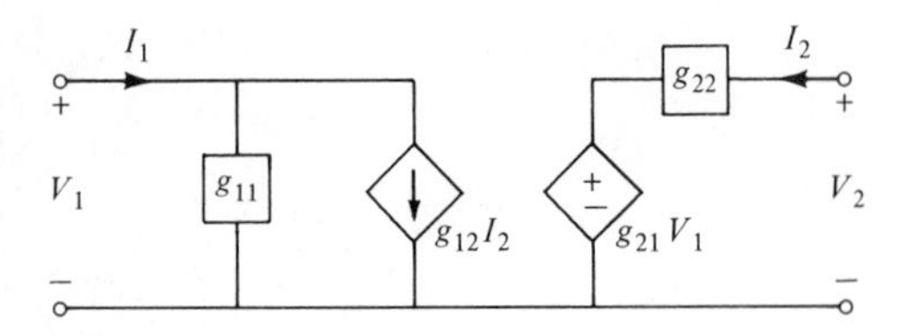

Figure 15.19 A 2-port representation using the inverse hybrid g-parameters.

and

$$V_2 = g_{21}V_1 + g_{22}I_2 \tag{15.41}$$

and are known as the inverse hybrid or g-parameters. The relationship of the h- and g-parameters to each other and to the y- and z-parameters is given in Table 15.1. Figure 15.19 shows a 2-port representation using inverse hybrid g-parameters.

EXAMPLE 15.2.3

The network shown in Figure 15.20 contains one current-controlled voltage source. Making use of Equations 15.35 to 15.38, we find that

$$h_{11} = \frac{R_aR_b + R_aR_c + R_bR_c(1-\alpha)}{R_b + R_c}, \tag{15.42}$$

$$h_{12} = \frac{R_c}{R_b + R_c}, \tag{15.43}$$

$$h_{21} = -\frac{\alpha R_b + R_c}{R_b + R_c}, \tag{15.44}$$

and

$$h_{22} = \frac{1}{R_b + R_c}. \tag{15.45}$$

Two other 2-port representations relating V_1, I_1, V_2, and I_2 are possible. One representation is in terms of *transmission parameters* where V_1 and I_1 are expressed

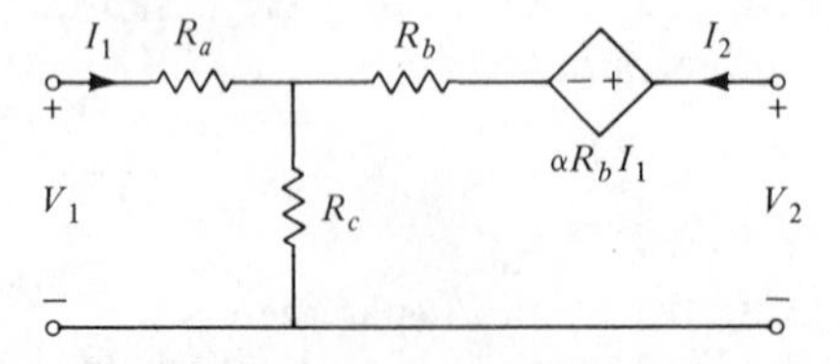

Figure 15.20 A network with a controlled source which is analyzed in Example 15.2.3 for the hybrid h-parameters.

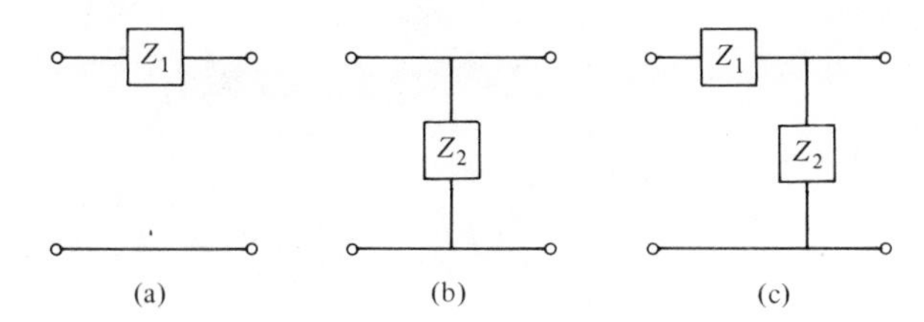

Figure 15.21

in terms of V_2 and I_2 (see Problem 15.5 for details), and another representation is in terms of *inverse transmission parameters* with V_2 and I_2 expressed in terms of V_1 and I_1 (see Problem 15.21 for details). These last two representations are especially useful when dealing with cascades of several 2-ports.

ERCISES

15.2.1 For each of the three networks shown in Figure 15.21, determine the y-, z-, h-, and g-parameters.

15.2.2 Repeat Exercise 15.2.1 for the two resistive networks of Figure 15.22.

15.2.3 For the RC network shown in Figure 15.23, show that

$$y_{11} = \frac{(s+1)(s+3)}{(s+2)(s+4)} \qquad \text{and} \qquad -y_{21} = \frac{k_{21}(s+1)}{(s+2)(s+4)}$$

and determine the numerical value of k_{21}.

15.2.4 Figure 15.24 shows a 2-port network embedded in another resistive network. The network is described by the short-circuit parameters $y_{11} = y_{22} = 2$ mhos, $y_{21} = 2$ mhos, and $y_{12} = 1$ mho. If $I_o = 1$ ampere, determine V_1 and V_2.

15.2.5 Consider the network of Figure 15.20 which is analyzed in Example 15.2.3. (a) Does $h_{11} \neq h_{22}$ imply that the network is nonsymmetrical? Why? (b) Does $h_{21} \neq h_{12}$ imply that the network is nonreciprocal? Why?

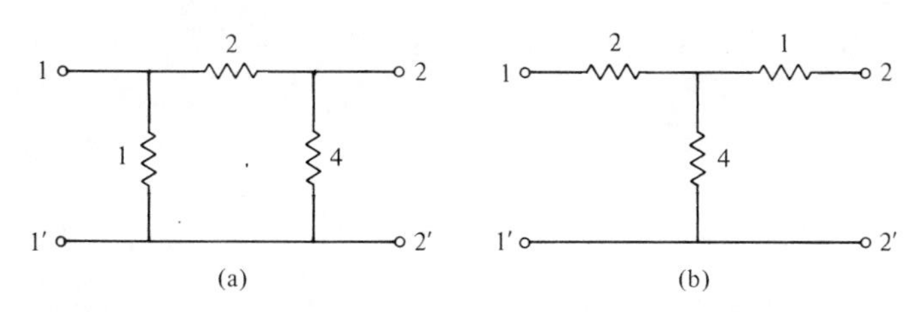

Figure 15.22

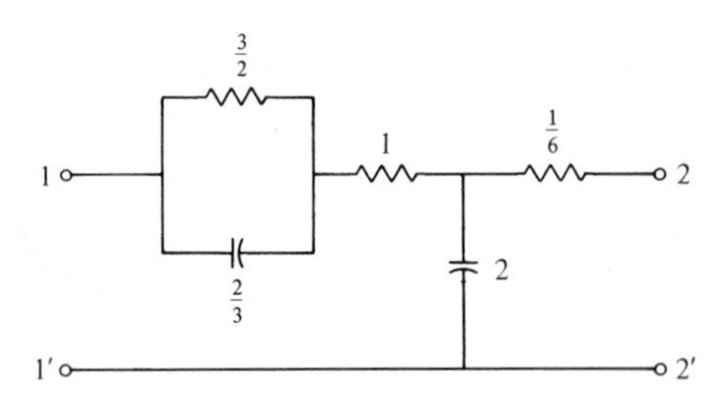

Figure 15.23

15.3 Two-port reciprocity

The principle of reciprocity applies to many physical systems. In antenna theory, for example, it is well known that all the receiving properties of an antenna can be deduced from the known transmitting properties of the same antenna.* Thus the directional radiation pattern for a transmitting antenna is determined by measuring its pattern as a receiving antenna. The reciprocity principle has interesting and important consequences when applied to the 2-port network, as we shall see in this section.

Our derivation relating to 2-port reciprocity stems from Tellegen's theorem which was studied in Section 6.5. Although the theorem was given there in terms of instantaneous quantities, it relies upon KVL and KCL and so applies to phasor quantities representing voltage and current as well as for $V(s)$ and $I(s)$ for exponential signals since we are considering only linear time-invariant networks.

Consider the Network N_1, two views of which are shown in Figure 15.25. While the networks of (a) and (b) are identical, the signals for the two networks are assumed to be different with the voltages and currents distinguished by an addi-

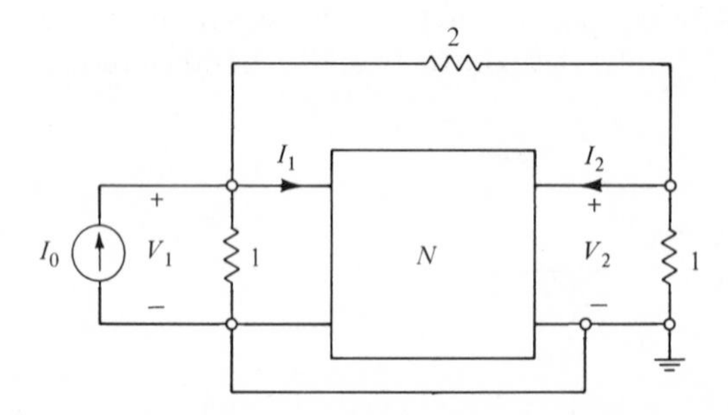

Figure 15.24

*E. C. Jordan and Keith G. Balmain, *Electromagnetic Waves and Radiating Systems,* 2nd edition (Englewood Cliffs, N.J.: Prentice-Hall, Inc., 1968), pp. 346–349, 479–485.

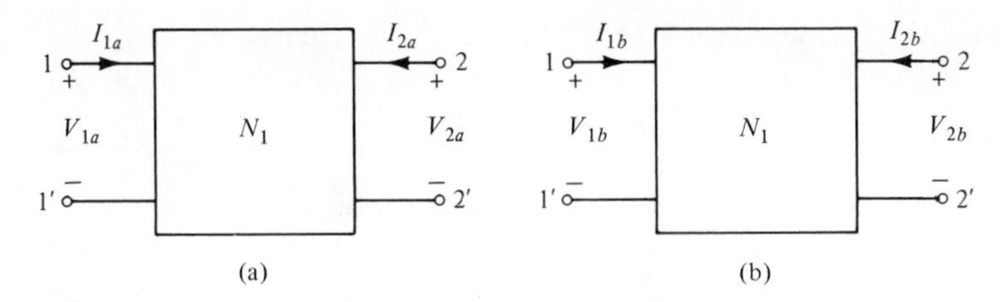

re 15.25 The same network N_1, shown twice, for voltages and currents which are assumed to be different for the two cases.

tional subscript a or b. If the signals are assumed to be sinusoidal or exponential, then Tellegen's theorem of Equation 6.36 may be written in the form

$$V_1 I_1 + V_2 I_2 = \sum_{\substack{k \text{ internal} \\ \text{elements}}} V_k I_k. \tag{15.46}$$

The first two terms in this equation have positive signs because the voltages at the ports have opposite reference conventions than those within the network. Now the amazing thing we discover from Chapter 6 is that it is necessary only that all V's in Equation 15.46 satisfy KVL and that all I's satisfy KCL. We exploit this fact by using voltages from one of the networks of Figure 15.25 and currents from the other, in Equation 15.46. This appears to be a most unusual thing to do, but wait until you see the results! Using the voltages of the network of (a) in Figure 15.25 and the currents from (b), we have the following form of Equation 15.46:

$$V_{1a} I_{1b} + V_{2a} I_{2b} = \sum_k V_{ka} I_{kb}. \tag{15.47}$$

If instead we use the currents of the network of (a) in Figure 15.25 and the voltages from (b), then Equation 15.46 has a different form which is

$$V_{1b} I_{1a} + V_{2b} I_{2a} = \sum_k V_{kb} I_{ka}. \tag{15.48}$$

At this point, we assume that the network contains no independent sources and that each branch is described by

$$V_{ka} = Z_k I_{ka} \tag{15.49}$$

and

$$V_{kb} = Z_k I_{kb}. \tag{15.50}$$

Using these relationships, we may modify the summation of Equation 15.47 as follows:

$$\sum_k V_{ka} I_{kb} = \sum_k (Z_k I_{ka}) I_{kb} = \sum_k (Z_k I_{kb}) I_{ka} = \sum_k V_{kb} I_{ka}, \tag{15.51}$$

which is identical with the summation of Equation 15.48. Hence, we have the important result

$$V_{1a}I_{1b} + V_{2a}I_{2b} = V_{1b}I_{1a} + V_{2b}I_{2a}. \tag{15.52}$$

Any 2-port network for which this identity holds is said to be a *reciprocal* 2-port. The reciprocity formula of Equation 15.52 has been derived, assuming that the 2-port is linear and time-invariant, and that there are no independent sources inside the network. Furthermore, it has been assumed that the internal branches are 1-ports, such that Equations 15.49 and 15.50 hold. Note that our branches can be 1-port subnetworks that may contain elements like gyrators, controlled sources, coupled coils, negative converters, and ideal transformers, in addition to resistors, inductors, and capacitors, as long as Equations 15.49 and 15.50 hold. For a slightly more general situation, see Problem 15.14.

As the first application of the result of Equation 15.52, consider the networks of Figure 15.26 with one port of each network shorted such that $V_{1b} = 0$ and $V_{2a} = 0$. For this condition, Equation 15.52 reduces to

$$V_{1a}I_{1b} = V_{2b}I_{2a}. \tag{15.53}$$

Rearranging this equation, we have

$$\left.\frac{I_{2a}}{V_{1a}}\right|_{V_{2a}=0} = \left.\frac{I_{1b}}{V_{2b}}\right|_{V_{1b}=0}. \tag{15.54}$$

Now the first term is recognized as the short-circuit admittance of N_1, y_{21}, while

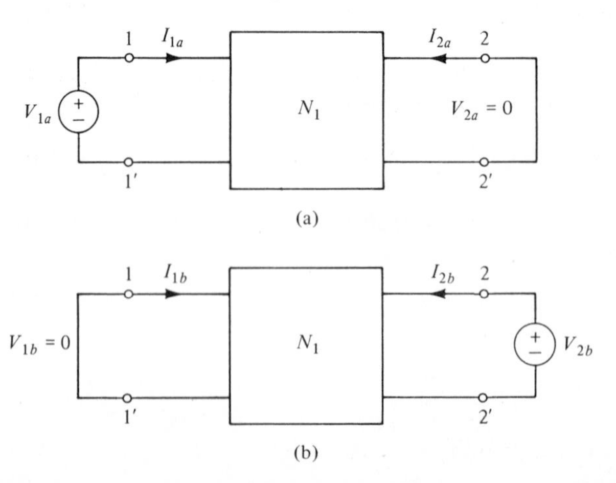

Figure 15.26 The networks of Figure 15.25 with one port of each network shorted. Analysis of these two networks leads to the conclusion that $y_{21} = y_{12}$ under the assumptions of the derivation.

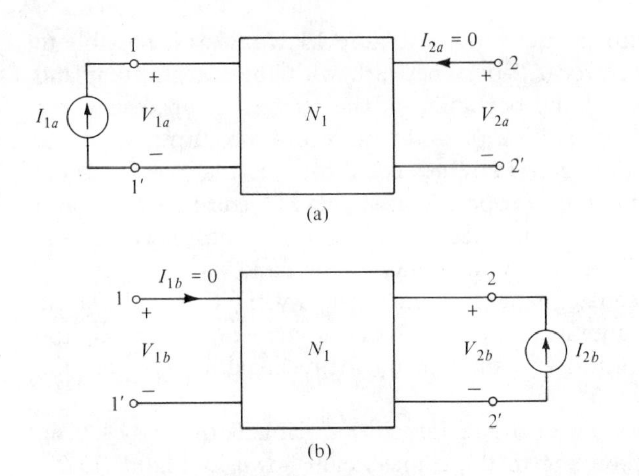

e 15.27 The networks of Figure 15.25 with one port of each network open. These networks are used in the derivation of Equation 15.57 or 15.58.

the second is the short-circuit admittance of N_1 y_{12}. From this result, we see that reciprocity in a 2-port network implies

$$y_{21} = y_{12} \tag{15.55}$$

as stated in the last section.

From Equation 15.54, we also see what might be regarded as the "classical" statement of the reciprocity theorem: In any linear reciprocal network, if a voltage V applied between any two terminals causes a current I in any branch as measured by an ammeter, then the same V and I will be obtained if the positions of the generator and ammeter are interchanged.

As a second application, we let $I_{2a} = 0$ and $I_{1b} = 0$ by the mechanism of opening the ports as shown in Figure 15.27. Under these conditions, Equation 15.52 becomes

$$V_{2a}I_{2b} = V_{1b}I_{1a}. \tag{15.56}$$

This equation may be arranged as

$$\left.\frac{V_{2a}}{I_{1a}}\right|_{I_{2a}=0} = \left.\frac{V_{1b}}{I_{2b}}\right|_{I_{1b}=0}. \tag{15.57}$$

Comparing these equations with those defining the z-parameters, we see that reciprocity implies that

$$z_{21} = z_{12}, \tag{15.58}$$

again a condition stated in the last section.

The analysis associated with the networks in Figure 15.27 makes it possible to compare a property of a reciprocal 2-port network with the transmitting and receiving antennas mentioned at the beginning of the section. Suppose that we connect a source at port 1 in Figure 15.27(a) and then record the output at port 2. The fact that the ratios of Equation 15.57 are equal tells us that for the same network we could connect our source at port 2, as shown in Figure 15.27(b), and then record the output at port 1. If the Network N_1 is a reciprocal network, the outputs for the two cases will be exactly the same. In the same way, the network of (a) may be considered the analog of a receiving antenna while that of (b) is the analog of the transmitting antenna. If the antenna system is reciprocal, the characteristics of the transmitting system will be identical with those for the receiving system.

An interesting third application of the reciprocity principle is made by letting $I_{2a} = 0$ and $V_{1b} = 0$ corresponding to the connections shown in Figure 15.28. Then Equation 15.52 reduces to

$$V_{1a}I_{1b} + V_{2a}I_{2b} = 0, \tag{15.59}$$

which may be written in the following form:

$$\left.\frac{V_{2a}}{V_{1a}}\right|_{I_{2a}=0} = \left.\frac{-I_{1b}}{I_{2b}}\right|_{V_{1b}=0}. \tag{15.60}$$

This result shows that the voltage ratio for one connection of the network, that of (a), is equal to the negative of the inverse current ratio for the connection of the network shown in (b). We see that a Network N_1 driven by a voltage source at

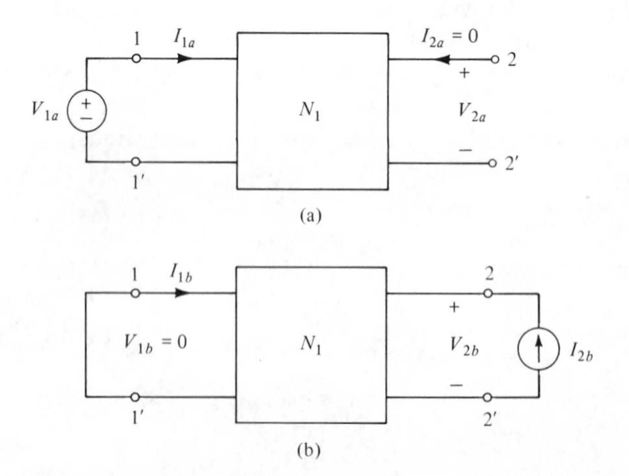

Figure 15.28 The modified networks derived from those of Figure 15.25, from which Equation 15.60 is found.

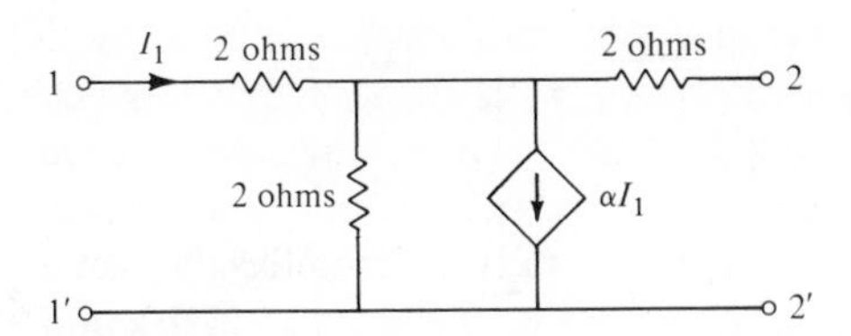

Figure 15.29

port 1 has an open-circuit voltage response at port 2, which is the same as the negative of the short-circuit current response at port 1, if the Network N_1 is driven by a current source at port 2, provided that the excitations are numerically equal, that is, $I_{2b} = V_{1a}$.

RCISES

15.3.1 Investigate the consequences of $I_{1a} = 0$ and $V_{2b} = 0$ in Equation 15.52 interpreted in terms of the networks of Figure 15.25. Compare your conclusions with those reached from Equation 15.60.

15.3.2 Starting with the equations

$$V_1 = z_{11}I_1 + z_{12}I_2 \qquad \text{and} \qquad V_2 = z_{21}I_1 + z_{22}I_2,$$

derive Equation 15.60 for the networks of Figure 15.25 under the condition that $z_{12} = z_{21}$.

15.3.3 We have seen that $y_{12} = y_{21}$ and $z_{12} = z_{21}$ for reciprocal networks. What is the corresponding condition for a reciprocal network in terms of the h-parameters?

15.3.4 The network of Figure 15.29 contains a current-controlled current source. For what range of values of α will the 2-port network be reciprocal?

15.3.5 We have shown in this section that all *RLC* 2-ports (with linear time-invariant elements) are reciprocal. What can we say about 2-port networks containing negative *R*'s?

15.3.6 Give an example of a reciprocal 2-port which contains a controlled source. Your network should remain reciprocal, regardless of the values of the parameters in the controlled source.

.4 Linear incremental models for some electronic 2-ports

The development of useful circuit models for new electronic devices is based on an understanding of the physical and quantum electronics of the device itself, coupled with the proper interpretation of measurements made at the terminals.

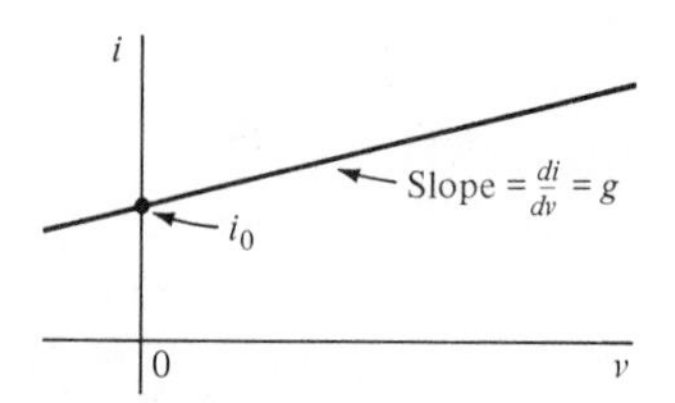

Figure 15.30 A straight-line v–i characteristic.

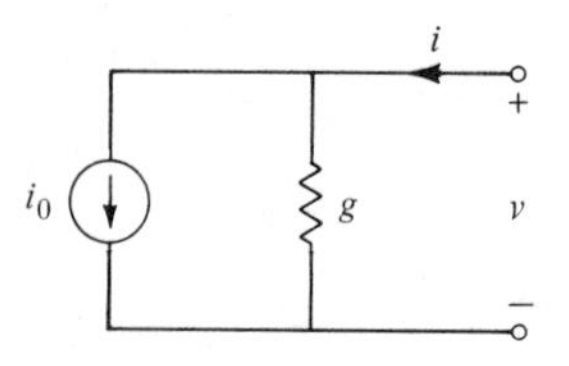

Figure 15.31 Circuit whose v–i characteristic is that shown in Figure 15.30.

The development of useful circuit models follows the invention phase, and is the result of cycles of proposing a model and testing the adequacy of this model in representing measurements — a process in which only successful models survive. Success is often based on simplicity in terms of the number of circuit elements required. In this section, we will consider some simple linear circuit models. In doing so, we should stress that some applications will require more complex models than will be treated here. The ability of the electrical engineer to select the model to match the needs of a particular problem is important, and is the result of experience in the laboratory coupled with computer simulation.

The linear models to be considered are intended for use with small signals superimposed on some time-invariant bias voltage or current. Consider a hypothetical 1-port device with a straight-line v–i characteristic as shown in Figure 15.30. The equation of this line is

$$i = gv + i_0, \tag{15.61}$$

where g is the slope and i_0 is the vertical-axis intercept. The 1-port network shown in Figure 15.31 with a resistor of conductance g and a current source i_0 in parallel is described by the v–i characteristic of Equation 15.61, and hence it is a circuit model for the v–i characteristics of Figure 15.30. Suppose that the i_0 intercept depends on another variable, say v_1, as given by the linear relationship

$$i_0 = g_m v_1, \tag{15.62}$$

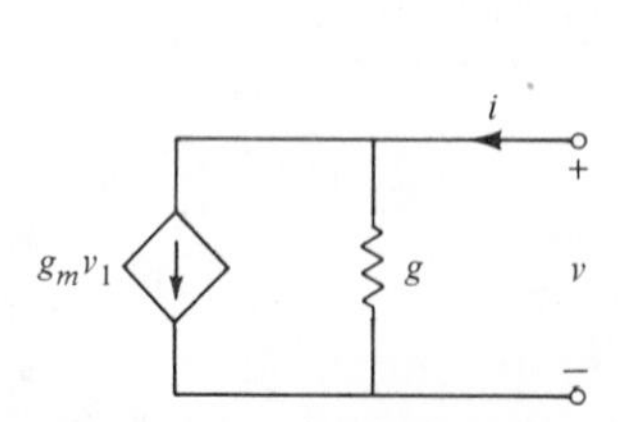

Figure 15.32 Circuit whose v–i characteristic is that shown in Figure 15.33.

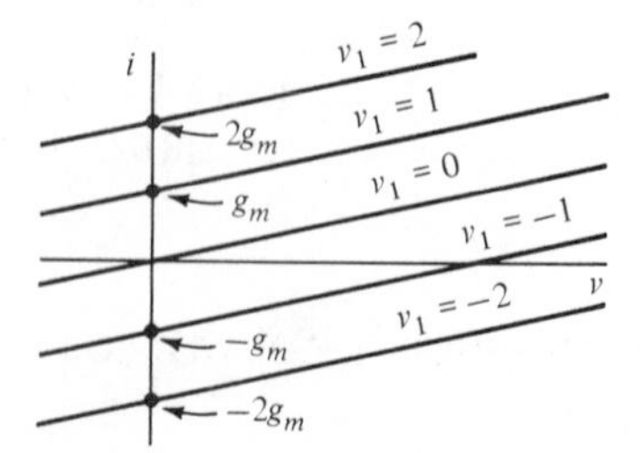

Figure 15.33 A family of v–i characteristics corresponding to Equation 15.63.

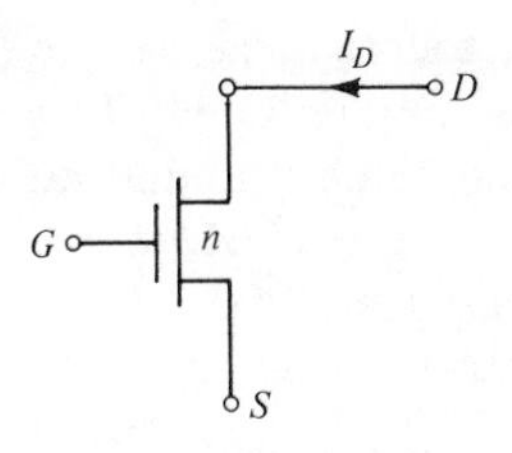

Figure 15.34 The symbol for n-channel metal-oxide semiconductor field effect transistor (MOSFET), where G, D, and S stand for gate, drain, and source terminals.

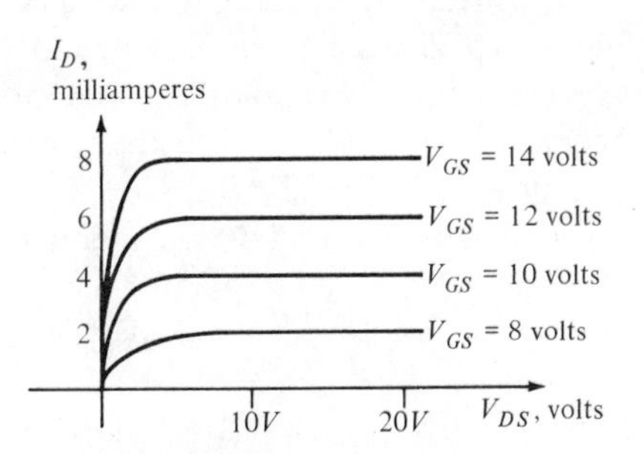

Figure 15.35 Typical v–i characteristics for a MOSFET.

where g_m is a constant. The source i_0 becomes a voltage-controlled current source as shown in Figure 15.32 and Equation 15.61 becomes

$$i = gv + g_m v_1. \tag{15.63}$$

Figure 15.33 shows a family of straight lines for different values of v_1. The coefficient g_m equals the change in i per unit change in v_1, holding v constant.

Three-terminal 2-port semiconductor devices are adequately described for some applications at low frequencies by static v–i characteristics at the ports. Typically, these characteristics also vary with temperature, so that several v–i characteristics and a range of temperatures must be given. For higher-frequency applications, capacitive effects are not negligible. These capacitive effects are described by terminal charge characteristics. Figure 15.35 shows a typical characteristic for a metal-oxide semiconductor field-effect transistor (MOSFET) whose schematic diagram is given in Figure 15.34. The current i_G is essentially zero. This MOSFET characteristic is similar to that for a field-effect transistor (FET).

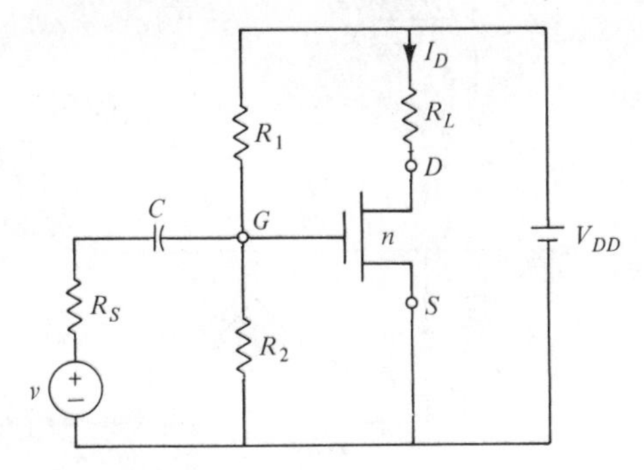

Figure 15.36 A MOSFET used in an amplifier network.

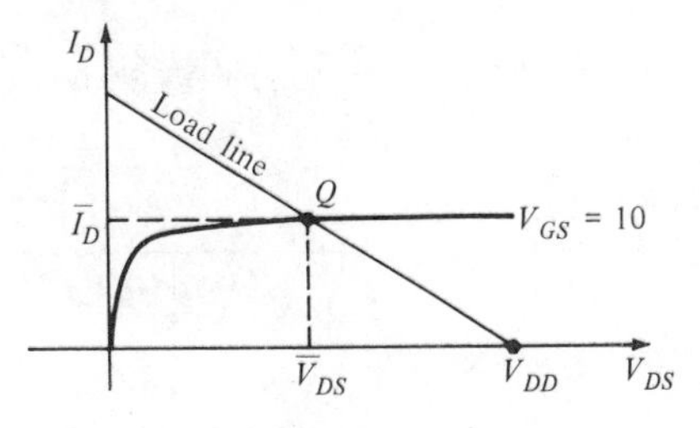

Figure 15.37 Illustrating the determination of the operating point for the network of Figure 15.36.

For operation beyond the so-called knee of the curves, the characteristic in Figure 15.35 may be approximated by straight lines, as in Figure 15.33. This approximation is valid if the device is operated with small signals around an operating point Q. To illustrate, consider the network of Figure 15.36, which includes a MOSFET. Suppose that v is zero. In the steady-state, there is no capacitor current, and V_{GS}, I_D, and V_{DS} are constant. Since $I_G = 0$,

$$V_{GS} = \frac{R_2}{R_1 + R_2} V_{DD}. \tag{15.64}$$

For a given V_{DD}, R_1 and R_2 may be chosen so that a specified or desired V_{GS} is achieved. For example, for $V_{DD} = 20$ volts, $R_1 = R_2$, then $V_{GS} = 10$ volts, by Equation 15.64. I_G and V_{DS} must be on the specific curve with $V_{GS} = 10$ volts, as shown in Figure 15.37. Furthermore, from the circuit diagram of Figure 15.36, we see that I_D and V_{DS} must satisfy the equation

$$I_D = \frac{V_{DD} - V_{DS}}{R_L} = -\frac{1}{R_L} V_{DS} + \frac{V_{DD}}{R_L}. \tag{15.65}$$

This is the equation of a straight line, and is shown in Figure 15.37 labeled as the *load line*. It has a slope of $-1/R_L$ and horizontal intercept of V_{DD} (or vertical intercept of V_{DD}/R_L). Since I_D and V_{DS} must be on the load line as well as on the characteristic curve corresponding to $V_{GS} = 10$, the only possibility is for I_D and V_{DS} to be at the intersection labeled Q in Figure 15.37, yielding $I_D = \bar{I}_D$ and $V_{DS} = \bar{V}_{DS}$. R_L and V_{DD} may be chosen to achieve the desired intersection.

For small v, the quantities I_D and V_{DS} will be near $\bar{I}_D$ and $\bar{V}_{DS}$. If the characteristic is given by the function

$$I_D = I_D(V_{DS}, V_{GS}), \tag{15.66}$$

then the approximation may be written

$$\begin{aligned} I_D &= I_D(\bar{V}_{DS}, \bar{V}_{GS}) + \frac{\partial I_D}{\partial V_{DS}}(\bar{V}_{DS}, \bar{V}_{GS})(V_{DS} - \bar{V}_{DS}) + \frac{\partial I_D}{\partial V_{GS}}(\bar{V}_{DS}, \bar{V}_{GS})(V_{GS} - \bar{V}_{GS}) \\ &= \bar{I}_D + g(V_{DS} - \bar{V}_{DS}) + g_m(V_{GS} - \bar{V}_{GS}), \end{aligned} \tag{15.67}$$

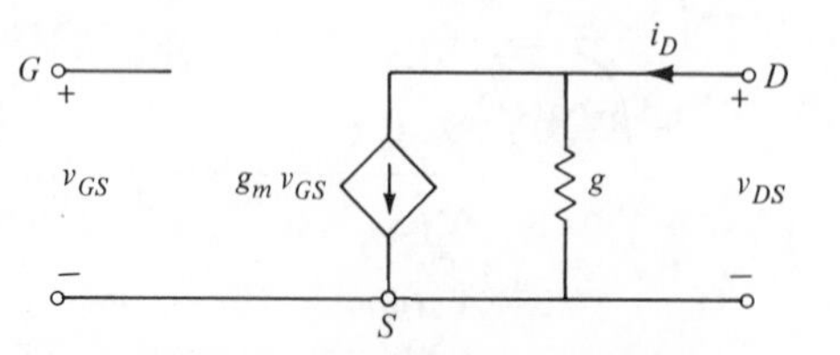

Figure 15.38 Linear incremental circuit model for MOSFET.

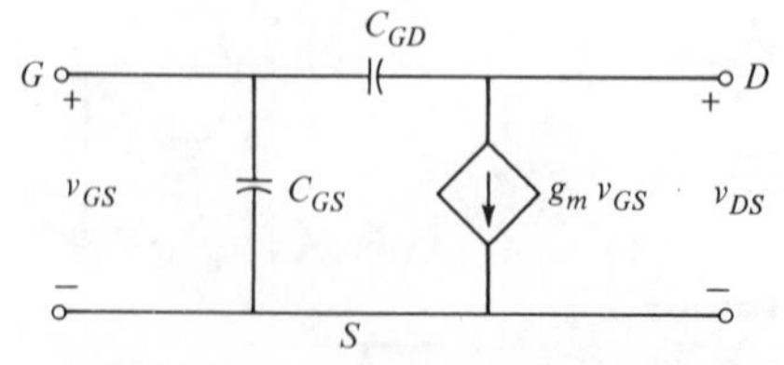

Figure 15.39 Linear incremental circuit model for the FET.

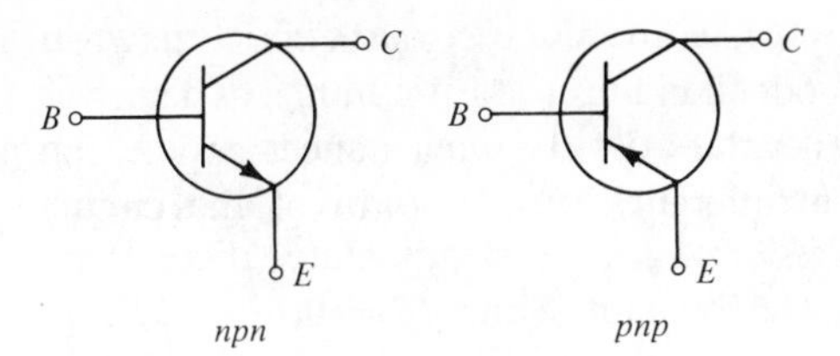

re 15.40 Symbols for *npn* and *pnp* bipolar transistors where *B*, *C*, and *E* stand for base, collector, and emitter terminals.

which is a straight line tangent to the nonlinear characteristic curve at the operating point Q. Using the deviation or incremental variables defined as

$$i_D = I_D - \bar{I}_D, \tag{15.68}$$

$$v_{DS} = V_{DS} - \bar{V}_{DS}, \tag{15.69}$$

and

$$v_{GS} = V_{GS} - \bar{V}_{GS}, \tag{15.70}$$

then we find that Equation 15.67 becomes

$$i_D = g v_{DS} + g_m v_{GS}. \tag{15.71}$$

The corresponding incremental circuit model is shown in Figure 15.38. Since the model is linearized, superposition applies. The network is analyzed first setting the input signal to zero and computing the operating point due to the bias voltage alone. Then the incremental circuit is analyzed, due to the input alone, setting the bias voltage to zero. The actual signals are the superposition of the incremental signals and the constant signals from the operating-point analysis.

For field-effect transistors (FETs), the shunt conductance is practically negligible. However, for higher frequencies, capacitive effects are important. Figure 15.39 shows a suitable FET linear incremental model.

Another type of widely used transistor is the bipolar junction transistor whose symbol is shown in Figure 15.40. A suitable linear incremental model is shown in

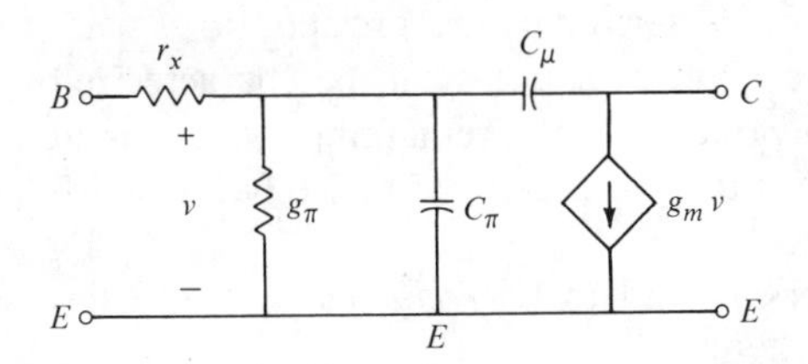

ure 15.41 Hybrid π linear incremental model for the bipolar transistors.

Figure 15.41. This model is known as a hybrid π because it consists of a π-network with a resistor r_x added. This hybrid π model includes the FET model as a special case when r_x and g_π are zero. Numerical values for the capacitances and g_m for these different classes of transistors are different, but the form of the circuit models is similar.

EXERCISES

15.4.1 Suppose that a FET device is modeled by characteristics shown in Figure 15.35, where i is the current I_D in milliamperes, v is V_{DS} in volts, and v_1 is V_{GS} in volts. Suppose that the vertical intercepts are 1 milliampere apart and that for a constant V_{GS} the increase in I_D per unit volt increase of V_{DS} is 0.1 milliampere. Find an equivalent circuit as in Figure 15.38.

15.4.2 Suppose that in Figure 15.36 $v = V_0 \sin \omega_0 t$ and that the impedance of C is negligible compared to R_s. If $g_m = 5 \times 10^{-3}$ mho, $g = 0$, $R_s = 1$ kilohm, $R_1 = R_2 = 50$ kilohms, $R_L = 3$ kilohms, then determine the incremental voltage v_{DS} using the linear incremental model of a FET in Figure 15.38.

15.4.3 Suppose that for the circuit in Figure 15.36, it is desired to locate the operating point Q at $\bar{I}_D = 4$ milliamperes and $\bar{V}_{DS} = 10$ volts. Suppose that V_{GS} is chosen so that this point is on the characteristic curve and suppose that V_{DD} is 24 volts. Determine the value of R_L that is needed.

15.4.4 Determine the inverse hybrid parameters for the model shown in Figure 15.38. What about its hybrid h-parameters?

15.5 Summary

1. An n-port network is one in which the $2n$ terminals are grouped in n pairs called ports. A 2-port with no internal independent sources can be characterized in six different ways, relating two terminal variables to the remaining two terminal variables. Using the reference convention shown in Figure 15.6, we have the information shown at the top of page 423.

2. A linear time-invariant 2-port network is said to be *reciprocal* if it satisfies

$$V_{1a}I_{1b} + V_{2a}I_{2b} = V_{1b}I_{1a} + V_{2b}I_{2a},$$

Parameters	Equations
(a) Open-circuit impedance	$V_1 = z_{11}I_1 + z_{12}I_2$ $V_2 = z_{21}I_1 + z_{22}I_2$
(b) Short-circuit admittance	$I_1 = y_{11}V_1 + y_{12}V_2$ $I_2 = y_{21}V_1 + y_{22}V_2$
(c) Hybrid	$V_1 = h_{11}I_1 + h_{12}V_2$ $I_2 = h_{21}I_1 + h_{22}V_2$
(d) Inverse hybrid	$I_1 = g_{11}V_1 + g_{12}I_2$ $V_2 = g_{21}V_1 + g_{22}I_2$
(e) Transmission	$V_1 = AV_2 - BI_2$ $I_1 = CV_2 - DI_2$
(f) Inverse transmission	$V_2 = A'V_1 - B'I_1$ $I_2 = C'V_1 - D'I_1$

where the subscripts a and b refer to two different conditions for the same network. In particular, any 2-port network which consists of an interconnection of linear time-invariant 1-port networks is reciprocal. When a network is reciprocal, $z_{12} = z_{21}$ and $y_{12} = y_{21}$ provided the parameters exist.

3. Three-terminal semiconductor devices, particularly field-effect transistors and bipolar transistors can be biased and operated in a linear mode, when the signals are small. The total signal may be considered as a superposition of dc signals obtained from an operating-point analysis, and small signals obtained from an incremental circuit analysis. An adequate linear incremental model for a transistor is a hybrid π consisting of two resistors, two capacitors, and a voltage-controlled current source. For a FET, the controlled source alone is sufficient as a model for low frequencies. For a bipolar transistor, the shunt resistance r_π together with the controlled source form a sufficiently good approximation for modeling at low frequencies.

OBLEMS

15.1 The network shown in Figure 15.42 contains coupled inductors L_1 and L_2 of mutual inductance M. (a) For this network, find the z-parameters. (b) From the results of (a), show that if $R_1 = R_2$ and $L_1 = L_2$, then $z_{11} = z_{22}$.

15.2 Given a 2-port network composed of an interconnection of linear time-invariant 1-port networks for which $z_{12} = z_{21}$ (and $y_{12} = y_{21}$): (a) Show that the T-network in Figure 15.43 is an equivalent network with respect

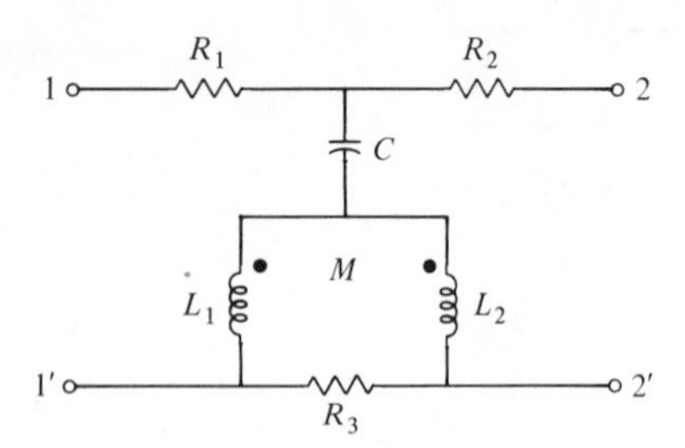

Figure 15.42

to measurements made at the ports (but no other place). (b) Show that the π-network of Figure 15.43 is equivalent to the 2-port network for measurements made at the ports.

15.3 Given the 2-port network of Figure 15.44, in which port 2 is terminated in Z_L: (a) Find the driving-point impedance at port 1 in terms of the z-functions and Z_L. (b) Repeat part (a) but in terms of the y-functions and $Y_L = 1/Z_L$.

15.4 For the capacitive network shown in Figure 15.45(a), determine the T-section equivalent network shown in (b) of the figure which is the equivalent 2-port network for an excitation.

15.5 For the series of problems that follow, we will consider another set of parameters for characterizing 2-port networks known as the *transmission parameters*. These parameters serve to relate parameters at one port to those parameters for the other port. We have

$$V_1 = AV_2 - BI_2 \qquad \text{and} \qquad I_1 = CV_2 - DI_2.$$

(a) Define the A, B, C, and D parameters in terms of short-circuit or open-circuit measurements made on the 2-port network as in Equations 15.5 to 15.8. (b) Express each of the transmission parameters in terms of the z-functions; the y-functions.

15.6 Find the transmission parameters for the network shown in Figure 15.46 if $\alpha = 2$ and $\beta = 2$.

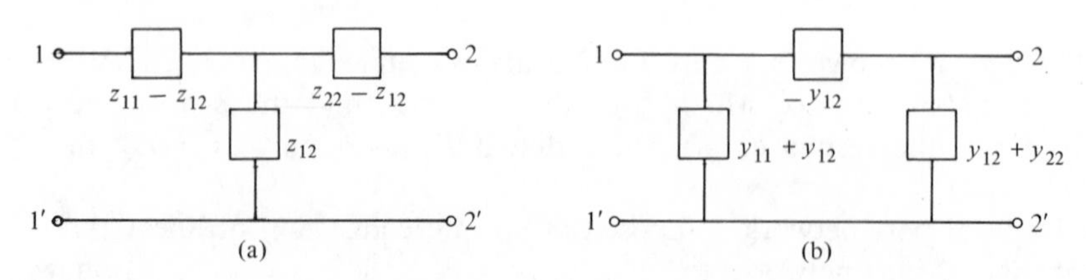

Figure 15.43

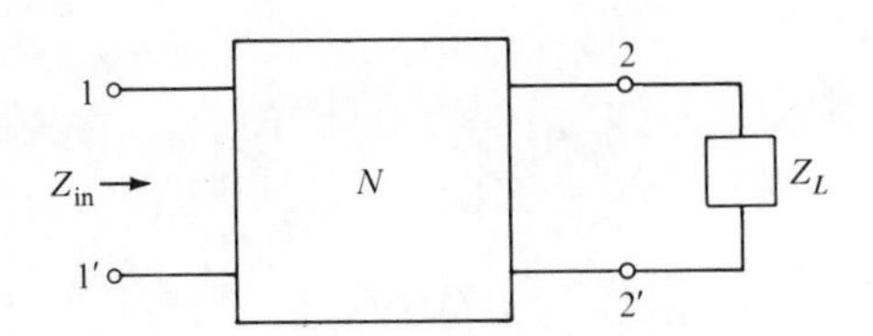

Figure 15.44

15.7 Show that for a 2-port network to be reciprocal, it is necessary that $AD - BC = 1$.

15.8 Repeat Problem 15.3 expressing Z_{in} in terms of the A-, B-, C-, and D-parameters and Z_L.

15.9 Two 2-port networks are connected in cascade as shown in Figure 15.47. Let the transmission parameters of the two networks be distinguished by subscripts a and b. Determine the driving-point impedance Z_{in} in terms of the eight transmission parameters and Z_L.

15.10 Given the network of Figure 15.48 with $R_x = 2$ ohms. (a) Determine the value of g (if one exists) for which the 2-port is reciprocal. (b) Determine the value of g (if one exists) for which the 2-port is electrically symmetrical.

15.11 Repeat Problem 15.10 with $R_x = 1$ ohm.

15.12 Consider the network of Figure 15.46. Does there exist a combination of values of α and β for which the 2-port network is reciprocal? Express in the form of an equation if possible.

15.13 We are given a resistive 2-port network for which an equivalent T- or π-section is desired for further analysis under different load conditions. We are permitted only driving-point measurements at port 1 and port 2. How many and what kinds of measurements must be made to determine the equivalent T- or π-section?

15.14 In our discussion of 2-port reciprocity, we have assumed that the 2-port network is linear, time-invariant, with no independent sources inside, and that it is an interconnection of 1-ports. Consider the 4-port network shown

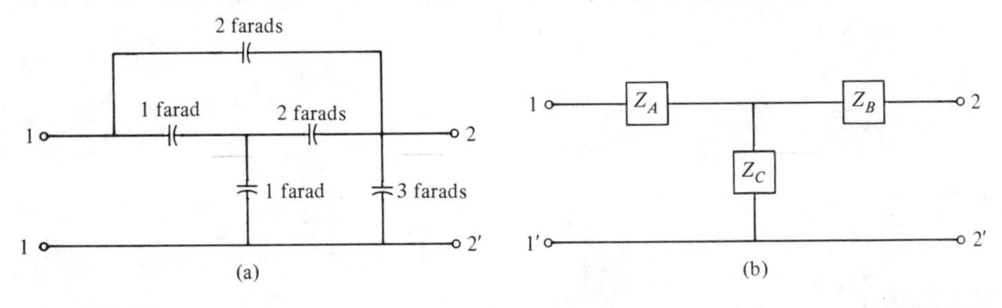

Figure 15.45

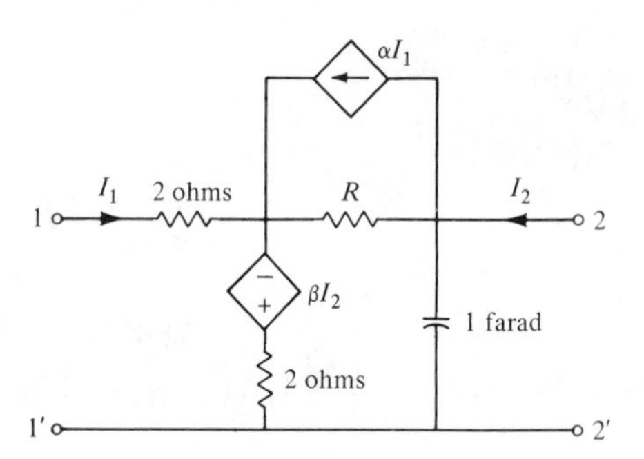

Figure 15.46

in Figure 15.49 in which Network N contains only 1-ports satisfying Equations 15.49 and 15.50. Using Tellegen's theorem of Chapter 6, show that this network exhibits 2-port reciprocity with respect to ports 1 and 2. Generalize this result to show that a 2-port network with any number of coupled inductors is reciprocal.

★15.15 A textbook published in 1948 states the reciprocity theorem in the following manner:

If at any particular frequency the only generator of a linear, bilateral network, located in the ith *branch, and an impedance in the* kth *branch equal to the internal impedance of the generator are interchanged, the current of the* kth *branch before the interchange is the same as the current in the* ith *branch after the interchange.*

Using the approach in Section 15.3, prove (or disprove) this theorem.

15.16 Repeat Problem 15.14, with the coupled inductors replaced by an ideal transformer.

★15.17 Instead of Equations 15.49 and 15.50, assume that

$$V_k = \sum_j Z_{kj} I_j$$

and $Z_{kj} = Z_{jk}$ for all k and j. Derive the reciprocity formula of Equation 15.52.

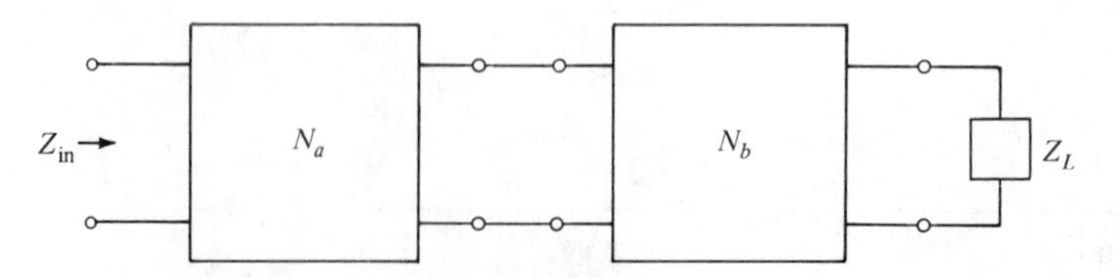

Figure 15.47

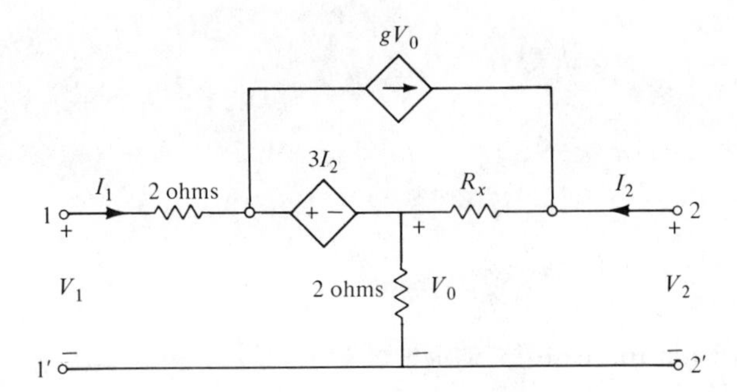

Figure 15.48

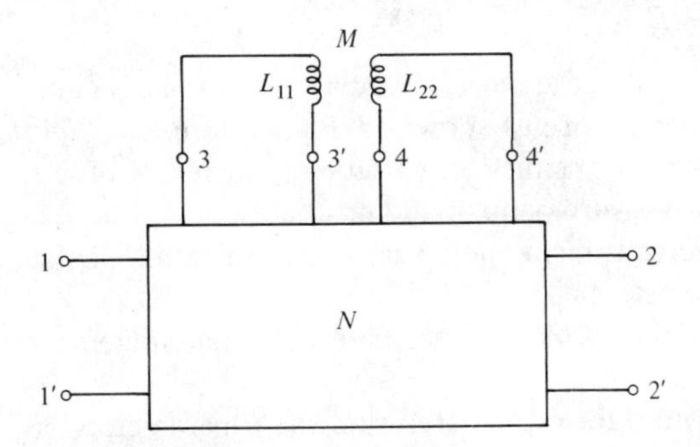

Figure 15.49

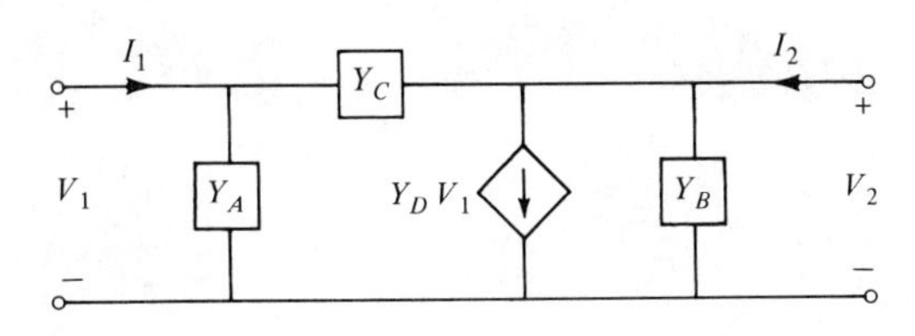

Figure 15.50

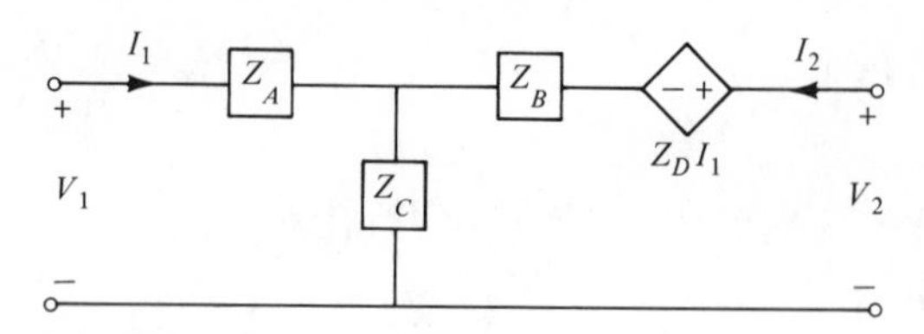

Figure 15.51

15.18 Repeat Problem 15.12 for the case in which C is replaced by a 1-ohm resistor in Figure 15.46.

15.19 Express Y_A, Y_B, Y_C, and Y_D in terms of the short-circuit y-parameters for the network in Figure 15.50.

15.20 Express Z_A, Z_B, Z_C, and Z_D in terms of the open-circuit z-parameters for the network in Figure 15.51.

15.21 A 2-port representation called the inverse transmission parameter representation expresses V_2 and I_2 in terms of V_1 and I_1 as follows:

$$V_2 = A'V_1 - B'I_1, \qquad I_2 = C'V_1 - D'I_1,$$

where A', B', C', and D' are called the *inverse transmission parameters* and the voltage and current references are as shown in Figure 15.6.
(a) Express each of the inverse transmission parameters in terms of the short-circuit or open-circuit measurements made on the 2-port.
(b) Express each of the inverse transmission parameters in terms of the open-circuit impedance parameters.

15.22 Using exponential signals of the form e^{st}, determine the h-parameters of the hybrid π model of Figure 15.41.

15.23 Using exponential signals of the form e^{st}, determine the y-parameters of the hybrid π model of Figure 15.41.

Chapter **16**

ɿree-Phase Power Networks

One of the most important electrical systems in most countries is the electric power system for generation, transmission, distribution, and consumption of electric energy. In terms of installed capital investment, it is the largest single industry in the United States. In terms of complexity, most power networks have thousands of nodes (buses), and these networks are themselves interconnected into a super power network in the whole country. With a few exceptions of high voltage dc transmissions, the electrical network consists of three-phase synchronous generators, transformers, transmission lines, distribution lines, control and telemetering equipment, and loads (end users). The primary sources of energy for the electric generators are fossil fuels, nuclear fuels, and hydroenergy. The synchronous generators are driven by steam turbines in the case of fossil and nuclear fuels and by hydroturbines in the case of hydroelectric plants.

The virtues of sinusoidal waveforms have been extolled in previous chapters, and the voltages generated by these giant synchronous machines are practically sinusoidal with a frequency of 60 Hz in the United States, or 50 Hz in many other countries. Furthermore, a set of three voltages is generated, where the amplitudes are equal and displaced 120° from each other. The three-phase generators have significant advantages over the single-phase machines in that for the same power to be transmitted and the same amount of conductor material, the energy loss in the transmission is less for three-phase than it is for single-phase machines. Moreover, as will be demonstrated below, although voltages and currents are sinusoidal, the power $p(t)$ delivered to a balanced load is constant for a three-phase system, but sinusoidal for a single-phase system. In this chapter we will discuss some basic concepts underlying three-phase network operations.

16.1 Balanced *Y*-*Y* systems

The simplest model for a three-phase synchronous generator (or alternator) is a set of three voltage sources:

$$\begin{aligned} v_a(t) &= V_p \sin \omega t, \\ v_b(t) &= V_p \sin (\omega t - 120°), \\ v_c(t) &= V_p \sin (\omega t - 240°), \end{aligned} \tag{16.1}$$

where V_p is a positive constant and ω is a fixed frequency, usually 60 Hz (see Section 2.4). Almost all alternators are connected in a Y structure where the voltages v_a, v_b, and v_c, called the *phase voltages*, are arranged as shown in Figure 16.1(a). When v_b lags v_a by 120° and v_c lags v_b by 120° as in Equation 16.1, we say that the *phase sequence* is *abc*. Otherwise, if v_c lags v_a by 120° and if v_b lags v_c by 120°, we say that the phase sequence is *acb*.

With the representation

$$\begin{aligned} v_a(t) &= \text{Im}\,(V_p e^{j\omega t}) = \text{Im}\,(V_a e^{j\omega t}), \\ v_b(t) &= \text{Im}\,(V_p e^{j\omega t} e^{-j120°}) = \text{Im}\,(V_b e^{j\omega t}), \end{aligned} \tag{16.2}$$

and

$$v_c(t) = \text{Im}\,(V_p e^{j\omega t} e^{-j240°}) = \text{Im}\,(V_c e^{j\omega t}),$$

the phasors V_a, V_b, and V_c are shown in Figure 16.1(b). The voltages v_{12}, v_{23}, and v_{31} are called *line voltages*. They are easily obtained by first determining V_{12}, V_{23}, and V_{31} by phasor analysis. By inspection of Figure 16.1(a) it is clear that

$$\begin{aligned} V_{12} &= V_a - V_b, \\ V_{23} &= V_b - V_c, \\ V_{31} &= V_c - V_a. \end{aligned} \tag{16.3}$$

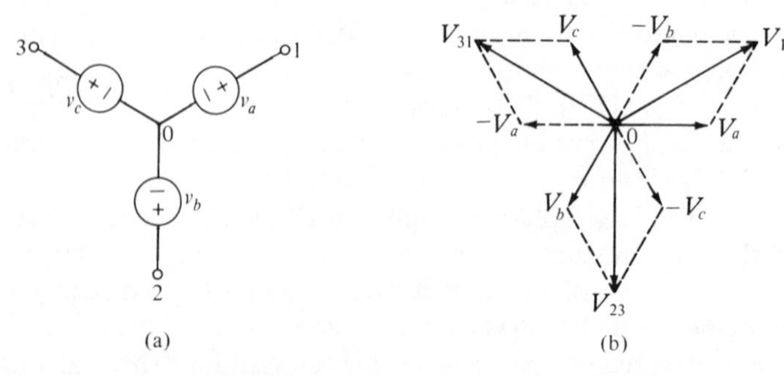

Figure 16.1 (a) Balanced Y-connected three-phase source. (b) Phasor diagram for three-phase source of (a).

From the geometrical construction in Figure 16.1(b) it is seen that V_{12}, V_{23}, and V_{31} are equal in magnitude and displaced 120° from each other. Since V_{12} leads V_a by 30°, inspection of the phasor diagram reveals that

$$|V_{12}| = 2|V_a| \cos 30° = \sqrt{3}\,|V_a| = \sqrt{3}\,V_p. \tag{16.4}$$

Thus

$$v_{12}(t) = \sqrt{3}\,V_p \sin(\omega t + 30°),$$
$$v_{23}(t) = \sqrt{3}\,V_p \sin(\omega t - 90°), \tag{16.5}$$

and

$$v_{31}(t) = \sqrt{3}\,V_p \sin(\omega t - 210°).$$

The simplest type of three-phase system is shown in Figure 16.2, where a Y-connected three-phase generator is connected to a Y-connection of three identical impedances. Furthermore, the common node 0 is connected to the common node 0′. This connection is called the *neutral wire.* The set of three impedances is called a *balanced Y-connected load.* Examination of the circuit diagram in Figure 16.2 verifies that

$$I_1 = \frac{V_a}{Z},$$
$$I_2 = \frac{V_b}{Z}, \tag{16.6}$$

and

$$I_3 = \frac{V_c}{Z}.$$

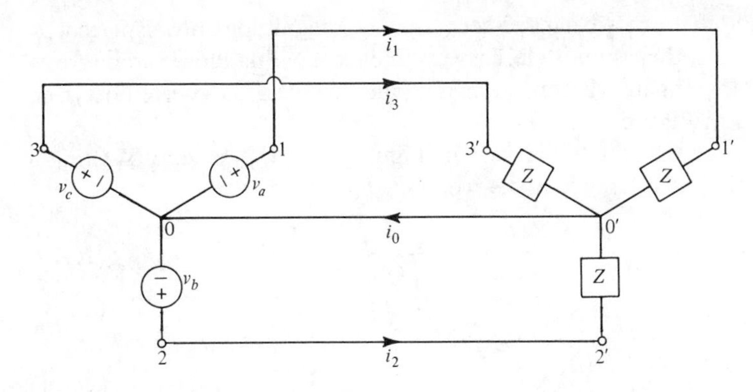

ure 16.2 Y-Y balanced three-phase system.

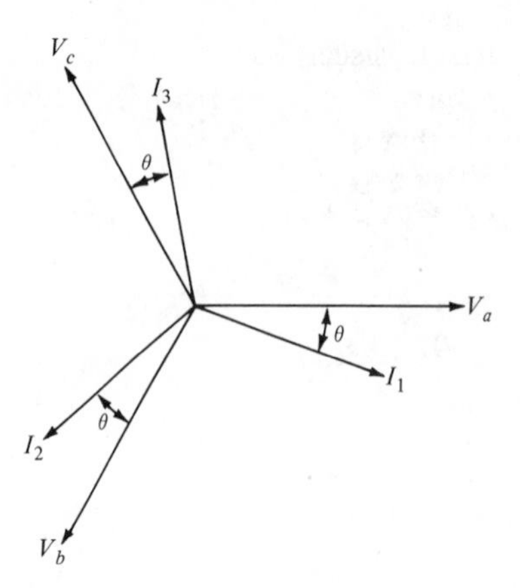

Figure 16.3 Phasor diagram for three-phase system shown in Figure 16.2.

Thus if Z is written in exponential form

$$Z = |Z|e^{j\theta}, \tag{16.7}$$

then

$$I_1 = \frac{V_p}{|Z|} e^{-j\theta} \equiv |I_1|e^{-j\theta}. \tag{16.8}$$

Since V_a, V_b, and V_c are symmetrically displaced from each other, so are I_1, I_2, and I_3. Thus I_1, I_2, and I_3 are equal in magnitude, I_2 lags I_1 by 120°, and I_3 lags I_2 by 120°. The phasor diagram is shown in Figure 16.3.

Writing KCL at node 0′

$$I_0 = I_1 + I_2 + I_3. \tag{16.9}$$

Because I_1, I_2, and I_3 form a balanced set (equal in magnitude and displaced by 120° from each other), their sum is clearly zero. Hence $I_0 = 0$. This is an interesting result. Let us investigate what happens if an impedance Z_0 is inserted between 0 and 0′ as shown in Figure 16.4.

Let us demonstrate that for the system in Figure 16.4, $V_{0'0}$ is zero. Writing a node equation with 0 as reference node, we have

$$\frac{V_{0'0} - V_a}{Z} + \frac{V_{0'0} - V_b}{Z} + \frac{V_{0'0} - V_c}{Z} + \frac{V_{0'0}}{Z_0} = 0. \tag{16.10}$$

Simplifying, we get

$$\left(\frac{3}{Z} + \frac{1}{Z_0}\right) V_{0'0} = \frac{1}{Z}(V_a + V_b + V_c). \tag{16.11}$$

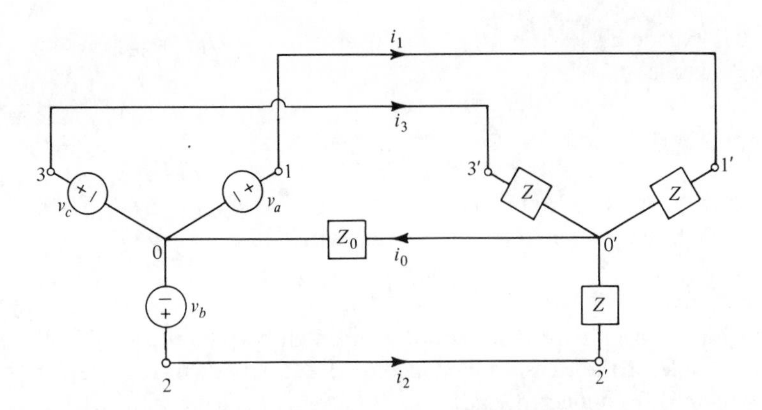

ure 16.4 Balanced three-phase Y-Y connected system with impedance in neutral line.

Since V_a, V_b, and V_c form a balanced system,

$$V_a + V_b + V_c = 0. \tag{16.12}$$

From (16.11) this implies that $V_{0'0}$ is also zero. But if $V_{0'0}$ is zero, then I_0 which is $V_{0'0}/Z_0$ is zero, regardless of the value of Z_0 (including $Z_0 = \infty$) provided that $Z_0 \neq 0$. The case of $Z_0 = 0$ was already analyzed earlier, and even for that case, $I_0 = 0$. We conclude that the neutral connection in Figure 16.4 has no consequence. Hence the currents in Figure 16.4 are equal to the corresponding currents in Figure 16.2.

Let us determine the average power delivered to Z in phase 1, using the formula in Chapter 10,

$$P_1 = \text{Re}\left[\frac{V_a}{\sqrt{2}}\left(\frac{I_1}{\sqrt{2}}\right)^*\right]. \tag{16.13}$$

Notice that the length of the phasors V_a and I_1 are peak values rather than rms values, so we have to divide each by $\sqrt{2}$. Thus

$$P_1 = \frac{|V_a|}{\sqrt{2}}\frac{|I_1|}{\sqrt{2}}\cos\theta. \tag{16.14}$$

Similarly, it is seen that

$$P_2 = \frac{|V_b|}{\sqrt{2}}\frac{|I_2|}{\sqrt{2}}\cos\theta$$

and

$$P_3 = \frac{|V_c|}{\sqrt{2}}\frac{|I_3|}{\sqrt{2}}\cos\theta. \tag{16.15}$$

Since the system is balanced, $|V_a| = |V_b| = |V_c|$ and $|I_1| = |I_2| = |I_c|$. Thus the total average power is

$$P_T = P_1 + P_2 + P_3 = 3 \frac{V_P}{\sqrt{2}} \frac{|I_1|}{\sqrt{2}} \cos \theta . \tag{16.16}$$

Denoting $|I_1|/\sqrt{2}$ by I_l (rms value of line current) and $\sqrt{3}\, V_P/\sqrt{2}$ by V_l (rms value of line voltage), Equation 16.16 can be written as

$$P_T = \sqrt{3}\, V_l I_l \cos \theta . \tag{16.17}$$

Equation 16.17 is an important formula for a balanced three-phase load. Recall that θ is the phase angle of the phase impedance load Z. As we will see later, it holds also for a balanced Δ-connected load.

EXAMPLE 16.1.1

Consider the balanced three-phase system in Figure 16.4 where the neutral line is disconnected ($Z_0 = \infty$). Let

$$v_a(t) = 141.4 \sin (377t + 30°) \text{ volts},$$

and let the phase sequence be *abc*; that is, v_b lags v_a by 120° and v_c lags v_b by 120°. Thus

$$v_b(t) = 141.4 \sin (377t - 90°)$$

and

$$v_c(t) = 141.4 \sin (377t - 210°) .$$

Let $Z = 5.6 + j1.8$ ohms. Using peak values for the lengths of our phasor representations, we have

$$V_a = 141.4e^{j30°}, \qquad V_b = 141.4e^{-j90°}, \qquad V_c = 141.4e^{-j210°}.$$

The current phasors are

$$I_1 = \frac{141.4e^{j30°}}{5.6 + j1.8} = \frac{141.4e^{j30°}}{5.9e^{j17.8°}} = 24e^{j12.2°},$$

$$I_2 = 24e^{-j107.8°},$$

$$I_3 = 24e^{j132.2°}.$$

Hence

$$i_1(t) = 24 \sin (377t + 12.2°) \text{ amperes},$$

$$i_2(t) = 24 \sin (377t - 107.8°) \text{ amperes},$$

$$i_3(t) = 24 \sin (377t + 132.2°) \text{ amperes}.$$

The total average power delivered to the load is

$$P_T = \sqrt{3}\left(\sqrt{3}\,\frac{141.4}{\sqrt{2}}\right)\frac{24}{\sqrt{2}}\cos 17.8^\circ = 4320 \text{ watts}.$$

Let us see how this compares with the total instantaneous power delivered to the load

$$p_T(t) = v_a(t)i_1(t) + v_b(t)i_2(t) + v_c(t)i_3(t).$$

Using previous computations, we have

$$\begin{aligned} p_T(t) = (141.4)(24)[&\sin(377t + 30^\circ)\sin(377t + 12.2^\circ) \\ &+ \sin(377t - 90^\circ)\sin(377t - 107.8^\circ) \\ &+ \sin(377t - 210^\circ)\sin(377t + 132.3^\circ)], \end{aligned}$$

$$\begin{aligned} p_T(t) = \frac{(141.4)(24)}{2}[&\cos 17.8^\circ - \cos(754t + 42.2^\circ) \\ &+ \cos 17.8^\circ - \cos(754t - 197.8^\circ) \\ &+ \cos 17.8^\circ - \cos(754t - 77.8^\circ)]. \end{aligned} \tag{16.18}$$

Notice that $\cos(754t + 42.2^\circ)$, $\cos(754t - 197.8^\circ)$, and $\cos(754t - 77.8^\circ)$ are displaced 120° from each other, so that their sum is zero. Thus

$$p_T(t) = \frac{3(141.4)(24)}{2}\cos 17.8^\circ = 4320 \text{ watts}.$$

The most important aspect of this answer is that it is a constant! That is, in spite of the sinusoidal character of the three voltages and three currents, the total instantaneous power delivered into the three-phase load is a constant. This constant equals the total average power.

The last result in the above example holds for any balanced system. If v_p and i_p denote the peak values of the phase voltages and currents, respectively, then

$$\begin{aligned} p_T(t) = v_p i_p[&\sin\omega t\sin(\omega t - \theta) + \sin(\omega t - 120^\circ)\sin(\omega t - 120 - \theta) \\ &+ \sin(\omega t - 240^\circ)\sin(\omega t - 240 - \theta)]. \end{aligned}$$

Thus

$$p_T(t) = \frac{3v_p i_p}{2}\cos\theta = \sqrt{3}\,V_l I_l\cos\theta. \tag{16.19}$$

ERCISES

16.1.1 A balanced three-phase Y-connected generator has a phase-sequence *abc*, and

$$v_a(t) = 1414\sin(377t + 25^\circ) \text{ volts}.$$

Determine the expressions for the other phase voltages v_b, v_c, and the line voltages v_{12}, v_{23}, and v_{31}. See Figure 16.1.

16.1.2 Repeat Exercise 16.1.1 if the phase sequence is *acb*.

16.1.3 A balanced three-phase Y-connected generator has a phase sequence *abc*, and the line voltage v_{12} is

$$v_{12}(t) = 2000 \sin 377t \text{ volts}$$

(see Figure 16.1). Determine the other line voltages v_{23} and v_{31} and the phase voltages v_a, v_b, and v_c.

16.1.4 A balanced three-phase Y-connected generator is connected to a balanced Y-connected load as in Figure 16.4, with the neutral line disconnected. The line voltage is 220 volts rms, and the line current is 10 amperes rms. The total average power delivered to the inductive load is 3000 watts. Determine the impedance in each phase of the load. Using a phase sequence of *abc* and assuming V_a as a reference (along the positive real axis), draw a phasor diagram showing all the voltages and currents.

16.2 Δ-*Y* transformations

We will find that the analysis of Δ-connected power networks is greatly simplified when the Δ-network is replaced by an equivalent Y-connected network. In this section, we derive the transformations for converting a *Δ-connected network*, as shown in Figure 16.5(b) to a Y-connected network and a Y-connected network to a Δ-connected network. These transformations are also useful in communication network problems.

Consider the Y- and Δ-networks in Figure 16.5. For equivalence, we impose the condition that the impedance between any two terminals for one network be equal to the corresponding impedance between the same terminals for the other network. Thus, for the Y-connected network the impedance between terminals 1 and 2 is

$$Z_{12} = Z_1 + Z_2.$$

For the Δ-connected network, Z_{12} is

$$Z_{12} = \frac{Z_a(Z_b + Z_c)}{Z_a + Z_b + Z_c}.$$

Thus we set

$$Z_1 + Z_2 = \frac{Z_a(Z_b + Z_c)}{Z_a + Z_b + Z_c}. \tag{16.20}$$

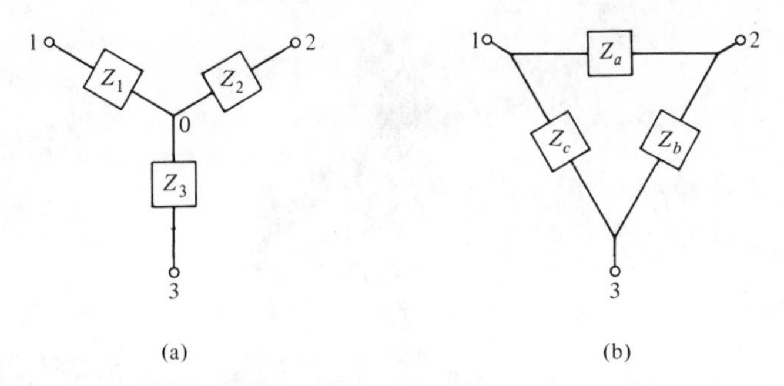

ure 16.5 (a) Y-connected network. (b) Δ-connected network. In communication networks the Y-network is known as a T-network, and the Δ-network is known as a π-network.

Similarly

$$Z_2 + Z_3 = \frac{Z_b(Z_a + Z_c)}{Z_a + Z_b + Z_c} \tag{16.21}$$

and

$$Z_3 + Z_1 = \frac{Z_c(Z_a + Z_b)}{Z_a + Z_b + Z_c}\cdot \tag{16.22}$$

Adding Equations 16.20 and 16.22, subtracting Equation 16.21 from the result, and dividing by 2, we have

$$Z_1 = \frac{Z_a Z_c}{Z_a + Z_b + Z_c}\cdot \tag{16.23}$$

Similarly, adding Equations 16.20 and 16.21, subtracting Equation 16.22, and dividing the result by 2, we obtain

$$Z_2 = \frac{Z_a Z_b}{Z_a + Z_b + Z_c}\cdot \tag{16.24}$$

Finally, adding Equations 16.21 and 16.22, subtracting Equation 16.20, and dividing by 2, we have

$$Z_3 = \frac{Z_b Z_c}{Z_a + Z_b + Z_c}\cdot \tag{16.25}$$

Equations 16.23, 16.24, and 16.25 are the formulas for converting from a Δ-connection to a Y-connection. Each leg of the Y-connection is obtained by multiplying the adjacent impedances and dividing by the sum of the impedances in the Δ. Verify this rule by examining Equations 16.23, 16.24, and 16.25.

We now develop formulas for the Δ impedances in terms of the Y-impedances.

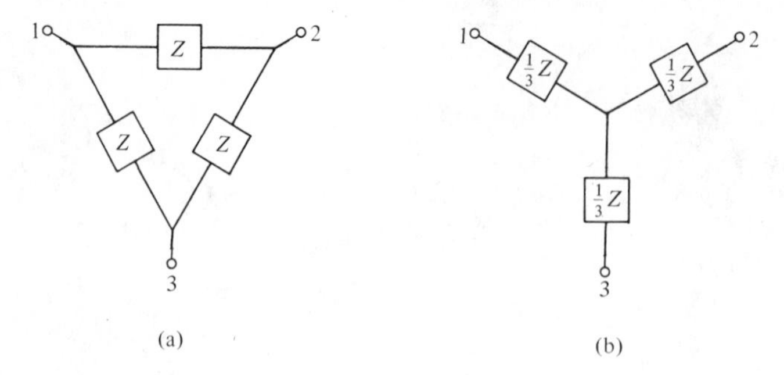

Figure 16.6 Two equivalent networks.

This can be done by solving for Z_a, Z_b, and Z_c in terms of Z_1, Z_2, and Z_3 in Equations 16.23, 16.24, and 16.25. It is readily verified that if we form $Z_1Z_2 + Z_2Z_3 + Z_3Z_1$ and divide by Z_3, using Equations 16.23, 16.24, and 16.25, we obtain Z_a:

$$Z_a = \frac{Z_1Z_2 + Z_2Z_3 + Z_3Z_1}{Z_3}. \tag{16.26}$$

Similarly

$$Z_b = \frac{Z_1Z_2 + Z_2Z_3 + Z_3Z_1}{Z_1} \tag{16.27}$$

and

$$Z_c = \frac{Z_1Z_2 + Z_2Z_3 + Z_3Z_1}{Z_2}. \tag{16.28}$$

These formulas are easily remembered if it is noted that the numerators are the same, and each is the sum of products of the Y-impedances taken two at a time. The denominator is the Y-impedance opposite the Δ-impedance that is being computed. Thus Z_3 in Figure 16.5(a) is opposite Z_a in Figure 16.5(b).

EXAMPLE 16.2.1

An important Δ-connected network is shown in Figure 16.6(a) where the individual impedances are equal. This is known as a balanced Δ-connected network. Applying Equations 16.23, 16.24, and 16.25, we obtain the Y-connected network in Figure 16.6(b). That is, each Y-impedance is $\frac{1}{3}$ of each Δ-impedance. Conversely, each Δ-impedance is 3 times each Y-impedance.

EXERCISES

16.2.1 Verify Equations 16.23, 16.24, and 16.25 by performing the details of the derivation suggested in the section.

16.2.2 Verify Equations 16.26, 16.27, and 16.28 by performing the details of the derivation suggested in the section.

16.2.3 A Δ-connected network has impedances $Z_a = 1 + j2$ ohms, $Z_b = 3 + j2$ ohms, and $Z_c = 5 + j1$ ohms, connected as in Figure 16.5(b). Determine its Y-connected equivalent network.

16.3 Balanced Y-Δ systems

A simple power system consisting of a balanced three-phase Y-connected generator and a balanced Δ-connected load is shown in Figure 16.7. All the material on the balanced 3-phase generator in Section 16.1 applies to the generator of this section. If the phase sequence is *abc*, the phasor diagram of Figure 16.1(b) applies. In this simple system the line voltages at the source side V_{12}, V_{23}, and V_{31} are equal to the corresponding line voltages $V_{1'3'}$, $V_{2'3'}$, and $V_{3'1'}$ at the load side. The phase load currents are easily computed as

$$I_{1'2'} = \frac{V_{12}}{Z},$$

$$I_{2'3'} = \frac{V_{23}}{Z},$$

and

$$I_{3'1'} = \frac{V_{31}}{Z}.$$

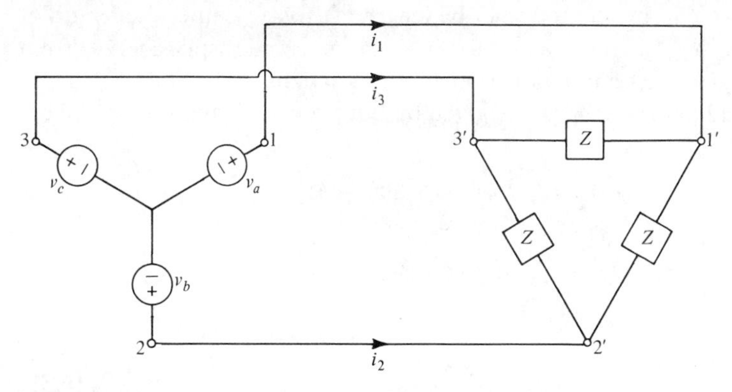

Figure 16.7 A simple balanced Y-Δ three-phase system.

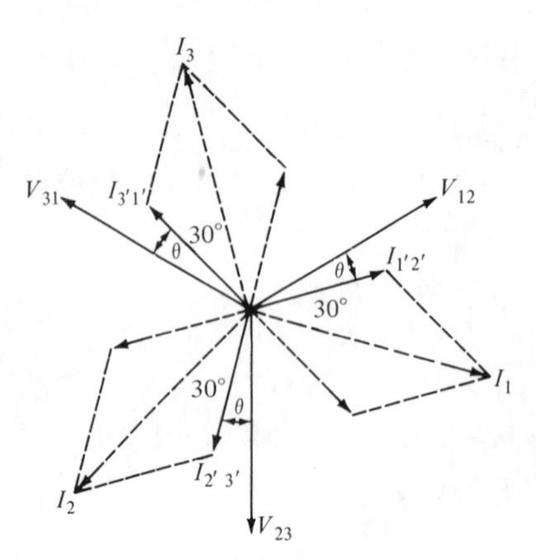

Figure 16.8 Phasor diagram for the Y-Δ three-phase system of Figure 16.7 with a load made up of an inductor and resistor in each phase.

Since V_{12}, V_{23}, and V_{31} are a set of balanced voltages, and since Z is the same for each phase, then $I_{1'2'}$, $I_{2'3'}$, and $I_{3'1'}$ are also a set of balanced currents. Figure 16.8 shows the phasor diagram, where the phase currents lag the phase voltages by θ (when the load is inductive so that the power factor is $\cos\theta$ lagging). From Figure 16.7 it is clear that the phasor line currents are

$$\begin{aligned} I_1 &= I_{1'2'} - I_{3'1'}, \\ I_2 &= I_{2'3'} - I_{1'2'}, \\ I_3 &= I_{3'1'} - I_{2'3'}. \end{aligned} \tag{16.29}$$

These phasors may be obtained geometrically as shown in Figure 16.8. It is readily verified that I_1 lags $I_{1'2'}$ by 30° and that the length of I_1 is $\sqrt{3}$ times the length of $I_{1'2'}$. Furthermore, I_1, I_2, and I_3 are equal in magnitude and displaced 120° from each other. Thus, if v_a, v_b, and v_c are as given in Equations 16.1, the line voltages (and hence the load phase voltages) are given by Equation 16.5, and the load phase currents are

$$\begin{aligned} i_{1'2'}(t) &= I_p \sin(\omega t + 30° - \theta), \\ i_{2'3'}(t) &= I_p \sin(\omega t - 90° - \theta), \\ i_{3'1'}(t) &= I_p \sin(\omega t - 210° - \theta), \end{aligned} \tag{16.30}$$

where

$$I_p = \frac{\sqrt{3}\, V_p}{|Z|}. \tag{16.31}$$

Observe that V_p is the peak value of the generator phase voltage. The line currents are

$$\begin{aligned} i_1(t) &= \sqrt{3}\, I_p \sin(\omega t - \theta), \\ i_2(t) &= \sqrt{3}\, I_p \sin(\omega t - 120° - \theta), \\ i_3(t) &= \sqrt{3}\, I_p \sin(\omega t - 240° - \theta). \end{aligned} \tag{16.32}$$

The average power per phase is

$$P_{\text{av}}\,(\text{per phase}) = \frac{\sqrt{3}\, V_p}{\sqrt{2}} \frac{I_p}{\sqrt{2}} \cos\theta,$$

so that the total average power delivered to the load is

$$P_{\text{av}} = 3P_{\text{av}}\,(\text{per phase}) = \frac{3\sqrt{3}\, V_p I_p}{2} \cos\theta. \tag{16.33}$$

Since the rms value of the line voltage V_l is $\sqrt{3}\, V_p/\sqrt{2}$ and the rms value of the line current I_l is $\sqrt{3}\, I_p/\sqrt{2}$, Equation 16.33 can be written as

$$P_{\text{av}} = \sqrt{3}\, V_l I_l \cos\theta. \tag{16.34}$$

This is exactly the same formula we obtained for the balanced Y-Y system.

The total instantaneous power $p_T(t)$ delivered to the load is

$$p_T(t) = v_{12}(t) i_{1'2'}(t) + v_{23}(t) i_{2'3'}(t) + v_{31}(t) i_{3'1'}(t). \tag{16.35}$$

By substitution of the expressions from Equations 16.5 and 16.30 into Equation 16.35 and simplifying, the expression reduces to

$$p_T(t) = \frac{3\sqrt{3}\, V_p I_p}{2} \cos\theta, \tag{16.36}$$

which is the same as Equation 16.33. Thus the total instantaneous power into a balanced 3-phase Δ-connected load is a constant!

From Equations 16.31 and 16.32 it is seen that the peak value of the line current is $3V_p/|Z|$ and that the line current lags the generator phase voltage by θ. This same result is also obtained by considering the Y equivalent of the Δ load. From Example 16.2.1 it is seen that the impedances of the equivalent Y-connected load are $\frac{1}{3}Z$ each. From the procedure of Section 16.1 we see that the peak value of the line current (and the peak value of the phase current of the equivalent Y-network) is $V_p/(\frac{1}{3}|Z|) = 3V_p/|Z|$, which we obtained earlier. Furthermore, the angle of $\frac{1}{3}Z$ is still θ.

Consider next the balanced three-phase Y-Δ system in Figure 16.9, which includes an impedance Z_l on each of the three lines connecting the generator to the load. These impedances Z_l represent the impedances of the transmission lines or distribution lines. The simplest way to analyze this network is to replace the balanced Δ-connected load by an equivalent Y-connected load. The impedance

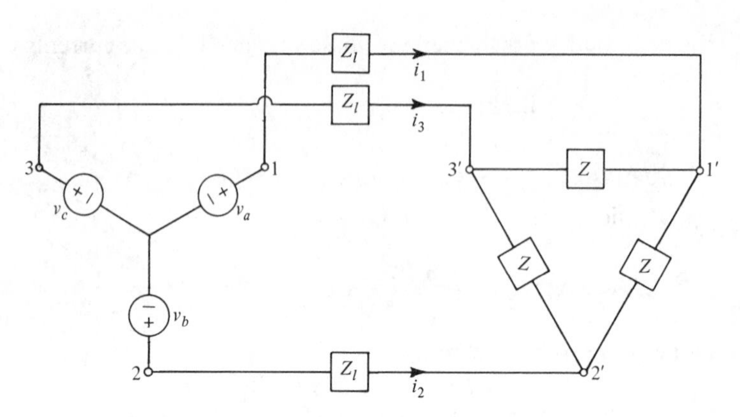

Figure 16.9 A balanced three-phase Y-Δ system with line impedances.

in each branch of the Y-connected load is $\frac{1}{3}Z$. The line impedance Z_l may be combined with this $\frac{1}{3}Z$ in series to form a composite Y-connected network. Thus the methods of Section 16.1 apply directly to obtain the line currents. Since these line currents are balanced, so are the phase currents, which are easily deduced by examining the phasor relationship between the line currents and the phase currents, as in Figure 16.8.

Proceeding with the above plan, we find that the line current I_1 is

$$I_1 = \frac{V_a}{Z_l + \frac{1}{3}Z}, \tag{16.37}$$

so that with an *abc* phase sequence, the two other line currents are

$$I_2 = I_1 e^{-j120°} \tag{16.38}$$

and

$$I_3 = I_1 e^{-j240°}. \tag{16.39}$$

The voltage across the line impedance on line 1 is $I_1 Z_l$, so that

$$V_{1'0} = V_a - I_1 Z_l = V_a - \frac{Z_l V_a}{Z_l + \frac{1}{3}Z}$$

and

$$V_{1'0} = \frac{\frac{1}{3}Z V_a}{Z_l + \frac{1}{3}Z}. \tag{16.40}$$

Again by symmetry and because the phase sequence is *abc*, the voltage $V_{2'0}$ lags $V_{1'0}$ by 120°, and $V_{3'0}$ lags $V_{2'0}$ by 120°. The line voltage at the load side $V_{1'2'}$, which is $V_{1'0} - V_{2'0}$, may be obtained by observing its phasor relationship with the phase voltages, similar to that in Figure 16.1(b). Thus

$$V_{1'2'} = \sqrt{3}\, e^{j30°} V_{1'0} = \sqrt{3}\, e^{j30°} V_a \frac{\frac{1}{3}Z}{Z_l + \frac{1}{3}Z}. \tag{16.41}$$

Notice that when $Z_l = 0$, Equation 16.41 yields $V_{1'2'} = \sqrt{3}\, e^{j30°} V_a$, which is the line voltage at the source side, as previously obtained.

The load phase currents are

$$\begin{aligned} I_{1'2'} &= \frac{V_{1'2'}}{Z} = \frac{(1/\sqrt{3})e^{j30°} V_a}{Z_l + \frac{1}{3}Z}, \\ I_{2'3'} &= I_{1'2'} e^{-j120°}, \\ I_{3'1'} &= I_{1'2'} e^{-j240°}. \end{aligned} \tag{16.42}$$

These phase currents can also be obtained from the line currents in Equations 16.37, 16.38, and 16.39 and noting the phasor relationship between the line currents and phase currents in Figure 16.8. That is, $I_{1'2'}$ leads I_1 by 30°, and its length is $1/\sqrt{3}$ times the length of the line current.

A more practical model for the three-phasor generator is to add a series impedance Z_g to each of the phase source voltages. The analysis is the same as above where we replace Z_l by $Z_l + Z_g$. Of course, the terminal phase voltage V_{10} will no longer be V_a, but instead it will be $V_{10} = V_a - Z_g I_1$.

ERCISES

16.3.1 Suppose that for the three-phase system in Figure 16.7 the phase sequence is *abc*. Draw a phasor diagram showing all the phase and line voltages and currents, where $Z = 1 + j1$ ohms and $v_a(t) = 141.4 \sin 377t$ volts.

16.3.2 Repeat Exercise 16.3.1 for a phase sequence *acb*.

16.3.3 For the system in Figure 16.9, suppose that the phase sequence is *abc*, $v_a(t) = 141.4 \sin 377t$ volts, $Z = 1 + j1$ ohms, and $Z_l = j0.5$ ohms. Draw a phasor diagram showing V_a, I_1, $V_{1'0}$, and $V_{1'2'}$.

16.3.4 Repeat Exercise 16.3.3 if the phase sequence is *acb*.

6.4 Unbalanced three-phase systems

As already mentioned earlier, the three-phase sources are manufactured as balanced generators. For large industrial users, the loads are generally balanced three-phase loads. However, smaller loads may be connected as single-phase loads. The loads are then distributed among the phases so that the total load is as close to balanced as possible. In this section we briefly consider the unbalanced case.

First consider the four-wire system in Figure 16.2 with the load impedance Z replaced by Z_1, Z_2, and Z_3 connected to 1′, 2′, and 3′, respectively. This is the simplest unbalanced system, and the line currents are easily computed as

$$I_1 = \frac{V_a}{Z_1},$$
$$I_2 = \frac{V_b}{Z_2}, \tag{16.43}$$
$$I_3 = \frac{V_c}{Z_3}.$$

The neutral current is still

$$I_0 = I_1 + I_2 + I_3, \tag{16.44}$$

but in general it will not be zero, unlike the balanced system.

Consider next the three-phase system in Figure 16.4 except that Z is replaced by Z_1, Z_2, and Z_3 connected to 1′, 2′, and 3′, respectively. A simple method for analyzing this network is to write the node equation

$$\left(\frac{1}{Z_0} + \frac{1}{Z_1} + \frac{1}{Z_2} + \frac{1}{Z_3}\right) V_{0'0} = \frac{V_a}{Z_1} + \frac{V_b}{Z_2} + \frac{V_c}{Z_3}. \tag{16.45}$$

When Z_1, Z_2, and Z_3 are different from each other, even if V_a, V_b, and V_c form a balanced set, the right-hand side of Equation 16.45 will generally not equal zero. Hence $V_{0'0}$ is generally not zero. The phase voltages at the load side are

$$V_{1'0'} = V_a - V_{0'0},$$
$$V_{2'0'} = V_b - V_{0'0}, \tag{16.46}$$
$$V_{3'0'} = V_c - V_{0'0}.$$

These are generally unbalanced voltages. The line currents are

$$I_1 = \frac{V_{1'0'}}{Z_1},$$
$$I_2 = \frac{V_{2'0'}}{Z_2}, \tag{16.47}$$
$$I_3 = \frac{V_{3'0'}}{Z_3}.$$

The formula for the neutral current I_0 is given by Equation 16.44, where I_1, I_2, and I_3 are obtained from Equation 16.47. When Z_0 is infinite (open circuit), Equations 16.45, 16.46, and 16.47 still hold, but this time $I_0 = 0$ even when I_1, I_2, and I_3 form an unbalanced set.

Finally, we consider the three-phase system of Figure 16.9, where the three-phase source is still balanced, the line impedances Z_l are still the same, but where the load impedances are unbalanced. The impedance between 1′ and 2′ is now Z_a, between 2′ and 3′ it is Z_b, and between 3′ and 1′ it is Z_c. It is possible to write three loop equations or three node equations, but the simplest method for analyzing this system is to replace the Δ load by an equivalent Y load using the transformation formulas in Section 16.2. This yields an equivalent Y-Y system so that there is only one node equation. Once this voltage $V_{0'0}$ is obtained, the line currents can be computed. Hence the voltages across the line impedances can be determined. The line voltages at the load are computed next, and the phase currents of the original Δ load are finally computed.

EXERCISES

16.4.1 Consider a three-phase system as in Figure 16.4 except that Z is replaced by

$$Z_1 = 3 + j1, \qquad Z_2 = 2 + j3, \qquad \text{and} \qquad Z_3 = 5 + j2$$

connected to 1′, 2′, and 3′, respectively, and Z_0 is disconnected (open circuit). The source is balanced with a phase sequence *abc*, and

$$v_a(t) = 1414 \sin 377t\,.$$

Draw a phasor diagram showing V_a, V_b, V_c, I_1, I_2, and I_3.

16.4.2 Repeat Exercise 16.4.1 if $Z_0 = 0 + j5$.

16.4.3 Consider the system in Figure 16.7 except that Z is replaced by $Z_a = 5 + j0$ connected between 1′ and 2′, $Z_b = 4 + j0$ connected between 2′ and 3′, and $Z_c = 3 + j0$ connected between 3′ and 1′. The source is balanced with a phase sequence of *abc*, and $v_a(t) = \sqrt{2}\,2000 \sin 377t$. Draw a phasor diagram showing V_a, V_b, V_c, V_{12}, V_{23}, V_{31}, $I_{1'2'}$, $I_{2'3'}$, $I_{3'1'}$, I_1, I_2, and I_3. Compute the total average power delivered to the load. Use V_a as a reference along the positive real axis.

16.4.4 Repeat Exercise 16.4.3 for a phase sequence of *acb*.

16.5 Two-wattmeter method

A wattmeter is an instrument with a current coil and potential coil. One end of each of these coils has a polarity mark ± or a dot symbol. When connected as shown in Figure 16.10, it reads

$$P = \frac{1}{T}\int_0^T v(t)i(t)\,dt\,. \tag{16.48}$$

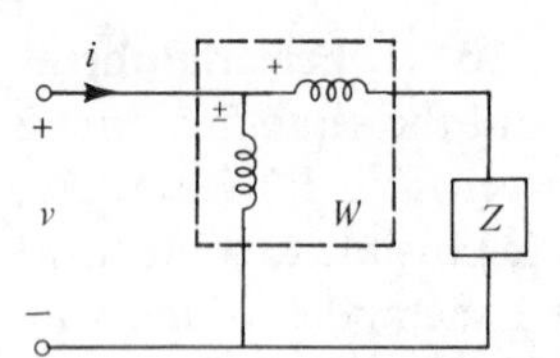

Figure 16.10 The wattmeter reads $P = (1/T) \int_0^T v(t)i(t)\,dt$ when its current coil and voltage coil are connected as shown.

Thus, if $v(t)$ and $i(t)$ are associated with an impedance Z as indicated in Figure 16.10, P is the average power delivered to Z. A wattmeter may be connected with each phase impedance to measure the average power delivered to each phase, and their sum is then the total average power to a three-phase load. In three-phase power systems, it may not be convenient to connect a wattmeter to each phase load. For example, a three-phase generator is connected to a three-phase load using three wires, and it may not even be possible to ascertain whether the load is Y-connected or Δ-connected, but it is important to measure the total power delivered by the generator. It is possible to measure this delivered power by using two wattmeters.

Consider a power network A connected to another power network B by three lines, as shown in Figure 16.11. It is desired to measure the total average power flow from Network A to Network B. Two wattmeters are connected as shown. Note that the current coil polarities are connected toward Network A and the potential coil polarities are connected to where the current coils are connected. The other ends of the potential coils are connected to the third line not containing current coils. The sum

$$P_T = P_1 + P_2 \tag{16.49}$$

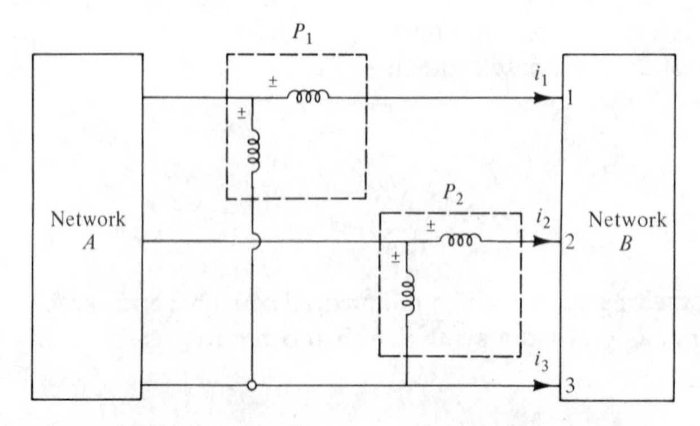

Figure 16.11 Two-wattmeter connections for measuring total average power flow, P_T, from Network A to Network B, $P_T = P_1 + P_2$.

is the total average power flow from Network A to Network B. Either P_1 or P_2 or both may be negative. If $P_1 + P_2$ is positive, then indeed the power flow is from A to B. If the sum is negative, the actual power flow is from B to A.

From Section 6.4, it is shown that the total energy flow from Network A to Network B from $t = 0$ to $t = T$ is

$$W_T = \int_0^T (v_{13}i_1 + v_{23}i_2)\, dt \tag{16.50}$$

(see Figures 6.6 and 6.8 and Equation 6.30). Thus the total average power is

$$P_T = \frac{1}{T}\int_0^T v_{13}i_1\, dt + \frac{1}{T}\int_0^T v_{23}i_2\, dt. \tag{16.51}$$

From the definition of the wattmeter and from the connection diagram of Figure 16.11, it is seen that

$$P_1 = \frac{1}{T}\int_0^T v_{13}i_1\, dt$$

and

$$P_2 = \frac{1}{T}\int_0^T v_{23}i_2\, dt.$$

Hence the sum $P_1 + P_2$ yields the desired total average power flow into Network B. As discussed in Section 6.4, it is evident that since the labeling of terminals 1, 2, and 3 is arbitrary, the same sum will be obtained by inserting the wattmeters in lines 2 and 3 as in lines 1 and 3. Of course, the end of the potential coil without a polarity mark must be connected to the line not containing current coils.

In terms of phasors, where the lengths of the phasors are equal to the peak values of the corresponding sinusoids, then

$$P_1 = \mathrm{Re}\left(\frac{V_{13}I_1^*}{2}\right) = \mathrm{Re}\left(\frac{V_{13}}{\sqrt{2}}\frac{I_1^*}{\sqrt{2}}\right) \tag{16.52}$$

and

$$P_2 = \mathrm{Re}\left(\frac{V_{23}I_2^*}{2}\right) = \mathrm{Re}\left(\frac{V_{23}}{\sqrt{2}}\frac{I_2^*}{\sqrt{2}}\right). \tag{16.53}$$

EXERCISES

16.5.1 Consider the balanced three-phase system in Figure 16.4, where the neutral wire is disconnected. Suppose a wattmeter is inserted in lines 1 and 2 in the manner of Figure 16.11. Assuming a phase sequence of *abc*, show that

$$P_1 = V_l I_l \cos(30° - \theta)$$

and

$$P_2 = V_l I_l \cos (30° + \theta),$$

where V_l and I_l are the rms values of the line voltage and line current and θ is the phase angle of the impedance load Z. Verify that

$$P_1 + P_2 = \sqrt{3}\, V_l I_l \cos \theta .$$

16.5.2 Using the result of Exercise 16.5.1, show that

$$\tan \theta = \sqrt{3}\,\frac{P_1 - P_2}{P_1 + P_2} .$$

16.5.3 Suppose that the phase sequence in the system of Exercise 16.5.1 is *acb*. Show that

$$P_1 = V_l I_l \cos (30° + \theta)$$

and

$$P_2 = V_l I_l \cos (30° - \theta) .$$

Verify that

$$P_1 + P_2 = \sqrt{3}\, V_l I_l \cos \theta$$

and

$$\tan \theta = \sqrt{3}\,\frac{P_2 - P_1}{P_1 + P_2} .$$

16.5.4 For the balanced three-phase system in Exercise 16.5.1, suppose that $P_1 = 4500$ watts, $P_2 = -1000$ watts, and $V_l = 240$ volts rms. (a) Determine the total power delivered to the three-phase load. (b) Using the result of Exercise 16.5.2, determine the phase angle θ of the load. (c) Determine the complex impedance Z.

16.6 Simple load-flow analysis

We have seen that when the sources are balanced and the loads are balanced, three-phase power network calculation may be carried out on a per-phase basis. Δ-connected loads may be replaced by equivalent Y-connected loads. The common or neutral terminals of the loads and the sources, even when no neutral wires are involved, are at the same voltage, and a neutral wire may be assumed to be connected for purposes of analysis. A network can then be drawn using only one phase

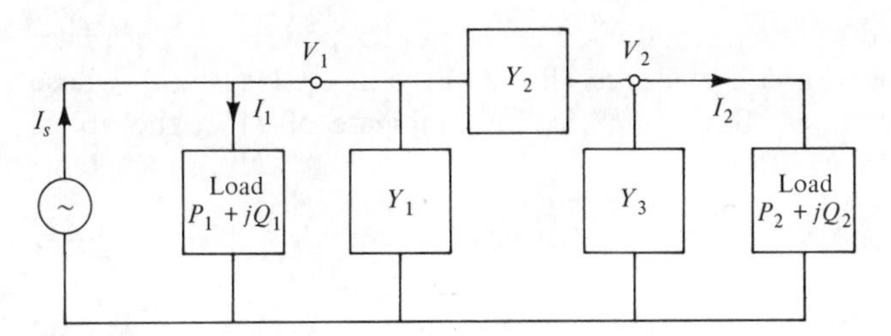

ure 16.12 Power network (per phase) where load complex powers are specified.

per load, one phase per source, and the neutral wire. Line impedances may be present, and transformers may be present. These transformers are represented by equivalent impedances, such as simple series impedances or π-networks.

Node equations can be written for such per-phase networks. The solution of these node equations for power networks is somewhat more complicated than those we have considered previously because of the special nature in which the unknown quantities appear in typical power-system problems. Instead of specifying the impedance of a load, the complex power $P + jQ$ into the load may be specified. Furthermore, in addition to voltage, the power and complex power from a generator are of interest. As we shall see below, this generally leads to a set of nonlinear equations. The analysis of such a network for various voltages and complex powers throughout the network is called a *load-flow analysis*.

Consider the simple network in Figure 16.12, where Y_1, Y_2, and Y_3 are known equivalent admittances representing transformers and transmission lines. Two loads are shown. However, instead of specifying their impedances Z_1 and Z_2, their power and reactive power consumptions P_1, Q_1, P_2, and Q_2 are specified. Likewise the generator complex power $P + jQ$ may be the desired quantity instead of the current I_s.

The node equations are

$$I_1 + (Y_1 + Y_2)V_1 - Y_2V_2 = I_s \tag{16.54}$$

and

$$-Y_2V_1 + (Y_2 + Y_3)V_2 + I_2 = 0. \tag{16.55}$$

Notice that I_1, I_2, and I_s are not known. Multiplying Equation 16.54 by V_1^* and Equation 16.55 by V_2^*, we have

$$I_1V_1^* + (Y_1 + Y_2)V_1V_1^* - Y_2V_2V_1^* = I_sV_1^* \tag{16.56}$$

and

$$-Y_2V_1V_2^* + (Y_2 + Y_3)V_2V_2^* + I_2V_2^* = 0. \tag{16.57}$$

If the lengths of the phasors correspond to rms values of corresponding sinusoidal waveforms, then $V_1I_1^* = P_1 + jQ_1$, $V_2I_2^* = P_2 + jQ_2$, and $V_1I_s^* = P_s + jQ_s$

(complex power delivered by the generator). These complex powers are per phase if the voltages are phasor phase voltages (for a Y-connected system) whose magnitudes are in rms values. Since VI^* is the conjugate of IV^*, the above equations may be written as

$$(Y_1 + Y_2)V_1V_1^* - Y_2V_2V_1^* = (P_s - P_1) - j(Q_s - Q_1) \qquad (16.58)$$

and

$$-Y_2V_1V_2^* + (Y_2 + Y_3)V_2V_2^* = -P_2 + jQ_2. \qquad (16.59)$$

Equations 16.58 and 16.59 are known as complex load-flow equations.* If Y_1, Y_2, and Y_3 are given complex constants of the system and if P_1, Q_1, P_2, and Q_2 are specified load quantities, the only remaining variables are V_1, V_2, P_s, and Q_s. Since V_1 and V_2 are complex, each is characterized by two real quantities. Thus there are six remaining real variables. The above two complex equations may be rewritten as four real equations. Of these six variables, two may be specified as controllable parameters (chosen at our disposal), and the other four are then solved by solving the four nonlinear equations. It is conventional in power-system analysis to specify the value of one of the generator node voltages. This specified node is called a *reference bus* or a *slack bus*. A *bus* in a power system is a set of all points having the same voltage. A node where no generators are connected is called a *load bus*, and a node where a generator is connected is called a *generator bus* unless it is already designated as a reference bus. In our simple power system in Figure 16.12, V_2 is associated with a load bus, and V_1 is associated with the reference bus. In a complex network with several generators, one of the nodes with a generator must be a slack bus.

Before proceeding further, we observe at this point that if we multiply Equations 16.58 and 16.59 by 3, then $3P_s$, $3Q_s$, $3P_1$, $3Q_1$, $3P_2$, and $3Q_2$ are the total three-phase power and reactive power associated with the source, load 1, and load 2. By writing 3 as $\sqrt{3}\cdot\sqrt{3}$, each voltage may be associated with $\sqrt{3}$. For example, the first term in Equation 16.58 becomes $(Y_1 + Y_2)(\sqrt{3}\,V_1)(\sqrt{3}\,V_1^*)$. Thus if all the V's in Equations 16.58 and 16.59 are interpreted to be phasors whose magnitudes are line voltage values in rms, then the P's and Q's may be interpreted to be total three-phase powers and reactive powers.

Using the following exponential notations, we find

$$\begin{aligned} V_1 &= |V_1|e^{j\alpha_1}, \\ V_2 &= |V_2|e^{j\alpha_2}, \\ Y_1 + Y_2 &= |Y_1 + Y_2|e^{j\theta_{12}}, \\ Y_2 &= |Y_2|e^{j\theta_2}, \\ Y_2 + Y_3 &= |Y_2 + Y_3|e^{j\theta_{23}}. \end{aligned} \qquad (16.60)$$

* The complex conjugates of these equations are also known as the complex load-flow equations.

Equations 16.58 and 16.59 become

$$|Y_1 + Y_2||V_1|^2 e^{j\theta_{12}} - |Y_2||V_2||V_1| e^{j(\theta_2 + \alpha_2 - \alpha_1)} = (P_s - P_1) - j(Q_s - Q_1) \tag{16.61}$$

and

$$-|Y_2||V_1||V_2| e^{j(\theta_2 + \alpha_1 - \alpha_2)} + |Y_2 + Y_3||V_2|^2 e^{j\theta_{23}} = -P_2 + jQ_2. \tag{16.62}$$

Separating the real and imaginary parts leads to the following four equations:

$$|Y_1 + Y_2||V_1|^2 \cos\theta_{12} - |Y_2||V_1||V_2| \cos(\theta_2 + \alpha_2 - \alpha_1) = P_s - P_1, \tag{16.63}$$

$$|Y_1 + Y_2||V_1|^2 \sin\theta_{12} - |Y_2||V_1||V_2| \sin(\theta_2 + \alpha_2 - \alpha_1) = -(Q_s - Q_1), \tag{16.64}$$

$$-|Y_2||V_1||V_2| \cos(\theta_2 + \alpha_1 - \alpha_2) + |Y_2 + Y_3||V_2|^2 \cos\theta_{23} = -P_2, \tag{16.65}$$

$$-|Y_2||V_1||V_2| \sin(\theta_2 + \alpha_1 - \alpha_2) + |Y_2 + Y_3||V_2|^2 \sin\theta_{23} = Q_2. \tag{16.66}$$

In the above four equations $|Y_1 + Y_2|$, $|Y_2|$, $|Y_2 + Y_3|$, θ_{12}, θ_2, θ_{23}, P_1, Q_1, P_2, and Q_2 are known. The unknown quantities are $|V_1|$, $|V_2|$, α_1, α_2, P_s, and Q_s. As we mentioned earlier, it is conventional to specify and choose the value of a reference voltage. We choose $|V_1|$ and α_1. For simplicity we choose $\alpha_1 = 0$, which is equivalent to saying that V_1 is chosen as a reference phasor. The four nonlinear equations have to be solved for $|V_2|$, α_2, P_s, and Q_s. Examining these four equations, we see that Equations 16.65 and 16.66 contain the two unknown quantities $|V_2|$ and α_2, but not P_s or Q_s. Hence, if there is a solution at all, $|V_2|$ and α_2 can be solved from these two equations. Once $|V_2|$ and α_2 are obtained, these can be substituted in Equation 16.63 to yield P_s, and into Equation 16.64 to yield Q_s. In general, these equations are solved numerically by using a digital computer. A powerful and widely used method is the Newton-Raphson method (see Appendix D).

The basic procedure for solving more complicated power networks is similar. The first step is to write n node equations. Multiplying the ith node equation (written for node i) by V_i^*, for all i, n complex-power flow equations are obtained. These are equivalent to $2n$ real equations. Load-flow analysis is used for planning studies, economic dispatch (scheduling which generators produce how much complex power to meet demands at least cost), and security (reliable operation, for example, by forecasting which lines or machines might become overloaded and preventing such occurrence). It is an important aspect of power-system analysis.

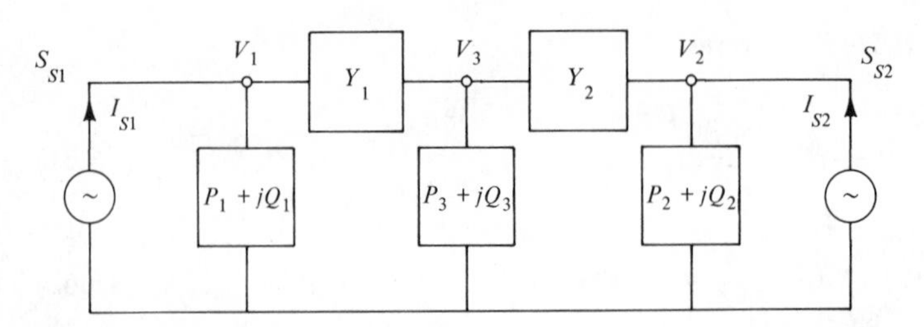

Figure 16.13 Power network for Exercise 16.6.3.

EXERCISES

16.6.1 From Equations 16.65 and 16.66, eliminate α_2 and show that $|V_2|$ can be obtained from

$$|Y_2 + Y_3|^2|V_2|^4 + 2|Y_2 + Y_3|(P_2 \cos \theta_{23} - Q_2 \sin \theta_{23})|V_2|^2 - |Y_2|^2|V_1|^2|V_2|^2 + (P_2^2 + Q_2^2) = 0.$$

Notice that this is quadratic in $|V_2|^2$.

16.6.2 Suppose that in the power network of Figure 16.12, $Y_1 = Y_3 = 0$ and $Y_2 = -j5$ mhos. Verify that $P_s = P_1 + P_2$.

16.6.3 Write the complex power flow equations for the network in Figure 16.13, using phasor voltages whose lengths are line values rms. The loads shown are total three-phase complex powers.

16.7 Summary

1. In a balanced three-phase system, a phase sequence *abc* means that v_a leads v_b by 120° and v_b leads v_c by 120°.
2. The voltage sources in a three-phase system are usually balanced and Y-connected. The load may be Y-connected or Δ-connected. A Y-Δ transformation is useful in converting a Δ-load to a Y-load and vice versa.
3. When the phasors V and I for sinusoidal voltages and currents have lengths equal to $1/\sqrt{2}$ times the corresponding peak values, the average power is

$$P_{\text{av}} = \text{Re}\,(VI^*) = |V|\,|I| \cos \theta$$

where θ, the power-factor angle, is the angle of V measured with respect to I. The reactive power is

$$Q = \text{Im}\,(VI^*),$$

and the complex power is

$$S = P_{av} + jQ = VI^*$$

4. In a balanced three-phase system, the total power is

$$P = \sqrt{3} V_l I_l \cos \theta$$

where V_l is the rms value of the line voltage, I_l is the rms value of the line current, and θ is the power-factor angle of the phase load. This formula is valid for both Δ- and Y-connected loads.
5. Power in a three-phase three-wire system can conveniently be measured by the two-wattmeter method.
6. The analysis of a balanced three-phase system is significantly simplified by exploiting symmetry and carrying out analysis on a per-phase basis.
7. The analysis for various voltages and complex powers in a power network is known as load-flow analysis. The complex load-flow equations are usually non-linear because of the nature in which the unknown quantities appear. These equations are obtained by multiplying each of the node equations by the conjugate of the corresponding node voltage.

ROBLEMS

16.1 Consider a single-phase power system with one source, one load, and two wires for transmitting power. It is required to deliver power at the load of P watts at a voltage of $|V|$ volts rms. The load is Z. If the resistive component of the impedance of each line is R_l, then the total losses in the line to deliver the required power are $P_l = 2|I|^2 R_l$. Now consider a balanced three-phase Y-Y connected system that uses three wires. It is required to deliver the same total average power P to the load at the same line-to-line voltage $|V|$ volts rms (power per phase is $P/3$). If the load has the same power factor as in the single-phase case, compute the rms values of the line currents. Determine what the resistive component of each line should be, in order that the total losses in the wires are the same as those for the single-phase case. Compare the relative sizes of the wires in the three-phase and the single-phase systems. (Note R_l is inversely proportional to cross-sectional area.) Compare the total amount of conducting material (mass) in the wires of the three-phase system and single-phase system.

16.2 A balanced three-phase three-wire system is connected to a balanced load. It is known that the load is inductive. Two wattmeters are connected as in Figure 16.11, and P_1 reads 1000 watts and P_2 reads 1200 watts. Is it possible to determine the phase sequence? Briefly show how or why not.

16.3 A balanced three-phase Y-Y three-wire system has a line voltage of 240 volts rms, and the impedance in each phase of the balanced load is $Z = 10 + j10$ ohms. Compute the total average power and reactive power delivered to the load. If the phase sequence is *abc*, draw and label the network diagram and draw a phasor diagram showing all voltage and current phasors. If wattmeters are connected as in Figure 16.11, what will P_1 and P_2 read?

16.4 Repeat Problem 16.3 except that the load is Δ-connected and the impedance in each phase is still $Z = 10 + j10$ ohms.

16.5 A balanced three-phase three-wire system consists of a balanced three-phase source with a line voltage of 2200 volts rms and a balanced load of 400 kilowatts at a power factor of 0.8 lagging. (a) Determine the line current rms value. (b) Determine the total reactive power delivered to the load. (c) Three identical capacitors are connected in a Y. Compute the reactance of each capacitor so that when connected to the same three-phase system at the load side (in parallel with the previous load), the total reactive power to the capacitors is the negative of the total reactive power to the inductive load. (d) Compute the new line currents from the generator. Is it greater or less than before?

16.6 A balanced Y-connected three-phase source with a line voltage of 220 volts rms is connected to a balanced Δ-connected load. The line impedance on each line is $Z_l = 0 + j0.5$ ohms, and the Δ-connected impedance in each phase is $Z = 3 + j2$ ohms. Compute the line voltage in rms at the load side.

16.7 A balanced Y-connected three-phase source is connected to a balanced Δ-connected load. The line impedances are $Z_l = 0 + j0.5$ ohms. It is desired to deliver a total average power of 3000 watts to the load, at 0.8 power factor lagging at 240 volts rms line voltage at the load. Determine (a) the required line voltage at the source, and (b) the total average power delivered by the source.

16.8 For the system in Problem 16.6, suppose that it is desired that the line voltage be 220 volts rms at the load side. What should be the line voltage at the source side?

16.9 Repeat Problem 16.8 except that instead of 220 volts rms at the load side, it is required to have $|V|$ volts. Express your answer in terms of $|V|$. Using this result, what is $|V|$ if the line voltage at the source side is 220 volts rms?

16.10 Repeat Problem 16.3 if the impedance on phase 1 is disconnected (open circuit).

16.11 Suppose that in Problem 16.6 the balanced Δ-connected load is replaced by an unbalanced Δ-connected load with $Z_a = 10 + j0$ ohms, $Z_b = 8 + j2$ ohms, and $Z_c = 7 - j3$ ohms. Compute all the line voltages in volts rms at the load side and all the phase currents in amperes rms.

16.12 For the power network in Figure 16.12, suppose $Y_1 = Y_3 = 0$ and $Y_2 = -j5$ mhos. The complex power demand $P_2 + jQ_2$ is $1000 + j300$ in watts and vars. The demand $P_1 + jQ_1$ is $800 + j100$ in watts and vars. The complex powers are total three-phase values. $|V_1|$ is chosen as 100 volts (line value in rms). Determine V_2, P_s, and Q_s. If there are several solutions, choose $|V_2|$ to be as close to $|V_1|$ as possible.

Chapter **17**

Fourier Series

17.1 Sums of exponential signals

In this chapter, we will study the representation of signals in the frequency domain. Although these words may sound imposing, the basic idea is very simple, as we shall see. In Chapter 2 in Equation 2.1, we expressed a signal in the following series form:

$$v(t) = a_1e^{s_1t} + a_2e^{s_2t} + a_3e^{s_3t} + \cdots. \tag{17.1}$$

Depending on the nature of the s_j's, there are several possible forms for the terms of this equation, such as

$$a_1e^{-2t}, \qquad a_2 \sin 3t\,, \qquad a_3e^{-t}\cos 2t\,, \qquad \text{etc.} \tag{17.2}$$

Our study is aimed at the discovery of how these terms can be used in combination to represent a wide class of signals. But before giving the specific details, we inquire as to motivation.

Linear network analysis is frequently carried out by the following steps:

Step 1. Given a signal $v(t)$ which may be described by straight-line segments, singularity functions, a graph, etc., we express $v(t)$ as a sum of exponential signals. With experience, the terms present and values of the coefficients tell us something about the signal.

Step 2. Complete a network analysis for each term in the series. Our past experience tells us that analysis with exponential signals, including the sinusoidal steady-state, is easy and routine. See Figure 17.1.

Step 3. Using the concept of superposition, we sum the individual solutions of (2) to determine the complete solution.

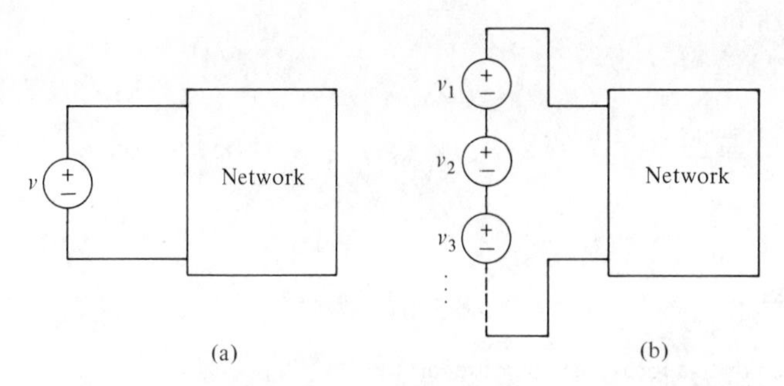

ure 17.1 The signal generator of (a) is equivalent to that of (b) if $v = v_1 + v_2 + v_3 + \cdots$. In the analysis of the network of (b), superposition is employed. This is a basic notion in the application of Fourier series to linear network analysis.

At first sight, this may appear to be a complicated way to solve a problem, compared with a direct solution, or solution using the ideas of convolution as explained in Chapters 7 and 8. It turns out that the method offers definite advantages.

In the discussion to follow, it is helpful to distinguish between a nonrecurring or nonperiodic signal and a periodic signal. A periodic signal is defined as one for which $v(t + T) = v(t)$, where T is the period. This means that the function is repeated exactly every T sec. If this relationship is not satisfied, then the signal is

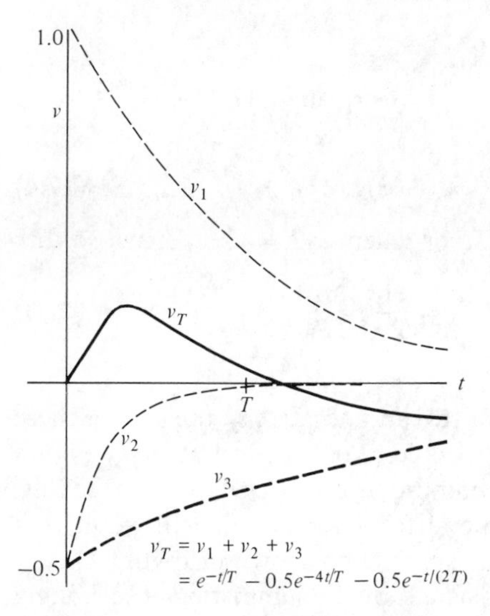

ure 17.2 The combination of three damped exponentials to give v_T, a nonperiodic signal.

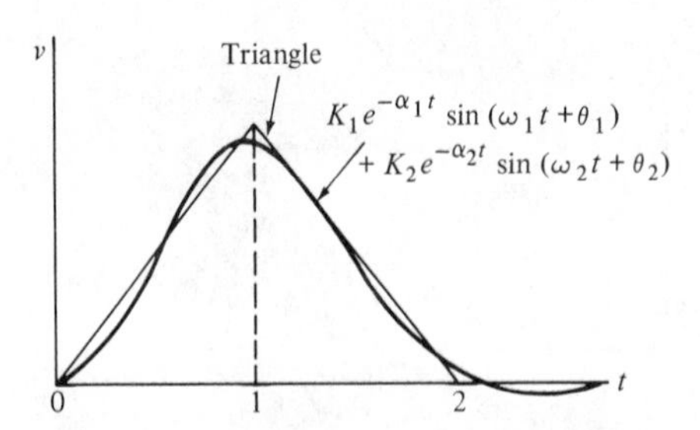

Figure 17.3 The sum of two damped sinusoids to approximate a nonrecurring triangular waveform.

nonperiodic. In terms of the general form of each component of the series representation of the signal in Equation 17.1, $a_j e^{s_j t}$, we see that:

1. When s_j is real, then the signal is nonperiodic, being an exponential function. When it is both real and negative, then the signal is a decaying exponential.
2. When s_j is complex and in conjugate pairs, then the signal is nonperiodic. When the real part of s_j is negative, then the signal is known as a damped sinusoid.
3. When s_j is imaginary and in conjugate pairs, then the signal is periodic. This follows from Euler's formula, which may be expressed as

$$e^{jx} + e^{-jx} = 2 \cos x, \tag{17.3}$$

which is a sinusoid, periodic for all time.

To study this periodic function further, let x in Equation 17.3 be $\omega_0 t + \phi$, so that the sinusoidal signal is

$$v_1(t) = K_j \cos(\omega_0 t + \phi). \tag{17.4}$$

This $v_1(t)$ begins repeating itself after $t = T$ or when $\omega_0 T = 2\pi$ radians, so that

$$\omega_0 = \frac{2\pi}{T} \quad \text{and} \quad f_0 = \frac{\omega_0}{2\pi} = \frac{1}{T}, \tag{17.5}$$

where ω_0 is in radians/sec and f_0 is in Hz.

Now if we add components resulting from real and negative s_j, the sum of these nonperiodic components will also be nonperiodic. This is illustrated by Figure 17.2, which shows a signal found by summing three decaying exponentials. Similarly, the addition of terms of the general form $Ke^{\sigma t} \sin \omega t$ will result in a nonperiodic signal. This is illustrated by the summation shown in Figure 17.3.

While we have discussed the formation of signals by the summation of seemingly arbitrary components, it is easy to visualize that it is possible to go the other way, to start with a given signal and then determine the components necessary to

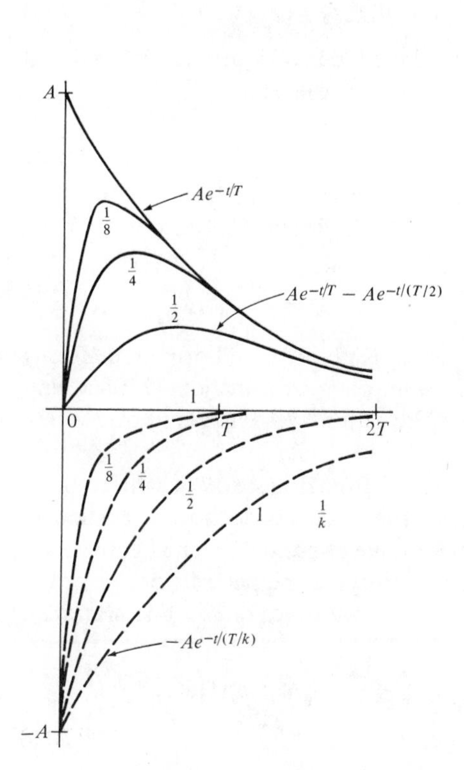

Figure 17.4 Plots for $\pm Ae^{-t/(T/k)}$ for several values of $1/k$. The linear combination of these exponentials for appropriate choices of A and k may be used to approximate a non-recurring signal.

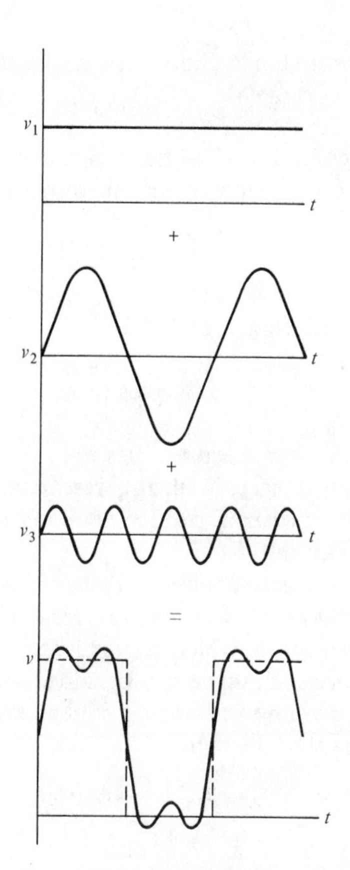

Figure 17.5 An example of a truncated Fourier series. The sum $v = v_1 + v_2 + v_3$ approximates the square wave for which the Fourier components were found.

represent the signal. This process is known as *approximation by exponentials.* Figure 17.4 shows the difference $Ae^{-t/T} - Ae^{-t/(T/k)}$ for several values of $1/k$ suggesting the components available for approximation by the use of different values of A and $1/k$.

Turning next to periodic signals, consider the component

$$v_1(t) = K_1 \cos \frac{2\pi}{T} t. \tag{17.6}$$

Here $v_1(t + T) = v_1(t)$, where T is the period, the smallest constant for which the relationship holds. Since the signal continues to repeat itself for all t, we know that

$$v_1(t) = v_1(t + nT), \qquad n = 0, \pm 1, \pm 2, \ldots. \tag{17.7}$$

We now inquire as to those signals that can be added to Equation 17.6 and still have the sum remain periodic of period T. Consider the signal

$$v_n(t) = K_n \cos n \frac{2\pi}{T} t. \tag{17.8}$$

In terms of this equation, we seek the values of n such that the sum $v = v_1 + v_n$ will have period T and $v(t) = v(t + T)$. The answer is simple. Since

$$\cos \left[n \frac{2\pi}{T} (t + T) \right] = \cos n \frac{2\pi}{T} t \tag{17.9}$$

holds only for integer values of n, then $v_1 + v_n$ is periodic and of period T only if n is an integer. With this restriction, the sum of v_1 in Equation 17.6 and any number of terms for different values of n in Equation 17.8 will result in a waveform of period T.

This integer requirement may be seen clearly from Figure 17.5, which shows the addition of two signals corresponding to $n = 1$ and 2 added to a constant to give a waveform which approximates a square wave of period T. The figure shows that more such terms can be added without destroying the periodic nature of the sum, providing that their frequencies are integer multiples of ω_0. The argument suggests that the sum

$$\begin{aligned} v(t) &= c_0 + c_1 \cos(\omega_0 t + \phi_1) + c_2 \cos(2\omega_0 t + \phi_2) + \cdots \\ &= c_0 + \sum c_j \cos(j\omega_0 t + \phi_j), \end{aligned} \tag{17.10}$$

which is of period $T = 2\pi/\omega_0$ is very useful in representing periodic signals.

The first person to make use of this series was the French mathematician J. B. J. Fourier (1758–1830), who did so in his study of problems in heat transfer. The series has come to be known as the Fourier series, and it plays an important role in electrical engineering as well as in other fields of engineering and science.

EXERCISES

17.1.1 For the following periodic functions, determine the numerical value for ω_0, f_0, and T. (a) $v_1(t) = 100 \sin 377t$. (b) $v_2(t) = 10 \cos 1000t$.

17.1.2 Plot each term of the following series and sum the components as in Figure 17.2:

$$v_1(t) = \pi/4 + \sin \omega_0 t + \tfrac{1}{2} \sin 2\omega_0 t + \tfrac{1}{3} \sin 3\omega_0 t.$$

17.1.3 Repeat Exercise 17.1.2 for the following signal:

$$v_2(t) = \cos t - \tfrac{1}{3} \cos 3t + \tfrac{1}{5} \cos 5t - \tfrac{1}{7} \cos 7t.$$

17.1.4 Find the period T for the waveform

$$v(t) = 2\cos 3t + 3\cos 2t\,.$$

17.1.5 Repeat Exercise 17.1.4 for the waveform

$$v(t) = 2\cos 2t + 3\cos(2+\pi)t\,.$$

17.1.6 Given the signal

$$v(t) = A_1\cos\omega_1 t + A_2\cos\omega_2 t\,,$$

show that $v(t)$ is periodic when ω_1/ω_2 is a rational number. Find the period T in terms of ω_1 and ω_2.

.2 The Fourier coefficients

Expressing a signal as a Fourier series is accomplished by the evaluation of the coefficients of the Fourier series for a specific signal. To begin our study of evaluation procedures, we will modify Equation 17.10 to gain simplicity in calculations. This is done by expanding terms like $c_1\cos(\omega_0 t + \phi_1)$ through a trigonometric identity for the sum of two angles as

$$\begin{aligned} c_1\cos(\omega_0 t+\phi_1) &= (c_1\cos\phi_1)\cos\omega_0 t + (-c_1\sin\phi_1)\sin\omega_0 t \\ &= a_1\cos\omega_0 t + b_1\sin\omega_0 t\,. \end{aligned} \tag{17.11}$$

In addition, if we let $c_0 = a_0$, then the Fourier series of Equation 17.10 becomes

$$\begin{aligned} v(t) = a_0 &+ a_1\cos\omega_0 t + a_2\cos 2\omega_0 t + a_3\cos 3\omega_0 t + \cdots \\ &+ b_1\sin\omega_0 t + b_2\sin 2\omega_0 t + \cdots\,. \end{aligned} \tag{17.12}$$

As a further simplification, we will make use of time scaling by letting $\omega_0 = 1$, making the variable t rather than $\omega_0 t$.* With this change of scale — or substitution for a new variable if you prefer — the period becomes 2π rather than $T = 1/f_0 = 2\pi/\omega_0$. We have lost no generality by making the change, of course, since we may employ inverse time scaling at any point we wish. It is usually more convenient to deal with times in seconds than in, say, nanoseconds with the attendant 10^{-9} factors.

With this change made, Equation 17.12 then has the simple form we will use:

$$\begin{aligned} v(t) = a_0 &+ a_1\cos t + a_2\cos 2t + a_3\cos 3t + \cdots \\ &+ b_1\sin t + b_2\sin 2t + b_3\sin 3t + \cdots\,. \end{aligned} \tag{17.13}$$

* For elaboration, see item 2 in Section 17.5.

From this equation, we wish to derive formulas to permit the evaluation of a_0, a_n, and b_n for a given $v(t)$. This will be accomplished by several simple steps:

1. Multiply Equation 17.13 by either $\sin mt$ or $\cos mt$, where m is an integer.
2. Integrate the resulting product over one complete period.
3. Make use of so-called orthogonality conditions involving the integrals of cosine and/or sine functions. It will be shown that most of the terms of (1) vanish when integrated over the period.
4. Rearrange the nonvanishing terms as solutions for a_0, a_n, or b_n.

The relationships to be used in step 2 above are the following. First, the sine and cosine functions have zero average value, and

$$\int_0^{2\pi} \cos nt \, dt = 0 \tag{17.14}$$

and

$$\int_0^{2\pi} \sin nt \, dt = 0. \tag{17.15}$$

As shown in Figure 17.6, the sum of area with a positive sign always equals that with a negative sign, for all n. Although less obvious, it turns out that

$$\int_0^{2\pi} \sin nt \cos nt \, dt = 0, \qquad \text{all } n; \tag{17.16}$$

$$\int_0^{2\pi} \cos mt \sin nt \, dt = 0, \qquad m \neq n; \tag{17.17}$$

$$\int_0^{2\pi} \cos mt \cos nt \, dt = 0, \qquad m \neq n; \tag{17.18}$$

and

$$\int_0^{2\pi} \sin mt \sin nt \, dt = 0, \qquad m \neq n. \tag{17.19}$$

Interpreted in terms of area in Figure 17.6(b), there are equal areas with positive and negative signs so that the sum is zero. Finally,

$$\int_0^{2\pi} \cos^2 nt \, dt = \pi \tag{17.20}$$

and

$$\int_0^{2\pi} \sin^2 nt \, dt = \pi, \tag{17.21}$$

for $n \neq 0$, as shown in Figure 17.6(c).

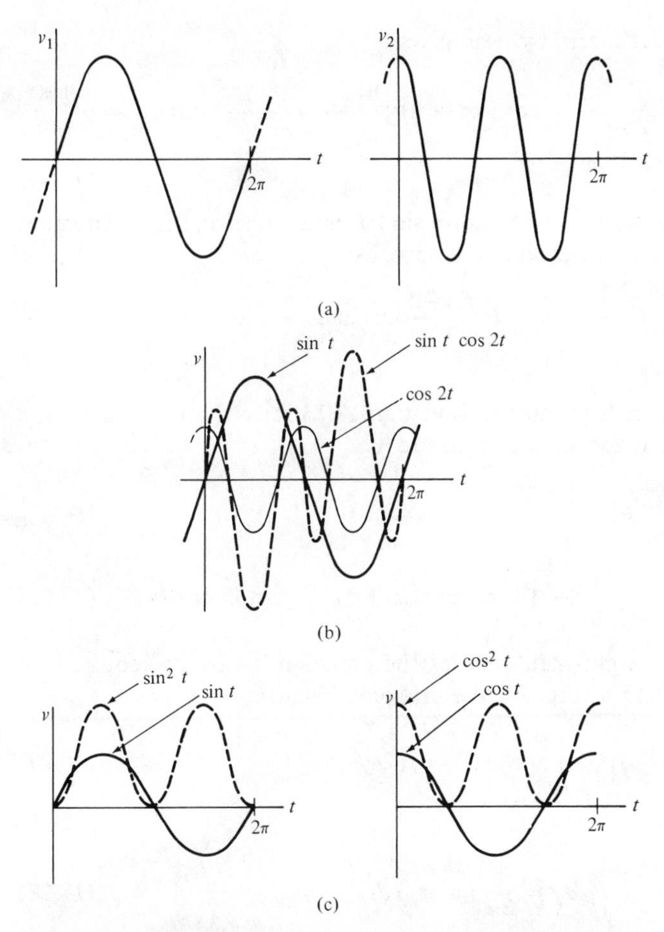

ure 17.6 (a) Two sinusoids. Over any number of periods, the sum of areas with a positive sign is equal to the sum of areas with a negative sign. (b) The product of sin t and cos $2t$, shown as a dashed line, has equal areas with positive and negative signs. (c) However, $\sin^2 t$ and $\cos^2 t$ enclose areas which are positive when summed over a period.

Any of these six equations is found through algebraic manipulation using the appropriate trigonometric identities. For example,

$$\int_0^{2\pi} \sin mt \cos nt \, dt \tag{17.22}$$

is first simplified making use of the equation

$$\sin \phi \cos \theta = \tfrac{1}{2}[\sin (\phi + \theta) + \sin (\phi - \theta)]. \tag{17.23}$$

Integrating Equation 17.22 term by term gives

$$\int_0^{2\pi} \sin mt \cos nt \, dt = \frac{-1}{2(m+n)} \cos [(m+n)t] \Big|_0^{2\pi} + \frac{-1}{2(m-n)} \cos (m-n) \Big|_0^{2\pi}$$

$$= 0, \qquad m \neq n. \tag{17.24}$$

If $m = n \neq 0$, then we use $2 \sin \theta \cos \theta = \sin 2\theta$, and substitute into Equation 17.22. Routine integration of the sine function gives

$$\int_0^{2\pi} \sin mt \cos nt \, dt = \frac{-1}{4m} \cos 2mt \Big|_0^{2\pi} = 0 \tag{17.25}$$

as stated.

To proceed with step 3, we return to Equation 17.13 for $v(t)$ and integrate the equation term by term. A few of the terms are

$$\int_0^{2\pi} v(t) \, dt = \int_0^{2\pi} a_0 \, dt + \int_0^{2\pi} a_1 \cos t \, dt + \cdots$$

$$+ \int_0^{2\pi} b_1 \sin t \, dt + \int_0^{2\pi} b_2 \sin 2t \, dt + \cdots. \tag{17.26}$$

All the integrals on the right-hand side of the equation vanish as required by Equations 17.14 and 17.15, except for the first one. Then

$$\int_0^{2\pi} v(t) \, dt = a_0 2\pi \tag{17.27}$$

or

$$a_0 = \frac{1}{2\pi} \int_0^{2\pi} v(t) \, dt, \tag{17.28}$$

which is the required integral form for the evaluation of a_0. Observe that this equation is simply the *average value* of $v(t)$, the area under the curve of $v(t)$ divided by the base. This is sometimes known as the dc value of the signal, and it is frequently possible to determine it by inspection of $v(t)$.

To find a_1, we simply multiply Equation 17.13 by $\cos t$ and integrate from 0 to 2π. This gives

$$\int_0^{2\pi} v(t) \, dt = \int_0^{2\pi} a_0 \cos t \, dt + \int_0^{2\pi} a_1 \cos^2 t \, dt + \int_0^{2\pi} a_2 \cos t \cos 2t \, dt + \cdots$$

$$+ \int_0^{2\pi} b_1 \sin t \cos t \, dt + \int_0^{2\pi} b_2 \sin 2t \cos t \, dt + \cdots. \tag{17.29}$$

All the integrals on the right-hand side of this equation vanish except for the second one, which involves a_1. Then

$$\int_0^{2\pi} v(t) \cos t \, dt = a_1 \pi \tag{17.30}$$

or

$$a_1 = \frac{1}{\pi} \int_0^{2\pi} v(t) \cos t \, dt. \tag{17.31}$$

The computation of a_1 is illustrated in Figure 17.7, which shows a representative $v(t)$ in (a), $\cos t$ in (b), and the product of $v(t)$ and $\cos t$ in (c). The integration of this product over a period, the cross-hatched area with due regard to sign divided by half the base, is the value of a_1. The procedure illustrated for a_1 applies to all other a_n and b_n, and the resulting equations are

$$a_n = \frac{1}{\pi} \int_0^{2\pi} v(t) \cos nt \, dt \tag{17.32}$$

and

$$b_n = \frac{1}{\pi} \int_0^{2\pi} v(t) \sin nt \, dt. \tag{17.33}$$

For completeness, we include the equation for a_0 from Equation 17.28,

$$a_0 = \frac{1}{2\pi} \int_0^{2\pi} v(t) \, dt. \tag{17.34}$$

These three equations are the means by which we calculate the Fourier coefficients.

In Equations 17.32, 17.33, and 17.34, the limits of integration have been 0 to 2π, representing one period. As first studied in Equation 3.5 of Chapter 3 in connection with the average value of a signal, the limits in general are t_1 to $t_1 + T$ for a signal

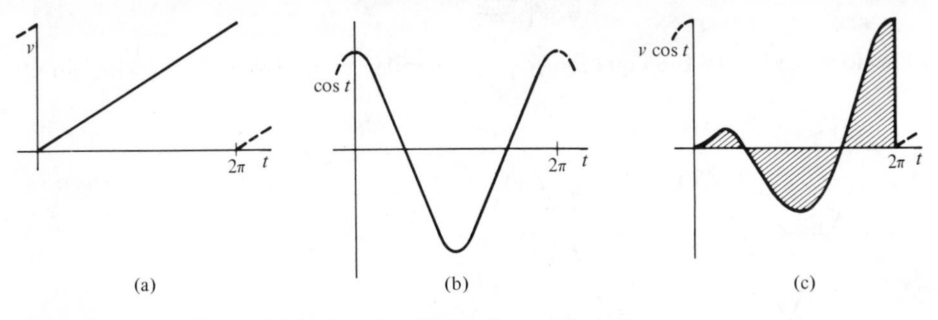

re 17.7 Plots of an example of $v(t)$ in Equation 17.31. The product of this $v(t)$ in (a) and $\cos t$ in (b) gives the curve of (c) for which the cross-hatched area divided by π is a_1.

of period T. The same notion applies to Equations 17.32 to 17.34, and the limits 0 to 2π may be replaced by $-\pi$ to $+\pi$, for example, or the complete period of duration 2π may be covered in any other way.

EXAMPLE 17.2.1

The signal shown in Figure 17.8 is triangular in form and representative of a waveform available in commercial generators. We wish to find the Fourier series which represents this $v(t)$. Now the waveform is made up of segments of straight lines. The general equation for a straight line, $y = mx + b$, is used where the y-intercept (in our case the v-intercept) is V_0, and the slopes are $\pm 2V_0/\pi$. Thus the signal may be represented from $-\pi$ to $+\pi$ by

$$v(t) = \begin{cases} V_0 - \dfrac{2V_0}{\pi} t, & 0 \le t < \pi, \\ V_0 + \dfrac{2V_0}{\pi} t, & -\pi < t \le 0. \end{cases} \tag{17.35}$$

Since the area under the curve of the signal is clearly zero, there is no need to evaluate a_0. The values of a_n are

$$a_n = \frac{V_0}{\pi}\left(\int_{-\pi}^{\pi} \cos nt\, dt + \int_{-\pi}^{0} \frac{2}{\pi} t \cos nt\, dt + \int_{0}^{\pi} \frac{-2}{\pi} t \cos nt\, dt\right). \tag{17.36}$$

Integrating and simplifying the resulting equation, we have

$$a_n = \frac{4V_0}{n^2\pi^2}(1 - \cos n\pi) \tag{17.37}$$

and from this

$$a_n = \begin{cases} 0, & n \text{ even}, \\ \dfrac{8V_0}{n^2\pi^2}, & n \text{ odd}. \end{cases} \tag{17.38}$$

The formulation of the equation for b_n is similar to Equation 17.36 with $\sin nt$

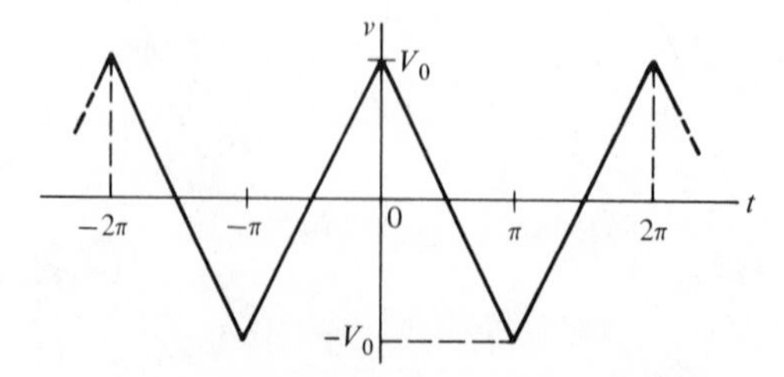

Figure 17.8 The triangular waveform used in Example 17.2.1.

terms replacing the cos nt terms. The result of the steps of integration and algebraic simplification is

$$b_n = 0. \tag{17.39}$$

Hence the Fourier series for the signal waveform of Figure 17.8 is

$$v(t) = \frac{8V_0}{\pi^2}\left(\cos t + \frac{1}{3^2}\cos 3t + \frac{1}{5^2}\cos 5t + \cdots\right). \tag{17.40}$$

This is a plausible result. The waveform shown in Figure 17.8 appears to have the general form of a cosine. The terms following the first in Equation 17.40 are just sufficient to change the cosine wave into the triangular form.

If the Fourier series found for a particular signal has both a_n and b_n terms for the same n, then we may wish to combine these terms by a procedure which is the inverse of that described at the beginning of the section. We wish to find c_n and ϕ_n in the equation

$$a_n \cos nt + b_n \sin nt = c_n \cos(nt + \phi_n). \tag{17.41}$$

This is accomplished by expanding the right-hand side of the equation and equating multipliers of cos nt and sin nt. Then

$$a_n = c_n \cos \phi_1 \tag{17.42}$$

and

$$b_n = -c_n \sin \phi_1. \tag{17.43}$$

Squaring and adding give

$$c_n = \sqrt{a_n^2 + b_n^2}, \tag{17.44}$$

and dividing Equation 17.43 by Equation 17.42 gives

$$\phi_n = \tan^{-1}\frac{-b_n}{a_n}. \tag{17.45}$$

With these results, we may use the equations of this section to define a number of terms that are used in describing signals expanded as a Fourier series. The value a_0 is known as the *average* or *dc value* of a signal. The frequency ω_0 is called the *fundamental frequency* and $n\omega_0$ is called the nth *harmonic* of the fundamental frequency. The value c_n is known as the *amplitude* of the nth harmonic, and ϕ_n is the *phase* of the nth harmonic. We will return to the use of these terms in describing signals represented as a Fourier series when we discuss signal spectra in a later section.

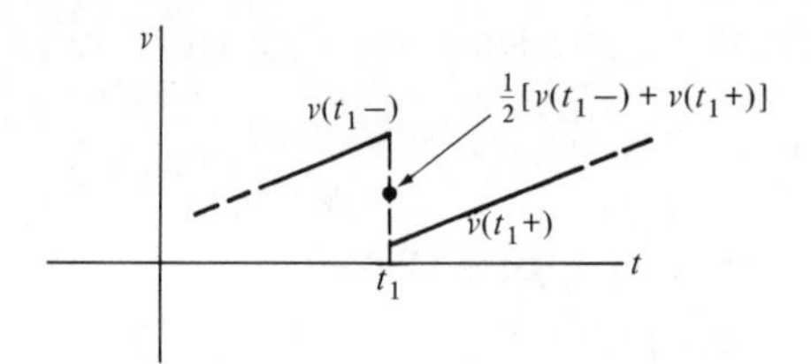

Figure 17.9 The signal $v(t)$ has a jump discontinuity at $t = t_1$, and the Fourier series converges to a point shown midway between the two segments of $v(t)$ at the jump.

In this section, we have concentrated on determining the Fourier coefficients assuming that $v(t)$ could be represented by a Fourier series. The conditions under which there is a Fourier-series representation of $v(t)$ are known as the *Dirichlet conditions*. The most important of these is that the signal $v(t)$ be absolutely integrable over a period so that

$$\int_0^{2\pi} |v(t)|\, dt = \text{finite} < \infty\,. \tag{17.46}$$

Note that this requirement does not exclude the impulse function for $v(t)$. The other two conditions are stated as follows:

(a) The function $v(t)$ has a finite number of discontinuities in one period, and

(b) the function $v(t)$ has a finite number of maxima and minima in one period.

The $v(t)$ we will deal with are *piecewise continuous* with jump discontinuities of the type shown in Figure 17.9. At these discontinuities, the Fourier series converges to a point midway between the two continuous curves of value

$$\tfrac{1}{2}[v(t_1-) + v(t_1+)], \tag{17.47}$$

where the quantities are as defined in the figure.

EXERCISES

17.2.1 The signal shown in Figure 17.10 is known as a *square wave*. For this waveform, determine the Fourier coefficients a_0, a_n, and b_n.

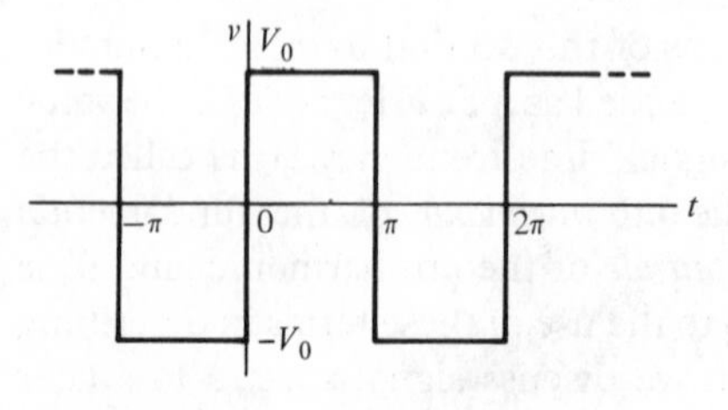

Figure 17.10

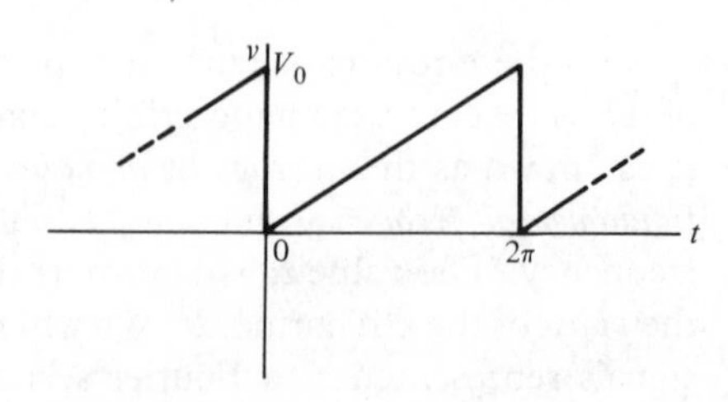

Figure 17.11

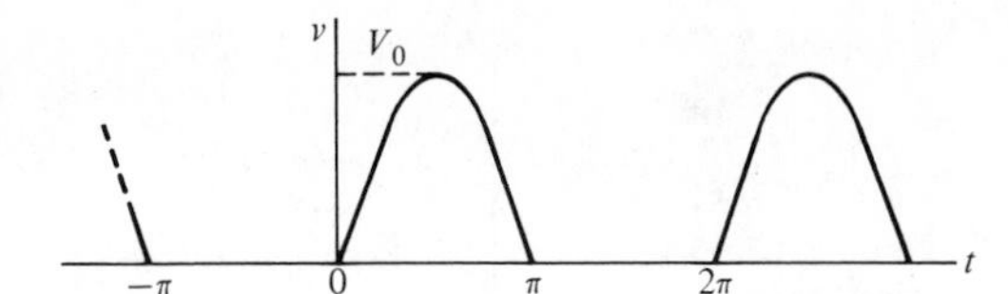

Figure 17.12

17.2.2 The signal of Figure 17.11 is triangular in form and is known as a *sweep* signal (from its use in a cathode-ray oscilloscope). For the waveform of the figure, determine the Fourier coefficients a_0, a_n, and b_n.

17.2.3 The signal shown in Figure 17.12 is a segment of a sine wave formed by removing the negative-going portion, and is known as a *half-wave-rectified sine wave*. For this waveform, determine the Fourier coefficients a_0, a_n, and b_n.

17.2.4 The signal of Figure 17.13 is similar to that of Exercise 17.2.3 except that the negative-going portions of the sine wave are replaced by positive-going parts. The equation is $v(t) = V_0|\sin t|$, and it is known as a *full-wave-rectified sine wave*. For this waveform, determine the Fourier coefficients a_0, a_n, and b_n.

3 Signal symmetry

In Example 17.2.1, it was found that $b_n = 0$ for all n through the evaluation of the integral equation for b_n. This result would be anticipated from a knowledge of symmetry of the signal waveform with respect to $t = 0$, and the implication of each class of symmetry with respect to the values of coefficients. The three most common forms of symmetry are (a) even symmetry, (b) odd symmetry, and (c) half-wave symmetry.

A signal $v(t)$ is said to be *even* with respect to $t = 0$ when

$$v(-t) = v(t) \tag{17.48}$$

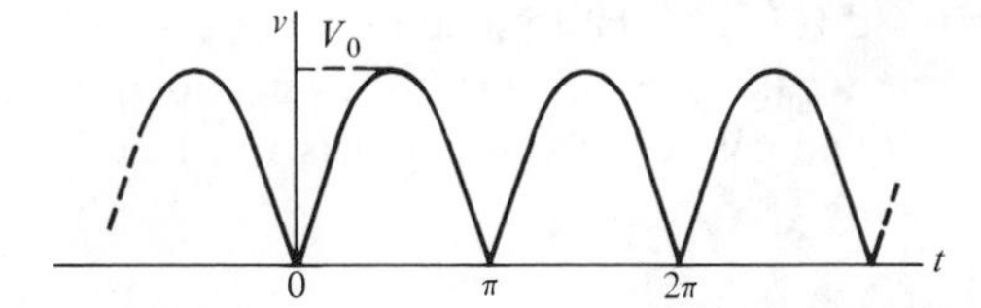

Figure 17.13

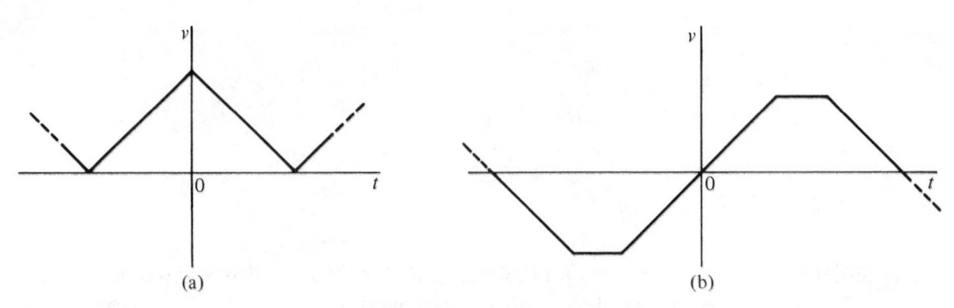

Figure 17.14 Signal waveforms which illustrate symmetry with respect to $t = 0$: (a) even symmetry; (b) odd symmetry.

and *odd* when

$$v(-t) = -v(t). \tag{17.49}$$

An even signal is shown in (a) of Figure 17.14, and an odd signal in (b) of that figure. Observe that an even signal has mirror symmetry with respect to $t = 0$, while an odd signal has a sign reversal about the line $t = 0$.

Now it seems clear that the sum of even functions is itself even while the sum of odd functions is itself odd. However, the sum of a number of even functions and one or more odd functions is neither even nor odd, but has components of each. We express this fact by noting that any signal has these two parts, either of which may vanish,

$$v(t) = v_e(t) + v_o(t), \tag{17.50}$$

where v_e is even and v_o is odd. Substituting $-t$ for t in this equation,

$$v(-t) = v_e(-t) + v_o(-t). \tag{17.51}$$

Substituting Equations 17.48 and 17.49 into Equation 17.51 gives

$$v(-t) = v_e(t) - v_o(t). \tag{17.52}$$

Adding Equation 17.52 to 17.50 and then subtracting it gives

$$v_e(t) = \tfrac{1}{2}[v(t) + v(-t)], \tag{17.53}$$

$$v_o(t) = \tfrac{1}{2}[v(t) - v(-t)]. \tag{17.54}$$

We will also make use of the fact that an odd function multiplied by an odd function gives an even function, an even function multiplied by another even function is itself even, and an even function multiplied by an odd function is itself odd. Observe also that

$$\int_{-\pi}^{\pi} v_e(t)\, dt = 2\int_0^{\pi} v_e(t)\, dt \tag{17.55}$$

and

$$\int_{-\pi}^{\pi} v_o(t)\,dt = 0. \tag{17.56}$$

Finally, we substitute $v = v_e + v_o$ into Equations 17.32 to 17.34:

$$a_n = \frac{1}{\pi}\left[\int_{-\pi}^{\pi} v_e(t)\cos nt\,dt + \int_{-\pi}^{\pi} v_o(t)\cos nt\,dt\right], \tag{17.57}$$

$$b_n = \frac{1}{\pi}\left[\int_{-\pi}^{\pi} v_e(t)\sin nt\,dt + \int_{-\pi}^{\pi} v_o(t)\sin nt\,dt\right], \tag{17.58}$$

$$a_0 = \frac{1}{2\pi}\left[\int_{-\pi}^{\pi} v_e(t)\,dt + \int_{-\pi}^{\pi} v_o(t)\,dt\right]. \tag{17.59}$$

With these equations, we make the following conclusions:

1. If $v(t)$ is an even periodic function, then from Equations 17.58 and 17.56 we see that

$$b_n = 0, \tag{17.60}$$

$$a_n = \frac{2}{\pi}\int_{0}^{\pi} v(t)\cos nt\,dt, \tag{17.61}$$

$$a_0 = \frac{1}{\pi}\int_{0}^{\pi} v(t)\,dt. \tag{17.62}$$

Hence an even periodic function contains only cosine terms in its Fourier-series representation.

2. If $v(t)$ is an odd periodic function, then we see that

$$a_n = a_0 = 0, \tag{17.63}$$

$$b_n = \frac{2}{\pi}\int_{0}^{\pi} v(t)\sin nt\,dt. \tag{17.64}$$

Then we conclude that an odd periodic function contains only sine terms and no constant (dc) term.

Another interesting symmetry is known as *half-wave symmetry* for which

$$v(t) = -v(t - \pi). \tag{17.65}$$

An example of a signal waveform with half-wave symmetry is shown in Figure 17.15. In general, such waveshapes are identical, but reversed in sign on alternate half-cycles. With this symmetry, $a_0 = 0$. But $v(t)$ may be neither even nor odd, and both sine and cosine terms may be present. Substituting Equation 17.65 into

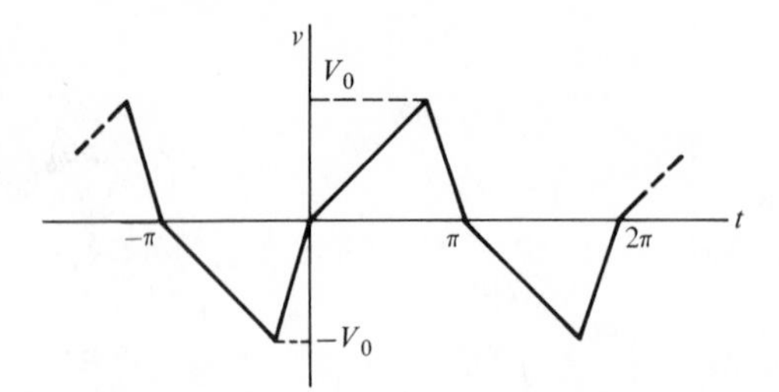

Figure 17.15

the definition for a_n of Equation 17.32, we find

$$a_n = \frac{1}{\pi}\left[\int_{-\pi}^{0} v(t)\cos nt\,dt + \int_{0}^{\pi} v(t)\cos nt\,dt\right], \tag{17.66}$$

$$a_n = \frac{1}{\pi}\left[\int_{0}^{\pi} v(t-\pi)\cos n(t-\pi)\,dt + \int_{0}^{\pi} v(t)\cos nt\,dt\right]. \tag{17.67}$$

Expanding the cosine function and making use of the fact that $\sin n\pi = 0$, we obtain

$$a_n = \frac{1}{\pi}\left[1 - (-1)^n\right]\int_{0}^{\pi} v(t)\cos nt\,dt. \tag{17.68}$$

From this, we see that

$$a_n = \begin{cases} 0, & n \text{ even}, \\ \dfrac{2}{\pi}\displaystyle\int_{0}^{\pi} v(t)\cos nt\,dt, & n \text{ odd}. \end{cases} \tag{17.69}$$

Similarly, for the b_n coefficients, we find that

$$b_n = \begin{cases} 0, & n \text{ even}, \\ \dfrac{2}{\pi}\displaystyle\int_{0}^{\pi} v(t)\sin nt\,dt, & n \text{ odd}. \end{cases} \tag{17.70}$$

Thus, we see that half-wave symmetry results in a Fourier series in which only odd harmonics are present! This class of symmetry is also known as *rotation symmetry*, since signals of this type are generated by rotating electrical machinery.

EXERCISES

17.3.1 Figure 17.16 shows four different plots of $v(t)$ from $-\pi$ to $+\pi$. Since these functions are neither even nor odd, separate them into their even part and their odd part, showing your results as graphs.

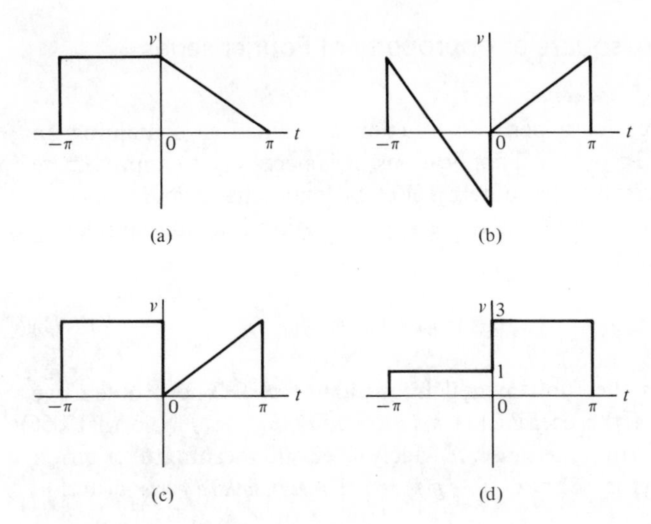

Figure 17.16

17.3.2 Sketch four different waveforms having half-wave symmetry.

17.3.3 Figure 17.17 shows $v(t)$ from $t = 0$ to $t = \pi$, where 2π is the period of $v(t)$. You are asked to construct four $v(t)$ to fill in the waveform from π to 2π such that the periodic function meets the following specifications:
(a) $a_0 = 0$.
(b) $b_n = 0$.
(c) $a_n = 0$, all n.
(d) a_n and b_n are present for odd n only.
Remember that you are asked to satisfy the specifications for each part of the problem, and *not* provide one waveform to simultaneously satisfy all four requirements.

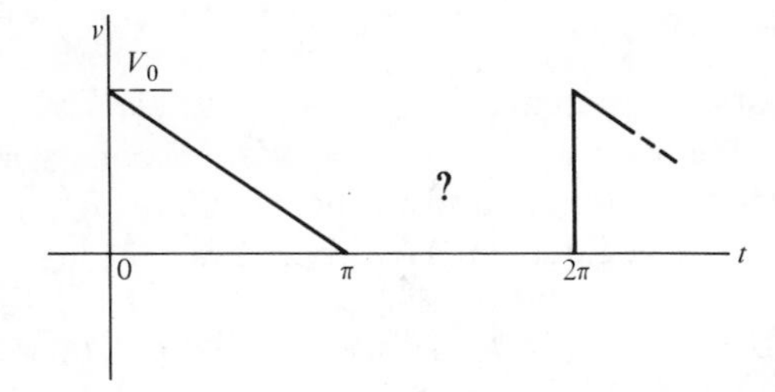

Figure 17.17

17.4 Truncation and least mean-square error property of Fourier series

The general Fourier series for a periodic function is defined as containing an infinite number of terms. In practical applications, it is necessary to approximate a periodic function by a *finite* number of terms. Let us denote the periodic function that we wish to approximate by $v(t)$ and the approximating function containing $2N + 1$ terms as $g(t)$, where

$$g(t) = a_0 + \sum_{n=1}^{N} (a_n \cos n\omega_0 t + b_n \sin n\omega_0 t), \tag{17.71}$$

for which, as usual, ω_0 is the fundamental frequency of $v(t)$. The problem we consider is the selection of the coefficients $a_0, a_1, \ldots, a_N, b_1, \ldots, b_N$, so that the error $v(t) - g(t)$ is "small" in some sense. A widely-used and meaningful criterion for the smallness of the error $e(t) \equiv v(t) - g(t)$ is the *mean-square error* criterion

$$E^2 = \frac{1}{T}\int_0^T e^2(t)\,dt. \tag{17.72}$$

Recall from Chapter 3 that the square root of E^2 is E, the rms value of $e(t)$.

The approximating function $g(t)$ has $2N + 1$ coefficients to be determined. Actually, it is not necessary that all $2N + 1$ coefficients be optimized. For example, it may be given that $a_0 = 0$ (we do not want a dc term present) or $a_1 = 0$ and $b_1 = 0$ (we do not want the fundamental frequency component to be present). The important fact about the approximating function $g(t)$ is that there are a finite number of terms and that the highest frequency allowed is $N\omega_0$. If the kth frequency component is not desired to be present, then a_k and b_k are set equal to zero. The remaining undetermined coefficients are now to be chosen in such a way that E^2 is as small as possible. This is accomplished as follows:

$$\frac{\partial E^2}{\partial a_0} = 0, \qquad \text{if } a_0 \text{ is present;} \tag{17.73}$$

$$\frac{\partial E^2}{\partial a_k} = 0, \qquad \text{if } a_k \text{ is present;} \tag{17.74}$$

$$\frac{\partial E^2}{\partial b_k} = 0, \qquad \text{if } b_k \text{ is present.} \tag{17.75}$$

Substituting Equation 17.71 with the definition of $e(t)$ into Equation 17.72 for E^2, we find that the minimizing condition for a_0 of Equation 17.73 becomes

$$\frac{\partial E^2}{\partial a_0} = \frac{1}{2\pi}\int_0^{2\pi} 2\left[v(t) - a_0 - \sum_{n=1}^{N} (a_n \cos nt + b_n \sin nt)\right] dt = 0. \tag{17.76}$$

Here, as in earlier sections, we have normalized time by letting $\omega_0 = 1$, and in such a way have replaced T by 2π. This minimization is accomplished when

$$a_0 = \frac{1}{2\pi}\int_0^{2\pi} v(t)\,dt. \tag{17.77}$$

Similarly, from Equation 17.74,

$$\frac{\partial E^2}{\partial a_k} = \frac{1}{2\pi}\int_0^{2\pi} 2\left[v(t) - a_0 - \sum_{n=1}^{N}(a_n \cos nt + b_n \sin nt)\right](-\cos kt)\,dt = 0. \tag{17.78}$$

Since

$$\left.\begin{aligned} \int_0^{2\pi} \sin nt \cos kt\,dt &= 0, \qquad k = n \\ \int_0^{2\pi} \cos nt \cos kt\,dt &= 0, \qquad k \neq n \end{aligned}\right\} \tag{17.79}$$

and

$$\left.\begin{aligned} \int_0^{2\pi} \cos nt \cos kt\,dt &= \pi, \qquad k = n \\ \int_0^{2\pi} \sin nt \cos kt\,dt &= 0, \qquad k \neq n \end{aligned}\right\}, \tag{17.80}$$

Equation 17.78 simplifies to

$$a_k = \frac{1}{\pi}\int_0^{2\pi} v(t) \cos kt\,dt. \tag{17.81}$$

Carrying out similar operations for $\partial E^2/\partial b_k$ gives

$$b_k = \frac{1}{\pi}\int_0^{2\pi} v(t) \sin kt\,dt. \tag{17.82}$$

It can be shown that the appropriate second partial derivatives are positive so that the solutions we have found lead to a minimum of E^2.

The result of our analysis now becomes clear. *If a periodic function* v(t) *is to be approximated by a function* g(t) *as in Equation 17.71, the best choice for the allowed coefficients which minimizes the mean-square error in Equation 17.72 is the set of Fourier coefficients!* Terms which are not allowed have zero coefficients. The implications of this statement will be shown through a number of examples.

AMPLE 17.4.1

Consider the full-wave-rectified sine wave

$$v(t) = |\sin \pi t|, \tag{17.83}$$

which has a Fourier series

$$(vt) = \frac{2}{\pi} - \frac{4}{\pi}\left(\tfrac{1}{3}\cos 2\pi t + \tfrac{1}{15}\cos 4\pi t + \tfrac{1}{35}\cos 6\pi t + \cdots + \frac{1}{4n^2-1}\cos 2n\pi t\right)$$

$$= \frac{2}{\pi} - \frac{4}{\pi}\left(\sum_{n=1}^{\infty}\frac{1}{4n^2-1}\cos 2n\pi t\right). \tag{17.84}$$

Notice that the fundamental frequency ω_0 of $v(t)$ is 2π. Suppose that we wish to approximate $v(t)$ by

$$g_1(t) = c_1 \cos 2\pi t + c_2 \cos 4\pi t\,. \tag{17.85}$$

The best choices for c_1 and c_2 which yield a minimum mean-square error,

$$E^2 = \frac{1}{T}\int_0^T [v(t) - g_1(t)]^2\,dt\,, \tag{17.86}$$

are the ordinary Fourier coefficients from Equation 17.84:

$$c_1 = \frac{-4}{3\pi} \quad \text{and} \quad c_2 = \frac{-4}{15\pi}. \tag{17.87}$$

If, instead, we wish to use as our approximation

$$g_2(t) = c_3 + c_4 \cos 6\pi t\,, \tag{17.88}$$

then the solution may be written directly from the coefficients of Equation 17.84. Thus the choice of

$$c_3 = \frac{2}{\pi} \quad \text{and} \quad c_4 = \frac{-4}{35\pi} \tag{17.89}$$

minimizes E^2. Finally, if we desire that

$$g_3(t) = c_5 \cos 2\pi t + c_6 \sin 2\pi t \tag{17.90}$$

be used as the approximating signal, the best choices for c_5 and c_6 which yield minimum mean-square error, E^2, are obtained directly from Equation 17.84 and are

$$c_5 = \frac{-4}{3\pi} \quad \text{and} \quad c_6 = 0\,. \tag{17.91}$$

A second important property of the Fourier-series approximation is that *the best choice for each coefficient is independent of the choices for the other coefficients, no matter how many terms are used in the approximation.* A practical implication of this property is that additional terms may be considered for improving the approximation without having to recompute the coefficients of the terms previously used.

MPLE 17.4.2

For the full-wave-rectified sine wave of the last example, suppose that $g_1(t)$ as in Equation 17.85 is used as an approximation. The minimum mean-square error is attained by the choices of c_1 and c_2 as found in Equation 17.87. Suppose that for a specific application the approximation using $g_1(t)$ is not good enough and that an additional term is to be considered:

$$g_4(t) = c_7 \cos 2\pi t + c_8 \cos 4\pi t + c_9 \cos 6\pi t. \tag{17.92}$$

Then the best choices for c_7 and c_8 remain the same as found in the previous approximation with $c_7 = c_1$ and $c_8 = c_2$. The best choice for the coefficient of the additional term is

$$c_9 = \frac{-4}{35\pi}, \tag{17.93}$$

which is the value taken directly from the Fourier series for $v(t)$ in Equation 17.84.

We next turn to the important matter of deriving a formula for the mean-square error if the best choices for the coefficients are used. We begin by substituting the expressions previously derived for a_0, a_k, and b_k in Equations 17.77, 17.81, and 17.82 with the defined $g(t)$ and $e(t)$ into Equation 17.72 for E^2. Thus

$$\begin{aligned} E^2 &= \frac{1}{2\pi}\int_0^{2\pi} [v^2(t) - 2v(t)g(t) + g^2(t)]\, dt \\ &= \frac{1}{2\pi}\int_0^{2\pi} v^2(t)\, dt - \frac{1}{\pi}\int_0^{2\pi} v(t)\left[a_0 + \sum_{n=1}^{N} (a_n \cos nt + b_n \sin nt)\right] dt \\ &\quad + \frac{1}{2\pi}\int_0^{2\pi}\left[a_0 + \sum_{n=1}^{N} (a_n \cos nt + b_n \sin nt)\right]^2 dt. \end{aligned} \tag{17.94}$$

Making use of the orthogonality conditions of Equations 17.79 and 17.80, we find that

$$\begin{aligned} E^2 &= \frac{1}{2\pi}\int_0^{2\pi}\left\{v^2(t) - 2\left[a_0^2 + \frac{1}{2}\sum_{n=1}^{N} (a_n^2 + b_n^2)\right] + a_0^2 + \frac{1}{2}\sum_{n=1}^{N} (a_n^2 + b_n^2)\right\} dt \\ &= \frac{1}{2\pi}\int_0^{2\pi} v^2(t)\, dt - \left[a_0^2 + \frac{1}{2}\sum_{n=1}^{N} (a_n^2 + b_n^2)\right]. \end{aligned} \tag{17.95}$$

It should be remembered that only the allowed coefficients are present in this result, Equation 17.95, and that the unallowed coefficients corresponding to terms absent from $g(t)$ are zero. Furthermore, the allowed coefficients, a_k and b_k, are obtained from equations already derived, Equations 17.77, 17.81, and 17.82.

Finally, it should be noted that E^2 cannot be negative, so that

$$a_0^2 + \frac{1}{2}\sum_{n=1}^{N}(a_n^2 + b_n^2) \leq \frac{1}{2\pi}\int_0^{2\pi} v^2(t)\,dt. \tag{17.96}$$

As more terms are admitted by permitting N to become larger, then

$$a_0^2 + \frac{1}{2}\sum_{n=1}^{N}(a_n^2 + b_n^2) \equiv M^2 \tag{17.97}$$

may increase but never decrease. Hence the expression is a monotonically non-decreasing function of N which has

$$\frac{1}{2\pi}\int_0^{2\pi} v^2(t)\,dt \tag{17.98}$$

as an upper bound. This concept will be illustrated with several examples.

EXAMPLE 17.4.3

Let us compute the minimum mean-square errors for the approximations of Examples 17.4.1 and 17.4.2. For those examples, $\omega_0 = 2\pi$ and $T = 2\pi/\omega_0 = 1$, so we will use the more general form for the integral expressions. Then

$$\frac{1}{T}\int_0^T v^2(t)\,dt = \frac{1}{T}\int_0^T \sin^2 \pi t\,dt = \frac{1}{T}\int_0^T \frac{1-\cos 2\pi t}{2}\,dt = \tfrac{1}{2}. \tag{17.99}$$

We now compute the minimum mean-square error E^2 for each of the approximations g_1, g_2, g_3, and g_4; these are

$$E_1^2 = \frac{1}{2} - \frac{1}{2}\left[\left(\frac{4}{3\pi}\right)^2 + \left(\frac{4}{15\pi}\right)^2\right] = 0.40633, \tag{17.100}$$

$$E_2^2 = \frac{1}{2} - \left[\left(\frac{2}{\pi}\right)^2 + \frac{1}{2}\left(\frac{4}{35\pi}\right)^2\right] = 0.094053, \tag{17.101}$$

$$E_3^2 = \frac{1}{2} - \frac{1}{2}\left(\frac{4}{3\pi}\right)^2 = 0.40994, \tag{17.102}$$

$$E_4^2 = \frac{1}{2} - \frac{1}{2}\left[\left(\frac{4}{3\pi}\right)^2 + \left(\frac{4}{15\pi}\right)^2 + \left(\frac{4}{35\pi}\right)^2\right] = 0.40567. \tag{17.103}$$

We now state a third important property of Fourier series. Although it is true that whenever a finite number of terms is used to approximate a periodic function, the least mean-square error choices for the coefficients are the Fourier coefficients, *it is not always true that the first* N *terms should be used if only* N *terms are allowed. Instead, the* N *terms with the largest coefficients should be used.* This will be shown by a number of examples.

AMPLE 17.4.4

Consider the periodic function

$$v(t) = 2.5 + 3\cos t + 3\sin t + 4\cos 2t + \sin 2t + \cos 3t - 2.5\sin 3t. \quad (17.104)$$

Observe that $v(t)$ is in Fourier-series form and that the fundamental frequency is $\omega_0 = 1$. Suppose that it is desired to use a one-term approximation and that the choices are

$$g(t) = c_1 \quad \text{or} \quad g(t) = c_2 \cos nt \quad \text{or} \quad g(t) = c_3 \sin nt. \quad (17.105)$$

From the equation for E^2 in Equation 17.95, it is clear that we should choose the largest of a_0^2, $a_n^2/2$, or $b_n^2/2$ from the harmonics present in $v(t)$ in Equation 17.104. By considering all of the possibilities, we conclude that the best choice is

$$g(t) = 4\cos 2t. \quad (17.106)$$

If $g(t)$ is allowed to have two terms, where $\cos nt$ and $\sin nt$ are considered as separate terms, then the best approximation is

$$g(t) = 2.5 + 4\cos 2t. \quad (17.107)$$

There are two best three-term approximations:

$$g(t) = 2.5 + 4\cos 2t + 3\sin t \quad (17.108)$$

and

$$g(t) = 2.5 + 4\cos 2t + 3\cos t. \quad (17.109)$$

AMPLE 17.4.5

Consider the same periodic function for $v(t)$ as in the last example. Let us determine the best approximation containing only one harmonic. Notice that $a_k \cos kt$ and $b_k \sin kt$ is actually one term (one frequency). This time we consider the largest value from the choices of a_0^2 and $(a_n^2 + b_n^2)/2$, for all possible n. By direct inspection of $v(t)$, we see that the best single-harmonic approximation is

$$g(t) = 2\cos t + 3\sin t. \quad (17.110)$$

Similarly, the best two-harmonic approximation is

$$g(t) = 3\cos t + 3\sin t + 4\cos 2t + \sin 2t, \quad (17.111)$$

and the best three-harmonic approximation is

$$g(t) = 2.5 + 3\cos t + 3\sin t + 4\cos 2t + \sin 2t. \quad (17.112)$$

ERCISES

17.4.1 For the full-wave-rectified sine wave of Example 17.4.1, determine the least mean-square error approximation of the form

$$g(t) = b_1 \sin 2\pi t + b_2 \sin 4\pi t.$$

17.4.2 For the signal $v(t)$ in Example 17.4.1, choose the least mean-square error approximation containing only two terms.

17.4.3 For $g(t)$ given by Equation 17.71, show that, with $\omega_0 = 1$,

$$\frac{1}{2\pi}\int_0^{2\pi} g^2(t)\,dt = a_0^2 + \frac{1}{2}\sum_{n=1}^{N}(a_n^2 + b_n^2).$$

Noting that:
(a) $a_n \cos nt + b_n \sin nt = \sqrt{a_n^2 + b_n^2}\cos(nt - \tan^{-1} b_n/a_n)$,
(b) the rms value of $c_n \cos(nt - \theta_n)$ is $|c_n|/\sqrt{2}$, and
(c) the rms value of a_0 is $|a_0|$,
the above relationship says that the mean-square value of a periodic signal $g(t)$, or $(1/2\pi)\int_0^{2\pi} g^2(t)\,dt$, which has only a finite number of harmonics, is equal to the sum of the squares of the rms values of the harmonics.

17.4.4 For the periodic signal $v(t)$ in Example 17.4.4, compute

$$\frac{1}{T}\int_0^T v^2(t)\,dt$$

by applying the result of Exercise 17.4.3.

17.5 Summary

1. A signal is said to be periodic if for all t and a constant T.

$$v(t) = v(t + T),$$

The smallest T which satisfies the equation is called the period.

2. The period T is related to frequencies ω_0 and f_0 by

$$\omega_0 = \frac{2\pi}{T}\text{ radians/sec}, \qquad f_0 = \frac{1}{T}\text{ Hz}.$$

By using the change of variable,

$$\omega_0 t = \tau,$$

we can write $v(t)$ as

$$v(t) = v\left(\frac{\tau}{\omega_0}\right) = \hat{v}(\tau),$$

where $\hat{v}(\tau)$ is the new function of τ. From

$$\hat{v}(\tau) = v(t) = v(t + T) = v\left(\frac{\tau}{\omega_0} + T\right)$$

$$= v\left[\frac{1}{\omega_0}(\tau + \omega_0 T)\right] = v\left[\frac{1}{\omega_0}(\tau + 2\pi)\right] = \hat{v}(\tau + 2\pi),$$

we see that a period of T in the t-domain means a period of 2π in the τ-domain.

3. The Fourier-series expansion of $v(t)$ is

$$v(t) = a_0 + \sum_{n=1}^{\infty} (a_n \cos nt + b_n \sin nt)$$

with time scaled so that $\omega_0 = 1$; otherwise nt is replaced by $n\omega_0 t$.

4. The Fourier coefficients are obtained from the relationships

$$a_0 = \frac{1}{T}\int_{t_0}^{t_0+T} v(t)\, dt,$$

$$a_n = \frac{2}{T}\int_{t_0}^{t_0+T} v(t) \cos n\omega_0 t\, dt,$$

$$b_n = \frac{2}{T}\int_{t_0}^{t_0+T} v(t) \sin n\omega_0 t\, dt.$$

We usually select $t_0 = 0$ or $t_0 = -T/2$. When time is scaled so that $\omega_0 = 1$, then T is replaced by 2π.

5. Several definitions are important:

ω_0 is the fundamental frequency,
$n\omega_0$ is the nth harmonic frequency,
a_0 is the average or dc value of $v(t)$.

6. When a periodic signal $v(t)$ is even,

$$v(-t) = v(t),$$

then

$$b_n = 0,$$

$$a_0 = \frac{2}{T}\int_0^{T/2} v(t)\, dt,$$

$$a_n = \frac{1}{T}\int_0^{T/2} v(t) \cos n\omega_0 t\, dt.$$

When a periodic signal $v(t)$ is odd, then

$$v(-t) = -v(t),$$

$$a_0 = a_n = 0,$$

$$b_n = \frac{1}{T}\int_0^{T/2} v(t) \sin n\omega_0 t \, dt.$$

When a periodic signal $v(t)$ has half-wave symmetry

$$v(t) = -v(t - T/2),$$

then it contains only odd harmonics.

7. When the Fourier-series representation of a periodic signal $v(t)$ is approximated by a finite trigonometric sum $g(t)$, an error generally results,

$$e(t) = v(t) - g(t).$$

The mean-square error

$$E^2 = \frac{1}{T}\int_0^T e^2(t) \, dt$$

is minimized by choosing the coefficients in $g(t)$ to be the Fourier coefficients of $v(t)$. This holds regardless of the number of terms present.

The minimum mean-square error is

$$E^2 = \frac{1}{T}\int_0^T v^2(t) \, dt - \left[a_0^2 + \frac{1}{2}\sum_{n=1}^{N} (a_n^2 + b_n^2)\right].$$

8. If there is a choice of which harmonics are to be included in the finite sum approximation $g(t)$, E^2 is minimized by including those with the largest amplitudes.

PROBLEMS

17.1 Given the following signal

$$v(t) = \sin t + \sin k_1 t + \sin k_2 t,$$

where $k_1 = 0.25$ and $k_2 = 0.55$, what is the period T of $v(t)$?

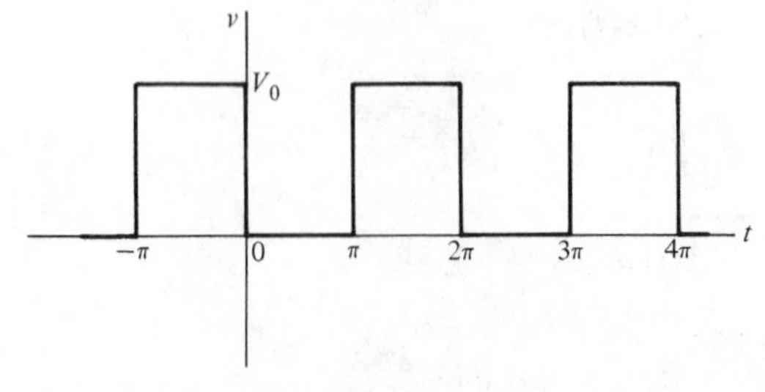

Figure 17.18

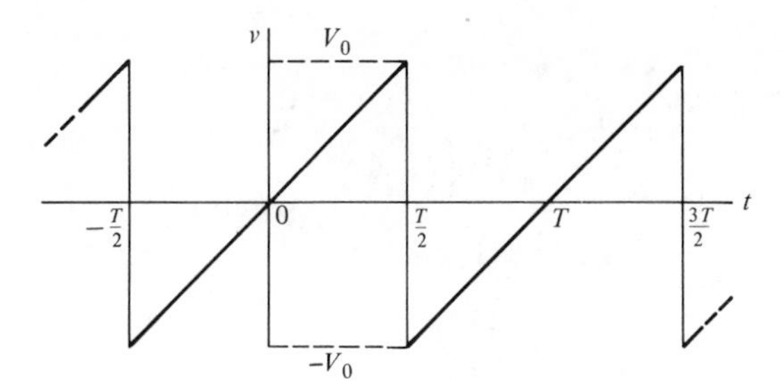

Figure 17.19

17.2 It is given that $v(t)$ is a periodic signal of period T. Determine the period of $v(kt)$ where k is a real positive constant and $k \neq 0$.

17.3 Find the period of the signal

$$v(t) = (3 \cos 2t)^2.$$

17.4 The signal shown in Figure 17.18 is a square wave of amplitude V_0. Determine the Fourier coefficients for this signal.

17.5 The signal shown in Figure 17.19 is a sweep voltage. Take advantage of any symmetry in computing the Fourier coefficients for $v(t)$.

17.6 The signal shown in Figure 17.20 is similar to that of Problem 17.5, except that the slope of the sweep is negative. Take advantage of any symmetry in computing the Fourier series for this $v(t)$.

17.7 Repeat Problem 17.5 for the signal waveform shown in Figure 17.21.

17.8 The signal shown in Figure 17.22 consists of one cycle of a sine function, followed by a period of equal duration of the cycle of sine wave for which the signal has zero value. Determine the coefficients for the Fourier series for this signal.

17.9 Figure 17.23 shows a signal which consists of a negative-going pulse, followed by a positive-going pulse. The signal has a period 2π. Taking advantage of any symmetries present, determine the coefficients for the Fourier-series representation of the signal.

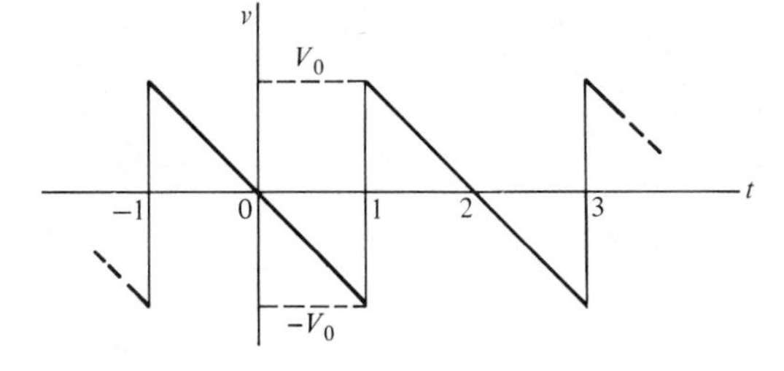

Figure 17.20

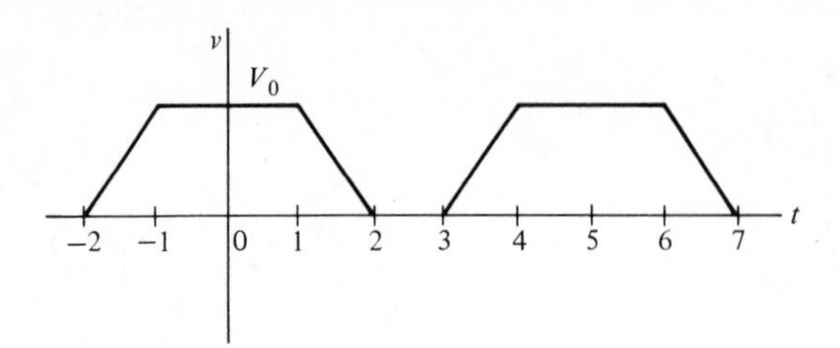

Figure 17.21

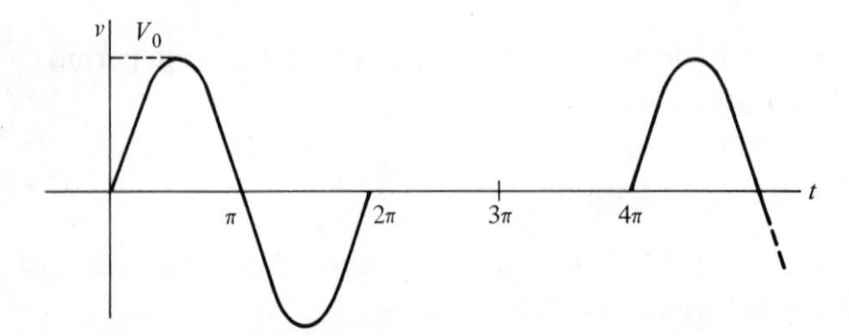

Figure 17.22

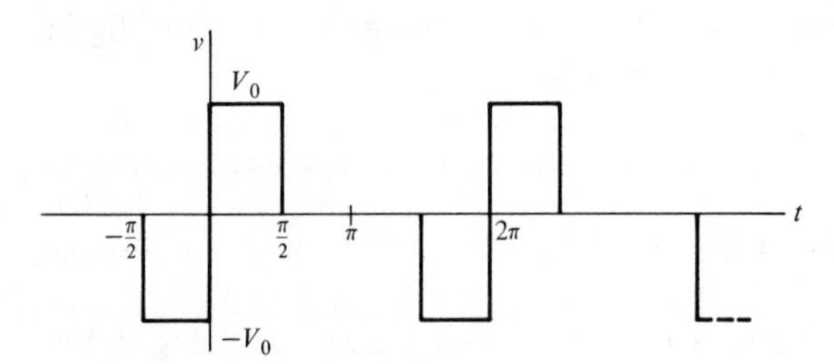

Figure 17.23

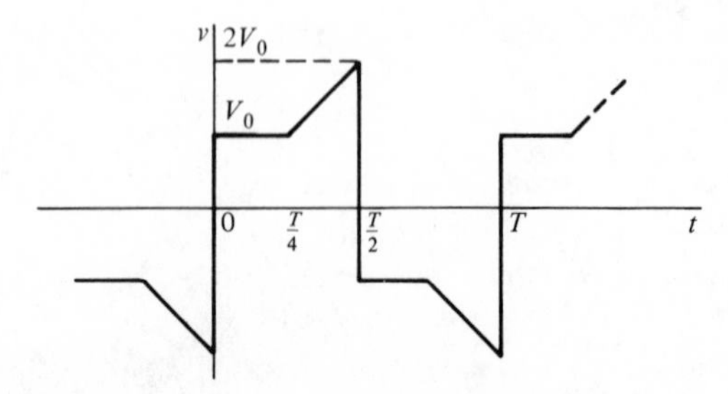

Figure 17.24

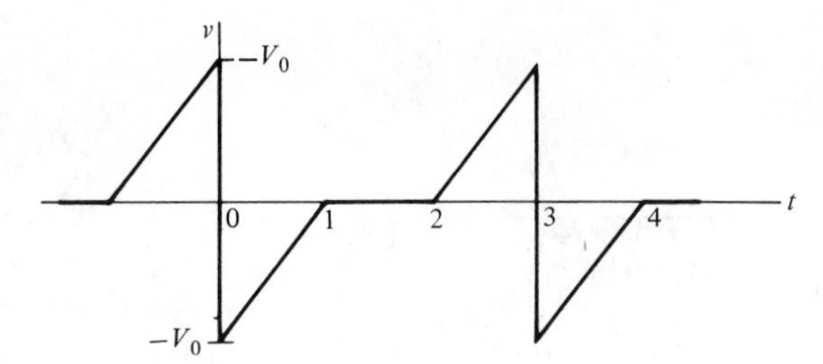

Figure 17.25

17.10 The signal shown in Figure 17.24 is made up of a number of straight-line segments and has a period of T sec. (a) Discuss the symmetry of this signal. (b) Determine the Fourier coefficients for this signal.

17.11 Repeat Problem 17.10 for the signal shown in Figure 17.25.

17.12 A signal is described by the equation

$$v(t) = Kt^2, \qquad -\pi < t < \pi,$$

and $v(t + 2\pi n) = v(t)$ for integer values of n. (a) Sketch this waveform and determine its period. (b) Determine the Fourier coefficients for this signal.

17.13 The signal shown in Figure 17.26 is a segment of a sine function for the first quarter of the cycle. For other values of t in the cycle, $v(t) = 0$. For this waveform, determine the Fourier coefficients.

17.14 The signal shown in Figure 17.27 is made up of two positive-going sine wave segments, followed by two negative-going sine wave segments, the period of the signal being 4π. Determine the Fourier coefficients for this signal.

17.15 Figure 17.28 shows a period function which is sinusoidal and of amplitude V_0 from $t = 0$ to $t = T/2$. It is also sinusoidal from $t = T/2$ to $t = T$, but of half amplitude, $V_0/2$. Such a waveform might be generated by a nonlinear device. For this signal, determine a_0, a_n, and b_n.

17.16 Inspection of the signal waveforms shown in the chapter shows that symmetry is influenced by the choice of the time reference $t = 0$. If that

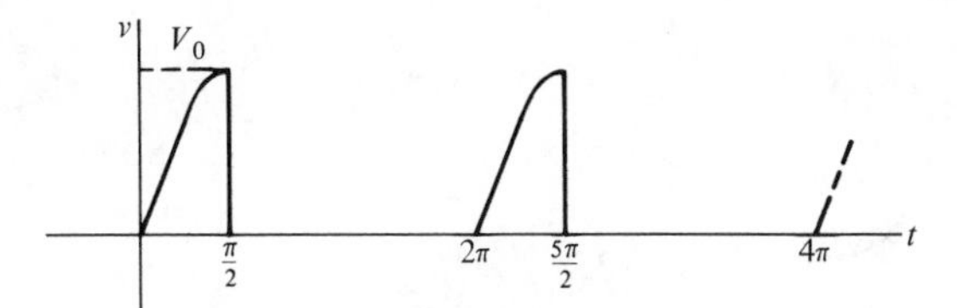

Figure 17.26

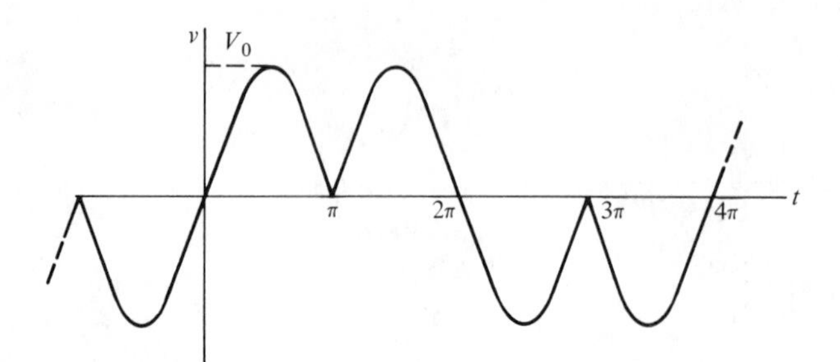

Figure 17.27

reference could be shifted, then it would be possible to take advantage of symmetries. Show that (a) shifting the time axis does not affect the value of $|c_n| = \sqrt{a_n^2 + b_n^2}$ and (b) delaying a signal by τ sec causes a shift in the phase characteristic of $-n\omega_0\tau$ radians.

17.17 Given two signals $v_1(t)$ and $v_2(t)$ having the same period T. Let the signals be piecewise continuous and the Fourier coefficients for v_1 be a_n and b_n while those for v_2 are α_n and β_n. Show that

$$\frac{2}{T}\int_{-T/2}^{T/2} v_1(t)v_2(t)\,dt = 2a_0\alpha_0 + \sum_{n=1}^{\infty}(a_n\alpha_n + b_n\beta_n).$$

17.18 Consider the periodic signal of Equation 17.104. This signal is to be passed through a bandpass filter which has the characteristic that two adjacent frequencies may be passed with a given setting and all others will be totally rejected. Which setting of the filter will result in a $g(t)$ which is the best approximation to $v(t)$ in a least mean-square error sense?

17.19 For the signal of Exercise 17.2.3, determine a least mean-square approximation $g_1(t)$ containing only (a) one harmonic, (b) two harmonics, and (c) three harmonics.

17.20 For the signal of Exercise 17.2.1, determine a minimum mean-square approximation $g(t)$ containing enough terms so that the mean-square of $g(t)$ is at least 90 percent of the mean-square of the original signal.

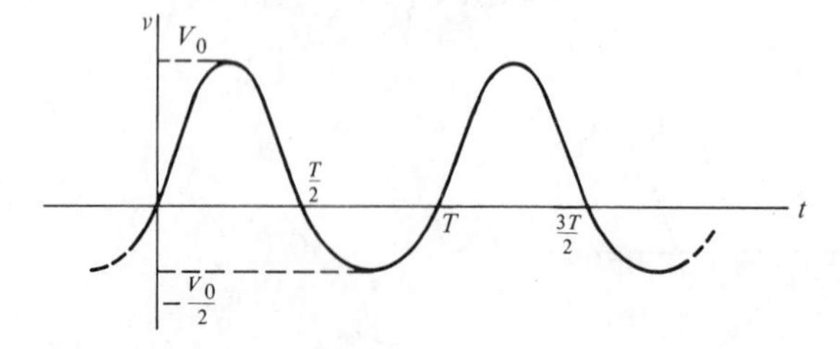

Figure 17.28

Chapter **18**

ransforms and Spectra

.1 Exponential Fourier series

The notions of the transform and signal spectra to be treated in this chapter stem from studies started in the last chapter: they are natural extensions of properties of Fourier series. Few topics have more importance in electrical engineering, for reasons that will be discussed as our study progresses.

We begin with the expression for the Fourier series of a periodic function from Chapter 17,

$$v(t) = a_0 + \sum_{n=1}^{\infty} (a_n \cos n\omega_0 t + b_n \sin n\omega_0 t), \tag{18.1}$$

where $\omega_0 = 2\pi/T$. To convert this equation to exponential form, we make use of equations for the sine and cosine functions in terms of exponentials,

$$\cos n\omega_0 t = \tfrac{1}{2}(e^{jn\omega_0 t} + e^{-jn\omega_0 t}) \tag{18.2}$$

and

$$\sin n\omega_0 t = \frac{1}{2j}(e^{jn\omega_0 t} - e^{-jn\omega_0 t}). \tag{18.3}$$

Substituting these two equations into Equation 18.1 and collecting coefficients of the two different exponentials, we have

$$v(t) = a_0 + \sum_{n=1}^{\infty} [\tfrac{1}{2}(a_n - jb_n)e^{jn\omega_0 t} + \tfrac{1}{2}(a_n + jb_n)e^{-jn\omega_0 t}]. \tag{18.4}$$

If we make the following definitions:

$$F_0 = a_0, \qquad F_n = \tfrac{1}{2}(a_n - jb_n), \qquad F_{-n} = \tfrac{1}{2}(a_n + jb_n), \tag{18.5}$$

then we may write $v(t)$ in Equation 18.4 in the following forms:

$$v(t) = F_0 + \sum_{n=1}^{\infty} (F_n e^{jn\omega_0 t} + F_{-n} e^{-jn\omega_0 t})$$

$$= F_0 + \sum_{n=1}^{\infty} F_n e^{jn\omega_0 t} + \sum_{n=-\infty}^{-1} F_n e^{+jn\omega_0 t}$$

$$= \sum_{n=-\infty}^{\infty} F_n e^{jn\omega_0 t}. \tag{18.6}$$

This last form is particularly compact. It is known as the exponential form of the Fourier series of $v(t)$.

The Fourier coefficients F_n and F_{-n} may be determined from Equation 18.4. For example,

$$F_n = \tfrac{1}{2}(a_n - jb_n)$$

$$= \frac{1}{T}\left[\int_{-T/2}^{T/2} v(t)\cos n\omega_0 t\,dt - j\int_{-T/2}^{T/2} v(t)\sin n\omega_0 t\,dt\right]$$

$$= \frac{1}{T}\int_{-T/2}^{T/2} v(t)(\cos n\omega_0 t - j\sin n\omega_0 t)\,dt \tag{18.7}$$

or

$$F_n = \frac{1}{T}\int_{-T/2}^{T/2} v(t)e^{-jn\omega_0 t}\,dt. \tag{18.8}$$

By similar operations, we find that

$$F_{-n} = \frac{1}{T}\int_{-T/2}^{T/2} v(t)e^{jn\omega_0 t}\,dt. \tag{18.9}$$

These two equations may be combined into a single equation, which also includes $F_0 = a_0$,

$$F_n = \frac{1}{T}\int_{-T/2}^{T/2} v(t)e^{-jn\omega_0 t}\,dt, \qquad n = 0, \pm 1, \pm 2, \ldots. \tag{18.10}$$

The complex Fourier coefficients, F_n, are useful in building the concept of a spectrum. They are expressed in rectangular form in Equations 18.5. We normally express them in polar form involving magnitude and phase, as

$$F_n = |F_n|e^{j\phi_n} = |F_n|\underline{/\phi_n} \tag{18.11}$$

and

$$F_{-n} = F_n^* = |F_n|e^{-j\phi_n} = |F_n|\underline{/-\phi_n}, \tag{18.12}$$

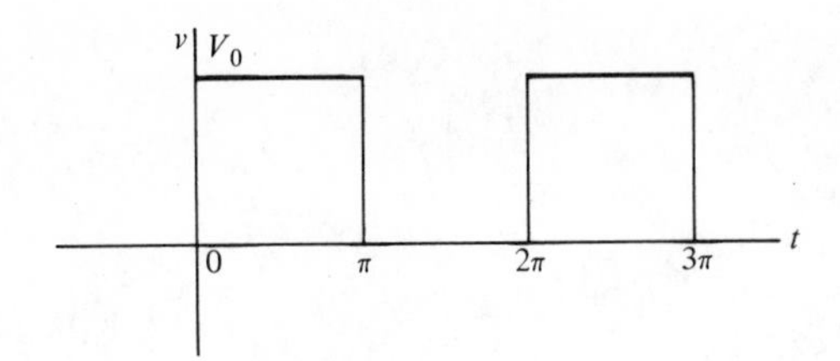

Figure 18.1 The signal waveform for which the exponential Fourier series is found in Example 18.1.1.

where, as usual, the asterisk designates complex conjugate. The magnitude and phase are related to the a_n and b_n coefficients through the equations

$$|F_n| = \tfrac{1}{2}\sqrt{a_n^2 + b_n^2} \tag{18.13}$$

and

$$\phi_n = \tan^{-1}\frac{-b_n}{a_n}. \tag{18.14}$$

EXAMPLE 18.1.1

The waveform shown in Figure 18.1 is known as a square wave. The time scale is chosen so that $\omega_0 = 1$, and thus $T = 2\pi$. The coefficients given by Equation 18.10 are to be found. We first observe that $F_0 = V_0/2$ by inspection of the figure; the remaining coefficients are

$$F_n = \frac{1}{2\pi}\int_0^{2\pi} v(t)e^{-jnt}\,dt = \frac{V_0}{2\pi}\left[\frac{e^{-jnt}}{-jn}\right]_0^{\pi} \tag{18.15}$$

or

$$F_n = \begin{cases} \dfrac{2}{jn}\dfrac{V_0}{2\pi}, & n \text{ odd}, \\ 0, & n \text{ even}, \quad n \neq 0, \\ \dfrac{V_0}{2}, & n = 0, \end{cases} \tag{18.16}$$

and so we see that the Fourier series in exponential form is

$$v(t) = V_0\left(\cdots\frac{-1}{j3\pi}e^{-j3t} - \frac{1}{j\pi}e^{-jt} + \frac{1}{2} + \frac{1}{j\pi}e^{jt} + \cdots\right). \tag{18.17}$$

EXAMPLE 18.1.2

The waveform of Figure 18.2 is known as a full-wave-rectified sine wave, described by the equation $v(t) = V_0|\sin t|$, where $\omega_0 = 2$, and so $T = \pi$. F_0 is seen by

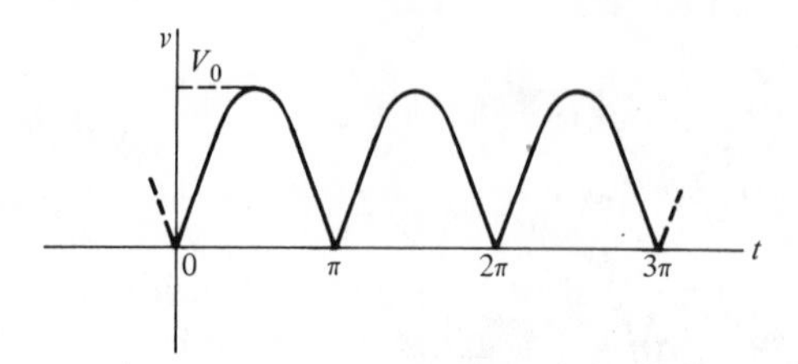

Figure 18.2 The signal waveform considered in Example 18.1.2, known as a full-wave-rectified sine wave. Note that the period of the signal is π, while that of the sinusoid from which it is derived is 2π.

inspection to be $2V_0/\pi$. The remaining terms are found from Equation 18.10:

$$F_n = \frac{1}{\pi}\int_0^{\pi} V_0|\sin t|e^{-j2nt}\,dt \tag{18.18}$$

$$= \frac{V_0}{\pi 2j}\left(\frac{e^{-j(2n-1)t}}{-j(2n-1)} - \frac{e^{-j(2n+1)t}}{-j(2n+1)}\right)\Bigg|_0^{\pi}. \tag{18.19}$$

Finally,

$$F_n = \frac{-2V_0}{(4n^2-1)\pi}. \tag{18.20}$$

Then the Fourier series in exponential form is

$$v(t) = \frac{-2V_0}{\pi}\cdot\sum_{n=-\infty}^{\infty}\frac{1}{4n^2-1}e^{jnt}. \tag{18.21}$$

In expanded form

$$v(t) = \frac{2V_0}{\pi}\left(\cdots + \tfrac{1}{35}e^{-j3t} + \tfrac{1}{15}e^{-j2t} + \tfrac{1}{3}e^{-jt} + 1 + \tfrac{1}{3}e^{jt} + \tfrac{1}{15}e^{j2t} + \cdots\right). \tag{18.22}$$

EXERCISES

18.1.1 Verify that

$$a_n = F_n + F_{-n} = 2\,\mathrm{Re}\,(F_n),$$
$$b_n = j(F_n - F_{-n}) = -2\,\mathrm{Im}\,(F_n).$$

18.1.2 Verify that F_n in Equation 18.8 may be written as

$$F_n = \frac{1}{T}\int_{t_0}^{t_0+T} v(t)e^{-jn\omega_0 t}\,dt.$$

.2 Discrete signal spectra

The magnitude spectrum of a signal $v(t)$ is the plot of the complex Fourier coefficient magnitude $|F_n|$ as a function of $n\omega_0$ (or simply n by frequency scaling). The phase spectrum is a similar plot of the phase function ϕ_n associated with the complex Fourier coefficient F_n as a function of $n\omega_0$ (or simply n). In both plots, n is an integer, and so the plot is made for discrete values of frequency only; for this reason, the spectrum is known as the *discrete frequency spectrum.* Actually, the plot is of points only, but for emphasis a line is drawn from zero to the point.

(E)XAMPLE 18.2.1

The spectrum for the signal shown in Figure 18.1 is given by Equation 18.16. Since the coefficients are of the form $V_0/(jn\pi)$ for n odd, the magnitudes are $V_0/(n\pi)$, and the phase is that associated with $1/(nj) = -j/n$. For n negative, the phase is 90°, while for n positive it is $-90°$. The spectrum for the square wave $v(t)$ is that shown in Figure 18.3.

(E)XAMPLE 18.2.2

The spectrum for the full-wave-rectified sine wave is found from the complex Fourier series of Equation 18.22. The magnitude functions are found directly from Equation 18.20. The phase of each term for $n \neq 0$ is that of a negative number which is $\pm 180°$. The spectrum is shown in Figure 18.4, with a choice of the sign of $\pm 180°$ made to make the phase spectrum an odd function.

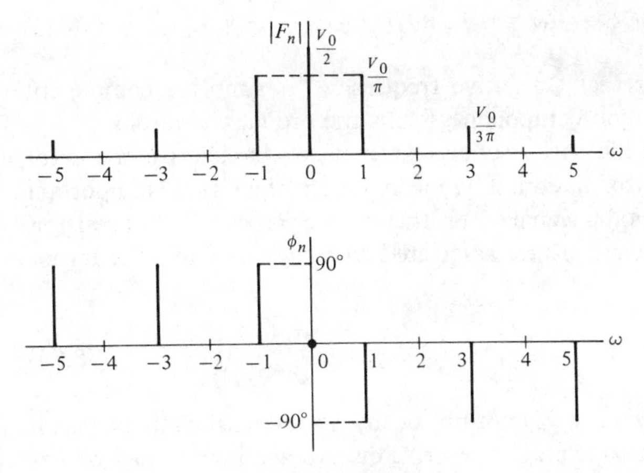

(Fi)gure 18.3 The amplitude and phase line spectra for Example 18.2.1.

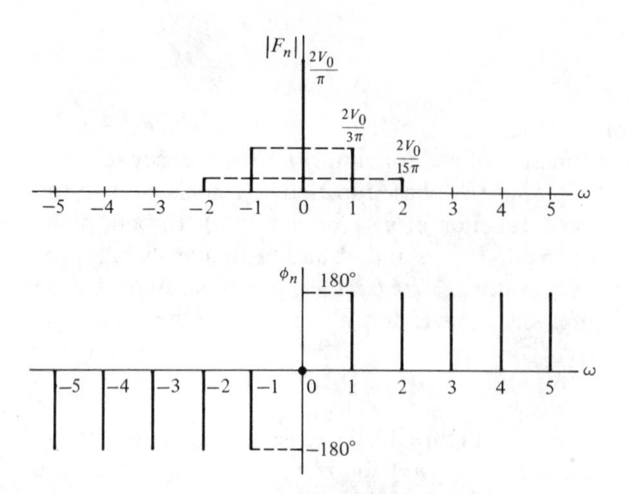

Figure 18.4 The amplitude and phase line spectra for the full-wave-rectified sine wave of Example 18.2.2.

Plots of magnitude and phase spectra are made for both positive and negative $n\omega_0$ since the infinite-series expansion of $v(t)$ extends from $-\infty$ to $+\infty$, and such plots always display the fact that the magnitude function is an even function, while the phase function is an odd function. It is natural that the question of the meaning of "negative frequency" should arise, since a physical interpretation seems difficult. Observe that the components of F_n occur in pairs, one for a negative frequency and one for a positive frequency, and further that $F_{-n} = F_n^*$. Hence terms can always be grouped together in pairs as

$$|F_n|e^{j\phi_n}e^{jn\omega_0 t} + |F_n|e^{-j\phi_n}e^{-jn\omega_0 t} = 2|F_n| \cos(n\omega_0 t + \phi_n). \tag{18.23}$$

From this, it becomes clear that "negative frequency" is simply a component that, together with a companion component, forms the ordinary sinusoid.

In Chapter 17, we used the mean-square concept to characterize the error between a signal and its approximation. The mean-square of a signal is important in its own right. When a periodic signal $v(t)$ is the voltage across a 1-ohm resistor, then $v^2(t)$ is also the *instantaneous power* associated with the signal and the mean-square of $v(t)$, or

$$V^2 = \frac{1}{T}\int_0^T v^2(t)\, dt, \tag{18.24}$$

is the average power. Even when V^2 is not physically nor dimensionally power, it is conventional to refer to it as average power. This average power may be expressed in terms of the discrete spectrum F_n of $v(t)$, as we will next show.

If we express $v(t)$ in its exponential Fourier-series form given as Equation 18.6, then it is

$$v(t) = \sum_{n=-\infty}^{\infty} F_n e^{jn\omega_0 t}. \tag{18.25}$$

It is not immediately obvious how we should square this infinite series to obtain the required $v^2(t)$. It turns out that this is best accomplished by using two indices, n and k, such that the substitution of $v(t)$ into the equation for V^2 (Equation 18.24) gives

$$V^2 = \frac{1}{T}\int_0^T \left(\sum_{n=-\infty}^{\infty} F_n e^{jn\omega_0 t}\right)\left(\sum_{k=-\infty}^{\infty} F_k e^{jk\omega_0 t}\right) dt. \tag{18.26}$$

The task of multiplying required by this equation is not as formidable as it may appear. Only two situations need be considered. The first is $n = -k$, in which case we have

$$V^2 = \frac{1}{T}\int_0^T \sum_{n=-\infty}^{\infty} F_n F_{-n}\, dt. \tag{18.27}$$

Since we know that $F_{-n} = F_n^*$ and that $F_n F_n^* = |F_n|^2$, then the summation becomes

$$\frac{1}{T}\int_0^T \sum_{n=-\infty}^{\infty} F_n F_{-n}\, dt = \sum_{n=-\infty}^{\infty} |F_n|^2. \tag{18.28}$$

The second situation is $n \neq -k$, the total of all other possibilities. In this case

$$\frac{1}{T}\int_0^T \sum_{\substack{n=-\infty \\ n\neq -k}}^{\infty} \sum_{k=-\infty}^{\infty} F_n F_k e^{j(n+k)\omega_0 t}\, dt = \frac{1}{T}\left[\sum_{\substack{n=-\infty \\ n\neq -k}}^{\infty} \sum_{k=-\infty}^{\infty} \frac{F_n F_k}{j(n+k)\omega_0} e^{j(n+k)\omega_0 t}\right]_0^T$$
$$= 0. \tag{18.29}$$

This sums to zero since $\omega_0 T = 2\pi$ and $n + k$ has integer values only. Substituting Equation 18.28 into Equation 18.29 and equating the result to Equation 18.24, we have

$$\frac{1}{T}\int_0^T v^2(t)\, dt = \sum_{n=-\infty}^{\infty} |F_n|^2. \tag{18.30}$$

This equation tells us that the mean-square of a signal is equal to the sum of the squares of the magnitude of the discrete spectrum F_n. We may think of $|F_n|^2$ as the average power associated with the frequency $n\omega_0$. The function $|F_n|^2$ plotted as a function of n (or by scaling $n\omega_0$) is known as the *discrete power spectrum* of $v(t)$. The relationship we have found and expressed in Equation 18.30 is known as *Parseval's theorem*.

EXAMPLE 18.2.3

The square-wave signal in Figure 18.1 considered in Example 18.2.1 has a mean-square or average power of

$$V^2 = \frac{1}{2\pi}\int_0^{2\pi} v^2(t)\,dt = \frac{1}{2\pi}\int_0^{\pi} V_0^2\,dt = \frac{V_0^2}{2}. \tag{18.31}$$

From the discrete spectrum of Equation 18.16, it is seen that the power spectrum is

$$|F_n|^2 = \begin{cases} \dfrac{V_0^2}{n^2\pi^2}, & n \text{ odd}, \\ 0, & n \text{ even}, \\ \dfrac{V_0^2}{4}, & n = 0. \end{cases} \tag{18.32}$$

From Parseval's theorem, we have

$$\begin{aligned} V^2 &= \frac{V_0^2}{4} + \sum_{\substack{n=1 \\ n \text{ odd only}}}^{\infty} \frac{2V_0^2}{n^2\pi^2} \\ &= V_0^2\left[\tfrac{1}{4} + \frac{2}{\pi^2}\left(1 + \tfrac{1}{9} + \tfrac{1}{25} + \tfrac{1}{49} + \cdots\right)\right]. \end{aligned} \tag{18.33}$$

An interesting mathematical curiosity follows from this result. Since V^2 must equal $V_0^2/2$, it must be true that

$$\frac{2}{\pi^2}\left(1 + \tfrac{1}{9} + \tfrac{1}{25} + \tfrac{1}{49} + \cdots\right) = \tfrac{1}{4}, \tag{18.34}$$

which may be verified from a mathematics table. Using an approximation function,

$$g(t) = \frac{V_0}{2} + \frac{V_0}{j\pi}e^{jt} + \frac{V_0}{-j\pi}e^{-jt} + \frac{V_0}{j3\pi}e^{j3t} + \frac{V_0}{-j3\pi}e^{-j3t} + \frac{V_0}{j5\pi}e^{j5t} + \frac{V_0}{-j5\pi}e^{-j5t}, \tag{18.35}$$

we discover that the mean-square G^2 is, by Parseval's theorem,

$$G^2 = \frac{V_0^2}{4} + \frac{2V_0^2}{\pi^2} + \frac{2V_0^2}{9\pi^2} + \frac{2V_0^2}{25\pi^2} = 0.483V_0^2, \tag{18.36}$$

and this is within 3.5 percent of $V^2 = V_0^2/2$. The discrete power spectrum plot, V^2, is shown in Figure 18.5.

EXERCISES

18.2.1 Determine the discrete magnitude and phase spectrum of

$$v(t) = 5 + 4\sin 2t + 6\cos 3t + 8\sin 3t.$$

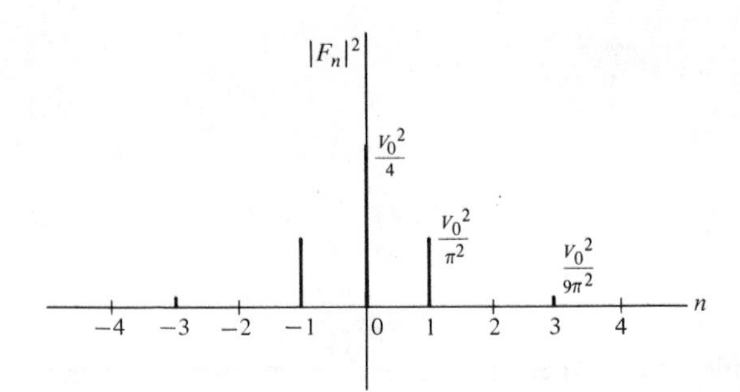

Figure 18.5 The discrete power spectrum of the square wave of Figure 18.1.

18.2.2 Determine and sketch the discrete power spectrum of the signal in Exercise 18.2.1.

18.2.3 Determine the discrete spectrum of $v(t + \pi/4)$ if $v(t)$ is the signal given in Exercise 18.2.1.

18.2.4 Determine the discrete power spectrum of $v(t + \pi/4)$ if $v(t)$ is the signal given in Exercise 18.2.1.

18.2.5 Using

$$g(t) = a_0 + b_1 \sin t + b_3 \sin 3t + b_5 \sin 5t + b_7 \sin 7t,$$

choose a_0, b_1, b_3, b_5, and b_7 so that the mean-square error

$$E^2 = \int_0^T (f - g)^2 \, dt$$

is minimized where f is the square wave in Example 18.2.1. Compute the discrete power spectrum of $g(t)$ and the average power of $g(t)$. Compare this average power with the average power of the square wave.

18.3 Steady-state response to periodic signals

From previous chapters, we have learned that if a linear time-invariant network or system is stable, then the steady-state response to a sinusoidal signal is also a sinusoidal signal of the same frequency. The magnitude and phase of the steady-state sinusoidal output signals are conveniently obtained using phasor analysis. The output phasor is expressed as the product of the input phasor and the transfer function of the network or system. This phasor analysis is useful in expressing the discrete spectrum of the output in terms of the discrete spectrum of the input when the input is periodic, but not necessarily a single-frequency sinusoid. We

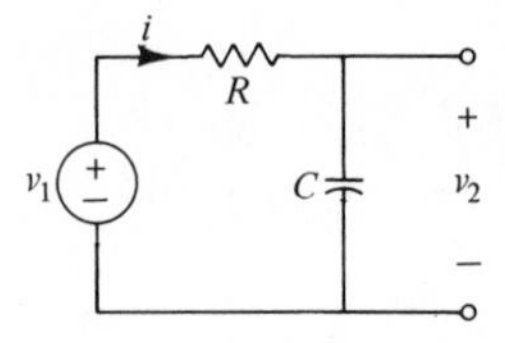

Figure 18.6 Series RC network for Example 18.3.1.

begin by considering an example of this spectrum analysis when the input is a simple sinusoid.

EXAMPLE 18.3.1

Consider the network shown in Figure 18.6, where

$$v_1(t) = V_1 \cos(\omega_0 t + \phi_1), \tag{18.37}$$

for which V_1 and ω_0 are positive constants and ϕ_1 is a real constant. Let us determine the spectrum of this input and the spectrum of the steady-state output. Expressing $v_1(t)$ in exponential form, we have

$$\begin{aligned} v_1(t) &= V_1 \frac{e^{j(\omega_0 t + \phi_1)} + e^{-j(\omega_0 t + \phi_1)}}{2} \\ &= \frac{V_1}{2} e^{j\phi_1} e^{j\omega_0 t} + \frac{V_1}{2} e^{-j\phi_1} e^{-j\omega_0 t}. \end{aligned} \tag{18.38}$$

Denoting the spectrum of $v_1(t)$ by X_n, we have

$$X_1 = \frac{V_1}{2} e^{j\phi_1} \quad \text{and} \quad X_{-1} = \frac{V_1}{2} e^{-j\phi_1}, \tag{18.39}$$

so that

$$v_1(t) = X_1 e^{j\omega_0 t} + X_{-1} e^{-j\omega_0 t}. \tag{18.40}$$

Now let us consider the input v_1 as the superposition of the two exponentials of Equation 18.40. For the input $X_1 e^{j\omega_0 t}$, the current is

$$i = \frac{X_1 e^{j\omega_0 t}}{R + 1/(j\omega_0 C)} = \frac{j\omega_0 C X_1 e^{j\omega_0 t}}{1 + j\omega_0 RC} \cdot \tag{18.41}$$

Then the voltage is

$$v_2' = \frac{1}{j\omega_0 C} i = \frac{X_1 e^{j\omega_0 t}}{1 + j\omega_0 RC} \cdot \tag{18.42}$$

Similarly, for an input $X_{-1}e^{-j\omega_0 t}$, the voltage is v_2'':

$$v_2'' = \frac{X_{-1}e^{-j\omega_0 t}}{1 - j\omega_0 RC}. \tag{18.43}$$

By superposition of v_2' and v_2'', the total voltage is

$$v_2 = v_2' + v_2'' = \frac{X_1}{1 + j\omega_0 RC} e^{j\omega_0 t} + \frac{X_{-1}}{1 - j\omega_0 RC} e^{-j\omega_0 t}. \tag{18.44}$$

If we denote the spectrum of $v_2(t)$ by Y_n, then

$$v_2 = Y_1 e^{j\omega_0 t} + Y_{-1}e^{-j\omega_0 t}, \tag{18.45}$$

where

$$Y_1 = \frac{X_1}{1 + j\omega_0 RC}, \qquad Y_{-1} = \frac{X_{-1}}{1 - j\omega_0 RC}. \tag{18.46}$$

Observe that X_1 and X_{-1} are conjugates of each other and also that Y_1 and Y_{-1} are conjugates of each other. Furthermore, $Y_1 e^{j\omega_0 t}$ and $Y_{-1}e^{j\omega_0 t}$ are also conjugates, so that v_2 in Equation 18.45 may be written

$$v_2(t) = 2 \operatorname{Re} (Y_1 e^{j\omega_0 t}) = 2 \operatorname{Re} \left(\frac{V_1 e^{j\phi_1} e^{-j\theta}}{2\sqrt{1 + (\omega_0 RC)^2}} e^{j\omega_0 t} \right), \tag{18.47}$$

where

$$\theta = \tan^{-1} \omega_0 RC. \tag{18.48}$$

Finally, we have

$$v_2(t) = \frac{V_1}{\sqrt{1 + (\omega_0 RC)^2}} \cos (\omega_0 t + \phi_1 - \theta). \tag{18.49}$$

Thus, for this simple network, the input $v_1(t)$ has a spectrum given by Equations 18.39 and 18.40, and the spectrum of the output is given by Equations 18.45 and 18.46. Finally, we note that for an input $e^{j\omega t}$, the transfer function V_2/V_1 of the network is

$$H(j\omega) = \frac{V_2}{V_1} = \frac{1/(j\omega C)}{R + 1/(j\omega C)} = \frac{1}{1 + j\omega RC}. \tag{18.50}$$

From Equation 18.46, we see that

$$Y_1 = H(j\omega_0)X_1 \quad \text{and} \quad Y_{-1} = H(-j\omega_0)X_{-1}. \tag{18.51}$$

The spectrum analysis of the network in the example we have just completed can be extended to any linear time-invariant network or system with a general periodic input. Denote the periodic input by

$$x(t) = V_0 + \sum_{n=1}^{\infty} V_n \cos(n\omega_0 t + \phi_n), \tag{18.52}$$

where ω_0, V_0, and the V_n's are positive numbers and the ϕ_n's are real numbers. In terms of the exponential Fourier series, which we have written in the form

$$x(t) = \sum_{n=-\infty}^{\infty} X_n e^{jn\omega_0 t}, \tag{18.53}$$

we have the identification

$$X_n = \begin{cases} V_0, & n = 0, \\ \dfrac{V_n}{2} e^{j\phi_n}, & n = 1, 2, 3, \ldots, \\ X_{-n}^*, & n = -1, -2, -3, \ldots. \end{cases} \tag{18.54}$$

Denoting the steady-state periodic output by

$$y(t) = \sum_{n=-\infty}^{\infty} Y_n e^{jn\omega_0 t}, \tag{18.55}$$

we obtain the transfer function $H(s)$ of the network or system. Then the discrete output spectrum is

$$Y_n = H(jn\omega_0)X_n = |H(jn\omega_0)|e^{j\theta_n}X_n, \tag{18.56}$$

where θ is the angle of $H(jn\omega_0)$. The conjugate pairs in Equation 18.55 may be combined, yielding

$$y(t) = H(0)V_0 + \sum_{n=1}^{\infty} |H(nj\omega_0)| V_n \cos(n\omega_0 t + \phi_n + \theta_n). \tag{18.57}$$

Since the discrete output power spectrum is $|Y_n|^2$, then using Equation 18.56, we have

$$|Y_n|^2 = |H(nj\omega_0)|^2 |X_n|^2. \tag{18.58}$$

Thus we can see that the discrete output power spectrum is equal to the discrete input power spectrum multiplied by the square of the magnitude of the transfer function evaluated at the corresponding harmonic frequency.

EXAMPLE 18.3.2

For the network of Figure 18.6 in Example 18.3.1, suppose that the input is the square wave of Figure 18.1 whose discrete spectrum is given in Equation 18.16.

The equation for the input signal is given by $v(t)$ in Equation 18.17. Since

$$H(s) = \frac{1}{1 + RCs}, \tag{18.59}$$

we have

$$Y_n = \begin{cases} \dfrac{V_0}{2}, & n = 0, \\ \dfrac{V_0}{jn\pi} \dfrac{1}{1 + jnRC}, & n \text{ odd}, \\ 0, & n \text{ even}. \end{cases} \tag{18.60}$$

The expression in this equation for n odd may be expressed in polar form as

$$\frac{V_0}{n\pi\sqrt{1 + (nRC)^2}} \bigg/ -\frac{\pi}{2} - \tan^{-1} nRC. \tag{18.61}$$

Then the discrete output power spectrum is

$$|Y_n|^2 = \begin{cases} \dfrac{V_0^2}{4}, & n = 0, \\ \dfrac{V_0}{n^2\pi^2[1 + (nRC)^2]}, & n \text{ odd}, \\ 0, & n \text{ even}. \end{cases} \tag{18.62}$$

The steady-state analysis described in this section is attractive for computation when there are only a few harmonics present in the input. However, an important point to remember is that regardless of the number of harmonics present, the output spectrum is related to the input spectrum by a simple multiplication by the transfer function evaluated at the appropriate harmonic frequency, as indicated by Equations 18.56 and 18.58. This is an important result.

EXERCISES

18.3.1 For the network in Figure 18.6, suppose that $RC = 1$ sec. Compute the output spectrum Y_n for $n = 0, \pm 1, \pm 3, \pm 5$ for the square-wave input of Figure 18.1 (see Example 18.3.2). Notice the relative reduction of higher harmonics because of the low-pass filter characteristic of the network.

18.3.2 For the network shown in Figure 18.1, suppose that $RC = 1$ sec. If the input is

$$v_1(t) = 10 + \cos t,$$

compute the discrete power spectrum of $v_2(t)$ and the average power (mean-square) of $v_2(t)$.

18.3.3 Repeat Exercise 18.3.2 for

$$v_1(t) = 10 + \cos t + \sin 2t.$$

18.4 Introduction to the Fourier transform

For periodic signals, the exponential Fourier-series pair of equations for $v(t)$ and F_n, Equations 18.6 and 18.10, provide a transformation for relating the discrete spectrum F_n to the time-domain signal $v(t)$. Equation 18.6 analyzes the harmonic content of $v(t)$ in the sense that it picks out the amplitude and phase of the nth harmonic component of $v(t)$. Equation 18.10 synthesizes the signal $v(t)$ as a superposition of elementary exponentials, each with a complex weighting coefficient F_n. A signal which is not periodic may be approximated by a periodic signal $v_T(t)$ which has a sufficiently large but finite period T such that

$$v_T(t) = v(t) \qquad \text{for} \quad 0 \leq t \leq T. \tag{18.63}$$

Now $\omega_0 = 2\pi/T$ is the spacing of the discrete spectrum, that is, the interval between the lines of the spectrum. This spacing becomes arbitrarily small as T becomes arbitrarily large. Thus the spectrum becomes arbitrarily dense and the number of terms arbitrarily large. For many practical applications in which the signals are to be examined only over a finite time interval, this type of approximation is adequate, and the methods of the previous sections are sufficient. However, even when an infinite time horizon is to be investigated, a frequency-domain theory is available. In this section, we will give a heuristic introduction to the Fourier transform pair which generalizes the spectrum F_n to $F(j\omega)$ and generalizes the Fourier series to the Fourier integral.

We denote the exponential Fourier-series representation of the periodic signal $v_T(t)$ in Equation 18.63 by

$$v_T(t) = \sum_{n=-\infty}^{\infty} F_n e^{jn\omega_0 t}, \tag{18.64}$$

where

$$F_n = \frac{1}{T} \int_{-T/2}^{T/2} v_T(t) e^{-jn\omega_0 t}\, dt. \tag{18.65}$$

When we substitute this equation for F_n into the previous equation for v_T and replace $1/T$ by $\omega_0/(2\pi)$, Equation 18.64 becomes

$$v_T(t) = \sum_{n=-\infty}^{\infty} \frac{\omega_0}{2\pi} \left(\int_{-T/2}^{T/2} v_T(\tau) e^{-jn\omega_0 \tau}\, d\tau \right) e^{jn\omega_0 t}. \tag{18.66}$$

As $T \to \infty$, the fundamental frequency $\omega_0 = 2\pi/T$ approaches a differential $d\omega$, $n\omega_0 \to \omega$, and the sum of the last equation approaches an integral. Assuming that the limit

$$F(j\omega) = \lim_{T \to \infty} \int_{-T/2}^{T/2} v_T(\tau) e^{-j\omega\tau}\, d\tau \tag{18.67}$$

exists, Equation 18.66 then assumes the form

$$v(t) = \lim_{T \to \infty} v_T(t) = \frac{1}{2\pi} \int_{-\infty}^{\infty} F(j\omega) e^{j\omega t}\, d\omega . \tag{18.68}$$

The integral of Equation 18.67 if it exists is called the *Fourier transform* of $v(t)$, and the integral of Equation 18.68 is called the *inverse Fourier transform* of $F(j\omega)$. This inverse transform is also called the *Fourier integral* which generalizes the Fourier series. Instead of the discrete spectrum F_n for the exponential $e^{jn\omega_0 t}$ at the frequency $n\omega_0$, we have a continuous spectrum $F(j\omega)\, d\omega/(2\pi)$ for the exponential $e^{j\omega t}$ at the frequency ω. Instead of a discrete distribution of frequencies, we now have a continuous distribution of frequencies.

EXAMPLE 18.4.1

Let us calculate the Fourier transform of

$$v(t) = \begin{cases} e^{-t}, & t \geq 0, \\ 0, & t < 0. \end{cases} \tag{18.69}$$

Using the equations just derived, we find that it is

$$F(j\omega) = \int_0^{\infty} e^{-t} e^{-j\omega t}\, dt = \left. \frac{e^{-(j\omega+1)t}}{-(j\omega+1)} \right|_0^{\infty} = \frac{1}{j\omega + 1} \cdot \tag{18.70}$$

This complex $F(j\omega)$ is routinely expressed in terms of its magnitude and phase,

$$F(j\omega) = |F(j\omega)| e^{j\phi(\omega)} = \frac{1}{\sqrt{1+\omega^2}}\, e^{-j \tan^{-1} \omega}. \tag{18.71}$$

Plots of the magnitude and phase functions associated with this $F(j\omega)$ are shown in Figure 18.7. These are known as the *continuous magnitude spectrum* and the *continuous phase spectrum* associated with the given signal $v(t)$.

This example illustrates the techniques for determining the transform of a signal $v(t)$ and displaying its magnitude and phase as continuous spectra. It is more difficult to give a physical explanation for this transform than it was for the discrete spectra associated with the exponential Fourier series. Here $F(j\omega)$ is a continuous function, defined for all ω. It suggests that if an exceedingly large

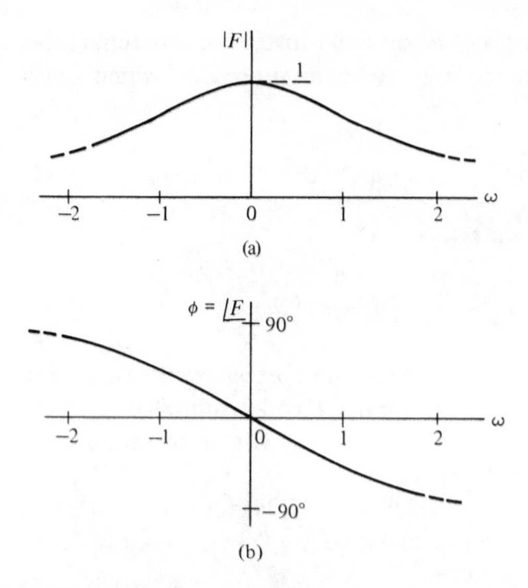

Figure 18.7 The continuous magnitude and phase spectra for $v(t) = e^{-t}$ considered in Example 18.4.1.

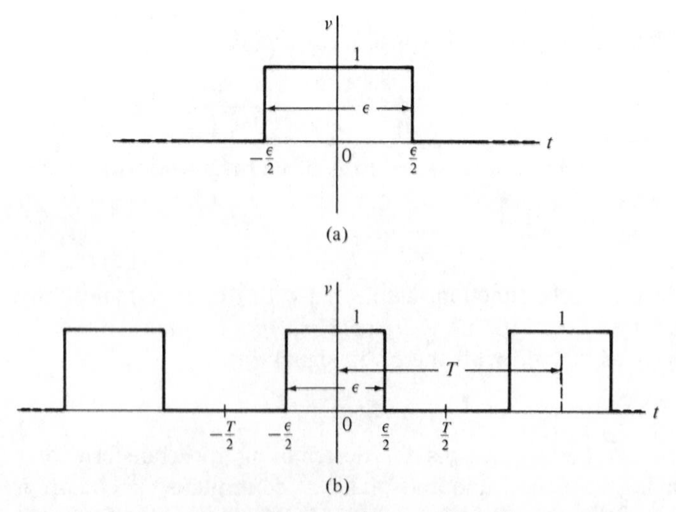

Figure 18.8 (a) A nonrecurring pulse of unit amplitude and width ϵ. The waveform of (a) may be considered as a limiting case for the periodic waveform of (b) as the period T approaches ∞.

number of signals are combined, with proper magnitude and phase characteristics, then the prescribed time signal $v(t)$ will result, occurring only once in all time. And to make the concept more difficult, we must remember that although $F(j\omega)$ is a finite function, the actual spectrum was $[1/(2\pi)]F(j\omega)\,d\omega$, meaning that the magnitude of the individual signals that combine to give $v(t)$ become incrementally small. It may be reassuring to know that, in spite of these visualization difficulties, the Fourier transform is a useful tool in electrical engineering and that there exist instruments by which it may be measured. Our next example is intended to clarify the meaning of the Fourier transform.

XAMPLE 18.4.2

The pulse is a signal which finds frequent use. It is shown in Figure 18.8(a) and is defined by the equation

$$v(t) = \begin{cases} 1, & -\epsilon/2 \le t \le \epsilon/2, \\ 0, & |t| > \epsilon/2. \end{cases} \tag{18.72}$$

The Fourier transform of $v(t)$ is

$$\begin{aligned} F(j\omega) &= \int_{-\epsilon/2}^{\epsilon/2} e^{-j\omega t}\,dt = \left.\frac{e^{-j\omega t}}{-j\omega}\right|_{-\epsilon/2}^{\epsilon/2} \\ &= \frac{e^{j\omega\epsilon/2} - e^{-j\omega\epsilon/2}}{j\omega} = \frac{2}{\omega}\,\frac{e^{j\omega\epsilon/2} - e^{-j\omega\epsilon/2}}{2j} \\ &= \frac{2}{\omega}\sin\frac{\omega\epsilon}{2} = \epsilon\,\frac{\sin(\omega\epsilon/2)}{\omega\epsilon/2}. \end{aligned} \tag{18.73}$$

This final form for the transform is of the general form of $(\sin x)/x$, which occurs frequently in mathematics and its applications, and is sometimes known as the *sampling function*. Its form is shown in Figure 18.9. The continuous magnitude and phase spectra associated with the pulse of unit magnitude are obtained directly

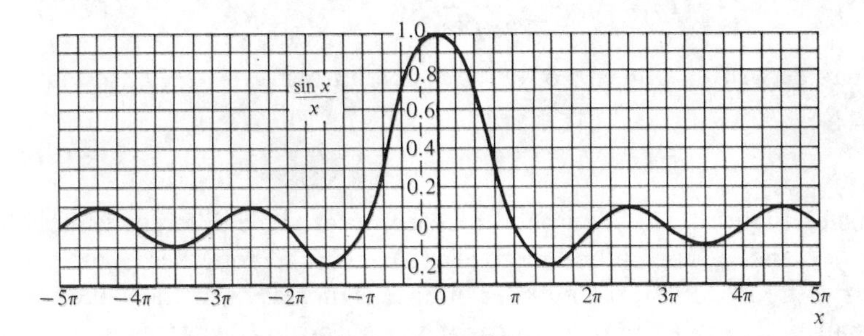

igure 18.9 Plot of $(\sin x)/x$ as a function of x.

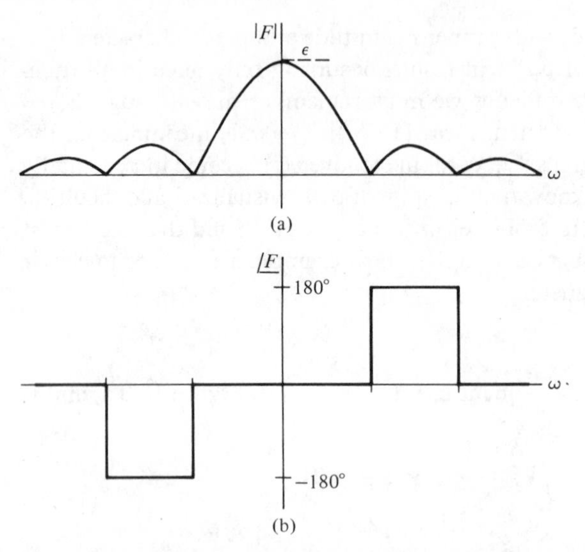

Figure 18.10 The continuous magnitude and phase spectra for a nonrecurring pulse of width ϵ.

from the $(\sin x)/x$ plot, and are as shown in Figure 18.10 with the magnitude spectrum in (a) and the phase spectrum in (b). Observe that the only values of phase found are 0° or $\pm$ 180°.

This example provides us with the opportunity to reexamine the limit process used in deriving the Fourier transform from the exponential form of the Fourier series, beginning with Equation 18.66. Rather than the nonrecurring pulse shown in (a) of Figure 18.8, consider the periodic pulse shown in Figure 18.8(b) of width ϵ and period T. The coefficients F_n for this recurring waveform are determined from Equation 18.65. The steps parallel those already given, and the result is

$$F_n = \frac{\epsilon}{T} \frac{\sin (n\omega_0\epsilon/2)}{n\omega_0\epsilon/2}. \tag{18.74}$$

This equation may be written in terms of ϵ/T by substituting $\omega_0 = 2\pi/T$, so that

$$F_n = \frac{\epsilon}{T} \frac{\sin (n\pi\epsilon/T)}{n\pi\epsilon/T}. \tag{18.75}$$

The three graphs of Figure 18.11 show the line spectra for three different values of ϵ/T, namely $\frac{1}{5}$, $\frac{1}{10}$, and $\frac{1}{20}$, corresponding to (a), (b), and (c). From these figures, we may observe two trends: First, the number of lines is increasing as the value of ϵ/T becomes smaller. Second, the magnitude of the spectrum decreases as ϵ/T becomes smaller. Since the pulse width ϵ remains constant, the ratio ϵ/T becomes

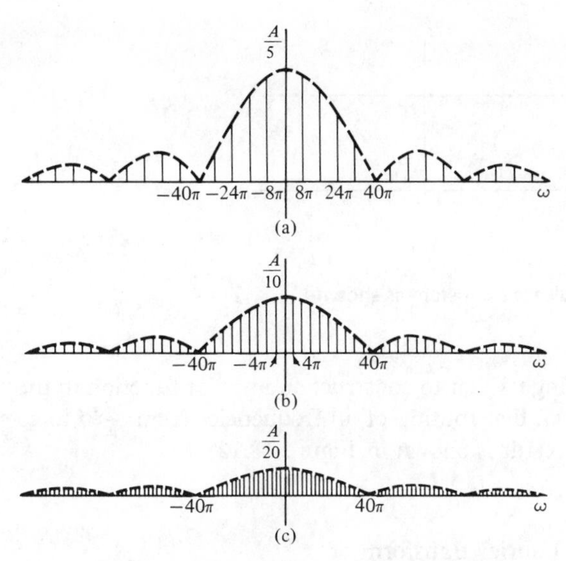

ure 18.11 Plots of the magnitude of F_n of Equation 18.75 for $\epsilon/T = \frac{1}{5}, \frac{1}{10}$, and $\frac{1}{20}$. In the limit as $\epsilon/T \to 0$, the spectrum will contain an infinite number of lines, each of amplitude approaching zero.

smaller by T becoming larger. Thus, we see the trend that in the limit the number of lines in the spectrum becomes infinite, and the magnitude of the spectrum approaches zero. We do not plot this spectrum $[1/(2\pi)]F(j\omega)\,d\omega$, of course, but only $F(j\omega)$.

We have assumed that the Fourier transform of a signal exists. This is ensured if $v(t)$ satisfies the Dirichlet conditions, which are:

(a) $\int_{-\infty}^{\infty} |v(t)|\,dt < \infty$,

(b) $v(t)$ has a finite number of discontinuities, and

(c) $v(t)$ has a finite number of maxima and minima.

These requirements do not, of course, exclude impulse functions.

AMPLE 18.4.3

The Fourier transform of

$$v(t) = \delta(t) \tag{18.76}$$

is

$$F(j\omega) = \int_{-\infty}^{\infty} \delta(t)e^{-j\omega t}\,dt = 1. \tag{18.77}$$

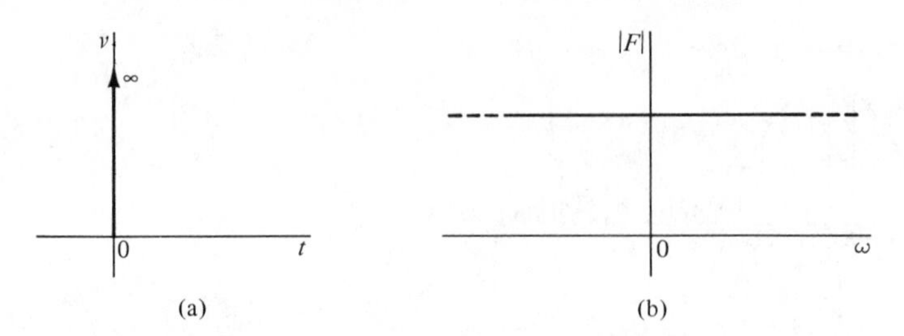

Figure 18.12 The Fourier transform of the unit impulse is a constant as shown by Equation 18.77.

This is an interesting result, telling us that to construct an impulse function in the time domain will require an equal distribution of all frequencies from $-\infty$ to ∞ in the frequency domain! This result is shown in Figure 18.12.

EXAMPLE 18.4.4

We next determine the inverse Fourier transform of

$$F(j\omega) = \delta(\omega). \tag{18.78}$$

It is

$$v(t) = \frac{1}{2\pi}\int_{-\infty}^{\infty} \delta(\omega)e^{j\omega t}\,d\omega = \frac{1}{2\pi}\cdot \tag{18.79}$$

This result is shown in Figure 18.13. It tells us that the Fourier transform of a constant, $1/(2\pi)$, is an impulse function $\delta(\omega)$.

EXAMPLE 18.4.5

As our final example, we consider the convolution integral which was studied extensively in an earlier chapter,

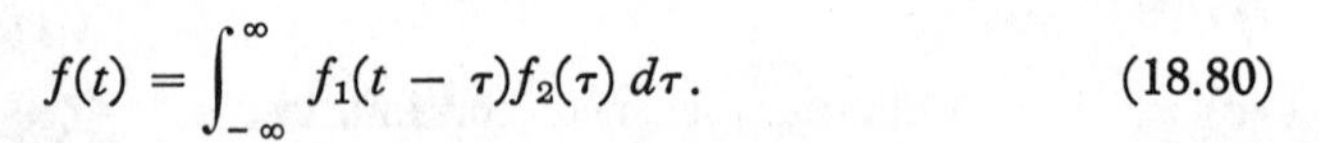

$$f(t) = \int_{-\infty}^{\infty} f_1(t-\tau)f_2(\tau)\,d\tau. \tag{18.80}$$

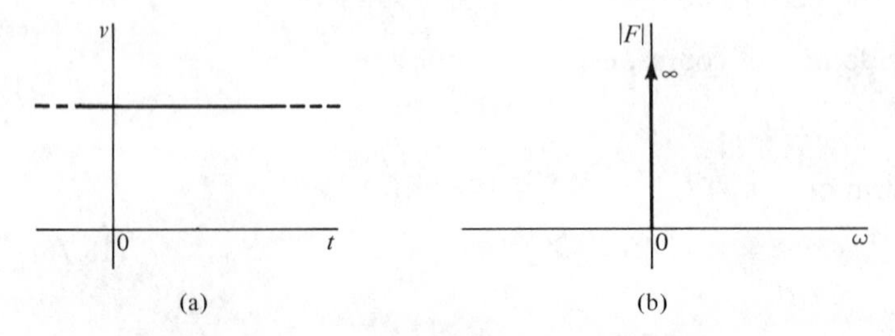

Figure 18.13 The Fourier transform of a constant is an impulse function at $\omega = 0$.

The Fourier transform is

$$F(j\omega) = \int_{-\infty}^{\infty} f(t)e^{-j\omega t}\,dt = \int_{-\infty}^{\infty}\int_{-\infty}^{\infty} f_1(t-\tau)f_2(\tau)e^{-j\omega t}\,d\tau\,dt\,. \quad (18.81)$$

Interchanging the order of integration and substituting ξ for $t - \tau$, we have

$$\begin{aligned} F(j\omega) &= \int_{-\infty}^{\infty}\int_{-\infty}^{\infty} f_1(\xi)e^{-j\omega(\xi+\tau)}f_2(\tau)\,d\xi\,d\tau \\ &= \int_{-\infty}^{\infty} F_1(j\omega)f_2(\tau)e^{-j\omega\tau}\,d\tau \\ &= F_1(j\omega)F_2(j\omega)\,. \end{aligned} \quad (18.82)$$

This tells us that multiplication in the frequency domain is equivalent to convolving two functions in the time domain.

The transform operation pairs of Table 18.1 are easily verified. These may be applied in combination to derive transform pairs for more complicated functions.

Table 18.1 Fourier transform pairs of operations

Operation	$v(t)$	$F(j\omega)$
1. Linearity (k_1 and k_2 are constants)	$k_1v_1(t) + k_2v_2(t)$	$k_1F_1(j\omega) + k_2F_2(j\omega)$
2. Time shifting (T is a real constant)	$v(t - T)$	$e^{-j\omega T}F(j\omega)$
3. Frequency shifting (ω_0 is a real constant)	$e^{j\omega_0 t}v(t)$	$F[j(\omega - \omega_0)]$
4. Time differentiation	$dv(t)/dt$	$j\omega F(j\omega)$
5. Frequency differentiation	$-jtv(t)$	$(d/d\omega)F(j\omega)$
6. Time integration	$\int_{-\infty}^{t} v(\tau)\,d\tau$	$(1/j\omega)F(j\omega)$
7. Time convolution	$v_1(t) * v_2(t)$	$F_1(j\omega)F_2(j\omega)$

EXERCISES

18.4.1 Verify entry 1 in Table 18.1.

18.4.2 Verify entry 2 in Table 18.1. [*Hint:* Determine the Fourier transform of $v(t - T)$.]

18.4.3 Verify entry 3 in Table 18.1. [*Hint:* Determine the Fourier transform of $e^{j\omega_0 t}f(t)$.]

18.4.4 Verify entry 4 in Table 18.1. [*Hint:* Differentiate with respect to time both sides of the inverse-transform equation.]

18.4.5 Verify entry 5 in Table 18.1. [*Hint:* Differentiate both sides of the transform equation with respect to ω.]

18.4.6 Verify entry 6 in Table 18.1. [*Hint:* Denote the time integral of $g(t)$ and note that $v(t) = g'(t)$. Then apply the time-differentiation property.]

18.4.7 Find the Fourier transform of

$$v(t) = \begin{cases} 1, & 0 \le t \le 1, \\ 0, & t < 0 \text{ and } t > 1. \end{cases}$$

Sketch the magnitude and phase spectra.

18.5 Spectra and steady-state response again

In our study of convolution in Chapter 8, we found that*

$$y(t) = \int_{-\infty}^{\infty} x(\tau)h(t - \tau)\, d\tau = \int_{-\infty}^{\infty} x(t - \tau)h(\tau)\, d\tau, \qquad (18.83)$$

where $x(t)$ is the input, $h(t)$ is the unit-impulse response, and $y(t)$ is the zero-state response. From Equation 18.82 of the last section, the corresponding Fourier transform equation is

$$Y(j\omega) = H(j\omega)X(j\omega), \qquad (18.84)$$

where $Y(j\omega)$ is the Fourier transform of the zero-state response, $H(j\omega)$ is the Fourier transform of the unit-impulse response, and $X(j\omega)$ is the Fourier transform of the input.

Let us now show that the Fourier transform $H(j\omega)$ may be obtained from sinusoidal steady-state measurements. Consider a periodic input $x(t) = \cos \omega_0 t$ applied to a stable linear system starting at $t = -\infty$. We let $\xi = t - \tau$ in the second form of Equation 18.83 and complete the steps of integration, noting that

* For causal systems, $h(\xi) = 0$ for $\xi < 0$ so that the upper limit of the first integral may be replaced by t and the lower limit of the second integral may be replaced by zero. Furthermore, if $x(\xi) = 0$ for $\xi < 0$, the lower limit of the first integral may be replaced by zero and the upper limit of the second integral may be replaced by t.

$\cos \xi = \frac{1}{2}(e^{j\xi} + e^{-j\xi})$. The result is

$$\begin{aligned} y(t) &= \tfrac{1}{2}[e^{j\omega_0 t}H(j\omega_0) + e^{-j\omega_0 t}H(-j\omega_0)] \\ &= \text{Re}\,[H(j\omega_0)e^{j\omega_0 t}] \\ &= \text{Re}\,[|H(j\omega_0)|e^{j\theta}e^{j\omega_0 t}] \\ &= |H(j\omega_0)| \cos(\omega_0 t + \theta), \end{aligned} \tag{18.85}$$

where the complex number $H(j\omega_0)$ is expressed in polar form in terms of magnitude $|H(j\omega_0)|$ and phase θ.

Thus we see that for the input $x(t) = \cos \omega_0 t$, the steady-state output has an amplitude $|H(j\omega_0)|$ and a phase θ. Thus the Fourier transform $H(j\omega)$ of $h(t)$ can be obtained from steady-state measurements made over a range of frequencies. These characterizing amplitude and phase quantities are those obtained by phasor analysis, and $H(j\omega_0)$ is indeed the transform function at the frequency $\omega = \omega_0$.

Now there are two ways in which we might make use of this result, shown as Equation 18.84. If $X(j\omega)$ and $H(j\omega)$ are known, then $Y(j\omega)$ is determined and $y(t)$ may be found from the inverse Fourier transform which is Equation 18.68. In many cases, we are as interested in a plot of $Y(j\omega)$ as we are in the analytical determination of $y(t)$, and $Y(j\omega)$ may be plotted simply by multiplying the plots of $H(j\omega) = G(j\omega)$ and $X(j\omega)$.

We see that $H(j\omega) = G(j\omega)$ relates to the network or system itself, and it is known as the *frequency response* of the system. It is simply the sinusoidal steady-state response which has been found routinely in earlier studies by the use of phasors. On the other hand, $X(j\omega)$ relates to the signal which is applied to the network, and it is called the *continuous frequency spectrum* of the signal. This spectrum is displayed in two plots: $|X(j\omega)|$ is called the *continuous amplitude spectrum*, and the angle of $X(j\omega)$ is called the *continuous phase spectrum*. The product of the frequency response and the input frequency spectrum is the frequency spectrum of the zero-state response. This will be illustrated by an example.

AMPLE 18.5.1

Consider the pulse of Example 18.4.2 as the input voltage to the network of Figure 18.6. The transfer function of this network is

$$H(j\omega) = \frac{1}{1 + j\omega RC}. \tag{18.86}$$

If the amplitude spectrum of the output signal is desired, then using this $H(j\omega)$ and the $V_1(j\omega)$ given by Equation 18.73, we have

$$|V_2(j\omega)| = \epsilon \left|\frac{\sin(\omega\epsilon/2)}{\omega\epsilon/2}\right| \frac{1}{\sqrt{1 + (\omega RC)^2}}. \tag{18.87}$$

This multiplication is illustrated in Figure 18.14, which shows the two spectra

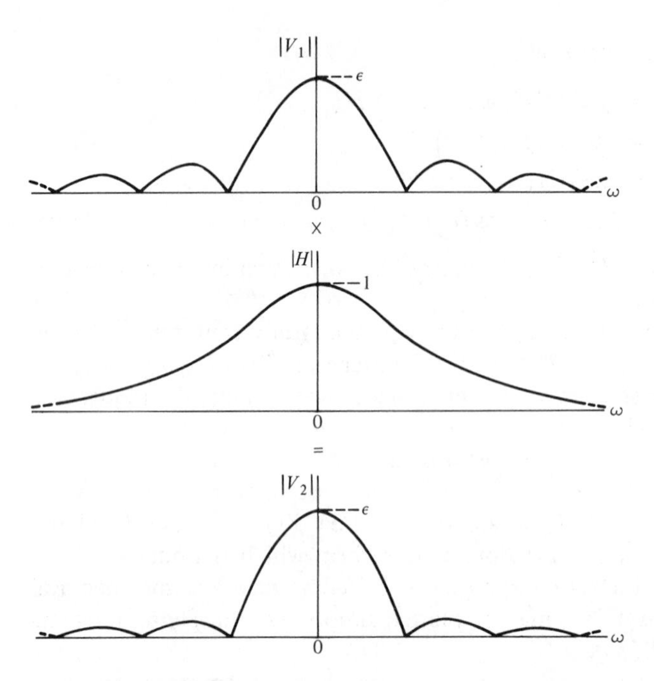

Figure 18.14 The product of $|V_1|$ and $|H|$, which are continuous functions of ω, is the amplitude spectrum $|V_2|$ of the output.

which are multiplied and the resulting spectrum of V_2. We see that the network is acting as a low-pass filter and reducing the amplitude of the $(\sin x)/x$ response at high frequencies.

We next turn to a study of the relationship of the integral of the square of the signal to the spectrum of the signal. If $v(t)$ is the voltage across a 1-ohm resistor, then the total energy delivered to the resistor is

$$W = \int_{-\infty}^{\infty} v^2(t)\, dt \,. \tag{18.88}$$

It is conventional to refer to the integral of the square of any signal as its *energy*, regardless of whether indeed it is the actual physical energy. This is analogous to the convention of using average power for a periodic signal to stand for the mean-square of the periodic signal. Assuming that the integral for W in Equation 18.88 exists and yields a finite answer, we write it as

$$W = \int_{-\infty}^{\infty} v(t)\left[\frac{1}{2\pi}\int_{-\infty}^{\infty} F(j\omega)e^{j\omega t}\, d\omega\right] dt \,. \tag{18.89}$$

If the integral converges absolutely, it is permissible to interchange the order of integration and rewrite it as

$$W = \frac{1}{2\pi}\int_{-\infty}^{\infty} F(j\omega)\left[\int_{-\infty}^{\infty} v(t)e^{j\omega t}\,dt\right] d\omega . \tag{18.90}$$

From the definition of the Fourier transform $F(j\omega)$,

$$\int_{-\infty}^{\infty} v(t)e^{j\omega t}\,dt = F(-j\omega) = F^*(j\omega). \tag{18.91}$$

Hence Equation 18.90 becomes

$$W = \frac{1}{2\pi}\int_{-\infty}^{\infty} F(j\omega)F^*(j\omega)\,d\omega = \frac{1}{2\pi}\int_{-\infty}^{\infty} |F(j\omega)|^2\,d\omega . \tag{18.92}$$

This result tells us that the total energy associated with a signal may be obtained as an integration of the frequency-domain quantity, $|F(j\omega)|^2$. This relationship is known as *Parseval's theorem,* which we have encountered earlier in this chapter, and the quantity

$$\frac{1}{2\pi}|F(j\omega)|^2 \tag{18.93}$$

is called the *energy-density spectrum.*

The energy-density spectrum depicts the distribution of energy over the entire frequency range. The energy-density spectrum of a signal may be modified by passing the signal through a network or system to yield another energy-density spectrum which has a modified shape. It is computed from the relationship

$$\frac{1}{2\pi}|Y(j\omega)|^2 = \frac{1}{2\pi}|H(j\omega)|^2|X(j\omega)|^2, \tag{18.94}$$

which is found from Equation 18.84 in combination with the quantity of Equation 18.93.

RCISES

18.5.1 The frequency response of a linear time-invariant network is

$$H(j\omega) = \frac{1}{j\omega + 3}.$$

Determine the steady-state response to an input $x(t) = \cos t$.

18.5.2 A signal $v(t)$ is given by

$$v(t) = \begin{cases} e^{-t}, & t \geq 0, \\ 0, & t < 0. \end{cases}$$

Determine (a) its energy and (b) its energy-density spectrum.

18.5.3 For the network of Exercise 18.5.1 and the input signal of Exercise 18.5.2, determine the amplitude and phase spectra of the output.

18.5.4 The impulse response of a linear time-invariant network is

$$h(t) = \begin{cases} e^{-kt}, & t \geq 0, \\ 0, & t < 0. \end{cases}$$

Using the methods of this section, determine the transfer function $H(j\omega)$ of the network.

18.6 An introduction to the Laplace transform

In this section, we give a brief introduction to the single-sided Laplace transform, which we will find is similar to the Fourier transform. It is more general than the Fourier transform in the sense that functions which are not absolutely integrable may have Laplace transforms. However, the single-sided Laplace transform is restricted to signals which have zero value for $t < 0$ (the two-sided Laplace transform covering the range $-\infty$ to ∞). The Laplace transform of a signal $v(t)$ is defined as

$$\mathscr{L}[v(t)] = V(s) = \int_0^\infty v(t)e^{-st}\,dt, \tag{18.95}$$

where s is a complex number

$$s = \sigma + j\omega. \tag{18.96}$$

The Laplace transform exists when

$$\int_{0-}^\infty |v(t)|e^{-\sigma_0 t}\,dt < \infty. \tag{18.97}$$

The smallest σ_0 for which the condition of Equation 18.97 holds is called *the abscissa* σ *of absolute convergence*. The Laplace transform is defined for all Re $(s) = \sigma \geq \sigma_0$. The lower limit is taken as $0-$ in Equation 18.97 to include impulse functions at $t = 0$.

EXAMPLE 18.6.1

We will first find the Laplace transform of

$$v(t) = e^{-t}u(t), \tag{18.98}$$

where $u(t)$ is the unit step function. The transform is

$$V(s) = \int_{0-}^{\infty} e^{-t}e^{-st}\,dt = \frac{e^{-(s+1)t}}{-(s+1)}\bigg|_{0-}^{\infty} = \frac{1}{s+1}. \tag{18.99}$$

The integral in this equation converges for all Re $(s) = \sigma \geq -1$.

In the last section of this chapter, we found that the Fourier transform of the $v(t)$ of the last example was

$$F(j\omega) = \frac{1}{j\omega + 1}, \tag{18.100}$$

which is identical with the Laplace transform of Equation 18.99 but with s replacing $j\omega$. This very close relationship holds for all signals which are zero for negative t for which the Fourier transform exists. In this case, the abscissa of absolute convergence σ_0 must be less than or equal to zero so that the Laplace transform exists for all Re $(s) \geq 0$. For such functions, the Laplace transform is the Fourier transform with $j\omega$ replaced by s. Provided that the resulting time functions of the operations given in Table 18.1 are Fourier transformable, and that the functions are zero for negative time, all the pairs of this table apply to the Laplace transformation with $j\omega$ replaced by s.

AMPLE 18.6.2

Consider the derivative $v'(t)$ of the signal of the previous example. Using entry 4 of Table 18.1, we have

$$\mathscr{L}[v'(t)] = \frac{s}{s+1}. \tag{18.101}$$

It should be noted that $v(t)$ has a jump discontinuity at $t = 0$, so that $v'(t)$ has an impulse function at $t = 0$:

$$\frac{dv(t)}{dt} = \delta(t) - e^{-t}. \tag{18.102}$$

Thus

$$\mathscr{L}[\delta(t) - e^{-t}] = \int_{0-}^{\infty} [\delta(t) - e^{-t}]e^{-st}\,dt$$

$$= 1 - \left[\frac{e^{-(s+1)t}}{-(s+1)}\right]_0^{\infty} = 1 - \frac{1}{s+1} = \frac{s}{s+1}, \tag{18.103}$$

which verifies the answer given in Equation 18.101.

EXAMPLE 18.6.3

The Laplace transform of $e^{j\omega_0 t}u(t)$ is

$$\mathscr{L}[e^{j\omega_0 t}u(t)] = \int_{0-}^{\infty} e^{j\omega_0 t}e^{-st}\,dt$$

$$= \left.\frac{e^{-(s-j\omega_0)t}}{-(s-j\omega_0)}\right|_0^{\infty} = \frac{1}{s-j\omega_0}, \qquad \text{for} \quad \mathrm{Re}\,(s) > 0. \quad (18.104)$$

Similarly,

$$\mathscr{L}[e^{-j\omega_0 t}u(t)] = \frac{1}{s+j\omega_0}, \qquad \mathrm{Re}\,(s) > 0. \quad (18.105)$$

Using the linearity property of entry 1, Table 18.1, we have

$$\mathscr{L}[\cos(\omega_0 t)u(t)] = \mathscr{L}\left[\frac{e^{j\omega_0 t} + e^{-j\omega_0 t}}{2}\right]u(t)$$

$$= \frac{1}{2}\left(\frac{1}{s-j\omega_0} + \frac{1}{s+j\omega_0}\right) = \frac{s}{s^2+\omega_0^2}. \quad (18.106)$$

EXAMPLE 18.6.4

The Laplace transform of the unit step function is

$$\mathscr{L}[u(t)] = \int_{0-}^{\infty} e^{-st}\,dt = \left.\frac{e^{-st}}{-s}\right|_{0-}^{\infty} = \frac{1}{s}, \qquad \mathrm{Re}\,(s) > 0. \quad (18.107)$$

The computation of transforms may be continued for additional functions. Once computed, the information is stored in a table of transforms. A short table of transforms is given in Table 18.2. Extensive tables are available, often as part of tables of mathematical functions.

The reverse process of recovering $v(t)$ given $V(s)$ is called the inverse Laplace transform,

$$v(t) = \mathscr{L}^{-1}[V(s)] = \frac{1}{2\pi j}\int_{\sigma_0 - j\infty}^{\sigma_0 + j\infty} V(s)e^{st}\,ds, \quad (18.108)$$

where σ_0 is any real number greater than the abscissa of absolute convergence of $V(s)$. In many engineering problems, it is simple to determine $v(t)$ using a table of Laplace transform pairs in conjunction with the partial-fraction expansion of $v(t)$.

The basic notion of the partial-fraction expansion is that $V(s)$ can be written as a sum of simpler functions, each of which may be found in a table of transforms.

Table 18.2 Short table of Laplace transforms

	$v(t)$	$V(s)$
1.	$u(t)$	$\frac{1}{s}$
2.	$e^{-at}u(t)$	$\frac{1}{s+a}$
3.	$(\cos \omega_0 t)u(t)$	$\frac{s}{s^2+\omega_0^2}$
4.	$(\sin \omega_0 t)u(t)$	$\frac{\omega_0}{s^2+\omega_0^2}$
5.	$\delta(t)$	1
6.	$tu(t)$	$\frac{1}{s^2}$
7.	$\delta'(t)$	s

Once this is accomplished, the inverse Laplace transform for each term in the sum is found from the table, rather than the inversion integral of Equation 18.108. The expansion and associated use of tables will be illustrated by several examples.

MPLE 18.6.5

We are required to determine the inverse transform of

$$V(s) = \frac{s}{(s+1)(s+2)}. \tag{18.109}$$

It is easily verified that

$$V(s) = \frac{-1}{s+1} + \frac{2}{s+2}. \tag{18.110}$$

Using the linearity property and transform pair 2 in Table 18.2, we have

$$\begin{aligned} \mathscr{L}^{-1}[V(s)] &= -\mathscr{L}^{-1}\left[\frac{1}{s+1}\right] + 2\mathscr{L}^{-1}\left[\frac{1}{s+2}\right] \\ &= (-e^{-t} + 2e^{-2t})u(t). \end{aligned} \tag{18.111}$$

Carrying out the partial-fraction expansion is simple. The method applies to rational functions, i.e., to ratios of polynomials in s. The first step in expansion is to simplify the quotient of polynomials until the degree of the numerator is less than the degree of the denominator. This is accomplished by long division. The next step is to factor the denominator polynomial,

$$q(s) = b_0(s + p_1)(s + p_2)\cdots(s + p_n), \tag{18.112}$$

where $-p_1, -p_2, \ldots, -p_n$ are the poles of $V(s)$. If the poles are distinct, they are called simple. A nondistinct pole is called a multiple pole. Suppose that $V(s)$ has only distinct poles. Then the steps we have described result in the following:

$$V(s) = \text{polynomial in } s + \sum_{i=1}^{n} \frac{K_i}{s + p_i}, \tag{18.113}$$

where

$$K_i = [(s + p_i)V(s)]_{s=-p_i}. \tag{18.114}$$

The formula for multiple poles is more complicated than this, but straightforward.

EXAMPLE 18.6.6

Let us expand

$$V(s) = \frac{s^3 + 3s^2 + 4}{s^2 + 2s}. \tag{18.115}$$

The first step is to divide the denominator into the numerator and continue until the remainder is a quotient of polynomials where the numerator is of smaller degree than the denominator. The result is

$$V(s) = s + 1 + \frac{-2s + 4}{s(s + 2)} = s + 1 + \frac{K_1}{s} + \frac{K_2}{s + 2}, \tag{18.116}$$

where

$$K_1 = \left[\frac{s(s^3 + 3s^2 + 4)}{s(s + 2)}\right]_{s=0} = 2, \tag{18.117}$$

$$K_2 = \left[\frac{(s + 2)(s^3 + 3s^2 + 4)}{s(s + 2)}\right]_{s=-2} = -4. \tag{18.118}$$

Thus

$$\frac{s^3 + 3s^2 + 4}{s(s + 2)} = s + 1 + \frac{2}{s} + \frac{-4}{s + 2}. \tag{18.119}$$

From Table 18.2, the inverse transform is

$$v(t) = \mathscr{L}^{-1}\left[\frac{s^3 + 3s^2 + 4}{s(s+2)}\right] = \delta'(t) + \delta(t) + 2u(t) - 4e^{-2t}u(t), \quad (18.120)$$

which completes the solution of the problem.

For signals which have zero value for $t < 0$, the Laplace transform of the derivative of the signal is simply s times the Fourier transform of the signal with $j\omega$ replaced by s. However, if $v(0-) \neq 0$, this is no longer true. Keeping this restriction in mind, we next obtain the Laplace transform of derivatives that are useful in circuit analysis. From the defining equation of the Laplace transform

$$\mathscr{L}\left[\frac{dv}{dt}\right] = \int_{0-}^{\infty} \frac{dv}{dt} e^{-st}\, dt. \quad (18.121)$$

Integrating by parts, we have

$$\mathscr{L}\left[\frac{dv}{dt}\right] = v(t)e^{-st}\Big|_{0-}^{\infty} - (-s)\int_{0-}^{\infty} v(t)e^{-st}\, dt = -v(0-) + s\int_{0-}^{\infty} v(t)e^{-st}\, dt.$$

Hence

$$\mathscr{L}\left[\frac{dv}{dt}\right] = sV(s) - v(0-). \quad (18.122)$$

It can be shown that the general result is

$$\mathscr{L}\left[\frac{d^n v}{dt^n}\right] = s^n V(s) - s^{n-1}v(0-) - s^{n-2}v'(0-) - \cdots - v^{(n-1)}(0-). \quad (18.123)$$

MPLE 18.6.7

Suppose that the capacitor in the network of Figure 18.6 has an initial voltage $v_2(0-) = 3$ volts, and that the input voltage is $v_1(t) = 5u(t)$ and $RC = 1$ sec. We will make use of the Laplace transform method to obtain $v_2(t)$. Writing KCL at the node common to R and C, we have

$$C\frac{dv_2}{dt} + \frac{v_2 - v_1}{R} = 0. \quad (18.124)$$

The Laplace transform of this equation is

$$C[sV_2(s) - v_2(0-)] + \frac{V_2(s) - 5/s}{R} = 0. \quad (18.125)$$

Dividing by C, substituting $RC = 1$, and solving for $V_2(s)$ yield

$$V_2(s) = \frac{5/s + 3}{s+1} = \frac{3s+5}{s(s+1)} = \frac{5}{s} + \frac{-2}{s+1}. \quad (18.126)$$

Using Table 18.2, we obtain

$$v_2(t) = \mathscr{L}^{-1}[V_2(s)] = (5 - 2e^{-t})u(t). \tag{18.127}$$

Finally, observe that if $v_2(0-) = 0$, then

$$V_2(s) = \frac{5}{s(s+1)} = \frac{5}{s}\left(\frac{1}{s+1}\right), \tag{18.128}$$

which we may interpret as being the product of the Laplace transform of the input signal and the Laplace transform of the unit-impulse response. This result holds for all linear time-invariant networks.

EXERCISES

18.6.1 Find the Laplace transform of

$$v(t) = u(t) - u(t-1).$$

18.6.2 Find the Laplace transform of

$$v(t) = e^{-t}u(t)\cos t.$$

18.6.3 Find the Laplace transform of

$$v(t) = e^{t}u(t).$$

18.6.4 Find the inverse Laplace transform of

$$V(s) = \frac{(s+1)(s+2)}{(s+3)(s+4)(s+5)}.$$

18.6.5 Repeat Exercise 18.6.4 for

$$V(s) = \frac{(s+1)(s+2)}{(s+3)(s+4)}.$$

18.6.6 Repeat Exercise 18.6.4 for

$$V(s) = \frac{s+1}{s(s^2+2)}.$$

18.7 Summary

1. For a periodic function $v(t)$ of period $T = 2\pi/\omega_0$, the exponential Fourier series is

$$v(t) = \sum_{n=-\infty}^{\infty} F_n e^{jn\omega_0 t},$$

where

$$F_n = \frac{1}{T}\int_{-T/2}^{T/2} v(t)e^{-jn\omega_0 t}\,dt\,.$$

F_n is called the discrete spectrum of $v(t)$. $|F_n|$ and $\underline{/F_n}$ are called the discrete amplitude and discrete phase spectra.

2. The average power of a periodic signal is

$$\frac{1}{T}\int_0^T v^2(t)\,dt = \sum_{n=-\infty}^{\infty} |F_n|^2.$$

This is known as Parseval's theorem for periodic signals, and $|F_n|^2$ is called the discrete power spectrum.

3. If $x(t)$ is a periodic input signal for a linear time-invariant stable system whose transfer function is $H(j\omega)$, and if $y(t)$ denotes the steady-state component of the output, then the output discrete spectrum Y_n is related to the input discrete spectrum by

$$Y_n = H(jn\omega_0)X_n\,.$$

Hence

$$|Y_n|^2 = |H(jn\omega_0)|^2|X_n|^2.$$

4. The Fourier transform of a signal $v(t)$ is

$$\mathscr{F}[v(t)] = F(j\omega) = \int_{-\infty}^{\infty} v(t)e^{-j\omega t}\,dt\,.$$

The signal $v(t)$ can be recovered from $F(j\omega)$ by means of the Fourier integral

$$v(t) = \frac{1}{2\pi}\int_{-\infty}^{\infty} F(j\omega)e^{j\omega t}\,d\omega\,,$$

which is also called the inverse Fourier transform of $F(j\omega)$. Furthermore, $F(j\omega)$ is known as the continuous spectrum of $v(t)$, $|F(j\omega)|$ is known as the continuous amplitude spectrum, and $\underline{/F(j\omega)}$ is known as the continuous phase spectrum.

5. The energy of a signal $v(t)$ is

$$\int_{-\infty}^{\infty} v^2(t)\,dt = \frac{1}{2\pi}\int_{-\infty}^{\infty} |F(j\omega)|^2\,d\omega\,.$$

This equation is known as Parseval's theorem. The quantity $[1/(2\pi)]|F(j\omega)|^2$ is known as the energy-density spectrum of $v(t)$.

6. If $x(t)$ is the input to a linear time-invariant stable system whose unit-impulse response is $h(t)$, and if $y(t)$ is the zero-state response, then

$$Y(j\omega) = H(j\omega)X(j\omega),$$

where $Y(j\omega)$, $H(j\omega)$, and $X(j\omega)$ are the corresponding Fourier transforms of $y(t)$, $h(t)$, and $x(t)$.

7. For a linear time-invariant stable system, the Fourier transform $H(j\omega)$ of the unit-impulse response $h(t)$ is equal to the transfer function. It is also equal to the frequency response of the system. If the input is $|A| \cos(\omega_0 t + \phi)$, then the steady-state output is

$$y(t) = |H(j\omega_0)||A| \cos(\omega_0 t + \phi + \theta_0),$$

where $\theta_0 = \underline{/H(j\omega_0)}$. Thus the frequency response $H(j\omega)$ is obtainable by repeated steady-state measurements at different frequencies.

8. The single-sided Laplace transform of $v(t)$ is defined as

$$\mathscr{L}[v(t)] = V(s) = \int_{0-}^{\infty} v(t)e^{-st}\, dt$$

and $v(t)$ may be recovered using the inverse Laplace transform

$$v(t) = \frac{1}{2\pi j}\int_{\sigma - j\infty}^{\sigma + j\infty} V(s)e^{st}\, ds.$$

When $V(s)$ is a rational function, i.e., the ratio of polynomials, the inverse transform is more conveniently obtained by partial-fraction expansion of $V(s)$ followed by the use of a table of transforms.

9. For signals which have zero value for negative time and are Fourier transformable, the Laplace transform is obtained from the Fourier transform by replacing $j\omega$ by s. Similarly, the Fourier transform is obtained from the Laplace transform by replacing s by $j\omega$.

PROBLEMS

18.1 Determine the class of all periodic signals which has a discrete power spectrum equal to that of the signal

$$v(t) = 5 + 10 \sin 2t + 6 \sin 3t + 8 \cos 3t.$$

18.2 If $v(t)$ is a periodic signal and $g(t)$ is a finite sum approximation,

$$g(t) = a_0 + \sum_{n=1}^{N} (a_n \cos n\omega_0 t + b_n \sin n\omega_0 t),$$

then the mean-square error

$$E^2 = \frac{1}{T}\int_0^T (v - g)^2\, dt$$

is minimized by choosing the a_n's and b_n's of $g(t)$ to be the corresponding

Fourier coefficients of $v(t)$. See Section 17.4. Furthermore, the minimum mean-square error is

$$E^2 = \frac{1}{T}\int_0^T v^2\,dt - \left[a_0^2 + \frac{1}{2}\sum_{n=1}^{N}(a_n^2 + b_n^2)\right].$$

(See Equation 17.95.) Show that this minimum mean-square error can be written as

$$E^2 = \frac{1}{T}\int_0^T v^2\,dt - \sum_{n=-N}^{N}|F_n|^2$$

and alternatively as

$$E^2 = 2\sum_{n=N+1}^{\infty}|F_n|^2,$$

where F_n is the discrete spectrum of $v(t)$.

18.3 A periodic signal $v(t)$ of frequency ω_0 has a discrete spectrum V_n. Determine the discrete spectrum and the discrete power spectrum of the signal $v(t + T_1)$, where T_1 is known and fixed.

★18.4 Show that if $f(t)$ and $g(t)$ are periodic and have the same period, where

$$f(t) = \sum_{n=-\infty}^{\infty} F_n e^{jn\omega_0 t}$$

and

$$g(t) = \sum_{n=-\infty}^{\infty} G_n e^{jn\omega_0 t},$$

then

$$\frac{1}{T}\int_0^T f(t)g(t)\,dt = \sum_{n=-\infty}^{\infty} F_n G_n^* = \sum_{n=-\infty}^{\infty} F_n^* G_n.$$

18.5 The transfer function of a linear time-invariant stable system is

$$H(j\omega) = \frac{1}{(j\omega + 1)(j\omega + 2)}.$$

If the input is $x(t) = \cos t + \sin 2t$:

(a) Determine the discrete spectrum of the input and the discrete spectrum of the steady-state component of the output.

(b) Determine the steady-state component of the output.

18.6 The unit-impulse response of a linear time-invariant network is

$$h(t) = \begin{cases} e^{-2t}, & t \geq 0, \\ 0, & t < 0. \end{cases}$$

(a) Determine the frequency response of the network.
(b) Determine the steady-state response to an input $x(t) = \sin t$.
(c) Repeat (a) and (b) if

$$h(t) = \begin{cases} 1, & 0 \le t \le 1, \\ 0, & t < 0 \text{ and } t > 1. \end{cases}$$

18.7 A signal $f(t)$ is given by

$$f(t) = \begin{cases} e^{-2t}, & t \ge 0, \\ 0, & t < 0. \end{cases}$$

(a) Determine its energy.
(b) Determine the fraction of its energy contained in all frequencies in the range $|\omega| < 1$.

18.8 An ideal low-pass filter is characterized by the transfer function

$$H(j\omega) = \begin{cases} 1, & |\omega| \le \Omega, \\ 0, & |\omega| > \Omega, \end{cases}$$

where Ω is a fixed real positive number. Determine the unit-impulse response of this filter. Is it physically realizable?

18.9 Show that if $f(t)$ and $g(t)$ are real-valued functions of time, then

$$\int_{-\infty}^{\infty} f(t)g(t)\,dt = \frac{1}{2\pi}\int_{-\infty}^{\infty} F(j\omega)G(-j\omega)\,d\omega = \frac{1}{2\pi}\int_{-\infty}^{\infty} F(-j\omega)G(j\omega)\,d\omega,$$

where $F(j\omega)$ is the Fourier transform of $f(t)$ and $G(j\omega)$ is the Fourier transform of $g(t)$.

18.10 We have shown that the following constitute a Fourier transform pair:

$$\frac{1}{2\pi} \leftrightarrow \delta(\omega).$$

Make use of the definition of the Fourier transform to show that the following also constitute a Fourier transform pair:

$$\frac{1}{2\pi} e^{\pm j\omega_0 t} \leftrightarrow \delta(\omega \pm \omega_0).$$

Make use of this result to show:
(a) $\mathscr{F}(\cos \omega_0 t) = \pi[\delta(\omega - \omega_0) + \delta(\omega + \omega_0)]$, and
(b) $\mathscr{F}(\sin \omega_0 t) = j\pi[\delta(\omega - \omega_0) - \delta(\omega + \omega_0)]$.

18.11 A real-valued signal $f(t)$ with a continuous spectrum in $F(j\omega)$ is used to modulate the amplitude of a carrier signal $f_c(t) = \cos \omega_c t$, where the constant ω_c is called the carrier frequency. Ideally, this amplitude modulated signal (AM signal) is the product

$$f_{AM}(t) = f(t) \cos \omega_c t$$

and this ideal model is widely used for analyzing AM communication systems. Determine the continuous spectrum of $f_{AM}(t)$. Briefly describe the effect of the modulation on the continuous spectrum of $f(t)$.*

18.12 The AM signal in Problem 18.11 is transmitted and propagated (assuming that ω_c is in the radio frequency range) and received at a distant station. The received signal is demodulated by passing it through a device whose output is modeled by

$$f_{AMD}(t) = f_{AM}(t) \cos \omega_c t,$$

where ω_c is the same carrier frequency used by the transmitter. Determine the continuous spectrum of $f_{AMD}(t)$. Suppose that the energy-density spectrum of $f(t)$ is essentially zero for $|\omega| > \Omega$, where Ω is a given finite positive constant $\ll \omega_c$, and the signal $f_{AMD}(t)$ is passed through an ideal low-pass filter whose transfer function is

$$H(j\omega) = \begin{cases} 1, & \text{for } |\omega| \le \Omega, \\ 0, & \text{for } |\omega| > \Omega. \end{cases}$$

Determine the continuous spectrum of the output of the low-pass filter. The demodulation-filter combination is a simplified ideal model for an AM receiver.*

18.13 Find the Fourier transform of

$$f(t) = e^{-|t|} \quad \text{for all } t.$$

18.14 Find the Fourier transform* of

$$f(t) = \begin{cases} \cos \omega_c t, & |t| \le 10\pi/\omega_c, \\ 0, & |t| > 10\pi/\omega_c. \end{cases}$$

18.15 A device that is useful in the study of sampled data systems is an impulse modulator described by

$$y(t) = x(t) \sum_{n=0}^{\infty} \delta(t - nT),$$

where $x(t)$ is the input to the impulse modulator, $y(t)$ is the output of the impulse modulator, and T is a constant called the sampling interval of the modulator. Determine the Fourier transform of $y(t)$.

* Make use of entry 3 from Table 18.1.

18.16 Determine the unit-impulse response of the system of Problem 18.5.

18.17 Make use of Tables 18.1 and 18.2 to evaluate

$$\mathscr{L}^{-1}\left[\frac{s+1}{(s+1)^2+1}\right].$$

18.18 (a) Determine the Laplace transform of

$$v(t) = e^t, \qquad t \geq 0,$$

and determine the region of convergence.
(b) Determine the Fourier transform of

$$v(t) = \begin{cases} -e^t, & t \leq 0, \\ 0, & t > 0. \end{cases}$$

(c) Is the Fourier transform of the signal in (a) obtained by replacing s by $j\omega$ in the answer of part (a)? State why or why not.

18.19 Show that if

$$y(t) = \int_{-\infty}^{\infty} x(\tau)h(t-\tau)\,d\tau,$$

then

$$Y(s) = H(s)X(s),$$

where Y, H, and X are the Laplace transforms of $y(t)$, $h(t)$, and $x(t)$, respectively.

18.20 The frequency response of a linear time-invariant system is given as

$$|H(j\omega)| = 1, \qquad \text{for all } \omega,$$
$$\underline{/H(j\omega)} = -k\omega \text{ radians}, \qquad \text{for all } \omega,$$

where k is a positive constant.
(a) Determine the impulse response of the system.
(b) Determine the zero-state response for an input, $x(t) = f(t)u(t)$.

18.21 Use Tables 18.1 and 18.2 to find the Fourier transform of

$$f(t) = e^{-2t}u(t)\cos 3t.$$

18.22 Use Laplace transforms to solve the differential equation

$$\frac{d^2y}{dt^2} + 3\frac{dy}{dt} + 2y = 5\frac{dx}{dt} + 4x$$

for $y(t)$ for $t \geq 0$ if $x(t) = u(t)$, $y(0-) = 0$, and $y'(0-) = 0$.

Appendix A

ꟷetwork Models for Mechanical ꟷevices and Systems

1 Linear mechanical elements

As in electrical networks, the analysis and design of mechanical systems or networks are often based on simple and idealized models for components. In this section, we will present the basic elements for mechanical systems which translate along one direction only (rectilinear motion) or rotate in one direction only (rotational motion). For rectilinear motion, the elements are mass, spring, and damper, and for rotational motion the elements are rotational inertia, rotational spring, and rotational damper.

Figure A.1 shows a pictorial representation for a block of material undergoing translational motion under the influence of a force f. The material is assumed to slide horizontally with no frictional resistance. Newton's law states that the mass will undergo a change in velocity. Specifically,

$$f = \frac{d}{dt}(M\hat{v}) = M\frac{d\hat{v}}{dt}, \tag{A.1}$$

where $\hat{v}$ is the velocity along the same direction as f, and M is a constant called *mass*. In the MKS system, if M is in kilograms and $\hat{v}$ is in meters per second (m/sec), then f is in newtons. The distance x is measured from an inertial reference frame. The quantity Mv is called *momentum*. Note that if the force variable f is thought of as a current variable in an electrical element, and velocity $\hat{v}$ as voltage,

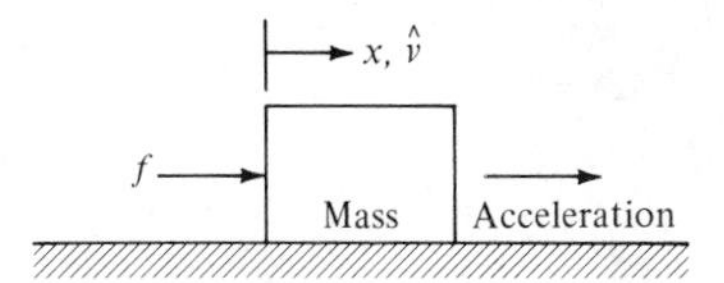

ure A.1 Force on mass causing acceleration.

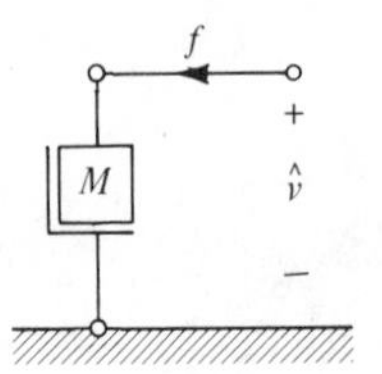

Figure A.2 Symbol for mass.

then Equation A.1 is similar to the defining equation for the linear time-invariant capacitor. On the other hand, if we think of force as analogous to voltage and velocity to current, then Equation A.1 is like the defining equation for the linear time-invariant inductor. This similar character of electrical and mechanical networks can be exploited to advantage. To facilitate the analogies, we will adopt symbols for the mechanical elements which emphasize their two-terminal nature. Figure A.2 shows the network symbol for mass, with a rectangle connected to a terminal point and an "L" to another terminal point. The L is tied to the inertial reference frame. The reference directions of the variables f and $\hat{v}$ (force and velocity) in Figure A.2 have the same significance as those for current and voltage in electrical elements. That is, if f is positive, the actual force is in the direction of the arrow in the pictorial representation of Figure A.1, and if f is negative, the actual force is in the opposite direction to that indicated in the pictorial representation of Figure A.1. The plus (+) reference mark for velocity denotes positive velocity reference with respect to the minus (−) terminal.

The second two-terminal linear time-invariant mechanical translational element we define is a spring shown in pictorial representation in Figure A.3(a) with a network symbol representation in Figure A.3(b). The defining equation is

$$\hat{v} = K\frac{df}{dt}, \tag{A.2}$$

where the coefficient K is a constant called *compliance*. The reciprocal of compliance is known as the *coefficient of stiffness*. The velocity $\hat{v}$ is the relative velocity of

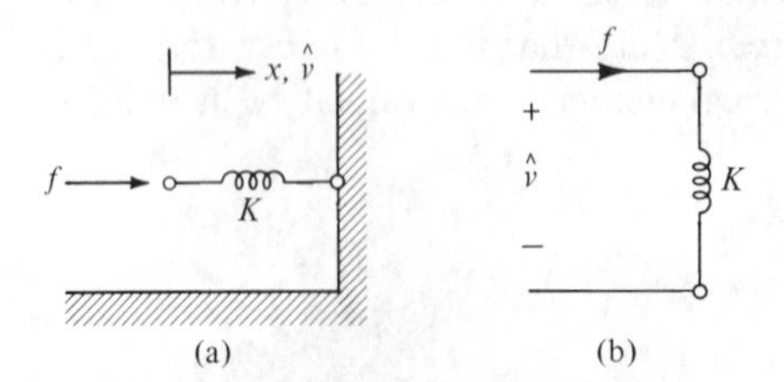

Figure A.3 (a) Force on a spring causing displacement. (b) The symbol for spring.

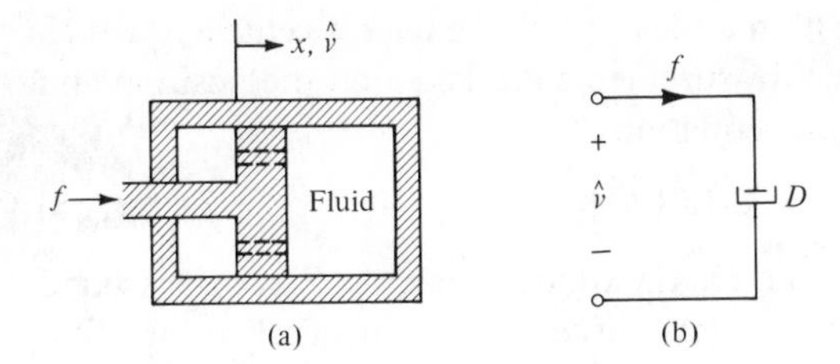

re A.4 (a) Force on a dashpot causing velocity. (b) The symbol for damper or dashpot.

compression of the spring. That is, in Figure A.3(a), if the left terminal has a velocity $\hat{v}_1$ to the right and the right terminal a velocity $\hat{v}_2$ to the right, then

$$\hat{v} = \hat{v}_1 - \hat{v}_2. \tag{A.3}$$

In the MKS system, compliance K is in meters per newton (m/N).

The third linear mechanical translational element is the *damper* shown symbolized by the dashpot in Figure A.4. It relates force and velocity by

$$f = D\hat{v}, \tag{A.4}$$

where the constant D is the *damping coefficient* in newton-seconds per meter (newton-sec/m) in the MKS system. The dashpot is similar to the shock absorber in an automobile, and, in our model, the force is linearly proportional to the velocity of the piston. This mechanical model is analogous to the resistor for electrical networks. If force is analogous to current and velocity to voltage, then the damping coefficient is analogous to conductance G.

Force sources and *velocity sources* are defined in exactly analogous manner as current sources and voltage sources. The symbol for force is that used for current source, and the symbol for velocity is the same as that for voltage source, as shown in Figure A.5. Mechanical systems consisting of M, K, D, velocity, and force sources are called translational mechanical networks or systems.

Let us consider a very simple numerical example for obtaining the parameter

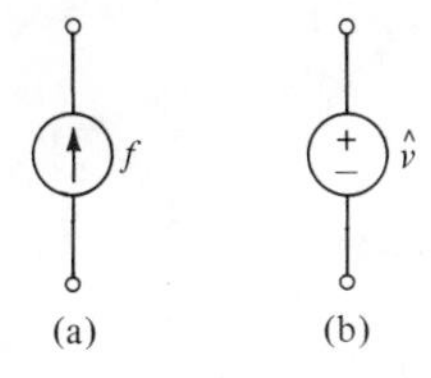

ure A.5 (a) The symbol for a force source. (b) The symbol for a velocity source.

for a mechanical model. Suppose that on a linear track we have a vehicle to which a constant force of 100 lb is applied. Measurements are taken on the position and can be described approximately by the equation

$$x = 0.1t^2, \tag{A.5}$$

where x is the position in meters (m) and t is the time in seconds. Since the second time derivative of the displacement x is a constant, and since the applied force is a constant, the vehicle may be modeled by a simple mass element. Since for a mass model Equation A.1 applies, all we have to do to obtain M is convert the given force and displacement data to a common system of units. If we choose the MKS system, we first convert the 100-lb force to newtons. Since 1 newton = 0.2248 lb, this conversion is simple. Then

$$\frac{100}{0.2248} = M\frac{d^2x}{dt^2} = M(2)(0.1). \tag{A.6}$$

Solving for M, we find

$$M = \frac{100}{(0.2248)(2)(0.1)} = 2224 \text{ kilograms}. \tag{A.7}$$

For rotational motion, the variables are torque τ (analogous to force f) and angular velocity ω (analogous to velocity v). The symbols for sources are the same as those in Figure A.5, with f replaced by τ for the torque source and $\hat{v}$ replaced by ω for the angular velocity source. We next define three elements analogous to M, K, and D.

First, we define rotational inertia by means of the equation

$$\tau = \frac{d}{dt}(J\omega) = J\frac{d\omega}{dt}, \tag{A.8}$$

where J is a constant called the *moment of inertia*. Figure A.6 shows the pictorial representation and the network symbol for rotational inertia. The angular velocity ω is measured with respect to the inertial reference frame, and $J\omega$ is called angular

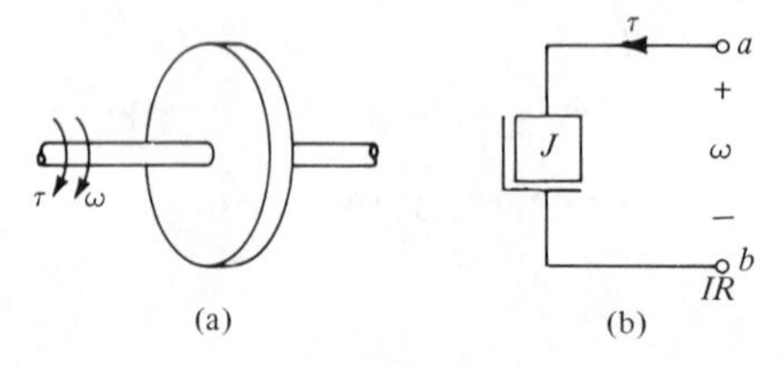

Figure A.6 (a) Pictorial diagram for rotational inertia. (b) The network symbol for J is the same as M except that the variables are τ and ω rather than f and v.

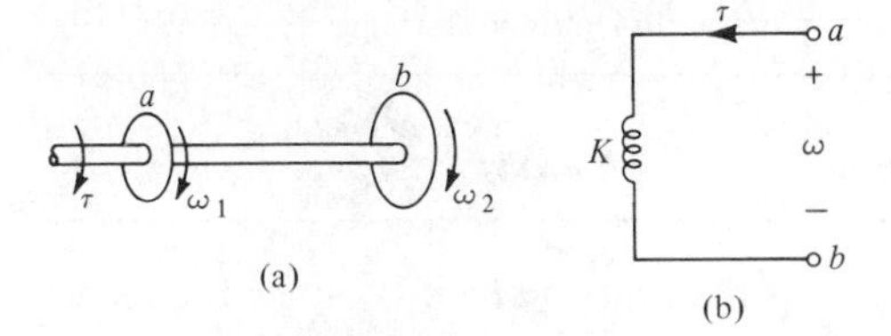

Figure A.7 (a) Pictorial diagram for rotational or torsional spring. (b) The network symbol for rotational spring or torsional spring.

momentum. It is assumed that the rotation is about the center of mass. Like mass, one terminal of the moment-of-inertia element is always connected to the inertial reference frame. In the MKS system, ω is in radians per second (radians/sec), τ is in newton-meters (newton-m), and J is in newton-meter-second^2 (newton-msec^2) or equivalently in kilogram-meters^2 (kilogram-m^2).

The rotational spring shown in Figure A.7 is modeled by

$$\omega = K\frac{d\tau}{dt}, \tag{A.9}$$

where

$$\omega = \omega_1 - \omega_2 \tag{A.10}$$

is the angular velocity of the shaft at a with respect to the shaft at b and K is a constant called *rotational compliance* measured in radians per newton-meter (radians/newton-m). The reciprocal of K is called *rotational coefficient of stiffness.*

Finally, we have a rotational damper shown in Figure A.8 analogous to the damper. It is modeled by

$$\tau = D\omega, \tag{A.11}$$

where the constant D is called *rotational damping coefficient* and ω is the relative angular velocity $\omega_1 - \omega_2$. D is measured in newton-meter-seconds (newton-msec).

Tables A.1 and A.2 summarize the defining equations for the mechanical elements and analogies to electrical elements.

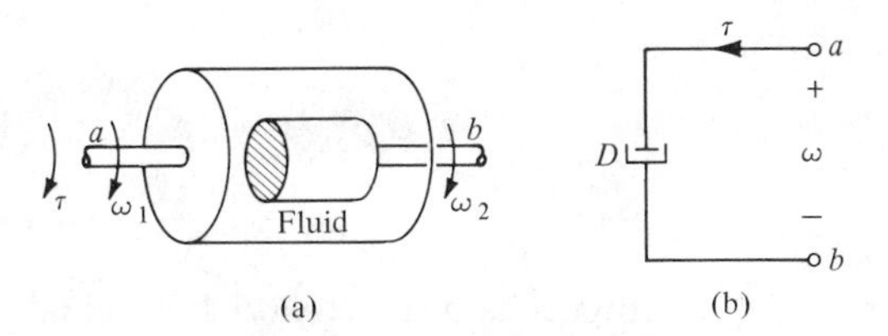

Figure A.8 (a) Pictorial diagram for torsional or rotational damping. (b) The network symbol for this damping.

Table A.1 Linear translational mechanical elements and electrical analogs

Element	*Translational Symbol*	*Equation*	*Electric analogy (force ~ current)*	*Electric analogy (force ~ voltage)*
Mass M	f, M, IR, $\hat{v}$	$f = M\dfrac{d\hat{v}}{dt}$	$M \sim C$	$M \sim L$
Translational spring K	f, K, $\hat{v}$	$\dfrac{df}{dt} = \dfrac{1}{K}\hat{v}$	$K \sim L$	$K \sim C$
Translational damper D	f, D, $\hat{v}$	$f = D\hat{v}$	$D \sim \frac{1}{R}$	$D \sim R$
Force source f	f, $\hat{v}$	$f = f(t)$ independent of $\hat{v}$	$f \sim i$, v	$f \sim v$, i
Velocity source $\hat{v}$	f, $\hat{v}$	$\hat{v} = \hat{v}(t)$ independent of f	$\hat{v} \sim v$, i	$\hat{v} \sim i$, v

EXERCISES

A.1.1 A 1-lb weight is resting on a helical spring. When the weight is removed, the spring elongates by 0.1 inch. What is the compliance K of the spring in MKS units?

A.1.2 A mass shown in Figure A.1 is initially at rest. A force of 1 newton is applied at $t = 0$ and removed at the end of 1 sec. What is the displacement at $t = 2$ sec if $M = 1$ kilogram?

A.2 D'Alembert's Principle

The equilibrium law for forces and torques in mechanical networks is called *D'Alembert's principle.* Its application is quite simple, and it is facilitated by drawing the mechanical network diagram, based on the mechanical pictorial or

Table A.2 Linear rotational mechanical elements and electrical analogs

Element	Rotational Symbol	Equation	Electric analogy (torque ~ current)	Electric analogy (torque ~ voltage)
Rotational inertia J	τ, J, ω, IR, $+$, $-$	$\tau = J\dfrac{d\omega}{dt}$	$J \sim C$	$J \sim L$
Torsional spring K	τ, K, ω, $+$, $-$	$\dfrac{d\tau}{dt} = \dfrac{1}{K}\omega$	$K \sim L$	$K \sim C$
Torsional damper D	τ, D, ω, $+$, $-$	$\tau = D\omega$	$D \sim \frac{1}{R}$	$D \sim R$
Torque source τ	τ, ω, $+$, $-$	$\tau = \tau(t)$ independent of ω	$\tau \sim i$, v, $+$, $-$	$\tau \sim v$, i
Angular velocity source ω	τ, ω	$\omega = \omega(t)$ independent of τ	$\omega \sim v$, i	$\omega \sim i$, v, $+$, $-$

schematic diagram. The first step is to assign a common positive reference direction in the pictorial diagram. Velocity (or angular velocity) variables are next assigned to each rigid part that moves. These rigid parts become nodes of the mechanical network diagram. The masses (or moments of inertia) are then drawn with one terminal from each mass connected to the inertial reference. The other terminals are associated with the assigned nodes. Dampers, springs, and sources are then drawn. Once the network is drawn, D'Alembert's principle is applied in a manner exactly analogous to Kirchhoff's current law.

(E)XAMPLE A.2.1

A force f is applied to a mass M and restrained by a spring K and frictional damping D, as shown pictorially in Figure A.9. Before drawing a mechanical network representation, we assign a positive reference direction for the velocity $\hat{v}$ of the mass M with respect to the inertial reference. If the force and velocity references are in the same direction in the pictorial diagram, then the reference direction of

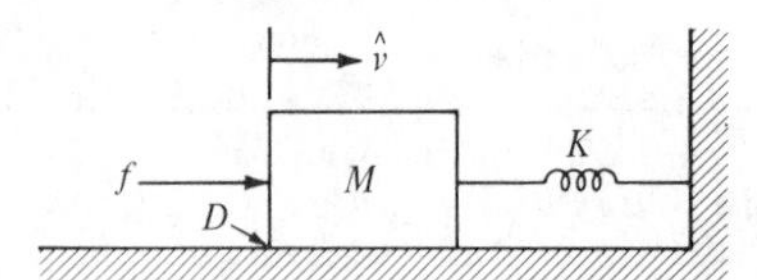

Figure A.9 A pictorial representation of a simple mechanical system.

f in the network diagram is toward the node $\hat{v}$, as shown in Figure A.10. Moreover, one terminal of M is connected to the inertial reference (IR). Next, we draw the damping D which produces a resisting force proportional to the velocity $\hat{v}$. Thus one terminal is connected to $\hat{v}$ and one terminal is connected to IR. Finally, the spring is connected between M and IR so that in the network diagram, one terminal is connected to the node $\hat{v}$ and one to the node IR. D'Alembert's principle is applied to the network in Figure A.10 just like Kirchhoff's current law:

$$M\frac{d\hat{v}}{dt} + D\hat{v} + \frac{1}{K}\int_{-\infty}^{t} \hat{v}(\xi)\, d\xi - f = 0. \tag{A.12}$$

For a force-to-current analogy where force is analogous to current and velocity is analogous to voltage, M is analogous to C, D is analogous to G, and K is analogous to L. The electrical analog is shown in Figure A.11. Note that the topologies of the mechanical and electrical network analog are identical. Kirchhoff's current law applied to the network in Figure A.11 yields

$$C\frac{dv}{dt} + Gv + \frac{1}{L}\int_{-\infty}^{t} v(\xi)\, d\xi - i = 0. \tag{A.13}$$

EXAMPLE A.2.2

Figure A.12 shows a torque source τ applied to a flywheel of moment of inertia J_1, rotating at angular velocity ω_1, restrained by a rotational damping D_1, and a torsional rod of compliance K. The other end of the rod is attached to another

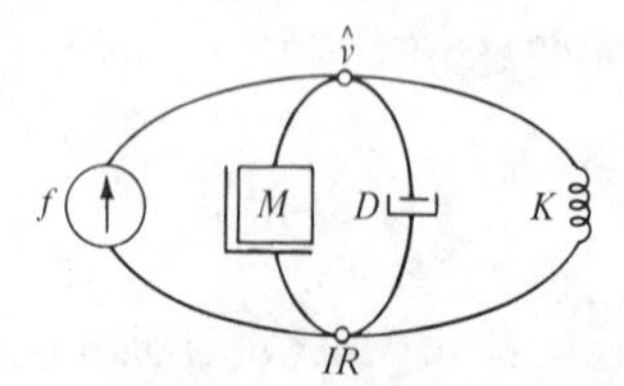

Figure A.10 Mechanical network diagram for the mechanical system shown in Figure A.9.

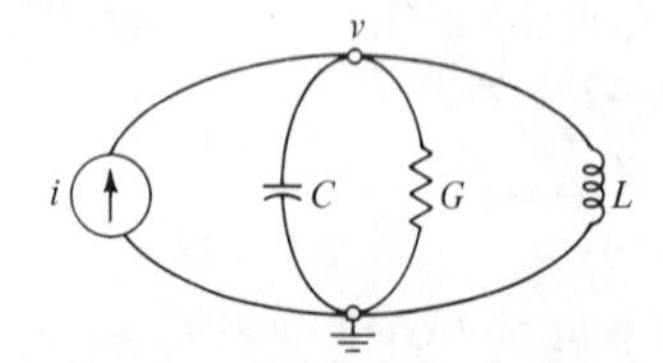

Figure A.11 Electrical network analog of the mechanical network of Figure A.10, using a force-to-current analogy.

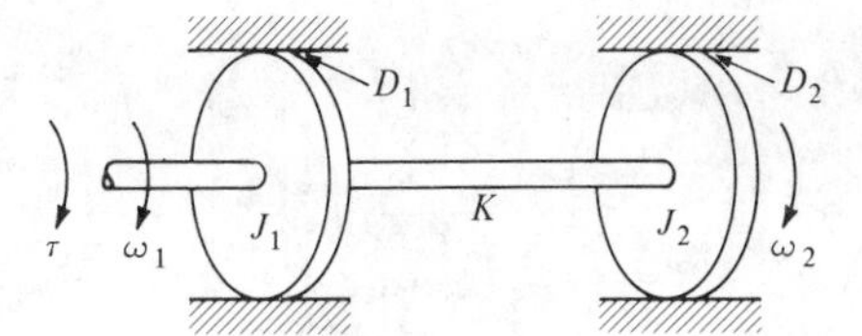

gure A.12 Pictorial representation of the mechanical system of Example A.2.2.

flywheel of moment of inertia J_2, rotating at angular velocity ω_2, restrained by a rotational damping D_2. Notice that in Figure A.12 we have assigned τ, ω_1, and ω_2 in the same direction. To draw the mechanical network, we first draw J_1 and J_2 with one terminal from each connected to the inertial reference IR. One terminal of J_1 is connected to node ω_1, and one terminal of J_2 is connected to node ω_2. The dampers are connected next. Finally, the spring K is connected, noting that one end rotates at ω_1 and the other end rotates at ω_2. The complete network is shown in Figure A.13. Once the network is drawn, D'Alembert's principle is easily applied, just like Kirchhoff's current law. Thus

$$J_1 \frac{d\omega_1}{dt} + D_1\omega_1 - \tau + \frac{1}{K}\int_{-\infty}^{t} (\omega_1 - \omega_2)\, d\xi = 0 \qquad \text{(A.14)}$$

and

$$J_2 \frac{d\omega_2}{dt} + D_2\omega_2 + \frac{1}{K}\int_{-\infty}^{t} (\omega_2 - \omega_1)\, d\xi = 0. \qquad \text{(A.15)}$$

These equations may be solved in exactly the same way as node equations for electrical networks are solved.

ERCISES

A.2.1 Draw the mechanical network diagram for the system shown in Figure A.14 and draw the electrical analog using a force-to-current analogy.

A.2.2 Draw the mechanical network diagram for the clutch system shown in Figure A.15. Apply D'Alembert's principle.

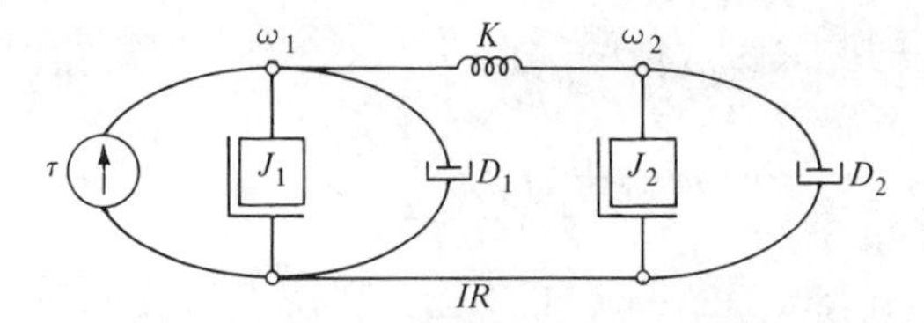

gure A.13 Mechanical network diagram for the mechanical system shown in Figure A.12.

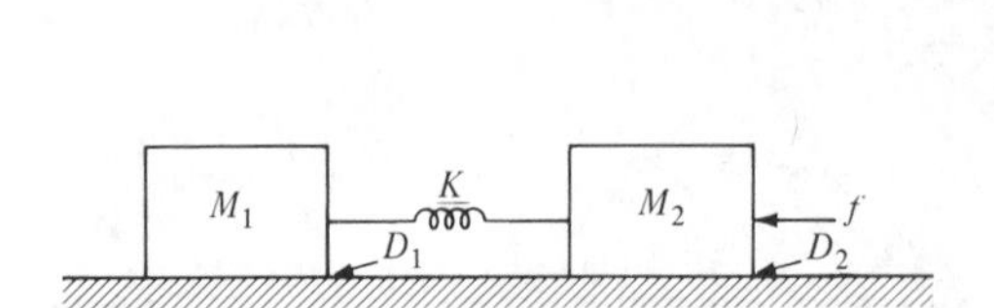

Figure A.14

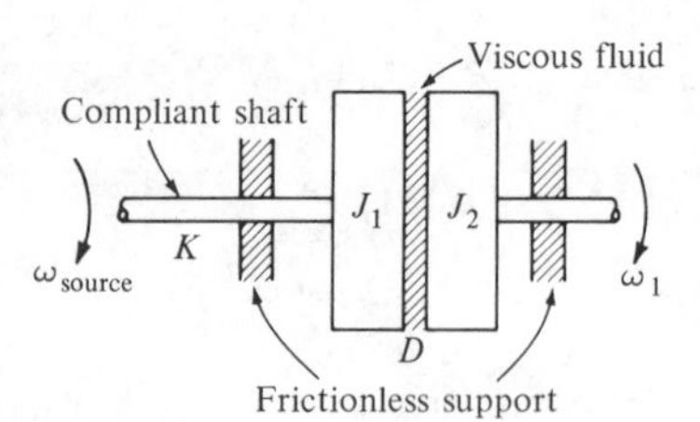

Figure A.15

A.2.3 For the mechanical system shown in Figure A.9 whose equilibrium equation is given in Equation A.12, suppose that a force-to-voltage analogy is used, so that velocity is analogous to current, mass is analogous to inductance, D is analogous to resistance, and compliance K is analogous to capacitance. Using these analogies, write the corresponding electrical equilibrium equation. Draw an electrical network for which this is Kirchhoff's voltage law. The resulting network is called the *dual* of the network of Figure A.11.

A.3 Summary

1. Linear mechanical elements for systems in translation or rotational motion may be represented by two-terminal network elements. Tables A.1 and A.2 summarize the defining equations, network symbols, and analogies with electrical elements.
2. Mass and inertial elements have one terminal always connected to the inertial reference.
3. D'Alembert's principle is an equilibrium law for forces or torques. It is applied like Kirchhoff's current law once the mechanical network is drawn.

A.4 Collateral reading references

Cannon, Robert H., Jr., *Dynamics of Physical Systems* (New York: McGraw-Hill Book Company, 1967), 904 pp.

Perkins, William R. and J. B. Cruz, Jr., *Engineering of Dynamic Systems* (New York: John Wiley & Sons, Inc., 1969), 568 pp.

Shearer, J. Lowen, Arthur T. Murphy, and Herbert H. Richardson, *Introduction to Systems Dynamics* (Reading, Mass.: Addison-Wesley Publishing Company, Inc., 1967), 420 pp.

Appendix B

Digital Computer Exercises

The first course in linear circuits is sometimes accompanied by a supporting software laboratory made up of a sequence of digital-computer exercises. The libraries of most computer centers contain the subroutines required for these exercises, or programs can be found in available manuals or textbooks. The exercises that are most appropriate to the topics of this book are given in this Appendix, along with suggested references to be consulted if detailed information is desired.

Note that the suggested exercises are divided into two parts. The first part is a listing of topics that might have been encountered by the student in an earlier course on the use of the computer which included an elementary treatment of topics in numerical analysis. The second is a listing of topics that arise in the specific circuit studies of the book, where it has been found that analysis can be greatly aided by the use of the computer.

Depending on the library available and the interests of the instructor, the digital-computer exercises may take such forms as the following: (1) Study the problem or method and draw a flowchart for computer solution. (2) Write a program to carry out the operations indicated in the flowchart. (3) Use the computer center library of subroutines to carry out the operations required, applying the program of subroutines to the solution of a problem for a specified network or signal.

B.1 Basic computational problems

Topic	Applicable in Chapter(s)	Suggested references
1. Solve a set of simultaneous algebraic equations. Use one of the following: Gauss elimination method, Gauss-Jordan method, Gauss-Seidel method.	14	1, 6, 13, 17
2. Study matrix inversion. Some topics include matrix operations by computer, construction of the inverse of a square matrix, eigenvalue problems.	5, 13, 14	7, 12, 13
3. Find the roots of an algebraic equation. Study interval halving, secant method, Newton-Raphson method, Lin-Bairstow method.	7, 11	1, 11, 9, 19
4. Solve one or a set of simultaneous nonlinear algebraic equations. Some methods: interval halving, Newton-Raphson.	1, 5, 13, 14	8, 13, 19
5. Carry out numerical integration. Make use of Newton's formula, the trapezoidal formula, or Simpson's integration formula.	1, 3, 4, 8	1, 7, 13, 19
6. Solve a set of simultaneous first-order differential equations. See Runge-Kutta method.	5, 7, 8, 13, 14	8, 9, 13

B.2 Network and system-oriented computational problems

Topic	Applicable in Chapter(s)	Suggested references
1. Solve a set of simultaneous algebraic equations for the steady-state solution. Apply to network equations formulated on the node or loop bases.	13	7, 9, 19
2. Study convolution of two functions by computer. See Section B.1.5.	8	7, 9, 11
3. Determine the frequency response of a network. For a specified network, plot magnitude and phase as a function of frequency, or plot magnitude vs. phase with frequency as the parameter.	9, 11, 17, 18	9, 14, 19

(Continued)

Topic	*Applicable in Chapter(s)*	*Suggested references*
4. Carry out the partial-fraction expansion of a quotient of polynomials. This will involve finding the roots of a polynomial; see Section B.1.3 and the computation of residues.	18, C	8, 11, 12, 14, 19
5. Solve a set of state equations representing a network. See Section B.1.6.	3, 6, 13	9, 11, 12, 14
6. Analyze a ladder network. Prepare a program applicable only to the ladder network case, and solve for a resistive network, a network operating in the sinusoidal steady-state, or the general case.	5, 7, 10	7, 19
7. Determine the root locus for the characteristic equation of some network, as a single parameter varies from 0 to ∞.	7, 11	8, 14
8. Fourier analysis. Determine the Fourier coefficients for a specified periodic signal.	17, 18	8, 9, 12
9. Fourier transform. Determine the magnitude and phase spectra for a specified nonperiodic signal. Display a plot of the spectra; see Section B.2.3.	18	9, 11
10. Fast Fourier transform (FFT). Use a computer program package for the FFT. For a given nonperiodic signal, determine the magnitude and phase spectra.	18	6, 11, 18
11. Analysis of a large network. Use a computer program package for network analysis, such as ECAP, SCEPTRE, NET-1, etc. Apply to a passive network (a filter, for example), to a network containing a transistor (using some simple model), or to a network with some one parameter assuming a range of values (as in design).	12, 13, 15	2, 10
12. Look up a program in your computer center library for solving the matrix equation, $\mathscr{I} = \mathscr{Y}\mathscr{V}$, where $\mathscr{Y}$ is the node admittance matrix which is sparse. Use L-U factorization. Apply the program to the analysis of a large resistive ladder network. (Such analysis programs find wide application in the power industry. See William F. Tinney and John W. Walker, "Direct solutions of sparse network equations by optimally ordered triangular factorization," *Proc. IEEE,* vol. 55, pp. 1801–1809, November, 1967.)	13, 14	6, 18

B.3 References

1. Beckett, Royce and James Hurt, *Numerical Calculations and Algorithms* (New York: McGraw-Hill Book Company, 1967), 298 pp.
2. Bowers, James C. and Stephen R. Sedore, *SCEPTRE: A Computer Program for Circuit and System Analysis* (Englewood Cliffs, N.J.: Prentice-Hall, Inc., 1971), 455 pp.
3. Calahan, Donald A., *Computer-Aided Network Design* (New York: McGraw-Hill Book Company, 1972), 350 pp.
4. Calahan, Donald A., Alan B. Macnee, and E. L. McMahon, *Computer-Oriented Circuit Analysis* (New York: Holt, Rinehart and Winston, Inc., 1973).
5. Dertouzos, Michael L., Michael Athans, Richard N. Spann, and Samuel J. Mason, *Systems, Networks, and Computation: Basic Concepts* (New York: McGraw-Hill Book Company, 1972), 514 pp.
6. Director, S. W., *Circuit Theory — The Computational Approach* (New York: John Wiley & Sons, Inc., 1974).
7. Gupta, Someshwar C., Jon W. Bayless, and Behrouz Peikari, *Circuit Analysis with Computer Applications to Problem Solving* (Scranton, Pa.: Intex Educational Publishers, 1972), 546 pp.
8. Huelsman, Lawrence P., *Basic Circuit Theory with Digital Computations* (Englewood Cliffs, N.J.: Prentice-Hall, Inc., 1972), 896 pp.
9. Huelsman, Lawrence P., *Digital Computations in Basic Circuit Theory* (New York: McGraw-Hill Book Company, 1968), 203 pp.
10. Jensen, Randall W. and Mark D. Lieberman, *IBM Electronic Circuit Analysis Program: Techniques and Applications* (Englewood Cliffs, N.J.: Prentice-Hall, Inc., 1968), 401 pp.
11. Kinariwala, B. K., Franklin F. Kuo, and Nai-kuan Tsao, *Linear Circuits and Computation* (New York: John Wiley & Sons, Inc., 1973).
12. Ley, B. James, *Computer-Aided Analysis and Design for Electrical Engineers* (New York: Holt, Rinehart and Winston, Inc., 1970), 852 pp.
13. McCracken, Daniel D., *FORTRAN with Engineering Applications* (New York: John Wiley & Sons, Inc., 1967), 237 pp.
14. Melsa, James L., and Stephen K. Jones, *Computer Programs for Computational Assistance in the Study of Linear Control Theory*, 2nd edition (New York: McGraw-Hill Book Company, 1973).
15. Ramey, Robert L. and Edward J. White, *Matrices and Computers in Electronic Circuit Analysis* (New York: McGraw-Hill Book Company, 1971).
16. Seely, Samuel, Norman H. Tarnoff, and David Holstein, *Digital Computers in Engineering* (New York: Holt, Rinehart and Winston, Inc., 1970), 329 pp.

17. Steiglitz, Kenneth, *An Introduction to Discrete Systems* (New York: John Wiley & Sons, Inc., 1973).
18. Tinney, William F. and John W. Walker, "Direct solutions of sparse network equations by optimally ordered triangular factorization," *Proc. IEEE*, vol. 55, pp. 1801–1809, November, 1967.
19. Wing, Omar, *Circuit Theory with Computer Methods* (New York: Holt, Rinehart and Winston, Inc., 1972), 529 pp.

Appendix C

The Laplace Transform

C.1 Definition

The single-sided Laplace transform of the time function $v(t)$ is defined as

$$\mathscr{L}[v(t)] = V(s) = \int_{0-}^{\infty} v(t)e^{-st}\,dt, \tag{C.1}$$

where the complex variable s equals $\sigma + j\omega$. The Laplace transform exists when

$$\int_{0-}^{\infty} |v(t)|e^{-\sigma_0 t}\,dt < \infty. \tag{C.2}$$

The smallest value of σ_0 for which this condition holds is the abscissa of absolute convergence. The lower limit of integration is $0-$, so that impulse functions at $t = 0$ are included. The inverse transform of $V(s)$ is $v(t)$, defined as

$$v(t) = \frac{1}{2\pi j}\int_{\sigma_1 - j\infty}^{\sigma_1 + j\infty} V(s)e^{st}\,ds, \tag{C.3}$$

where σ_1 is greater than the σ_0 for $V(s)$. For the single-sided Laplace transform case, the inverse transform of $V(s)$ is unique. This is an important property that makes it possible to avoid the use of Equation C.3 in favor of tables.

C.2 Table of transforms

A few signal waveforms $v(t)$ occur frequently in the study of circuits. One of these is the unit step function $u(t)$. The Laplace transform of $u(t)$ is found from Equation C.1:

$$V(s) = \int_{0-}^{\infty} u(t)e^{-st}\,dt = -\frac{e^{-st}}{s}\bigg|_0^{\infty} = 0 - \left(\frac{-1}{s}\right) = \frac{1}{s}. \tag{C.4}$$

In Example 18.5.1, it was found that if $v(t) = e^{-t}u(t)$, then

$$V(s) = \frac{1}{s+1}. \tag{C.5}$$

Using the same procedure as in that example, we find that

$$\mathscr{L}[Ae^{-at}u(t)] = \frac{A}{s+a}. \tag{C.6}$$

Finally, in Example 18.6.3 the following transform pair was established:

$$\mathscr{L}[(\cos \omega_0 t)u(t)] = \frac{s}{s^2 + \omega_0^2}. \tag{C.7}$$

The compilation of functions and their transforms, suggested by these examples, leads to a table of transforms having the following form:

$v(t)$	$V(s)$
1. $u(t)$	$\frac{1}{s}$
2. $Ae^{-at}u(t)$	$\frac{A}{s+a}$
3. $(\cos \omega_0 t)u(t)$	$\frac{s}{s^2 + \omega_0^2}$

Some of the transforms that are found most useful in the study of linear circuits are shown in Table C.1.

Table C.1 Table of Laplace transforms

$v(t) = \mathscr{L}^{-1}[V(s)]$	$V(s) = \mathscr{L}[v(t)]$	$v(t) = \mathscr{L}^{-1}[V(s)]$	$V(s) = \mathscr{L}[v(t)]$
1. $u(t)$	$\frac{1}{s}$	8. $(\sin \omega_0 t)u(t)$	$\frac{\omega_0}{s^2 + \omega_0^2}$
2. $\delta(t)$	1	9. $(\cos \omega_0 t)u(t)$	$\frac{s}{s^2 + \omega_0^2}$
3. $tu(t)$	$\frac{1}{s^2}$	10. $e^{-at}(\sin \omega_0 t)u(t)$	$\frac{\omega_0}{(s+a)^2 + \omega_0^2}$
4. $t^n u(t)$	$\frac{n!}{s^{n+1}}$	11. $e^{-at}(\cos \omega_0 t)u(t)$	$\frac{s+a}{(s+a)^2 + \omega_0^2}$
5. $e^{-at}u(t)$	$\frac{1}{s+a}$	12. $\sinh atu(t)$	$\frac{a}{s^2 - a^2}$
6. $te^{-at}u(t)$	$\frac{1}{(s+a)^2}$	13. $\cosh atu(t)$	$\frac{s}{s^2 - a^2}$
7. $t^n e^{-at}u(t)$	$\frac{n!}{(s+a)^{n+1}}$		

C.3 Transforms of integro-differential equations

We make use of three properties of the Laplace transform in obtaining the transform equation from an integro-differential equation. These are:

(1) The linearity property. For the linear combination of two variables,

$$\mathscr{L}[a_1 v_1(t) + a_2 v_2(t)] = a_1 V_1(s) + a_2 V_2(s). \tag{C.8}$$

This result is easily established by substituting the linear combination of v_1 and v_2 into Equation C.1. It may be generalized for the linear combination of any number of time functions.

(2) The transform of derivatives of functions. From the defining equation, Equation C.1, we write

$$\mathscr{L}\left[\frac{dv(t)}{dt}\right] = \int_{0-}^{\infty} e^{-st} \frac{dv(t)}{dt}\, dt. \tag{C.9}$$

This integration is accomplished by integrating by parts:

$$\mathscr{L}\left[\frac{dv(t)}{dt}\right] = e^{-st}v(t)\Big|_{0-}^{\infty} - \int_{0-}^{\infty} v(t)(-se^{-st}\, dt)$$

$$= e^{-s\infty}v(\infty) - e^{-s0}v(0-) + s\int_{0-}^{\infty} v(t)e^{-st}\, dt.$$

Since we are considering the class of functions for which $e^{-st}v(t) = 0$ as $t \to \infty$, the first term vanishes and the last term is $sV(s)$, so that

$$\mathscr{L}\left[\frac{dv(t)}{dt}\right] = sV(s) - v(0-). \tag{C.10}$$

By defining a new function $g_1(t) = dv(t)/dt$ and then applying the result just obtained as Equation C.10, we may show that

$$\left[\frac{d^2v(t)}{dt^2}\right] = s^2V(s) - sv(0-) - \frac{dv(t)}{dt}\Big|_{t=0-}. \tag{C.11}$$

In general,

$$\mathscr{L}\left[\frac{d^nv(t)}{dt^n}\right] = s^nV(s) - s^{n-1}v(0-) - s^{n-2}\frac{dv}{dt}\Big|_{t=0-} - \cdots$$

$$- s\frac{d^{n-2}v}{dt^{n-2}}\Big|_{t=0-} - \frac{d^{n-1}v}{dt^{n-1}}\Big|_{t=0-}. \tag{C.12}$$

(3) The transform of the integral of a function. To find the Laplace transform of the integral, let

$$g_2(t) = \mathscr{L}^{-1}[G_2(s)] = \int_{0-}^{t} v(\tau)\, d\tau. \tag{C.13}$$

From this equation, observe that $g_2(0-) = 0$ and $dg_2(t)/dt = v(t)$. Next, we apply the result found in Equation C.10:

$$\mathscr{L}\left[\frac{dg_2}{dt}\right] = sG_2(s) - g_2(0-) = V(s). \tag{C.10a}$$

Since $g_2(0-) = 0$, then

$$\mathscr{L}\left[\int_{0-}^{t} v(\tau)\, d\tau\right] = \mathscr{L}[g_2(t)] = G_2(s) = \frac{V(s)}{s}, \tag{C.14}$$

which is the desired result.

XAMPLE C.3.1

Consider the differential equation

$$\frac{d^2v}{dt^2} + 3\frac{dv}{dt} + 2v = u(t) + 2e^{-3t} \tag{C.15}$$

with zero initial conditions. Taking the Laplace transform of the equation term by term, and factoring the common $V(s)$, we have

$$V(s)(s^2 + 3s + 2) = \frac{1}{s} + \frac{2}{s+3}. \tag{C.16}$$

Solving for $V(s)$, we find

$$V(s) = \frac{1}{s(s^2 + 3s + 2)} + \frac{2}{(s+3)(s^2 + 3s + 2)}. \tag{C.17}$$

XAMPLE C.3.2

The following integro-differential equation describes an *RLC* series circuit:

$$L\frac{di}{dt} + Ri + \frac{1}{C}\int_{0}^{t} i(\tau)\, d\tau = V_1 \cos 2t. \tag{C.18}$$

Taking the Laplace transform of the equation term by term, and substituting the numerical values $V_1 = 1$, $R = 1$, $L = 1$, and $C = \frac{1}{2}$, we have, for zero initial conditions,

$$V(s)\left(s + 1 + \frac{2}{s}\right) = \frac{s}{s^2 + 4}. \tag{C.19}$$

Solving this equation for $V(s)$, we get

$$V(s) = \frac{s^2}{(s^2 + 4)(s^2 + s + 2)} \cdot \tag{C.20}$$

Equations like C.17 and C.20 will be considered further after we introduce the partial-fraction expansion.

C.4 Additional properties of the Laplace transform

A number of properties of the Laplace transform find application in circuit analysis. These properties follow from the defining equation, Equation 6.1:

(1) The time-shifting property. If the signal $v(t)$ is shifted by t_0, a positive number, then

$$\mathscr{L}[v(t - t_0)u(t - t_0)] = V(s)e^{-st_0}. \tag{C.21}$$

(2) The frequency-shifting property. A result that is similar to that just given is the following:

$$\mathscr{L}[v(t)e^{s_0 t}] = F(s - s_0). \tag{C.22}$$

(3) The scaling property. The relationship between time scaling and frequency scaling is expressed by the result

$$\mathscr{L}[v(at)] = \frac{1}{a} V\left(\frac{s}{a}\right), \tag{C.23}$$

where a is a positive real constant.

(4) The Laplace transform of periodic signals.* If $v(t)$ is a periodic function of period T, then making use of the result given in Equation C.21, we find that

$$\mathscr{L}[v(t)u(t)] = \frac{1}{1 - e^{-sT}} \int_0^T v(t)e^{-st}\, dt. \tag{C.24}$$

EXAMPLE C.4.1

Consider a unit pulse given by the equation

$$v(t) = u(t) - u(t - 1). \tag{C.25}$$

Making use of Equation C.21, we see that

$$\mathscr{L}[v(t)] = V(s) = \frac{1}{s} - \frac{e^{-s}}{s} = \frac{1}{s}(1 - e^{-s}). \tag{C.26}$$

* Since $v(t) = 0$ for $t < 0$, v is strictly speaking not periodic; we are considering only $v(t)u(t)$ where $v(t)$ is periodic.

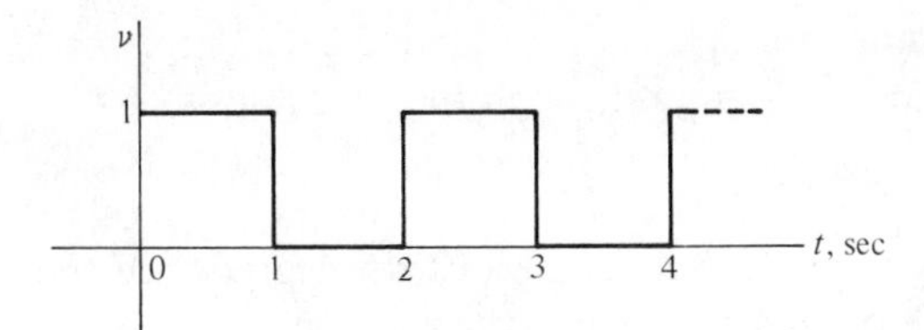

Figure C.1 Signal waveform for Example C.4.2.

EXAMPLE C.4.2

Consider a square wave shown in Figure C.1, in which the period is $T = 2$ and the width of each pulse is 1. The waveform may be described by the infinite series of step functions,

$$v(t) = u(t) - u(t-1) + u(t-2) - u(t-3) + u(t-4) - u(t-5) + \cdots, \tag{C.27}$$

and the Laplace transform of each term taken using Equation C.21. Alternatively, we may use Equation C.24 and find only the Laplace transform for the first cycle of waveform. Between $t = 0$ and $t = 2$, the function is described by

$$v_1(t) = u(t) - u(t-1), \tag{C.28}$$

and the transform of this function is given in Equation C.26. Then, from Equation C.24,

$$\mathscr{L}[v(t)] = \frac{1 - e^{-s}}{s(1 - e^{-2s})} = \frac{1}{s(1 + e^{-s})}. \tag{C.29}$$

This is an elegant result, being in closed form, but in the solution of a network problem it is probably more convenient to make use of the idea of superposition by using the sum given by Equation C.27.

C.5 Partial-fraction expansion

An important step in circuit analysis by the Laplace transform method is the simplification of the transform function $V(s)$ by a partial-fraction expansion. In carrying out this expansion, we first carry out steps of preparation which we initially explain. Given a ratio of polynomials in s,

$$V(s) = \frac{N_1(s)}{D(s)}, \tag{C.30}$$

we first examine the degree of $N_1(s)$ and $D(s)$. For the methods we will describe it is necessary that the degree of the numerator be less than that of the denominator. If this requirement is not satisfied, then we first go through the preparation of the given $V(s)$.

Step 1. Expand $V(s)$ by ordinary long division. For example, if the degree of $N_1(s)$ is 2 greater than the degree of $D(s)$, long division gives

$$\frac{N_1(s)}{D(s)} = \alpha_1 s^2 + \alpha_2 s + \alpha_3 + \frac{N(s)}{D(s)}, \tag{C.31}$$

and we may proceed to expand $N(s)/D(s)$.

Step 2. Factor the denominator polynomial $D(s)$ into its roots:

$$D(s) = b_0(s - s_1)(s - s_2)\cdots(s - s_m). \tag{C.32}$$

Since the coefficients of the polynomial $D(s)$ are real, there are only a few possibilities for the roots. In terms of s-plane locations, the roots may be real, conjugate complex, or conjugate imaginary. And the roots may be either simple or repeated. We consider several important possibilities.

Simple Real Roots. To illustrate the general method of partial-fraction expansion for simple real roots, consider a quotient of polynomials with a denominator of degree 3

$$V(s) = \frac{N(s)}{(s - s_1)(s - s_2)(s - s_3)}, \tag{C.33}$$

where the degree of $N(s)$ is less than 3 and s_1, s_2, and s_3 are distinct. The expansion of $V(s)$ is

$$V(s) = \frac{K_1}{s - s_1} + \frac{K_2}{s - s_2} + \frac{K_3}{s - s_3}. \tag{C.34}$$

The partial-fraction expansion of $V(s)$ consists of determining the values of K_1, K_2, and K_3. The manner in which this is accomplished will be illustrated for K_1. We first multiply $V(s)$ by $s - s_1$:

$$(s - s_1)V(s) = K_1 + K_2\frac{s - s_1}{s - s_2} + K_3\frac{s - s_1}{s - s_3}. \tag{C.35}$$

If we now let $s = s_1$, then we have determined K_1:

$$K_1 = (s - s_1)V(s)|_{s=s_1}. \tag{C.36}$$

Generalizing this result, we find

$$K_i = (s - s_i)V(s)|_{s=s_i}. \tag{C.37}$$

EXAMPLE C.5.1

Consider the polynomial quotient

$$V(s) = \frac{s^4 + 5s^3 + 9s^2 + 12s + 3}{s^3 + 4s^2 + 3s}. \tag{C.38}$$

Since the degree of the numerator exceeds that of the denominator, it is first necessary to simplify Equation C.38 by long division. The result is

$$V(s) = s + 1 + \frac{2s^2 + 9s + 3}{s^3 + 4s^2 + 3s}. \tag{C.39}$$

The partial-fraction expansion of $V(s)$ is

$$V(s) = s + 1 + \frac{K_1}{s} + \frac{K_2}{s+1} + \frac{K_3}{s+3}. \tag{C.40}$$

Using Equation C.37, we obtain K_1, K_2, and K_3:

$$K_1 = \left.\frac{2s^2 + 9s + 3}{(s+1)(s+3)}\right|_{s=0} = 1,$$

$$K_2 = \left.\frac{2s^2 + 9s + 3}{s(s+3)}\right|_{s=-1} = 2, \tag{C.41}$$

and

$$K_3 = \left.\frac{2s^2 + 9s + 3}{s(s+1)}\right|_{s=-3} = -1.$$

Simple Complex Roots. Equation C.37 is applicable to the case of simple complex roots as well as the simple real roots illustrated by the last example. There is a slight simplification in this case, however. If the denominator of $V(s)$ is

$$D(s) = D_1(s)(s - a - jb)(s - a + jb), \tag{C.42}$$

then the partial-fraction expansion of $V(s)$ is

$$V(s) = \frac{K_1}{s - a - jb} + \frac{K_2}{s - a + jb} + \frac{N_1(s)}{D_1(s)}. \tag{C.43}$$

It is a simple matter to show that $K_2 = K_1^*$; that is, that for complex conjugate roots, the K's of the partial-fraction expansion are themselves complex conjugate.

EXAMPLE C.5.2

The partial-fraction expansion for a $D(s)$ like Equation C.42 is

$$V(s) = \frac{1}{[(s+1)^2 + 1](s+2)} = \frac{K_1}{s + 1 - j1} + \frac{K_1^*}{s + 1 + j1} + \frac{K_2}{s+2}. \tag{C.44}$$

Using Equation C.37, we find

$$K_1 = V(s)(s + 1 - j1)|_{s=-1+j1} = \frac{-1 - j1}{4} = \frac{1}{2\sqrt{2}} \angle 225^\circ, \tag{C.45}$$

and, of course, K_1^* has the same magnitude but an angle of -225°.

Repeated Roots. If $D(s)$ contains a root which is repeated n times,

$$D(s) = D_1(s)(s - s_i)^n, \tag{C.46}$$

then the partial-fraction expansion has the form

$$V(s) = \frac{K_1}{(s - s_i)^n} + \frac{K_2}{(s - s_i)^{n-1}} + \cdots + \frac{K_n}{(s - s_i)} + \frac{N_1(s)}{D_1(s)}. \tag{C.47}$$

We next form $V_1(s)$ by multiplying by $(s - s_i)^n$:

$$V_1(s) = K_1 + K_2(s - s_i) + K_3(s - s_i)^2 + \cdots + K_n(s - s_i)^{n-1} + R(s)(s - s_i)^n, \tag{C.48}$$

where $R(s)$ represents the remaining terms, $N_1(s)/D_1(s)$. From this equation, note that K_1 is evaluated by letting $s = s_i$, as in Equation C.37. To find K_2, we differentiate V_1 with respect to s and let $s = s_i$:

$$\frac{d}{ds} V_1(s) = K_2 + 2K_3(s - s_i) + \cdots + K_n(n - 1)(s - s_i)^{n-2} + \cdots. \tag{C.49}$$

That is,

$$K_2 = \frac{d}{ds} V_1(s)\bigg|_{s=s_i}. \tag{C.50}$$

In general,

$$K_j = \frac{1}{(j - 1)!} \frac{d^{j-1}}{ds^{j-1}} V_1(s)\bigg|_{s=s_i}, \qquad j = 1, \ldots, n. \tag{C.51}$$

EXAMPLE C.5.3

If we are given the $V(s)$,

$$V(s) = \frac{2s^3 + 4s^2 + 5s + 1}{s(s + 1)^3}, \tag{C.52}$$

the required form of the partial-fraction expansion is

$$V(s) = \frac{K_1}{(s + 1)^3} + \frac{K_2}{(s + 1)^2} + \frac{K_3}{(s + 1)} + \frac{K_4}{s}. \tag{C.53}$$

We first form $V_1(s)$ by multiplying $V(s)$ by $(s + 1)^3$:

$$V_1(s) = \frac{2s^3 + 4s^2 + 5s + 1}{s}. \tag{C.54}$$

K_1 is determined directly from $V_1(-1)$ and is $K_1 = 2$. We next differentiate $V_1(s)$ with respect to s, getting

$$\frac{dV_1}{ds} = \frac{4s^3 + 4s^2 - 1}{s^2}. \tag{C.55}$$

K_2 is found by letting $s = -1$ in dV_1/ds and is $K_2 = -1$. Differentiating once more and making use of Equation C.51, we find that $K_3 = 1$ and $K_4 = 1$, so that the expansion is

$$V(s) = \frac{2}{(s + 1)^3} + \frac{-1}{(s + 1)^2} + \frac{1}{s + 1} + \frac{1}{s}. \tag{C.56}$$

C.6 Network analysis by the Laplace transform method

The steps that have been described thus far are all useful in using the Laplace transform method for the analysis of networks. This will be illustrated by example.

EXAMPLE C.6.1

The RC network shown in Figure C.2 is described by two differential equations involving the node-to-datum voltages v_1 and v_2. Using the numerical values shown, these equations are

$$\frac{dv_1}{dt} + v_1 - \frac{dv_2}{dt} = u(t)$$

and (C.57)

$$-\frac{dv_1}{dt} + \frac{3}{2}\frac{dv_2}{dt} + 2v_2 = 0.$$

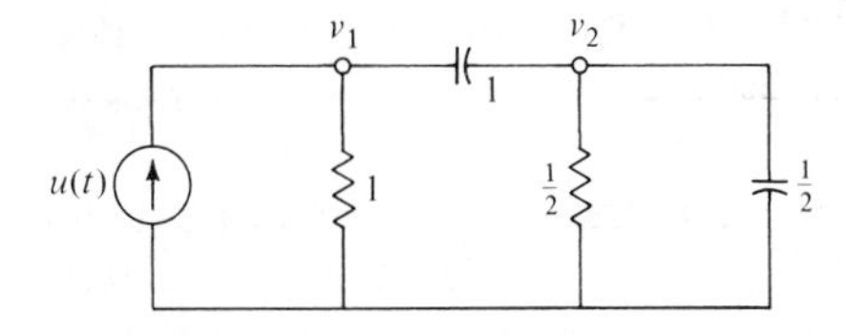

Figure C.2 *RC* network for Example C.6.1.

Assuming that both capacitors are uncharged at $t = 0$, we see that $v_1(0-) = v_2(0-) = 0$, and the Laplace transform equations corresponding to the differential equations of Equation C.57 are

$$(s + 1)V_1 - sV_2 = \frac{1}{s}$$

and (C.58)

$$-sV_1 + (\tfrac{3}{2}s + 2)V_2 = 0.$$

Suppose that we are interested in the voltage $v_2(t)$. Then we solve Equation C.58 for $V_2(s)$, which is

$$V_2(s) = \frac{2}{s^2 + 7s + 4}. \tag{C.59}$$

The partial-fraction expansion for $V_2(s)$ is

$$V_2(s) = \frac{K_1}{s + 0.625} + \frac{K_2}{s + 6.375}. \tag{C.60}$$

K_1 and K_2 are found by using Equation C.37; there results

$$V_2(s) = \frac{0.348}{s + 0.625} + \frac{-0.348}{s + 6.375}. \tag{C.61}$$

Making use of a table of Laplace transforms, we find that

$$v_2(t) = 0.348(e^{-0.625t} - e^{-6.375t})u(t). \tag{C.62}$$

Observe that $v_2(0-) = 0$ as required.

C.7 Summary

1. Definition: $\mathcal{L}[v(t)] = V(s) = \int_{0-}^{\infty} v(t)e^{-st}\,dt.$
$V(s)$ exists if: $\int_0^{\infty} |v(t)|e^{\sigma_0 t}\,dt < \infty$ for some σ_0.
Inverse transform: $v(t) = \mathcal{L}^{-1}[V(s)]$.
2. Uniqueness: $\mathcal{L}[v(t)]$ is unique, and $\mathcal{L}^{-1}[V(s)]$ is unique. This uniqueness makes it practical to use tables of transforms in the Laplace transform method. A short table of transforms is given as Table C.1.
3. Some of the properties of the Laplace transform that are important in circuit analysis are given under these titles: linearity property, transforms of derivatives, transforms of integrals, time-shifting property, frequency-shifting property, time and frequency scaling, and transforms of repeated signals defined for $t > 0$.

4. Partial-fraction expansion rules are given for simple roots and repeated roots. (a) For m simple (distinct) roots in $D(s)$ of $V(s) = N(s)/D(s)$,

$$V(s) = \sum_{j=1}^{m} \frac{K_j}{s - s_j},$$

provided that $\deg N < \deg D$. (b) For a root s_i of $D(s)$ repeated n times, the partial-fraction expansion is

$$V(s) = \sum_{j=1}^{n} \frac{K_j}{(s - s_i)^j} + \text{terms not containing } s_i \text{ as a pole},$$

again provided that $\deg N < \deg D$. If $D(s)$ has both simple and repeated roots, then both rules are used in the partial-fraction expansion. The expansion is complete when all of the K's are determined.

5. In circuit analysis by the Laplace transform method, integro-differential equations are transformed into algebraic equations in s. Analysis is completed by finding the transform of the desired variable and then expanding by partial fractions before applying the inverse Laplace transformation. Note that the initial conditions needed to solve the equation are specified in the transformed equations.

C.8 Collateral reading references

1. Lathi, B. P., *Signals, Systems, and Communication* (New York: John Wiley & Sons, Inc., 1965), 607 pp.
2. Strum, Robert D. and John R. Ward, *Laplace Transform Solution of Differential Equations (A Programmed Text)* (Englewood Cliffs, N.J.: Prentice-Hall, Inc., 1968), 197 pp.
3. Van Valkenburg, M. E., *Network Analysis*, 3rd edition (Englewood Cliffs, N.J.: Prentice-Hall, Inc., 1974), 492 pp.

EXERCISES

C.1 Determine the Laplace transforms of the following functions:

(a) $v(t) = K \sin(\omega t + \theta)$,
(b) $v(t) = \cos(\omega t + \theta)$,
(c) $v(t) = e^{-at} \cos(\omega t + \theta)$,
(d) $v(t) = \sinh bt$,
(e) $v(t) = \cosh bt$,
(f) $(t - 1)u(t - 1)$,

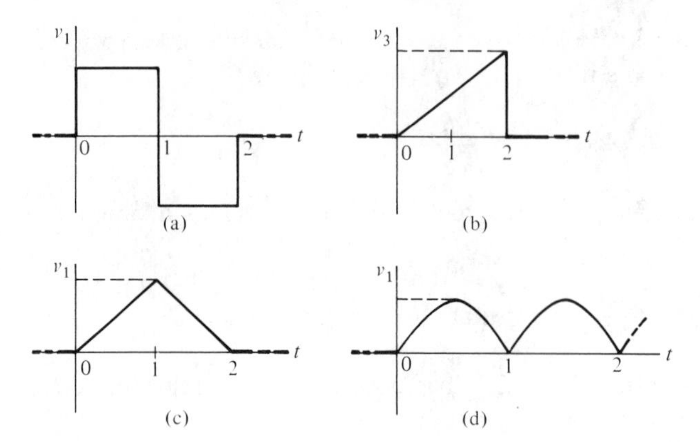

Figure C.3

(g) $v(t) = tu(t-1)$,
(h) $v(t) = K^t$, K real and $K > 0$.

C.2 Determine the Laplace transforms of each of the signal waveforms shown in Figure C.3.

C.3 Determine the partial-fraction expansion of the following functions:

(a) $V(s) = \dfrac{(s+1)(s+2)}{(s+3)(s+4)(s+5)}$,

(b) $V(s) = \dfrac{(s+1)(s+2)}{(s+3)(s+4)}$,

(c) $V(s) = \dfrac{s+1}{s(s^2+2)}$,

(d) $V(s) = \dfrac{s+2}{(s+1)^2(s+3)^2}$,

(e) $V(s) = \dfrac{1}{s^2(s+1)^3}$.

C.4 Solve the following differential equations, using the Laplace transform method. For each problem, let $v(0-) = 0$ and $dv/dt(0-) = 0$.

(a) $\dfrac{d^2v}{dt^2} + 2\dfrac{dv}{dt} + v = u(t)$,

(b) $\dfrac{d^2v}{dt^2} + 2\dfrac{dv}{dt} + 1 = (e^{-t} - 2e^{-2t})u(t)$,

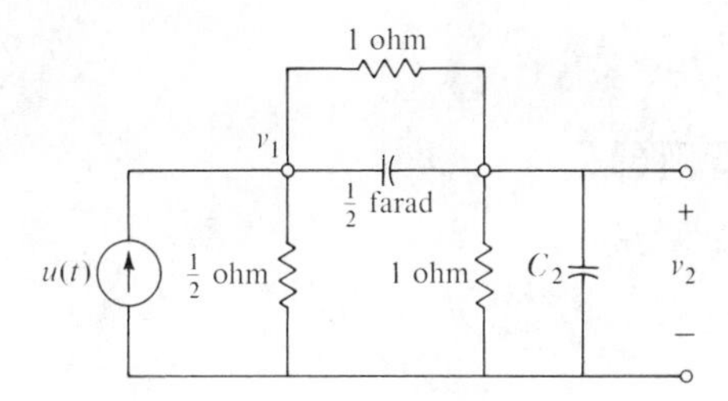

Figure C.4

(c) $\dfrac{d^2v}{dt^2} + 2\dfrac{dv}{dt} + 2 = u(t),$

(d) $\dfrac{d^2v}{dt^2} + 2\dfrac{dv}{dt} + 2v = u(t) + 2e^{-t}u(t),$

(e) $\dfrac{d^3v}{dt^3} + 3\dfrac{d^2v}{dt^2} + 3\dfrac{dv}{dt} = v = te^{-t}u(t).$

C.5 Repeat the solution of the differential equations of Exercise C.4 when $v(0-) = 1$ and $dv/dt(0-) = -1$.

C.6 In the network shown in Figure C.4, let $C_2 = 0$ and $v_2(0-) = 0$. Using the Laplace transform method and the element values given on the network, find $v_2(t)$.

C.7 This exercise makes use of the network given for Exercise C.6, but with $C_2 = 1$ farad and $v_2(0-) = 1$ volt. Using the Laplace transform method and the element values given on the network, find $v_2(t)$.

Appendix D

The Newton-Raphson Method

Most of the tools developed in the text pertain to the analysis of linear networks and systems. In the simplest case the network analysis may lead to the formulation of a system of linear algebraic equations (as in the analysis of linear resistive networks and in the steady-state analysis of linear time-invariant networks). In this Appendix we broaden the applicability of such tools by considering the solution of nonlinear equations by successive linearizations. All the linear network and system concepts we have developed apply at each stage of the linearization. This linearization process that we will describe is called the *Newton-Raphson method.*

First we consider a nonlinear equation

$$f(x) = 0 \tag{D.1}$$

in one variable x, where f is a continuous differentiable function. Let x_0 be an initial guess for the solution of Equation D.1 and assume that x_0 is not the correct solution. Refer to Figure D.1, which shows a plot of $y = f(x)$ vs. x. We now approximate the curve $f(x)$ by a straight line L_1 which is tangent to $f(x)$ at $x = x_0$. This line is given by

$$y = f(x_0) + f'(x_0)(x - x_0)\,. \tag{D.2}$$

This line has a slope $f'(x_0)$, and at $x = x_0$ it has the value $y = f(x_0)$ as required. An approximate solution of Equation D.1 is obtained by setting y in Equation D.2 to zero, obtaining

$$x = x_0 - \frac{f(x_0)}{f'(x_0)}\,. \tag{D.3}$$

This value of x is denoted by x_1 (see Figure D.1). In general x_1 is not the exact solution of Equation D.1, but (hopefully) it is better than x_0. This process is

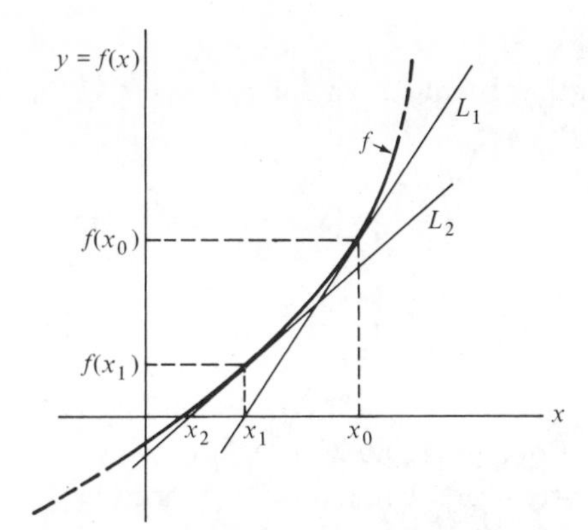

Figure D.1 Illustration of the basic Newton-Raphson method.

repeated by approximating $y = f(x)$ by a straight line tangent to $f(x)$ at $x = x_1$. This is indicated as L_2 in Figure D.1:

$$y = f(x_1) + f'(x_1)(x - x_1). \tag{D.4}$$

Setting this equal to zero yields

$$x_2 = x_1 - \frac{f(x_1)}{f'(x_1)}. \tag{D.5}$$

This linearization may be continued, and a sequence $x_0, x_1, x_2, \ldots$ may be generated by

$$x_n = x_{n-1} - \frac{f(x_{n-1})}{f'(x_{n-1})}. \tag{D.6}$$

The process is repeated until $f(x_n) \approx 0$. As a practical matter, the process is stopped when the magnitude of the difference $x_n - x_{n-1}$ is less than a predetermined small number or when the magnitude of $f(x_n)$ is less than a predetermined small number. The process is also stopped if $f(x_n)$ progressively diverges away from zero or if the number of iterations exceeds a predetermined number.

The line L_1 is also obtained by expanding $f(x)$ in Taylor's series at x_0 and retaining only the linear terms. Similarly, L_2 is a Taylor-series expansion of $f(x)$ at x_1 retaining only the linear terms. This basic linearization may be extended to several variables.

Consider a system of two nonlinear equations in two unknowns:

$$f_1(x_1, x_2) = 0, \tag{D.7}$$

$$f_2(x_1, x_2) = 0. \tag{D.8}$$

Suppose that $x_1 = (x_1)_0$ and $x_2 = (x_2)_0$ are initial guesses for the solution. Let us linearize these functions f_1 and f_2 by expanding them in Taylor's series at $(x_1)_0$ and $(x_2)_0$ and retaining only the linear terms:

$$f_1(x_1, x_2) \approx f_1[(x_1)_0, (x_2)_0] + \left(\frac{\partial f_1}{\partial x_1}\right)_0 [x_1 - (x_1)_0] + \left(\frac{\partial f_1}{\partial x_2}\right)_0 [x_2 - (x_2)_0], \quad \text{(D.9)}$$

$$f_2(x_1, x_2) \approx f_2[(x_1)_0, (x_2)_0] + \left(\frac{\partial f_2}{\partial x_1}\right)_0 [x_1 - (x_1)_0] + \left(\frac{\partial f_2}{\partial x_2}\right)_0 [x_2 - (x_2)_0]. \quad \text{(D.10)}$$

The partial derivatives are evaluated at $x_1 = (x_1)_0$ and $x_2 = (x_2)_0$. Equations D.7 and D.8 are solved approximately by setting Equations D.9 and D.10 to zero and solving for x_1 and x_2. These approximate solutions are denoted by $(x_1)_1$ and $(x_2)_1$:

$$\begin{bmatrix} \left(\frac{\partial f_1}{\partial x_1}\right)_0 & \left(\frac{\partial f_1}{\partial x_2}\right)_0 \\ \left(\frac{\partial f_2}{\partial x_1}\right)_0 & \left(\frac{\partial f_2}{\partial x_2}\right)_0 \end{bmatrix} \begin{bmatrix} (x_1)_1 - (x_1)_0 \\ (x_2)_1 - (x_2)_0 \end{bmatrix} = - \begin{bmatrix} (f_1)_0 \\ (f_2)_0 \end{bmatrix}. \quad \text{(D.11)}$$

Equation D.11 is a system of two *linear* equations which could be solved for $(x_1)_1 - (x_1)_0$ and $(x_2)_1 - (x_2)_0$. The procedure may be repeated several times, yielding

$$\begin{bmatrix} \left(\frac{\partial f_1}{\partial x_1}\right)_n & \left(\frac{\partial f_1}{\partial x_2}\right)_n \\ \left(\frac{\partial f_2}{\partial x_1}\right)_n & \left(\frac{\partial f_2}{\partial x_2}\right)_n \end{bmatrix} \begin{bmatrix} (x_1)_{n+1} - (x_1)_n \\ (x_2)_{n+1} - (x_2)_n \end{bmatrix} = - \begin{bmatrix} (f_1)_n \\ (f_2)_n \end{bmatrix}. \quad \text{(D.12)}$$

EXAMPLE D.1

Let us obtain an approximate solution for the nonlinear equations

$$x_1^2 + 3x_1x_2^2 - 1 = 0, \quad \text{(D.13)}$$

$$x_1^3x_2 + x_2^2 - 1 = 0. \quad \text{(D.14)}$$

Suppose that $(x_1)_0 = 0.32$ and $(x_2)_0 = 0.98$ are initial guesses and we wish to use the Newton-Raphson method to obtain a new approximation $(x_1)_1$ and $(x_2)_1$. Expanding Equations D.13 and D.14 in Taylor's series at $(x_1)_0$ and $(x_2)_0$ and retaining only the linear terms, we have

$$(x_1)_0^2 + 3(x_1)_0(x_2)_0^2 - 1 + [2(x_1)_0 + 3(x_2)_0^2][x_1 - (x_1)_0] + 6(x_1)_0(x_2)_0[x_2 - (x_2)_0] = 0, \quad \text{(D.15)}$$

$$(x_1)_0^3(x_2)_0 + (x_2)_0^2 - 1 + 3(x_1)_0^2(x_2)_0[x_1 - (x_1)_0] + [(x_1)_0^3 + 2(x_2)_0][x_2 - (x_2)_0] = 0. \quad \text{(D.16)}$$

Denoting $x_1 - (x_1)_0$ by Δx_1 and $x_2 - (x_2)_0$ by Δx_2 and substituting $(x_1)_0 = 0.32$ and $(x_2)_0 = 0.98$, we have

$$\begin{bmatrix} 3.52 & 1.88 \\ 0.301 & 3.798 \end{bmatrix} \begin{bmatrix} \Delta x_1 \\ \Delta x_2 \end{bmatrix} = \begin{bmatrix} -0.0245 \\ 0.046 \end{bmatrix}. \tag{D.17}$$

This linear matrix equation is solved to yield $\Delta x_1 = -0.0138$ and $\Delta x_2 = 0.013$. Thus $(x_1)_1 = 0.3062$ and $(x_2)_1 = 0.993$.

The Newton-Raphson method above may be extended to the case of k nonlinear equations in k unknowns:

$$\begin{aligned} f_1(x_1, x_2, \ldots, x_k) &= 0, \\ f_2(x_1, x_2, \ldots, x_k) &= 0, \\ &\vdots \\ f_k(x_1, x_2, \ldots, x_k) &= 0. \end{aligned} \tag{D.18}$$

Linearization yields

$$\begin{bmatrix} \left(\dfrac{\partial f_1}{\partial x_1}\right)_n & \left(\dfrac{\partial f_1}{\partial x_2}\right)_n & \cdots & \left(\dfrac{\partial f_1}{\partial x_k}\right)_n \\ \left(\dfrac{\partial f_2}{\partial x_1}\right)_n & \left(\dfrac{\partial f_2}{\partial x_2}\right)_n & \cdots & \left(\dfrac{\partial f_2}{\partial x_k}\right)_k \\ \cdots & \cdots & \cdots & \cdots \\ \left(\dfrac{\partial f_k}{\partial x_1}\right)_n & \left(\dfrac{\partial f_k}{\partial x_2}\right)_n & \cdots & \left(\dfrac{\partial f_k}{\partial x_k}\right)_n \end{bmatrix} \begin{bmatrix} (x_1)_{n+1} - (x_1)_n \\ (x_2)_{n+1} - (x_2)_n \\ \vdots \\ (x_k)_{n+1} - (x_k)_n \end{bmatrix} = - \begin{bmatrix} (f_1)_n \\ (f_2)_n \\ \vdots \\ (f_k)_n \end{bmatrix}. \tag{D.19}$$

Starting from initial guesses $(x_1)_0, (x_2)_0, \ldots, (x_k)_0$, Equation D.19 may be used to generate a sequence until a satisfactory approximate solution for Equation D.18 is obtained.

The linear matrix equation above may be solved using the Gauss elimination method described in Chapter 14 or by some triangular reduction scheme. Furthermore, if the coefficient matrix in Equation D.19 has many zero entries, sparse matrix techniques could be exploited. See Appendices B and E for references.

Appendix E

References

E.1 Collateral reading

Balabanian, Norman, *Fundamentals of Circuit Theory* (Boston: Allyn and Bacon, Inc., 1961), 548 pp.

Bose, Amar G. and Kenneth N. Stevens, *Introductory Network Theory* (New York: Harper & Row, Publishers, Inc., 1965), 357 pp.

Close, Charles M., *The Analysis of Linear Circuits* (New York: Harcourt, Brace & World, Inc., 1966), 716 pp.

Desoer, Charles A. and Ernest S. Kuh, *Basic Circuit Theory* (New York: McGraw-Hill Book Company, 1969), 876 pp.

Friedland, B., Omar Wing, and R. B. Ash, *Principles of Linear Networks* (New York: McGraw-Hill Book Company, 1961), 270 pp.

Hayt, William H., Jr., and Jack E. Kemmerly, *Engineering Circuit Analysis*, 2nd edition (New York: McGraw-Hill Book Company, 1971), 653 pp.

Huang, Thomas S. and Ronald R. Parker, *Network Theory: An Introductory Course* (Reading, Mass.: Addison-Wesley Publishing Company, Inc., 1971), 653 pp.

Leon, Benjamin J. and Paul A. Wintz, *Basic Linear Networks for Electrical and Electronics Engineers* (New York: Holt, Rinehart and Winston, Inc., 1970), 479 pp.

Scott, Ronald E., *Linear Circuits* (Reading, Mass.: Addison-Wesley Publishing Company, Inc., 1960), 928 pp.

Skilling, Hugh H., *Electrical Engineering Circuits*, 2nd edition (New York: John Wiley & Sons, Inc., 1965), 783 pp.

Smith, Ralph J., *Circuits, Devices and Systems*, 2nd edition (New York: John Wiley & Sons, Inc., 1971), 743 pp.

.2 Intermediate-level textbooks

Chua, Leon O., *Introduction to Nonlinear Network Theory* (New York: McGraw-Hill Book Company, 1969), 987 pp.

Dertouzos, Michael L., Michael Athans, Richard N. Spann, and Samuel J. Mason, *Systems, Networks, and Computation: Basic Concepts* (New York: McGraw-Hill Book Company, 1972), 514 pp.

Huelsman, Lawrence P., *Theory and Design of Active RC Circuits* (New York: McGraw-Hill Book Company, 1968), 297 pp.

Karni, Shlomo, *Intermediate Network Analysis* (Boston: Allyn and Bacon, Inc., 1971), 377 pp.

Kim, Wan H. and Henry E. Meadows, Jr., *Modern Network Analysis* (New York: John Wiley & Sons, Inc., 1971), 431 pp.

Kuo, Franklin F., *Network Analysis and Synthesis*, 2nd edition (New York: John Wiley & Sons, Inc., 1966), 515 pp.

Lathi, B. P., *Signals, Systems, and Communication* (New York: John Wiley & Sons, Inc., 1965), 607 pp.

Mitra, Sanjit K., *Analysis and Synthesis of Linear Active Networks* (New York: John Wiley & Sons, Inc., 1969), 565 pp.

Perkins, William R. and Jose B. Cruz, Jr., *Engineering of Dynamic Systems* (New York: John Wiley & Sons, Inc., 1969), 568 pp.

Rohrer, Ronald A., *Circuit Theory: An Introduction to the State Variable Approach* (New York: McGraw-Hill Book Company, 1970), 314 pp.

Truxal, John G., *Introductory Systems Engineering* (New York: McGraw-Hill Book Company, 1972), 596 pp.

Van Valkenburg, M. E., *Network Analysis*, 3rd edition (Englewood Cliffs, N.J.: Prentice-Hall, Inc., 1974).

.3 Device modeling

Gray, Paul E. and Campbell L. Searle, *Electronic Principles* (New York: John Wiley & Sons, Inc., 1969), 1016 pp.

Hamilton, D. J., F. A. Lindholm, and A. H. Marshak, *Principles and Applications of Semiconductor Device Modeling* (New York: Holt, Rinehart and Winston, Inc., 1971), 485 pp.

Millman, Jacob, *Electronic Devices and Models* (New York: McGraw-Hill Book Company), in preparation.

Millman, Jacob and Christos C. Halkias, *Integrated Electronics: Analog and Digital Circuits and Systems* (New York: McGraw-Hill Book Company, 1972), 911 pp.

E.4 Historical interest

Bode, H. W., *Network Analysis and Feedback Amplifier Design* (New York: Van Nostrand Reinhold, 1945), 551 pp.

Gardner, Murray F. and J. L. Barnes, *Transients in Linear Systems* (New York: John Wiley & Sons, Inc., 1942), 389 pp.

Guillemin, E. A., *Introductory Circuit Theory* (New York: John Wiley & Sons, Inc., 1953), 550 pp.

E.5 Computer-oriented analysis

Bowers, James C. and Stephen R. Sedore, *SCEPTRE: A Computer Program for Circuit and System Analysis* (Englewood Cliffs, N.J.: Prentice-Hall, Inc., 1971), 455 pp.

Calahan, Donald A., *Computer-Aided Network Design* (New York: McGraw-Hill Book Company, 1972), 350 pp.

Gupta, Someshwar C., Jon W. Bayless, and Behrouz Peikari, *Circuit Analysis with Computer Applications to Problem Solving* (Scranton, Pa.: Intex Educational Publishers, 1972), 546 pp.

Huelsman, Lawrence P., *Basic Circuit Theory with Digital Computations* (Englewood Cliffs, N.J.: Prentice-Hall, Inc., 1972), 896 pp.

Huelsman, Lawrence P., *Digital Computations in Basic Circuit Theory* (New York: McGraw-Hill Book Company, 1968), 203 pp.

Jensen, Randall W. and Mark D. Lieberman, *IBM Electronic Circuit Analysis Program: Techniques and Applications* (Englewood Cliffs, N.J.: Prentice-Hall, Inc., 1968), 401 pp.

Kinariwala, B. K., Franklin F. Kuo, and Nai-kuan Tsao, *Linear Circuits and Computation* (New York: John Wiley & Sons, Inc., 1973).

Ley, B. James, *Computer-Aided Analysis and Design for Electrical Engineers* (New York: Holt, Rinehart and Winston, Inc., 1970), 852 pp.

McCracken, Daniel D., *FORTRAN with Engineering Applications* (New York: John Wiley & Sons, Inc., 1967), 237 pp.

Steiglitz, Kenneth, *An Introduction to Discrete Systems* (New York: John Wiley & Sons, Inc., 1973), 320 pp.

Wing, Omar, *Circuit Theory with Computer Methods* (New York: Holt, Rinehart and Winston, Inc., 1972), 529 pp.

Answers to Exercises

CHAPTER 1

1.1.1 At $t = \pi/2$, $v_1(t) = 5 \sin [2(\pi/2) + \pi/6] = -2.5$ volts. It is negative. Hence the plus reference terminal is negative with respect to the other terminal.

1.1.2 (a) At $t = \pi/2$, $i_{12}(t) = 5 \sin [2(\pi/2) + \pi/6] = -2.5$. The current is 2.5 amperes from 2 to 1. Electron flow is opposite, from 1 to 2.
(b) For $t = 0$, $i_{12}(0) = 2.5$, so the current is 2.5 amperes from 1 to 2; electron flow is from 2 to 1.

1.1.3 The algebraic sum is zero. The two readings are negatives of each other.

1.2.1 $f(x) = e^{-ax} \sin bx$; $f(\xi) = e^{-a\xi} \sin b\xi$; $f(\xi + 1) = e^{-a(\xi+1)} \sin b(\xi + 1)$.

1.2.2 $v(x + 2\pi)$ is shifted to the left by 2π compared to $v(x)$, and $v(x - \pi)$ is shifted to the right by π compared to $v(x)$.

1.2.3 $p(t) = \frac{1}{2}(\sin t)^2$; $w(t) = \frac{1}{4}t - \frac{1}{8} \sin 2t$.

1.2.4 (a) and (b) are shifted to the left by 5, and (c) and (d) are shifted to the right by 5.

1.2.5 Abscissas for (a) and (b) are compressed by 10 and abscissas for (c) and (d) are expanded by 10.

1.2.6 (a) Shift to right and compress scale; (b) shift to left and expand scale.

1.2.7 (a) $I_1 = 25$; (b) $I_2 = 25$; (c) $I_3 = 25$; (d) $I_4 = 25$. Yes.

1.2.8 $I = 0$ for $t < 0$; $I = \frac{1}{4}t^2$ for $0 \leq t \leq 10$; $I = 25$ for $t > 10$.

1.2.9 The waveform in Figure 1.2(b) is rotated by 180° about the time axis to obtain the new waveform.

1.3.1 $v_{out}(t) = (1/T) \int_0^T |v_{in}(\tau)| \, d\tau$.

1.3.2 $$v_1(t) = \frac{dv_3(t) - bv_4(t)}{ad - bc}.$$

1.3.3 $$LC \frac{d^2v_2}{dt^2} + RC \frac{dv_2}{dt} + v_2 = v_1.$$

1.3.4 Node a: $i_1 + i_2 + i_6 = 0.$
Node b: $-i_2 - i_3 - i_4 = 0.$
Node c: $-i_5 - i_6 = 0.$

1.3.5 Loop 1: $-v_1 - v_3 + v_2 = 0.$
Loop 2: $-v_2 - v_5 - v_6 = 0.$
Loop 3: $v_3 + v_4 + v_5 = 0.$

HAPTER 2

2.1.1 Yes.
2.1.2 $v_3 = -25e^{-t} + 55e^{-3t}.$

2.1.3 $|K| \leq 1 \Big/ \left| \left(\frac{\alpha}{\beta}\right)^{-\alpha/(\alpha-\beta)} - \left(\frac{\alpha}{\beta}\right)^{\beta/(\alpha-\beta)} \right|.$

2.1.4 (a) $y = \sigma x + \ln K$; (b) $\ln (d^n v/dt^n) = \ln (\sigma^n K) + \sigma t$. Since $\ln (\sigma^n K) = \ln K + n \ln \sigma$, the spacing is constant and equal to $\ln \sigma$.

2.1.5 $b = \dfrac{2\pi}{0.8} = 2.5\pi$; $K = 4$; $\sigma = \frac{5}{3}$.

2.2.1 (a) volt-sec; (b) volt/sec.
2.2.2 Delayed step function $v_2(t) = 3.4u(t - 2)$.
2.2.3 $v_2(t) = 3\delta(t - t_0)$.
2.2.4 Maximum of $v_2(t)$ is at T_0, equal to $V_0 T_0$. Minimum of $v_2(t)$ is $-V_0$.
2.2.5 $v_2(t) = 2i(t) + u'(t - 1) - 3u(t - 2)$.
2.2.6 $v_2(t) = 2.34\delta(t - 1) + 2.06\delta(t - 2) - 4.4\delta(t - 3)$.
2.2.7 $u_{-1}(t)$ – unit step; $u_{-2}(t)$ – unit ramp.
2.3.1 (a) $\omega = 380$ radians/sec, (b) $T = 16.64$ millisec,
(c) $f = 60$ Hz, (d) B lags A by 45°, C lags A by 112.5°,
(e) A: 7.6 sin 380t,
B: 170 sin (380t – 45°),
C: 12 sin (380t – 112.5°).
2.3.2 $v_1(t) = 2e^{1/4}e^{-2t} \sin 4\pi t.$
2.3.3 $v_3(t) = v_1 v_2 = 2 \sin 3t \sin 2t = \cos t - \cos 5t.$
2.3.4 $v_2(t) = \sqrt{5} \sin (2t + 116.5°).$
2.6.1 $v_2(t) = v_1(t - 4).$
2.6.2 $v_2(t) = K_1 e^{-\sigma_1(t-a)} \sin [\omega_1(t - a) + \phi_1]u(t - a).$
2.6.3 $v_2(t) = A_2 \sin [\omega_2(t - a) + \phi_2]$. Negative of phase shift due to delay is $\omega_2 a$, $d(\omega_2 a)/d\omega_2 = a$.
2.6.7 $v_2(t) = \sin 2t.$
2.6.9 $v(t) = \sum_{k=-\infty}^{\infty} v_1(t - kT).$
2.6.10 $v_2(t) = u(t) + u(t - a) + u(t - 2a) - 3u(t - 3a).$

HAPTER 3

3.1.1 Period = T_0.
3.1.2 $\bar{v} = 1/\pi$.
3.1.3 No, does not satisfy condition for periodicity.

3.1.4 $\bar{v} = \dfrac{\sqrt{2} - 1}{2\pi}.$

3.1.5 $\bar{v} = \dfrac{K(T - a)}{T}$.

3.1.6 Average of v_{21} is $-V_0$.

3.2.1 (a) Peak value = 25, peak-to-peak = 25 + 21 = 46; (b) peak value = 2, peak-to-peak = 4; (c) peak value = 3, peak-to-peak = 4.

3.3.1 $V_{\text{rms}} = \dfrac{\sqrt{10}}{2}$.

3.3.2 (a) 0.648 volts, (b) 0.648 volts, (c) 0.648 volts, (d) $0.648/\sqrt{2}$, (e) $0.648/\sqrt{2}$, (f) $0.648/\sqrt{2}$.

3.3.3 (a) 1 watt; (b) 1 watt; (c) 1 watt; (d) 2 watts; (e) 4 watts.

3.3.4 $V_{\text{rms}} = \sqrt{\dfrac{11}{15}}$.

3.3.5 $V_{\text{rms}} = \dfrac{1}{\sqrt{2}}$.

3.3.6 $V_{\text{rms}} = \dfrac{V_m}{\sqrt{2}}$.

3.3.7 $V_{\text{rms}} = \sqrt{27}$.

3.3.8 $V_{\text{rms}} = \sqrt{\dfrac{21}{2}}$.

3.3.9 rms value of v_{21} is V.

3.3.10 $I_{\text{rms}} = \dfrac{1}{2}\sqrt{\dfrac{1}{2} - \dfrac{1}{\pi}}$.

3.4.1 Overshoot = 5; percent overshoot = $\frac{5}{100}(100) = 5\%$.

3.4.2 Overshoot = $e^{-(1/\omega)[\pi - \tan^{-1}(1/\omega)]} \dfrac{\omega}{\sqrt{1 + \omega^2}}$.

3.4.3 $t_2 - t_1 = \dfrac{1}{\sigma_1} \ln 9 = \dfrac{2.2}{\sigma_1}$.

3.4.4 Rise time = 2.4 − 0.45 = 1.95 millisec. Time delay = 1.4 millisec.

3.4.5 $t_1 = \dfrac{\ln 2}{\sigma_1} = \dfrac{0.692}{\sigma_1}$.

3.4.6 $t_s = 2.99$.

CHAPTER 4

4.1.1 Reference directions for v and i provide the clue.

4.1.2 Capacitor is the model for the component; capacitance is the parameter describing the model.

4.1.3 $i = 1000v, v > 0; i = v, v \leq 0.$

4.1.4 $i = \psi^3.$

4.1.9 $\psi = (1 - e^{-t})i.$

4.1.10 $dq/dt = i = c(dv/dt) + v(dc/dt).$

4.2.2 $v_2 = M(di_s/dt).$

4.2.3 $v_2 = -M(di_s/dt).$

4.2.4 $v_2 = M_{12}(di_s/dt).$

4.2.5 $v_1 = -L_1(di_a/dt) + M(di_b/dt); v_2 = L_2(di_b/dt) - M(di_a/dt).$

4.2.6 $v_a = -L_1(di_1/dt) - M(di_2/dt); v_2 = L_2(di_2/dt) + M(di_1/dt).$

4.2.7 $v_1 = L_1(di_1/dt) - M(di_2/dt); v_2 = L_2(di_2/dt) - M(di_1/dt).$

4.3.1 (a) $v_1 = (1/n^2)L(di_1/dt).$ (b) $i_1 = Cn^2(dv_1/dt).$

4.3.2 (a) $v_1 = (K^2/R)i_1.$ (b) $i_1 = (L/K^2)(dv_1/dt).$
(a) $v_2 = (K^2/R)i_2.$ (b) $i_2 = (L/K^2)(dv_2/dt).$

4.3.3 (a) $v_1 = -L(di_1/dt). \therefore L_{eq} = -L.$
(b) $i_1 = i_2 = C(dv_2/dt) = -C(dv_1/dt), C_{eq} = -C.$

4.3.4 $v_1 = K_1(v_2/K_2). \therefore v_2 = (K_2/K_1)v_1.$
$i_2 = (1/K_2)v_4 = (1/K_2)(-K_1i_1) = -(K_1/K_2)i_1.$
Therefore, relations are like those of transformer with $n = K_2/K_1$.

4.4.1 $v_{in} = v_O; i_{in} = 0$. The terminal characteristics are identical to those of the isolator of Example 4.4.1.

4.4.2 $v_O = -v_{in}; i_{in} = v_{in}/R$. The terminal characteristics are identical to those of the inverter of Example 4.4.2.

4.4.3 $v_O = -(v_a + v_b); i_a = v_a/R; i_b = v_b/R.$

4.4.4 $v_O = (v_a + v_b); i_a = v_a/R; i_b = v_b/R$. The terminal characteristics are identical to those of the summer of Example 4.4.3.

HAPTER 5

5.1.1 (a) *f, ad, be, bcd, ace.*
(b) *acd, ae, kf, amf, abf, khg, amhg, abhg.*

5.1.2 (a) *acd, ae, amf, abf, amhg, abhg, kf, khg, kmcd, kme.*
(b) *amk, abk, mb, mcdf, bcdf, cde, fgh, mef, bef, acdfk, acdghk, mcdgh, bcdgh, megh, begh, aefk, aeghk.*

5.2.1 $i_a + i_b + i_f = 0$ (1). $-i_a - i_c + i_d = 0$ (2).
$-i_d - i_e - i_f = 0$ (3). $-i_b + i_c + i_e = 0$ (4).

5.2.2 $i_{12} + i_{14} + i_{13} = 0$ (1). $i_{21} + i_{23} + i_{24} = 0$ (2).
$i_{32} + i_{34} + i_{31} = 0$ (3). $i_{41} + i_{42} + i_{43} = 0$ (4).

5.2.3 Add (1) and (2) from Exercise 5.2.1 : $i_f + i_b - i_c + i_d = 0.$

5.2.4 $a, b, d, e; i_a + i_b - i_e - i_d = 0.$

5.2.5 $a, c, e, f; i_a + i_f + i_c + i_e = 0.$

5.2.6 From Exercise 5.2.1, add (2), (3), (4): $-i_a - i_b - i_f = 0 \Rightarrow i_a + i_b + i_f = 0$(1).

5.2.7 Add (1), (3), (4): $i_a + i_c - i_d = 0$ or $-i_a - i_c + i_d = 0$ (2).

5.2.8 Add (1), (2), (4): $i_d + i_e + i_f = 0$ (4).

5.2.9 Add (1), (2), (3): $i_b - i_c - i_e = 0$ or $-i_b + i_c + i_e = 0$ (4).

5.3.1 (a) $v_a + v_d + v_e - v_b = 0,$ (b) $v_a - v_c - v_b = 0,$
(c) $v_d + v_e + v_c = 0,$ (d) $v_b - v_e - v_f = 0.$

5.3.2 (a) $v_{12} + v_{23} + v_{34} + v_{41} = 0$, (b) $v_{12} + v_{24} + v_{41} = 0$,
(c) $v_{23} + v_{34} + v_{42} = 0$, (d) $v_{14} + v_{43} + v_{31} = 0$.

5.3.3 (a) $v_{13} + v_{32} + v_{21} = 0$, (b) $v_{34} + v_{45} + v_{53} = 0$,
(c) $v_{34} + v_{45} + v_{52} + v_{23} = 0$, (d) $v_{56} + v_{62} + v_{25} = 0$.

5.3.4 $-v_a + v_b + v_c - v_d = 0$; $5 + 3 + v_c + 6 = 0$; $v_c = -14$.

5.3.5 $v_{32} = -4$; $v_{14} = -3$; $v_{43} = 9$.

5.4.1 $v_R = Ri = i$ if $R = 1$; then i is as shown in Figure 5.13(a).
$v_L = L(di/dt) = di/dt$, as shown in Figure 5.13(b).

5.4.2 $v_C = \int_0^t (\tau - 1)\, d\tau = \frac{1}{2}t^2 - t$ for $0 < t < 2$, as plotted in Figure 5.14(a).
$v_C = -t^2/2 + 3t - 4$ for $2 \le t \le 4$, as shown in Figure 5.14(b).
$v_C(t)$ in $4 \le t \le 8$ is the same as that in $0 \le t \le 4$ in Figure 5.14(c).

5.4.3 $v = v_L + v_C + v_R$, as in Figure 5.15.

5.4.4 $v_R = 5 \sin 5t$; $v_L = 25 \cos 5t$;

$$v = \begin{cases} 1 + 24 \cos 5t + 5 \sin 5t & \text{for } 0 \le t \le 2\pi/5, \\ 0, & \text{elsewhere.} \end{cases}$$

5.4.5 $$i = \begin{cases} \sin 10t + 4 \cos 10t + 1 & \text{for } t \ge 0, \\ 0 & \text{for } t < 0. \end{cases}$$

5.4.6 $$i = \begin{cases} i_R + i_C + i_L = 10 & \text{for } t > 2, \\ t + \frac{1}{2} + 5t^2 & \text{for } 0 \le t \le 1, \\ -5t^2 + 19t - \frac{17}{2} & \text{for } 1 \le t \le 2. \end{cases}$$

5.5.1 Node 1: $C_4[d(v_1 - v_6)/dt] + (1/R_7)(v_1 - v_2) - i_1 = 0$.
Node 2: $(1/R_7)(v_2 - v_1) + C_1[d(v_2 - v_3)/dt] + (1/R_4)(v_2 - v_4) = 0$.
Node 4: $C_2[d(v_4 - v_3)/dt] + C_3[d(v_4 - v_5)/dt] + (1/R_4)(v_4 - v_2) + (1/R_2)v_4 = 0$.
Node 5: $C_3[d(v_5 - v_4)/dt] + (1/R_5)(v_5 - v_6) + (1/R_3)v_5 - i_2 = 0$.
Node 6: $C_4[d(v_6 - v_1)/dt] + (1/R_6)(v_6 - v_3) + (1/R_5)(v_6 - v_5) = 0$.

5.5.2 Node 1: $(1/R_4)v_{14} + (1/R_5)(v_{14} - v_{54}) + (1/R_1)(v_{14} + v_1 - v_{24}) = 0$.
Node 2: $(1/R_1)(v_{24} - v_1 - v_{14}) + (1/R_6)(v_{24} - v_{56}) + (1/R_2)(v_{24} - v_{34}) = 0$.
Node 3: $(1/R_2)(v_{34} - v_{24}) + (1/R_7)(v_{34} - v_{54}) + (1/R_3)v_{34} = 0$.
Node 5: $(1/R_5)(v_{54} - v_{14}) + (1/R_6)(v_{54} - v_{24}) + (1/R_7)(v_{54} - v_{34}) + (1/R_8)v_{54} = 0$.

5.5.3 For loop a: $R_1i_a + R_6(i_a - i_b) + R_5(i_a - i_c) - v_1 = 0$.
For loop b: $R_6(i_b - i_a) + R_2i_6 + R_7(i_b - i_d) = 0$.
For loop c: $R_8(i_d - i_c) + R_7(i_d - i_b) + R_3i_d = 0$.

5.5.4 (a) loops 41534, 4534, 2512, 23512; tree $R_1R_5R_7R_3$.
(b) loops 12351, 2532, 5415, 43514; tree $R_5R_7R_2R_4$.
(c) loops 1521, 14521, 23452, 5345; tree $R_1R_6R_8R_3$.
(d) loops 12541, 1541, 5325, 34523; tree $R_6R_8R_4R_2$.

5.5.5 $R_7C_1R_1R_2R_3R_6$ is a tree. The following link currents are loop currents: i_1, i_{16}, i_{65}, i_{24}, i_{34}, i_{45}, i_3.

CHAPTER 6

6.1.1 $p = +vi$.

6.1.2 $p = -vi$.

6.1.3 $p = +vi.$

6.1.4 $p = -vi.$

6.2.1 $P_{av} = \dfrac{L\omega_0 I_0^2}{2T}\displaystyle\int_0^T \sin 2\omega_0 t\, dt = 0.$

6.2.2 $P_{av} = \dfrac{1}{T}\displaystyle\int_0^T 3V_0^4\omega_0 \sin^3 \omega_0 t \cos \omega_0 t\, dt = \dfrac{3}{T} V_0^4 \tfrac{1}{4}(\sin \omega_0 t)^4 \Big|_0^T = 0.$

6.2.3 $P_{av} = (1/T)\int_0^T \omega_0 I_0^2 \cos \omega_0 t(1 - \sin^2 \omega_0 t)\, dt = 0.$ Storing.

6.2.4 $P_{av} = (1/T)\int_0^T I_0 \cos \omega_0 t(-2\omega_0 I_0 \sin \omega_0 t + \omega_0 I_0 \sin^2 \omega_0 t - I_0 \omega \cos^2 \omega_0 t)\, dt = 0.$ Storing.

6.2.5 No. It depends on the periodicity of R compared to that of v.

6.2.6 No.

6.2.7 $P_{av} = \dfrac{10}{T}\displaystyle\int_0^T i^2\, dt = \dfrac{10}{2}[(I_1 \cos \alpha_1 + I_2 \cos \alpha_2)^2 + (I_1 \sin \alpha_1 + I_2 \sin \alpha_2)^2].$

6.3.4 $P_{av} = \dfrac{1}{T}\displaystyle\int_0^T 0\, dt = 0.$ Lossless.

6.3.5 Passive.

6.3.6 Yes.

6.4.1 W_{13} from source $= \displaystyle\int_0^t \frac{10}{12}(10)\, d\tau,$

$W_{R_1} = \displaystyle\int_0^t 10\left(\frac{10}{12}\right)^2 d\tau, \qquad W_{R_2} = \int_0^t 2\left(\frac{10}{12}\right)^2 d\tau.$

6.4.2 $W_{13} = \displaystyle\int_0^t 10\left(\frac{-5 \pm \sqrt{45}}{2}\right) d\tau, \qquad W_{R_1} = \int_0^t 10\left(\frac{-5 \pm \sqrt{45}}{2}\right)^2 d\tau,$

$W_{R_2} = \displaystyle\int_0^t 2\left(\frac{-5 \pm \sqrt{45}}{2}\right)^3 d\tau, \qquad W_{R_1} + W_{R_2} = W_{13}.$

6.4.3 $W_{13} = \displaystyle\int_0^t 10\left(\frac{10}{8}\right) d\tau, \qquad W_{R_1} = \int_0^t 10\left(\frac{10}{8}\right)^2 d\tau,$

$W_{R_2} = \displaystyle\int_0^t -2\left(\frac{10}{8}\right)^2 d\tau, \qquad W_{13} = W_{R_1} + W_{R_2}.$

6.4.4 $W_R = 10\left(t + \dfrac{2L}{R}e^{-(R/L)t} - \dfrac{L}{2R}e^{-(2R/L)t} - 2\dfrac{L}{R} + \dfrac{L}{2R}\right),$

$W_L = \tfrac{1}{2}L(1 - 2e^{-(R/L)t} + e^{-(2R/L)t}),$

$W_{13} = 10\left(t + \dfrac{L}{R}e^{-(R/L)t} - \dfrac{L}{R}\right). \qquad W_{13} = W_L + W_R.$

6.4.5 $P_R = 5;\ P_L = 0.$

6.4.6 $P_a = \left(\dfrac{3}{2}\right)^2(2), \qquad P_b = \left(\dfrac{5}{2}\right)^2(2), \qquad P_c = \left(\dfrac{8}{2}\right)^2(2),$

$P_{12}(\text{out}) = 33/2, \qquad P_{12}(\text{out}) = 65/2.$

6.5.1 (a) $v_5 = v_1 + v_2$, $i_5 = i_1 = i_2$, $v_1 i_1 + v_2 i_2 - v_s i_s = 0$.
(b) If v and i satisfy branch equations, then $v_1 i_1$ is instantaneous power into box 1, $v_2 i_2$ is instantaneous power into box 2, and $v_s i_s$ is instantaneous power out of source; $v_s i_s = v_1 i_1 + v_2 i_2$.

6.5.2 $v_1 = 3$, $i_1 = e^{-t}$; $\quad v_2 = 8$, $i_2 = e^{-t}$; $\quad v_s = 11$, $i_s = e^{-t}$.

6.5.3 By Tellegen's theorem,

$$v_{11'} i_1 + v_{22'} i_2 + \cdots + v_{nn'} i_n - (v_1 i_1 + v_2 i_2 + \cdots + v_n i_n) = 0.$$

$$\therefore \underbrace{\textstyle\sum_{k=1}^{n} v_{kk'} i_k}_{\substack{\text{power into}\\ n\text{-port}}} = \underbrace{\textstyle\sum_{k=1}^{n} v_k i_k}_{\substack{\text{power out}\\ \text{of sources}}}.$$

6.6.1 (a) $W_C = \dfrac{1}{2C} q^2 = \dfrac{1}{2}(2t^2) \quad$ for $\quad 0 \le t \le 1$,

$W_C = \frac{1}{2}(2) \quad$ for $\quad t > 1$;
(b) $W_C = \frac{1}{2}t^2 \quad$ for $\quad 0 \le t \le 1$,
$W_C = \frac{1}{2}(-t^2 + 4t - 3)^2 \quad$ for $\quad 1 < t < 2$.
For $t > 2$, W_C for (a) and W_C for (b) are equal. Reversing terminals has no effect on W_C.

6.6.2 (a) $W_R = \begin{cases} 8t & \text{for } 0 \le t \le 1, \\ 8 & \text{for } t > 1; \end{cases}$

(b) $W_R = \begin{cases} \frac{8}{3}t^3 & \text{for } 0 < t < 1, \\ \frac{8}{3}t^3 - 16t^2 + 32t - 16 & \text{for } 1 \le t \le 2, \\ \frac{16}{3} & \text{for } t > 2. \end{cases}$

Reversing switch has no effect.

6.6.3 $v = \begin{cases} 2i, & i \ge 0, \\ 10i, & i < 0. \end{cases} \qquad W_R = \begin{cases} 8t, & 0 \le t \le 1, \\ 8, & t > 1. \end{cases}$

If terminals are reversed,

$$W_R = \begin{cases} 40t, & 0 \le t \le 1, \\ 40, & t > 1. \end{cases}$$

6.6.4 $W_R = \begin{cases} \frac{8}{3}t^3 & \text{for } 0 \le t \le 1, \\ \frac{8}{3} - 16t^2 + 32t - 16 & \text{for } 1 \le t \le 2, \\ \frac{16}{3} & \text{for } t > 2. \end{cases}$

If terminals are reversed,

$$W_R = \begin{cases} \frac{40}{3}t^3, & 0 \le t \le 1, \\ 5(\frac{8}{3}t^3 - 16t^2 + 32t - 16), & 1 \le t \le 2, \\ \frac{80}{3}, & t > 2. \end{cases}$$

6.6.5 $W_L = \frac{2}{3}$, $t > 1$.

6.6.6 $W_L = \dfrac{1}{2L}\psi^2 = \begin{cases} \frac{1}{6}t^2, & 0 \le t \le 1, \\ \frac{1}{6}(-t^2 + 4t - 2), & 1 \le t \le 2, \\ \frac{1}{6}(2), & t > 2. \end{cases}$

Reversing terminals has no effect on W_L.

6.6.7 $W_L = \frac{1}{3}\psi^3 - \frac{1}{2}\psi^2 = -\frac{1}{6}$.
6.6.8 $W_L = -\frac{1}{6}$. If terminals are reversed, $\psi(1) = -1$, $W_L = -\frac{1}{3} - \frac{1}{2} = -\frac{5}{6}$.
6.6.9 $W_L = \frac{23}{6}$.

HAPTER 7

7.1.1 $v = R(t)i$. If $\left.\begin{matrix} v_1 = R(t)i_1 \\ v_2 = R(t)i_2 \end{matrix}\right\}$ then $K_1v_1 + K_2v_2 = R(t)(K_1i_1 + K_2i_2)$.

7.1.2 $\psi = Li = \int v\,d\tau$, $i = (1/L)\int_0^t v(\tau)\,d\tau + i(0)$. For $v \equiv 0$, $i = i(0)$ zero-input linear. For $i(0) \equiv 0$, $i_1 = (1/L)\int_0^t v_1(\tau)\,d\tau \Rightarrow i_2 = Ki_1 = (1/L)\int_0^t Kv_1\,d\tau$, and if $v_3 = v_1 + v_2$, $i_3 = i_1 + i_2$. Therefore, 1-port is zero-state linear, and since the element satisfies the decomposition property, it is linear.

7.1.3 $Ri + L(di/dt) = v$. If $Ri_1 + L(di_1/dt) = v_1$, then $R(Ki_1) + L(d/dt)(Ki_1) = Kv_1$ (homogeneous); and if $Ri_2 + L(di_2/dt) = v_2$, then $R(i_1 + i_2) + L(d/dt)(i_1 + i_2) = v_1 + v_2$ (additive). Therefore, the system is zero-state linear.

7.1.4 $Ri + L(di/dt) = 0$. Let $i_1(0)$ be an initial state. And, since $i = Ke^{-(R/L)t}$, then $i(t) = i(0)e^{-(R/L)t}$. If $i_2(0) = K_1i_1(0)$, then $i_2(t) = K_1i_1(0)e^{-(R/L)t}$ (homogeneous). If $i_3(0) = i_2(0) + i_1(0)$, then $i_3(t) = [i_2(0) + i_1(0)]e^{-(R/L)t}$ (additive). Therefore, the system is zero-input linear.

7.1.5 Using the method of I.F., we have

$$\frac{d}{dt}\, e^{(R/L)t} i = v(t)e^{(R/L)t}$$

$$e^{(R/L)t} i(t) - i(0) = \int_0^t e^{(R/L)\tau} v(\tau)\,d\tau$$

$$i(t) = \underbrace{e^{-(R/L)t} i(0)}_{\substack{\text{zero-input}\\ \text{response}}} + \underbrace{\int_0^t e^{-(R/L)(t-\tau)} v(\tau)\,d\tau}_{\substack{\text{zero-state}\\ \text{response}}}$$

Therefore, the system satisfies the decomposition property.

7.1.6 $v \neq f(q)$. Assuming that i is input, $v = f(\int i\,d\tau)$. If f is a nonlinear function, then $f(Kq) \neq Kf(q)$. Nonlinear.

7.1.7 The system is nonlinear and zero-state homogeneous.

7.1.8 $x_1 \to y_1$, $x_2 \to y_2$, $x_1 + x_2 \to y_1 + y_2$. Therefore, if $x_2 = x_1$, $2x_1 \to 2y_1$. Repeat for $x_2 = x_1$, $2x_1 + x_2 \to 2y_1 + y_2$ or $3x_1 \to 3y_1$, etc. Therefore, $Cx_1 \to Cy_1$ for C = integer.

7.1.9 $x_1 \to y_1$, $-x_1 \to -y_1$, $\therefore x_1 - x_1 = 0 \to y_1 - y_1 = 0$.

7.1.10 $v(t) = e^{-t}(\frac{41}{44}\cos 2t + \frac{6}{22}\sin 2t)u(t)$.

7.1.11 $v = e^{-t}(\frac{-5}{22}\cos 2t + \frac{20}{22}\sin 2t)u(t)$.

7.1.12 $v = e^{-t}(\frac{125}{44}\cos 2t + \frac{380}{44}\sin 2t)u(t)$.

7.1.13 $v(t) = -\frac{1}{2}(10e^{-t}\cos 2t)u(t) + 3(1 - e^{-t}\cos 2t)u(t)$.

7.2.1 $R_{eq} = R_2$; $v_{eq} = v$.

7.2.2 $R_{eq} = R_2$; $i_{eq} = v/R_2$.

7.2.3 $v_{eq} = 0$; $R_{eq} = 2R$.

7.2.4 (a) $v_{\text{open}} = (1/C)\int_0^t u(\tau)\,d\tau = (1/C)r(t)$ (ramp); (b) $v_{oc} = (1/C)u(t)$.
7.2.5 $i_{sc} = C(dv/dt)$.
7.2.6 $v = L(di/dt)$.
7.2.7 $i = (1/L)\int_0^t v\,d\tau$.
7.3.1 No.
7.3.2 $i = dq/dt = (df/dv)(dv/dt)$. Let $i_1(t)$ be the output due to $v_1(t)$. Let $v_2(t) = v_1(t - \Delta)$. Then

$$i_2(t) = \left.\frac{df}{dv}\right|_{v_2(t)} \frac{dv_2}{dt} = \left.\frac{df}{dv}\right|_{v_1(t-\Delta)} \frac{d}{dt} v_1(t - \Delta) = i_1(t - \Delta).$$

7.3.3 $v = d\psi/dt = (dg/di)(di/dt)$. So if $v_1(t)$ is the output due to $i_1(t)$,

$$v_1(t) = \left.\frac{dg}{dt}\right|_{i_1(t)} \frac{di_1(t)}{dt}.$$

Now for $i_2(t) = i_1(t - \Delta)$, we have

$$v_2(t) = \left.\frac{dg}{di}\right|_{i_1(t-\Delta)} \frac{di_1(t - \Delta)}{dt} = v_1(t - \Delta).$$

7.4.1 $\dfrac{R_1}{R_L}(R_2 + R_L)C\dfrac{dv_L}{dt} + \dfrac{R_1 + R_2 + R_L}{R_L} v_L = v.$

7.4.2 $v = -KC(dv_{ab}/dt)$.
7.4.3 $y = c_1 e^{jt} + c_2 e^{-jt}$.
7.4.4 $y = C_1 e^{-t} + C_2 e^{jt} + C_3 e^{-jt}$.
7.4.5 $y = C_1 e^{-3t} + C_2 e^{-2t} + C_3 t e^{-2t} + C_4 t^2 e^{-2t} + C_5 t^3 e^{-2t}$.

7.4.6 $y = \dfrac{1 - j}{2} e^{jt} + \dfrac{1 + j}{2} e^{-jt} = \cos t + \sin t.$

7.5.1 Stable.
7.5.2 Unstable, two roots in right half of s-plane.
7.5.3 Stable.
7.5.4 Unstable, two roots in right half of s-plane.
7.5.5 Unstable, oscillatory, two roots on j-axis.

CHAPTER 8

8.1.2 $i = 2u(t - 1) - 2r(t - 1) + 2r(t - 2)$;
$y = 2v(t - 1)u(t - 1) - 2\int_0^t v(\tau - 1)u(\tau - 1)\,d\tau + 2\int_0^t v(\tau - 2)u(\tau - 2)\,d\tau$.
8.1.3 $y = 2\int_0^t v(\tau - 1)u(\tau - 1)\,d\tau - 2\int_0^t\int_0^\xi v(\lambda - 1)u(\lambda - 1)\,d\lambda\,d\xi$
$+ 2\int_0^t\int_0^\xi v(\lambda - 2)u(\lambda - 2)\,d\lambda\,d\xi$.
8.1.4 $y = 2u(t - 1) - 2\delta(t - 2) - 2u(t - 2)$.
8.1.5 $y = 2\delta(t - 1) - 2u(t - 1) + 2u(t - 2)$.
8.2.1 $y = (1 - e^{-t})u(t)$.
8.2.2 $y = (1 - e^{-t})u(t) - (1 - e^{-(t-1)})u(t - 1)$.
8.2.3 Answer $= 2(t - 1)u(t - 1) - 2(1 - e^{-(t-1)})u(t - 1) - 2(t - 2)u(t - 2)$.
8.2.4 It can be shown that if $x_1(t)$ produces $y_1(t)$, $Kx_1(t)$ produces $Ky_1(t)$; and if $x_1(t)$ produces $y_1(t)$ and $x_2(t)$ produces $y_2(t)$, then $x_1(t) + x_2(t)$ produces $y_1(t) + y_2(t)$.

8.2.5 $y = \frac{1}{2}e^{-t} + \frac{1}{2}\sin t - \frac{1}{2}\cos t \quad \text{for} \quad t \geq 0.$

8.3.1 $y(1) = 1 - 1/e.$

8.3.2 $$y = \begin{cases} 0 & \text{for} \quad t < 1, \\ -t^2 + 4t - 3 & \text{for} \quad 1 \leq t < 2, \\ t^2 - 6t + 9 & \text{for} \quad 2 \leq t < 3, \\ 0 & \text{for} \quad t > 3. \end{cases}$$

8.3.3 $$y = \begin{cases} 0 & \text{for} \quad t < 0, \\ t & \text{for} \quad 0 \leq t < 1, \\ -t + 2 & \text{for} \quad 1 \leq t < 2, \\ 0 & \text{for} \quad t \geq 2. \end{cases}$$

8.3.4 From $h(t)$, form $h(\tau)$; then form $h(t + \tau)$ which is shifted to the left by t, compared to $h(\tau)$. Finally, fold $h(t + \tau)$ vs. τ, with respect to the vertical axis. This means replacing τ by $-\tau$, so that we have $h(t - \tau)$.

8.3.5 $$y = \begin{cases} 0 & \text{for} \quad t < 2, \\ \frac{2}{3}t^3 - 2t + \frac{4}{3} & \text{for} \quad 2 \leq t < 3, \\ -\frac{2}{3}t^3 + 4t^2 - 4t - \frac{16}{3} & \text{for} \quad 3 \leq t < 4, \\ 0 & \text{for} \quad t \geq 4. \end{cases}$$

8.3.6 $$y(t) = \begin{cases} t & \text{for} \quad 0 \leq t < 1, \\ 1 & \text{for} \quad t \geq 1, \\ 0 & \text{for} \quad t < 0. \end{cases}$$

8.4.1 $v_{23}(t) = [1/(RC)]e^{-t/(RC)}u(t).$

8.4.2 $v_{12}(t) = \delta(t) - [1/(RC)]e^{-t/(RC)}u(t).$

8.4.3 $h(t) = (e^{-t} + 2te^{-t})u(t).$

8.4.4 $h(t) = (-\frac{1}{8}e^{-4t} + \frac{1}{8} + \frac{1}{2}t)u(t).$

8.4.5 Substitute y of Equation 8.54 in left-hand side of Equation 8.43 and this comes out to be equal to 1. Hence Equation 8.43 is satisfied.

8.4.6 [*Hint:* Use Euler's identity for trigonometric functions.]

8.5.1 $y_{zi}(t) = \frac{1}{3}e^{-t} + \frac{2}{3}e^{-t/4}. \qquad y_{zs}(t) = 1 - \frac{1}{3}e^{-t} - \frac{2}{3}e^{-t/4}.$

$v_0(t) = y_{zi}(t) + y_{zs}(t) = 1.$

8.5.2 $y_{zi}(t) = \frac{1}{3}e^{-t} + \frac{2}{3}e^{-t/4}, \qquad t \geq 0.$

$y_{zs}(t) = (1 - \frac{1}{3}e^{-t} - \frac{2}{3}e^{-t/4})u(t) - (1 - \frac{1}{3}e^{-(t-1)} - \frac{2}{3}e^{-(t-1)/4})u(t - 1).$

$v_0(t) = y_{zi}(t) + y_{zs}(t).$

$$v_0(t) = \begin{cases} 1 & \text{for} \quad 0 \leq t \leq 1, \\ \frac{1}{3}e^{-(t-1)} + \frac{2}{3}e^{-(t-1)/4} & \text{for} \quad t > 1. \end{cases}$$

8.5.3 $$\frac{d^2v_{C_2}}{dt^2} + \frac{5}{4}\frac{dv_{C_2}}{dt} + \frac{1}{4}v_{C_2} = \frac{1}{4}v_{\text{in}}.$$

8.6.1 $$y_p(t) = \frac{b_0K}{2a_0}t^2.$$

8.6.2 $$y_p(t) = \frac{b_0K}{(k + 1)!\, a_{k+1}}t^{k+1}.$$

8.6.3 $y_p(t) = \dfrac{b_m(j\omega)^m + \cdots + b_0}{(j\omega)^n + \cdots + a_0}$.

8.6.4 $y_p = 1, y_c = k_1e^{-t} + k_2e^{-t/4}$. $y = 1 + k_1e^{-t} + k_2e^{-t/4}$.
$v_{C_2}(t) = 1 + \frac{1}{3}e^{-t} + \frac{2}{3}e^{-t/4}$.

CHAPTER 9

9.1.1 (a) $i_1 = 5 \sin (t + \tan^{-1} \frac{4}{3})$,
(b) $v_2 = \sqrt{15^2 + 7^2} \sin (3t + \tan^{-1} \frac{7}{15})$,
(c) $i_3 = \sqrt{(3 - 2\sqrt{2})^2 + 8} \sin \left(t + \tan^{-1} \dfrac{2\sqrt{2}}{3 - 2\sqrt{2}}\right)$,
(d) $q_1 = 18 \sin (t - \pi/4)$, (e) $v_1 = 3 \sin (t + \pi/2)$.

9.1.2 (a) $i_1 = 5 \cos (t + \tan^{-1} \frac{4}{3} - \pi/2)$,
(b) $v_2 = \sqrt{274} \cos (3t + \tan^{-1} \frac{7}{15} - \pi/2)$,
(c) $i_3 = \sqrt{(3 - 2\sqrt{2})^2 + 8} \cos \left(t + \tan^{-1} \dfrac{2\sqrt{2}}{3 - 2\sqrt{2}} - \dfrac{\pi}{2}\right)$,
(d) $q_1 = 18 \cos (t - 3\pi/4)$, (e) $v_1 = 3 \cos t$.

9.2.1 (a) $5\underline{/-53.2^\circ}$, (b) $5\underline{/233.2^\circ}$, (c) $5\underline{/126.8^\circ}$,
(d) $6.8\underline{/17.1^\circ}$, (e) $13.09\underline{/83.4^\circ}$, (f) $13\underline{/157.3^\circ}$,
(g) $5\underline{/180^\circ}$, (h) $8\underline{/90^\circ}$, (i) $8\underline{/-90^\circ}$,
(j) $15\underline{/-1.9^\circ}$.

9.2.2 (a) $0 + j5$, (b) $\sqrt{2} + j\sqrt{2}$, (c) $0 - j10$,
(d) $-4\sqrt{3} + j4$, (e) $-3 + j0$, (f) $3\sqrt{3} + j3$,
(g) $3\sqrt{3} - j3$, (h) $-12 \cos 75^\circ + j12 \sin 75^\circ$,
(i) $-\sqrt{2} + j\sqrt{2}$, (j) $4.5 + j0$.

9.2.3 (a) $14.5 + j15$,
(b) $(-12 \cos 75^\circ + 3\sqrt{3}) + j(12 \sin 75^\circ + 3)$,
(c) $-4\sqrt{3} + j4 + 3\sqrt{3} - j3 = -\sqrt{3} + j1$.

9.2.4 (a) $25e^{j\pi}$, (b) $170.2e^{j240.7^\circ}$, (c) $45e^{j178.1^\circ}$.

9.2.5 $(e^{j\theta})^* = (\cos \theta + j \sin \theta)^* = \cos \theta - j \sin \theta = e^{-j\theta}$.

9.2.6 $(c_1c_2)^* = [(a_1 + jb_1)(a_2 + jb_2)]^* = [(a_1a_2 - b_1b_2) + j(b_1a_2 + b_2a_1)]^*$
$= a_1a_2 - b_1b_2 - j(b_1a_2 + b_2a_1)$;
$c_1^*c_2^* = (a_1 - jb_1)(a_2 - jb_2) = a_1a_2 - b_1b_2 - j(b_1a_2 + a_1b_2)$; $\therefore (c_1c_2)^* = c_1^*c_2^*$.

9.3.1 $6.65\underline{/114.9^\circ}$.

9.3.2 (a) jA leads A by 90°.
(b) $-2jA$ lags A by 90° and it is twice as long as A.
(c) $(1 - 2j)A$ lags A by 63.5° and it is 2.24 times as long.
(d) A^* is the mirror image of A with respect to the real axis.

9.3.3 $9.95 \sin (377t + 59^\circ)$.

9.3.4 $5.33 \sin (377t + 26^\circ)$; $5.33 \sin (377t - 14^\circ)$; yes.

9.3.5 $V_0 = I_0\sqrt{R^2 + (\omega L)^2}$, $\theta_V = \theta_0 + \tan^{-1} \dfrac{\omega L}{R}$.

9.4.1 $\dfrac{d^2v_{12}}{dt^2} + \dfrac{R}{L}\dfrac{dv_{12}}{dt} + \dfrac{1}{LC}v_{12} = \dfrac{1}{C}\dfrac{di}{dt} + \dfrac{R}{LC}i$, $v_{12} = 2 \sin (10^4t + 119.4^\circ)$.

9.4.2 $i(t) = \frac{1}{2} \times 10^{-3} \sin (10^4 t - 89.4°)$.

9.4.3 $\frac{d^2 v_{13}}{dt^2} + \frac{3dv_{13}}{dt} + v_{13} = \frac{dv}{dt} + v$, $v_{13}(t) = 1.05 \cos (t - 71.5°)$.

9.4.4 $\frac{d^2 v_{23}}{dt^2} + \frac{3dv_{23}}{dt} + v_{23} = v$, $v_{23} = 0.745 \cos (t - 116.5°)$.

CHAPTER 10

10.1.1 $i = 5\sqrt{2} \sin (3t + 15°)$.

10.1.2 $G = \frac{R}{R^2 + X^2}$. $B = \frac{-X}{R^2 + X^2}$.

10.1.3 $G = \frac{1}{|Z|} \cos \theta$, $B = \frac{-1}{|Z|} \sin \theta$.

10.2.1 Capacitor. $C = 1/(40\pi)$.
10.2.2 Inductor. $L = 1/(2\pi)$.
10.2.3 Resistor. $R = \frac{10}{3}$.
10.3.1 $v_R(t) + v_L(t) + v_C(t) = v(t)$. Since we are considering only the sinusoidal steady-state, the voltages are of the form

$$v_R(t) = |V_R| \sin (\omega t + \theta_R) = \text{Im} (|V_R| e^{j(\omega t + \theta_R)}),$$
$$v_L(t) = |V_L| \sin (\omega t + \theta_L) = \text{Im} (|V_L| e^{j(\omega t + \theta_L)}),$$
$$v_C(t) = |V_C| \sin (\omega t + \theta_C) = \text{Im} (|V_C| e^{j(\omega t + \theta_C)}),$$
$$v(t) = |V| \sin (\omega t + \theta) = \text{Im} (|V| e^{j(\omega t + \theta)}),$$
$$v_R(t) + v_L(t) + v_C(t) = \text{Im} [(V_R + V_L + V_C) e^{j\omega t}] = v = \text{Im} (V e^{j\omega t}),$$

where $V_R = |V_R| e^{j\theta_R}$, $V_L = |V_L| e^{j\theta_L}$, $V_C = |V_C| e^{j\theta_C}$, and $V = |V| e^{j\theta}$.

Since the time-domain equation holds for all t, it holds for time equal to $t + \pi/2$. But if $\sin (\omega t + \theta + \pi/2) = \cos (\omega t + \theta)$, then
$v_R(t + \pi/2) + v_L(t + \pi/2) + v_C(t + \pi/2)$

$$= |V_R| \cos (\omega t + \theta_R) + |V_L| \cos (\omega t + \theta_L) + |V_C| \cos (\omega t + \theta_C)$$
$$= v(t + \pi/2) = |V| \cos (\omega t + \theta)$$
$$= \text{Re} [(V_R + V_L + V_C) e^{j\omega t}] = \text{Re} (V e^{j\omega t}).$$

Since the real parts and imaginary parts of the above rotating phasors are equal for all time, the rotating phasors are equal: $(V_R + V_L + V_C) e^{j\omega t} = V e^{j\omega t}$. Hence $V_R + V_L + V_C = V$, which verifies Equation 10.24.

10.3.2 Using a procedure similar to that for Exercise 10.3.1, show that the imaginary parts of the corresponding rotating phasors are equal. Similarly, show that the real parts of the rotating phasors match. Hence the corresponding rotating phasors are equal, thus verifying Equation 10.41.
10.3.3 $V_{ai} = 6.3$ volts; $V_{cg} = 1$ volt.
10.3.4 $|V_2| = 76.9$ volts.
10.3.5 $|V_L| = 2$ volts; $|V_R| = 0.707$ volt; $V_C = 1.293$ volts.
10.3.6 $|V_R| = 12$; $|V_L| = 16$; $V_C = 7$ or 25.
10.3.7 For $\omega = 1$, $Z = 1 - j\frac{1}{4}$; for $\omega = 2$, $Z = 1 - j\frac{1}{2}$.

10.3.8 $Z = \frac{7}{5} - j\frac{13}{10}$.

10.3.9 $|V_1| = 130$ volts.

10.4.1 $v = 2\sin(t - 75°)$.

10.4.2 $v_1(t) = 7.2\sin(2t + 101.3°)$.

10.4.3 $V_{I_{rms}} = 3$ volts.

10.4.4 (a) $v_1(t) = 2\sqrt{2}\sin(\frac{1}{2}t + 15°)$, (b) $v_1(t) = 11.65\sin(2t + 120.9°)$, (c) $v_1(t) = \sqrt{2}\sin(0.1t + 45°)$, (d) $v_1(t) = 2\sqrt{5}\sin(0.5t - 33.4°)$, (e) $v_1(t) = 2.91\sin(0.5t + 14.1°)$, (f) $v_1(t) = 3.09\sin(t + 30.9°)$, (g) $v_1(t) = 20\sqrt{2}\sin(0.5t - 45°)$, (h) $v_1(t) = 2\sin(2t)$, (i) $v_1(t) = 4\sin(4t - 7.5°)$, (j) $v_1(t) = 150\sqrt{2}\sin(t - 0.6°)$, (k) $v_1(t) = \frac{5}{3}\sin(\frac{1}{16}t - 33.1°)$, (l) $v_1(t) = 4\sin(0.5t + 120°)$, (m) $v_1(t) = \frac{11}{12}\sin(t + 240°)$, (n) $v_1(t) = (7/\sqrt{2})\sin(27t + 713°)$, (o) $v_1(t) = 5.54\sin(0.25t + 51.2°)$, (p) $v_1(t) = \sqrt{2}\sin(t - 180°)$, (q) $v_1(t) = 6.18\sin(t - 45.9°)$, (r) $v_1(t) = (3/\sqrt{2})\sin(t + 135°)$, (s) $v_1(t) = 7.28\sin(2t - 75.9°)$, (t) $v_1(t) = (3\sqrt{10}/2)\sin(0.5t - 1.6°)$, (u) $v_1(t) = 8\sin(t - 45°)$, (v) $v_1(t) = 1.95\sin(0.5t + 68.6°)$, (w) $v_1(t) = 152.5\sin(t - 101.3°)$, (x) $v_1(t) = 1.21\sin(0.5t + 39.1°)$, (y) $v_1(t) = (15/\sqrt{2})\sin(t - 90°)$.

10.5.1 $I_{rms} = 1 + j1 = \sqrt{2}\,\underline{/45°}$ amperes. $P_{av} = I_R{}^2R = 1$ watt.
$S = V_{rms}I^*_{rms} = 1 - j1$ va. $\theta = -45°$.

10.5.2 $I_{rms} = 2 + j1 = \sqrt{5}\,\underline{/26.5°}$ amperes. $P_{av} = I_R{}^2R = 2$ watts.
$S = V_{rms}I^*_{rms} = 2 - j1$ va. $\theta = -26.5°$.

10.5.3 $I_4 = 1, I_3 = 1, I_2 = -1 + j2, I_1 = j2$ amperes.
$V_1 = -1 + j\frac{5}{2}$ volts.
$S_4 = 1, S_3 = j/2, S_2 = -j\frac{5}{2}, S_1 = 4 + j4$ va.
$S_{source} = V_1I_1^* = 5 + j2 = S_1 + S_2 + S_3 + S_4$ va.

10.5.4 $I_4 = 2, I_3 = 2, I_2 = -2 + j2, I_1 = j2$ amperes.
$V_1 = -1 + j3$ volts.
$S_4 = 2, S_3 = j2, S_2 = -j4, S_1 = 4 + j4$ va.
$S_{source} = V_1I_1^* = 6 + j2 = S_1 + S_2 + S_3 + S_4$ va.

10.6.1
$$G_{21} = \frac{2000s + \frac{8}{3}\times 10^6}{s^2 + 4\times 10^3 s + \frac{8}{3}\times 10^6}.$$
$$v_2(t) = \frac{10[(\frac{8}{3}\times 10^6)^2 + (2000\omega)^2]^{1/2}}{[(\frac{8}{3}\times 10^6 - \omega^2)^2 + (4000\omega)^2]^{1/2}} \times \sin\left(\omega t + 45° + \tan^{-1}\frac{3\omega}{4000} - \tan^{-1}\frac{4000\omega}{\frac{8}{3}\times 10^6 - \omega^2}\right).$$

10.6.2
$$G_{21}(j\omega) = \frac{1}{(1 - \omega^2 + \omega^4)^{1/2}}\underline{\Big/ -\tan^{-1}\frac{\omega}{1-\omega^2}}.$$

10.6.3 (a) $Z_{21} = V_2/I_1 = \frac{2}{5} + j\frac{1}{5}$; (b) $Y_{21} = I_2/V_1 = j\frac{1}{3}$; (c) $Z_{21} \neq 1/Y_{21}$.

10.6.4 $\alpha_{21} = I_2/I_1 = \frac{2}{5} + j\frac{1}{5}$.

10.6.5 $G_{21} = s^2/(s^2 + 3s + 1)$.

10.6.6 $\alpha_{21} = s/(2s + 1)$.

10.7.1 $I = I_R + I_L + I_C = V/R + V/(Ls) + CsV$. Hence, $I/V = Y = 1/R + 1/(Ls) + Cs$.

10.7.2 (a) $Z_{21}(s) = \dfrac{s}{2s + 1}$, (c) $G_{21}(s) = \dfrac{s^2}{s^2 + 3s + 1}$,

(b) $Y_{21}(s) = \dfrac{s^2}{s^2 + 3s + 1}$, (d) $\alpha_{21}(s) = \dfrac{s}{2s + 1}$.

For $s = -1$, $Z_{21} = 1$, $Y_{21} = -1$, $G_{21} = -1$, $\alpha_{21} = 1$.

10.7.3 $G_{21}(s) = \dfrac{2s^2}{2s^2 + s + 1}$.

10.7.4 $G_{21}(s) = 1/(s^2 + s + 1)$; $v_2(t) = \frac{10}{13}e^{3t}$.

10.8.1 $V_{\text{Th}} = \frac{1}{2}(-1 - j3)$ volts. $Z_{\text{Th}} = \frac{1}{2}(-1 + j3)$ ohms.

10.8.2 $V_{\text{Th}} = V_1 \dfrac{1 + jk\omega C_1}{1 + j\omega R_1 C_1}$ volts. $Z_{\text{Th}} = R_2 + \dfrac{k - R_1}{1 + j\omega R_1 C_1}$ ohms.

10.8.3 $V_{\text{Th}} = \dfrac{1 + k_1}{1 + j\omega R_1 C_1}$ volts. $Z_{\text{Th}} = R_2$ ohms.

HAPTER 11

11.1.1 Time constant $= RC$; bandwidth $= \omega_c = 1/(\text{time constant})$.

11.1.2 $v_0(t) = 100 + 0.531 \sin(377t - 178.5°)$.

11.1.3 $v_0(t) = 20 \cos(377t - 178.5°)$.

11.1.4 $\phi = -\tan^{-1} \omega RC$; $\phi_c = -45°$.

11.1.5 $\phi = 90° - \tan^{-1} \omega RC$; $\phi_c = 45°$.

11.2.1 (a) Replace L by $1/(\omega_r^2 C)$ in Equation 11.35 and obtain $Q = 1/(\omega_r RC)$; (b) replace ω_r by $1/\sqrt{LC}$ in Equation 11.35 and obtain $Q = (1/R)\sqrt{L/C}$.

11.2.2 From Equation 11.35, $V_{C_m} = V_m/(\omega_r RC)$. But from Equation 11.40, $V_{L_m} = \omega_r L V_m/R$. Since $1/(\omega_r RC) = Q = \omega_r L/R$, then $V_{L_m} = V_{C_m}$.

11.2.3 Let $\omega \to \infty$ in Equation 11.40 and obtain $V_{L_m} \to V_m$.

11.2.4 $|Y| = 1/\sqrt{R^2 + [\omega L - 1/(\omega C)]^2}$.

11.2.5 Selectivity $= \omega_r/\Delta\omega = \omega_r/(R/L) = \omega_r L/R$. But for the series RLC, $\omega_r L/R = Q$. Hence the Q expressions in Exercise 11.2.1 are also equal to S for the series RLC circuit.

11.3.1 $\omega_r \approx 10^6$ radians/sec; $S \approx 10$; $\Delta\omega = 10^5$ radians/sec; $Q = 10$.

11.3.2 $|Z|$ at resonance $\approx L/(RC) = 10^5$ ohms.
At $\omega = 0$, $|Z| = R = 1000$.
At $\omega = \infty$, $|Z| = 0$.
At $\omega = \omega_r - \frac{1}{4}\Delta\omega = 10^6(0.975)$, $|Z| = 0.865 \times 10^5$ ohms.
At $\omega = \omega_r + \frac{1}{4}\Delta\omega = 10^6(1.025)$, $|Z| = 0.913 \times 10^5$ ohms.
At $\omega = \frac{1}{2}\omega_r = 0.5 \times 10^6$, $|Z| = 0.675 \times 10^4$.
At $\omega = 2\omega_r = 2 \times 10^6$, $|Z| = 0.667 \times 10^4$.

11.3.3 $v(t) = 10^4 \cos \omega_r t + 0.667 \times 10^3 \sin 2\omega_r t$. There is a filtering effect.

11.3.4 $\omega_r = 1/\sqrt{LC}$.

11.3.5 $\Delta\omega = 1/(RC)$; $S = R\sqrt{C/L}$; $Q = R/(\omega_r L)$.

11.3.6 $\omega_r = 1/\sqrt{LC}$.

11.4.1 From the locus diagram, form a right triangle with the hypotenuse equal to the maximum value of I, and one side corresponding to an angle of 45°. It is clear that the sides are $1/\sqrt{2}$ times the length of the hypotenuse.

11.4.2 Plot the values of I in the complex plane corresponding to the three values of C. Draw a circle passing through the three points. Then compute the value of I corresponding to the fourth value of C. Verify that the point lies on the circle that was drawn.

11.4.3 The diameter of the circular locus is 1. For $C \to \infty$, $I = 1/(1 + jl)$ so that the locus consists of $\frac{3}{4}$ of the circle starting at I lagging V by 45° ($C \to \infty$) up to I leading V by 90° ($C = 0$).

11.4.4 (a) $I_L = V_m/(R_L + j\omega L)$; (b) the locus of I_C with C as parameter is similar to that in Figure 11.17; (c) the locus of I with C as parameter is similar to that in Figure 11.20.

11.4.5 The locus is similar to that in Figure 11.23.

11.4.6 This applies only to a series RLC situation.

11.5.1 There is a single root on the negative real axis which varies from 0 to $-\infty$ as R varies from ∞ to zero.

11.5.2 There are two roots. The locus consists of the segment of the negative real axis from -1 to zero and the vertical line $-\frac{1}{2} + jx$, where x is any real number.

11.5.3 There are two roots. The locus consists of the segment of the negative real axis from -100 to 0 and the vertical line $-50 + jx$, where x is any real number.

11.5.4 The characteristic equation is $s^2 + Rs + 1 = 0$. The locus consists of the negative real axis and the semicircle, as in Figure 11.26, with $\omega_r = 1$ and $2\zeta = R$.

11.6.1 Poles are at $-500 \pm j500\sqrt{3}$; zero at the origin.

11.6.2 Compare your result with that in Exercise 11.3.2.

11.6.3 (a) Pole is at $s = -1/(RC)$; no finite zero. (b) Pole at $s = -1/(RC)$; zero at $s = 0$.

11.6.4 $Y = \dfrac{1}{R} + Cs + \dfrac{1}{Ls}$,

$$Z = 1\Big/\left(Cs + \frac{1}{R} + \frac{1}{Ls}\right) = Ls\Big/\left(LCs^2 + \frac{Ls}{R} + 1\right)$$

$$Z = \frac{1}{C}s\Big/\left(s^2 + \frac{1}{RC}s + \frac{1}{LC}\right).$$

Zero is at $s = 0$. If $[1/(RC)]^2 - 4/(LC) > 0$, there are two real poles at

$$s = \left[-\frac{1}{RC} \pm \sqrt{\left(\frac{1}{RC}\right)^2 - \frac{4}{LC}}\right]\Big/2.$$

If $[1/(RC)]^2 - 4/(LC) = 0$, there is a double pole at $s = -1/(2RC)$. If $[1/(RC)]^2 - 4/(LC) < 0$, there are two complex poles at

$$s = \left[-\frac{1}{RC} \pm j\sqrt{\frac{4}{LC} - \left(\frac{1}{RC}\right)^2}\right]\Big/2.$$

CHAPTER 12

12.1.1 Sets (b) and (d) are redundant.

12.1.2 At (a): $i_{af} + i_1 + i_5 = 0$. At (c): $-i_3 + i_4 - i_5 = 0$.
At (b): $-i_1 + i_2 + i_3 = 0$. At (e): $-i_{af} - i_2 - i_4 = 0$.
The four equations are not linearly independent.

12.1.3 1. $v_1 + v_2 - v_s = 0$. 2. $-v_2 + v_3 + v_4 = 0$.
3. $-v_1 + v_5 - v_3 = 0$. 4. $v_1 + v_3 + v_4 - v_s = 0$.
5. $-v_s + v_5 + v_4 = 0$. 6. $-v_s + v_5 - v_3 + v_2 = 0$.
These six equations are not linearly independent.

12.1.4 $v_1 = v_a - v_b$; $v_2 = v_b$; $v_3 = v_b - v_c$; $v_4 = v_c$; $v_5 = v_a - v_c$; $v_s = v_a$.

12.1.5 $i_2 = i_1 - i_3$; $i_4 = i_3 + i_5$; $i_{af} = -i_1 - i_5$.

12.2.1 (a) *cad*; (b) *efd*; (c) *eda*; (d) *dcb*; (e) *cef*. There are many more.

12.2.2 (a) *bfe*; (b) *abc*; (c) *bfc*; (d) *aef*; (e) *abd*.

12.2.3 *c*, *ab*, *de*, *afe*, *dfb*.

12.2.4 1673, 5678, 5724, 1683, 1237, 1238, 1235, 2347, 2456, 1268.

12.2.5 2458, 1234, 1368, 2457, 4568, 4567, 4678, 1568, 1378, 3457, respectively.

12.3.1 (a) 3; (b) 3.

12.3.2 (a) 3; (b) 3.

12.3.3 (a) 4; (b) 4.

12.3.4 (a) 4; (b) 4.

12.3.5 1. $i_{ed} + i_{eg} = 0$. 2. $i_{kd} + i_{kb} = 0$.
3. $i_{dc} + i_{kb} + i_{eg} = 0$. 4. $i_{ca} + i_{cf} + i_{kb} + i_{eg} = 0$.
5. $i_{ab} + i_{cf} + i_{kb} + i_{eg} = 0$. 6. $i_{bh} + i_{cf} + i_{eg} = 0$.
7. $i_{hg} + i_{eg} = 0$. 8. $i_{fh} + i_{cf} = 0$.

12.3.6 1. $v_{ac} + v_{cf} - v_{hf} - v_{bh} - v_{ab} = 0$. 2. $v_{kd} + v_{dc} - v_{ac} + v_{ab} - v_{kb} = 0$.
3. $v_{ed} + v_{dc} - v_{ac} + v_{ab} + v_{bh} + v_{hg} - v_{eg} = 0$.

12.4.1 *acb*, *cde*, *afd*.

12.4.2 *adf*, *bfe*, *cde*.

12.4.3 $i_{12} + i_{13} + i_{14} = 0$; $-i_{12} + i_{24} + i_{23} = 0$; $-i_{23} - i_{13} + i_{34} = 0$.

12.4.4 $v_{12} + v_{24} - v_{14} = 0$; $v_{23} + v_{34} - v_{24} = 0$; $v_{13} - v_{23} - v_{12} = 0$.

12.4.5 $i_{12} + i_{13} + i_{14} + i_{15} = 0$; $-i_{12} + i_{23} + i_{24} + i_{25} = 0$;
$-i_{13} - i_{23} + i_{34} + i_{35} = 0$; $-i_{14} - i_{24} - i_{34} + i_{45} = 0$.

12.4.6 (a) No, because it is nonplanar. (b) Yes, since it is planar, but first it has to be redrawn in planar form.

12.5.1 $i_{14} = -i_{13} + i_{21}$; $i_{34} = i_{13} + i_{23}$; $i_{24} = -i_{21} - i_{23}$.

12.5.2 i_{12}, i_{23}, i_{34}, i_{45}, i_{56}, $i_{6,12}$, $i_{12,18}$, $i_{18,24}$, $i_{24,23}$, $i_{23,22}$, $i_{22,21}$, $i_{21,20}$, $i_{20,19}$, $i_{19,13}$, $i_{13,7}$, i_{71}.

12.5.3 Choose a tree such as (15, 54, 43, 32); then the chords are appropriate variables: 12, 24, 14, 25, 35, and 13.

12.5.4 No. It is nonplanar.

12.5.5 v_{15}, v_{25}, v_{35}, v_{45}.

12.5.6 v_{15}, v_{54}, v_{43}, v_{32}.

CHAPTER 13

13.1.1 Let v_1, v_2, v_4, and v_5 be the node voltages of nodes 1, 2, 4, and 5 with respect to node 3. Then

$$C_1 \frac{dv_1}{dt} + \frac{1}{R_1} v_1 - \frac{1}{R_1} v_2 = i,$$

$$-\frac{1}{R_1} v_1 + C_3 \frac{dv_2}{dt} + \left(\frac{1}{R_1} + \frac{1}{R_2}\right) v_2 - \frac{1}{R_2} v_4 - C_3 \frac{dv_5}{dt} = 0,$$

$$-\frac{1}{R_2} v_2 + C_2 \frac{dv_4}{dt} + \left(\frac{1}{R_2} + \frac{1}{R_4}\right) v_4 - \frac{1}{R_4} v_5 = 0,$$

$$-C_3 \frac{dv_2}{dt} - \frac{1}{R_4} v_4 + C_3 \frac{dv_5}{dt} + \left(\frac{1}{R_3} + \frac{1}{R_4}\right) v_5 = -i.$$

13.1.2 Let v_1, v_2, v_3, v_5 be the node voltages of nodes 1, 2, 3, and 5 with respect to node 4. Then

$$C_1 \frac{dv_1}{dt} + \frac{1}{R_1} v_1 - \frac{1}{R_1} v_2 - C_1 \frac{dv_3}{dt} = i,$$

$$-\frac{1}{R_1} v_1 + C_3 \frac{dv_2}{dt} + \left(\frac{1}{R_1} + \frac{1}{R_2}\right) v_2 - C_3 \frac{dv_5}{dt} = 0,$$

$$-C_1 \frac{dv_1}{dt} + (C_1 + C_2) \frac{dv_3}{dt} + \frac{1}{R_3} v_3 - \frac{1}{R_3} v_5 = 0,$$

$$-C_3 \frac{dv_2}{dt} - \frac{1}{R_3} v_3 + C_3 \frac{dv_5}{dt} + \left(\frac{1}{R_3} + \frac{1}{R_4}\right) v_5 = -i.$$

13.1.3 Let V_1, V_2, V_4, V_5, V_6, and V_7 be the node voltages with respect to node 3. Then

$$\left(\frac{1}{R_5} + j\omega C_1\right) V_1 - j\omega C_1 V_2 - \frac{1}{R_5} V_7 = I_1,$$

$$-j\omega C_1 V_1 + \left(j\omega C_1 + \frac{1}{R_2} + \frac{1}{j\omega L_1}\right) V_2 - \frac{1}{R_2} V_4 = 0,$$

$$-\frac{1}{R_2} V_2 + \left(\frac{1}{R_2} + \frac{1}{R_6} + j\omega C_2\right) V_4 - j\omega C_2 V_5 - \frac{1}{R_6} V_6 = 0,$$

$$-j\omega C_2 V_4 + \left(j\omega C_2 + \frac{1}{R_3}\right) V_5 - \frac{1}{R_3} V_7 = 0,$$

$$-\frac{1}{R_6} V_4 + \left(\frac{1}{R_6} + \frac{1}{R_4}\right) V_6 - \frac{1}{R_4} V_7 = 0.$$

$$-\frac{1}{R_5} V_1 - \frac{1}{R_3} V_5 - \frac{1}{R_4} V_6 + \left(\frac{1}{R_5} + \frac{1}{R_1} + \frac{1}{R_3} + \frac{1}{R_4}\right) V_7 = -I_1.$$

13.1.4

$$\left(\frac{1}{R_5} + j\omega C_1\right) V_1 - j\omega C_2 V_2 - \frac{1}{R_5} V_7 = I_1,$$

$$-j\omega C_1 V_1 + \left(j\omega C_1 + \frac{1}{R_2} + \frac{1}{j\omega L_1}\right) V_2 - \frac{1}{j\omega L_1} V_3 - \frac{1}{R_2} V_4 = 0,$$

$$-\frac{1}{j\omega L_1} V_2 + \left(\frac{1}{R_1} + \frac{1}{j\omega L_1}\right) V_3 - \frac{1}{R_1} V_7 = 0,$$

$$-\frac{1}{R_2} V_2 + \left(\frac{1}{R_2} + \frac{1}{R_6} + j\omega C_2\right) V_4 - \frac{1}{R_6} V_6 = 0,$$

$$-\frac{1}{R_6} V_4 + \left(\frac{1}{R_6} + \frac{1}{R_4}\right) V_6 - \frac{1}{R_4} V_7 = 0,$$

$$-\frac{1}{R_5} V_1 - \frac{1}{R_4} V_6 + \left(\frac{1}{R_5} + \frac{1}{R_1} + \frac{1}{R_3} + \frac{1}{R_4}\right) V_7 = -I_1.$$

13.1.5 $\left(\frac{1}{R_5} + j\omega C_1\right) V_1 - \frac{1}{R_5} V_7 = I_1,$

$$\left(\frac{1}{R_1} + \frac{1}{j\omega L_1}\right) V_3 - \frac{1}{R_1} V_7 = 0,$$

$$\left(j\omega C_2 + \frac{1}{R_2} + \frac{1}{R_6}\right) V_4 - j\omega C_2 V_5 - \frac{1}{R_6} V_6 = 0,$$

$$-j\omega C_2 V_4 + \left(j\omega C_2 + \frac{1}{R_3}\right) V_5 - \frac{1}{R_3} V_7 = 0,$$

$$-\frac{1}{R_6} V_4 + \left(\frac{1}{R_6} + \frac{1}{R_4}\right) V_6 - \frac{1}{R_4} V_7 = 0,$$

$$-\frac{1}{R_5} V_1 - \frac{1}{R_1} V_3 - \frac{1}{R_3} V_5 - \frac{1}{R_4} V_6 + \left(\frac{1}{R_5} + \frac{1}{R_1} + \frac{1}{R_3} + \frac{1}{R_4}\right) V_7 = -I_1.$$

13.1.6 $\left(C_1 s + \frac{1}{R_5}\right) V_1 - C_1 s V_2 - \frac{1}{R_5} V_2 = I_1,$

$$-C_1 s V_1 + \left(C_1 s + \frac{1}{R_2} + \frac{1}{L_1 s}\right) V_2 - \frac{1}{L_1 s} V_3 = 0,$$

$$-\frac{1}{L_1 s} V_2 + \left(\frac{1}{L_1 s} + \frac{1}{R_1}\right) V_3 - \frac{1}{R_1} V_7 = 0,$$

$$\left(C_2 s + \frac{1}{R_3}\right) V_5 - \frac{1}{R_3} V_7 = 0,$$

$$\left(\frac{1}{R_6} + \frac{1}{R_4}\right) V_6 - \frac{1}{R_4} V_7 = -I_1,$$

$$-\frac{1}{R_5} V_1 - \frac{1}{R_1} V_3 - \frac{1}{R_3} V_5 - \frac{1}{R_4} V_6 + \left(\frac{1}{R_5} + \frac{1}{R_1} + \frac{1}{R_3} + \frac{1}{R_4}\right) V_7 = 0.$$

13.1.7 $\left(C_1 s + \frac{1}{R_5}\right) V_1 - C_1 s V_2 - \frac{1}{R_5} V_7 = I_1,$

$$-C_1 s V_1 + \left(C_1 s + \frac{1}{R_2} + \frac{1}{L_1 s}\right) V_2 - \frac{1}{L_1 s} V_3 - \frac{1}{R_2} V_4 = 0,$$

$$-\frac{1}{L_1 s} V_2 + \left(\frac{1}{L_1 s} + \frac{1}{R_1}\right) V_3 - \frac{1}{R_1} V_7 = 0,$$

$$-\frac{1}{R_2} V_2 + \left(\frac{1}{R_2} + \frac{1}{R_6} + C_2 s\right) V_4 - C_2 s V_5 = 0,$$

$$-C_2 s V_4 + \left(C_2 s + \frac{1}{R_3}\right) V_5 - \frac{1}{R_3} V_7 = 0,$$

$$-\frac{1}{R_5} V_1 - \frac{1}{R_1} V_3 - \frac{1}{R_3} V_5 + \left(\frac{1}{R_5} + \frac{1}{R_1} + \frac{1}{R_3} + \frac{1}{R_4}\right) V_7 = -I_1.$$

13.1.8 $$\left(C_1 s + \frac{1}{L_1 s} + \frac{1}{R_2}\right) V_2 - \frac{1}{L_1 s} V_3 - \frac{1}{R_2} V_4 = 0,$$

$$-\frac{1}{L_1 s} V_2 + \left(\frac{1}{L_1 s} + \frac{1}{R_1}\right) V_3 - \frac{1}{R_1} V_7 = 0,$$

$$-\frac{1}{R_2} V_2 + \left(C_2 s + \frac{1}{R_2} + \frac{1}{R_6}\right) V_4 - C_2 s V_5 - \frac{1}{R_6} V_6 = 0,$$

$$-C_2 s V_4 + \left(C_2 s + \frac{1}{R_3}\right) V_5 - \frac{1}{R_3} V_7 = 0,$$

$$-\frac{1}{R_6} V_4 + \left(\frac{1}{R_6} + \frac{1}{R_4}\right) V_6 - \frac{1}{R_4} V_7 = 0,$$

$$-\frac{1}{R_1} V_3 - \frac{1}{R_3} V_5 - \frac{1}{R_4} V_6 + \left(\frac{1}{R_5} + \frac{1}{R_1} + \frac{1}{R_3} + \frac{1}{R_4}\right) V_7 = -I_1.$$

13.1.9 Let v_1, v_3, and v_4 be the voltages of nodes 1, 3, and 4 with respect to node 2. Then

$$C_1 \frac{dv_1}{dt} + \frac{1}{R_1} v_1 - \frac{1}{R_1} v_4 = i_1,$$

$$\frac{1}{R_3} v_3 + \frac{1}{L_1} \int_{-\infty}^{t} v_3(\tau)\, d\tau - \frac{1}{R_3} v_4 = i_2,$$

$$-\frac{1}{R_1} v_1 - \frac{1}{R_3} v_3 + \left(\frac{1}{R_1} + \frac{1}{R_2} + \frac{1}{R_3}\right) v_4 = -i_1 - i_2.$$

The inductor current $i_{32}(t)$ may be written as

$$i_{32}(t) = \frac{1}{L_1} \int_{t_1}^{t} v_3(\tau)\, d\tau + i_{32}(t_1),$$

where $i_{32}(t_1)$ is the initial current at the instant t_1.

13.1.10 $$C_1 \frac{dv_1}{dt} + \frac{1}{R_1} v_1 - C_1 \frac{dv_2}{dt} - \frac{1}{R_1} v_4 = i_1,$$

$$-C_1 \frac{dv_1}{dt} + C_1 \frac{dv_2}{dt} + \frac{1}{R_2} v_2 + \frac{1}{L_1} \int_{t_1}^{t} v_3(\tau)\, d\tau + i_{23}(t_1) - \frac{1}{R_2} v_4 = 0,$$

$$-\frac{1}{R_1} v_1 - \frac{1}{R_2} v_2 + \left(\frac{1}{R_1} + \frac{1}{R_2} + \frac{1}{R_3}\right) v_4 = -i_1 - i_2.$$

13.1.11

$$\begin{bmatrix} \left(\frac{1}{R_1}+\frac{1}{R_2}\right) & -\frac{1}{R_2} & 0 & 0 \\ -\frac{1}{R_1} & \left(\frac{1}{R_2}+\frac{1}{R_3}+\frac{1}{R_4}\right) & -\frac{1}{R_4} & 0 \\ 0 & -\frac{1}{R_4} & \left(\frac{1}{R_4}+\frac{1}{R_5}+\frac{1}{R_6}\right) & -\frac{1}{R_6} \\ 0 & 0 & \left(-\frac{1}{R_6}-G\right) & \left(\frac{1}{R_6}+\frac{1}{R_7}\right) \end{bmatrix} = y.$$

No, y is not symmetric.

13.1.12

$$\left(C_1s+\frac{1}{R_5}\right)V_1 - C_1sV_2 = I_1,$$

$$-C_1sV_1 + \left(C_1s+\frac{1}{L_1s}+\frac{1}{R_2}\right)V_2 - \frac{1}{L_1s}V_3 - \frac{1}{R_2}V_4 = 0,$$

$$-\frac{1}{L_1s}V_2 + \left(\frac{1}{L_1s}+\frac{1}{R_1}\right)V_3 = 0,$$

$$-\frac{1}{R_2}V_2 + \left(\frac{1}{R_2}+\frac{1}{R_6}+C_2s\right)V_4 - C_2sV_5 - \frac{1}{R_6}V_6 = 0,$$

$$-\frac{K}{L_1s}V_2 + \frac{K}{L_1s}V_3 - C_2sV_4 + \left(C_2s+\frac{1}{R_3}\right)V_5 = 0,$$

$$-\frac{1}{R_6}V_4 + \left(\frac{1}{R_6}+\frac{1}{R_4}\right)V_6 = 0.$$

13.2.1

$$\left(R_1+j\omega L_1+\frac{1}{j\omega C}\right)I_a + \frac{1}{j\omega C}I_b - (R_1+j\omega L_1)I_c = V_1,$$

$$\frac{1}{j\omega C}I_a + \left(R_2+R_3+j\omega L_2+\frac{1}{j\omega C}\right)I_b + R_2I_c = 0,$$

$$-(R_1+j\omega L_1)I_a + R_2I_b + (R_1+R_2+R_4+j\omega L_1)I_c = 0,$$

$$Z = \begin{bmatrix} \left(R_1+j\omega L_1+\frac{1}{j\omega C}\right) & \frac{1}{j\omega C} & -(R_1+j\omega L_1) \\ \frac{1}{j\omega C} & \left(R_2+R_3+j\omega L_2+\frac{1}{j\omega C}\right) & R_2 \\ -(R_1+j\omega L_1) & R_2 & (R_1+R_2+R_4+j\omega L_1) \end{bmatrix}$$

13.2.2 $$\begin{bmatrix} \left(R_1+j\omega L_1+\dfrac{1}{j\omega C}\right) & -\dfrac{1}{j\omega C} & (R_1+j\omega L_1) \\ -\dfrac{1}{j\omega C} & \left(R_2+R_3+j\omega L_2+\dfrac{1}{j\omega C}\right) & R_2 \\ (R_1+j\omega L_1) & R_2 & (R_1+R_2+R_4+j\omega L_1) \end{bmatrix} \times \begin{bmatrix} I_a \\ I_b \\ I_c \end{bmatrix} = \begin{bmatrix} V_1 \\ 0 \\ 0 \end{bmatrix}.$$

13.2.3 $$(R_4 + R_7)i_a + (R_4 + R_7)i_b + R_4 i_c - R_4 i_d - R_7 i_e = v_1,$$
$$(R_4 + R_7)i_a + (R_1 + R_4 + R_6 + R_7)i_b + (R_1 + R_4)i_c - (R_1 + R_4)i_d - R_7 i_e = v_1,$$
$$R_4 i_a + (R_1 + R_4)i_b + (R_1 + R_2 + R_4)i_c - (R_4 + R_1)i_e = 0,$$
$$-R_4 i_a - (R_4 + R_1)i_b - (R_4 + R_1)i_c + (R_1 + R_3 + R_4 + R_5)i_d - R_3 i_e = 0,$$
$$-R_7 i_a - R_7 i_b + 0 - R_3 i_d + (R_3 + R_7 + R_8)i_d = 0.$$

13.2.4 $$(R_1 + R_6)i_a - R_6 i_b - R_6 i_e = v_1,$$
$$-R_6 i_a + (R_2 + R_6 + R_7)i_b - R_7 i_c + R_7 i_d + R_6 i_e = 0,$$
$$-R_7 i_b + (R_3 + R_7 + R_8)i_c - R_7 i_d + R_8 i_e = 0,$$
$$R_7 i_b - R_7 i_c + (R_4 + R_7)i_d = v_1,$$
$$-R_6 i_a + R_6 i_b + R_8 i_c + (R_5 + R_6 + R_8)i_e = 0.$$

13.2.5 (1) $$R_7 i_0 + \frac{1}{C_2}\int_{t_1}^{t} i_0(\tau)\,d\tau - R_7 i_1 - \frac{1}{C_2}\int_{t_1}^{t} i_1(\tau)\,d\tau + v_{cf}(t_1) + \frac{1}{C_2}\int_{t_1}^{t} i_2(\tau)\,d\tau - \frac{1}{C_2}\int_{t_1}^{t} i_3(\tau)\,d\tau - \frac{1}{C_2}\int_{t_1}^{t} i_5(\tau)\,d\tau = v_1.$$

(2) $$-R_7 i_0 - \frac{1}{C_2}\int_{t_1}^{t} i_0(\tau)\,d\tau + (R_1 + R_7)i_1 + \left(\frac{1}{C_1} + \frac{1}{C_2}\right)\int_{t_1}^{t} i_1(\tau)\,d\tau + v_{bf}(t_1) - v_{cf}(t_1) + \frac{1}{C_2}\int_{t_1}^{t} i_3(\tau)\,d\tau - \left(\frac{1}{C_1} + \frac{1}{C_2}\right)\int_{t_1}^{t} i_2(\tau)\,d\tau + \frac{1}{C_2}\int_{t_1}^{t} i_5(\tau)\,d\tau - \frac{1}{C_1}\int_{t_1}^{t} i_6(\tau)\,d\tau = 0.$$

(3) $$-\frac{1}{C_2}\int_{t_1}^{t} i_0(\tau)\,d\tau + \frac{1}{C_2}\int_{t_1}^{t} i_1(\tau)\,d\tau - \frac{1}{C_2}\int_{t_1}^{t} i_2(\tau)\,d\tau + R_3 i_3 + \left(\frac{1}{C_2} + \frac{1}{C_3}\right)\int_{t_1}^{t} i_3(\tau)\,d\tau + v_{df}(t_1) - v_{cf}(t_1) - \frac{1}{C_3}\int_{t_1}^{t} i_4(\tau)\,d\tau + \frac{1}{C_2}\int_{t_1}^{t} i_5(\tau)\,d\tau + \frac{1}{C_3}\int_{t_1}^{t} i_6(\tau)\,d\tau = 0.$$

(4) $$R_2 i_2 + \frac{1}{C_2}\int_{t_1}^{t} [i_2(\tau) + i_0(\tau) - i_1(\tau) - i_3(\tau) - i_5(\tau)]\,d\tau + v_{cf}(t_1) + \frac{1}{C_1}\int_{t_1}^{t} [-i_1(\tau) + i_2(\tau) + i_6(\tau)]\,d\tau - v_{bf}(t_1) = 0.$$

(5) $R_4 i_4 + v_2 + \frac{1}{C_3}\int_{t_1}^{t}[-i_3(\tau) + i_4(\tau) - i_6(\tau)]\,d\tau - v_{af}(t_1) = 0.$

(6) $R_5 i_5 + v_2 + \frac{1}{C_2}\int_{t_1}^{t}[-i_0(\tau) + i_1(\tau) - i_2(\tau) + i_3(\tau) + i_5(\tau)]\,d\tau - v_{cf}(t_1) = 0.$

(7) $R_6 i_6 + \frac{1}{C_3}\int_{t_1}^{t}[i_3(\tau) - i_4(\tau) + i_6(\tau)]\,d\tau + v_{af}(t_1)$

$$+ \frac{1}{C_1}\int_{t_1}^{t}[-i_1(\tau) + i_2(\tau) + i_6(\tau)]\,d\tau - v_{bf}(t_1) = 0.$$

13.2.6
$$\begin{bmatrix} (R_1 + R_2) & -R_2 & 0 \\ -R_2 & \left(R_2 + R_3 + \frac{1}{Cs}\right) & -\frac{1}{Cs} \\ 0 & \left(-\frac{1}{Cs} - R\right) & \left(Ls + R_4 + \frac{1}{Cs}\right) \end{bmatrix} \begin{bmatrix} I_a \\ I_b \\ I_c \end{bmatrix} = \begin{bmatrix} V_1 \\ 0 \\ 0 \end{bmatrix}.$$

No, Z is not symmetric.

13.2.7 $(10 - j100)I_1 + j100I_2 = 10,$ $j100I_1 + (40 - j90)I_2 - 30I_3 = 0,$ $-30I_2 + 50I_3 = -5e^{-j30°}.$
Note that the phasor $V_1 = 10$ corresponds to $v_1 = 10 \sin 1000t$.

13.2.8 (1) $(R_1 + L_1 s)(I_a - I_c) - MsI_b + \frac{1}{Cs}(I_a - I_b) = V_1.$

(2) $\frac{1}{Cs}(I_b - I_a) + R_2(I_b - I_c) + (R_3 + L_2 s)I_b + MsI_c = 0.$

(3) $(R_1 + L_1 s)(I_c - I_a) + MsI_b + R_4 I_c - R(I_a - I_b) + R_2(I_c - I_b) = 0.$

13.3.1 $I_c = (V_b - V_c - V_2)j\omega C$. This is not the same as the current through C in the transformed network.

13.3.2 Yes. If a source transformation is performed with respect to a branch (voltage source in series is converted to current source in parallel), then the current through the transformed branch is not the same as that in the original branch.

13.3.3 Two node equations. Use node d as reference. Denote the unknown node voltages by V_b and V_c. Then

$$\frac{V_b - V_1}{R_1} + j\omega C V_b + \frac{V_b - V_c}{j\omega L} = 0,$$

$$\frac{V_c - V_b}{j\omega L} + \frac{V_c}{R_2} = I_1.$$

13.3.4 Two node equations:

$$\left(j\omega C + \frac{1}{R_1} + \frac{1}{j\omega L}\right)V_b - \frac{1}{j\omega L}V_c = \frac{V_1}{R_1}$$

and

$$-\frac{1}{j\omega L}V_b + \left(\frac{1}{j\omega L} + \frac{1}{R_2}\right)V_c = I_1 + I_2.$$

13.3.5 Three loop equations. Assign mesh currents directed clockwise, I_a, I_b, and I_c, for the first three meshes from left to right. The fourth mesh current is I_2.

$$R_1I_a - R_1I_b = V_1, \qquad -R_1I_a + (R_1 + R_2 + R_3 + j\omega L)I_b - (R_3 + j\omega L)I_c = 0,$$

$$-(R_3 + j\omega L)I_b + \left(R_3 + R_4 + j\omega L + \frac{1}{j\omega C}\right)I_c = -R_4I_2.$$

13.4.1 $C_1R_1R_3$ is a proper tree. Note that i_1 is part of the R_1 branch and that i_2 is part of the R_3 branch. There are only two proper trees.

13.4.2 R_3CR_4 is a proper tree. Note that i_2 is part of the R_4 branch, R_1 may be eliminated, and R_2 in series with v_1 is a link branch. There are only two proper trees.

13.4.3 $dv_C/dt = [1/(RC)]v_c + (1/C)i_L$. $\qquad di_L/dt = (1/L)v_C + (1/L)v$.

13.4.4 $dv_2/dt = -2v_2 + 2v_3$. $\qquad dv_3/dt = 2v_2 - 4v_3 + 2v_1$.

CHAPTER 14

14.1.1 (a) The solution is the same as before; (b) the solutions are equal.

14.1.2 $v_1 = 1$, $v_2 = \frac{3}{4}$, $v_3 = \frac{5}{4}$. The answer is twice the answer in Example 14.1.1.

14.1.3 (a) $v_1 = .371$, $v_2 = .174$, $v_3 = .293$.
(b) $v_1 = .371 \sin 1000t$, $v_2 = .174 \sin 1000t$, $v_3 = .293 \sin 1000t$.
(c) $v_1 = .146 \sin 3000t + .225 \sin 500t$, $v_2 = .106 \sin 3000t + .068 \sin 500t$, $v_3 = .220 \sin 3000t + .073 \sin 500t$.

14.1.4 The answer is the same. There are four others, but the answers are all the same.

14.1.5 $v_1(t) = .471 \sin(\omega t + 7.7°)$, $v_2(t) = .389 \sin(\omega t + 28.9°)$, $v_3(t) = .612 \sin(\omega t + 24.2°)$.

14.2.1 (a) $\det A = -a_{12}(a_{21}a_{33} - a_{23}a_{31}) + a_{22}(a_{11}a_{33} - a_{13}a_{31}) - a_{32}(a_{11}a_{23} - a_{13}a_{21})$.
(b) $\det A = a_{13}(a_{21}a_{32} - a_{22}a_{31}) - a_{23}(a_{11}a_{32} - a_{12}a_{31}) + a_{33}(a_{11}a_{22} - a_{12}a_{21})$.

14.2.2 (a) $\det A = a_{11}(a_{22}a_{33} - a_{23}a_{32}) - a_{12}(a_{21}a_{33} - a_{23}a_{31}) + a_{13}(a_{21}a_{32} - a_{22}a_{31})$.
(b) $\det A = -a_{21}(a_{12}a_{33} - a_{13}a_{32}) + a_{22}(a_{11}a_{33} - a_{13}a_{31}) - a_{23}(a_{11}a_{32} - a_{12}a_{31})$.
(c) $\det A = a_{31}(a_{12}a_{23} - a_{13}a_{22}) - a_{32}(a_{11}a_{23} - a_{13}a_{21}) + a_{33}(a_{11}a_{22} - a_{12}a_{21})$.

14.2.3 (a) $x_2 = [-b_1(a_{21}a_{33} - a_{23}a_{31}) + b_2(a_{11}a_{33} - a_{13}a_{31}) - b_3(a_{11}a_{23} - a_{13}a_{21})]/\det A$.
(b) $x_3 = [b_1(a_{21}a_{32} - a_{22}a_{31}) - b_2(a_{11}a_{32} - a_{12}a_{31}) + b_3(a_{11}a_{22} - a_{12}a_{21})]/\det A$.

14.2.4

$$\begin{vmatrix} a_{11} & a_{12} \\ a_{21} & a_{22} \end{vmatrix} = \begin{vmatrix} a_{11} & a_{12} \\ 0 & a_{22} - \dfrac{a_{21}a_{12}}{a_{11}} \end{vmatrix} = \begin{vmatrix} a_{11} & 0 \\ 0 & a_{22} - \dfrac{a_{21}a_{12}}{a_{11}} \end{vmatrix}$$

$$= a_{11}\left(a_{22} - \frac{a_{21}a_{12}}{a_{11}}\right)\begin{vmatrix} 1 & 0 \\ 0 & 1 \end{vmatrix} = a_{11}a_{22} - a_{12}a_{21}.$$

14.2.5 (a) From Cramer's rule: six multiplications, two divisions, three subtractions.
(b) From Gauss's elimination: three multiplications, three divisions, three subtractions.

14.3.1

$$3\frac{d^3x_1}{dt^3} + 18\frac{d^2x_1}{dt^2} + 29\frac{dx_1}{dt} + 7x_1 = 4\frac{d^2u_1}{dt^2} + 18\frac{du_1}{dt} + 18u_1 + \frac{d^2u_2}{dt^2}$$

$$+ 4\frac{du_2}{dt} + 2u_2 + 4\frac{d^2u_3}{dt^2} + 28\frac{du_3}{dt} + 40u_3.$$

Three initial conditions are needed.

14.3.2 $$3\frac{d^3x_2}{dt^3} + 18\frac{d^2x_2}{dt^2} + 29\frac{dx_2}{dt} + 7x_2 = 2\frac{d^2u_1}{dt^2} + 8\frac{du_1}{dt} + 4u_1 + 2\frac{d^2u_2}{dt^2} + 7\frac{du_2}{dt} + 2u_2 + 8\frac{d^2u_3}{dt^2} + 28\frac{du_3}{dt} + 12u_3.$$

Three initial conditions are required.

14.3.3 $$3\frac{d^3x_3}{dt^3} + 18\frac{d^2x_3}{dt^2} + 29\frac{dx_3}{dt} + 7x_3 = 2\frac{d^2u_1}{dt^2} + 14\frac{du_1}{dt} + 20u_1 + 2\frac{d^2u_2}{dt^2} + 7\frac{du_2}{dt} + 3u_2 + 20\frac{d^2u_3}{dt^2} + 76\frac{du_3}{dt} + 60u_3.$$

Three initial conditions are needed.

14.3.4 $-2x_1 - (p + 1)x_2 + (p + 2)x_3 = 4u_3,$
$(7p + 17)x_2 - (p + 2)x_3 = 4u_1 + 4u_2 + 12u_3,$
$(3p^3 + 18p^2 + 29p + 7)x_3 = (p^2 + 7p + 10)2u_1 + (2p^2 + 7p + 3)u_2 + (20p^2 + 76p + 60)u_3.$

14.3.5 (a) $C\frac{d}{dt}(v_2 - v_1) + \frac{v_2}{R_2} + \frac{v_2 - v_3}{R_3} = 0,$

$$\frac{v_3 - v_2}{R_3} + \frac{1}{L}\int_{t_1}^{t} v_3(\tau)\, d\tau + i_L(0) = 0;$$

(b) $\left[p^2 + \left(\frac{R_3}{L} + \frac{1}{R_2C}\right)p + \frac{1}{LC}\left(1 + \frac{R_3}{R_2}\right)\right]v_2 = \left(p^2 + \frac{R_3}{L}p\right)v_1;$

(c) $v_2(t) = -1.049e^{-.869\times 10^6 t} - 0.049e^{-.23\times 10^6 t}.$

CHAPTER 15

15.1.1 (a) An LC 2-port terminated in R_1 and R_2; (b) a resistive 4-port terminated in L_1, L_2, C_1, and C_2.

15.1.2 (a) An LC 4-port terminated in R_1, R_2, R_3, and R_4; (b) a resistive 5-port terminated in C_1, C_2, C_3, L_1, and L_2.

15.1.3 $$R_1 = \frac{R_aR_b}{R_a + R_b + R_c}, \quad R_2 = \frac{R_aR_c}{R_a + R_b + R_c}, \quad R_3 = \frac{R_bR_c}{R_a + R_b + R_c}.$$

15.2.1 (a) $y_{11} = 1/Z_1, y_{22} = 1/Z_1, y_{12} = y_{21} = -1/Z_1.$
Z-parameters do not exist.
$h_{11} = Z_1, h_{12} = 1, h_{21} = -1, h_{22} = 0.$
$g_{11} = 0, g_{12} = -1, g_{21} = 1, g_{22} = Z_1.$

(b) y-parameters do not exist.
$z_{11} = z_{12} = z_{21} = z_{22} = Z_2.$
$h_{11} = 0, h_{12} = 1, h_{21} = -1, h_{22} = 1/Z_2.$
$g_{11} = 1/Z_2, g_{12} = -1, g_{21} = 1, g_{22} = 0.$

(c) $y_{11} = 1/Z_1, y_{12} = -1/Z_1, y_{21} = -1/Z_1, y_{22} = 1/Z_1 + 1/Z_2.$
$z_{11} = Z_1 + Z_2, z_{12} = z_{21} = Z_2, z_{22} = Z_2.$
$h_{11} = Z_1, h_{12} = 1, h_{21} = -1, h_{22} = 1/Z_2.$
$g_{11} = 1/(Z_1 + Z_2), g_{12} = -Z_2/(Z_1 + Z_2), g_{21} = 1, g_{22} = Z_1Z_2/(Z_1 + Z_2).$

15.2.2 (a) $y_{11} = \frac{3}{2}, y_{12} = y_{21} = -\frac{1}{2}, y_{22} = \frac{3}{4}$.
$z_{11} = \frac{6}{7}, z_{12} = z_{21} = \frac{4}{7}, z_{22} = \frac{12}{7}$.
$h_{11} = \frac{2}{3}, h_{12} = \frac{1}{3}, h_{21} = -\frac{1}{3}, h_{22} = \frac{7}{12}$.
$g_{11} = \frac{7}{6}, g_{12} = -\frac{2}{3}, g_{21} = \frac{2}{3}, g_{22} = \frac{4}{3}$.
(b) $y_{11} = \frac{5}{14}, y_{12} = y_{21} = -\frac{2}{7}, y_{22} = \frac{3}{7}$.
$z_{11} = 6, z_{12} = z_{21} = 4, z_{22} = 5$.
$h_{11} = \frac{14}{5}, h_{12} = \frac{4}{5}, h_{21} = -\frac{4}{5}, h_{22} = \frac{1}{5}$.
$g_{11} = \frac{1}{6}, g_{12} = -\frac{2}{3}, g_{21} = \frac{2}{3}, g_{22} = \frac{7}{3}$.

15.2.3 $k_{21} = 3$.

15.2.4 Assume that the 2-port measurements are made with the two bottom nodes of the two ports connected to ground. Then $V_1 = \frac{7}{23}$, $V_2 = -\frac{3}{23}$.

15.2.5 (a) No. h_{11} is a short-circuit driving-point impedance, but h_{22} is an open-circuit driving-point admittance. (b) Not necessarily. If $h_{12} = -h_{21}$, the network is reciprocal. (See Table 15.1.)

15.3.1 $$\left.\frac{I_{2b}}{I_{1b}}\right|_{V_{2b}=0} = -\left.\frac{V_{1a}}{V_{2a}}\right|_{I_{1a}=0}.$$

This is the same result as in Equation 15.60, except that the roles of ports 1 and 2 are interchanged.

15.3.2 $$\left.\frac{V_2}{V_1}\right|_{I_2=0} = \frac{z_{21}}{z_{11}}, \qquad \left.\frac{I_1}{I_2}\right|_{V_1=0} = -\frac{z_{12}}{z_{11}}.$$

Hence, if $z_{12} = z_{21}$, Equation 15.60 is verified.

15.3.3 $h_{12} = -h_{21}$.

15.3.4 $z_{21} = 2(1 - \alpha)$; $z_{12} = 2$. For the network to be reciprocal, $\alpha = 0$.

15.3.5 As long as the 2-port can be represented as an interconnection of 1-ports, it is reciprocal.

15.3.6 Replace the 2-ohm resistor in Figure 15.22(a) by a 1-port, using port 1 of the network in Figure 15.29, and leaving port 2 of the network of Figure 15.29 unused.

15.4.1 $g = 10^{-4}$ mho, $g_m = 10^{-3}$ mho.

15.4.2 $v_{DS} = (-15 \times 25/26)V_0 \sin \omega_0 t$ volts.

15.4.3 $R_L = 3.5$ kilohms.

15.4.4 $g_{11} = 0, g_{12} = 0, g_{21} = -g_m/g, g_{22} = 1/g$. The h-parameters do not exist for this model.

CHAPTER 16

16.1.1 $v_b(t) = 1414 \sin(377t - 95°)$, $v_c(t) = 1414 \sin(377t - 215°)$,
$v_{12}(t) = \sqrt{3}\, 1414 \sin(377t - 65°)$, $v_{23}(t) = \sqrt{3}\, 1414 \sin(377t - 185°)$,
$v_{31}(t) = \sqrt{3}\, 1414 \sin(377t - 305°)$.

16.1.2 $v_b(t) = 1414 \sin(377t - 215°)$, $v_c(t) = 1414 \sin(377t - 95°)$,
$v_{12}(t) = \sqrt{3}\, 1414 \sin(377t - 5°)$, $v_{23}(t) = \sqrt{3}\, 1414 \sin(377t + 115°)$,
$v_{31}(t) = \sqrt{3}\, 1414 \sin(377t - 125°)$.

16.1.3 $v_{23}(t) = 2000 \sin(377t - 120°)$, $v_{31}(t) = 2000 \sin(377t - 240°)$,
$v_a(t) = (2000/\sqrt{3}) \sin(377t - 30°)$, $v_b(t) = (2000/\sqrt{3}) \sin(377t - 150°)$,
$v_c(t) = (2000/\sqrt{3}) \sin(377t - 270°)$.

16.1.4 $Z = 10 + j(2\sqrt{46}/3)$.

16.2.1 Perform the operations as indicated.

16.2.2 Perform the operations as indicated.

16.2.3 $Z_1 = (67 + j57)/106$, $Z_2 = (31 + j77)/106$, $Z_3 = (182 + j52)/106$.

16.3.1 $v_b(t) = 141.4 \sin (377t - 120°)$, $v_c(t) = 141.4 \sin (377t - 240°)$, $v_{12}(t) = 200 \sin (377t + 30°)$, $v_{23}(t) = 200 \sin (377t - 90°)$, $v_{31}(t) = 200 \sin (377t - 210°)$, $i_1(t) = 300 \sin (377t - 45°)$, $i_2(t) = 300 \sin (377t - 165°)$, $i_3(t) = 300 \sin (377t - 285°)$.

16.3.2 $v_b(t) = 141.4 \sin (377t - 240°)$, $v_c(t) = 141.4 \sin (377t - 120°)$, $v_{12}(t) = 200 \sin (377t - 30°)$, $v_{23}(t) = 200 \sin (377t + 90°)$, $v_{31}(t) = 200 \sin (377t + 210°)$, $i_1(t) = 300 \sin (377t - 45°)$, $i_2(t) = 300 \sin (377t + 75°)$, $i_3(t) = 300 \sin (377t + 195°)$.

16.3.3 $V_a = 141.4$, $I_1 = 3(141.4)/(1 + j2.5)$, $V_{1'0} = 141.4(1 + j1)/(1 + j2.5)$. Other phasors are obtained using *abc* sequence information.

16.3.4 Same as in Exercise 16.3.3 except that phases b and c are interchanged.

16.4.1 Use Equations 16.45, 16.46, and 16.47.

16.4.2 Use same procedure as in Exercise 16.4.1.

16.4.3 V_a, V_b, V_c, V_{12}, V_{23}, and V_{31} have phase relationships as shown in Figure 16.1(b). The phase currents are computed as

$$I_{1'2'} = V_{12}/Z_a, \qquad I_{2'3'} = V_{23}/Z_b, \qquad I_{3'1'} = V_{31}/Z_c.$$

Line currents are computed using Equation 16.29. $P_T = 9.4$ megawatts.

16.4.4 Same procedure as in Exercise 16.4.3. $P_T = 9.4$ megawatts.

16.5.1 V_{13} lags V_a by 30° and I_1 lags V_a by θ (note that θ may be negative). Hence the phase difference between V_{13} and I_1 is $30° - \theta$. Similarly, V_{23} leads V_b by 30° and I_2 lags V_b by $\theta°$. Hence the phase difference between V_{23} and I_2 is $30° + \theta°$.

16.5.2 Expand $\cos (30° - \theta)$ and $\cos (30° + \theta)$ and form $(P_1 - P_2)/(P_1 + P_2)$.

16.5.3 V_{13} leads V_a by 30° and I_1 lags V_a by $\theta°$. Hence the phase difference between V_{13} and I_1 is $30° + \theta$. Similarly, V_{23} lags V_b by 30° and I_2 lags V_b by θ. Hence the phase difference between V_{23} and I_2 is $30° - \theta$.

16.5.4 (a) 3500 watts, (b) $\tan \theta = 11\sqrt{3}/7$, (c) $|Z| = 240^2(7)/(3500\sqrt{412})$, angle θ obtained from (b).

16.6.1 In general and for more complicated networks, a closed solution is not possible.

16.6.2 This is in accordance with conservation of energy.

16.6.3 $S_{S_1}^* = P_1 - jQ_1 + V_1^*(Y_1V_1 - Y_1V_3)$,
$S_{S_2}^* = P_2 - jQ_2 + V_2^*(Y_2V_2 - Y_3V_3)$,
$0 = P_3 - jQ_3 + V_3^*(-Y_1V_1 - Y_2V_2)$.

CHAPTER 17

17.1.1 (a) $\omega_0 = 377$ radians/sec, $f_0 = 60$ Hz, $T = 1/60$ sec.
(b) $\omega_0 = 1000$ radians/sec, $f_0 = 1000/(2\pi)$ Hz, $T = 2\pi/1000$ sec.

17.1.2 This is an approximation for a negative-going sawtooth wave with a period of $T = 2\pi$, starting at $3\pi/4$ at $t = 0$, going down linearly to $-\pi/4$ at $t = 2\pi$, and then repeating.

17.1.3 This is an approximation for a square wave which has a period of 2π,

$$v(t) = \begin{cases} \pi/4, & 0 \le t \le \pi/2, \\ -\pi/4, & \pi/2 \le t \le 3\pi/2, \\ \pi/4, & 3\pi/2 \le t \le 2\pi. \end{cases}$$

17.1.4 $T = 2\pi$ sec.

17.1.5 $v(t)$ is not periodic since the frequencies are not harmonically related.

17.1.6 If ω_2/ω_1 is a rational number, it is a ratio of integers $= m/n$, where m and n have no common factors. Set $\omega_2 = \omega_0 m$. Then $\omega_0 = \omega_2/m$ and $T = 2\pi/\omega_0 = 2\pi m/\omega_2$.

17.2.1 $v(t) = (4V_0/\pi)(\sin t + \frac{1}{3}\sin 3t + \frac{1}{5}\sin 5t + \cdots)$.

17.2.2 $v(t) = V_0/2 - (V_0/\pi)(\sin t + \frac{1}{2}\sin 2t + \frac{1}{3}\sin 3t + \cdots)$.

17.2.3 $v(t) = V_0/\pi + (V_0/2)\sin t - (2V_0/\pi)(\frac{1}{3}\cos 2t + \frac{1}{15}\cos 4t + \cdots)$.

17.2.4 $v(t) = \dfrac{2V_0}{\pi} + \dfrac{4V_0}{\pi}\sum_{n=1}^{\infty}\dfrac{1}{1-4n^2}\cos 2nt.$

17.3.1 Apply Equations 16.53 and 16.54 to the signals.

17.3.2 See Figure 17.10, Figure 17.14(b), Figure 17.24, and $\sin t$.

17.3.3 (a) Continue downward sloping line until $t = 2\pi$.
(b) Use straight line such that $v = 0$ at $t = \pi$ and $v = V_0$ at $t = 2\pi$.
(c) Sketch of (a) is satisfactory.
(d) Use straight line such that $v = -V_0$ at $t = \pi$ and $v = 0$ at $t = 2\pi$.

17.4.1 $b_1 = 0$ and $b_2 = 0$.

17.4.2 $g(t) = 2/\pi - [4/(3\pi)]\cos 2\pi t$.

17.4.3 Perform the indicated integration, making use of the orthogonality relations in Equations 17.79 and 17.80.

17.4.4 $V^2 = (2.5)^2 + \frac{1}{2}[3^2 + 3^2 + 4^2 + 1^1 + 1^2 + (-2.5)^2]$.

CHAPTER 18

18.1.1 $a_n = (a_n - jb_n)/2 + (a_n + jb_n)/2 = 2\,\mathrm{Re}\,(F_n)$.

18.1.2 Since $v(t)$ has period T, the integral is not affected by choice of t_0 as long as the interval is T.

18.2.1 $F_0 = 5$, $F_2 = 2\underline{/-90^\circ}$, $F_3 = 5\underline{/-53.2^\circ}$, $F_{-2} = 2\underline{/90^\circ}$, $F_{-3} = 5\underline{/53.2^\circ}$.

18.2.2 $|F_0|^2 = 25$, $|F_2|^2 = 4$, $|F_3|^2 = 25$, $|F_{-2}|^2 = 4$, $|F_{-3}|^2 = 25$.

18.2.3 $F_0 = 5$, $F_2 = 2\underline{/0^\circ}$, $F_3 = 5\underline{/126.8^\circ}$, $F_{-2} = 2\underline{/0^\circ}$, $F_{-3} = 5\underline{/-126.8^\circ}$.

18.2.4 Same as in Exercise 18.2.2.

18.2.5 $a_0 = V_0/2$, $b_1 = 2V_0/\pi$, $b_3 = 2V_0/(3\pi)$, $b_5 = 2V_0/(5\pi)$, $b_7 = 2V_0/(7\pi)$.
$G_0 = (V_0/2)^2$, $G_1 = (V_0/\pi)^2$, $G_3 = [V_0/(2\pi)]^2$, $G_5 = [V_0/(5\pi)]^2$, $G_7 = [V_0/(7\pi)]^2$.
Average power of $g(t) = G_0 + G_1 + G_{-1} + G_3 + G_{-3} + G_5 + G_{-5} + G_7 + G_{-7} = 0.488V_0^2$. Average power of $v(t) = 0.5V_0^2$.

18.3.1 $Y_0 = V_0/2$, $Y_1 = [V_0/(\pi\sqrt{2})]\underline{/-3\pi/4}$, $Y_3 = [V_0/(3\sqrt{10}\,\pi)]\underline{/-\pi/2 - \tan^{-1}3}$, $[V_0/(5\sqrt{26}\,\pi)]\underline{/\pi/2 - \tan^{-1}5}$, $Y_{-1} = Y_1^*$, $Y_{-3} = Y_3^*$, $Y_{-5} = Y_5^*$.

18.3.2 $|Y_0|^2 = 100$, $|Y_1|^2 = \frac{1}{8}$, $|Y_{-1}|^2 = \frac{1}{8}$. Average power of $v_2(t) = 100.25$.

18.3.3 $|Y_0|^2 = 100$, $|Y_1|^2 = \frac{1}{8}$, $|Y_{-1}|^2 = \frac{1}{8}$, $|Y_2|^2 = \frac{1}{20}$, $|Y_{-2}|^2 = \frac{1}{20}$. Average power of $v_2(t) = 100.35$.

18.4.1 Apply the Fourier transform definition to $k_1v_1(t) + k_2v_2'(t)$.
18.4.2 Apply the Fourier transform definition to $v(t - T)$ and replace $t - T$ by τ.
18.4.3 Apply the definition of Fourier transform.
18.4.4 Differentiate the inverse transform formula for $v(t)$ with respect to t.
18.4.5 Differentiate expression for $F(j\omega)$ with respect to ω.
18.4.6 Follow the hint.
18.4.7 $F(j\omega) = e^{-j\omega/2}(\sin \omega/2)/(\omega/2)$.
18.5.1 $y(t) = (1/\sqrt{10}) \cos (t - \tan^{-1} \frac{1}{3})$.
18.5.2 (a) $\frac{1}{2}$; (b) $[1/(2\pi)][1/(1 + \omega^2)]$.
18.5.3 $|Y(j\omega)| = 1/\sqrt{(1 + \omega^2)(9 + \omega^2)}$, $\underline{/Y(j\omega)} = -\tan^{-1} \omega - \tan^{-1} (\omega/3)$.
18.5.4 $H(j\omega) = 1/(j\omega + 1)$.
18.6.1 $F(s) = (1 - e^{-s})/s$.
18.6.2 $F(s) = (s + 1)/[(s + 1)^2 + 1]$.
18.6.3 $F(s) = 1/(s - 1)$.
18.6.4 $f(t) = (e^{-3t} - 6e^{-4t} + 6e^{-5t})u(t)$.
18.6.5 $f(t) = \delta(t) + (2e^{-3t} - 6e^{-4t})u(t)$.
18.6.6 $f(t) = [\frac{1}{2} - \frac{1}{2} \cos \sqrt{2}\, t + (1/\sqrt{2}) \sin \sqrt{2}\, t]u(t)$.

Index

BCDEFGHIJ–VB–7987654